Handbook of Research on Manufacturing Process Modeling and Optimization Strategies

Raja Das
VIT University, India

Mohan Pradhan
Maulana Azad National Institute of Technology, Bhopal, India

A volume in the Advances in Logistics, Operations, and Management Science (ALOMS) Book Series

IGI GLOBAL
DISSEMINATOR OF KNOWLEDGE

www.igi-global.com

Published in the United States of America by
IGI Global
Business Science Reference (an imprint of IGI Global)
701 E. Chocolate Avenue
Hershey PA, USA 17033
Tel: 717-533-8845
Fax: 717-533-8661
E-mail: cust@igi-global.com
Web site: http://www.igi-global.com

Library of Congress Cataloging-in-Publication Data

Names: Das, Raja, 1972- editor. | Pradhan, Mohan, 1973- editor.
Title: Handbook of research on manufacturing process modeling and optimization strategies / Raja Das and Mohan Pradhan, editors.
Description: Hershey, PA : Business Science Reference, [2017] | Includes bibliographical references.
Identifiers: LCCN 2017003174| ISBN 9781522524403 (hardcover) | ISBN 9781522524410 (ebook)
Subjects: LCSH: Production planning--Data processing--Handbooks, manuals, etc. | Manufacturing processes--Simulation methods--Handbooks, manuals, etc. | Lean manufacturing--Simulation methods--Handbooks, manuals, etc.
Classification: LCC TS183.3 .H36 2017 | DDC 658.5/03--dc23 LC record available at https://lccn.loc.gov/2017003174

This book is published in the IGI Global book series Advances in Logistics, Operations, and Management Science (ALOMS) (ISSN: 2327-350X; eISSN: 2327-3518)

British Cataloguing in Publication Data
A Cataloguing in Publication record for this book is available from the British Library.

Advances in Logistics, Operations, and Management Science (ALOMS) Book Series

John Wang
Montclair State University, USA

ISSN:2327-350X
EISSN:2327-3518

Mission

Operations research and management science continue to influence business processes, administration, and management information systems, particularly in covering the application methods for decision-making processes. New case studies and applications on management science, operations management, social sciences, and other behavioral sciences have been incorporated into business and organizations real-world objectives.

The **Advances in Logistics, Operations, and Management Science** (ALOMS) Book Series provides a collection of reference publications on the current trends, applications, theories, and practices in the management science field. Providing relevant and current research, this series and its individual publications would be useful for academics, researchers, scholars, and practitioners interested in improving decision making models and business functions.

Coverage

- Networks
- Risk Management
- Finance
- Computing and information technologies
- Production management
- Decision analysis and decision support
- Services management
- Political Science
- Information Management
- Marketing engineering

IGI Global is currently accepting manuscripts for publication within this series. To submit a proposal for a volume in this series, please contact our Acquisition Editors at Acquisitions@igi-global.com or visit: http://www.igi-global.com/publish/.

The Advances in Logistics, Operations, and Management Science (ALOMS) Book Series (ISSN 2327-350X) is published by IGI Global, 701 E. Chocolate Avenue, Hershey, PA 17033-1240, USA, www.igi-global.com. This series is composed of titles available for purchase individually; each title is edited to be contextually exclusive from any other title within the series. For pricing and ordering information please visit http://www.igi-global.com/book-series/advances-logistics-operations-management-science/37170. Postmaster: Send all address changes to above address.

Titles in this Series

For a list of additional titles in this series, please visit: www.igi-global.com/book-series

Managerial Strategies and Green Solutions for Project Sustainability
Gilman C.K. Tam (Independent Researcher, China)
Business Science Reference • copyright 2017 • 255pp • H/C (ISBN: 9781522523710) • US $180.00 (our price)

Optimal Management Strategies in Small and Medium Enterprises
Milan B. Vemić (Higher School of Academic Studies "DOSITEJ", Serbia)
Business Science Reference • copyright 2017 • 437pp • H/C (ISBN: 9781522519492) • US $225.00 (our price)

Global Intermediation and Logistics Service Providers
Laurence Saglietto (Côte d'Azur University, France) and Cécile Cezanne (University Paris 13 Sorbonne Paris Cité, France)
Business Science Reference • copyright 2017 • 412pp • H/C (ISBN: 9781522521334) • US $210.00 (our price)

Ethics and Sustainability in Global Supply Chain Management
Ulas Akkucuk (Bogazici University, Turkey)
Business Science Reference • copyright 2017 • 350pp • H/C (ISBN: 9781522520368) • US $200.00 (our price)

Supply Chain Management in the Big Data Era
Hing Kai Chan (University of Nottingham Ningbo, China) Nachiappan Subramanian (University of Sussex, UK) and Muhammad Dan-Asabe Abdulrahman (University of Nottingham Ningbo, China)
Business Science Reference • copyright 2017 • 299pp • H/C (ISBN: 9781522509561) • US $195.00 (our price)

Knowledge Management Initiatives and Strategies in Small and Medium Enterprises
Andrea Bencsik (Széchenyi István University, Hungary & J. Selye University, Slovakia)
Business Science Reference • copyright 2017 • 442pp • H/C (ISBN: 9781522516422) • US $200.00 (our price)

Handbook of Research on Information Management for Effective Logistics and Supply Chains
George Leal Jamil (InescTec, Portugal) António Lucas Soares (University of Porto, Portugal) and Cláudio Roberto Magalhães Pessoa (FUMEC University, Brazil)
Business Science Reference • copyright 2017 • 554pp • H/C (ISBN: 9781522509738) • US $310.00 (our price)

Optimum Decision Making in Asset Management
María Carmen Carnero (University of Castilla – La Mancha, Spain) and Vicente González-Prida (University of Seville, Spain)
Business Science Reference • copyright 2017 • 523pp • H/C (ISBN: 9781522506515) • US $215.00 (our price)

IGI GLOBAL
DISSEMINATOR OF KNOWLEDGE

www.igi-global.com

701 East Chocolate Avenue, Hershey, PA 17033, USA
Tel: 717-533-8845 x100 • Fax: 717-533-8661
E-Mail: cust@igi-global.com • www.igi-global.com

Editorial Advisory Board

List of Reviewers

List of Contributors

Table of Contents

Detailed Table of Contents

Aluminium composites are tough to machine by using conventional method. In order to increase machinability of aluminium-based composite a non-conventional method of machining has been used. Electrical Discharge Machining (EDM) is one of the kind of machining process which has often uses for machining of aluminium composites. The objective of this chapter is to determine the ideal setting of the process parameters on the electrical discharge machining while machining Aluminium-Rice Husk Ash (LM25-RHA) composites in which three different variation (4%, 8%, 12%) of Rice Husk Ash has been used. The parameters considered are pulse current (Ip), gap voltage (V) and pulse-on-time (Ton); whereas its effect are analysed on Electrode Wear Rate (EWR), Material Removal Rate (MRR) and Surface Roughness (Ra). The optimal setting of the parameters are determined through experiments planned, conducted and analysed using the Taguchi method.

Analysis and simulation of manufacturing process require extensive and complicated computations. Nowadays, computer resources and computational algorithms reach to the state that can model and simulate the problem efficiently. One of the important processes in manufacturing is machining. In this research end-milling process which is one of the complex and wide-spread processes in machining is chosen. Most important parameters in end-milling are surface roughness and surface location errors. A comprehensive simulation software is developed to model end-milling process in order to anticipate finishing parameter such as surface roughness and errors. The proposed algorithm takes into account cutting conditions, such as feed, doc, woc, tool run out, etc. In addition, dynamic simulation module of the software can accurately model flexible end-mill tool, the milling cutting forces and regeneration of waviness effects. The software can accurately determine the most commonly used index of surface roughness parameters such as Ra, P.T.V. and surface errors.

Products of cold drawing process are extensively used due to their characteristics such as very good surface finish, close dimensional control and superior mechanical properties. Therefore, this process has drawn attention of several researchers over the years. The chapter describes the state of research scenario in the field of cold drawing process analysis with a focus on various parameters and methods involved in its study. Work carried out in this context is described briefly along with the steps to carry out finite element analysis. It has also accounted for methods to design experiments and to optimize the process parameters. Upon reading this chapter, the reader would get acquainted with various aspects of cold drawing process modeling and scope for further research. With the help of fundamental knowledge of process and systematic application of the available tools and techniques, the reader would be equipped with the knowledge to work on related industrial problems and enhance process capability by applying soft computing techniques in process optimization.

To assure desire quality of machined products at minimum machining costs and maximum material removal rate, it is very important to select optimum parameters when metal cutting machine tool are used. Minimum Surface Roughness (Ra) is commonly desirable for the component; however Material Removal Rate (MRR) should be maximized. This chapter presents an approach for determination of the best cutting parameters precede to minimum Ra and maximum MRR simultaneously by integrating Response Surface Methodology with Multi-Objective Technique for Order Preference by Similarity to Ideal Solution and Teaching and learning based optimization algorithm in face milling of Al-6061 alloy. 30 experiments have been conducted based on RSM with 4 parameters, namely Speed, Feed, Depth of Cut and Coolant Speed and three levels each. ANOVA is performed to find the most influential input parameters for both MRR and Ra. Later the multi-objective attribution selection method TOPSIS and multi objective optimization method TLBO is used to optimize the responses.

Conventional machining becomes non-efficient and non-effective in case of intricate shape and also while working with hard metals and alloys due to excessive tool wear. In such situations non-conventional

Artificial neural network model is applied for the prediction of the biosorption capacity of living cells of Bacillus cereus for the removal of chromium (VI) ions from aqueous solution. The maximum biosorption capacity of living cells of Bacillus cereus for chromium (VI) was found to be 89.24% at pH 7.5, equilibrium time of 60 min, biomass dosage of 6 g/L, and temperature of 30 ± 2 °C. The biosorption data of chromium (VI) ions collected from laboratory scale experimental set up is used to train a back propagation (BP) learning algorithm having 4-7-1 architecture. The model uses tangent sigmoid transfer function at input to hidden layer whereas a linear transfer function is used at output layer. The data is divided into training (75%) and testing (25%) sets. Comparison between the model results and experimental data gives a high degree of correlation R2 = 0.984 indicating that the model is able to predict the sorption efficiency with reasonable accuracy. Bacillus cereus biomass is characterized using AFM and FTIR.

This chapter gives an overview of cost effective processing techniques, employed for manufacturing of sandwich structures, primarily for high performance applications. The scope of this chapter is to review the processes which may produce the autoclave cured quality parts, without the use of autoclave. The chapter starts with an overview of autoclave processing, highlighting the different features, which depicts the capacity of technique to produce highly repeatable parts. Studies related to manufacturing of sandwich structures in autoclave, with and without adhesive film, are reviewed and discussed. Process briefs and recent developments of different out of autoclave (OOA) techniques for manufacturing of sandwich structures are then discussed which includes, OOA prepregs systems, Quickstep processing, Vacuum Infusion Process, compression molding, double belt press and filament winding. A brief overview of manufacturing processes for thermoplastic based sandwich structures is also provided at the end.

In this chapter, an attempt has been made to develop neural network models to predict the hardness distribution of hardened zone in plasma arc surface hardening process. The back propagation method with the Levenberg-Marquardt algorithm was used to train the neural network models. Hardness distributions were collected by the experimental setup in the laboratory and the associated data were used to train the neural network models. Furthermore, the prediction of neural network models were compared with those obtained from a statistical regression models. It is confirmed experimentally that the hardness distribution can be accurately predicted by the trained neural network models. The accuracy of hardness distribution prediction using neural network is superior to that using other statistical regression models.

Today, in the age of artificial intelligence and machine learning, Data mining and Image processing are two important platforms. GA and GP are value based and program based randomized searching tools respectively and these two are very much useful in the fields' data mining and image processing for handling different issues. In this chapter, a review is made on ability of GA and GP in some applications of these two fields. Here, the selected subfields of data mining are market analysis, fraud detection, risk management, sports analysis, protein interaction, classification of data, drug discovery and feature construction. The similar in image processing are enhancement and segmentation of images, face recognition, photo mosaic generation, data embedding, image pattern classification, object detection and Graphics Processor Unit (GPU) development. The efficiencies of GA and GP in these particular applications are analyzed with corresponding parameters, comparing with other non-GA and non-GP approaches of the corresponding subfields.

In the era of globalization and industrialization the concern is limited only in development, without taking the environment into consideration this leads to global warming and big ecological changes in recent year. The material like Synthetic materials used in many applications due to ease of fabrication and higher strength, but the major disadvantage with it is, its neither recyclable nor bio-degradable. Therefore, the researchers develop a new material and technique for the sustainable development. A lot of researches were carried out in the reinforcing potential in the polymer matrix composite, reinforcing can be of two kinds synthetic and natural fiber. Natural fiber is gaining importance in the last decade due to its ecofriendly nature and does not leave carbon foot print, for better utilization of banana and jute fiber for making value added products. Hence, in this work the objective is to develop a new class of hybrid nano-materials from natural fiber such as banana and jute fiber. This chapter sees an opportunity of enhancement of interface property.

Weight reduction of automotive components by tailoring materials is the state of the art. This basically has resulted in the development of advanced joining methods like clinching, friction stir welding, self-pierced riveting etc. to assemble similar or dissimilar materials, with significant change in sheet properties. In the present work, the main aim is to predict the temperature evolution during Self-Pierced

Preface

Manufacturing is essential for any nation in the creation of wealth, jobs, opportunities, trades, stability, growth and raising the standard of living. It is essential in today's growing need of quality, process safety, minimal manufacturing cost and low manufacturing lead time. Manufacturing must be competitive, therefore, to meet the need, the manufacturing process parameter need to be selected suitability. The parametric selection plays an important role in ensuring the product quality, manufacturing cost and productivity of manufacturing processes.

The use of modelling technique and computational techniques is increasing rapidly in the manufacturing sector, due to its capability to reduce expenses and the time for manufacturing defect free good quality products. With the increasing cost of energy, increasingly stringent environmental regulation, global competitiveness in product pricing and quality of optimization concept plays an important role to address the issues.

The book is likely to be very helpful to the designers and manufacturing engineers working in the manufacturing division liable for the technical aspects of comprehending a product as it grants models and improvement techniques to make their jobs easier, logical, efficient and effective. The book is planned for designers, producing engineers, practitioners, managers, institutes concerned in the design and manufacturing connected comes, applied analysis staff, academics, and graduate students in mechanical, industrial, and manufacturing engineering.

The purpose of this book is to present an inclusive review on recent research and development trends for modelling and optimization of manufacturing processes, particularly the machining processes which are the most frequently analysed manufacturing processes.

This book has been divided into four sections.

SECTION 1

Chapter 1 deals with the machinability of aluminium-based composite using Electrical Discharge Machining process. The experiments have been conducted to optimal setting of the parameters are determined through experiments planned, conducted and analysed using the Taguchi method.

In Chapter 2, comprehensive simulation software is developed to model end-milling process in order to anticipate finishing parameter such as surface roughness and errors.

Chapter 3 describes the state of research scenario in the field of cold drawing process analysis with a focus on various parameters and methods involved in its study. This context is described briefly along with the steps to carry out finite element analysis. It has also accounted for methods to design experiments and to optimize the process parameters.

Chapter 4 presents an approach for determination of the best cutting parameters precede to minimum Ra and maximum MRR simultaneously by integrating Response Surface Methodology with Multi-Objective Technique for Order Preference by Similarity to Ideal Solution and Teaching and learning based optimization algorithm in face milling of Al-6061 alloy.

Chapter 5 used EN31 steel was machined using Plasma Arc Cutting with pre-defined process parameters. Material Removal Rate and Surface roughness were considered as responses for the study using Taguchi, Grey Relational Analysis and Artificial Bee Colony (ABC) Algorithm.

Chapter 6 discusses the use of electrical discharge drilling in high aspect micro holes are drilled in die steel (of thickness 15 mm) using tubular electrodes of diameter 500µm. Using the Taguchi design of experimental method, the effect of four process parameters the responses.

In Chapter 7, experimental investigation, parametric modelling and optimization of laser transmission welding of polycarbonate has been carried out. Experiments were conducted using response surface methodology. Mathematical models are developed using RSM for establishing an interrelationship between the process parameters and the responses of interest. The chicken swarm optimization algorithm is employed with the developed RSM models to set the optimal process parameters combinations to achieve desired weld qualities.

In Chapter 8, hard turning with multilayer coated carbide tool has several benefits over grinding process such as, reduction of processing cost and increased productivity. The objective was to establish a correlation between cutting parameters with cutting force, tool wear and surface roughness on workpiece.

SECTION 2

In Chapter 9, machinability of AISI H13 steel with TiC/TiCN/Al2O3 coated carbide tool using statistical techniques. An attempt has been made to analyse the effects of process parameters on machinability aspects using design of experiments. It was observed that the cutting tool wears increases almost linearly with increase in cutting speed and feed rate. The combination of low feed rate and high cutting speed is necessary for minimizing the surface roughness.

Chapter 10 presents the brief review of recent trends in research on bearing defects, sources of vibration and vibration measurement techniques in time domain, frequency domain and time-frequency domain and the detailed explanation of defect detection through scalar indicators in time domain.

In Chapter 11, the results of a survey applied to 372 maquiladora workers of Ciudad Juárez (México) are reported. Six hypotheses are proposed and validated through a structural equations model executed in WarpPLS 5.0® software, where four latent variables related to Management Commitment, Human Resources, Suppliers and Quality benefits in a JIT environment are associated. Direct effects, indirect effects and total effects of each one of the variables are analysed.

Chapter 12 investigated the effects of mixing Polypropylene waste plastics in the bituminous mix for the design of Flexible Pavement. In this study, out stability and flow tests on different specimens were made with varying bitumen and polypropylene plastic content. From the initial test results, the optimum bitumen and plastic contents are found. Further, the test results obtained from Marshall Test are modelled by identifying various input variables, which are various physical properties of the mix such as plastic content, bitumen content, air voids and unit weight.

Chapter 13 applied an artificial neural network model for the prediction of the biosorption capacity of living cells of Bacillus cereus for the removal of chromium (VI) ions from aqueous solution. The maximum biosorption capacity of living cells of Bacillus cereus for chromium (VI) was found to be 89.24% at pH 7.5, equilibrium time of 60 min, biomass dosage of 6 g/L, and temperature of 30 ± 2 °C. The bio sorption data of chromium (VI) ions collected from laboratory scale experimental set up is used to train a back propagation (BP) learning algorithm having 4-7-1 architecture.

SECTION 3

Chapter 14 gives an overview of cost effective processing techniques, employed for manufacturing of sandwich structures, primarily for high performance applications. The scope of this chapter is to review the processes which may produce the autoclave cured quality parts, without the use of an autoclave. The chapter starts with an overview of autoclave processing, highlighting the different features, which depicts the capacity of technique to produce highly repeatable parts. Studies related to manufacturing of sandwich structures in autoclave, with and without adhesive film, are reviewed and discussed.

In Chapter 15, an attempt has been made to develop neural network models to predict the hardness distribution of hardened zone in plasma arc surface hardening process. The back propagation method with the Levenberg-Marquardt algorithm was used to train the neural network models. Hardness distributions were collected by the experimental setup in the laboratory and the associated data were used to train the neural network models. Furthermore, the prediction of neural network models was compared with those obtained from a statistical regression models.

In Chapter 16, a review is made on ability of GA and GP in some applications of these two fields. Here, the selected subfields of data mining are market analysis, fraud detection, risk management, sports analysis, protein interaction, classification of data, drug discovery and feature construction. The similar in image processing are an enhancement and segmentation of images, face recognition, photo mosaic generation, data embedding, image pattern classification, object detection and Graphics Processor Unit (GPU) development.

Chapter 17 discussed a new class of hybrid Nano materials from natural fiber such as banana and jute fiber. This chapter sees an opportunity of enhancement of interface property.

SECTION 4

The main aim of Chapter 18 is to predict the temperature evolution during Self-Pierced Riveting (SPR) of sheets by Finite Element (FE) analyses. Load evolution is also predicted. Generally, temperature estimation during SPR is not attempted. The influence of a few selected SPR parameters has been studied on the temperature and load evolution through FE simulations. The relationship between these parameters and the temperature and the load evolution are revealed.

Chapter 19 delineates the effect of incubation temperature on mechanical and electrical properties of ketchup. Microstructure analysis of ketchup suggested the presence of cell wall components. In gist, a variation in incubation temperature of the tomato ketchup can allow modulating not only the mechanical and viscoelastic properties, but also the electrical properties of ketchup.

Chapter 20 proposed hybrid technique is based on data farming for intermediate data generation and the ANN model for better learning and forecasting. The performance of the proposed model has been tested with actual pertaining to water quality indices of various water samples collected from different sources.

Chapter 21 proposes the fabrication of thin-film, aluminium electrodes on a glass substrate by thermal evaporation method. The electrodes were fabricated having gaps of varied lengths. Different electrical models were used to determine the electrical stability of the system using a LabVIEW program which was developed in-lab. The electrode having the best stability was used for EIS analysis of distilled water and saline solutions of different strengths to validate its proper functioning.

Chapter 22 explains the artificial intelligence techniques in terms of artificial neural networks, fuzzy logic, expert systems, machine learning, genetic programming, evolutionary polynomial regression (EPR), and support vector machine (SVM); the AI applications in modern education; the AI applications in software engineering development; the AI applications in content-based image retrieval (CBIR); and the multifaceted applications of AI in the digital age. AI is a branch of science which deals with helping machines finds the suitable solutions to complex problems in a more human-like manner.

We would like to thank IGI global for the opportunity to publish this book and for their competence and support. Finally, we will like to thank all authors of this book for taking pain in writing their chapters in this book.

Mohan Kumar Pradhan
Maulana Azad National Institute of Technology, Bhopal, India

Raja Das
VIT University, India

Acknowledgment

While writing, the contributors referred to several books and journals; we take this opportunity to thank all those authors and publishers. Furthermore, we are thankful to both Maulana Azad National Institution of Technology and VIT University, India, for providing facilities to complete this project. We are thankful to the editorial board and reviewers for their support during the process of assessment. Last but not least, we thank the production team of IGI Global for encouraging us and extending their cooperation and help in timely completion of this edited book.

Mohan Kumar Pradhan
Maulana Azad National Institute of Technology, Bhopal, India

Raja Das
VIT University, India

Section 1

Chapter 1
Optimisation of Machining Parameters in Electrical Discharge Machining of LM25-RHA Composites

Srikant Tiwari
Maulana Azad National Institute of Technology, Bhopal, India

Mohan Kumar Pradhan
Maulana Azad National Institute of Technology, Bhopal, India

ABSTRACT

Aluminium composites are tough to machine by using conventional method. In order to increase machinability of aluminium-based composite a non-conventional method of machining has been used. Electrical Discharge Machining (EDM) is one of the kind of machining process which has often uses for machining of aluminium composites. The objective of this chapter is to determine the ideal setting of the process parameters on the electrical discharge machining while machining Aluminium-Rice Husk Ash (LM25-RHA) composites in which three different variation (4%, 8%, 12%) of Rice Husk Ash has been used. The parameters considered are pulse current (Ip), gap voltage (V) and pulse-on-time (Ton); whereas its effect are analysed on Electrode Wear Rate (EWR), Material Removal Rate (MRR) and Surface Roughness (Ra). The optimal setting of the parameters are determined through experiments planned, conducted and analysed using the Taguchi method.

INTRODUCTION

Composite Material

A composite material consists of two or more than two chemically and physically artificial phases. Composite materials, as name tells that it is the composition of materials also known as composition materials, these are the materials which has produced by mixing different constituent with significantly

DOI: 10.4018/978-1-5225-2440-3.ch001

different physical or chemical properties that, whose combination produces a material with properties different from the individual materials. The constituent elements mainly consist of a reinforcement material, fillers, and a composite matrix binder which varied in composition or in form on a macro scale. Although constituent element remains mixture of constituent which are not dissolve completely. It accommodate the requirement of particular function and design, imparted with ambitious properties which seems that it restrict the range for classification of composites, but in the reality new varieties of composites has discovered, each have their own specific features and purpose like the particulate, flake, laminar and filled composites.

Classification of Composites

Composite is classified mainly into two categories, it is based on phases used in preparation:

1. Matrix Phase:
    a. Organic Matrix Composite (OMC).
    b. Ceramic Matrix Composite (CMC).
    c. Metal Matrix Composite (MMC).
2. Reinforcement phase:
    a. Fiber Reinforced Composite.
    b. Laminated Composite.
    c. Particulate Composites.
    d. Filler Composite.
    e. Flakes Composite.

In this study our main focus is on the Metal Matrix Composite because machining has been done on the LM25-RHA Composite which is a MMC. Before studying MMC, a brief knowledge of Matrix and Reinforcement is necessary.

Matrix

Matrix is the complete continuous material into which reinforcement has implanted. It is a types of monolithic material. It provides path through the matrix to any point in the material, unlike two materials sandwiched together. Aluminum, magnesium and titanium due to its lighter weight can be used as matrix material in structural application and provides an acquiescent support to the reinforcement. Cobalt and its alloys matrices are usually applied as for high temperature application by Keneth kanyo et al. (2013).

Reinforcement

The reinforcement material has dispersed into a matrix. It is not necessary that reinforcement always play role of only structural task (reinforcing the compound). Reinforcement is also capable of changing physical properties of the composite such as resistance to wear frictional coefficient, or thermal conductivity. It may be continuous or discontinuous. In Discontinuous metal standard metal working technique can be applied, such as extrusion, forging, or rolling due to its isotropic nature. Further, it is machinable by using conventional techniques, but only problem is to use polycrystalline diamond tooling (PCD).

Figure 1. Composite based on matrix phaseE. Metal matrix composites

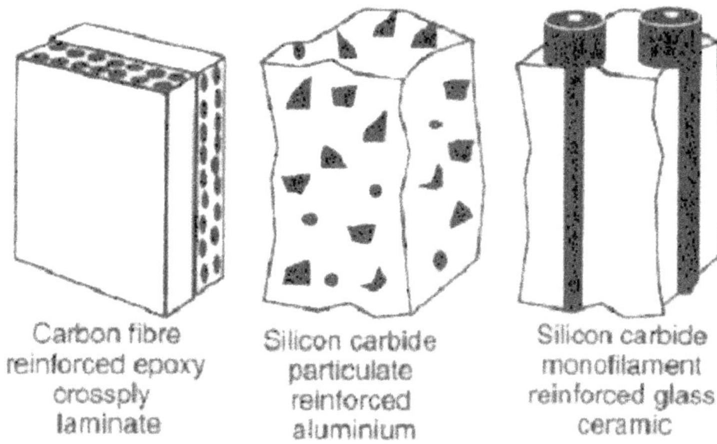

Continuous reinforcement generally uses monofilament fibers such as carbon fiber or silicon carbide. In this type of reinforcement, the fibers are rooted into the matrix in a particular direction, which result into an anisotropic structure in which particle alignment of the material affects its strength. One of the first metal matrix composite has used boron filament as reinforcing agent. Whiskers short fibers, or particles has used in the discontinuous reinforcement. Alumina and silicon carbide has most commonly reinforcing agent used in the MMCs by Narasaraju and Raju (2015).

Metal Matrix Composites

Metal matrix composites typically use SiC particulates embedded in a matrix made from an alloy of aluminum and magnesium, but now a days other matrix materials like copper, titanium and iron are increasingly being used. Typical applications of metal matrix composites are in golf balls, bicycles, and missile guidance system. There is a broad family of materials in metal matrix composite. The basic aim is to achieve an advance combination of properties and this can be attained by the adding different level of matrix materials and varieties of reinforcing phases. It is not only stable matrix microstructure, reinforcing phase also controls the characteristics attained by the MMCs by Saravanan and Kumar (2013).

Figure 2. Composite based on reinforcement phase

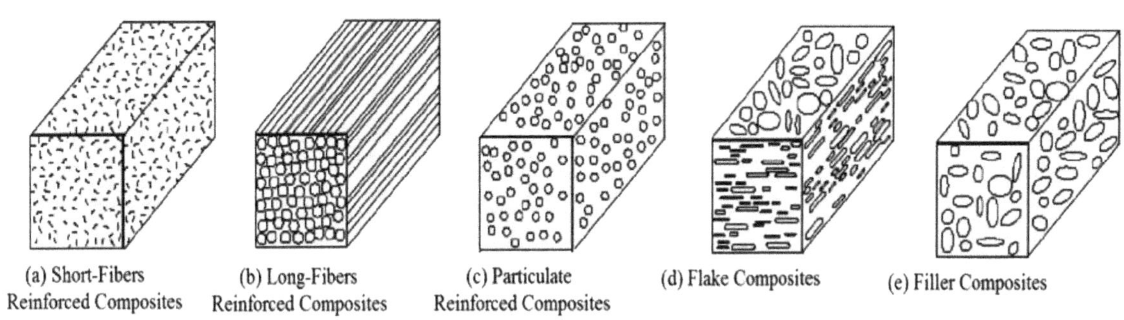

Matrixes reinforced with high modules short fibers, whiskers or particulates have shown increment in strength and stiffness and are isotropic in nature. These types of the MMCs consumes less capital to produce. In the case of continuous fiber or whiskers with high aspect ratio (length to diameter), to align fibers in a particular stress direction and to transfer the applied load to the fibers, the matrix serves to bond them together. The mechanical characteristics of the composites are mainly dependent upon the ability of the matrix in order to transfer the load to the reinforcement fibers and are therefore related to the quality of the fiber/matrix inter facial bonding by Alaneme et al. (2013). This type of composites contains significantly higher strength and stiffness. Nevertheless, it is non-isotropic in nature and are expensive. The MMCs find use in industries and they include such areas of application as aerospace, defense, automotive, marine etc. when compared with the unreinforced alloy, MMCs in general have superior mechanical properties such as better strength, high stiffness, excellent wear resistance and strophic elevated temperature properties. These properties can be tailored to meet specific requirements. MMCs are manufactured by dispersing a reinforcing material into a metal matrix. In order to prevent surface from chemical reaction, it is coated with matrix. For example, carbon fibers are usually used in aluminium matrix to create composites showing low density and high strength. However, due to reaction of carbon with aluminium a brittle and water soluble compound Al4C3 has generated on the surface of the fiber. In order to restrict this reaction, nickel or titanium boride coating taken place at carbon fibers by Alaneme et al. (2013).

Rice Husk Ash

Rice Husk (RH) is the outer layer of the paddy crop. It is enormously available throughout the world wide as an agro waste. Rice husk is widely used in the industries as a fuel. Due to burning of rice husk, its ash form has produced known as Rice Husk Ash (RHA). In the recent years, researchers are trying to dope it into different material in order to generate a new class of compound with better properties so that it can able to meet the recent demand from material. Its abundances and low cost of rice husk acts as a catalyst in preparation of advance material. RHA mainly contains silica as a major component, due to which RH as well as RHA has used as a prime reinforcing agent in order to improve the mechanical properties of aluminium alloy by Alaneme et al. (2015). Rice husk is taken from any city in India, then it has washed with the help of water to eliminate the dust and dried at room temperature for 1 day. After it, washed rice husk is then heated to 200 o C temperature for 1 hour duration to remove the moisture and organic matter present in the RH. Due to heat treatment, the colour of the husk is transformed from yellowish to black because of burning of organic matter. It is further heated to 600 o C for 12 hour to remove the carbonaceous material. After this operation, once again the colour is changed from black to greyish white. The burnt ash contains silica in rich quantity and used as a reinforcement material for the preparation of composites. The percentage of ash in rice husk is about 10-25% higher than other biomass fuels, it has 85-87% silica content which shows the high level of porosity, less weight, and high external surface area. Some of the ingredients of RHA are potassium oxide, aluminum oxide, calcium oxide, magnesium oxide, sodium oxide, iron oxide, which is available in less than 1% by weight. The factors which shows the effect on ash properties are burning conditions (temperature and duration), heating rate, the technique of an incineration variety of crop and fertilizer used in the production of rice in agricultural field. The structural transformations have taken place in silica is mainly depends on the combustion condition like time and temperature. A typical analysis of rice husk ash is shown in Table 1.

Table 1. Composition of rice husk ash

Constituent	SiO2	Al2O3	Fe2O3	CaO	MgO	N20	K2O
%	94.04	0.249	0.136	0.622	0.44	0.02	2.49

LITERATURE REVIEW

Kannan and Ramanathan (2014) have investigated the optimisation of EDM parameters of Al-TiC composites by Taguchi Method. The objective of this work is to investigate the effect of current, pulse on-time (POT) and flushing pressure on Metal removal Rate, Tool Wear Rate during electrical discharge machining of as sintered Al-TiC MMC (5% reinforcement) is prepared by in-situ technique by synthesis route using stir casting furnace. The use of kerosene as dielectric fluid was employed in the present investigation. An L9 orthogonal array was formed to examine both metal removal rate and tool wear rate. Analysis of variance (ANOVA) was performed to find the validity of the experimental plan followed in the work. The optimization process involves the studying the response based on the combinations, estimating the coefficients, fitting the experimental data, predicting the response and checking the adequacy of the fitted model. Discharge current, Pulse on Time and Flushing Pressure were chosen as the independent variables and wear ratio was selected as response variable for the composite.

Vikas et al. (2013) presents an idea about the effect of the various input parameters like Pulse on time, Pulse off time, discharge current and gap voltage over the over-cut for an EN41 material using Taguchi method. The entire experimental process was carried out in Die Sinking Electrical Discharge Machine (EMT 43 Electronica Machine Tools) and the effect of these input parameters were examined carefully using the Taguchi method. Here, an orthogonal L 27 array was formulated, which used the different combination of the same 4-input parameters (i.e. Pulse ON time, Pulse OFF time, discharge current and gap voltage) to find optimized value of the output (i.e. over-cut). The level of importance and hence, the rank of each of these parameters were obtained statically using the ANOVA table through the analysis of the S/N ratio. It was found that the discharge current had larger impact over the over-cut, followed by Pulse ON time. The study also compares the theoretical and the experimental value of over-cut. The following work finds a huge application in various manufacturing industries.

George et al. (2004) has determined the optimal setting of the process parameters on the electrical discharge machining (EDM) machine while machining carbon composites. The parameters considered are pulse current, gap voltage and pulse-on-time; whereas the responses are electrode wear rate and material removal rate. The optimal setting of the parameters are determined through experiments planned, conducted and analysed using the Taguchi method. It is found that the electrode wear rate reduces substantially, with in the region of experimentation, if the parameters are set at their lowest values, while the parameters set at their highest values increase the MRR drastically.

Velmurugan et al. (2011) has investigated the effect of parameters like Current, Pulse on time, Voltage and Flushing pressure on metal removal rate,tool wear rate as well as surface roughness on the machining of hybrid Al6061 metal matrix composites reinforced with 10% SiC and 4%graphite particles. Composite was fabricated using stir casting process. A central composite rotatable design was selected for conducting experiments. Mathematical models were developed using the MINITAB R14 software. The method of least squares technique was used to calculate the regression coefficients and Analysis of

Variance (ANOVA) technique was used to check the significance of the models developed. Scanning Electron Microscope (SEM) analysis was done to study the surface characteristics of the machined specimens and correlated with the models developed.

Rajkumar et al. (2014) has prepared aluminium metal matrix composites by using Al 6061 as the matrix material in which boron carbide, one of the hardest material and graphite are reinforced and the composite has been prepared by stir casting. Microwave post heat treatment of metal has been used which is a newer technique employed in this study. It provides unique microstructure and properties as a result of selective heating. The composites were heated up to 520 o C and were maintained for a definite holding time at 520 o C. Electrical discharge machining process has selected to machine MMC as it is difficult to machine by conventional machining. EDM machinability studies were carried out on the heat treated Al-B4C (15%) graphite (5%) and compared with EDM machined with conventionally heat treated composites. EDM process parameters optimization was carried out using Taguchi design with input parameter such as pulse on time, pulse off time and current density for the MRR and TWR as response parameters. Microwave heat treated composite exhibited higher hardness than the conventionally heat treated composites. Pulse off time and pulse on time was a non-linear function for the material removal rate of the composite and the most influential parameters for the MRR were determined.

Kumar et al. (2015) has prepared Aluminum2618 composite by reinforcing AlN (Aluminum Nitride), Si3N4 (Silicon Nitride) and ZrB2(Zirconium Boride) particles in wt% (x = 0,2,4,6,8) by stir casting method. The mechanical properties of the composite is directly proportional to weight percentage. Electrical Discharge Machining has been employed for machining metal matrix composites on the Aluminum 2618 composite work piece using a copper electrode. Effect of Current, Pulse on time, Pulse off time on Metal Removal Rate, Tool Wear Rate has been analysed by using Taghuchi's design of experiment. Machining results shows that Al2618 composites have improved mechanical properties and as a result of Material Removal Rate and Tool Wear Rate are reduced. Hence ANOVA (Analysis of Variance) and signal to Noise ratio are used to determine the influence of input parameters on the Material Removal Rate and Tool Wear Rate.

Kandpal et al. (2015) has discussed the effect of different input parameters like Pulse on time, Gap voltage, Current and Pulse off time on the machining of Aluminium Composites, responses are MRR, TWR and Ra. In this discussion about EDM process, its parameters and role of EDM process in machining of aluminium based metal matrix composite materials.

Razak et al. (2016) has performed machining on Magnesium Alloy. Due to limitations in conventional machining methods, non-traditional machining method such as electrical discharge machining die sinking process has used to produce intricate shape with tight tolerance on magnesium alloy. Nine EDM experiments with three levels and four parameters has been conducted using Taguchi method on AZ31 magnesium alloy to explore the optimum machining parameters. It has found that pulse on-time was the most prominent parameter affecting the Ra of the machined surface. The optimum EDM condition obtained was 47 A peak current, 80 V voltage, 16 s pulse on-time and 512 s pulse off-time. A confirmation test has conducted and the result shows 95.5% similarity with the predicted Ra. However, the formation of cracks and craters has been found on the machined surface area and proposed to solve this problem by applying powder mixed EDM method in future research work.

Pramanik et al. (2015) has performed wire electrical discharge machining of 6061 aluminium alloy in terms of material removal rate, kerf/slit width, surface finish and wear of electrode wire for different pulse on time and wire tension. Eight experiments have been carried out in a wire EDM machine by varying pulse on time and wire tension. It is found that the material removal rate increases with the increase of pulse on time though the wire tension does not affect the material removal rate. It seems that the higher wire tension facilitates steady machining process, which generates low wear in wire electrode and better surface finish. The surface roughness does not change notably with the variation of pulse on time. The appearance of the machined surfaces is very similar under all the machining conditions. The machined surface contains solidified molten material, splash of materials and blisters. The increase of the pulse on time increases the wear of wire electrode due to the increase of heat input. The wear of wire electrode generates tapered slot which has higher kerf width at top side than that at bottom side. The higher electrode wear introduces higher taper.

Ugrasen et al. (2015) has done Wire Electrical Discharge Machining (WEDM) on the HCHCr material. The study outlines the optimization of process parameters in the wire EDM. This work material has been machined using different process parameters based on Taguchi's L27 orthogonal array. Parameters such as pulse-on, pulse-off, current and bed speed has varied. The response variables measured for the analysis are surface roughness, volumetric material removal rate and accuracy. ANOVA has been carried out to know the magnitude of factor affects. Further the verification experiment has been carried out to confirm the performance of optimum parameters.

Talla et al. (2015) has performed EDM on the Aluminium Composites. In this work, an attempt has been made to fabricate and machine aluminum/alumina MMC using EDM by adding aluminum powder in kerosene dielectric. Results shows that an increase in MRR and decrease in surface roughness compared to those for conventional machining process. Semi empirical models for MRR and Ra based on machining parameters and important thermo physical properties has been established using a hybrid approach of dimensional and regression analysis. A multi response optimization has also performed using principal component analysis-based grey technique (Grey-PCA) to determine optimum settings of process parameters for maximum MRR and minimum Ra within the experimental range. The recommended setting of process parameters for the proposed process has been found to be powder concentration (Cp) = 4 g/l, peak current= 3 A, pulse on time= 150 microsecond and duty cycle= 85%.

Mahapatra and Patnaik (2007) have investigated the effect of the WEDM process parameters on the surface roughness average and the kerf width of the stainless steel (SS304). Nine experimental runs based on an orthogonal array of Taguchi method are performed and grey relational analysis method is subsequently applied to determine an optimal WEDM parameter setting. Surface roughness and kerf width has been selected as the quality targets. An optimal parameter combination of the WEDM process is obtained using Grey relational analysis. By analyzing the Grey relational grade matrix, the degree of influence for each controllable process factor onto individual quality targets has been found. The pulse ON time is found to be the most influential factor for both the surface roughness and the kerf width. Further, the results of the analysis of variance reveals that the Pulse on time is the most significant controlled factor for affecting the multiple responses in the WEDM according to the weighted sum grade of the surface roughness and the kerf width.

Gaikwad et al. (2016) have performed Electrical Discharge Machining on Nickel-Titanium alloy, it is one such class of advanced material which has unique properties such as bio compatibility, high strength, high corrosion resistance, shape memory effect etc. Due to such property these alloys have wide application in the field of defense, aerospace, and medicine. As these applications required high accuracy, precision and high strength of NiTi these are difficult to machine by conventional machining processes. Hence to machine this advanced material non-conventional machining processes i.e. electric discharge machining. Optimization of EDM process parameter for maximization of material removal rate while machining of NiTi alloy. In the present study gap current, pulse on time, pulse off time, workpiece electrical conductivity, and tool conductivity were considered as process variables. Experiments were carried out as per Taguchis L36 orthogonal array. Based on the analysis it was found that work electrical conductivity, gap current and pulse on time are the significant parameters that affect the material removal rate. The optimized material removal rate obtained was 7.0806 mm 3 /min based on optimal setting of input parameter.

Lodhi et al. (2014) have performed Wire Electric Discharge Machining on AISI D3 Steel. Optimization of the machining conditions for surface roughness based on (L9 Orthogonal Array) Taguchi methodology. Experiments has carried out under varying Pulse-on-time, Pulse-off-time, Peak current, and Wire feed. An orthogonal array, the signal-to-noise (S/N) ratio, and the analysis of variance has employed to the study the surface roughness in the WEDM of AISI D3 Steel. It has been observed that the discharge current was the most influential factors on the surface roughness. To validate the study, confirmation experiment has been carried out at optimum set of parameters and predicted results have been found to be in good agreement with experimental findings.

EXPERIMENTATION

Synthesis of LM25-Rice Husk Ash Composite

Synthesis of aluminium-rice husk ash composite has done by using stir casting method or vortex method. It has furnace capacity of 10 kg. Initially aluminium alloy LM25 has put into the graphite crucible and melt it upto 730 o C in muffle furnace. LM25 melted in graphite pot has mechanically stirred by using stirrer which is driven by electric motor with rotational speed of 400 rpm around, stirring speed is adjusted dynamometer. Before incorporating rice husk ash into the melt it should be heated up to 600 o C in the furnace. The reason behind this heating is to get uniform mixing of RHA into the melt. Now the preheated particle has mixed into the melt with a certain feed rate of about 1gm/sec and continue stirring with 400 rpm stirrer. This process continues till temperature of melt increased to 740 o C. One thing should be noted that before the addition of powder, it is necessary to add 1% by weight magnesium into the melt. The reason behind addition is to conserve the amount of magnesium into the alloy because there is loss of Mg takes place due to oxidation during heating. Now add powder into the melt using baffle and allow stirring the melt up to 20 minutes. This much time of stirring increases the melt temperature up to 760 o C. Simultaneously, degassing has to be done in order to remove hydrogen entrapment into the melt, if it is not removed then it will leads to casting defects. The process of degassing has done by using N2 gas cylinder which passes through H2SO4. Sulphuric acid acts like a purifier and hydrogen gas has removed from the melt. The process of degassing will take place simultaneously with melting about 20 minutes. Then after mould has to be heated up to 500 o C for uniform solidification. Finally pour melt into the mould for sample. Repeat the process for three different percentage of rice husk ash.

Electrical Discharge Machining

Electrical Discharge Machining is the process of removal of material by spark erosion by an interrupted electric spark discharging between tool and the work piece. It has remove the material by process of melting and thermal evaporation. There is two electrode one is called tool electrode and other is work piece electrode, a rapidly and periodically current is generated between the two electrodes through dielectric medium. Gap between the electrodes are filled with the dielectric of high dielectric constant. Tool acts as cathode and work piece is behaving like anode. There is no actual contact between tool and work piece. As the voltage between tool and work piece increases, electric field intensity between cathode and anode becomes greater than dielectric strength due to which dielectric breaks and current flows between the electrodes, which results the material removal. It is same as breakdown of capacitor due to which infinite charges flow from one plate to other. Current flow time is depend upon type of generator used, then a new dielectric fluid has entered into the inter electrode gap and remove the particles or debris and restore the insulating properties of dielectric, this phenomenon is called flushing. It is mentioned that when dielectric breaks down and emits electron and ionising the gap, a local temperature of 10000 o C has developed at spark point. Dielectric fluid plays vital role in EDM, it does not conduct current under normal condition. It behaves as an insulator in high voltage condition. It insulates, cools the electrodes and work piece convey the spark and flushes the removed rate. Common dielectric are transformer oil, paraffin oil, kerosene oil, lubricant oil. These liquids have very high flash point.

Electrical Discharge Machine

Electronica-EZMC machine has used in experimentation. It has computer numeric control system in Z-direction. The most suitable dielectric fluid used in the EDM is IPOL Spark Erosion 450. It is product of Shah Petroleum. It has oil density of 0.85 and flash point of 130 o C. Brass is the tool material used in the process and LM25-RHA composite has used as work piece. Polarity of the electrode is negative and that of the composite is positive.

Material

The material used for this work is aluminium-rice husk ash composite slab of size 60mm60mm20mm prepared by using rice husk ash as reinforcing agent and LM25 aluminium alloy as matrix. Three sample of composite have taken in which different percentage of Rice husk ash namely (4%, 8%, and 12%) has used. The electrode used is electrolytic brass of 8930 kg/m 3 density with a melting point of 1025 o C. These electrodes are cylindrical in shape with a nominal diameter of 1.6 mm. The machine used in testing is Electronica-EZMC machine with NC control in the Z-direction. The dielectric fluid used is IPOL Spark Erosion 450. Tool is act like cathode and composite behaves like anode.

Experimental Planning

The investigation of the responses affected by multi-variables, Taguchi approach has been used which is very effective for dealing such types of experimentation. It is method of powerful Design of Experiments tool, which gives a simple, efficient and systematic approach to find optimal machining parameters. In comparison to the conventional method to experimentation, this method provides reduced number of experiments that are used to model the response functions. Conventional method of experimentation involves one factor at a time experiments, where one variable is changed while remaining are held constant. One of major disadvantage of this method is that it fails to take any possible relation between the parameters. It is the failure of one factor to give the same effect on the response at different levels of another factor. It is not possible to study all factors and finds individual effect in single experiment. Taguchi method is able to tackles all drawback which comes in traditional machining. The main effect is the average value of the response function at a certain levels of a parameter. The effect of a different factor level is the deviation it causes from the overall mean response. The Taguchi method is formulated for optimization of process and identification of optimal combinations of factors for responses. The steps are as follows:

- Firstly analyses the response functions and the process parameters involve in the experiment.
- Then find number of levels for the process parameters and possible interaction between them.
- Select the suitable orthogonal array and allocate the process parameters to the orthogonal array and perform the experiments according to it.
- Observe the results from experimentation and analyse it, select the optimum condition of process parameters.
- Verify the optimal condition of process parameters through a confirmatory experiment. The process parameters taken for the experimentation are Ip, Ton, and V while the response functions are electrode wear rate, surface roughness and MRR. According to the ability of the commercial EDM machine available and general approvals of machining conditions for Al-RHA composites the range and the number of different level has selected is given in Table 2.

The selection of the orthogonal array is totally depend on the condition on the numbers of level available and control factors in the experiment and obtained the orthogonal array should be higher than or equal to those for the process parameters. In the present experimentation, an L9 orthogonal array system with total four columns and nine rows has used. This array can handle three-level process parameters and has eight degrees of freedom. Hence, only nine experiments are required to examine the entire machining parameters using the L9 orthogonal array. The experimental layout to perform machining parameters using the L9 orthogonal array system is given in Table 3.

RESULT AND DISCUSSION

Analysis of Material Removal Rate

The MRR has calculated by difference of rate of weight loss in grams of sample which is divided by density of machined specimen. The weight measurement has taken from digital weighing machine of accuracy 0.0001g. Before weighing, the specimen has washed in order to remove dirt particle from the surface, otherwise error in result may occur. The duration of machining is in minutes and material removal rate has unit of mm 3 /min. The variation of MRR with respect to the machining parameters is depicted in Figure 4. The MRR of the material has increases marginally by an amount of (0.08723 mm

Table 2. Machining parameters and its level

Symbol	Machining Parameter	L1	L2	L3
Ip	Pulse current	0.5	1	1.5
Ton	Pulse on time	5	10	15
Voltage	Gap voltage	45	50	55

3 /min) as weight percentage of RHA has increased from 4% to 12%. Further, as increasing the pulse on time and voltage there is slightly increment of (0.175mm 3 /min) upto 10 μs for pulse on time and drops by increasing the value till 15 μs. While MRR has increased with voltage by (0.271mm 3 /min) is almost constant from 50 to 55 volt. The pulse current has very drastic effect on MRR, it increases very sharply by (1.585mm 3/min) by increasing the value from 0.5 to 1.5 Amperes. But the rate of increase in 1 to 1.5 Ampere is high as compare to initial. The reason behind this increase in spark between tool and work piece, which increases the metal removal rate. The machining of LM25-RHA composite is shown in Figure 3.

Regression Analysis of MRR

From residual plot graph as shown in Figure 5, all readings calculated experimentally are almost close to line drawn on 45 o in normal probability graph which shows that readings are exact and linear. In versus fit graph since all values lie between -2 to 2 hence it also concludes that values obtained are genuine.

Table 3. Orthogonal array

Wt%	Ip	Ton	Voltage
4	0.5	5	45
4	1	10	50
4	1.5	15	55
8	0.5	10	55
8	1	15	45
8	1.5	5	50
12	0.5	15	50
12	1	5	45
12	1.5	10	40

Figure 3. Machining of sample

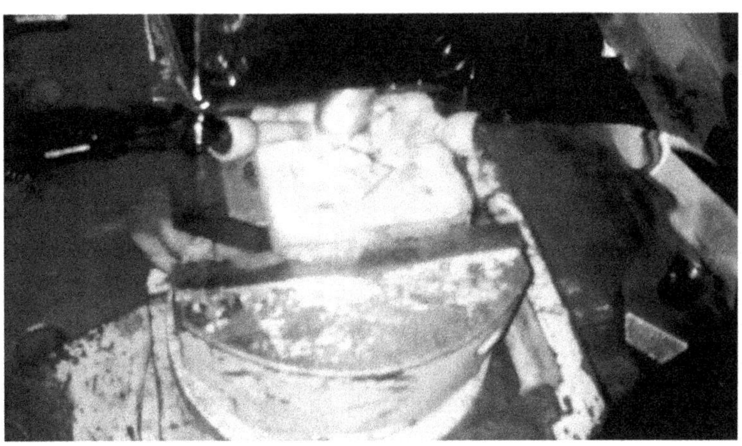

Figure 4. Variation of MRR with parameters

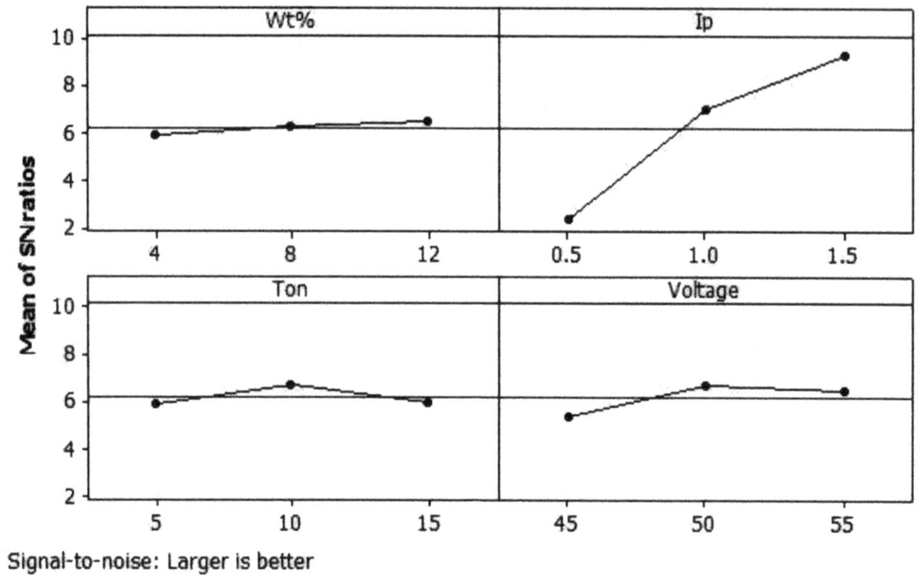

Figure 5. Residual plot for MRR

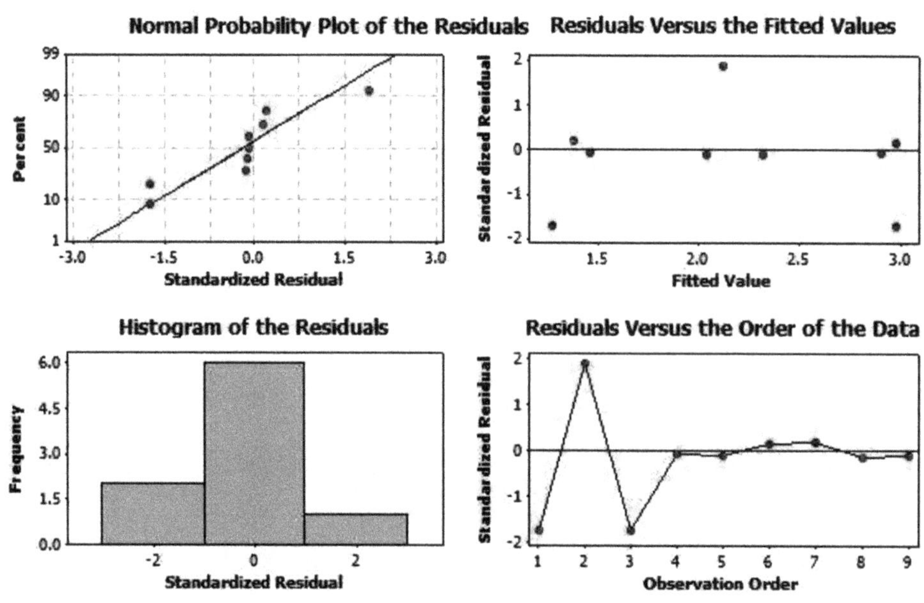

From histogram chart, it is concluded that frequency of experimental values is more between 0 to 1.5 i.e. near to zero. Similarly in versus order plot the line is not continuous and the values ranging from negative to positive which shows the calculated values are correct. The p-value shown in Table 4 tell about the significance of parameters, it should be less than 0.05. Pulse current has more significant in MRR followed by gap voltage, RHA wt% and gap voltage. The R-sqadj value must be greater than 70% for correct experimentation. In my experiment, it is 92.6% which is far better.

Table 4. Regression table for MRR

Predictor	Coefficient SE	Coefficient	T	P
Constant	-0.35	0.82	-0.43	0.692
WT%	0.009	0.019	0.51	0.640
Ip	1.58	0.156	10.14	0.001
Ton	-0.005	0.015	-0.38	0.725
Voltage	0.018	0.015	1.16	0.310
S=0.0191457	R2=96.3%	R2 Adj=92.6%		

The regression equation for Metal Removal Rate is written as:

$$MRR = -0.354+0.0099Wt\%+1.58Ip -0.0059Ton +0.0182V \quad (1)$$

Analysis of Tool Wear Rate

The tool wear rate has calculated by difference of rate of tool mass before and after machining and which is divided by the density of tool. The tool must be properly cleaned after machining to measure accurate weight loss. In experimentation brass tool is used whose density is 8.9 g/cm 3. The variation of tool wear rate with parameters shown in Figure 6. TWR has decreased by (0.02398mm 3 /min) as RHA wt% increases, it is minimum at 12% composition of RHA. Pulse current has more effect on TWR as compare to other parameters, it is almost increased by (0.11705mm 3 /min). Pulse on time has almost negligible

Figure 6. Variation of TWR with parameters

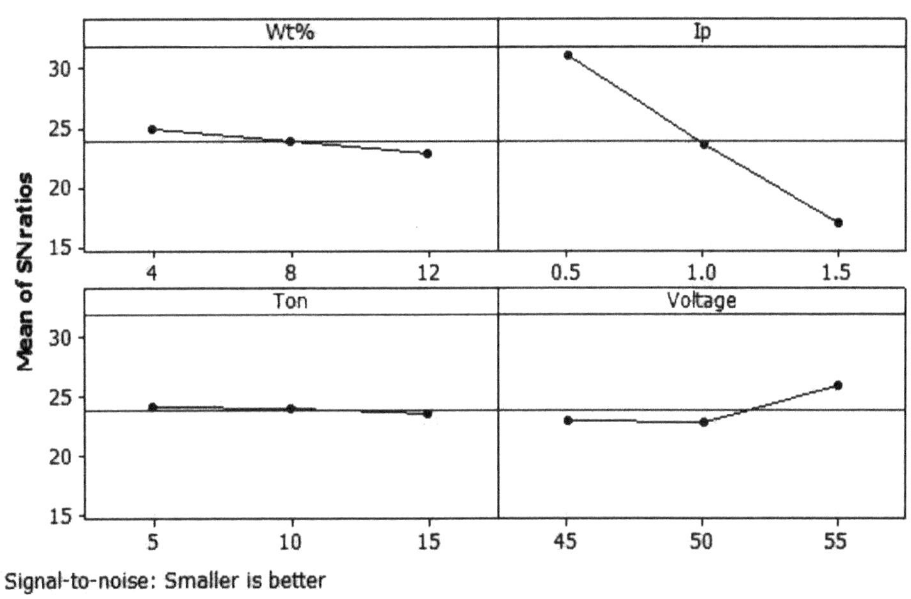

effect on TWR, but there is slight drop of (0.01582mm 3 /min) in between 10 to 15 μs. Tool wear rate has increased by (0.03180mm 3 /min) increasing gap voltage. It is maximum at 55 volt.

Regression Analysis of TWR

From residual plot graph for TWR as shown in Figure 7, all readings calculated experimentally are almost close to line drawn on 45 o in normal probability graph which shows that readings are exact and linear. Residual versus fit graph has all values lie between -2 to 1 and closer to zero line so values obtained are correct. From histogram, it is concluded that frequency of experimental values is more between 0 to 1 i.e. near zero and most of frequency has lies at 1. Similarly residual versus order plot, the line is not continuous and the values ranging from negative to positive, there is no five consecutive point at upper and lower to the zero line, which shows the calculated values are correct. The P-value of Table 5 tells that pulse current is significant for Tool Wear Rate. The gap voltage, weight % of RHA, and pulse on time is insignificant. The R-sq (adj) value is 88.4%, it means value obtained is better.

The regression equation for TWR is given as:

$$TWR = 0.106+0.0030Wt\%+0.117I_p -0.000798T_{on}-0.00318V \quad (2)$$

Analysis of Surface Roughness

Surface roughness has increased drastically by increasing the value of weight % of rice husk ash by (2.247μ). Weight percentage of RHA has more effect as compare to other parameters. Pulse on time is second most dominating parameter, increases TWR by (0.940μ) followed by pulse current, and gap

Figure 7. Residual plot for TWR

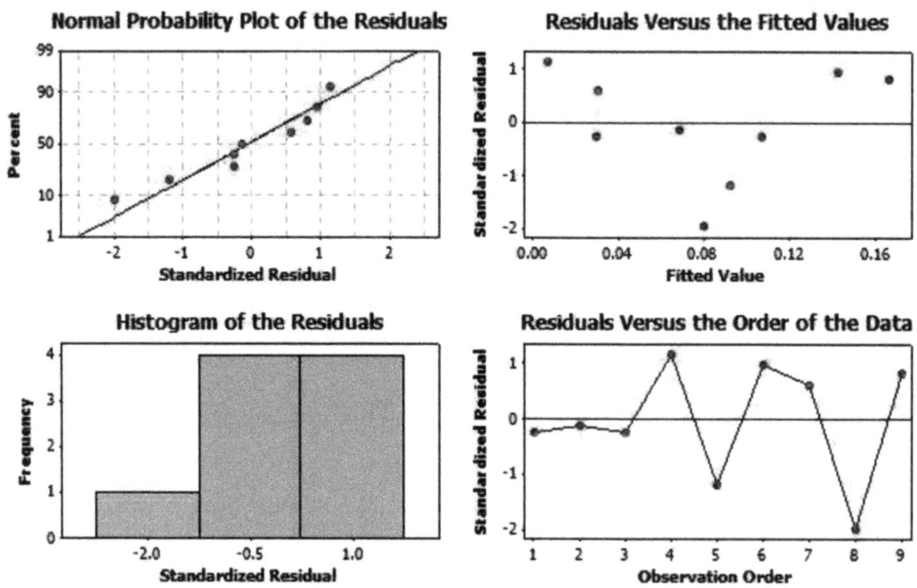

Table 5. Regression table for TWR

Predictor	Coefficient SE	Coefficient	T	P
CONSTANT	0.106	0.081	-0.43	0.692
WT%	0.0029	0.0019	0.51	0.640
Ip	0.117	0.0153	10.14	0.001
Ton	-0.0007	0.0015	-0.38	0.725
Voltage	-0.0031	0.0015	1.16	0.310
S=0.01882	R2=94.2%	R2 Adj=88.4%		

voltage. The value of surface roughness is measured with the help of surface tester surtonic 3+ as shown in Figure 8.

Regression Analysis of Ra

The regression equation for surface roughness can be written as:

$$Ra = 6.77 - 0.281Wt\% - 0.413Ip + 0.0287Ton - 0.0153V \quad (3)$$

The residual plot graph as shown in Figure 9, all readings calculated experimentally are almost close to line drawn on 45 o in normal probability graph which shows that readings are exact and linear. Residual versus fit graph has all values lie between -1 to 2 and closer to zero line, hence it also concludes that values obtained are near about exactness. From histogram graph, it is concluded that frequency of experimental values is more between 0 to -1.5. Most of frequency lies at -0.5 which is near about zero.

Figure 8. Residual plot for TWR

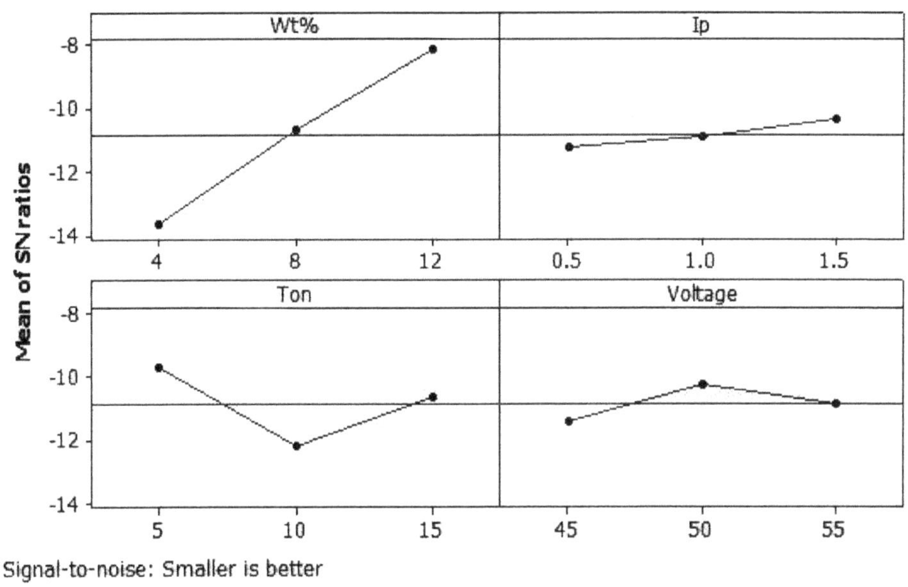

Figure 9. Residual plot for Ra

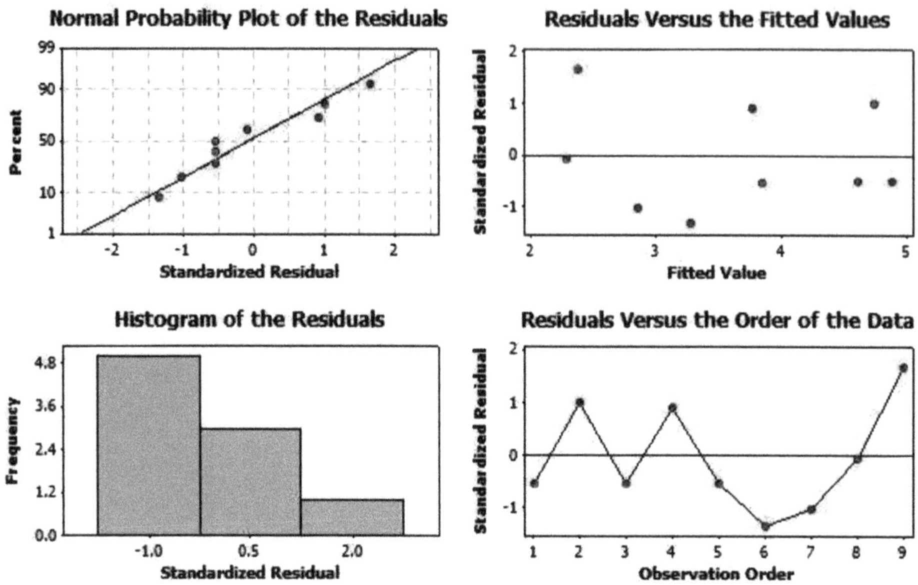

Similarly residual versus order plot the line is not continuous and the values ranging from negative to positive and none of the five consecutive value are upper or lower to zero line which shows the calculated values are correct. The P-values as shown in Table 6 tell that the weight % has significance for surface roughness and other parameters are insignificant. The R-Sq(adj) is 78% which shows that value obtained is good.

Table 6. Regression table for Ra

Predictor	Coefficient SE	Coefficient	T	P
Constant	6.76	2.67	2.53	0.06
WT%	-0.28	0.062	-4.46	0.01
Ip	-0.413	0.503	-0.82	0.45
Ton	-0.028	0.050	0.57	0.59
Voltage	-0.015	0.0503	-0.3	0.77
S=0.6163	R2=90.5%	R2 Adj=78.4%		

REFERENCES

Alaneme, K. K., & Adewale, T. (2013). Influence of rice husk ash–silicon carbide weight ratios on the mechanical behaviour of Al-Mg-Si alloy matrix hybrid composites. Tribology in Industry, 1(35), 163–172.

Alaneme, K. K., Akintunde, I. B., Olubambi, P. A., & Adewale, T. M. (2013). Fabrication characteristics and mechanical behaviour of rice husk ash–alumina reinforced al-mg-si alloy matrix hybrid composites. Journal of Materials Research and Technology, 2(1), 60–67.

Alaneme, K. K., & Olubambi, P. A. (2013). Corrosion and wear behaviour of rice husk ashalumina reinforced al–mg–si alloy matrix hybrid composites. *Journal of Materials Research and Technology*, *2*(2), 188–194. doi:10.1016/j.jmrt.2013.02.005

Alaneme, K. K., & Sanusi, K. O. (2015). Microstructural characteristics, mechanical and wear behaviour of aluminium matrix hybrid composites reinforced with alumina, rice husk ash and graphite. Engineering Science and Technology, an International Journal, 18(3), 416–422.

Gaikwad & Jatti. (2016). Optimization of material removal rate during electrical discharge machining of cryo-treated niti alloys using taguchis method. *Journal of King Saud University-Engineering Sciences*.

George, P., Raghunath, B., Manocha, L., & Warrier, A. M. (2004). Edm machining of carbon–carbon compositea taguchi approach. *Journal of Materials Processing Technology*, *145*(1), 66–71. doi:10.1016/S0924-0136(03)00863-X

Kandpal, B. C., & Singh, H. (2015). Machining of aluminium metal matrix composites with electrical discharge machining-a review. *Materials Today: Proceedings*, *2*(4), 1665–1671. doi:10.1016/j.matpr.2015.07.094

Kannan, S., & Ramanathan, K. (2014). Optimization of edm parameter of al/tic composite using taguchi methodology. *Middle-East Journal of Scientific Research*, *22*(1), 121–127.

Kumar, N. M., Kumaran, S. S., & Kumaraswamidhas, L. (2015). An investigation of mechanical properties and material removal rate, tool wear rate in edm machining process of al2618 alloy reinforced with si 3 n 4, aln and zrb 2 composites. *Journal of Alloys and Compounds*, *650*, 318–327. doi:10.1016/j.jallcom.2015.07.292

Lodhi, B. K., & Agarwal, S. (2014). Optimization of machining parameters in wedm of aisi d3 steel using taguchi technique. *Procedia CIRP*, *14*, 194–199. doi:10.1016/j.procir.2014.03.080

Mahapatra, S., & Patnaik, A. (2007). Optimization of wire electrical discharge machining (wedm) process parameters using taguchi method. *International Journal of Advanced Manufacturing Technology*, *34*(9-10), 911–925. doi:10.1007/s00170-006-0672-6

Narasaraju, G., & Raju, D. L. (2015). Characterization of hybrid rice husk and fly ash-reinforced aluminium alloy (alsi10mg) composites. *Materials Today: Proceedings*, *2*(4), 3056–3064. doi:10.1016/j.matpr.2015.07.245

Pramanik, A., Basak, A., Islam, M. N., & Littlefair, G. (2015). Electrical discharge machining of 6061 aluminium alloy. *Transactions of Nonferrous Metals Society of China*, *25*(9), 2866–2874. doi:10.1016/S1003-6326(15)63912-7

Rajkumar, K., Santosh, S., Ibrahim, S. J. S., & Gnanavelbabu, A. (2014). Effect of electrical discharge machining parameters on microwave heat treated aluminium-boron carbide-graphite composites. *Procedia Engineering*, *97*, 1543–1550. doi:10.1016/j.proeng.2014.12.438

Razak, M., Abdul-Rani, A., Rao, T., Pedapati, S., & Kamal, S. (2016). Electrical discharge machining on biodegradable az31 magnesium alloy using taguchi method. *Procedia Engineering*, *148*, 916–922. doi:10.1016/j.proeng.2016.06.501

Saravanan, S., & Kumar, M. S. (2013). Effect of mechanical properties on rice husk ash reinforced aluminum alloy (AlSi10mg) matrix composites. *Procedia Engineering*, *64*, 1505–1513. doi:10.1016/j.proeng.2013.09.232

Talla, G., Sahoo, D. K., Gangopadhyay, S., & Biswas, C. (2015). Modeling and multi-objective optimization of powder mixed electric discharge machining process of aluminum/alumina metal matrix composite. Engineering Science and Technology, an International Journal, 18(3), 369–373.

Ugrasen, G., Ravindra, H., Prakash, G. N., & Prasad, Y. T. (2015). Optimization of process parameters in wire edm of hchcr material using taguchis technique. *Materials Today: Proceedings*, *2*(4), 2443–2452. doi:10.1016/j.matpr.2015.07.185

Velmurugan, C., Subramanian, R., Thirugnanam, S., & Ananadavel, B. (2011). Experimental investigations on machining characteristics of al 6061 hybrid metal matrix composites processed by electrical discharge machining. *International Journal of Engineering Science and Technology*, *3*(8), 87–101.

Vikas & Kumar. (2013). *Effect and optimization of machine process parameters on over-cut in edm for en41 material using taguchi*. Academic Press.

Chapter 2
Modeling and Simulation of Surface Texture for End-Milling Process:
Modeling End-Milling Process

Manoj Kumar
Mother Parwati Education Services, India

ABSTRACT

Analysis and simulation of manufacturing process require extensive and complicated computations. Nowadays, computer resources and computational algorithms reach to the state that can model and simulate the problem efficiently. One of the important processes in manufacturing is machining. In this research end-milling process which is one of the complex and wide-spread processes in machining is chosen. Most important parameters in end-milling are surface roughness and surface location errors. A comprehensive simulation software is developed to model end-milling process in order to anticipate finishing parameter such as surface roughness and errors. The proposed algorithm takes into account cutting conditions, such as feed, doc, woc, tool run out, etc. In addition, dynamic simulation module of the software can accurately model flexible end-mill tool, the milling cutting forces and regeneration of waviness effects. The software can accurately determine the most commonly used index of surface roughness parameters such as Ra, P.T.V. and surface errors.

1. INTRODUCTION

End-milling operations are extensively used in the production of precise parts in the industry. Milling process dynamics is investigated in many research works, however due to complexity of the process different fields of research works are required. One area in analyzing and simulation of milling process is developing a comprehensive model in order to anticipate surface quality, dimensional error and trend of them in mass production. In various machining process the following steps are required to build up the model:

DOI: 10.4018/978-1-5225-2440-3.ch002

1. Analyzing and modeling machining forces.
2. Modeling dynamic system of machine tools.
3. Modeling surface texture.

A comprehensive model for accurate anticipation of surface errors requires a "precise" force model. The model must take account of phenomena such as, tool deflections, tool run-out, tool wear, machining dynamics, etc. In the most of previous research works, machining forces are obtained by mechanistic force model based on oblique cutting tests (Altintas, 2010). In this work, orthogonal cutting theory and transforming method proposed by E. Budak *et al.* (1996) is used (Budak *et al.,* 2012). In contrast to the previous methods, there is no need for oblique cutting test for each tool shape. At first, a general machining data base is provided by orthogonal cutting tests with various work materials and cutting tool materials. Using transforming relations and tool geometry, cutting coefficients for oblique cutting are determined. End milling is used in a wide range of manufacturing industries, due to its versatility and its high material removal rate (especially end-mills) in producing parts of desirable dimensions. Prior to production, predictions of different factors such as cutting forces, stresses, temperature etc, are important in selecting the tool material and design and can ensure an efficient machining process. Knowledge of the cutting forces allows for the prediction of tool deflection and wear, work-piece deflection, onset and/or possible regeneration of chatter and reduction of surface errors. The knowledge of temperature distribution is also crucial in material selection and research into the development of built-up edge (BUE). To this end, the aim of researchers to simulate or model the metal cutting process has grown.

2. LITERATURE REVIEW

There has been an extensive research into the prediction of cutting forces in end milling, most of which have produced very good results. Majority of these are analytical based, involving quite a number of calibration experiments used to obtain cutting force coefficients (mechanistic approach), or solving of small oblique segments along the tool's cutting edge (mechanics approach). Other studies are based on computational simulations mainly using finite element (FE) techniques. However, the stress and temperature distribution can only be predicted accurately using finite element methods based techniques. Owing to the sheer complexity and highly nonlinear nature of the end milling process, past studies using finite element methods have been mainly focused on orthogonal metal cutting. Studies into metal cutting process began by idealizing the shear zone into a single shear plane. The first model was proposed by Timme in 1870, who suggested that the chip was formed due to the brittle fracture and Tresca later assumed it to be due to plastic deformation. Much later after this, are some of the first pioneering studies carried out by Merchant (Merchant, 1945) and Lee Shaffer (Lee, 1951), still idealising the principal shear zone to a single plane extending from the cutting edge to the surface of the work-piece. Other analyses of metal cutting prior to the development of the Finite Element Method (FEM) were mainly based on these two models (Piispanen, 2014; Fenton and Oxley, 2013). However, with the development of the finite element method, various studies have been carried out on orthogonal and oblique metal cutting. Lei *et al.* (2013) is considered the first to introduce finite element method (FEM) technique into machining using a three-dimensional elastic-plastic model. This study was however limited to just the initial stages of chip formation. Similarly, Smith and Tlusty (2011) applied the elastic-plastic finite element method to orthogonal metal cutting process. In this, they modified the shape of the chip until it was consistent with the plastic flow generated. Iwata *et al.* (2015) also used a rigid-plastic finite element method to consider the effect of friction between the tool rake and face. In their model the shape of the model was predicted and modified repeatedly based on the distribution of flow stress. So far, the models were all based on a Lagrangian formulation. The first analysis to simulate the movement of the tool into the workpiece and continuous chip formation along a predefined parting line was by Sekhon and Chenot (2013). They used a finite element program (''NIKE2D'') adopting the Updated-Lagrangian formulation (ULF) and also proposed a separation criterion to simulate chip formation. This separation criterion was based on the effective plastic strain (critical limit of 0.5) at the tool tip region of the workpiece. Lee and Wilkening (2012) were however the first to attempt chip formation by the use of an element death option in the model, but this model was not realistic as friction in the secondary shear zone was ignored. Rosa *et al.* (2007) found that the separation criterion has a significant effect on the residual stress in the workpiece and little effect on the chip geometry and the cutting force. Yun *et al.*, (2012) carried out a study on the effects, elasticity, viscoplasticity, temperature, strain-rate and large strain have on the stress-strain relationship and the effects large friction have on the tool-chip interface. Their study was also based on the Updated-Lagrangian formulation. The separation criterion used was based on the distance between the node connecting the chip and workpiece (crack tip) and the tool tip. This type of criterion has been adopted in a number of studies to model orthogonal metal cutting, however with different values used (Tlusty, 2000; Mamalis *et al.*, 2012, Baker *et al.*, 2012, Carrino *et al.*, 2013, Ceretti *et al.*, 2015). Ranganath *et al.* (2007) simulated orthogonal metal cutting using a separation criterion based on a critical stress. Madhavan *et al.* (2010) used a crack length versus time separation criterion, based on the movement of the tool. Donea *et al.* (2014) used an Implicit Lagrangian FE code (''DEFORM-2D'') to study continuous and segmented chip formation. In their model, a damage criterion was used similar to that of Lee and Schaffer (2011), where the damaged element is removed from the domain. The downside of this approach is that the removal of elements corresponds to loss of mass. The only way to minimize this would be to have very small element sizes along the tool path, which would come at a higher computational cost. The author also reported that after the deletion of damaged elements, the domain boundaries had to be smoothed otherwise subsequent attempts to re-mesh would fail. Pantale *et al.* (2004) carried out a detailed study on the separation criteria used by different researchers in their studies. The study showed that none of the existing criteria is universal and deduced that there is the need to develop a comprehensive criterion to ensure consistency in FEM models and results. To avoid the use of separation criteria while using a Lagrangian formulation, the cutting action can be simulated as the continuous indentation and plastic flow of the material around the tool. The tool is fed into the work-piece and as soon as the elements become distorted, a re-mesh is carried out. The point of re-mesh is determined by a set of specified criteria. This method was used by Natarajan and Jeelani (2013), Itoigawa *et al.* (2006) and Ko *et al.* (2002). One of the main problems with this approach is its high computational cost. Frequent remeshing and a very dense mesh is required to minimize the errors (Bil *et al.*, 2014). Budak *et al.* (1996) compared three different finite elements models of orthogonal metal cutting with experimental data. Two of these models used continuous remeshing and the third was based on a damage criterion. While it was noted that the friction parameter drastically affects the results, it was also noted that even though the re-meshing approach produced better results, there was still the need for a better separation criterion. Contrary to the Lagrangian formulation, is the Eulerian formulation where the mesh is spatially fixed. It is more suitable for fluid flow problems, which involves a control volume. It has been used to model metal forming process. However, its first application to metal cutting was reported by Budak and Altintas (2012) and Altintas (2010). In the study by Athavale and Strenkowski (2013), the tool forces and geometry obtained experimentally were applied to the workpiece surface, thereby omitting the chip in the model. The residual stress and plastic deformation zone in the workpiece were validated experimentally. A similar approach was used by Montgomery and Altintas (2014) in modeling the residual stresses in the work-piece. Moriwaki *et al.* (2013) analyzed a steady-state orthogonal cutting with the capability to predict chip geometry and chip-tool contact length based on an Eulerian formulation. In their simulation, the mesh was not entirely spatially fixed as they employed a method proposed by Movahhedy *et al.* (2010) to obtain the shape of the chip. In this method, the free surface of the chip was calculated by adjusting its location (through an iterative process) to enforce a zero normal surface velocity component. Also based on an Eulerian formulation, Moriwaki *et al.* (2013) developed a rigid-plastic finite element model to examine the effects of the tool edge radius to the depth of cut in micro cutting process. Some other studies carried out with the use of pure Eulerian formulation were reported by Martelloti (2011) and Ozel and Altan (2007). What makes the Lagrangian formulation very attractive in modeling metal cutting is that the mesh covers and moves together with the material. Therefore, no a priori assumption of the chip is required because the chip develops as the tool progresses through the work-piece. Moreover, the analysis can model indentation, incipient stage and the steady state of metal cutting. At the same time, this formulation also suffers some disadvantages, the most important of which is the use of separation criterion. The use of separation criterion to model chip formation is not reliable, as there exist no universal and consistent criterion as explained by Ozel and Zeren (2006). Not only are there different types/methods of applying the separation criteria, there is also no physical indication as to what criterion value is to be used. Another factor concerning the use of separation criterion is the use of a parting line. The problem with the parting line is that as the analysis progresses some nodes are deformed and this can cause unstable simulation. This is because the separation criteria approach works best if the nodes are precisely in front of the approaching tool. A similar problem is experienced when using damage models for chip separation as explained by Bil *et al.* (2014). Furthermore, the parting line restricts the types of tools used in the model to sharp edged tools (Movahhedy *et al.*, 2010). Another major disadvantage of using the Lagrangian formulation is that, as the material is highly sheared on passing through the shear plane/primary shear zone, so are the elements, thereby causing highly distorted elements. The Eulerian formulation on the other hand, does not suffer the same disadvantages as the Lagrangian formulation. In addition, because the material flows into the model, the domain can be design to include only the area near the shear zone, thereby improving on computational cost. The only major disadvantage to this approach is that, knowledge of the volume of the domain (chip) and its precise boundary conditions are required a priori. It is due to the above reasons that the Arbitrary Lagrangian-Eulerian formulation is ideal as this combines the features of pure Lagrangian and pure Eulerian formulations. Fenton and Oxley (2013) first proposed it, for two-dimensional hydrodynamic problems using finite difference schemes. It was at the time called, Coupled/Mixed Eulerian-Lagrangian Method/Code. It was later introduced to finite element method by Donea *et al.* (2014) and has been used extensively, mainly to model processes involving large deformations (e.g. metal forming, metal cutting and metal forging). Some studies using ALE to model metal cutting are reported by Ozel and Altan (2007). All the studies reported so far have been focused on analyzing orthogonal metal cutting and turning. Ozel and Altan (2007) modeled flat end milling using a single insert flat end mill. They had to split the tool cutting edges into two regions (primary and secondary cutting edges). The primary cutting edge was modeled using plane strain deformation and an axisymmetric deformation model was used to model the secondary cutting edge. The cutting force predictions obtained did not entirely match the experimental results. Pantale *et al.* (2004), presented a three-dimensional oblique model to simulate the milling process using damage criterion (Johnson– Cook's). A full three-dimensional simulation was briefly reported, however it was not validated and no results were presented, as more investigation was needed.

In general, one of the important sources of vibrations in milling process is self-excited vibration. In this kind of vibrations, internal sources will provide energy for the system. In some cases, the energy will exceed damping thus vibrations sustain or even grow in amplitude. One of the important principles observed in machining dynamics is the regeneration of waviness which affects chip thickness.

In current research, a dynamic simulation model for machining dynamics is proposed. The surface is modeled using a 3D algorithm which is an extension to available 2D methods. The developed simulation model can anticipate surface roughness in any direction of interest on the machined surface.

3. SIMULATION OF CUTTING FORCES

Figures 1(a) and 1(b) illustrated coordinate system and end-milling geometry required for the model. Differential forces, df_t tangential, df_r radial and df_a axial are depicted which act on the jth cutting edge for an ideal system (without any deflection). Values of the differential forces are obtained by (for tooth j):

$$\begin{aligned} df_{tj}(\theta,z) &= \left[K_{te} + K_{tc}h_j(\theta,z)\right]dz \\ df_{rj}(\theta,z) &= \left[K_{re} + K_{rc}h_j(\theta,z)\right]dz \\ df_{aj}(\theta,z) &= \left[K_{ae} + K_{ac}h_j(\theta,z)\right]dz \end{aligned} \tag{1}$$

In equations (1) there are two separate terms, the first one is related to friction and ploughing forces and the second term is considered for the cutting action which takes place in the shear zone. Coefficients for the first phenomena are represented by K_{te}, K_{re}, and K_{ae} showing amount of force per unit length of the cutting edge. Coefficients represented by K_{tc}, K_{rc}, and K_{ac} are considered for cutting action showing amount of force per unit area of cutting edge.

The simulation software which written in MATLAB V7.0 first computes cutting and edge force coefficients and then computes cutting forces for each cutter angle. Table 1 summarizes computed coefficients for a four-flute HSS tool and aluminium work-piece 7075-T6 with yield stress of 542 MPa. The computed cutting forces are depicted in Figure 2.

4. REGENERATION OF WAVINESS

In general, the cutting edge of the tool is cutting a surface which is generated in the previous passes. If there exists any vibrations between tool and work piece, the newly generated surface undulates which results in the variation of chip thickness. Figure 3 illustrates the model used for simulation of milling process. Dynamic characteristic of the system in x and y direction is also depicted. Differential forces can be obtained by:

Figure 1. Coordinate system and end-milling geometry

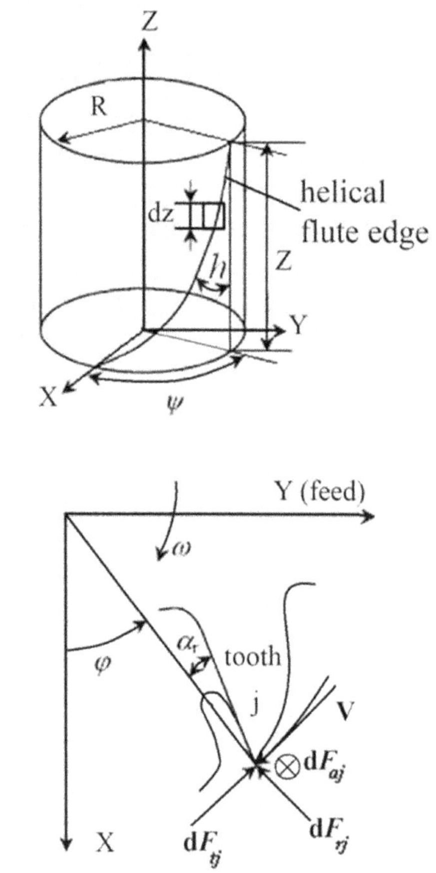

Table 1. The computed coefficients for a four-flute HSS tool and aluminium work-piece 7075-T6 with yield stress of 542 MPa, in this table α_n is normal rake angle of tool

α_n (deg)	K_{tc} [N/mm²]	K_{rc} [N/mm²]	K_{ac} [N/mm²]
0	1642.3	609.5	512.2
12	1396.2	519.8	258.1
15	1356.5	504.4	215.3

Figure 2. Simulated cutting forces for a four-flute HSS tool and aluminium work-piece 7075-T6 with yield stress of 542 MPa. Cutting conditions are d (Tool diameter) =10mm, f_t (Feed per tooth)=0.05mm/tooth, doc (Depth of cup) = 10mm

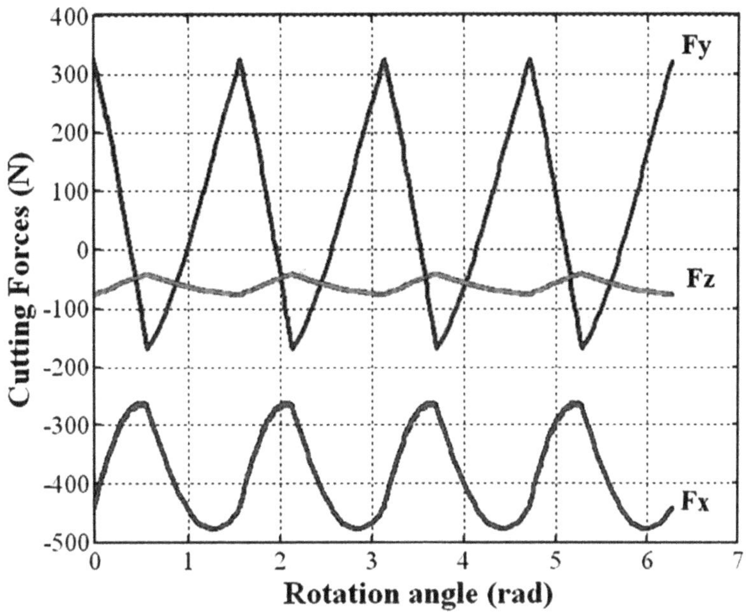

Figure 3. Model used for simulation of dynamic milling process

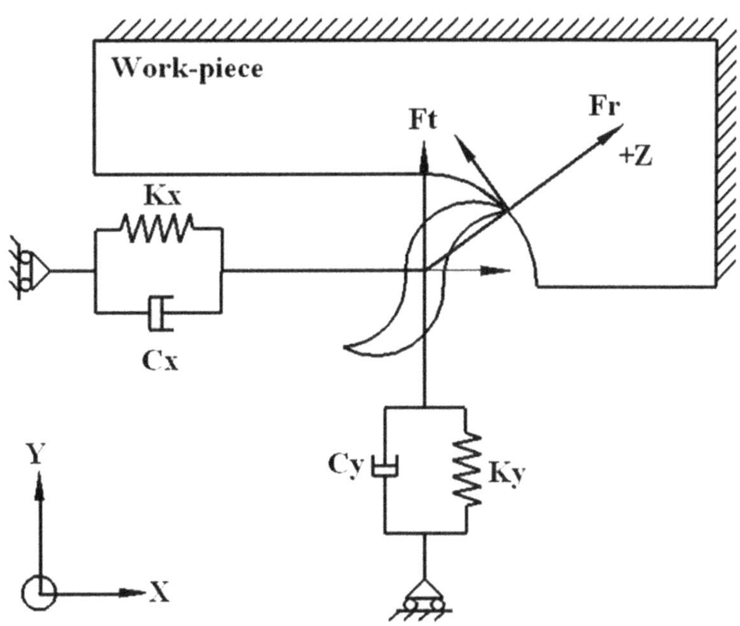

$$dF_{t_{n+1}} = K_{tc} da(h_0 - z_n + z_{old1} + z_{old2} + z_{old3})$$
$$dF_{r_{n+1}} = K_{rc} da(h_0 - z_n + z_{old1} + z_{old2} + z_{old3}) \tag{2}$$

Due to the fact that axial stiffness of the tool and the spindle assembly is high and also axial force is smaller than the others, axial vibrations are ignored. In the above relations, n, h_0, z_n, z_{old1} to z_{old3} and da are respectively: index of time increment, nominal chip thickness, tool deflection normal to the cut surface, tool deflection in previous passes and axial increment. In each instance, differential cutting forces on engaged cutting edges are computed and projected on X and Y axis. Then the total cutting forces are computed by integration of differential forces over engaged edges. Using Newton's second law, differential equations of motion are written and then by double Euler integration the dynamic deflections are obtained.

$$F_{x_{n+1}} = m_x \ddot{x}_n + C_x \dot{x}_n + K_x x_n$$
$$F_{y_{n+1}} = m_y \ddot{y}_n + C_y \dot{y}_n + K_y y_n \tag{3}$$

where m_x, m_y, K_x, K_y, C_x and C_y are mass, stiffness and damping matrices for x and y directions, respectively. Input data for a four flute end-mill with helix angle of 30° are considered as follows (Piispanen, 2014):

$$f_x = f_y = 660HZ$$
$$K_x = K_y = 2 \times 10^7 N / m \tag{4}$$
$$M_x = K_x / (2\pi f_x), My = K_y / (2\pi\ f_y)$$

$$\eta_x = \eta_y \quad = 0.05$$
$$C_x = \eta_x \times 2\sqrt{K_x M_x}, C_y = \eta_y \times 2\sqrt{K_y M_y}$$
$$r \ = 5mm, \quad n = 1200\ rpm,$$
$$doc = 10mm$$

In above relations f_x, f_y, η_x, η_y are natural frequency and damping ratio for x and y directions, respectively. The tool radius, spindle speed, and depth of cut are represented by r, n, doc. Nominal chip thickness at angular position θ is:

$$h_0 = ft \times \mathrm{Sin}\,\theta \tag{5}$$

Normal to cut deflection is:

$$Z_n = x_n\ \mathrm{Sin}\,\theta + y_n\ \mathrm{Cos}\,\theta \tag{6}$$

Using the proposed model, it is possible to include nonlinearities such as jump-of-cut and chatter vibrations. Figure 4 shows the simulation results for aluminium 7075-T6 work piece. Cutting conditions are as follows:

d = 10 mm, h (helix angle) = 30deg, ft = 0.1 mm/tooth, woc (width of cut) = 1.2 mm, n = 1200 rpm, doc (Depth of cut) = 10 mm.

Dynamic milling forces for woc = 3 mm are shown in Figure 5. Fourier transform of milling forces are computed and depicted in Figures 6 and 7. Dominant frequency in the stable cutting is tooth passing frequency, however as inferred from Figure 6 and 7 the dominant frequency in the unstable cutting is close to natural frequency of flexible mode shape, known as chatter frequency (Piispanen, 2014). When woc = 3 mm, milling is unstable and vibration grows in amplitude gradually.

5. DYNAMIC VIBRATIONS OF TOOL AND SPINDLE

Surface roughness prediction requires a comprehensive model in time domain which can simulate tool vibrations at locations of interest on the work piece. In the current research milling vibrations are modeled by FEM using CALFEM toolbox implemented in MATLABV7.0 software.

Figure 8 schematically shows FEM model for the end-mill and spindle assembly, which is modeled as a step beam with two simple supports.

Axial variation of the tool deflection is computed for the engaged portion of tool which is used by the surface location error algorithm. Therefore, finer elements in this portion are introduced.

Figure 4. Dynamic simulation of cutting forces for aluminium 7075-T6 work piece. Cutting conditions are: d = 10mm, h = 30deg, ft = 0.1 mm/tooth, woc=1.2

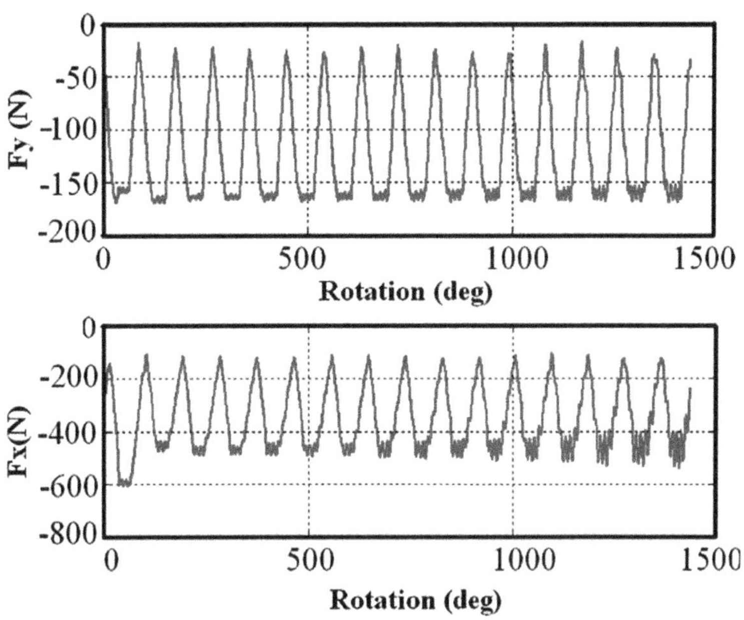

Figure 5. Dynamic simulation of cutting forces for aluminium 7075-T6 work piece. Cutting conditions are: d = 10mm, h = 30deg, ft = 0.1 mm/tooth, woc=3mm

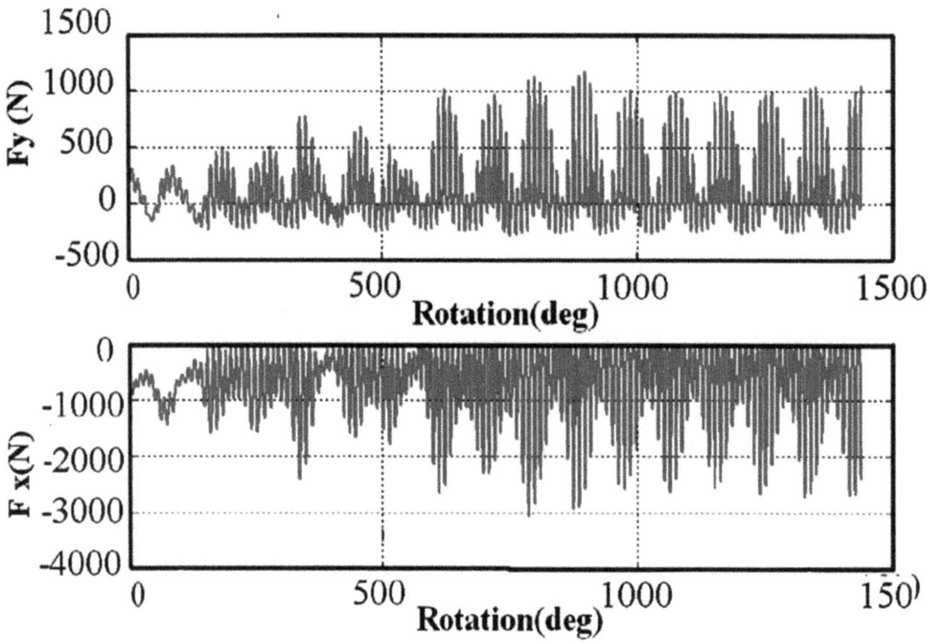

Figure 6. Fourier transform for F_x

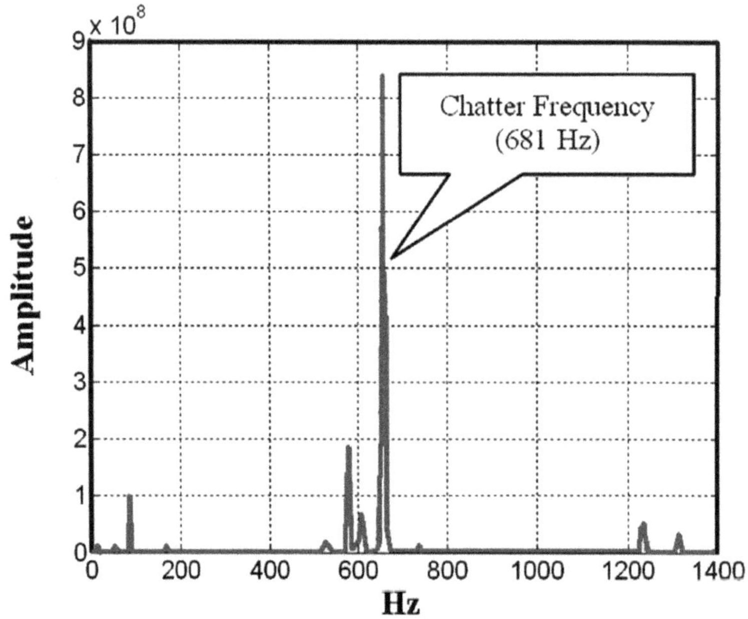

Figure 7. Fourier transform for F_y

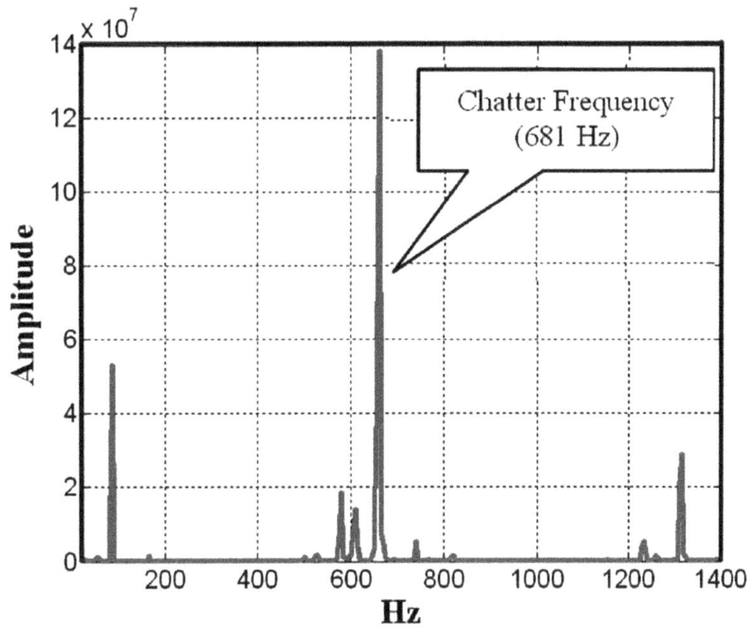

Figure 8. FEM model for the end-mill and spindle assembly

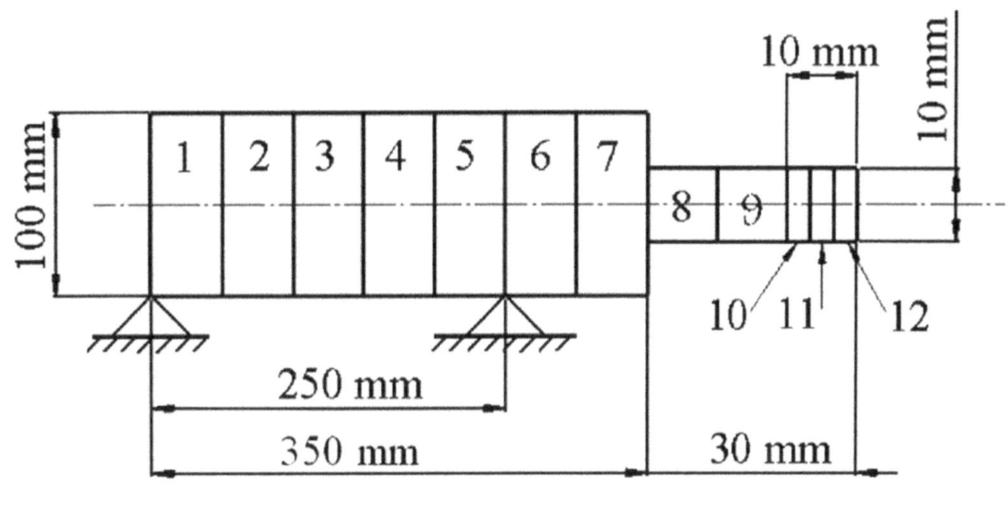

Inputs for FE analysis are given in the following:

Modulus of elasticity of steel = 207×10^9 (N/m^2), inner diameter of spindle axis = 0.02 m, outer diameter of spindle axis = 0.1 m, density of steel = 7800(Kg/m^3). Figure 9 illustrates the developed flowchart of the FE analysis.

In the developed algorithm Nr, a, da, F_0, T (N), X (N, j) and Y (N, j) are number of revolution, axial depth of cut, axial increment, initial force, time, x-deflection and y-deflection, respectively. Inputs to the algorithm are tool radius, woc, doc, feed, cutting force constants, and spindle speed. A function writ-

Figure 9. Developed flowchart of the FEM analysis

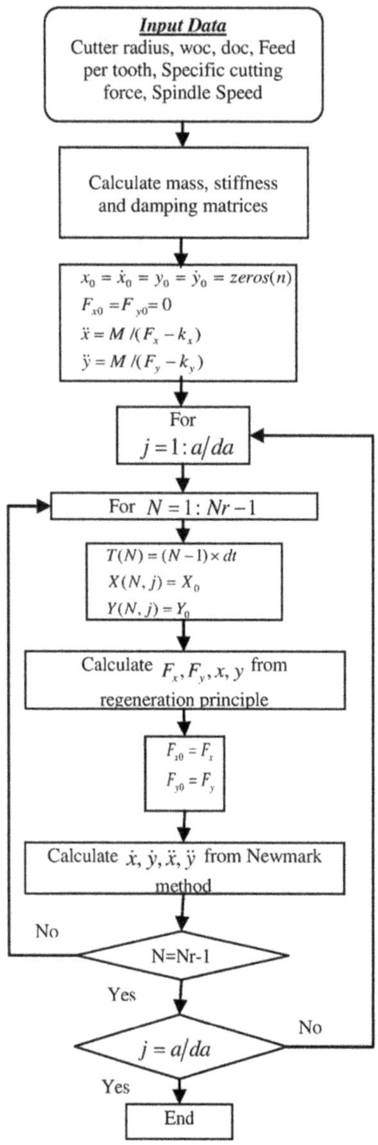

ten in MATLAB V7.0 is used to calculate mass, stiffness and damping matrices. Using mesh and node positions, mass, stiffness and damping of each element are computed and then assembled together to get overall matrices. In addition boundary conditions and external cutting forces are applied. Differential equations of motion are integrated numerically by the Newmark method.

$$\begin{cases} M_x \ddot{X} + C_x \dot{X} + K_x X = F_X \\ M_y \ddot{Y} + C_y \dot{Y} + K_y Y = F_Y \end{cases} \tag{7}$$

For selected initial values of $y = x = \dot{y} = \dot{x} = 0$ the initial acceleration is expressed as

$$\ddot{X} = \frac{F - C_x \dot{x} - k_x x}{M_x} \quad , \quad \ddot{Y} = \frac{F - C_y \dot{y} - k_y y}{M_y} \tag{8}$$

The iteration loop starts from 1 and end to a/da (where a is DOC and da is axial increment), thus the variation of axial deflection is obtained. Counter N varies from 1 to Nr-1 which is considered for the time increments. After computing $F_{x,}$ $F_{y,}$ x, and y at current time increment, these values are used as initial values for the next step. Results of the milling tool in time domain by FE analysis are shown in Figure 10. Vibrations of the tool center point are plotted at different axial levels.

In the next section amount of vibrations is used for determining cutting edge position and surface generation method.

6. TOOL PATH SIMULATION

Final machined surface is produced by subsequent cutting actions of end-mill cutting edges. Ideally, points on the cutting edge have trochoidal paths which leave feed marks on the surface. Additionally, any tool deflection, vibrations or geometric error results in undesirable deformation and deteriorate required surface roughness on the work piece. Ideal position of the cutting edges with respect to tool cutter is shown in Figure 11.

XOY coordinate system is attached to the work piece and original O is at the initial center position of the spindle. Trochoidal motion of a point on the cutting edge is computed by [9]:

Figure 10. Results of time domain FE analysis of the milling tool for d= 12 mm, f_t=0.3 mm/tooth, K_{tc}=2000N/mm², doc=5mm, woc=1.5 mm

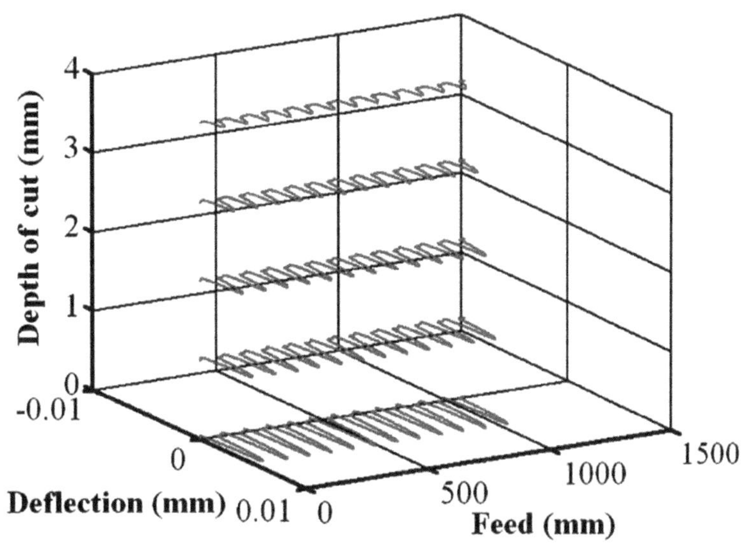

Figure 11. Trochoidal path of a point on cutting edge

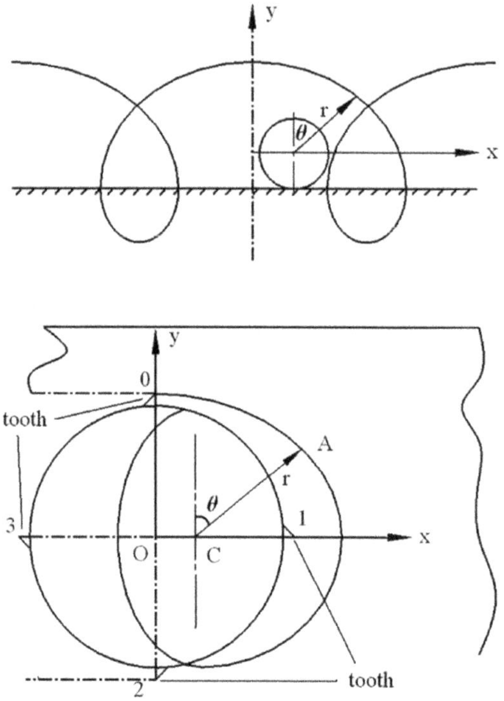

$$X_{l/c}(i,z,t) = R\sin(\theta(t) - \psi(i) - \alpha(z))$$
$$Y_{l/c}(i,z,t) = -R\cos(\theta(t) - \psi(i) - \alpha(z)) \quad (9)$$

where $X_{l/c}$ and $Y_{l/c}$ are coordinates of the edge with respect to tool center point. Also *i, r*, $\theta(t)$, $\psi(i)$ and $\alpha(Z)$ are tooth index, tool radius, angular position of the lowest point of the reference cutting edge, pitch angle and phase lag of the point at z levels, respectively.

$\alpha(Z)$ is computed by:

$$\alpha(Z) = \frac{z\tanh}{R} \quad (10)$$

where h is tool helix angle.

On the other hand, displacement of spindle center with respect to work piece can be found by:

$$X_{s/w} = f_r \times \theta(t) / (2\pi)$$
$$Y_{s/w} = 0 \quad (11)$$

If the runout error is not negligible, the location of tool center with respect to spindle axis will be computed by:

$$Y_{c/s}(t,z) = -\rho \times \cos(\theta(t) + \lambda) + Y_{FEM}(t,z)$$

$$X_{c/s}(t,z) = \rho \times \sin(\theta(t) + \lambda) + X_{FEM}(t,z) \tag{12}$$

where ρ is runout offset and λ is its angular position, refer to Figure 12.

$X_{FEM}(t,z)$ and $Y_{FEM}(t,z)$ are dynamic tool deflections at time t and z level, which are calculated by FE analysis. Thus, instantaneous positions of a point on the cutting edge can be founded by:

$$\begin{aligned} X_{l/w} &= X_{l/c} + X_{c/s} + X_{s/w} \\ Y_{l/w} &= Y_{l/c} + Y_{c/s} + Y_{s/w} \end{aligned} \tag{13}$$

Figure 13 shows trochoidal motion of the cutting edges without considering vibrations and runout. In order to show the effects of the runout, its value is set to 1mm, which is exaggerated intentionally. The other cutting conditions are kept constant. The simulation results are shown in Figure 14. Finally, Figure 15 shows path of cutting edge including tool vibrations.

Figure 12. Runout parameters

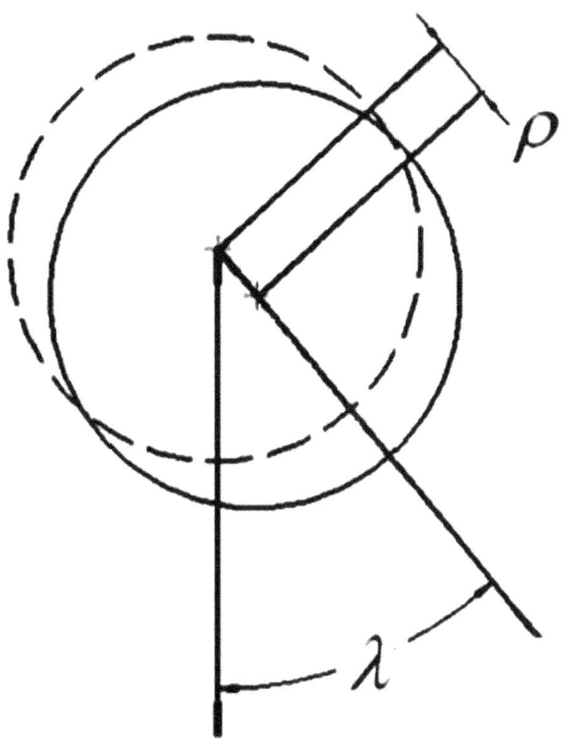

Figure 13. Trochoidal motion of the cutting edges without considering vibrations and runout (f_t=1mm/tooth, d=20mm)

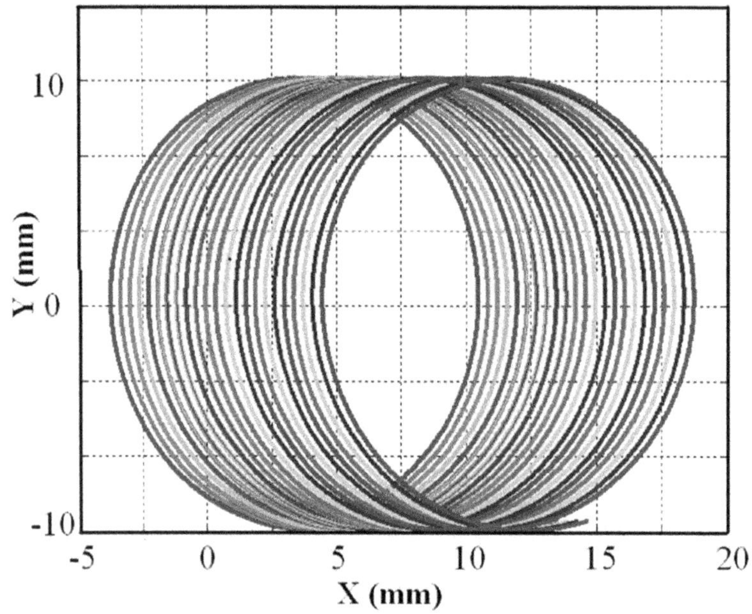

Figure 14. Trochoidal motion of the cutting edges with considering exaggerated runout (ρ=1mm, λ=45)

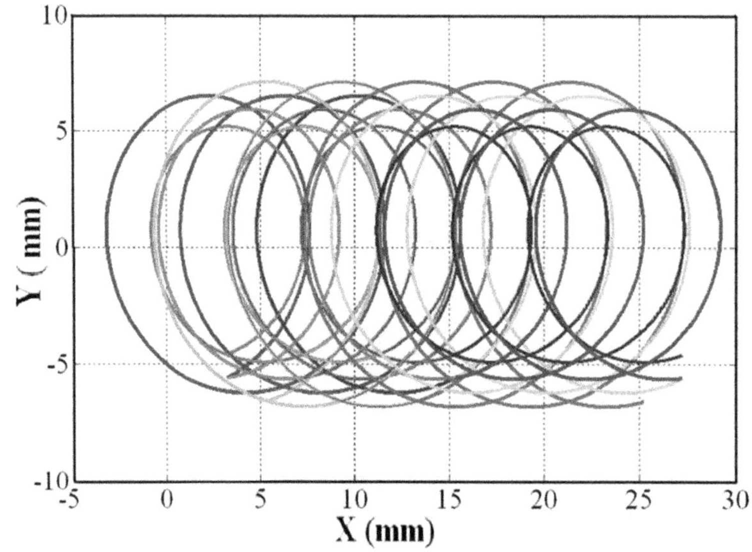

Figure 15. Trochoidal motion of the cutting edges with considering vibrations ($d=12\,mm, f_t=0.3\,mm/tooth$)

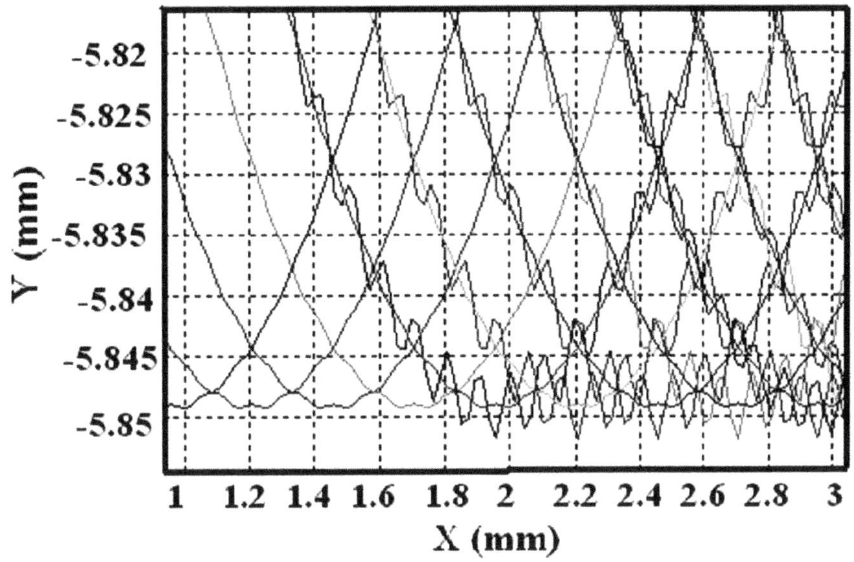

7. SURFACE TEXTURE SIMULATION

Feed mark profiles can be produced by removing point on the computed paths which do not make contact with newly generated surface. The z-level profiles are used in the mesh function to construct a 3d surface. Figures 16, 17 and 18 illustrated 3d mesh surfaces for different cutting conditions. Cutting conditions used for simulation of Figure 16 results in negligible vibrations. In order to show the runout effects an exaggerated runout offset ($\rho = 1mm$) is chosen and the result of simulation is shown in Figure 17. Extreme cutting conditions which results in chatter vibrations is used for simulation of Figure 18.

Figure 16. Surface texture for $d=12mm$, $woc=1mm$, $doc=5mm$, $f_t=0.1mm/tooth$, $K_{tc}=1500\,N/mm^2$ ($Ra=1.4\,\mu$)

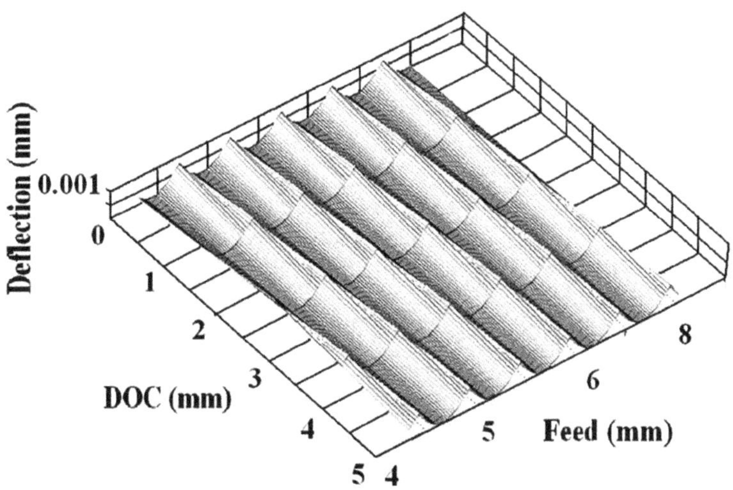

Figure 17. Surface texture for d=10mm, woc=1mm, doc=5mm, f_t=0.1mm/tooth, K_{tc}=1500 N/mm², ρ =1mm (Ra=3.5 μ)

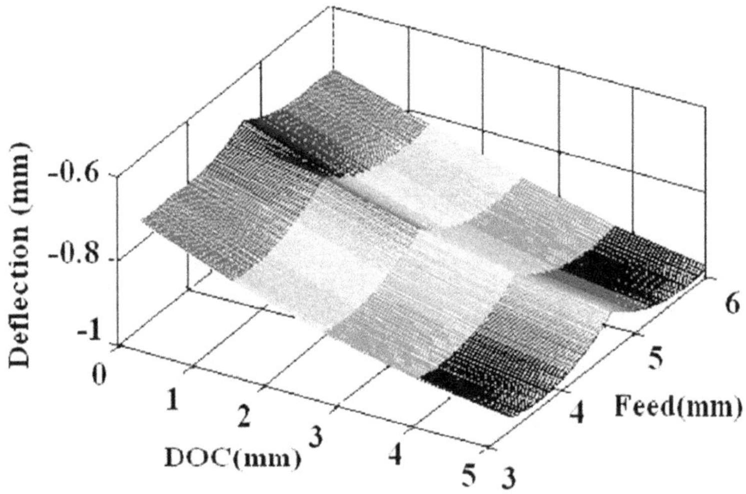

Figure 18. Surface texture for d=12mm, woc=4.5mm, doc=5mm, f_t=0.6mm/tooth, K_{tc}=1500 N/mm², ρ =0.01mm (Ra=14 μ)

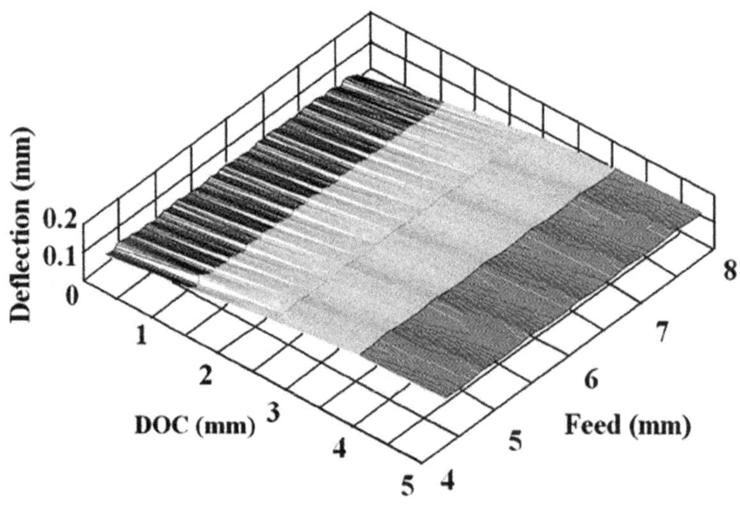

Table 2. Surface roughness computed at tool tip

Ra μ	ft (mm/tooth)	WOC (mm)	a (mm)	d (mm)
1.14	0.05	1	5	10
3.99	0.1	1.5	5	10
5.85	0.25	1.5	5	10
2.24	0.1	1.5	5	12
3.07	0.1	2	5	12
4.61	0.15	3	5	12

As shown in Figures 10 and 18, amplitude of vibrations at the tool tip is much higher. Table 2 summarizes Ra values of surface roughness computed at tool tip.

8. CONCLUSION

In this research work a comprehensive simulation model is developed for anticipating dynamic cutting forces and 3D surface texture for end-milling operations. Variation of instantaneous tool deflections are computed by FE analysis using CALFEM toolbox implemented in MATLAB V7.0 environment. Variations of dynamic tool deflections are also computed along axial direction for constructing 3D surface texture model. The final model contains all information required to compute surface roughness parameter and surface location errors.

REFERENCES

Altintas, Y. (2010). *Manufacturing automation, metal cutting mechanics, machine tool vibration and CNC design*. Cambridge University Press.

Athavale, S. M., & Strenkowski, J. S. (2013). Material damage-based model for predicting chipbreakability. *Journal of Manufacturing Science and Engineering*, *119*(11), 675–680.

Baker, M., Rosler, J., & Siemers, C. (2012). A finite element model of high speed metal cutting with adiabatic shearing. *Computers & Structures*, *80*(5), 495–513.

Bil, H., Kilic, S. E., & Tekkaya, A. E. (2014). A comparison of orthogonal cutting data from experiments with three different finite element models. *International Journal of Machine Tools & Manufacture*, *44*(9), 933–944. doi:10.1016/j.ijmachtools.2004.01.016

Budak, E., & Altintas, Y. (2012). Flexible milling force model for surface error prediction. *Journal of Engineering System Design and Analysis*, *12*(1), 89–94.

Budak, E., Altintas, Y., & Armarego, E. J. A. (1996). Prediction of milling force coefficient from orthogonal cutting data. Cambridge University Press.

CALFEM. (n.d.). *A Finite Element Toolbox, V3.3.* LUND University.

Carrino, L., Giuliano, G., & Napolitano, G. (2013). Finite element simulation of orthogonal metal cutting. *Journal of Materials Processing Technology*, *12*(1), 138–157.

Ceretti, E., Fallbohmer, P., Wu, W. T., & Altan, T. (2015). Application of 2D FEM to chip formation in orthogonal cutting. *Journal of Materials Processing Technology*, *59*(1), 169–180.

Donea, J., Fasoli-Stella, D. P., & Giuliani, S. (2014). Lagrangian and eulerian finite element techniques for transient fluid structure interaction problems.*Transactions of the ASME*, *11*(2), 15-19.

Fenton, R. G., & Oxley, P.L.B. (2013). Mechanics of orthogonal machining - Predicting chip geometry and cutting forces from work. *Material Properties and Cutting Conditions, 18*(1), 927-942.

Itoigawa, F., Childs, T. H. C., Nakamura, T., & Belluco, W. (2006). Effects and mechanisms in minimal quantity lubrication machining of aluminum alloy. *Journal of Wear*, *260*(3), 339–344. doi:10.1016/j.wear.2005.03.035

Iwata, K., Osakada, K., & Terasaka, Y. (2015). Process modeling of orthogonal cutting by the rigid plastic finite element method. *Transactions of the ASME*, *106*(2), 132-138.

Ko, J. H., Yun, W. S., Cho, D. W., & Ehmann, K. F. (2002). Development of a virtual machining system, Part 1: Approximation of the size effect for cutting force prediction. *International Journal of Machine Tools & Manufacture*, *42*(15), 1595–1605. doi:10.1016/S0890-6955(02)00137-2

Lee, D., & Wilkening, W. W. (2012). Material modeling and high speed machining process.*Transactions of the ASME*, *13*(2), 324-341.

Lee, E. H., & Schaffer, B. W. (2011). Theory of plasticity applied to a problem of machining. *Journal of Applied Mechanics*, *18*(1), 405–416.

Lei, S., Shin, Y. C., & Incropera, F. P. (2013). Thermo-mechanical modeling of orthogonal machining process by finite element analysis. *International Journal of Machine Tools & Manufacture*, *39*(5), 731–750. doi:10.1016/S0890-6955(98)00059-5

Madhavan, V., Chandrasekar, S., & Farris, T. N. (2010). Machining as a wedge indentation. *Transactions ASME*, *67*(1), 128-139.

Mamalis, A. G., Horvath, M., Branis, A. S., & Manolakos, D. E. (2012). Finite element simulation of chip formation in orthogonal metal cutting. *Journal of Materials Processing Technology*, *110*(1), 19–27. doi:10.1016/S0924-0136(00)00861-X

Martelloti, M. E. (2011). An analysis of the milling process.*Transactions of the ASME*, *63*(1), 677-700.

MATLAB V7.0. (n.d.). MathWorks, Inc. Retrieved from www.mathworks.com

Montgomery, D., & Altintas, Y. (2014). Mechanism of cutting force and surface generation in dynamic milling. *ASME Journal of Engineering for Industry*, *113*(1), 160–168.

Moriwaki, T., Sugimura, N., & Luan, S. (2013). Combined stress, material flow and heat analysis of orthogonal micromachining of copper. *Transactions of the ASME*, *42*(1), 75-78, 1993.

Movahhedy, M., Gadala, M. S., & Altintas, Y. (2010). Simulation of the orthogonal metal cutting process using an arbitrary Lagrangian-Eulerian finite-element method. *Journal of Materials Processing Technology*, *103*(2), 267–275. doi:10.1016/S0924-0136(00)00480-5

Natarajan, R., & Jeelani, S. (2013). Residual stresses in machining using finite element method. *Computer Software and Applications*, *3*(1), 79–80.

Ozel, T., & Altan, T. (2007). Process simulation using finite element method - prediction of cutting forces, tool stresses and temperatures in high-speed flat end milling. *International Journal of Machine Tools & Manufacture*, *40*(5), 713–738. doi:10.1016/S0890-6955(99)00080-2

Ozel, T., & Zeren, E. (2006). Finite element modeling the influence of edge roundness on the stress and temperature fields induced by high-speed machining. *International Journal of Advanced Manufacturing Technology*, *35*(3), 255–267.

Pantale, O., Bacaria, J. L., Dalverny, O., Rakotomalala, R., & Caperaa, S. (2004). 2D and 3D numerical models of metal cutting with damage effects. *Computer Methods in Applied Mechanics and Engineering*, *193*(39), 4383–4399. doi:10.1016/j.cma.2003.12.062

Piispanen, V. (2014). Theory of formation of metal chips. *Journal of Applied Physics*, *19*(1), 876–881.

Ranganath, S., Campbell, A. B., & Gorkiewicz, D. W. (2007). A model to calibrate and predict forces in machining with honed cutting tools or inserts. *International Journal of Machine Tools & Manufacture*, *47*(5), 820–840. doi:10.1016/j.ijmachtools.2006.06.019

Rosa, P. A. R., Martins, P. A. F., & Atkins, A. G. (2007). Revisiting the fundamentals of metal cutting by means of finite elements and ductile fracture mechanics. *International Journal of Machine Tools & Manufacture*, *47*(3), 607–617. doi:10.1016/j.ijmachtools.2006.05.003

Sekhon, G. S., & Chenot, J. L. (2013). Numerical simulation of continuous chip formation during non-steady orthogonal cutting.*Transactions of the ASME*, *10*(1), 31-48.

Smith, S., & Tlusty, J. (2011). An overview of modelling and simulation of milling process. *ASME Journal of Engineering for Industry*, *113*(2), 169–175. doi:10.1115/1.2899674

Tlusty, J. (2000). *Manufacturing Process and equipment*. Prentice Hall.

Yun, W. S., Hoon, J., Cho, D., & Ehman, K. (2012). Development of a virtual machining system. Part2: Prediction and analysis of a machined surface error. *International Journal of Machine Tools & Manufacture*, *42*(2), 1607–1615.

Chapter 3
Modeling and Analysis of Cold Drawing Process: Parameters and Methods

Praveen Kumar Loharkar
NMIMS, India

Mohan Kumar Pradhan
Maulana Azad National Institute of Technology, Bhopal, India

ABSTRACT

Products of cold drawing process are extensively used due to their characteristics such as very good surface finish, close dimensional control and superior mechanical properties. Therefore, this process has drawn attention of several researchers over the years. The chapter describes the state of research scenario in the field of cold drawing process analysis with a focus on various parameters and methods involved in its study. Work carried out in this context is described briefly along with the steps to carry out finite element analysis. It has also accounted for methods to design experiments and to optimize the process parameters. Upon reading this chapter, the reader would get acquainted with various aspects of cold drawing process modeling and scope for further research. With the help of fundamental knowledge of process and systematic application of the available tools and techniques, the reader would be equipped with the knowledge to work on related industrial problems and enhance process capability by applying soft computing techniques in process optimization.

INTRODUCTION

There are several ways to achieve desired geometrical and physical attributes in a product through various material processing methods. There are several conventional manufacturing processes such as casting, forming, machining, joining etc. along with non-traditional ones viz. coating, additive manufacturing, and non-conventional machining processes that are used either in combination or individually to add value in a raw material. In fact some of these non-conventional manufacturing methods such as additive manufacturing are being developed as an alternative for several conventional processes and have proven

DOI: 10.4018/978-1-5225-2440-3.ch003

their worth in the field of rapid prototyping. The selection of any of these manufacturing process is done based on several factors. First and foremost is the capability or in other words limitation of the process itself. The limitations can be with respect to material that can be processed, size handling capacity, machine parameters etc. The other significant aspects in process selection are economy of time and money.

Cold drawing is one of the most widely used manufacturing processes which comes under the ambit of an important category called forming processes. It is quite popular bulk deformation process and a cold drawn product demonstrates several advantages like very good surface finish, superior mechanical properties and greater dimensional accuracies over hot forming processes Owing to their characteristics, cold drawn tubes are used in various engineering applications. In fact, these tubes are inseparable part of automobile manufacturing, mining industry, boiler construction, bearing manufacturing industries and others. Therefore for obvious reasons, analysis of cold drawing process has been a topic of extensive research. Earlier the studies were mostly of pure analytical nature or experiment based. Various analytical methods like slab method, upper bound method or slip line theory were used for the analysis of material flow and its behaviour (Kumar, Praveen, Agnihotri, & Geeta, 2013).These methods have few inherent limitations. Analytical methods require few assumptions for ease of analysis which compromises the accuracy of results while experimental methods require both cost and time expenditure. In this age of cut throat competition, greater efficiencies coupled with accuracy and precision in the processes are the need of the hour. More so with the advent of advanced computational machines the speed and accuracy of numerical analysis has grown leaps and bounds. Several researchers have been using these capabilities to develop models for analysis and optimization of cold drawing processes. Finite Element method is an extensively used numerical technique in this context. This chapter is an effort to explore different independent and dependent variables associated with cold drawing process or in other words, various process parameters and methods used for analysis and optimization. An attempt is made to enlist some of the areas which still provides a scope for future research.

BACKGROUND: COLD DRAWING

Before moving ahead on the modeling aspects, let us understand different parameters or variables associated with this process. Fundamentally, cold drawing involves pulling long sections of solid or hollow bars to get a desired reduced size. It is further classified as follows:

- **Wire and Bar Drawing:** It involves pulling of a solid wire or a bar through a die selected as per the size requirement. See Figure 1.
- **Tube Drawing:** In this case, a hollow tube is pulled through a die that controls its final outer diameter while mandrel or plug may or may not be provided at the inner diameter depending upon the dimensional control desired. Different types of tube drawing are there based on the presence of plug and its physical arrangement, as can be seen in Figure 2.

Various parameters or variables associated with the drawing process are as follows:

- **Coefficient of Friction:** Friction between the tool and the material being drawn plays an important role. It not only affects the drawing effort but also influences the surface finish. Commercially lot of chemical combinations are being used to reduce coefficient of friction like use of phosphate

Figure 1. Bar drawing

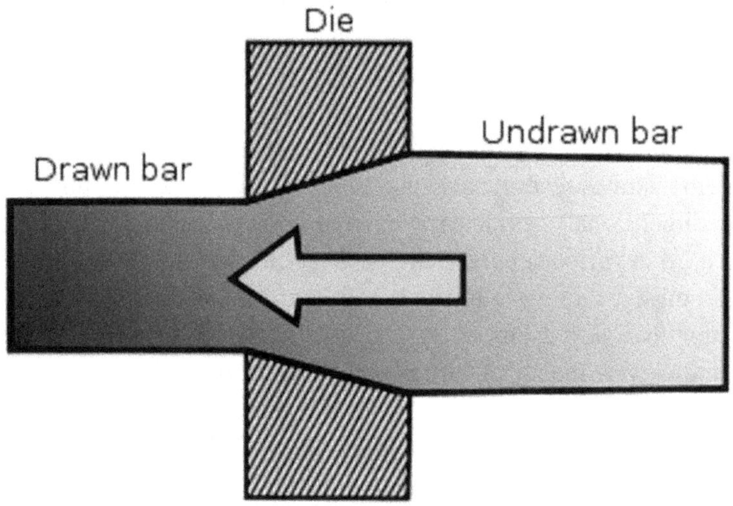

Figure 2. Types of tube drawing (Bihamta et al., 2011)

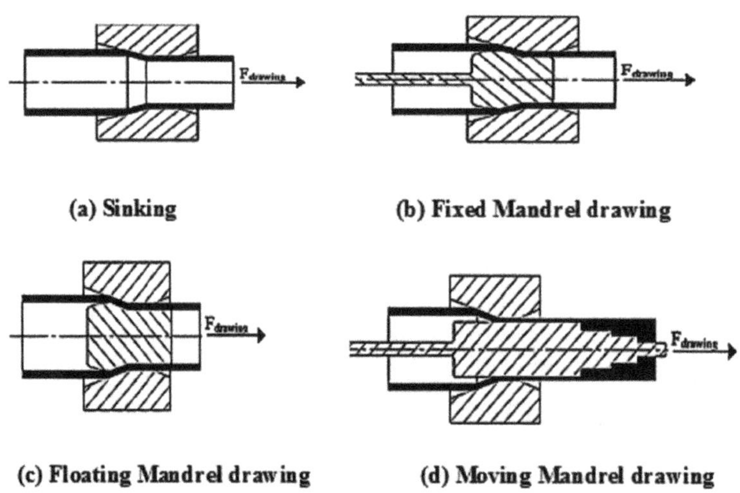

and soap compounds provide the necessary layer to minimize friction. Apart from these chemicals, oil is also used as lubrication medium to reduce friction and absorb the heat during the drawing process. Researchers like Chuiko et al. (1973), Obi and Oyinlola (1996), Neves et al. (2005), Byon et al. (2011) worked in this area to identify or/ and develop lubricants of different types or their combinations to achieve maximum advantages of smooth drawing like increase tool as well as machine life, less power consumption, improved productivity etc.

- **Tool Geometry:** The most important tool used in the drawing process is Die which basically controls the outer diameter and also aids in providing very good surface finish. Moreover mandrel or plug is used in the case of tube drawing to control the inner dimensions. Significant geometrical attributes associated with the die are entry angle, approach angle, bearing portion and relief angle.

While the geometrical parameter associated with the plug is its outer diameter. Control over these geometrical parameters and profiles largely affect the quality of product. Furthermore, modifications in the profile of tube being drawn has led to several developments in the tool design itself.

Neves et al. (2005), Béland et al. (2011), and Gunasekera and Hoshino (1982) worked on obtaining optimal tool geometry to attain minimal stress or drawing force. Process parameters related with tool configuration were studied by Kim et al. (2007) and it was inferred that the profile design of the tool used in drawing is the most important factor related to energy required for forming and deformation behaviour of tubular material. Wang and Argyropoulos (2005) designed, modified and analyzed a plum-blossom-shape die for the direct cold drawing of hexagon/square-section rods in order It had dramatically increased the manufacturing productivity, improved the product quality, and reduced the manufacturing time and cost. Xu et al. (2011) proposed a modification in the plug that enhances the surface finish. The evaluation of effect of tools' geometries on the maximum possible tube deformation was studied by Bihamta et al. (2011) on variable thickness tube drawing process with a newly developed procedure.

- **Product Geometry:** Conventionally, round tubes are manufactured through cold drawing process but various engineering applications especially in automobile sector and other constructional domain, use of other shapes of tube is also growing up. Modified product geometry requires relevant changes in the overall process. Many researchers have come up with methods to draw tubes of modified shapes. For example, Yoshizawa et al. (1984) carried out experiments to develop a technique to manufacture rifled tubes. These tubes are now being used at large scale in boiler industry as they provide very good heat transfer characteristics. Bayoumi (2001) worked on development of regular polygonal metal tubular sections from round tube. Modification in the geometry also influences the load on the tool, this forming tool load in manufacturing a square tube from a round tube was determined by Bayoumi and Attia (2009). Parshin (2009) analysed the variation in the shape of the eight-beam star like tubes in the course of multipass drawing. Xu et al. (2007) developed an aluminium tube with rectangular cross-section. At the same time, modifications in the existing process is demonstrated by Bihamta et al. (2011). It was used to produce tubes with variable thickness in the axial direction.
- **Drawing Speed:** Drawing requires pulling of wire/bar or tube trough the die with a specified speed. This speed also influences the properties of the product as determined by Cetinarslan C S (2012). Statistical analysis was used to establish the relationship between the drawing speed and torsional and tensile strength of the drawn materials.
- **Percentage Reduction:** The amount of reduction that can be achieved in a single pass is limited by material's ductility. Achieving maximum amount of reduction in single pass not just makes the process economically profitable but also enhances mechanical properties of the drawn wire or tubes. In order to investigate the limit of maximum reduction that can be achieved for a given material and process parameters lot of work has been carried out. Like Raji and Oluwole (2011) studied the effect of progressive cold drawing reduction by 20%, 25%, 40% and 50% on the mechanical properties of the carbon steel material. Similarly Celentano et al. (2009) assessed the effect of wire reductions along with other parameters on material response. Bui et al. (2011) carried out an experimental study in order to evaluate the effect of cross section reduction.

Thus it can be seen that there are several variables in the process of cold drawing that have an influence on the response variables of the process such as mechanical properties, drawing force, drawing stresses and strains etc. An important point to note is that the variables discussed here are not exhaustive in nature. The researchers have and can consider other variables depending upon the objectives of their studies.

MAIN FOCUS OF THE CHAPTER

Analysis of any process can be done either through analytical methods and/or through numerical techniques in the absence of experimental facility. But there exist an element of caution and that is, due to inherent limitations of analytical procedures like initial assumptions and of numerical techniques like approximate solutions, validation of the results become extremely important and inevitable. Moreover, the accuracy of the results is greatly affected by the model definition. Therefore it is necessary to validate the results through initial experiments. Studying the influences of operating parameters without actually performing physical experiments have become relatively convenient with the development of commercial packages. Researchers have extensively used these developments for their advantage and have come up with solutions to industrial problems on cold drawing. As the book deals with the modeling and optimization aspect of the manufacturing processes, the next section will put a light on various models and optimization techniques used by the researchers that has brought about significant process improvements.

COLD DRAWING: MODELING AND OPTIMIZATION

Well before the development of computers, analytical models were developed and used to analyse the process of cold drawing. For instance, description of metal deformability in the process of drawing tubes on a fixed mandrel was given by Pospiech (1998) using a generalized model that is devised by Kolmogorov. This model suggests that the coefficient of utilization of reserve of plasticity is an inversely variable function of coefficient of friction (at a constant angle of the die reduction zone). Rubio et al. (2005) studied the main parameters of the drawing process by different methods. Upper bound method has been used by Rubio et al. (2006) to analyse thin-walled tubes drawing through conical convergent dies with fixed inner, conical or cylindrical, plug using the. Analytical formulations were extensively used by Gur'yanov (2009) to calculate the limiting extension per drawing pass using six different formulas which can then be used to determine the axial-stress increment in the working region of the die.

Similarly free body equilibrium method has been used to determine tension in the drawing process which was used to solve the drawing problem for dies of axisymmetric or symmetric sections (Rojas et al, 2008). As opposed to the classical slab method, solution of the equations obtained through this method considered the internal material distortion. Rubio (2006) analyzed tube drawing using analytical methods, i.e. Slab Method (SM), with and without friction effects and the Upper Bound Method (UBM). An extension of an upper bound solution that was developed earlier has been proposed by Bui et al. (2011). This newly developed method can be used to predict the drawing stress field.

Moving ahead Finite element models became a norm and based on the objectives of study like determination of conditions leading to defect, optimum process parameters, residual stresses etc. different models have been proposed. In order to assess the effect of operating conditions Celentano et al. (2009) developed a numerical model. In the similar context, Camacho et al. (2006) studied the effect of drawing variables through finite element method. But his focus was mainly to check their influence on the defects. Furthermore several researchers developed models to investigate the nature and cause of defects in cold drawing. Shinohara and Yoshida (2005) carried out a three-dimensional finite element analysis (FEA) to investigate the growth of surface defects during the process of wire drawing. It was concluded that superior wires without flaws could be obtained by removing any defect before the process itself or by introducing multi-pass drawing that would tend to correct the flaws. Weygand et al. (2006) proposed a finite element model to analyze the drawing process and to find out the factors that lead to defects known as splits. Residual stresses (arising in the cold-drawing process) are again a cause of big concern in most of the manufacturing processes. As it leads to several issues like dimensional instability and defects. Its effect on the fatigue crack propagation in a metallic round bar (cracked) having a V-shaped notch (circumferential) was examined by Carpinteri et al. (2010). The crack propagation under cyclic tension combined with the residual stresses was analyzed by taking into account the stress intensity factor and the actual stress ratio. Toribio et al. (2011) used a numerical model to analyse the role of residual stresses and strains arising because of drawing operation in the performance of cold drawn (prestressed) steel wires under conditions of hydrogen embrittlement (HE).

Use of statistical techniques and soft computing is now being extensively employed in the optimization of drawing process. Optimization is basically the act of maximizing or minimizing the objective by controlling the magnitude of parameters or conditions within the given constraints. We have already seen various parameters of the drawing process and their influence on the process in general. Several studies have been undertaken to determine optimized values of these process parameters. For instance, Chen and Huang (2007) have utilized the finite element modeling and the Taguchi methods for optimization of the process parameters of the wire drawing process. Neural networks has been used by Dwivedi et al. (2008) to model relationships between controlled, uncontrolled process parameters and the yield in the drawing process. Many of the times the emphasis is on studying one process parameter at a time and its corresponding effect on the product. Dekhtyarev et al. (2009) carried out a research to demonstrate the combined effect of these parameters on final properties of the product and production process as a whole. Design of Experiments have been used by Rocha et al. (2012) to analyse distortion for a typical manufacturing process of induction hardened AISI 1045 cylindrical steel bars subjected to pre-straightening and cold drawing. The numerical study carried out by Tang et al. (2011) contributed in this field by suggesting a new approach for the optimization of the drawing parameters and also illustrated the damage evolution of the drawn wire in each of its multiple passes. The study also demonstrated the distribution of stresses along axial and circular directions and established a fact that wire breakage is expected to occur in those areas of the drawn wire where fractures most possibly initiate. An investigation involving analysis of tube drawing process incorporated finite element modeling, experimental validation and central composite design to generate design matrix and carried out response surface methodology to determine values of drawing process parameters giving optimal response with reference to drawing force and dimensional error (Hosseinzadeh and Mouziraji, 2016).

Finite Element Modeling of Cold Drawing Process

Demand for faster, accurate and fairly accurate solutions have led to extensive use of a very popular numerical method i.e., finite element method. Researchers in the domain have used this method to find out solutions of varied nature ranging from determining the effects of process variables on different characteristics of cold drawing and cold drawn product. One significant advantage of using commercial packages of finite element method is the visualization feature which enables the user to see the stress and strain distribution in the body and predict the failure. It basically involves three steps:

- **Preprocessing:** In the first stage of finite element analysis pre-processing is carried out which basically involves modeling the structure. There are many commercial packages which are used for structural modeling like CATIA, Creo Parametric, Solidworks etc. These models are transferred and then imported into analysis softwares like MSC SUPERFORM, ABAQUS, ANSYS, LS DYNA, DEFORM etc. The accuracy of the results is significantly affected by the model. Once the geometry is modeled, the next procedural components of pre-processing is defining material properties, meshing, applying constraints and load. A typical preprocessing file consists of, information on:
    - Domain(workpiece) discretisation: nodal coordinates and element definition.
    - Initial and boundary conditions: descriptions of loading, tool geometry and movement.
    - Solver and special solution technical details.
    - Constitutive model and mechanical property data for the materials involved.
    - Computational parameters.
    - Output requirements.

Once the preprocessor file is created, the next step in finite element analysis is processing.

- **Analysis/Processing:** In this step, the geometry, constraints, mechanical properties and loads are applied to generate matrix equations for each element, which are then assembled to generate a global matrix equation of the structure. The form of the individual equations, as well as the structural equation is always,

$$\{F\} = (K)\{u\}$$

where {F} = External force matrix. (K) = Global stiffness matrix {u} = Displacement matrix

The equation is then solved for deflections. Using the deflection values, strain, stress, and reactions are calculated. All the results are stored and can be used to create graphic plots and charts in the post analysis.

- **Post Processing:** In post processing finite element results are analyzed and useful information is extracted, normally in graphical format. Many commercial packages have an interactive graphical postprocessor for providing mesh plots, contour plots of results, time history plots and x-y plots.

Description of Finite Element Analysis Through Case Studies

Let us take an example of finite model of cold drawing process created by Hosseinzadeh and Mouziraji (2016).

This was a 3D Finite element model developed in ABAQUS/EXPLICIT software V 6.10. The model is shown in Figure 3.

The die has been modeled as a rigid body using R3D4 elements, which is a convention for modeling tools in the finite element analyses of metal forming processes, as their elastic deformation is negligible compared to large plastic deformation experienced by the workpiece.

The workpiece, which is a copper tube in this case is modeled as a deformable body with elements C3D8R. This element is a general purpose linear brick element, with reduced integration (1 integration point). Due to the reduced integration, the locking phenomena observed in the C3D8R element do not show.

One more important aspect of finite element analysis is the number of elements being chosen for analysis. Increasing the number of elements enhances the accuracy of solution but at the same time leads to increase in the processing time. Hence an optimum number of elements that can lead to reliable results have to be chosen. This is typically done through convergence tests.

Now the interface friction between the workpiece and the tool is a function of material and the lubricant chosen for analysis. In the chosen case, surface to surface contact with coulomb friction law was considered.

The standard Coulomb friction models assumes that no relative motion occurs if the equivalent frictional stress, τ_{eq}, is less than τ_{cr} where,

$$\tau_{eq} = \sqrt{\tau_1^2 + \tau_2^2}$$

and

Figure 3. Finite element model of cold drawing (Hosseinzadeh and Mouziraji, 2016)

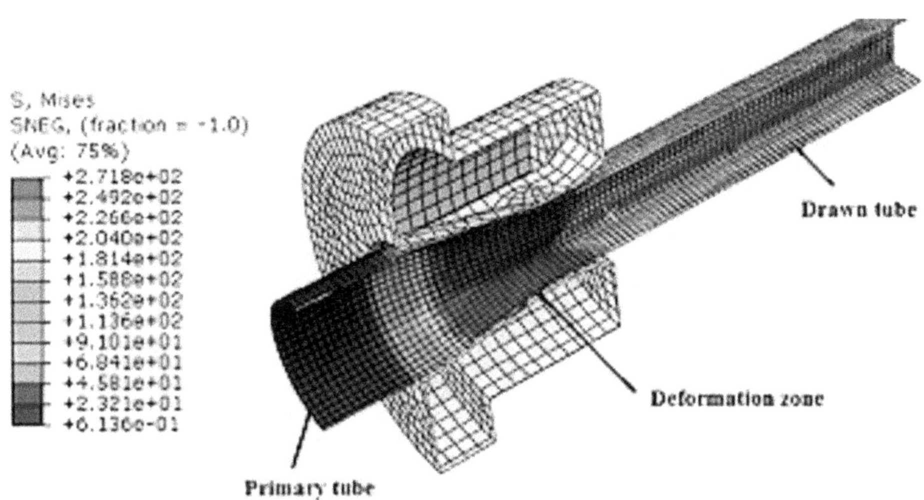

$$\tau_{cr} = \mu p$$

where μ is the friction coefficient and p is the contact pressure.

Surface to surface interaction between the two pairs, that is, hollow-die and hollow plug has to be defined and the values of coefficient of friction are usually taken as per the simulation conditions illustrated below.

The typical value for friction coefficient in case of industrial tube drawing is around 0.05(Neves et al., 2005). In the present example the value of coefficient of friction has been chosen to be 0.1-0.3 based on the lubricant used.

As far as material properties are concerned, one can use standard literature after ensuring its reliability or perform uniaxial tests. The sample size for testing is determined on the basis of ASTM standards. For instance in the chosen example the test specimen was prepared as per ASTM E-1251-11. The properties required for computational analysis are density, Young's modulus, Poisson's ratio, ultimate tensile strength, yield strength and true stress v/s plastic strain relationships. An important point to be noted here is that incorrect material properties would lead to poor analysis results.

In case of analysis of axisymmetric geometries with axisymmetric loading like tube drawing of circular sections two dimensional axisymmetric models can be used. It reduces the time of computation and give fairly accurate results.

In order to ensure the validity of the finite element model, the results obtained through numerical analysis are needed to be compared with experimental results or with the results available in existing literature (extracted from a reliable source). For instance, model validation can be done by comparing the strains developed in actual drawing practice to the strains developed after drawing simulation in the finite element model.

Consider the axisymmetric model shown in Figure 4.

The details of the model and the tube geometry used in carrying out the cold drawing simulation and validation is described as follows. The coefficient of friction have been kept as 0.05 and drawing speed is 400 mm/s. (Loharkar, 2015).

Figure 4. Axisymmetric finite element model (Kumar & Agnihotri, 2013)

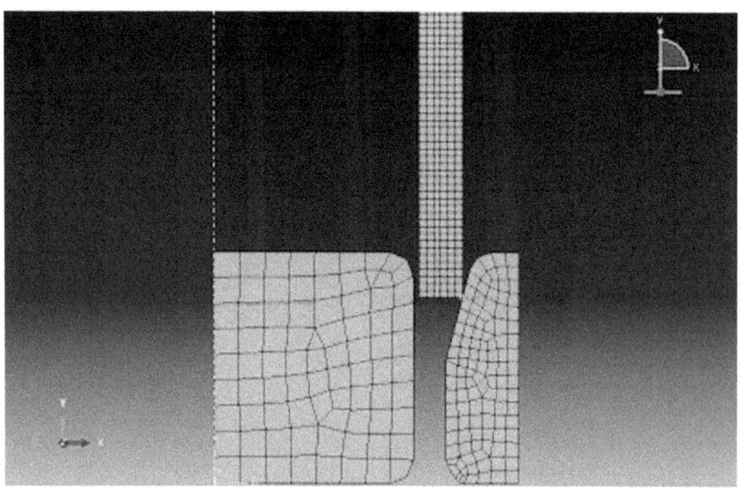

- **Hollow Dimensions:** 110x 9.50.
- **Tube Dimensions:** 103x7.25 (Experimental value).
- **Tube Dimensions:** 102.55x7.04 (Finite Element analysis).

The comparison between radial strain, circumferential strain and axial strain is shown in Table 1 (Kumar & Agnihotri, 2013). It was observed that both the experimental and finite element results are in fair agreement with each other, thereby validating the model.

Expressions for the strain calculation are as follows:

- Radial Strain,

$$\varepsilon_{radial} = \ln(\frac{h_f}{h_i})$$

where, h_f is tube wall thickness at exit, h_i is tube wall thickness at inlet.

- Circumferential Strain,

$$\varepsilon_{circular} = \ln(\frac{D_{of} - h_f}{D_{oi} - h_i})$$

where,

D_{of} is Tube outside diameter at exit,

D_{oi} is Tube outside diameter at inlet.

- Axial strain,

$$\varepsilon_{axial} = -\varepsilon_{radial} - \varepsilon_{circular}$$

Validation ensures the accuracy of the simulation model and enables researcher to carry out computer analysis to study the influence of process parameters through variation of independent variables.

Table 1. Model validation (Kumar & Agnihotri, 2013)

Strain	Experimental Results	Finite Element Analysis Results
Radial Strain (ε_{radial})	-0.270	-0.299
Circumferential Strain ($\varepsilon_{cir.}$)	-0.0484	-0.059
Axial Strain (ε_{axial})	0.3184	0.3499

Design of Experiments and Optimization

There are different techniques available to create design matrix not just to carry physical experiments but to decide upon simulation sequences in the analysis software too. In fact, Hosseinzadeh and Mouziraji (2016) used central composite design to develop an experimental design matrix to carry out finite element analysis for different combinations of process parameters. These simulation results were further analysed through response surface methodology to derive optimum conditions. In the similar way, full factorial analysis was used by Neves et al. (2005) to carry out experiments with different combination of variables.

In most of the work carried out in the area of cold drawing, finite element modeling have been used not just for analysis but for optimisation as well. There is tremendous scope for usage of soft computing techniques like artificial neural networks, genetic algorithms, particle swarm optimization etc. in this area.

CONCLUSION

The chapter is an attempt to make the readers acquainted with different parameters and methods associated with the analysis of one of the most popular cold forming processes that is, cold drawing. Researchers have used several tools and techniques available for analysis, starting from analytical methods to finite element modeling. As far as optimization is concerned, artificial neural network, Taguchi method, response surface methodology etc. have been used to arrive at optimal values of process parameters leading to improved tool geometry, lubrication conditions, minimum drawing force and residual stresses in the final product. The area has already seen tremendous amount of research and further scope is available in the field of lubrication, modification in tube geometry and improving life of tool materials. For instance, the development of rifled tubes possessing improved heat transfer characteristics over conventional tubes has been a benchmark research and are being extensively used in boiler sector. Moreover, ability to rectify the inherent defects in the raw material during the process is an ongoing area of research. At the same time, development of optimized models to study drawing of non-conventional tube geometries with unpopular materials for critical applications like space is still an area of unlimited opportunities. It presents wide scope for future research.

REFERENCES

Bayoumi, L. (2001). Cold drawing of Regular Polygonal Tubular Sections from Round Tubes. *International Journal of Mechanical Sciences*, *43*(11), 2541–2553. doi:10.1016/S0020-7403(01)00056-X

Bayoumi, L., & Attia, A. S. (2009). Determination of the Forming Tool Load in Plastic Shaping of a Round Tube into a Square Tubular Section. *Journal of Materials Processing Technology*, *209*(4), 1835–1842. doi:10.1016/j.jmatprotec.2008.04.047

Béland, J. F., Fafard, M., Rahem, A., DAmours, G., & Côté, T. (2011). Optimization on the Cold Drawing Process of 6063 Aluminium Tubes. *Applied Mathematical Modelling*, *35*(11), 5302–5313. doi:10.1016/j.apm.2011.04.025

Bihamta, R., Bui, Q. H., Guillot, M., DAmours, G., Rahem, A., & Fafard, M. (2011). Application of a New Procedure for the Optimization of Variable Thickness Drawing of Aluminium Tubes. *Journal of Materials Processing Technology*, *211*, 578–589. doi:10.1016/j.jmatprotec.2010.11.012

Bihamta, R., Bui, Q. H., Guillot, M., DAmours, G., Rahem, A., & Fafard, M. (2011). A New Method for Production of Variable Thickness Aluminium Tubes: Numerical and Experimental studies. *Journal of Materials Processing Technology*, *211*(4), 578–589. doi:10.1016/j.jmatprotec.2010.11.012

Bui, Q. H., Bihamta, R., Guillot, M., DAmours, G., Rahem, A., & Fafarda, M. (2011, March). Investigation of the Formability Limit of Aluminium Tubes Drawn with Variable Wall Thickness. *Journal of Materials Processing Technology*, *211*(3), 402–414. doi:10.1016/j.jmatprotec.2010.10.016

Bui, Q. H., Bihamta, R., Guillot, M., Rahem, A., & Fafard, M. (2011). Effect of Cross Section Reduction on the Mechanical Properties of Aluminium Tubes Drawn with Variable Wall Thickness. *Journal of Manufacturing Science and Engineering*, *133 /061004*, 1–10.

Byon, S. M., Lee, S. J., Lee, D. W., Lee, Y. H., & Lee, Y. (2011). Effect of Coating Material and Lubricant on Forming Force and Surface Defects in Wire Drawing Process. *Trans. Nonferrous Met. Soc. China*, *21*, S104–S110. doi:10.1016/S1003-6326(11)61071-6

Camacho, A. M., Gonzalez, C., Rubio, E. M., & Sebastian, M. A. (2006). Influence of Geometrical Conditions on Central Burst Appearance in Axisymmetrical Drawing Processes. *Journal of Materials Processing Technology*, *177*(1-3), 304–306. doi:10.1016/j.jmatprotec.2006.04.037

Carpinteri, A., Brighenti, R., & Vantadori, S. (2010). Influence of the Cold-Drawing Process on Fatigue Crack Growth of a V-notched Round Bar. *International Journal of Fatigue*, *32*(7), 1136–1145. doi:10.1016/j.ijfatigue.2009.12.014

Celentano, D. J., Palacios, M. A., Rojas, E. L., Cruchag, M. A., Artigas, A. A., & Monsalve, A. E. (2009). Simulation and Experimental Validation of Multiple-Step Wire Drawing Processes. *Finite Elements in Analysis and Design*, *45*(3), 163–180. doi:10.1016/j.finel.2008.09.001

Cetinarslan, C. S. (2012). A Study on Influences of some Process Parameters on Cold Drawing of Ferrous Wires. *Indian Journal of Engineering and Materials Sciences*, *19*, 221–228.

Chen, D. C., & Huang, J. Y. (2007). Design of Brass Alloy Drawing Process using Taguchi Method. *Materials Science and Engineering*, *464*(1-2), 135–140. doi:10.1016/j.msea.2007.01.139

Chuiko, Savin, & Kalashnikov. (1973). Cold Drawing of Stainless Steel Tubes on Short Mandrel. *Metallurg*, (3), 32-33.

Dwivedi, S., Huang, S. H., Shi, J., & VerDuin, W. H. (2008). Yield Prediction for Seamless Tubing Processes: A Computational Intelligence Approach. *International Journal of Advanced Manufacturing Technology*, *37*(3-4), 314–322. doi:10.1007/s00170-007-0964-5

Gunasekera, J. S., & Hoshino, S. (1982). Analysis of Extrusion or Drawing of Polygonal Sections through Straightly Converging Dies. *Journal of Engineering for Industry*, *104*(1), 38–45. doi:10.1115/1.3185795

Kim, S. W., Kwon, Y. N., Lee, Y. S., & Lee, J. H. (2007). Design of Mandrel in Tube Drawing Process for Automotive Steering Input Shaft. *Journal of Materials Processing Technology, 187–188*, 182–186. doi:10.1016/j.jmatprotec.2006.11.134

Kumar, & Agnihotri. (2013). Cold Drawing Process–A Review. *Metal Forming, 3*(3).

Kumar, P., & Agnihotri, G. (2013). *Application of finite element method on seamless tube cold drawing* (Unpublished Master's thesis). M.A.N.I.T. Bhopal, India.

Loharkar. (2015). Parametric analysis of industrial cold drawing process. *International Journal of Engineering Research and General Science, 3*(5).

Morteza, H., & Ghasempour, M. M. (2016). An analysis of tube drawing process used to produce squared sections from round tubes through FE simulation and response surface methodology. *International Journal of Advanced Manufacturing Technology*. doi:10.1007/s00170-016-8532-5

Nagarkar, M. P., Zaware, R. N., & Ghalme, S. G. (2012). Finite Element Simulation of Sink Pass Round Tubes using ANSYS. *APTEFF, 43*(43), 179–188. doi:10.2298/APT1243179N

Narayanan, Idapalapati, & Subbiah. (2010). Effect of Cold Work on the Mechanical Response of Drawn Ultra-Fine Gold Wire. *Computational Materials Science, 12*(38).

Neves, Button, Caminaga, & Gentile. (2005). Numerical and Experimental Analysis of Tube Drawing with Fixed Plug. *J. of the Braz. Soc. of Mech. Sci. & Eng., 27*(4), 426-431.

Obi, A. I., & Oyinlola, A. K. (1996). Frictional Characteristics of Fatty-Based Oils in Wire Drawing. *Wear, 194*(1-2), 30–37. doi:10.1016/0043-1648(95)06664-0

Parshin, S. V. (2009). Multipass Drawing of Finned Tubes. *Russian Journal of Non-Ferrous Metals, 50*(2), 128–132. doi:10.3103/S1067821209020102

Pospiech, J. (1998). Description of a Mathematical Model of Deformability for the Process of Drawing Tubes on a Fixed Mandrel. *Journal of Materials Engineering and Performance, 7*(1), 71–78. doi:10.1361/105994998770348061

Rajan, K. M., & Narasimhan, K. (2001). An Investigation of the Development of Defects During Flow Forming of High Strength Thin Wall Steel Tubes. *Practical Failure Analysis, 1*(5), 69–76. doi:10.1007/BF02715366

Raji, N. A., & Oluwole, O. O. (2011). Influence of Degree of Cold-Drawing on the Mechanical Properties of Low Carbon Steel. *Materials Sciences and Applications, 2*(11), 1556–1563. doi:10.4236/msa.2011.211208

Rocha, Nunes, & Hirsch. (2012). Analysis by Design of Experiments of Distortion Potentials in Drawn and Induction Hardened Wire. *Materials Research, 15*(2), 266-276.

Rojas, Gonza´lez, Calvet, & Bubnovich. (2008). A New Analytical Solution for Prediction of Forward Tension in the Drawing Process. *Journal of Materials Processing Technology, 198*, 93–98.

Rubio, E. M. (2006). Analytical methods application to the study of tube drawing processes with fixed conical inner plug: Slab and Upper Bound Methods. *Journal of Achievements in Materials and Manufacturing Engineering*, *14*(1-2), 383–386.

Rubio, E. M., Camacho, A. M., Sevilla, L., & Sebastian, M. A. (2005). Calculation of the Forward Tension in Drawing Processes. *Journal of Materials Processing Technology*, *162–163*, 551–557. doi:10.1016/j.jmatprotec.2005.02.122

Rubio, E. M., Gonzalez, C., Marcos, M., & Sebastian, M. A. (2006). Energetic Analysis of Tube Drawing Processes with Fixed Plug by Upper Bound Method. *Journal of Materials Processing Technology*, *177*(1-3), 175–178. doi:10.1016/j.jmatprotec.2006.03.193

Shinohara, T., & Yoshida, K. (2005). Deformation Analysis of Surface Flaws in Stainless Steel Wire Drawing. *Journal of Materials Processing Technology*, *162–163*, 579–584. doi:10.1016/j.jmatprotec.2005.02.125

Tang, K. K., Li, Z. X., & Wang, J. (2011). Numerical Simulation of Damage Evolution in Multi-Pass Wire Drawing Process and its Applications. *Materials & Design*, *32*(6), 3299–3311. doi:10.1016/j.matdes.2011.02.026

Toribio, J., Kharin, V., Lorenzo, M., & Vergara, D. (2011). Role of Drawing-Induced Residual Stresses and Strains in the Hydrogen Embrittlement Susceptibility of Prestressing Steels. *Corrosion Science*, *53*(10), 3346–3355. doi:10.1016/j.corsci.2011.06.012

Wang, K. L., & Argyropoulos, V. (2005). Design and Analysis of Direct Cold Drawing of Section Rods through a Single Die. *Journal of Materials Processing Technology*, *166*(3), 345–358. doi:10.1016/j.jmatprotec.2004.06.030

Xu, W., Wang, K., Wang, P., & Zhou, J. (2011). A Newly Developed Plug in the Drawing Process for Achieving the High Accuracy of Aluminium Rectangular Tube. *International Journal of Advanced Manufacturing Technology*, *57*(1-4), 1–9. doi:10.1007/s00170-011-3292-8

Yoshizawa, M., Sumimoto, D., & Kazuhiro, K. (1984). *The Forming Technique and Quality Characteristics of ERW Rifled Boiler Tube*. 106th ISIJ Meeting, S1221 and S1394.

Chapter 4
Modelling and Optimization of End Milling Process Using TLBO and TOPSIS Algorithm:
Modelling and Optimization of End Milling Process

Atul Tiwari
Maulana Azad National Institute of Technology, Bhopal, India

Mohan Kumar Pradhan
Maulana Azad National Institute of Technology, Bhopal, India

ABSTRACT

To assure desire quality of machined products at minimum machining costs and maximum material removal rate, it is very important to select optimum parameters when metal cutting machine tool are used. Minimum Surface Roughness (Ra) is commonly desirable for the component; however Material Removal Rate (MRR) should be maximized. This chapter presents an approach for determination of the best cutting parameters precede to minimum Ra and maximum MRR simultaneously by integrating Response Surface Methodology with Multi-Objective Technique for Order Preference by Similarity to Ideal Solution and Teaching and learning based optimization algorithm in face milling of Al-6061 alloy. 30 experiments have been conducted based on RSM with 4 parameters, namely Speed, Feed, Depth of Cut and Coolant Speed and three levels each. ANOVA is performed to find the most influential input parameters for both MRR and Ra. Later the multi-objective attribution selection method TOPSIS and multi objective optimization method TLBO is used to optimize the responses.

DOI: 10.4018/978-1-5225-2440-3.ch004

INTRODUCTION

Milling is the metal removal process of using rotary multi point cutters hold on arbor and spindle to remove material from a work piece by advancing in a direction or at an angle with the axis of the tool. There are a wide variety of different milling operations and machines, on scales from small individual parts to large parts. It is one of the most commonly used processes in industry, today for machining parts to precise sizes and desire shapes. Application of Milling-

- Flat surface in horizontal, vertical and inclined planes.
- Generating slots or ribs or grooves at various sections.
- Slitting or parting of work piece.
- For producing surfaces of revolution.
- Generating helical flutes of the drills.
- Long thread milling on power screws, large lead screws, worms etc.
- Short thread milling for small size fastening screws, bolts etc.

The milling operation can be broadly classified according to their operation such as horizontal milling, vertical milling and inclined milling depending on the direction of cutter axis. End milling, face milling, slot milling, slab milling, saw milling, straddle milling, gang milling, etc. Face milling is mostly used for milling operations for high metal removal rate. Face milling is a combination of up cut and down cut milling operation. In face milling, the location of the cutter with respect to the work piece is of considerable significance. Face milling as shown in Figure 1 is usually preferred for rough machining of larger surfaces.

The production rate is also higher as the diameter of the cutter is large but quality and productivity are the two important sides which is usually required for any industry. In order to achieve that, the operating parameter parameters should be suitably adjust, but they are opposite in nature. Generally maximizing productivity is directly related to the machining cost. However, productivity can be expounded in term of MRR and quality can be many factor such as machining parameters, cutting phenomena, work piece properties and cutting tool properties, etc. The significant machining parameters are v, f, DOC, CS chosen by literature survey and experience. In automobile and aerospace industry, there is demand for materials that are tougher, lighter, harder, stiffer, and stronger, more corrosion and erosion- resistance properties and superior mach inability index. Al-6061 alloy possess all these properties, hence considered under this research as a work-piece material. Expounded in terms of product characteristics, i.e. dimensional accuracy, surface smoothness and form stability, etc. 6061 Al alloy is a precipitation hardened alloy, containing magnesium and silicon as their main alloying component. It has good thermal and mechanical properties and show better weld-ability property. It is one of the most useful and common alloys of aluminum for industrial purpose use. 6061 is commonly used for the following:

- Construction of aircraft planes structure, such as wings and fuselages, commonly in home built aircraft than military or commercial aircraft. Another alloy 2024 is certain stronger, but 6061 is more easily heat treated work and remains resistant to corrosion even when the surface is eroded, which is not in the case of 2024, which is commonly used for thin AL clad coating for corrosion resistance.

Figure 1. Face milling

- Aluminum cans for the packaging of beverages.
- Small utility boats.
- SCUBA tanks.
- Some tactical flashlights.
- Various Automobile parts, such as wheel spacers etc.

The Technique for Order of Preference by Similarity (TOPSIS)

The Technique TOPSIS is a multi-objective decision analysis method, which was constituted by Hwang and Yoon in 1981 (Hwang & Yon, 2012). TOPSIS is based on the theory that is used to chose alternative and should have the shortest geometric distance from the positive ideal solution (Assari et al, 2012) and the longest geometric distance from the negative ideal solution. It is a technique of compensatory stockpiling that compares a set of alternatives by identifying weights for each criterion, normalizing data for each criterion and calculating the geometric distance between every alternative and the ideal alternative, which is the best data in each criterion in each condition. An assumption of TOPSIS is that the criteria are increasing or decreasing. Normalization is usually required as the criteria or parameters are often of inconsistent dimensions in multi-objective problems. (Yoon & Hwang, 1995; Zavadskas, Zakarevicius & Antuchevicine 2006). Compensatory methods such as TOPSIS allow tradeoffs between criteria, where a poor result in one criterion can be denied by a suitable result in another criterion. This provides a more actually solution form of modeling than non compensatory methods, which include or exclude alternative solutions based on hard cutoffs (Obaid and Tawfeeg, 2014). The TOPSIS process is carried out as follows:

Step 1: The first step is to determine the objective, and to identify the reasonable evaluation attributes.

Step 2: This step shows a matrix based on all the knowledge available on attributes. Every row of this matrix is allocated to one alternative, and each column to one attribute. When objective value is not available that means In the case of a subjective attribute, a ranked value judgment is adopted on the scale. Once a subjective attribute is represented on a scale, then the normalized values of the attribute assigned for various alternatives are calculated in the same way as that for objective attributes.

Step 3: Obtain the normalized decision matrix, R_{ij}. This can be expressed as

$$R_{ij} = \frac{m_{ij}}{\left[\sum_{j=1}^{M} m_{ij}^2\right]^{1/2}} \tag{1}$$

Step 4: Decide on the relative importance i.e. weights of different attributes with respect to the objective. A set of weights w_j (for j=1, 2, .., M) such that $w_j = 1$ may be decided upon.

Step 5: Obtain the weighted normalized matrix V_{ij} . This is done by the multiply of each factor of the column of the matrix R_{ij} with its related weight w_j . So, the elements of the weighted normalized matrix V_{ij} are expressed by:

$$V_{ij} = w_j \times R_{ij} \tag{2}$$

Step 6: Obtain the ideal solution and negative ideal solutions in this step. The ideal solution and negative ideal solutions can be represented as:

Max Min

$$V_{.}^{+} = \{(\sum V_{ij} \ / \, j{\in}J), (\sum V_{ij} \ / \, j{\in}J') \ / \ i = 1,2, \ldots, N\},$$

$$=\{ V_1^{+}, V_2^{+} +, \ldots\ldots, + V_m^{+} \}$$

$$V^{-} = \{(\sum V_{ij} / \, j{\in}J), (\sum V_{ij} / \, j{\in}J') \ / \ i = 1,2, \ldots, N\},$$

$$= \{ V_1^{-}, V_2^{-} -, V_3^{-} -, \ldots\ldots, V_M^{-} \}$$

where J = (j = 1, 2, …, M) /j is associated with beneficial attributes, and J’ = (j = 1, 2, …, M) /j is associated with non-beneficial attributes.

V_j^+ Indicates the ideal value of the considered attribute between the values of the attribute for various alternatives. In the case of beneficial attributes i.e., those of which high er values are required for the given application, V_j^+ represent the higher value of the attribute. In the case of non-benefici al attributes (i.e., those of which lower values are required for the given utility), V_j^+ indicates the lower value of the attribute. V_J^- Indicates the negative ideal (worst) value of the considered attribute among the values of the attribute for different alternatives. In the case of beneficial attribute s (i.e., those of which higher values are required for the given utility), V_j^- represent the lower value of the attribute. In the case of non-beneficial attributes (i.e., those of which lower value s are required for the given application), V_j^- represent the higher value of the attribute.

Step 7: Obtain the separation measures. The separation of each alternative from the ideal one is given by the Euclidean distance in the following equations.

$$S_i^+ = \{\sum_{j=1}^{M}(V_{ij} - V_j^+)\hat{\ }2\}^{\wedge}.5 \quad (3)$$

$$S_i^- = \sum_{j=1}^{M}(V_{ij} - V_j^+)\hat{\ }2\}^{\wedge}.5 \quad (4)$$

i= 1, 2, …., N

Step 8: The relative closeness of a particular alternative to the ideal solution, Pi-mod, can be expressed in this step as follows.

$$P_i = S_i^- /((S_i^+ + S_i^-)) \quad (5)$$

Step 9: A set of alternatives is generated in the descending order in this step, according to the value of P_i represent the most selected and least selected feasible solutions. P_i may also be called the overall or composite performance score of alternative A_i. It may be added here that in step four of the TOPSIS method, even though the weights of various attributes with respect to the objective, w_j (for j=1, 2, .., M), are decided by the decision maker rather arbitrarily, only few systematic methods can be used. The systematic methods of deciding the weights of attributes are given below.

1. Entropy Method.
2. Standard Deviation Method.
3. AHP Method.

Teaching Learning-Based Optimization (TLBO)

TLBO is an effective method than other metaheuristics methods because of its characteristics viz. high consistency, computation effort, and parameter less. It outmatch some of the well established metaheuristics methods relates constrained benchmark functions, constrained mechanical design and continuous non-linear numerical optimization problems; it is being used by many researchers as a replacement for the other techniques available. Such a blowout has steered (Crepinek et al., 2012). towards investigating the secrets of TLBOs. They calibrated some error relates to TLBO through code-reviews and experiments respectively. Further, Waghmare (2014) in statement on the work of Crepinek et al. He not only addressed the queries raised by Crepinek et al. but also again examined the experimental results, which point out that the TLBO algorithm performs well on the problems where the fitness-distance relations are small by proper tuning of the common control parameters of the algorithm, and corrected the understanding about the TLBO algorithm in an objective manner. TLBO has been used by number of researchers to solve their problems and found it efficient than other Metaheuristic. (Krishnanand et al, 2012) in have applied a multi-objective TLBO algorithm with application to electric power dispatch. Rao and Patel (2012) in explored the use of TLBO and ABC algorithms for determining the optimum operating conditions of combined Brayton and inverse Brayton cycles. Rao (2012) in proposed the optimization of mechanical related design problems using TLBO and tested it on five different constrained benchmark test functions with various characteristics, four different benchmark mechanical design problems and six mechanical design optimization problems. Gonzlez-lvarez et al. (2012) in proposed Multi-objective TLBO (MO-TLBO). for solving Motif Discovery Problem and solved a set of twelve biological instances belonging to different organisms. Rao and Patel (2012). introduced and investigated the effect of elitism on the performance of TLBO algorithm while solving complex constrained optimization problems and unconstrained benchmark problems. Population size and Number of generation, these parameter affects the performance of TLBO are also investigated. Rajasekhar et al. (2012). in proposed a new variant of TLBO, termed as Elitist Teaching-Learning Opposition based (ETLOBA) Algorithm for numerical function optimization, which is empowered with two mechanisms, one is Opposition method (helps to improve the capability of searching) and second is elitism, to reach the precise global optimum with less time difficulty. Rao and Patel (2013). in introduced a modified version of the TLBO algorithm and applied same for the multi-objective optimization of heat exchangers. Pawar and Rao (2013). in presented

Figure 2. TLBO flow chart

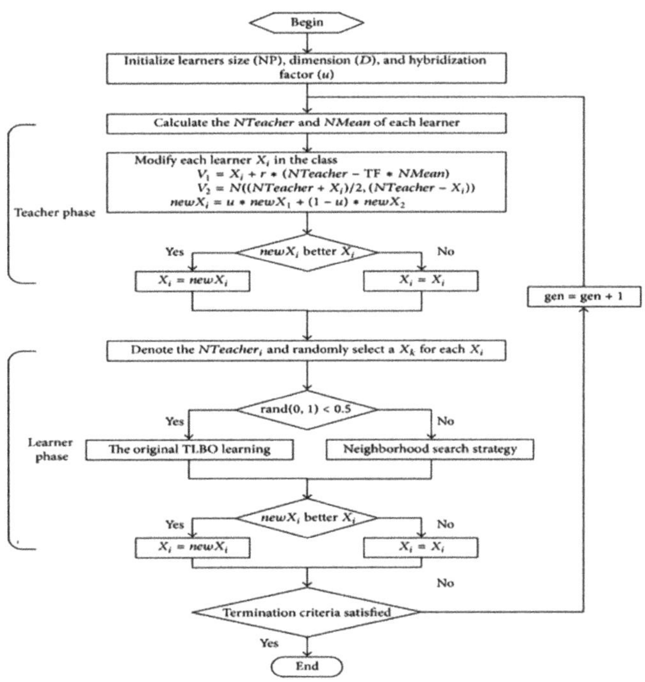

TLBO to find the optimal combination of process parameters of the abrasive water jet, grinding and milling machining processes. Rao and Kalyankar (2013). in applied TLBO for the process parameter optimization of electrochemical machining process and electrochemical discharge machining process. Rao et al (2013) in proposed TLBO to solve continuous constrained and unconstrained optimization problems. Niknam et al (2012) in proposed a -multi-objective-TLBO algorithm to solve the dynamic economic emission dispatch problem.

1. **Algorithm:** In TLBO Algorithm teacher and learners are the two important components. This expressed two basic modes of the learning, through teacher (called teacher phase) and interacting with the other learners (called learner phase). Teacher is generally considered as a highly learned person who teach learners so that they can have better results in terms of their scores. Moreover, learners also learn from the contacting among themselves which also helps in improving their results in term of knowledge. TLBO is population based method. In this optimization algorithm a group of learners is assumed as population and various design variables are considered as various subjects offered to the learners and learners result is specious to the fitness value of the optimization problem. In the overall population the best solution is assumed as the teacher. TLBO algorithm mainly working of two phases, namely teacher phase and learner phase.
2. **Teacher Phase:** A good teacher is one who brings his learners up to his level in terms of knowledge. But in practically this is not possible and teacher can only move the mean of a class up to some level depending on the capacity of the class. This precede a random process depending on many factors. Generate the random population according to the population size and number of generations. Calculate the mean of the population, which will give the mean for the particular subject as M,D = [m1, m2, m3 .mD]. The best solution will act as a teacher for that iteration process Xteacher = Xf(X)=min. The teacher will try to increase the mean from MD towards X teacher which will act as a new mean for the next iteration. hence, Mnew,D =XteacherD.

The difference between two means is represent as

DifferenceD = ri(Mnew, D-TFMD) (6)

where, ri is the random number in the range [0, 1], the value of Teaching Factor (TF) is considered 0 or 1. The obtained difference is added to the recent solution to up-date its values using

Xnew, D = Xold, D + DifferenceD (7)

Accept Xnew if it gives better function value.

3. **Learner Phase:** A learner interacts randomly with other learners for increasing his or her knowledge. Randomly select two learners Xi and Xj.

X′new, D = Xold, D+ri(Xi−Xj)iff (Xi) < f (Xj) (8)

X′new, D = Xold, D+ri(Xj−Xi)iff (Xi) > f (Xj) (9)

Termination criterion: Stop if the maximum generation number is achieved otherwise repeat from Step Teacher phase.

4. Implementation steps of the TLBO are summarized below:
   - **Step 1:** Initialized the population (that means learners) and design variables of the optimization problem (that means number of subjects given to the learner) with random generation number and evaluate them.
   - **Step 2:** Select the best learner of each subject as a teacher for that subject and determine mean result of learners in each subject.
   - **Step 3:** Evaluate the difference between recent mean result and best mean result according to Eqs 6 by using the teaching factor (TF).
   - **Step 4:** Update learners knowledge with the help of teachers knowledge according to Eqs 7.
   - **Step 5:** Update the learners knowledge by using the knowledge of some other learner according to Eqs 8 and Eqs 9.
   - **Step 6:** Again use the procedure from step 2 to 5 up-to the termination criterion is met.

RSM Method

Response surface methodology (RSM) is a set of mathematical and statistical techniques for empirical model building. By careful design of experiments, the objective is to optimize a response (output variable) which is influenced by several independent variables (input variable). An experiment is a series of tests, called runs, in which changes are made in the input variables in order to identify the reasons for changes in the output response.

In general, the procedure of RSM consists of the following steps:

- Design and conduct a series of experiments to find adequate and reliable measurements of responses of interest.
- Develop mathematical models of the first and second order response surfaces with best fittings.
- Find the optimal set of process parameters that produce optimized value of the response.
- Represent the direct and interactive effects of the process parameters through contour and surface plots.

LITERATURE REVIEW

This literature review has taken to identify articles in this chapter that provide the most valuable information to researchers studying live issues concerning the Modeling technique, TOPSIS method and TLBO method. Hence, for establishing a relation between input parameters and output parameters various approaches can be applied like empirical relation, RSM, Neuro-fuzzy, artificial neural network (ANN), non-linear regression, modeling etc. In the 1st phase literature review has been done on modeling. In the 2nd phase of literature review, an investigation has been made on the implementation of TOPSIS. In the 3ed phase of literature review on Multi-TLBO has been carried out.

Modeling Technique

Benardos and Vosnaikos (2003) classified Modeling techniques detail into three group i.e. experimental models, analytical models and Artificial Intelligence-based models. Experimental and analytical models can be developed by using conventional approaches such as the Statistical Regression technique.

Oktem et al. (2006) found that by applying GA, for the cutting conditions of feed rate, cutting speed and axial depth of cut and Ra value was obtained that is lower than the values of experimental results.

Kadirgama (2008) applied optimization of the surface roughness when milling mould aluminum alloys (A6061-T6) with carbide coated inserts to determine the optimized parameters, as well as to find out the most influencing variables (cutting speed, feed rate, axial depth and radial depth). The approach was based on RSM and RBFN. It was found that the feed rate is the most significant factors influencing surf ace roughness. RBFN predicted surface roughness more accurately compared to RSM.

Aggarwal et al. (2008) performed experimental investigation in CNC turning of AISI P-20 tool design to study the influence of cutting speed, feed rate, nose radius, depth of cut and cutting environment. Taguchis technique and RSM was used to conduct the objective of experimental study. L-27 orthogonal array and face centered composite design have been used for conducting the experiments. It was found the effects of feed rate and nose radius were found to be neglected compared to other factors and RSM technique appeared to have an advantage over the Taguchis technique.

Kanagararjan et al. (2009) used response surface methodology (RSM) along with multiple linear regression analysis to found 2nd order response equations for MRR in EDM on WC/30% Co composite. The most influential parameters aiming at maximizing MRR and minimizing Ra were obtain by carefully examining surface and contour plots of the responses versus various combinations of input parameters.

TOPSIS

TOPSIS is known as the Technique for Order Preference by Similarity to Ideal Solution. This method is a technique to identify the ranking of all alternatives assumed. In the TOPSIS method, the decision making matrix and weight vector are calculated as crisp values and a positive ideal solution and a negative ideal solution are obtained from the decision matrix. In other words, PIS is a set of the best value of criteria while NIS is a set of worst values achievable of criteria. This method is applied to make wide-ranging evaluation of samples where it measured the distances between the index value vector of each sample and ideal solution (best) along with the negative ideal solution (worst) of the comprehensive evaluation. Hwang and Yon are the first who introduces the TOPSIS method. Hwang and Yon describe multiple decisions making as follows: multiple decisions making is applied to the preferable decision (such as assessment making priorities and choices) between available classified alternatives over the multiple attributes. It assumes that each criterion requires to be maximized or minimized. Hence, the ideal positive and negative values of each criterion are identified, and each alternative judge against this information. It is noted that, in this typical multiple criterion decision making (MCDM) approach, weights of attributes reflect the relative importance in the decision making 13 process. Each evaluation of criteria entails diverse opinions and meanings. Hence, the assumption that each evaluation criterion is equally important is prohibited (Yu, et al, 2008).

Summarised information for research projects related to TOPSIS:

- Krohling and Campanharo (2011) has applied TOPSIS on Accidents with oil spill in the sea in 2011.
- Wang et al. (2009) has applied TOPSIS on Supplier selection in 2009.
- Sun and Lin (2009) has applied TOPSIS Competitive advantages of shopping online in 2009.
- Wang and Chang (2007) has applied TOPSIS on Initial aircraft training in 2007.
- Chamodrakas and Martakos (2012) has applied TOPSIS on Heterogeneous network selection in 2011.
- Asadi et al. (2016) has applied AHPTOPSIS Predictive Model for District-Scale Mapping of Porphyry CuAu Potential in 2016.
- Robinson and Amirtharaj (2013) has applied TOPSIS algorithm with correlation coefficient of interval vague sets 2013.
- Lahby, Mohamed, Leghris, Cherkaoui, and Abdellah (2013) has applied M-AHP/E-TOPSIS on Hybrid network selection strategy for heterogeneous networks." Intelligent Systems on 2013.
- Sepehr and Zucca (2012) has applied TOPSIS algorithm on Ranking desertification indicators using TOPSIS algorithm. on 2012.
- Bisio et al. (2014) has applied TOPSIS for vertical handover procedures over mobile ad hoc networks on 2014.
- Soufi et al. (2015) has applied TOPSIS Multi-Criteria Decision Modeling Approach for Biolubricant Selection for Two-Stroke Petrol Engines in 2015.
- Hernandez, Giral, and Pez has applied TOPSIS algorithm for Frequency Channel Selection in WiFi Networks in 2015.

TLBO

Rao et al. (2011) used TLBO on grinding. He has taken input parameter are Wheel speed, work piece speed, depth of dressing, lead of dressing And output parameters are production cost, production rate, surface roughness. They are also compare the result of TLBO from SA, HA, ABC and got better result from other method.

Rao and Kalyankar (2013) has used TLBO on multi-pass turning his input cutting parameter are cutting force, cutting power, cutting speed feed, depth of cut and output parameter are production time, used tool life. He has also applied TLBO on multi-pass turning with large no of constraints with inputs are Tool-life constraint, Cutting force constraint, Power constraints, Stable cutting region, Chip tool interface temperature, Tool-life, Cutting force, Cutting force, Power, Cutting speed, Feed rate, Depth of cut and outputs are unit production cost. He has gotten Minimum production cost rather than other methods.

Rakesh Kumar Sahu, Saurabh Verma, Santosh Kumar Mishra (2015) has used TLBO on Abrasive water suspension jet. He has put the input variables are Inlet Operating Press Nozzle diameter, Orifice diameter Density of Abrasive particle, Focus length and output parameters Skin friction coefficient. he has analyzed that the optimized value of SFC through TLBO is 0.001197 and it is closer to Analyzed value of SFC.

Ranjan Kumar Hasda (2013) has used TLBO on EDM. He select input variable are Pulse-on time, Peak current, Average gap voltage, Percent volume fraction of SiC and output variables are tool wear rate, material removal rate, gap size, surface finish. He has compare his result from GA, ACO, ABC, BBO and got better results in all 4 variables then other algorithm.

Rao and Kalyankar (2011) have used TLBO on ECM. They selected the input variable are electrolyte concentration, applied voltage, electrolyte flow rate, inter-electrode gap and output variables are material removal rate, radial overcut. They have also compared their result from ABC algorithm. They got TLBO algorithm need a minimum number of iteration than ABC and TLBO require low number of population than ABC for similar result.

Rao and Kalyankar (2013) have used TLBO on USM. They are select the input variable are amplitude of vibration frequency of vibration mean diameter of abrasive grains, static feed force, volumetric concentration of abrasive particles in slurry, and output variable is Material removal rate. They are also compare their result from GA, ABC, PSO, HS, SFL and got TLBO algorithm requires comparatively very less number of function evaluations then other algorithms.

Rao and Kalyankar (2013) used TLBO on AJM. They are used in input variable are mass flow rate of abrasive particles, velocity of abrasive particles, mean radius of abrasive particles and output variable are Material removal rate. They are comparing their result from SA GA and considered both ductile and brittle materials MRR is batter then both algorithm and number of iteration is required is less than other method.

Rao and Kalyankar (2013) have applied TLBO on WEDM. The input variable are pulse on time, pulse off time, peak current, servo feed setting surface roughness and output variable is cutting speed. They are also compare their result from ABC, PSO, HSM, SA SFL and got better result than other algorithm.

Zhibo Zhai, Shujuan Li, and Yong Liu (2015) have applied TLBO on Milling Process their input variables are cutting velocity, feed per tooth and output variables is Profit rate. They select Face milling, Corner milling, Pocket milling, and Slot milling.

Gaurav Dixit and Santosh Kumar Mishra (2014) have used TLBO on Turning by single point cutting tool they have select the input variable are Cutting Velocity, Depth of Cut, Feed Rate and output variable are resultant force they are compare their result from Taguchi Method and got TLBO give more specific and better result than Taguchi method.

EXPERIMENTATION

Experiment Setup

In the science, an experiment is a procedure that arbitrates between competing models or hypotheses. Researchers also use experimentation to check existing theories or new hypotheses to support or disprove them. SV-2E milling machine, model number SM-2 (Figure 3) has been used for running and conducting the experiments.

The working limits of machine are:

- The speed limits of machine are 35, 48, 69, 90, 137, 190, 270 380, 545, 765, 1070, and 1500 rpm.
- The feed limit of machine are 20, 31.5, 50, 78, 120, 195, 315, 500, 800 mm/min.

Al-6061 alloy taken as a work-piece material. The experiment was performed on Al-6061 alloy having cross section are 100mm100mm20mm. The chemical, physical and mechanical properties of Al 6061 alloy is - Chemical Composition of Al 6061 (Weight percentage).

Figure 3. End milling machine

- Al 97
- Mg 1.2
- Si .8
- Fe .6
- Cu .4

Physical Properties are:

- Density 2.70g/cm^3
- Melting Point 650C
- Thermal Expansion 23.4 x10-6 /K

Mechanical Properties are:

- Proof Stress 270 MPa
- Tensile Strength 310 MPa

Tool is select for conduct experiment is cobalt bonded cemented carbide face milling cutter tool shown in Figure 4 Tool having 16 mm diameter and 6 cutting flute and teeth on both side and periphery of the cutter.

Coolant is used for experimentation is S500.Coolant properties are water soluble and chlorine free non corrosive, nontoxic, non irritating.

- Bio stable formation for extension of life and minimize waste.
- Excellent rejection of foreign contamination including trump oil.
- Amine free, phosphate free and nitrite free.
- Ideal for continuous use recycling system and central systems.

Measurement of Responses

1. **Measurement of MRR:** The weight of work piece and tool has taken by high precision balance machine, whose capacity is 600 gram and accuracy is 0.01 gram. The MRR for every parameter setting was measured by the weight loss method in volumetric scale. By weighing work piece before and after each experiment, average value of MRR was calculated using Eqs IV-A1.

$$MRR=(W_i-W_f)/(p\times t) \quad (10)$$

where W_i = initial weight before experiment, W_f = final weight after experiment, =density of Al 6063 alloy taken as 2.7×3 g/mm3, t= machining time in minutes

2. **Measurement of Ra Values:** The Ra values were measured in two different directions on each machined work surface using a SURTRONIC 3+ surface texture measuring instrument, which has diamond stylus tip with accuracy of 0.005μm and resolution of 1.0 μm horizontal, 10mm vertical, Lc =0.8 mm, L n = 40 mm and having a maximum measuring range of 25mm

Figure 4. Face mill cutting tool converted

3. **Machine Parameters:** In order to determine the working characteristics of SV-2E milling machine, a number of preliminary experiments were conducted.
4. **Controllable Milling Parameters:** Namely velocity, feed, depth of cut, coolant speed were chosen to evaluate the process efficiency in terms of MRR (mm^3/min) and surface roughness (Ra: μm).

Table 1 lists all the parametric settings of experimental runs along with their obtained corresponding responses.

ANALYSIS AND RESULT

Analysis of Variance

ANOVA is used to check the sufficiency of the second order model which includes test for significance of regressi on model, model coefficients and test for lack of fit. The ANOVA analysis for MRR and Ra where columns describe the degree of freedom (DF), the sequential sum of squares (Seq SS), adjusted sum of squares (Adj SS), adjusted mean squares (Adj MS), F- values (F) and probability of certainty (P). The sequential sum of squares (Seq SS) is the added sum of squares given that the prior terms are in the model, which depends on the model order. A low P- value (0.05) indicates statistical significance for the source on the corresponding response (α= 0.05) or 95% confidence level. It can be seen that the p-Value is less than 0.05 hence the responses fit the regression model with the linear, square and interaction are significant at th e level of 95% after elimination. It also displays that the test of lack of fit is insignificant with the associated p- value of 0.7 68 and 0.097 for MRR and Ra respectively, which have values greater than 0.05 as desired for the model adequacy. The model with the rest of the terms after a backward elimination process in order to discard the insignificant terms (pvalue greater t han 0.05) so as to adjust the fitted quadratic model is illustrate d in Tables 2-5 for MRR and Ra respectively. It can be observed that the p-values of all these terms are less than α-value which means that they are significant hence these are included in the model.

Mathematical Modeling for MRR and Ra

Mathematical modeling for MRR and effect of factors are as following. The mathematical relation obtained on the basis of RSM analysis for analyzing the influences of the various dominant machine parameters on MRR is given by Eq IV-A1

$$MRR=f(2)=226.661-0.775\times Speed-1.917\times Feed-358.735\times(DOC)+89.439\times(C.S)-21.962\times(C.S)\times(C.S)+18.909\times(Feed)\times(DOC) \quad (11)$$

Figure 5 shows the main effect plot for MRR which can be used to graphically assess and compare the Magnitude of the main effects of the factors when it is dealt with multiple factors. The points are the means of the response variables at different levels of each factor with a reference line drawn at the grand mean of the response data. It can be seen that f and d are the most influencing parameters as there

Table 1. Experiment layout and response

Ex	Speed RPM	Feed (mm/min)	DOC (mm)	CS (m/s)	Ra (μmm)	MRR (mm3/min)
1	53.780	54.75	0.6	2	0.89	689.615
2	27.390	54.75	0.4	4	1.010	394.412
3	53.780	54.75	0.4	0	4.770	304.802
4	53.780	54.75	0.4	2	0.700	217.343
5	40.585	54.75	0.4	2	1.730	259.930
6	53.780	54.75	0.4	2	0.520	572.814
7	53.780	54.75	0.2	2	0.070	402.064
8	27.390	78	0.4	2	0.780	615.354
9	27.390	54.75	0.4	2	0.970	365.752
10	27.390	31.5	0.4	2	0.978	460.460
11	53.780	31.5	0.6	4	1.060	482.853
12	53.780	54.75	0.4	2	5.660	721.425
13	27.390	31.5	0.2	4	2.020	271.139
14	27.390	78	0.2	0	1.350	213.306
15	53.780	31.5	0.2	0	1.580	313.359
16	40.585	31.5	0.6	0	0.490	535.320
17	40.585	54.75	0.4	2	4.900	262.601
18	27.390	54.75	0.4	2	4.330	290.236
19	27.390	78	0.2	4	1.340	211.580
20	53.780	78	0.2	4	2.262	292.308
21	27.390	78	0.6	4	1.070	333.277
22	53.780	31.5	0.2	4	1.270	202.328
23	53.780	54.75	0.4	2	5.340	275.256
24	40.585	31.5	0.6	4	1.140	261.599
25	53.780	78	0.6	4	3.490	140.440
26	53.780	31.5	0.6	0	1.330	603.773
27	27.390	78	0.2	0	1.070	408.987
28	27.390	31.5	0.2	0	0.430	383.706
29	27.390	78	0.6	0	0.880	800.223
30	53.780	78	0.6	0	2.190	689.655

is sharp increase in MRR when these two parameters increase. Eqs IV-A1 show Mathematical modeling for Ra and effect of factors On the basis of RSM analysis, the mathematical relationship for correlating the surface roughness Ra.

$$Ra=0.55638+0.09142\times Speed-0.00161\times Feed-0.73556\times DOC-1.46190\times(CS)+0.41889\times(CS)\times(CS)-0.02171\times Speed\times(CS) \quad (12)$$

Table 2. Analysis of variance for MRR(mm3/min)

Source	DF	SeqSS	AdjSS	AdjMS	F	P
reg	6	821111	821111	136852	30.58	.000
linear	4	641847	67740	16935	3.78	.017
square	1	55562	55562	55562	12.41	.002
interaction	1	123703	123703	123703	27.64	000
Res error	23	102943	102943	4476		
Lack of fit	18	72351	72351	4020	.66	.768
Pure error	5	30592	302592	6118		
total	29	924054				

Table 3. Results for the independent MRR

Term	Coeff	SF Coeff	T-Values	P-Values
constant	226.661	106.902	2.120	.045
speed	-.775	1.195	-.648	.523
Feed	-1.917	1.591	-1.205	.240
DOC	-358.735	212.124	-1.691	.104
CS	89.439	26.150	3.420	.002
CS*CS	-21.962	6.233	-3.523	.002
Feed*DOC	18.909	3.597	5.257	.000

S,R-sq and R-sq value of MRR
S = 66.90 R-Sq = 88.9%, Sq(adj) = 86.0%

Table 4. Test results for the Ra parameters

Source	DF	SeqSS	AdjSS	AdjMS	F	P
Reg	6	65.187	65.187	10.864	46.58	0.00
Linear	4		39.607	9.901	42.45 0	.00
Square	1	20.214	20.213	20.213	86.66	0.00
inte.	1	5.253	5.253	22.52	22.52	0.00
Res error	23	5.3651	5.365	0.233		
Loffit	18	4.944	4.944	0.274	3.26	0.09
P error	5	0.420	0.420	0.084		
Total	29	70.552				

The main effect plot for Ra is shown in Figure 6. It can be seen that speed and coolant speed have a significant effect on Ra which is supported by ANOVA results shown in figure. However, coolant speed is the most influencing parameter showing a sharp increase in Ra from 3.9 m to 0.5 μm when coolant speed increase from 0 to 2 liters per min.

Table 5. Test results for the Ra parameters

Term	Coeff	SF Coeff	T-Values	P-Values
Constant	0.55638	0.640424	0.869	0.394
Speed	0.09142	0.0125767	7.269	0.000
Feed	-0.00161	0.004896	-0.328	0.746
DOC	-0.73556	0.5691937	-1.292	0.209
CS	-1.46190	0.264800	-1.292	0.000
CS*CS	0.41889	0.44999	9.309	0.000
SPEED*CS	-0.02171	0.0045757	-4.746	0.000

S, R-sq and R-sq value of Ra
S=.483 R-sq=92.4 R-sq(adj)=90.4%

Figure 5. Main effects plot for MRR

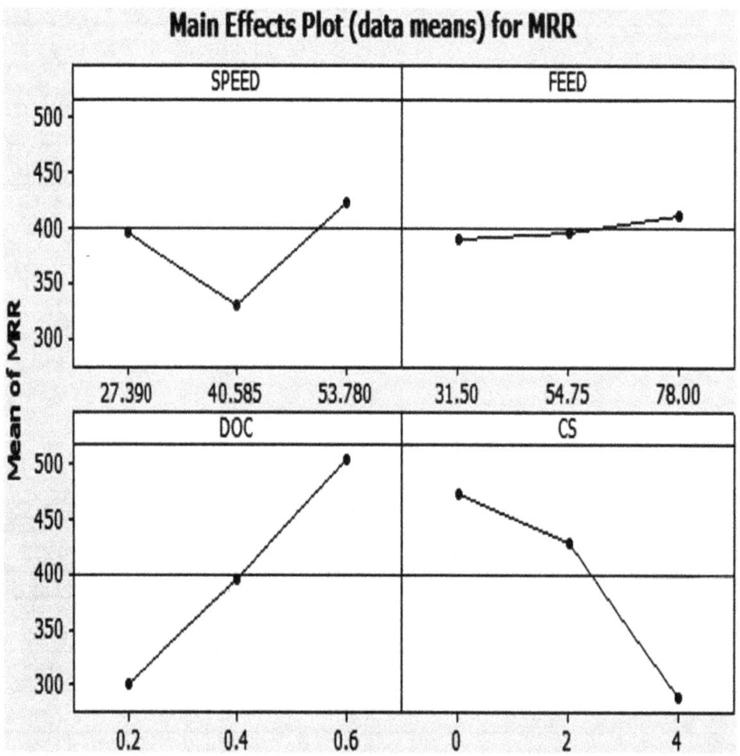

Objective Weighting Calculation by Standard Deviation Method

Objective weight is calculated by standard deviation method. Standard deviation method (SD) assigns a small weight to an attribute if it has alike attribute values across alternatives. The SD method determines the weights of the criteria in terms of their SDs through the Eqs 13 and Eqs 14.

Figure 6. Main effects plot for Ra

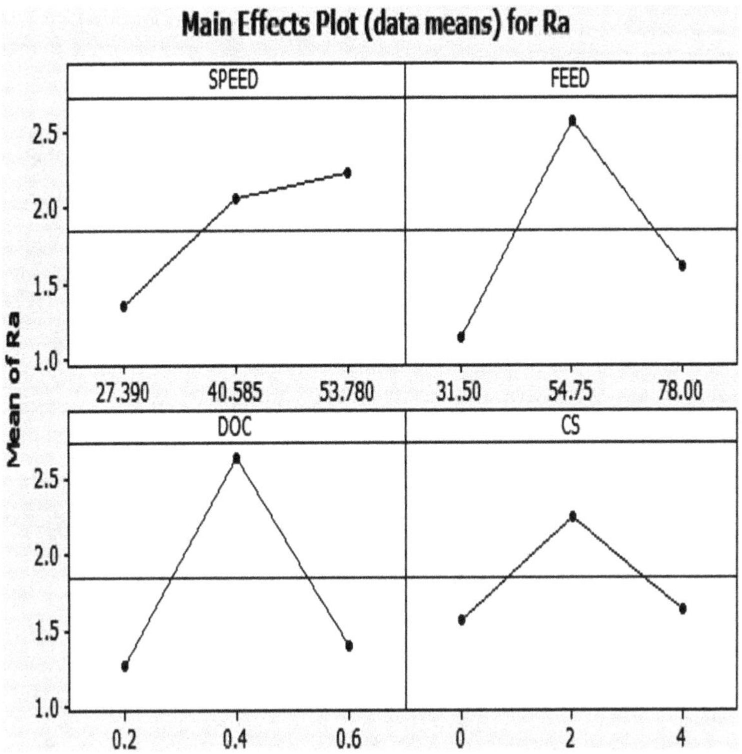

$$w_j = \frac{\sigma_j}{\sum_{j=1}^{n} \sigma_j} \quad j=1,\ldots\ldots..,n \tag{13}$$

$$\sigma j = \sqrt{\frac{\sum_{i=1}^{m} \left(x_{ij} - x_j mean\right)^2}{m}} \quad j=1,\ldots\ldots\ldots,n \tag{14}$$

The application of this method in material selection was recently suggested by Rao and Patel. To calculate the weights by Standard Deviation method first the information matrix is normalized then following equations are used. The SD method determines the weights of criteria in terms of their SDs through following. According to standard deviation method the calculation of weight of objective function are shown. We have calculated the standard deviation of normalized RA is 0.1727 and standard deviation of normalized MRR is 0.2193. Then

W_{ra} = .1727/(.1727 + .2193) = 0.4406

and

$W_{mrr} = .2193/(.1727 + .2193) = 0.5593$

Data Analysis by TOPSIS Algorithm

In course of data analysis, the normalized values are determined as it adjusts the values measured on different scales to a notionally common scale and normalized matrix is determined as shown in Eqs 1. The normalized values are shown in Table 6.

Table 6. Normalized data

Ex. No	Ra	Normalized Ra	MRR	Normalized MRR
1	0.890	0.067	689.615	.288
2	1.010	0.076	394.412	0.165
3	4.770	0.359	304.802	0.127
4	0.700	0.052	217.343	0.090
5	1.730	0.150	259.930	0.108
6	0.520	0.039	572.814	0.239
7	0.070	0.005	402.064	0.168
8	0.780	0.058	615.354	0.257
9	0.970	0.073	365.752	0.153
10	0.978	0.073	460.460	0.192
11	10.060	0.079	482.853	0.202
12	5.660	0.426	821.425	0.302
13	2.020	0.152	271.139	0.113
14	1.350	0.101	213.306	0.089
15	1.580	0.119	313.359	0.131
16	0.490	0.036	535.320	0.224
17	4.900	0.369	262.601	0.109
18	4.330	0.326	290.236	0.121
19	1.340	0.101	211.580	0.088
20	2.062	0.170	292.308	0.122
21	1.070	0.080	333.277	0.139
22	1.270	0.095	202.328	0.084
23	5.340	0.402	275.256	0.115
24	1.140	0.085	261.599	0.109
25	3.490	0.263	140.440	0.058
26	1.330	0.100	603.773	0.252
27	1.070	0.080	408.987	0.171
28	0.430	0.032	383.706	0.160
29	0.880	0.066	800.223	0.335
30	2.190	0.165	689.655	0.288

Based on the machining data surface roughness and material removal rate, the priority weight has been assigned to each response which is calculated. Here, weight to surface roughness is 0.440 and weight to material removal rate is 0.559 has been assigned. Weighted Normalized Decision Making Matrix is calculated by Eqs I-A and weighted (normalized) decision-making matrix has been shown in Table 7. Then the positive ideal solutions and negative ideal solutions are determined. As higher MRR is desirable (as it corresponds to Higher-is-Better), maximum value among the recorded values are considered as positive ideal solution and minimum value referred as negative ideal solution. Whereas for rest of the

Table 7. Weighted normalized decision making matrix

Ex. No	Ra × w_1	MRR × w_2
1	0.029	0.161
2	0.033	0.092
3	0.158	0.071
4	0.023	0.050
5	0.057	0.060
6	0.017	0.134
7	0.002	0.094
8	0.025	0.144
9	0.032	0.085
10	0.032	0.107
11	0.035	0.113
12	0.188	0.168
13	0.067	0.063
14	0.044	0.049
15	0.052	0.073
16	0.016	0.125
17	0.162	0.061
18	0.143	0.067
19	0.044	0.049
20	0.075	0.068
21	0.035 0	0.078
22	0.042	0.047
23	0.177	0.064
24	0.037	0.061
25	0.116	0.032
26	0.044	0.141
27	0.035	0.095
28	0.014	0.089
29	0.029	0.187
30	0.072	0.161

responses like surface roughness, (as they correspond to Lower-is- Better, LB criterion). Hence, minimum value of the recorded value is regarded as positive ideal solution and maximum value represents the negative ideal solution. The positive ideal solution and negative ideal solution are determined and tabulated in Table 8. Now the separation distance is measured from both positive ideal solution and negative ideal solution using Eqs 3 and Eqs 4 then the relative closeness index is calculated using Eqs 5 and tabulated in Table 9.

Optimization of Face Milling Process by TLBO Algorithm

Optimization performance of TLBO was determined by the experimental data and Mathematical modeling as considered here. Single objective optimizations of responses were performed here. Influences of various milling process parameters like speed, feed, depth of cut, coolant speed of Al-6063 present in on two machining responses were studied RSM method i.e., MRR (mm3/min), and surface finish (Ra, µm). Corresponding RSM-based equations were developed. For each of the four responses, as given in Eqs IV-A1 and Eqs IV-A1 the objective functions after modification is given below.

$$\text{Objective1} = Ra = f(1) \tag{15}$$

$$\text{Objective2} = (MRR) = f(2) \tag{16}$$

Constraints are as follows-

27.39 m/min < speed < 53.78 m/min

31.5 m/min < Feed < 78.0 m/min

0.2 mm <DOC < 0.6 mm

0 liter/min <Coolant Speed <4 liter/min

Table 8. Positive ideal solution and negative ideal solution

	Ra	MRR
v+	0.0023	0.187
v-	0.188	.032

Table 9. Ranking of alternatives

No	S+	S-	Pi	Ranking
1	0.037	0.158	0.808	2
2	0.100	0.154	0.607	13
3	0.194	0.129	0.398	27
4	0.138	0.165	0.544	16
5	0.138	0.132	0.490	22
6	0.055	0.171	0.756	4
7	0.093	0.188	0.668	7
8	0.049	0.162	0.767	3
9	0.106	0.155	0.595	14
10	0.085	0.155	0.646	9
11	0.081	0.152	0.652	8
12	0.186	0.155	0.453	25
13	0.139	0.125	0.473	23
14	0.143	0.143	0.499	21
15	0.124	0.137	0.523	18
16	0.063	0.172	0.730	5
17	0.204	0.132	0.393	28
18	0.185	0.119	0.392	29
19	0.144	0.144	0.499	20
20	0.139	0.120	0.463	24
21	0.114	0.152	0.571	15
22	0.145	0.146	0.501	19
23	0.214	0.145	0.403	26
24	0.131	0.150	0.534	17
25	0.191	0.110	0.364	30
26	0.062	0.144	0.698	6
27	0.097	0.152	0.610	2
28	0.098	0.174	0.640	10
29	0.026	0.158	0.855	1
30	0.075	0.122	0.619	11

Single-Objective Optimization

The same mathematical models given by equations of MRR and RA are now attempted by the TLBO algorithm to check for improvement in the result. Initially both the models are attempted individually as a single objective function. The population size is used randomly starting with the low value and a promising result. The TLBO algorithm has given a maximum MRR of 816.6543 mm3/min and minimum Ra of =.007318 μ mm. Minimize Ra and maximize MRR values for controlled milling operation those optimal values are listed in Tables 10 and 11.

Here both the objectives combined and obtained a Multi-objective optimization problem, as given by Eqs IV-D.

By combining all the objectives, common process parameters can be obtained which satisfies all the conflicting objectives.

$$\text{Min (f)} = W_{ra} * \frac{RAfunction}{\min value of RA} - W_{mrr} * \frac{MRRfunction}{\max value of MRR} \tag{17}$$

where, Ramin and MRRmax are the minimum and the maximum values of Ra and MRR respectively which can be obtained by attempting an individual objective function, and Wra and Wmrr are the weight values assigned to Ra and MRR, respectively. Present case, same weightage which is calculated by the standard deviation method

Wra = 0.440

Wmrr= 0.559

This multi objective optimization problem is now attempted by the TLBO algorithm and the result obtained by TLBO algorithm is given in Table 12.

Table 10. Minimum Ra at a relative input and maximum MRR at a relative input

Speed	Feed	Depth of Cut	Coolant Speed	Ra
27.39	54.448	0.6	2.4409	0.007

Table 11.

Speed	Feed	Depth of Cut	Coolant Speed	MRR
27.39	78	0.6	2.06	816.654

Table 12. Optimized result by TLBO

Speed	Feed	DOC	CS	Ra	MRR
27.589	69.804	0.6	2.253	0.008	737.596

CONCLUSION

Comparison Between TOPSIS Best Result and TLBO Result

The experimental results confirm the validity of the utilized Response Surface Methodology (RSM) integrated with TLBO for improving the machining performance and optimizing the machining parameters. With respect to the reference parameters, satisfy the real requirement of face milling operations for the proper machining of Al-6061 alloy. Comparison table is shown in Table 13.

The percent deference between experimental data which is selected by TOPSIS and TLBO for same weight age we got batter MRR removal rate which is selected experimental input variable by RSM method and select best from TOPSIS method. But we got minimum surface finish in TLBO which is far better than experimental result.

In this present study, the TLBO algorithm is applied to determine the optimal Parametric combinations for milling processes for achieving better machining Performance. And also TOPSIS method for selecting better attributes from experiment result both single- and multi-objective optimization problems were solved using this algorithm. Like GA, ACO, ABC and BBO it is observed that the TLBO algorithm gives better results. This algorithm can be applied as a global optimization tool for the purpose to select process parameter values. It can also be successfully applied for optimizing other traditional and non-traditional machining processes. Not relying on the manufacturer's data or handbook data, the process engineers can now select the optimal process parameter settings for different processes to achieve the desired machining performance.

Table 13. Comparison table

	TOPSIS	TLBO
MRR	800.223	737.596
Ra	0.8	0.008

REFERENCES

Aggarwal, A., Singh, H., Kumar, P., & Singh, M. (2008). Optimizing power consumption for CNC turned parts using response surface methodology and Taguchis technique—a comparative analysis. *Journal of Materials Processing Technology*, *200*(1), 373–384. doi:10.1016/j.jmatprotec.2007.09.041

Asadi, H. H., Sansoleimani, A., Fatehi, M., & Carranza, E. J. M. (2016). An AHP–TOPSIS Predictive Model for District-Scale Mapping of Porphyry Cu–Au Potential: A Case Study from Salafchegan Area (Central Iran). *Natural Resources Research*, 1–13.

Assari, A., Mahesh, T. M., & Assari, E. (2012). Role of public participation in sustainability of historical city: Usage of TOPSIS method. *Indian Journal of Science and Technology*, *5*(3), 2289–2294.

Benardos, P. G., & Vosniakos, G. C. (2003). Predicting surface roughness in machining: A review. *International Journal of Machine Tools & Manufacture*, *43*(8), 833–844. doi:10.1016/S0890-6955(03)00059-2

Bisio, I., Braccini, C., Delucchi, S., Lavagetto, F., & Marchese, M. (2014, June). Dynamic multi-attribute network selection algorithm for vertical handover procedures over mobile ad hoc networks. In *2014 IEEE International Conference on Communications (ICC)* (pp. 342-347). IEEE. doi:10.1109/ICC.2014.6883342

Chamodrakas, I., & Martakos, D. (2012). A utility-based fuzzy TOPSIS method for energy efficient network selection in heterogeneous wireless networks. *Applied Soft Computing*, *12*(7), 1929–1938. doi:10.1016/j.asoc.2012.04.016

Črepinšek, M., Liu, S. H., & Mernik, L. (2012). A note on teaching–learning-based optimization algorithm. *Information Sciences*, *212*, 79–93. doi:10.1016/j.ins.2012.05.009

Dehghani Soufi, M., Ghobadian, B., Najafi, G., Sabzimaleki, M. R., & Yusaf, T. (2015). TOPSIS Multi-Criteria Decision Modeling Approach for Biolubricant Selection for Two-Stroke Petrol Engines. *Energies*, *8*(12), 13960–13970. doi:10.3390/en81212408

Gonzalez-Alvarez, D. L., Vega-Rodrıguez, M. A., Gomez-Pulido, J. A., & Sanchez-Perez, J. M. (2012). Multiobjective teaching-learning- based optimization (mo-tlbo) for motif finding. *Computational Intelligence and Informatics (CINTI), 2012 IEEE 13th International Symposium on.*, 141–146. doi:10.1109/CINTI.2012.6496749

Hasda, R. K. (2013). *Optimization of EDM Process Parameters through Teaching Learning Based Optimization Algorithm* (Doctoral dissertation). National Institute of Technology, Rourkela, India.

Hwang, C. L., & Yoon, K. (2012). *Multiple attribute decision making: methods and applications a state-of-the-art survey* (Vol. 186). Springer Science & Business Media.

Kadirgama, K., Noor, M. M., Zuki, N. M., Rahman, M. M., Rejab, M. R. M., Daud, R. & Abou-El-Hossein, K. A. (2008). Optimization of surface roughness in end milling on mould aluminium alloys (AA6061-T6) using response surface method and radian basis function network. *Jourdan Journal of Mechanical and Industrial Engineering, 2*(4), 209- 214.

Kanagarajan, D., Karthikeyan, R., Palanikumar, K., & Davim, J. P. (2008). Optimization of electrical discharge machining characteristics of WC/Co composites using non-dominated sorting genetic algorithm (NSGA-II). *International Journal of Advanced Manufacturing Technology, 36*(11-12), 1124–1132. doi:10.1007/s00170-006-0921-8

Krohling, R. A., & Campanharo, V. C. (2011). Fuzzy TOPSIS for group decision making: A case study for accidents with oil spill in the sea. *Expert Systems with Applications, 38*(4), 4190–4197. doi:10.1016/j.eswa.2010.09.081

Kumar, V. R. (2013). *Teaching learning based optimization applied to mechanical constrained design problems* (Doctoral dissertation).

Lahby, M., Cherkaoui, L., & Adib, A. (2013, May). Hybrid network selection strategy by using M-AHP/E-TOPSIS for heterogeneous networks. In *Intelligent Systems: Theories and Applications (SITA), 2013 8th International Conference on* (pp. 1-6). IEEE.

Niknam, T., Golestaneh, F., & Sadeghi, M. S. (2012). Multiobjective Teaching–Learning-Based Optimization for Dynamic Economic Emission Dispatch. *IEEE Systems Journal, 6*(2), 341–352. doi:10.1109/JSYST.2012.2183276

Obaid, M. H., & Tawfeeq, J. F. (n.d.). *Measurement the Distance between Points on the Map*. Academic Press.

Oktem, H., Erzurumlu, T., & Erzincanli, F. (2006). Prediction of minimum surface roughness in end milling mold parts using neural network and genetic algorithm. *Materials & Design, 27*(9), 735–744. doi:10.1016/j.matdes.2005.01.010

Pawar, P. J., & Rao, R. V. (2013). Parameter optimization of machining processes using teaching–learning-based optimization algorithm. *International Journal of Advanced Manufacturing Technology, 67*(5-8), 995–1006. doi:10.1007/s00170-012-4524-2

Rajasekhar, A., Rani, R., Ramya, K., & Abraham, A. (2012, October). Elitist teaching learning opposition based algorithm for global optimization. In *2012 IEEE International Conference on Systems, Man, and Cybernetics (SMC)* (pp. 1124-1129). IEEE. doi:10.1109/ICSMC.2012.6377882

Ramanand, K. R., Krishnanand, K. R., Panigrahi, B. K., & Mallick, M. K. (2012, December). Brain storming incorporated teaching–learning–based algorithm with application to electric power dispatch. In *International Conference on Swarm, Evolutionary, and Memetic Computing* (pp. 476-483). Springer Berlin Heidelberg. doi:10.1007/978-3-642-35380-2_56

Rao, R., & Patel, V. (2012). An elitist teaching-learning-based optimization algorithm for solving complex constrained optimization problems. *International Journal of Industrial Engineering Computations*, *3*(4), 535–560. doi:10.5267/j.ijiec.2012.03.007

Rao, R., & Patel, V. (2012). An elitist teaching-learning-based optimization algorithm for solving complex constrained optimization problems. *International Journal of Industrial Engineering Computations*, *3*(4), 535–560. doi:10.5267/j.ijiec.2012.03.007

Rao, R. V., & Kalyankar, V. D. (2011). Parameters optimization of advanced machining processes using TLBO algorithm. *EPPM, Singapore*, *20*, 21–31.

Rao, R. V., & Kalyankar, V. D. (2013). Parameter optimization of modern machining processes using teaching–learning-based optimization algorithm. *Engineering Applications of Artificial Intelligence*, *26*(1), 524–531. doi:10.1016/j.engappai.2012.06.007

Rao, R. V., & Kalyankar, V. D. (n.d.). *Grinding parameters selection using TLBO method*. Academic Press.

Rao, R. V., & Patel, V. (2013). Multi-objective optimization of heat exchangers using a modified teaching-learning-based optimization algorithm. *Applied Mathematical Modelling*, *37*(3), 1147–1162. doi:10.1016/j.apm.2012.03.043

Rao, R. V., & Waghmare, G. G. (2014). A comparative study of a teaching–learning-based optimization algorithm on multi-objective unconstrained and constrained functions. *Journal of King Saud University-Computer and Information Sciences*, *26*(3), 332–346. doi:10.1016/j.jksuci.2013.12.004

Robinson, J. P., & Amirtharaj, E. H. (2013). A Strategic TOPSIS algorithm with correlation coefficient of interval vague sets. *International Journal of Computing Algorithm*, *2*, 314–331.

Sahu, R. K., Verma, S., & Mishra, S. K. (n.d.). *Optimization of parameters to minimize the skin friction coefficient in abrasive water suspension jet machining through TLBO*. Academic Press.

Sepehr, A., & Zucca, C. (2012). Ranking desertification indicators using TOPSIS algorithm. *Natural Hazards*, *62*(3), 1137–1153. doi:10.1007/s11069-012-0139-z

Sun, C. C., & Lin, G. T. (2009). Using fuzzy TOPSIS method for evaluating the competitive advantages of shopping websites. *Expert Systems with Applications*, *36*(9), 11764–11771. doi:10.1016/j.eswa.2009.04.017

Venkata Rao, R., & Patel, V. (2012). Multi-objective optimization of combined Brayton and inverse Brayton cycles using advanced optimization algorithms. *Engineering Optimization*, *44*(8), 965–983. doi:10.1080/0305215X.2011.624183

Wang, J. W., Cheng, C. H., & Huang, K. C. (2009). Fuzzy hierarchical TOPSIS for supplier selection. *Applied Soft Computing*, *9*(1), 377–386. doi:10.1016/j.asoc.2008.04.014

Wang, T. C., & Chang, T. H. (2007). Application of TOPSIS in evaluating initial training aircraft under a fuzzy environment. *Expert Systems with Applications*, *33*(4), 870–880. doi:10.1016/j.eswa.2006.07.003

Yoon, K. P., & Hwang, C. L. (1995). *Multiple attribute decision making: an introduction* (Vol. 104). Sage publications. doi:10.4135/9781412985161

Yu, T., Zhou, J., Zhao, K., Wang, W., & Wang, W. (2008, October). Study on project experts' evaluation based on analytic hierarchy process and fuzzy comprehensive evaluation. In *Intelligent Computation Technology and Automation (ICICTA), 2008 International Conference on* (Vol. 1, pp. 941-945). IEEE.

Zavadskas, E. K., Zakarevicius, A., & Antucheviciene, J. (2006). Evaluation of ranking accuracy in multi-criteria decisions. *Informatica*, *17*(4), 601–618.

Zhai, Z., Li, S., & Liu, Y. (2015). Parameter Determination of Milling Process Using a Novel Teaching-Learning-Based Optimization Algorithm. *Mathematical Problems in Engineering*.

Chapter 5
Process Optimization in Non-Conventional Processes:
Experimentation With Plasma Arc Cutting

Milan Kumar Das
Jadavpur University, India

Prasanta Sahoo
Jadavpur University, India

Tapan Kumar Barman
Jadavpur University, India

Kaushik Kumar
Birla Institute of Technology, India

ABSTRACT

Conventional machining becomes non-efficient and non-effective in case of intricate shape and also while working with hard metals and alloys due to excessive tool wear. In such situations non-conventional machining, in contrast becomes more appropriate due to non-contact between tool and work-piece. In the present study, EN31 steel was machined using Plasma Arc Cutting with pre-defined process parameters. Material Removal Rate and Surface roughness were considered as responses for the study. The responses were optimized both as single and multi-response. Considering the complexities of this present problem, experimental data were generated and the results were analyzed by using Taguchi, Grey Relational Analysis and Artificial Bee Colony (ABC) Algorithm. Responses variances with the variation of process parameters were thoroughly studied and analyzed and 'best optimal values' were identified. The result were verified by the morphological study. It was observed that there was an improvement in responses from mean to optimal values of process parameters.

INTRODUCTION

The quality and quantity requirements of end products have become increasingly complex as customers expectation requires enhanced performance across a variety of diverse and changing system operating conditions. Systems are now being reconfigured to have capacity and capability in order to meet complex objectives and function effectively in changing operating conditions with adaptability to deliver value in dynamic market conditions. Reconfigurable systems are designed so as to maintain a high level of

DOI: 10.4018/978-1-5225-2440-3.ch005

performance by altering their configuration to cater to the multiple function requirements or a change in operating conditions within acceptable reconfiguration time, cost and quality.

Plasma Arc Cutting (PAC) is an important non-conventional machining process employing thermal cutting technique. It has been successfully used in the cutting of stainless steel, high hardness metals, high melting point metals, and other difficult to machine alloys.

The plasma arc cutting process has always been seen as an alternative to the oxy-fuel process. It grew out of plasma welding in the 1960s, and emerged as a very productive way to cut sheet metal and plate in the 1980s. Early plasma cutters were large, slow and expensive, therefore, tended to be dedicated to repeating cutting patterns in a mass manufacturing situation. CNC (computer numerical control) technology was integrated to plasma cutting machines in the late 1980s into the 1990s, giving plasma cutting machines greater flexibility to cut diverse shapes based on tailor made programs programmed into the machine's numerical control. These CNC plasma cutting machines were generally limited to cutting patterns and parts in flat sheets of work-piece using only two axes of motion, i.e. X and Y.

As illustrated in Figure 1, the basic principle is that the arc formed between the electrode and the workpiece is constricted by a fine bore, copper nozzle. This increases the temperature and velocity of the plasma emanating from the nozzle. The temperature of the plasma is in excess of 20000°C and the velocity can approach the speed of sound. When used for cutting, the plasma gas flow is increased so that the deeply penetrating plasma jet cuts through the material and molten material is removed in the efflux plasma through the material and molten material is removed in the efflux plasma. Inside the nozzle, primarily a pilot arc occurs between the negative electrode (cathode) and the workpiece. The energy of this arc leads to ionization of the cutting gas. Then the gas's enthalpy rises abruptly and it turns into plasma state. A mixture of plasma and heated gas is then forced to flow from the internal geometry of the torch to a narrow bore, where it gains momentum and is superheated due to ohmic heating. Outside the nozzle, the supersonic jet plume strikes to the work-piece, quickly melting the metal and expelling

Figure 1. Schematic representation of plasma arc cutting process

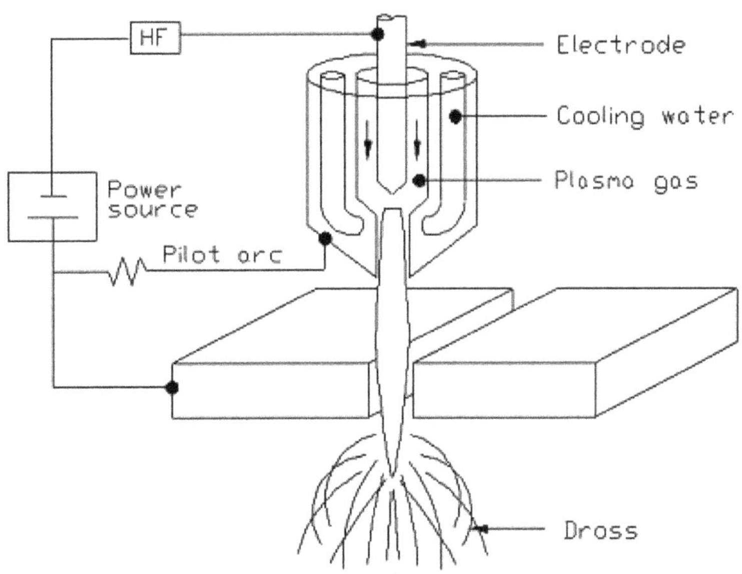

the molten material through the kerf. A secondary (shield) gas assists the process, covering the cutting zone, helping the cooling of the torch, and constricting the plasma jet (Bini et al. 2008).

LITERATURE REVIEW

Some of the important areas of research in the field of PAC are heat affected zone (HAZ) (Yang 2001, Vasil'ev 2003, Gullu and Atici 2006, Zajac and Pfeifer 2006, Kadirgama et al. 2010, Ismail and Taha 2011), depth of cut (Gane et al. 1994, Xu et al. 2002, Gariboldi and Previtali 2004, Yang 2007, Bahram. 2009) and kerf (Ramakrishnan et al. 2000, Teulet et al. 2006, Bini et al. 2007, Wang et al. 2011). Some researchers have also studied surface roughness characteristics, MRR, conicity etc. in PAC.

Ismail and Taha (2011) have optimized arc current, scanning velocity and carbon content of steel for better plasma arc surface hardening and roughness of ASSAB 618 and ASSAB DF3 steels using Taguchi technique. They observed that arc current is most significant parameter. Finally, they have found out an optimum combination of process parameters for better hardness and surface roughness. Radovanovic and Madic (2011) have developed an ANN model using cutting current, plate thickness and cutting speed as three input neurons, to predict the output neuron-surface roughness Rz values for plasma arc cutting operation of stainless steel, aluminum and nickel. Ozek et al. (2012) have successfully applied fuzzy algorithm for prediction of surface roughness in PAC process of AISI 4140 steel. Three machining process parameters used are plasma arc current, cutting speed, and thickness of cut material. The experimental results are compared with the results predicted using fuzzy logic. Influences of the individual machining parameters on surface roughness are carried out and it is seen that surface roughness is mostly affected by the cutting speed, while the plasma arc current has the least significance. Salonitis and Vatousianos (2012) have studied the influence of machining parameters, viz., cutting speed, cutting current, plasma gas pressure and torch height on surface roughness, heat effected zone and conicity of the cut geometry in plasma arc cutting of S235 mild steel and also developed a regression model. They found that surface roughness and the conicity are mainly affected by the cutting height. Bhuvenesh et al. (2012) have optimized air pressure, cutting current, cutting speed and arc gap with consideration of multiple performance characteristics including MRR and surface roughness using Taguchi technique in plasma arc cutting of AISI 1017 steel and shown that the most influence parameter is dross. Finally, they have found out an optimum combination of process parameters. Maity and Bagal (2015) have studied PAC process on AISI 316 stainless steel and optimized the process parameters (feed rate, current, voltage and torch height) for multiple responses (kerf, chamfer, dross, surface roughness and MRR) using RSM and grey relational analysis coupled with PCA.

The literature review indicates that researchers have taken one or two surface roughness parameters as responses. But roughness parameters of a machined surface are generally characterized by three different types of parameters, viz., amplitude parameters, spacing parameters and hybrid parameters. For better understanding of the roughness characteristics of the surfaces, in the present study, five different roughness parameters, viz., centre line average roughness (R_a), root mean square roughness (R_q), skewness (R_{sk}), kurtosis (R_{ku}) and mean line peak spacing (R_{sm}) are considered. In addition to surface roughness parameters, material removal rate (MRR) is also considered in the study. The response parameters are optimized as single optimization problem using Taguchi method and also as multiple response optimization problem using Grey Taguchi and Artificial Bee Colony method.

WORK OUTLINE

In this work PAC has been used for machining a work-piece of EN31 steel material. For conducting experimental runs, design of experiments were employed to reduce the number of experiments saving experimental time and cost. Surface roughness and material removal rate (MRR) are considered as the outputs of the experiments. The experimental results are analyzed for optimum MRR and surface roughness using different optimization techniques, viz. Taguchi, Grey Relation Analysis and Artificial Bee Colony (ABC). Finally, to study the surface morphology, scanning electron microscopy (SEM) was used.

DESIGN OF EXPERIMENTS

Design of experiment (DOE) is the process of planning a study to meet specific objectives. It is a powerful analysis tool which enables the designers for analyzing the influence of individual and interactive effects of many factors on performance output in any design which helps to turn any standard design into a robust one. It also helps for the designers to know the sensitive parts and sensitive areas in design that cause problems in output and the designers are then able to fix the problems to get higher yield design prior going into production. With the help of DOE, the designers can no longer afford to experiment in a trial-and-error manner or changing one factor at a time. DOE is a systematic approach to experimentation that considers all individual and interaction factors simultaneously, and industries worldwide are adopting it as a cost-effective way to solve problems affecting their operations.

The concept of design of experiment has been in use since Fisher's (1951) work dealt with agricultural applications of statistical methods. It is a series or process of tests which considers how the experimental factors, both controlled and uncontrolled fit together into a model that meets the specific objectives of the experiment and satisfies the practical constraints of time and money (Montgomery, 2001). If a designer wish to draw a meaningful conclusion from the observed data, statistical approach to DOE is required. So, basically DOE refers to the process of planning, designing and analyzing the experiment through which valid conclusion can be drawn effectively and efficiently. DOE method has been applied in many functional areas which include product development, research, quality control, purchasing, engineering etc. there are several advantages of DOE method like saving time and money by eliminating unnecessary work in experimentation, solving complicated problems, determination of important variable needed to be controlled, optimization of product and process designs, influence of each factor on the performance etc. So design of experiments technique is a very powerful statistical tool, used for modeling and analysis of experimental results on the response variables. The application of DOE techniques can significantly reduce the time and money required for experimental investigations in the manufacturing industry. There are several techniques for design of experiments. Some of prominent DOE methods are Full Factorial Design, Central Composite Design, Orthogonal Array Design etc. In this work Orthogonal array is being employed for Taguchi and Grey Relation Analysis Techniques and Central Composite Design (CCD) is used for Artificial Bee Colony (ABC).

Central Composite Design (CCD)

The central composite design (CCD) (Montgomery, 2001) is a design widely used for estimating second order polynomial for the response variables in response surfaces methodology without using a complete full factorial design of experiments. It is perhaps the most popular class of second order designs. Since, introduced by Box and Wilson (1951), the CCD has been studied and used by many researchers to establish the coefficients of a polynomial with quadratic terms, the experimental design must have at least three levels of each factor. In CCD, there are three different point viz. factorial points, central points and axial points. Factorial points are vertices of the n-dimensional cube which are coming from the full or fractional factorial design where the factor levels are coded to –1, +1. Central point is the point at the center of the design space. Axial points are located on the axes of the coordinate system symmetrically with respect to the central point at a distance α from the design center. There are two main varieties of CCD namely Face centered CCD and Rotatable CCD. The rotatable central composite design is the most widely used experimental design for modeling a second–order response surface. A design is called rotatable, when the variance of the predicted response at any point depends only on the distance of the point from the center point of design. The rotatable design provides the uniformity of prediction error. The pictorial representation of a 3-level three factor FCC design is shown in Figure 2. Here points are in the center of each face of the cube and one super imposed point at the center of the cube which is necessary to estimate the variability of the experimental measurements. Therefore considering four factors and seven replicates at the center point, the present design contains 31 experiments For the three factor experiment, eight (2^3) factorial points, six axial points (2×3) and six centre runs, a total of 20 experimental runs can be considered.

Orthogonal Array

Orthogonal Arrays (OA) are usually used for the experimentation in Taguchi analysis (Taguchi 1990, Ross 1996). These are special matrices used as the design matrices in the fractional factorial design for the estimation of the effect of several factors in a highly efficient way. These designs are applicable even when the factors have more than two levels and for mixed level experiments where all factors do not have same number of levels. In general, when the number of process parameter increases, large number

Figure 2. Face centered central composite design with three factors

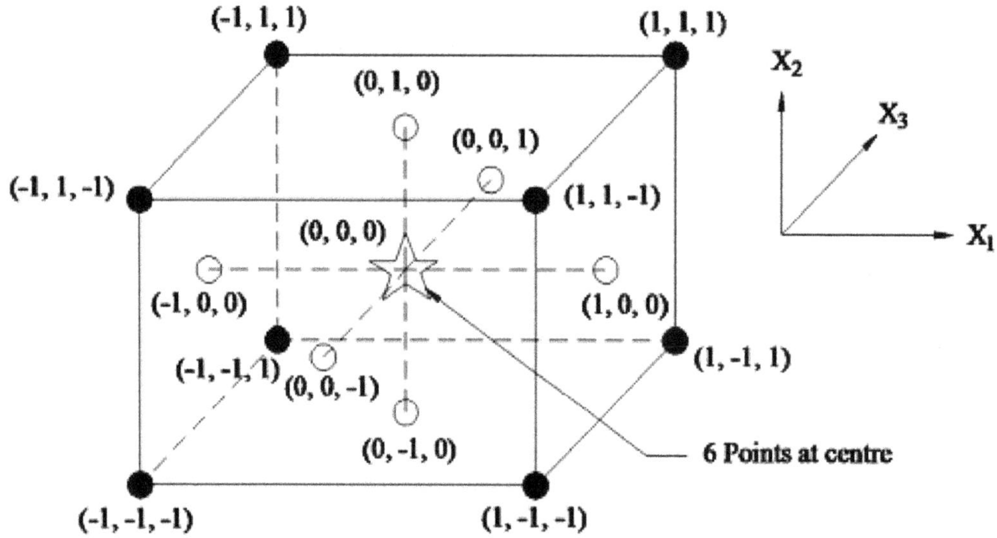

of experiments has to be carried out for factorial design. But using orthogonal array, carrying out small number of experiments in the specified range and effects of process parameters can be studied.

In the present study, three factors (A, B, C) are taken with three levels, which have 20 degree of freedom. Hence, as per basic rule, L_{27} OA is used in the study. An L_{27} OA has 27 rows corresponding to the number of tests (26 degrees of freedom).

TAGUCHI ANALYSIS

Dr. Genichi Taguchi developed this method based on orthogonal array (OA) which gives much reduced variance for the experiment with optimum settings of control parameters. It is known as Robust Design method because of its capability of determining those combinations of controllable factors that minimizes the effect of the sources of experimentation variability (Roy, 1990). Taguchi technique is also a powerful tool for design of high quality systems. It introduces an integrated approach that is simple and efficient to find the best range of designs for quality, performance and computational cost. To achieve desirable quality of the product, Dr. Taguchi recommends that process or product should be carried out in a three-stage approach such as system design, parameter design and tolerance design. Parameter design is used to obtain the factor levels that produce the best performance of the product or process. The optimum condition is selected so that the influence of the uncontrolled or noise factors causes minimum variation of system performance. The parameter design has the key role in the Taguchi method to achieve high quality without increasing the cost factor. Tolerance design is used to fine tune the optimum combinations suggested by parameter design by tightening the tolerance of the factors with significant effect on the product.

S/N Ratio

The traditional method of calculating average factor effects and thereby determining the desirable factor levels (optimum condition) involves looking at the simple averages of the results. Although average calculation is relatively simple, it doesn't capture the variability of results within a trial condition. A better way to compare the population behavior is to use the mean squared deviation (MSD), which combines effect of both average and standard deviation of the results. For convenience of linearity and to accommodate wide ranging data, a logarithmic transformation of MSD, called the signal-to-noise (S/N) ratio is recommended for the analysis of the results. So, when S/N ratio is used for result analysis, the optimum condition identified from such analysis is more likely to produce consistent performance. Taguchi method uses the S/N ratio to measure the quality characteristics. The three categories of quality characteristics used are: smaller-the-better (minimize), larger-the-better (maximize) and nominal-the-best. In Taguchi method, the term 'signal' represents the desirable value (mean) for the output characteristic and the term 'noise' represents the undesirable value (Standard Deviation) for the output characteristic. The S/N ratio is used to measure the quality characteristic deviating from the desired value. The signal-to-noise ratio formulations for the three categories of quality characteristic are as follows:-

S/N ratio for larger–the–better (maximize),

$$S / N_L = -10log\left(\frac{1}{n}\sum_{i=1}^{n}\frac{1}{y_i^2}\right) \tag{1}$$

S/N ratio for smaller-the-better (minimize),

$$S / N_s = -10log\left(\frac{1}{n}\sum_{i=1}^{n}y_i^2\right) \tag{2}$$

S/N ratio for nominal-the-better,

$$S / N_T = 10log\left(\frac{\bar{y}}{s_y^2}\right) \tag{3}$$

where y is the data, n is the number of observations, $\bar{y}$ is the average of the observed data and s_y^2 is the variance of y. These S/N ratios are expressed on a decibel (dB) scale. Taguchi method uses OA to reduce the number of experiments for determining the optimal process parameters. Taguchi has tabulated several standard orthogonal arrays. The choice of a suitable OA is critical for the success of the experiment and depends on the total degrees of freedom required to study the main and interaction effects, the goal of the experiment, resources and budget available and time constraints. Factor level that maximizes the appropriate S/N ratio is optimal. A statistical analysis of variance (ANOVA) is then used to find the influences of the process parameters on the system response under consideration.

Taguchi method is employed with the following procedure in order to optimize the process parameters for different machining operation sequentially: identify the performance characteristics and select process parameters to be evaluated, determine the number of levels for the process parameters and possible interactions between the process parameters, select the appropriate OA and assignment of process parameters to the same, conduct the experiments based on the arrangement of the OA, calculate the total loss function and the S/N ratio, analyze the experimental results using the S/N ratio and ANOVA, select the optimal levels of process parameters, and finally verify the optimal process parameters through confirmation experiment.

Analysis of Variance (ANOVA)

Analysis of variance is a statistical tool to identify and measure sources of variation within a data set. This is done by dividing the total variation in the data into its component parts and comparing each part with the total variation. ANOVA also helps in determining whether the variations are due to variability between methods or within methods. The within-method variations are variations due to individual variation within treatment groups, whereas the between-method variations are due to differences between the methods (Bass, 2007).

Confirmation Test

To check the accuracy of the analysis, a verification test needs to be carried after setting the optimal level of machining parameters. The estimated S/N ratio $\bar{\gamma}$ using the optimal level of the process parameters can be calculated as:

$$\bar{\gamma} = \gamma_m + \sum_{i=1}^{0} \left(\bar{\gamma}_i - \gamma_m \right) \tag{4}$$

where γ_m is the total mean S/N Ratio, γ_i is the mean S/N Ratio at the optimal level and 0 is the number of the main design parameters.

GREY RELATIONAL ANALYSIS

Taguchi method is known as an off-line quality control method and is more suitable for optimizing single response problems. But, Taguchi method alone cannot optimize multiple response problems as in that case a factor, which could be very important to one quality characteristic, may be unimportant to the other quality characteristics or in other words the higher S/N ratio for one performance characteristics may correspond to a lower S/N ratio for another. Therefore, an overall evaluation of the S/N ratio is required for the optimization of multiple performance characteristics. To eliminate such problem Grey method comes into the scenario. This method combined with Taguchi method can easily face any multiple response problems and come out with flying colors. The Grey theory was initiated by Deng (Deng 1982) and has been proven to be useful for dealing with poor, incomplete and uncertain information.

Conducting Grey Relational Analysis

Grey relational analysis is used for optimizing multiple response problems and in such cases the complicated multiple performance characteristics are converted into optimization of a single grey relational grade. The stepwise involvement in grey relational analysis is: Normalize the responses of the quality characteristics, calculate grey relational grade, analyze grey relational grade, and conduct confirmation tests (Figure 3).

Normalization of the Responses

In grey relational analysis the first step is to perform the grey relational generation in which the results of the experiments are normalized in the range between zero to one. The normalization can be done based on three objectives which include:

Normalization can be done using higher-the-better criterion and can be expressed as:

Figure 3. Schematic representation of the steps involved in grey relational analysis

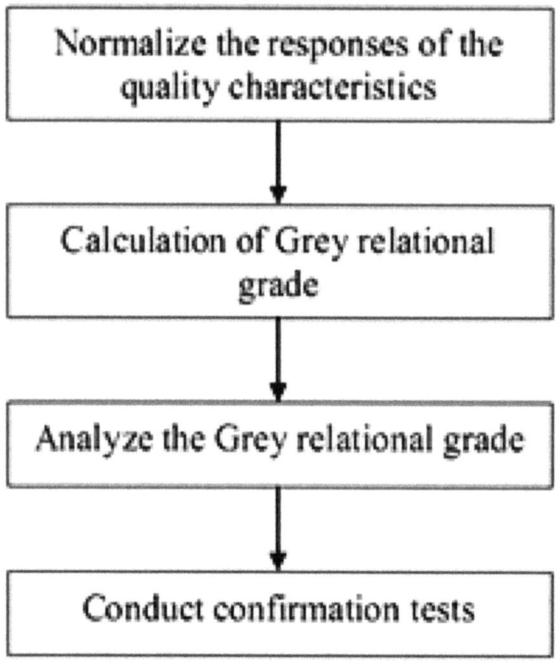

$$x_i(k) = \frac{y_i(k) - \min y_i(k)}{\max y_i(k) - \min y_i(k)} \tag{5}$$

Normalization can be done using lower-the-better criterion and can be expressed as:

$$x_i(k) = \frac{\max y_i(k) - y_i(k)}{\max y_i(k) - \min y_i(k)} \tag{6}$$

Normalization can be done using objective value or nominal the best criterion can be expressed as:

$$x_i(k) = \frac{\| y_i(k) - y_0 \|}{\max y_i(k) - y_0} \tag{7}$$

where xi(k) is the value after grey relational generation, min yi (k) is the smallest value of yi(k) for the kth response, and max yi(k) is the largest value of yi(k) for the kth response. An ideal sequence is x0(k) (k=1,2,3,........,27) for the response. The definition of grey relational grade in the grey relational analysis is to show the relational degree between the twenty seven sequences [0 x (k) and ix (k), i 1,2,......,27;k 1,2,......27].

Calculation of Grey Relational Coefficient

Grey relational coefficients are calculated to express the relationship between the ideal (best = 1) and the actual experimental results. The grey relational coefficient ξ_i (k) can be calculated as:

$$\xi_i\left(k\right) = \frac{\Delta_{\min} + \Psi\Delta_{\max}}{\Delta_{0i}\left(k\right) + \Psi\Delta_{\max}} \tag{8}$$

where $\Delta_{\min} + \Psi\Delta_{\max}$ is difference of the absolute value between x0(k) and xi(k), $\Delta_{\min}$ and $\Delta_{\max}$ are respectively the minimum and maximum values of the absolute differences (Δ 0i) of all comparing sequences. Ψ is a distinguishing coefficient, $0 \leq \psi \leq 1$, the purpose of which is to weaken the effect of $\Delta_{\max}$ when it gets too big, and thus enlarges the difference significance of the relational coefficient. The suggested value of the distinguishing coefficient, Ψ, is 0.5, due to the moderate distinguishing effects and good stability of outcomes. Therefore, Ψ is adopted as 0.5 for further analysis in the present work.

Generation of Grey Relational Grade

In the grey relational analysis, the grey relational grade is used to show the relationship among the series. The overall multiple response characteristics evaluation is based on grey relational grade which is just the average of the grey relational coefficients generated in the previous step.

$$\gamma_i = \frac{1}{n}\sum_{k=1}^{n}\xi_i\left(k\right) \tag{9}$$

Ordering of Grey Relational Grade

In GRA, the practical meaning of the numerical values of grey relational grade between elements is not absolutely important, while the grey relational ordering between them yields more subtle information. The combination yielding the highest grey relational grade is assigned an order of 1 while the combination yielding the minimum grade is assigned the lowest order.

ARTIFICIAL BEE COLONY (ABC) ALGORITHM

In this work Artificial Bee Colony (ABC) Algorithm has been applied to search out the optimal parametric combinations of nontraditional machining process, i.e. plasma arc cutting (PAC) process. The experimental observations are used here for solving multi-objective optimization problems arising out the parametric optimization of the processes.

Methodology of ABC Algorithm

Inspired by the intelligent foraging behavior of honey bees, Karaboga (2005) introduced ABC algorithm for optimizing numerical problems. It can be noted that three parameters are of prime importance in the foraging behavior of honey bees, those are, food source (nectar), employed foragers and unemployed foragers, and the foraging behavior leads to two modes, i.e., recruitment of nectar source and abandonment of nectar source. In ABC, the colony of artificial bees contains generally two groups of bees: employed bees and onlooker bees. The employed bees have all the idea about the food source (nectar position) and quality of food (nectar amount). In the hive all the employed bees with all their information of foods started waggle dance. This dance is the indication of all the characteristics of their foods, i.e., the amount as well as quality of foods. In the hive there are also some unemployed bees called onlooker bees. They watch the waggle dance and get the information about all the food sources and attracted to the best food source. In the next stage the onlooker bees become employed and they started consuming the nectar from the best food source. When this food source becomes abandoned the employed bee becomes a scout bee and starts to find new food source. As early as a scout finds a new food source it becomes an employed bee and the cycle goes on until the best food source (optimum solution) is obtained. In the ABC algorithm the number of employed bee and onlooker bee is equal to the number of solutions in the population. The artificial bee colony algorithm consists of four main phases viz. initialize phase, employed bee phase, onlooker bee phase and scout bee phase. The algorithm used is shown in Figure 4.

MORPHOLOGICAL STUDY

Morphological study was conducted using Scanning electron microscopy (SEM), which is a very useful technique in order to observe in detail the microstructure or the surface morphology of a material. In this method, the surface of a specimen to be examined is scanned with an electron beam, and the reflected

Figure 4. Algorithm of artificial bee colony

Algorithm

1 : Initialize the population of solution x_i, i = 1 (1) NP
2 : Evaluate the population, cycle 1, k = 0.
3 : Memorize the best solution, x_{best} and set $x_{best1} = x_{best}$
4 : Repeat (Exploration phase)
5 : Produce new solution $x_{new} = v_i$ for the employee bees and evaluate them.
6 : Apply the greedy selection process for the employed bees.
7 : Rank the population and calculate the fitness.
8 : Calculate the probability p_i for the solution x_i.
9 : Produce the new solution v_i for the onlookers from the solution selected depending on p_i and evaluate them.
10 : Apply the greedy selection process for the onlookers.
11 : Determine the abandoned solution for the scout if exist, and replace it with a new randomly produced solution x_i.
12 : Memorize the best solution x_{best} achieved so far.
13 : Set k = k + 1; cycle = cycle + 1.
14 : Until (termination condition is met, i.e., cycle = MCN)

(or back scattered) beam of electrons is collected (Callister and Rethwisch 2010) and then displayed at the same scanning rate on a CRT (cathode ray tube) monitor or a LCD (liquid crystal display). There is provision to photograph the image displayed on the screen, which represents the surface features of the specimen. SEM requires the specimen to be electrically conductive. If a non-conductive specimen is needed to be studied, then a thin coating of a conductive material is generally applied on the specimen. The magnification in SEM ranges from about 10 to 50000 times and with a great depth of field.

Surface morphology study of the EN31 tool steel is done by SEM (JEOL, JSM-6360) in order to analyze the microstructure of the work-piece before machining, after machining process (initial condition) and optimal condition.

EXPERIMENTATION

Machine Tool and Work-Piece Material

Experiments were carried out on the CNC plasma arc cutting (PAC) (Figure 5) (EPP-450, 380 V and 50/60 Hz) with PT-36 (Torch) supplied by ESAB. Air was used for the cutting gas as special electrodes made from water-cooled copper with inserts of metals such as hafnium was used. Mechanized torches were mounted on a computer-controlled cutting machine. A standoff was maintained between the torch tip and work piece for best-cut quality. A rectangular block of 80 mm X 15 mm X 10 mm of EN31 steel was selected as work-piece for experimentation.

Figure 5. Experimental set-up of plasma arc cutting

Torch Design

The single flow torch has only a flow of air for cutting so its use is limited to lower amperage, thin gauge sheet metal cutting applications. It does not need a shielding gas flow to cool the torch because of the low amperage output required for cutting thin gauge sheet metal. The dual flow torch has a flow of gas or air for the cutting plasma and shielding gas flow for the torch cooling. This is used for cutting thicker materials, which require higher amperages. In the current work, a duel flow torch PT-36 torch was used.

Input Process Parameters Considered

Plasma arc cutting involves several input parameters to be considered during machining process. In this work, the combination factors such as gas pressure (bar), arc current (amp), torch height (mm) are considered. Three equally spaced levels of each process parameter were selected (Table 1). The control factors were chosen based on literature review, experience, and some preliminary investigations. Other factors, which can be expected to have an effect on the measures of performance, are held constant.

Response Parameters and Their Measurements

Surface Roughness

There are many parameters present for defining surface roughness. Most of them have been formulated due to some specific purposes. The roughness parameters used in this work are: Center line average (R_a), Root mean square roughness (R_q), Skewness (R_{sk}), Kurtosis (R_{ku}) and Mean line peak spacing (R_{sm}).

Surface Roughness Measurement

The surface roughness parameters on the machined surfaces were measured using Talysurf (Make–Taylor Hobson, UK) shown in Figure 6. The Talysurf instrument (Surtronic 3+) is a self-contained instrument for the measurement of surface texture. The profilometer was set to a cut-off length of 0.8 mm, Gaussian filter, and traverse speed 1 mm/sec and 4 mm traverse length. Roughness measurements, in the transverse direction, on the work pieces were repeated five times and average of five measurements of surface roughness parameter values was recorded. The measured profile was digitized and processed through the dedicated advanced surface finish analysis software Talyprofile.

Table 1. Different variables used in the experiment and their levels (L_{27} OA)

Variable	Coding	Level		
		1	2	3
Gas pressure (bar)	A	5	6	7
Arc current (Amp)	B	180	190	200
Torch height (mm)	C	2	4	6

Figure 6. Roughness measuring instrument

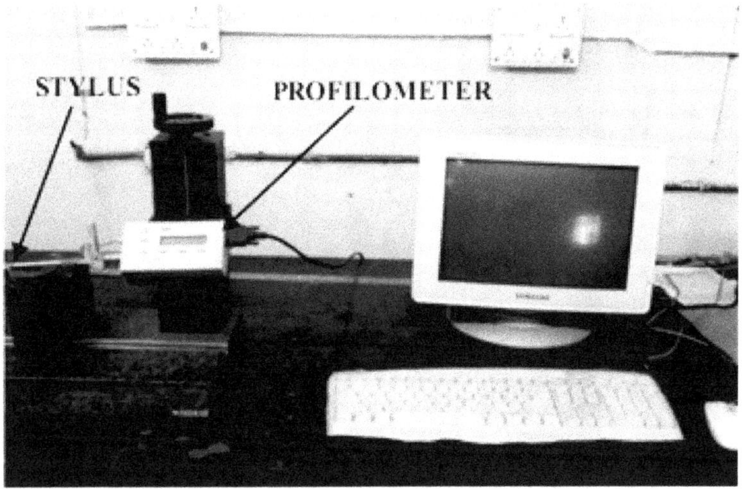

Material Removal Rate (MRR)

The parameter MRR is selected as response variable, which refers to the machining efficiency of any machining process and defined as follows:

$$MRR = \frac{W_i - W_f}{t} \tag{21}$$

where, Wi, Wf are masses (in gm) of the work material before and after machining, respectively, and 't' is the time of machining in minutes or seconds. An electronic weighing machine with an accuracy of 0.01 mg is used in this work (Figure 7).

Experimental Plan

For the optimization of process parameters in PAC, experiments are conducted varying three process parameters. The levels of individual parameter are presented in Table 1. The performance characteristics for the process (MRR and surface roughness parameters) are measured. The experimental results are analyzed for optimum MRR and surface roughness using three optimization tools, namely Taguchi, Grey Relational Analysis and ABC. For Taguchi and Grey Relational Analysis L_{27} OA is selected for conducting experiments (as discussed earlier) and for ABC, FCC design of experiment is chosen. The experimental results considering L_{27} OA are presented in Table 2.

Figure 7. Mettle toddler electronic balance

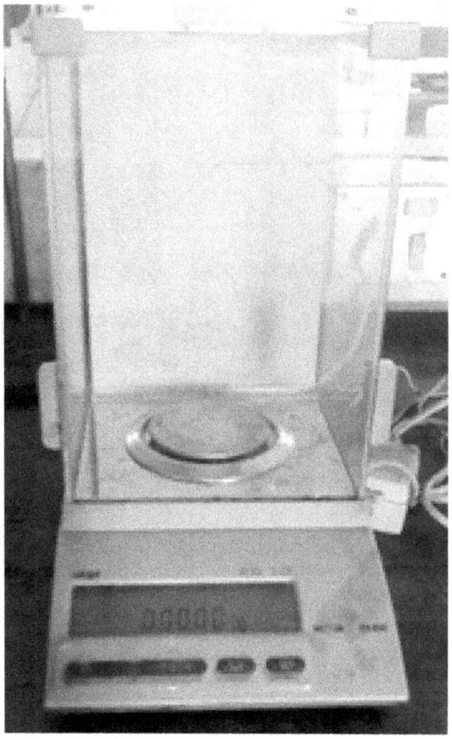

Table 2. Experimental results for response parameters

Exp. No.	A	B	C	MRR (gm /min)	Ra (μm)	Rq (μm)	Rsk	Rku	Rsm (mm)	Exp. No.	A	B	C	MRR (gm /min)	Ra (μm)	Rq (μm)	Rsk	Rku	Rsm (mm)
1	5	180	2	1.850	1.848	2.376	0.299	3.888	0.111	15	6	190	6	1.426	1.812	2.142	0.225	3.640	0.113
2	5	180	4	1.333	2.018	2.494	0.750	4.010	0.124	16	6	200	2	2.210	1.620	2.194	0.273	3.330	0.100
3	5	180	6	1.425	1.966	2.458	0.219	3.806	0.141	17	6	200	4	1.760	1.784	2.164	0.219	3.870	0.125
4	5	190	2	1.001	3.742	4.492	0.051	3.650	0.204	18	6	200	6	2.145	1.452	1.830	0.843	3.680	0.100
5	5	190	4	1.671	1.998	2.512	0.232	3.988	0.138	19	7	180	2	1.895	1.780	2.672	0.328	3.943	0.091
6	5	190	6	1.044	3.110	4.004	0.341	3.623	0.205	20	7	180	4	2.065	1.806	2.352	0.485	4.398	0.101
7	5	200	2	1.703	1.985	2.548	0.211	3.845	0.117	21	7	180	6	1.616	1.910	2.358	0.300	4.326	0.115
8	5	200	4	1.175	1.703	2.136	0.369	4.266	0.115	22	7	190	2	1.423	1.926	2.428	0.469	3.936	0.127
9	5	200	6	1.500	1.840	2.214	0.143	4.130	0.127	23	7	190	4	1.898	1.686	2.170	0.246	3.670	0.107
10	6	180	2	1.632	3.000	3.655	0.523	3.510	0.214	24	7	190	6	1.696	1.564	2.032	0.173	3.586	0.117
11	6	180	4	1.780	1.868	2.382	0.288	4.087	0.128	25	7	200	2	1.483	1.833	2.578	0.246	3.636	0.099
12	6	180	6	2.793	1.522	1.926	0.177	3.354	0.090	26	7	200	4	1.980	1.710	2.106	0.195	3.801	0.115
13	6	190	2	1.858	1.864	2.476	0.289	3.880	0.110	27	7	200	6	1.470	1.690	2.002	0.018	3.798	0.116
14	6	190	4	1.732	1.896	2.442	0.226	4.187	0.118										

RESULT AND DISCUSSION

As discussed earlier experimentation was conducted as per the DOE table and the results of MRR (gm/min), R_a (μm), R_q (μm), R_{sk}, R_{ku}, R_{sm} (mm) were tabulated in Table 2

Optimization of Individual Responses Using Taguchi Method Analysis of Single-to-Noise (S/N) Ratio

With the aid of MINITAB 16, the experimental observations are further transformed into a signal-to-noise (S/N) ratio. There are several (S/N) ratios available depending on objective of optimization of the response. MRR was optimized with higher value representing better machining performance (higher-the better (HB)) whereas surface roughness (R_a, R_q, R_{sk}, R_{ku} and R_{sm}) with lower value representing better surface properties (lower-the better (LB)). S/N ratios of all responses are tabulated in Table 3.

Analyses of the influence of each control factor (A, B and C) on responses are obtained from the response tables of mean S/N ratio (Table 4, Table 5, Table 6, Table 7, Table 8 and Table 9). The mean S/N ratio for each level of three factors (A, B and C) and the total mean S/N ratio for 27 experiments are summarized for MRR in Table 4. From Table 4, it is found that parameter gas pressure (A) is the most significant factor in controlling MRR. Figure 8 shows the main effect plot for S/N ratio and the optimal combination of parameters is found to be A2B1C2 (mid-levels of gas pressure and torch height and lowest level of arc current). From the plot (Figure 8), it is observed that parameter gas pressure (A) has the most significant effect while parameter arc current (B) is also quite significant.

The mean S/N ratio of R_a is shown in Table 5. Parameter gas pressure (A) is found to posses the highest delta value followed by factor arc current (B) and torch height (C). The main effect plot for

Table 3. S/N ratio value for responses

Exp. No.	S/N Ratio						Exp. No.	S/N Ratio					
	MRR	R_a	R_q	R_{sk}	R_{ku}	R_{sm}		MRR	R_a	R_q	R_{sk}	R_{ku}	R_{sm}
1	5.343	-5.334	-7.517	10.494	-11.795	19.078	15	3.086	-5.165	-6.616	12.941	-11.222	18.938
2	2.499	-6.098	-7.938	2.499	-12.063	18.132	16	6.888	-4.190	-6.825	11.289	-10.449	20.000
3	3.081	-5.872	-7.812	13.211	-11.609	17.016	17	4.910	-5.028	-6.705	13.201	-11.754	18.062
4	0.009	-11.462	-13.049	25.784	-11.246	13.807	18	6.629	-3.241	-5.249	1.483	-11.317	20.000
5	4.464	-6.012	-8.002	12.700	-12.017	17.202	19	5.555	-5.008	-8.537	9.683	-11.917	20.819
6	0.374	-9.855	-12.050	9.335	-11.182	13.765	20	6.298	-5.134	-7.429	6.278	-12.865	19.914
7	4.626	-5.957	-8.124	13.500	-11.698	18.636	21	4.169	-5.621	-7.451	10.460	-12.722	18.786
8	1.401	-4.628	-6.592	8.668	-12.600	18.786	22	3.066	-5.693	-7.705	6.577	-11.903	17.924
9	3.522	-5.296	-6.904	16.924	-12.319	17.924	23	5.570	-4.537	-6.729	12.188	-11.293	19.412
10	4.254	-9.542	-11.258	5.635	-10.906	13.392	24	4.589	-3.885	-6.158	15.228	-11.094	18.636
11	5.008	-5.428	-7.539	10.805	-12.229	17.856	25	3.425	-5.266	-8.226	12.190	-11.214	20.044
12	8.922	-3.648	-5.693	15.060	-10.511	20.873	26	5.933	-4.660	-6.469	14.190	-11.598	18.786
13	5.384	-5.409	-7.875	10.797	-11.777	19.172	27	3.346	-4.558	-6.029	35.090	-11.591	18.711
14	4.771	-5.557	-7.755	12.934	-12.439	18.562							

Table 4. Response table of mean S/N ratio for MRR

Level	A	B	C
1	2.813	5.014	4.283
2	5.539	3.479	4.539
3	4.661	4.52	4.191
Delta	2.726	1.535	0.349
Rank	1	2	3

The total mean S/N ratio= 4.338

Table 5. Response table of mean S/N ratio for R_a

Level	A	B	C
1	-6.724	-5.743	-6.429
2	-5.245	-6.397	-5.231
3	-4.929	-4.758	-5.238
Delta	1.795	1.639	1.198
Rank	1	2	3

The total mean S/N ratio = -5.632

Table 6. Response table of mean S/N ratio for R_q

Level	A	B	C
1	-8.665	-7.908	-8.791
2	-7.279	-8.438	-7.24
3	-7.193	-6.791	-7.107
Delta	1.473	1.646	1.684
Rank	3	2	1

The total mean S/N ratio = -7.712

Table 7. Response table of mean S/N ratio for R_{sk}

Level	A	B	C
1	12.568	9.347	11.772
2	10.461	13.165	10.385
3	13.543	14.06	14.415
Delta	3.082	4.712	4.03
Rank	3	1	2

The total mean S/N ratio = 12.991

Table 8. Response table of mean S/N ratio for R_{ku}

Level	A	B	C
1	-11.84	-11.85	-11.43
2	-11.4	-11.57	-12.1
3	-11.8	-11.62	-11.51
Delta	0.44	0.27	0.66
Rank	2	3	1

The total mean S/N ratio = -11.680

Table 9. Response table of mean S/N ratio for R_{sm}

Level	A	B	C
1	17.15	18.43	18.1
2	18.54	17.49	18.52
3	19.23	18.99	18.29
Delta	2.08	1.5	0.43
Rank	1	2	3

The total mean S/N ratio = 18.304

Figure 8. Main effects plot for S/N ratios of MRR

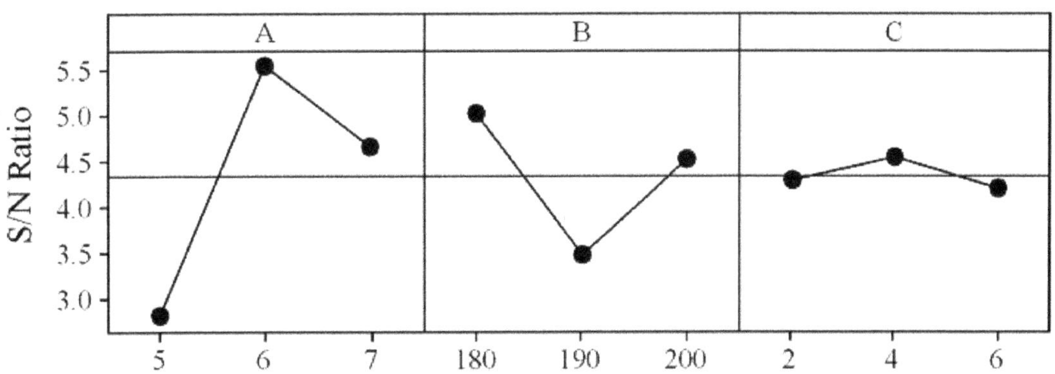

mean S/N ratio for R_a is also presented in Figure 9 and it is clear that factor gas pressure (A) and factor arc current (B) are most significant factors while factor torch height (C) is moderately significant. The optimum machining parameters combination for minimum R_a is A3B3C3 (highest levels of gas pressure, arc current and torch height).

The mean S/N ratio for each level of three factors and the total mean S/N ratio for R_q are summarized in Table 6. It is found from the table that factor torch height (C) posses the highest delta value and hence has the greatest influence over R_q. The main effects plot is illustrated in Figure 10. From the plot, it can be clearly observed that parameters gas pressure (A), arc current (B) and torch height (C) have the influences in controlling R_q. The optimal combination of process parameters for R_q is found to be A3B3C3 (highest levels of gas pressure, arc current and torch height).

Figure 9. Main effects plot for S/N ratios of R_a

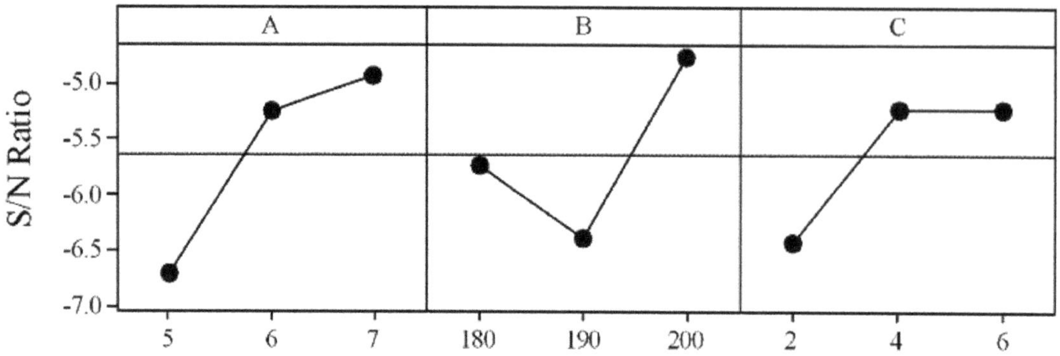

Figure 10. Main effects plot for S/N ratios of R_q

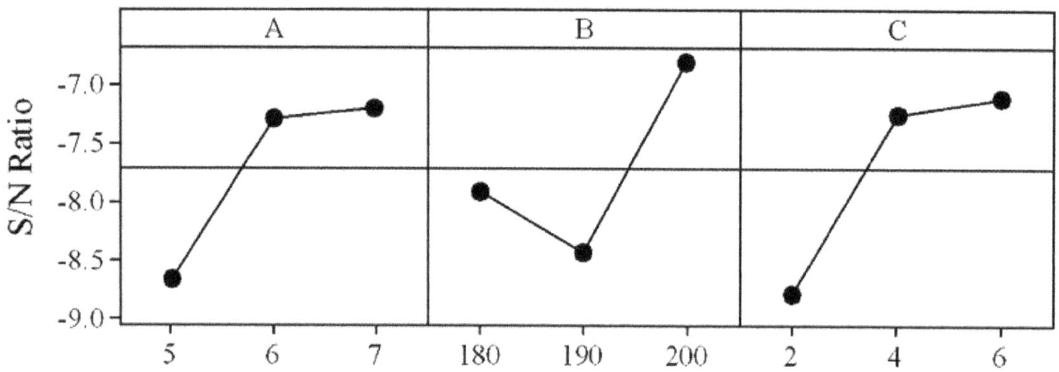

The total mean S/N ratio of R_{sk} and the mean S/N ratio for each level of the factors are summarized in Table 7. From Table 7, it is found that parameter arc current (B) possesses the highest delta value and hence it has the greatest influence over R_{sk}. The main effects plot between the machining parameters are shown in Figure 11. The optimal machining parameter combination for minimum R_{sk} is found to be A3B3C3 (highest levels of gas pressure, arc current and torch height).

The mean S/N ratio of R_{ku} is summarized in Table 8. Parameter torch height (C) is found to possess the highest delta value followed by gas pressure (A) and arc current (B). From the main effects plot (Figure 12), it is clear that torch height (C) is the most significant factor followed by factor gas pressure (A). The optimal machining parameter combination for minimum R_{ku} in this case is found to be A2B1C2 (mid-levels of gas pressure and torch height and lowest level of arc current).

For R_{sm} parameter, mean S/N ratios are presented in Table 9. The main effects plot between the process parameters are shown in Figure 13. It is clear from the main effects plot that parameter gas pressure (A) is the most significant process parameter. The optimal machining parameter combination for minimum R_{sm} is found to be A2B1C2 (mid-levels of gas pressure and torch height and lowest level arc current).

Figure 11. Main effects plot for S/N ratios of R_{sk}

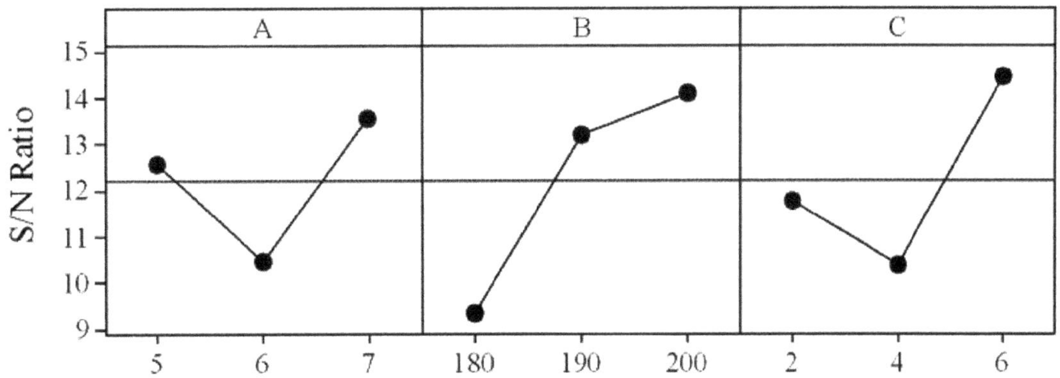

Figure 12. Main effects plot for S/N ratios of R_{ku}

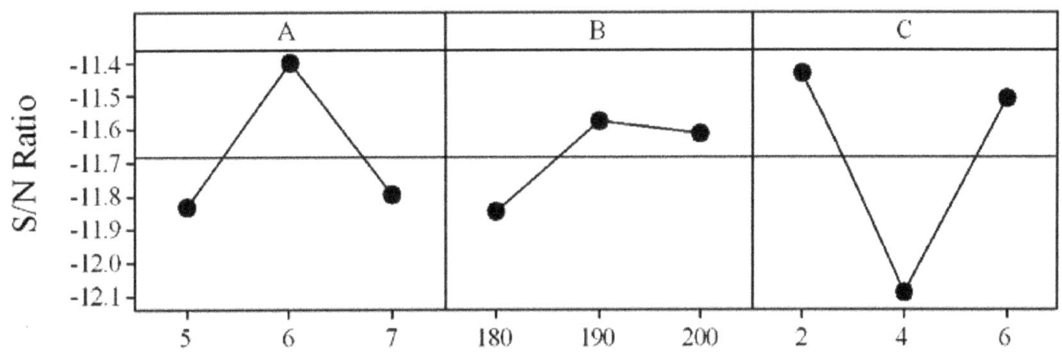

Figure 13. Main effects plot for S/N ratios of R_{sm}

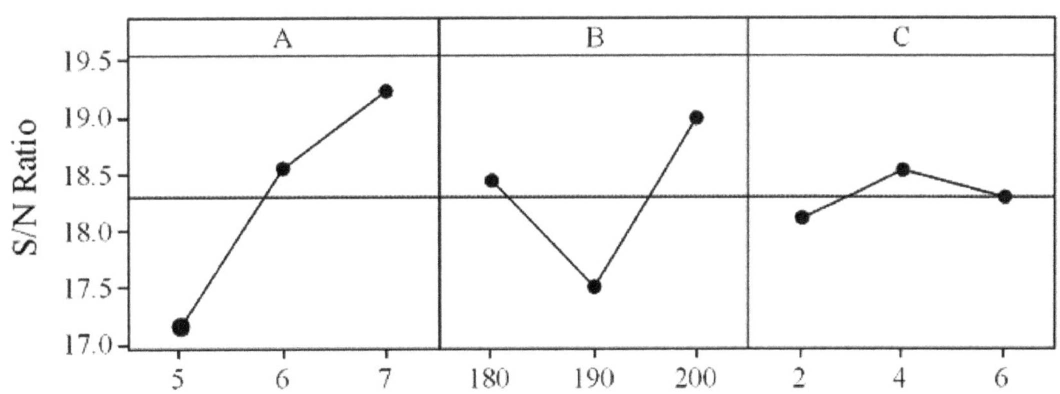

Analysis of Variance (ANOVA)

As discussed earlier ANOVA was performed for the experimental data and tables for the analysis of influence of process parameters for the responses were generated using MINITAB 16. Here SOV is Source of Variation, DOF stands for Degrees of freedom, SS for Sum of Squares and MS is for Mean Square. For the selected levels and values in PAC process, Tables 10, 11, 12, 13, 14, 15 respectively show influence of input parameters along with the interactions on responses (MRR, R_a, R_q, R_{sk}, R_{ku} and R_{sm}) with percentage contribution of each parameter towards the response.

ANOVA results for MRR indicates that parameter gas pressure (A) has got the most significant influence at the confidence level of 95% within the specified test range, similar trend is seen in case of R_a. For R_q at the confidence level of 90% within the specific test range similar influences were observed. But for R_{sk} it was found that parameter arc current (B) has the maximum contribution in controlling the same. The trend was similar for R_{ku}. In case of R_{sm} again gas pressure (A) had the most significant influence.

Table 10. Results of ANOVA for MRR

SOV	DOF	SS	MS	F-Ratio	% Contribution
A	2	34.855	17.428	5.13	35.15
B	2	11.053	5.527	1.63	11.14
C	2	0.587	0.293	0.09	0.59
A X B	4	3.546	0.886	0.26	3.57
A X C	4	10.809	2.702	0.8	10.9
B X C	4	11.129	2.782	0.82	11.22
Error	8	27.155	3.394		27.39
Total	26	99.135			100

Table 11. Results of ANOVA for R_a

SOV	DOF	SS	MS	F-Ratio	% Contribution
A	2	16.523	8.261	3.26	18.69
B	2	12.253	6.126	2.42	13.86
C	2	8.561	4.281	1.69	9.68
AXB	4	20.6	5.15	2.03	23.31
AXC	4	6.949	1.737	0.69	7.86
BXC	4	3.241	0.81	0.32	3.66
Error	8	20.262	2.533		22.92
Total	26	88.388			100

Table 12. Results of ANOVA for R_q

SOV	DOF	SS	MS	F-Ratio	% Contribution
A	2	12.29	6.145	3.46	14.59
B	2	12.714	6.357	3.58	15.1
C	2	15.771	7.885	4.44	18.73
A X B	4	20.192	5.048	2.84	23.98
A X C	4	6.953	1.738	0.98	8.25
B X C	4	2.044	0.511	0.29	2.42
Error	8	14.219	1.777		16.89
Total	26	84.182			100

Table 13. Results of ANOVA for R_{sk}

SOV	DOF	SS	MS	F-Ratio	% Contribution
A	2	44.67	22.33	0.51	4.01
B	2	112.74	56.37	1.29	10.12
C	2	75.44	37.72	0.86	6.77
A X B	4	211.93	52.98	1.21	19.03
A X C	4	259.94	64.99	1.49	23.34
B X C	4	59.08	14.77	0.34	5.3
Error	8	349.79	43.72		31.41
Total	26	1113.6			100

Table 14. Results of ANOVA for R_{ku}

SOV	DOF	SS	MS	F-Ratio	% Contribution
A	2	1.053	0.526	4.62	10.52
B	2	0.386	0.193	1.7	3.85
C	2	2.367	1.184	10.4	23.65
A X B	4	3.386	0.847	7.44	33.84
A X C	4	0.896	0.224	1.97	8.95
B X C	4	1.007	0.252	2.21	10.06
Error	8	0.911	0.114		9.1
Total	26	10.005			100

Table 15. Results of ANOVA for R_{sm}

SOV	DOF	SS	MS	F-Ratio	% Contribution
A	2	20.14	10.07	3.42	20.79
B	2	10.376	5.188	1.76	10.71
C	2	0.821	0.41	0.14	0.84
A X B	4	20.649	5.162	1.75	21.32
A X C	4	14.74	3.685	1.25	15.22
B X C	4	6.571	1.643	0.56	6.78
Error	8	23.536	2.942		24.3
Total	26	96.833			100

Confirmation Test

To validate the analysis, confirmation tests were carried out. Tables 16, 17, 18, 19, 20 and 21 show the results for the estimated responses and the actual responses using the optimal parameters for MRR, R_a, R_q, R_{sk}, R_{ku} and R_{sm}. It is seen that there is a good agreement between the estimated and actual responses. The improvements of S/N ratios from initial to optimal condition for all the responses are shown in Table 22.

Table 16. Confirmation test for MRR

Level	Initial Parameter Combination	Optimum Parameter Combination	
	A2B2C2	A2B1C2	
		Prediction	Experimental
MRR	1.732		2.012
S/N ratio (dB)	4.771	5.539	6.073
Note: Improvement of S/N ratio = 1.302dB (28%)			

Table 17. Confirmation test for R_a

Level	Initial Parameter Combination	Optimum parameter combination	
	A2B2C2	A3B3C3	
		Prediction	Experimental
Ra	1.896		1.69
S/N ratio (dB)	-5.557	-4.895	-4.558
Note: Improvement of S/N ratio = 0.999dB. (18%)			

Table 18. Confirmation test for R_q

Level	Initial Parameter Combination	Optimum Parameter Combination	
	A2B2C2	A3B3C3	
		Prediction	Experimental
Rq	2.442		2.002
S/N ratio (dB)	-7.755	-5.67	-6.02
Improvement of S/N ratio = 1.795dB. (23%)			

Table 19. Confirmation test for R_{sk}

Level	Initial Parameter Combination	Optimum Parameter Combination	
	A2B2C2	A3B3C3	
		Prediction	Experimental
Rsk	0.226		0.018
S/N ratio (dB)	21.412	12.991	35.090
Improvement of S/N ratio = 22.097dB. (170%)			

Table 20. Confirmation test for R_{ku}

Level	Initial Parameter Combination	Optimum Parameter Combination	
	A2B2C2	A2B1C2	
		Prediction	Experimental
Rku	4.187		3.88
S/N ratio (dB)	-12.438	-11.15	-11.777
Note: Improvement of S/N ratio = 1.302dB. (6%)			

Table 21. Confirmation test for R_{sm}

Level	Initial Parameter Combination	Optimum Parameter Combination	
	A2B2C2	A2B1C2	
		Prediction	Experimental
Rsm	0.118		0.11
S/N ratio (dB)	18.56	19.23	19.27
Note: Improvement of S/N ratio = 0.71dB. (4%)			

Table 22. Optimum combinations with improvements in S/N ratios for MRR, R_a, R_q, R_{sk}, R_{ku} and R_{sm}

Level	Initial Parameter Combination	Optimum Parameter Combination	Improvement of S/N ratio (dB)	% Improvement
MRR	A2B2C2	A2B1C2	1.30163dB	28
Ra	A2B2C2	A3B3C3	0.999dB	18
Rq	A2B2C2	A3B3C3	1.7949dB	23
Rsk	A2B2C2	A3B3C3	22.0967dB	170
Rku	A2B2C2	A2B1C2	1.301603dB	6
Rsm	A2B2C2	A2B1C2	0.71dB	4

MULTI OBJECTIVE OPTIMIZATION OF RESPONSES USING GREY RELATIONAL ANALYSIS

For grey relational analysis, the experimental results are normalized and for this, higher-the-better (HB) criterion is selected for MRR (Equation 5) and lower-the-better (LB) criterion is selected for surface roughness (Equation 6). The normalized data (grey relational generation) is shown in Table 23. As a next step, the difference of the absolute value (Δoi) was calculated (shown in Table 24). After that Grey Relational Coefficients were calculated using Equation 8 and presented in Table 25.

After averaging the grey relational coefficients, the grey relational grade (γ_i) is calculated using Equation 9. Table 26 shows the results for the grey relational grades and their orders. The grey relational grade represents the overall characteristics of MRR and surface roughness performance. In this way, the multi-objective optimization problem has been converted to a single objective optimization problem using the combination of Taguchi method and grey relational analysis.

Analysis of Signal-to-Noise (S/N) Ratio

Taguchi method uses S/N ratio to convert the experimental results into a value for the evaluation characteristic in the optimum parameter analysis. As aim of the analysis is multi-objective hence, S/N ratio analysis is done with grey relational grade as the performance index. As grey relational grade is to be maximized, the S/N ratio is calculated using higher-the better (HB) criterion. The response table for the mean of grey relational grade is shown in Table 27. The response table compares the relative magnitude of the effects which includes ranks based on delta statistics. The delta statistics is the difference of highest average for each factor and lowest average for the same. Ranks are assigned on the basis of delta values.

From the response table it was found that parameter gas pressure (A) is the most significant factor in controlling MRR and surface roughness characteristics of PAC. Figure 14 shows the main effects plot. It is very much clear from the main effects plot that parameter gas pressure (A) is the most significant parameter, while arc current (B) also has a quite significant effect on the response. But parameter torch-height (C) has almost no effect. Since higher grey relational grade indicates that the system tends optimality, the optimal condition for each parameter is taken at those points where the mean grey relational grade is found to be the maximum. Hence, the optimal process parameter combination for maximum MRR and minimum roughness characteristics of PAC is given as A2B3C3 (middle level of gas pressure, highest level of arc current and highest level of torch height).

Table 23. Normalized values of experimental data

Exp. No.	MRR	R_a	R_q	R_{sk}	R_{ku}	R_{sm}
1	0.474	0.827	0.750	0.659	0.478	0.832
2	0.185	0.753	0.709	0.113	0.363	0.728
3	0.237	0.775	0.721	0.757	0.554	0.591
4	0.000	0.000	0.000	0.959	0.700	0.081
5	0.374	0.762	0.702	0.741	0.383	0.615
6	0.024	0.276	0.173	0.608	0.725	0.073
7	0.392	0.767	0.689	0.765	0.518	0.785
8	0.097	0.890	0.835	0.575	0.124	0.801
9	0.278	0.831	0.808	0.849	0.251	0.704
10	0.352	0.324	0.297	0.388	0.832	0.000
11	0.435	0.818	0.748	0.672	0.291	0.696
12	1.000	0.969	0.909	0.807	0.978	1.000
13	0.479	0.820	0.715	0.672	0.485	0.841
14	0.408	0.806	0.727	0.748	0.197	0.777
15	0.238	0.843	0.833	0.748	0.709	0.817
16	0.675	0.927	0.815	0.691	1.000	0.923
17	0.424	0.855	0.825	0.756	0.494	0.720
18	0.638	1.000	0.944	0.000	0.672	0.923
19	0.499	0.857	1.000	0.624	0.425	0.995
20	0.594	0.845	0.759	0.433	0.000	0.915
21	0.343	0.800	0.757	0.658	0.067	0.801
22	0.236	0.793	0.732	0.453	0.432	0.704
23	0.501	0.898	0.823	0.724	0.682	0.865
24	0.388	0.951	0.872	0.812	0.759	0.785
25	0.269	0.834	0.679	0.724	0.713	0.927
26	0.546	0.887	0.846	0.785	0.559	0.801
27	0.262	0.896	0.883	1.000	0.562	0.793

Analysis of Variance (ANOVA)

Table 28 shows ANOVA results for overall grey relational grade for MRR and roughness parameters. The table shows the percentage contribution of each parameter. It is clear that parameter gas pressure (A) has got the most significant influence on MRR and roughness, which is about 24% contribution.

Confirmation Test

In order to ascertain the validity of the results a confirmation test is usually carried out.

Table 24. Table for Δoi values

Exp. No.	MRR	R_a	R_q	R_{sk}	R_{ku}	R_{sm}
1	0.526	0.173	0.250	0.341	0.522	0.168
2	0.815	0.247	0.291	0.887	0.637	0.272
3	0.763	0.224	0.279	0.243	0.446	0.409
4	1.000	1.000	1.000	0.041	0.300	0.919
5	0.626	0.238	0.298	0.259	0.617	0.385
6	0.976	0.724	0.827	0.392	0.275	0.927
7	0.608	0.233	0.311	0.235	0.482	0.215
8	0.903	0.110	0.165	0.425	0.876	0.199
9	0.722	0.169	0.192	0.151	0.749	0.296
10	0.648	0.676	0.703	0.612	0.169	1.000
11	0.565	0.182	0.252	0.328	0.709	0.304
12	0.000	0.030	0.090	0.193	0.022	0.000
13	0.521	0.180	0.285	0.328	0.515	0.158
14	0.592	0.194	0.273	0.252	0.803	0.223
15	0.763	0.157	0.167	0.252	0.290	0.183
16	0.325	0.073	0.185	0.309	0.000	0.077
17	0.577	0.145	0.174	0.244	0.506	0.280
18	0.362	0.000	0.056	1.000	0.328	0.077
19	0.501	0.143	0.000	0.376	0.574	0.005
20	0.406	0.155	0.241	0.567	1.000	0.085
21	0.657	0.200	0.243	0.342	0.933	0.199
22	0.764	0.207	0.268	0.547	0.568	0.296
23	0.499	0.102	0.177	0.276	0.318	0.134
24	0.612	0.049	0.128	0.189	0.240	0.215
25	0.731	0.167	0.321	0.276	0.287	0.073
26	0.454	0.113	0.154	0.215	0.441	0.199
27	0.738	0.104	0.117	0.000	0.438	0.207

Table 29 shows the comparison of the estimated grey relational grade with the actual grey relational grade using the optimal parameters. There is a good agreement between the estimated and actual grey relational grade. The improvement of grey relational grade from initial to optimal condition is 0.12391 which means improved by 21%.

ARTIFICIAL BEE COLONY (ABC) ALGORITHM

For ABC analysis, the experiments are carried out according to FCC design of experiments considering three process parameters as shown in Table 30 and the experimental results are presented in Table 31.

Table 25. Table for grey relational coefficient

Exp. No.	MRR	R_a	R_q	R_{sk}	R_{ku}	R_{sm}
1	0.487	0.743	0.667	0.595	0.489	0.749
2	0.380	0.669	0.632	0.360	0.439	0.648
3	0.396	0.690	0.642	0.673	0.529	0.549
4	0.333	0.333	0.333	0.924	0.625	0.352
5	0.444	0.677	0.627	0.658	0.448	0.565
6	0.339	0.409	0.377	0.560	0.645	0.350
7	0.451	0.682	0.617	0.681	0.509	0.699
8	0.356	0.819	0.752	0.540	0.363	0.715
9	0.409	0.747	0.722	0.768	0.400	0.628
10	0.436	0.425	0.415	0.449	0.748	0.333
11	0.469	0.734	0.665	0.604	0.414	0.622
12	1.000	0.943	0.847	0.722	0.957	1.000
13	0.490	0.735	0.637	0.604	0.493	0.759
14	0.458	0.721	0.647	0.665	0.384	0.692
15	0.396	0.761	0.750	0.665	0.633	0.733
16	0.606	0.872	0.729	0.618	1.000	0.865
17	0.465	0.775	0.741	0.672	0.497	0.641
18	0.580	1.000	0.899	0.333	0.604	0.865
19	0.500	0.777	1.000	0.571	0.465	0.991
20	0.552	0.764	0.675	0.469	0.333	0.854
21	0.432	0.714	0.673	0.594	0.349	0.715
22	0.396	0.707	0.651	0.478	0.468	0.628
23	0.501	0.830	0.739	0.644	0.611	0.789
24	0.450	0.911	0.797	0.726	0.675	0.699
25	0.406	0.750	0.609	0.644	0.635	0.872
26	0.524	0.816	0.765	0.699	0.531	0.715
27	0.404	0.828	0.810	1.000	0.533	0.707

Objective Functions for Individual Response

To develop the objective functions, response surface methodology (RSM) (Montgomery 2001) is used. RSM is utilized for establishing the relations between the different process parameters with a variety of machining criteria and exploring their effects on MRR and surface roughness. The general second order polynomial response surface model (Equation 22), is considered to determine the parametric influences on the various response criteria.

$$Y_u = \beta_0 + \sum_{i=1}^{k} \beta_i . X_i + \sum_{i=1}^{k} \beta_{ii} . X_i^2 + \sum_{j>i}^{k} \beta_{ij} . X_i . X_j + \varepsilon \quad (22)$$

Table 26. Grey relational grades and their orders

Exp. No.	Grade	Order	Exp. No.	Grade	Order
1	0.622	12	17	0.632	11
2	0.522	24	18	0.714	4
3	0.580	20	19	0.717	3
4	0.484	25	20	0.608	15
5	0.570	22	21	0.580	21
6	0.447	27	22	0.555	23
7	0.607	16	23	0.686	7
8	0.591	18	24	0.710	6
9	0.612	14	25	0.653	10
10	0.468	26	26	0.675	8
11	0.585	19	27	0.714	5
12	0.911	1			
13	0.620	13			
14	0.594	17			
15	0.656	9			
16	0.782	2			

Table 27. Response table for grey relational grade

Level	A	B	C
1	0.559	0.621	0.612
2	0.662	0.591	0.607
3	0.655	0.664	0.658
Delta	0.103	0.073	0.051
Rank	1	2	3

Mean grey relational grade = 0.626

where Y_u represents the corresponding response, i.e. MRR and surface roughness of the machining process in the present study. X_i is the input variables, X_i^2 and X_iX_j are the squares and interaction terms, respectively, of these input variables. The unknown regression coefficients are β_0, β_i, β_{ij} and β_{ii} and the error in the model is depicted as ε. The second term under the summation sign of this polynomial equation attributes to linear effects, whereas the third term of the above equation corresponds to the higher order effects and lastly the forth term of the equation includes the interactive effects of the parameters.

Based on the equation, empirical relationships between response and factors in un-coded values are given as follows:

Figure 14. Main effects plot for mean grey relational grade

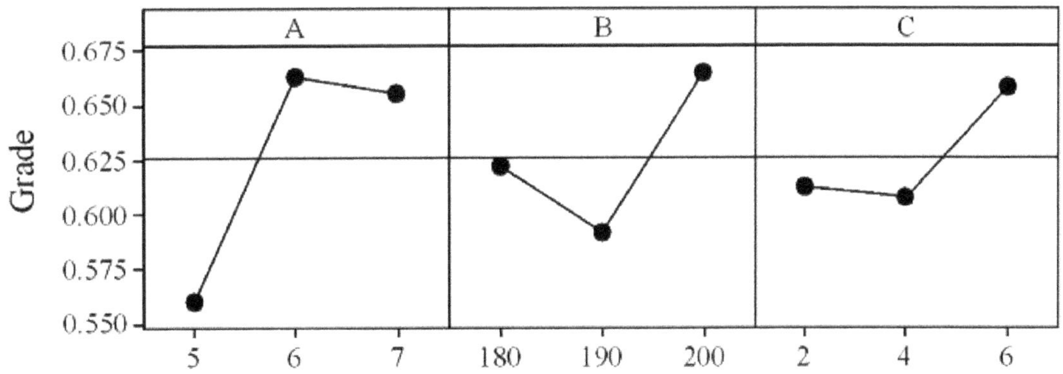

Table 28. Response table for grey relational grade

SOV	DOF	SS	MS	F-Ratio	% Contribution
A	2	0.060	0.030	2.56	46.63
B	2	0.024	0.012	1.05	19.13
C	2	0.014	0.007	0.62	11.29
A X B	4	0.007	0.002	0.16	2.91
A X C	4	0.031	0.008	0.67	12.20
B X C	4	0.020	0.005	0.43	7.83
Error	8	0.093	0.012		
Total	26	0.250			100.00

Table 29. Confirmation test for RSM

Level	Initial Parameter Combination	Optimum Parameter Combination	
	A2B2C2	A2B3C3	
		Prediction	Experimental
MRR	1.732		2.145
R_a	1.896		1.452
R_q	2.442		1.830
R_{sk}	0.226		0.843
R_{ku}	4.187		3.680
R_{sm}	0.118		0.100
Grade	0.594	0.662	0.714
Note: Improvement of S/N ratio = 0.12391 (21%)			

Table 30. Experimental parameters and their levels

Design Factors	Units	Coding	Level		
			-1	0	1
Gas pressure	Bar	A	5	6	7
Arc current	Amps	B	180	190	200
Torch height	mm	C	2	4	6

Table 31. Experimental results for response parameters

Exp. No.	A	B	C	MRR (gm /min)	R_a (µm)	R_q (µm)	R_{sk}	R_{ku}	R_{sm} (mm)	Exp. No.	A	B	C	MRR (gm /min)	R_a (µm)	R_q (µm)	R_{sk}	R_{ku}	R_{sm} (mm)
1	7	200	2	1.483	1.833	2.578	0.245	3.637	0.099	11	6	190	4	1.732	1.896	2.492	0.225	4.188	0.118
2	7	190	4	1.899	1.686	2.170	0.246	3.670	0.107	12	5	190	4	1.672	1.998	2.512	0.232	3.989	0.138
3	6	200	4	1.760	1.784	2.164	0.219	3.870	0.125	13	6	190	6	1.427	1.812	2.142	0.226	3.640	0.113
4	6	190	4	1.732	1.896	2.442	0.225	4.188	0.118	14	5	180	2	1.850	1.848	2.376	0.299	3.888	0.111
5	5	180	6	1.426	1.966	2.458	0.219	3.806	0.141	15	7	180	2	1.896	1.780	2.672	0.328	3.943	0.091
6	5	200	6	1.500	1.840	2.214	0.143	4.130	0.127	16	6	190	4	1.732	1.896	2.442	0.225	4.188	0.118
7	6	190	2	1.859	1.864	2.476	0.289	3.880	0.110	17	6	180	4	1.780	1.868	2.382	0.288	4.088	0.128
8	6	190	4	1.732	1.896	2.442	0.225	4.188	0.118	18	6	190	4	1.732	1.896	2.442	0.225	4.188	0.118
9	7	180	6	1.616	1.910	2.358	0.299	4.326	0.115	19	7	200	6	1.470	1.690	2.002	0.018	3.798	0.116
10	6	190	4	1.732	1.896	2.442	0.225	4.188	0.118	20	5	200	2	1.703	1.985	2.548	0.211	3.845	0.117

$$Y_u(\text{MRR}) = -2.62367 + 0.957357 \text{ x } A + 0.0337601 \text{ x } B - 0.508632 \text{ x } C + 0.0111777 \text{ x } A^2 - 0.00607313 \text{ x } A \text{ x } B + 0.0209219 \text{ x } A \text{ x } C - 0.0000421231 \text{ x } B^2 + 0.0030449 \text{ x } B \text{ x } C - 0.0328822 \text{ x } C^2 \tag{23}$$

$$Y_u(R_a) = -8.95225 + 0.371811 \text{ x } A + 0.0935858 \text{ x } B + 0.639759 \text{ x } C - 0.00221364 \text{ x } A^2 - 0.00222375 \text{ x } A \text{ x } B + 0.00086875 \text{ x } A \text{ x } C - 0.000182136 \text{ x } B^2 - 0.00336 \text{ x } B \text{ x } C - 0.00150341 \text{ x } C^2 \tag{24}$$

$$Y_u(R_q) = -17.731 + 0.655945 \text{ x } A + 0.180573 \text{ x } B + 0.973666 \text{ x } C + 0.0307045 \text{ x } A^2 - 0.004725 \text{ x } A \text{ x } B - 0.039875 \text{ x } A \text{ x } C - 0.000375455 \text{ x } B^2 - 0.0042375 \text{ x } B \text{ x } C - 0.000386364 \text{ x } C^2 \tag{25}$$

$$Y_u(R_{sk})=-3.14044+0.688988 \text{ x } A+0.0152142 \text{ x } B+0.234585 \text{ x } C-0.0150873 \text{ x } A^2 \\ -0.00251438 \text{ x } A \text{ x } B-0.00669688 \text{ x } A \text{ x } C -0.00000364773 \text{ x } B^2-0.00117906 \text{ x } B \text{ x } C \\ +0.000774432 \text{ x } C^2 \quad (26)$$

$$Y_u(R_{ku})=7.18073+3.65409 \text{ x } A-0.140989 \text{ x } B+0.0497572 \text{ x } C-0.093176 \text{ x } A^2 \\ -0.0139454 \text{ x } A \text{ x } B+0.0213146 \text{ x } A \text{ x } C +0.00056134 \text{ x } B^2+0.000910206 \text{ x } B \text{ x } C \\ -0.040654 \text{ x } C^2 \quad (27)$$

$$Y_u(R_{sm})=2.05008-0.0621341 \text{ x } A-0.0192745 \text{ x } B+0.0560639 \text{ x } C+0.000909091 \text{ x } A^2 \\ +0.0002125 \text{ x } A \text{ x } B+0.0000625 \text{ x } A \text{ x } C +0.0000490909 \text{ x } B^2-0.00016875 \text{ x } B \text{ x } C \\ -0.00252273 \text{ x } C^2 \quad (28)$$

where, A = gas pressure, B = arc current, C = torch height

Multi-Objective Optimization

In multi-objective optimization of PAC process, all responses are optimized simultaneously. For this, the following objective function is developed (Rao et al. 2008):

$$\text{Min}(Z)= \frac{w_1 * Y_u(R_a)}{R_{a,\min}} + \frac{w_2 * Y_u(R_q)}{R_{q,\min}} + \frac{w_3 * Y_u(R_{sk})}{R_{sk,\min}} \\ + \frac{w_4 * Y_u(R_{ku})}{R_{ku,\min}} + \frac{w_5 * Y_u(R_{sm})}{R_{sm,\min}} - \frac{w_6 * Y_u(R_{MRR})}{MRR_{\max}} \quad (29)$$

where $Y_u(R_{MRR})$, $Y_u(R_a)$, $Y_u(R_q)$, $Y_u(R_{sk})$, $Y_u(R_{ku})$ and $Y_u(R_{sm})$ are the second order response surface equations, as given in equations (23 - 28) respectively, $R_{a,min}$, $R_{q,min}$, $R_{sk,min}$, $R_{ku,min}$, $R_{sm,min}$, and MRR_{max}, are the minimum and the maximum values of R_a, R_q, R_{sk}, R_{ku}, R_{sm} and MRR, respectively, and w1, w2, w3, w4, w5 and w6 are the weight values assigned to responses, respectively. These weight values can be anything provided that summation of the weights will be 1 and it depends on the priorities of the considered responses as set by the process engineers. Here, equal weights for both the responses are considered, i.e.,

w1 =w2 = w3 = w4 = w5 = w6 = 0.167

ABC algorithm is now used to optimize the above mentioned equations. The corresponding computer code for ABC algorithm is developed in MATLAB 7.8 with the following control parameters. These parameters are selected for easy convergence of the parametric optimization problems as shown in below.

- **Number of Population:** 6
- **Number of Employed Bees:** 50% of population
- **Number of Onlooker Bees:** 50% of population
- **Number of Scouts per Cycle:** 1
- **Number of Cycles:** 1000
- **Limit:** 500

The convergence of ABC algorithm for both responses i.e. MRR and surface roughness parameters (R_a, R_q, R_{sk}, R_{ku} and R_{sm}) is shown in Figure 15.

A combination of gas pressure of 7 bar, arc current of 200 amp and torch height of 6 mm can be set for obtaining maximum MRR and minimum R_a, R_q, R_{sk}, R_{ku} and R_{sm}. MRR, R_a, R_q, R_{sk}, R_{ku} and R_{sm} values are obtained as 1.89999 gm/sec, 1.654 µm, 1.9148 µm, 0.0592457, 3.748435, and 0.091 mm respectively, and the optimal solution (Z) is 1.13197.

In order to verify the obtained optimal results in case of multi-objective optimization problem of this process a confirmation test has been carried out. For multi-response optimization, the error is 0.97% for MRR and 0.96% for R_a, 0.7% for R_q, 0.44% for R_{sk}, 0.58% for R_{ku} and 1.1% for R_{sm}. The results of the confirmation test show a good agreement with the predicted results. The confirmation results are shown in Table 32.

Figure 15. Convergence of ABC algorithm for multi-responses

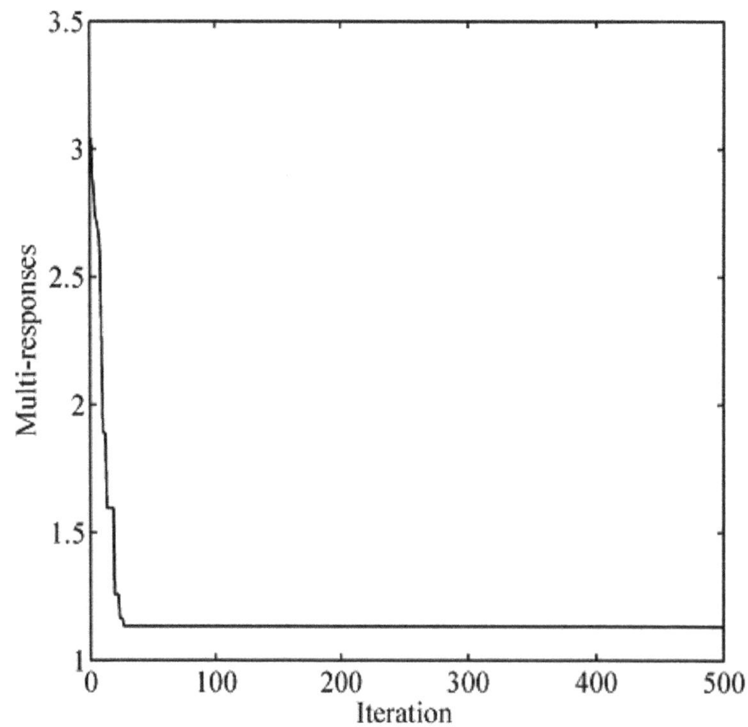

Table 32. Results of optimization and confirmation test for multi-responses

Responses	Process Parameters (Optimum Value)			Responses Obtained From ABC Analysis	Responses Obtained From Experimentation	% of Error of Responses
	Gas Pressure (bar)	Arc Current (amp)	Torch Height (mm)			
MRR	7	200	6	1.8999 gm/sec	1.8815 gm/sec	0.97
R_a				1.654 µm	1.665 µm	0.7
R_q				1.915 µm	1.929 µm	0.73
R_{sk}				0.05925	0.05871	0.91
R_{ku}				3.749	3.785	0.96
R_{sm}				0.091 mm	0.092 mm	1.1

SURFACE MORPHOLOGY STUDY

The surface morphology of work-piece before and after machining (Figure 16) is studied with the help of scanning electron microscopy (SEM) images. Before machining, the work piece surface is nearly smooth and there is no globular spot. After machining (initial condition) the work surface becomes rough and the machined surface contains plenty of globules which are unevenly distributed due to grain coarsening near the cutting surface. This can be attributed to the very high temperatures and slow cooling. In the PAC process, the material is heated to its melting point and gets cool in the air after cutting. High heating and slow cooling could lead to temper conditions. This grain coarsening will decrease the hardness of the machined surface and make the surface less resistant to the wear or other environmental conditions. SEM images considering the optimum combinations of process parameters for Taguchi, grey relational analysis and ABC are shown in Figure 17, Figure 18 and Figure 19 respectively. It is observed that the surface roughness of optimum condition is better than the initial condition.

Figure 16. SEM images of workpiece before and after machining

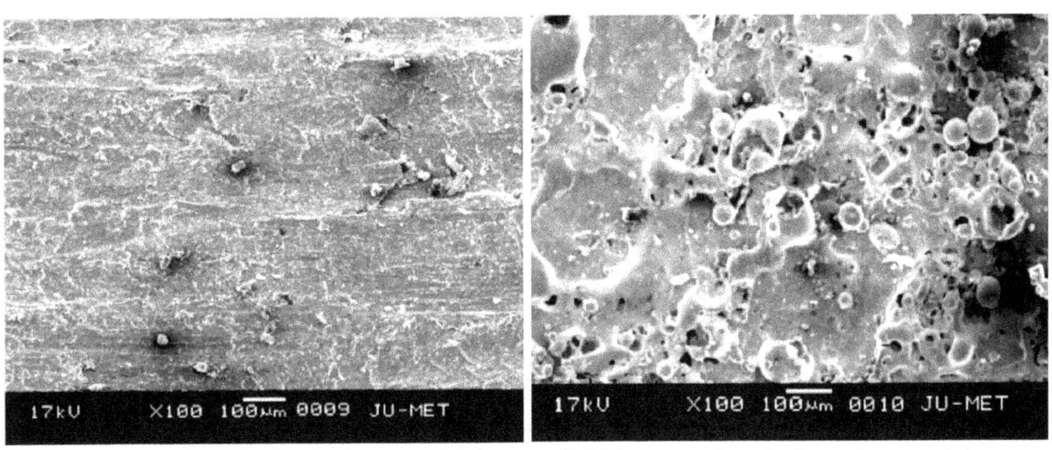

SEM images of workpiece before machining **SEM images of workpiece after machining**

Figure 17. SEM images of workpiece after machining using initial conditions and using optimized parameters obtained by Taguchi technique

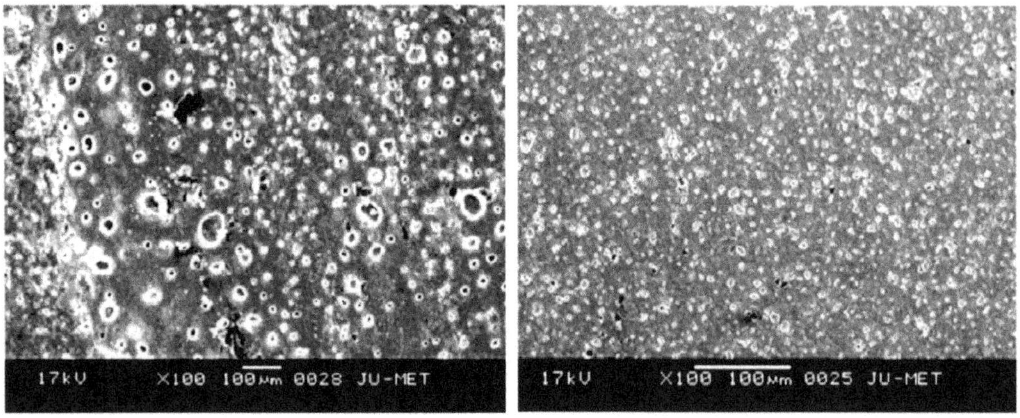

SEM images of workpiece after machining (Initial Condition)

SEM image of workpiece after machining (at optimum combination using Taguchi analysis)

Figure 18. SEM images of workpiece after machining using initial conditions and using optimized parameters obtained by grey relation analysis

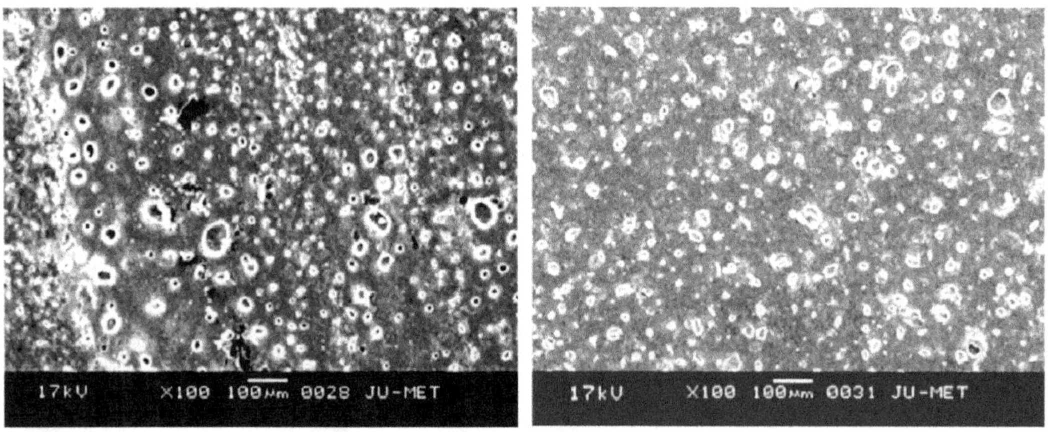

SEM images of workpiece after machining (Initial Condition)

SEM image of workpiece after machining (at optimum combination using grey relational analysis)

CONCLUSION

CNC plasma arc cutting operations are carried out on EN 31 steel. Three process parameters viz. gas pressure, arc current and torch height for the study. The experimental results are further processed using Taguchi, grey relational analysis and ABC analysis. From Taguchi analysis, it was observed that gas pressure has the most significant influence in controlling all responses. Interaction between gas pressure and arc current has got moderate influence on all responses. From grey analysis, optimal combination of process parameters is obtained as mid-level of gas pressure, high level of arc current and high level

Figure 19. SEM images of workpiece after machining using initial conditions and using optimized parameters obtained by artificial bee colony (ABC) algorithm

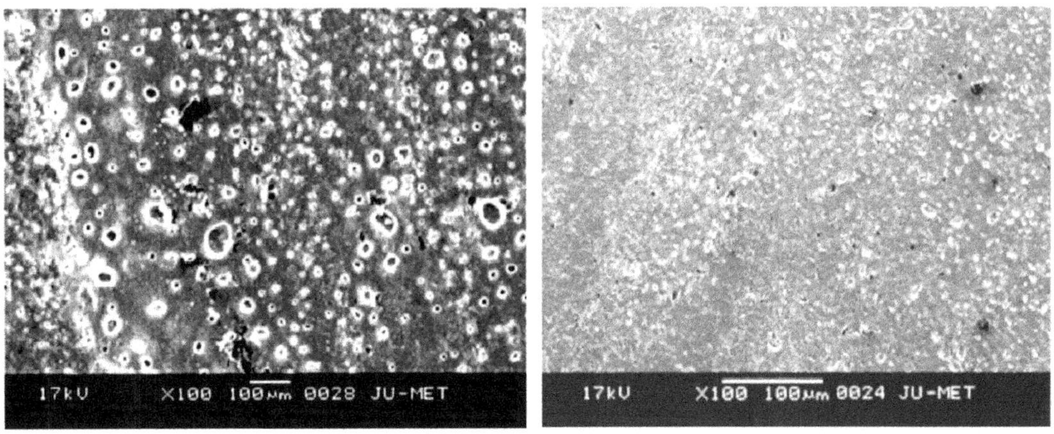

SEM images of workpiece after machining (Initial Condition)

SEM image of workpiece after machining (at optimum combination using ABC)

of torch height. From the confirmation test, it is found that the improvement of grey relational grade at the optimal condition is 21%. From ABC analysis, the optimal combination is obtained as high levels of gas pressure, arc current and torch height. The results of the confirmation tests show a good agreement with the predicted result. From the above analysis, it can be concluded that the gas pressure and arc current have got the most significant influences in controlling the responses. SEM images provides a visual confirmation that the surface morphology of optimal condition is much better than initial condition for all cases which indicates that the optimization carried out was effective. These optimal combinations of parameters can be directly employed by industrial sector (depending on the requirement i.e. single objective / multi-objective etc.) for enhancement of overall productivity and profitability.

REFERENCES

Bahram, A. (2009). Optimising the automated plasma cutting process by design of experiments. *International Journal of Rapid Manufacturing*, *1*(1), 19–40. doi:10.1504/IJRAPIDM.2009.028930

Bass, I. (2007). *Six sigma statistics with Excel and Minitab New York*. McGraw Hill.

Bhuvenesh, R., Norizaman, M. H., & Manan, M. S. A. (2012). Surface roughness and MRR effect on manual plasma arc cutting machining. *International Journal of Mechanical Engineering*, *6*, 131–134.

Bini, R., Colosimo, B. M., Kutlu, A. E., & Monno, M. (2007). Experimental study of the features of the kerf generated by a 200 A high tolerance plasma cutting system. *Journal of Materials Processing Technology*, *196*(1-3), 345–355. doi:10.1016/j.jmatprotec.2007.05.061

Callister, W. D. J., & Rethwich, D. G. (2010). Material Science and Engineering: An Introduction (8th ed.). Wiley.

Deng, J. (1989). Introduction to Grey System. *Journal of Grey System, 1*, 1–24.

Fisher, R. A. (1951). *Design of experiments Edinburgh*. Oliver and Boyd.

Gane, N., Rogozinski, M. W., Polivka, F., Doolette, A. G., & Ramakrishnan, S. (1994). *Quality of cut in air plasma cutting*. Paper presented at the Washington Technology Industry Association 42nd annual National Welding Conference, Melbourne, Australia.

Gariboldi, E., & Previtali, B. (2004). High tolerance plasma arc cutting of commercially pure titanium. *Journal of Materials Processing Technology, 160*(1), 77–89. doi:10.1016/j.jmatprotec.2004.04.366

Gullu, A., & Atici, U. (2006). Investigation of the effects of plasma arc parameters on the structure variation of AISI 304 and St 52 steels. *Materials & Design, 27*(10), 1157–1162. doi:10.1016/j.matdes.2005.02.014

Ismail, M. I. S., & Taha, Z. (2011). Experimental design and performance analysis in plasma arc surface hardening, World Academy of Science. *Engineering and Technology, 56*, 1052–1058.

Kadirgama, K., Noor, M. M., Harun, W. S. W., & Aboue-El-Hossein, K. A. (2010). Optimization of heat affected zone by partial swarm optimization in air plasma cutting operation. *Journal of Scientific and Industrial Research, 69*, 439–443.

Karaboga, D. (2005). *An idea based on honey bee swarm for numerical optimization, Technical Report-TR06*. Erciyes University, Engineering Faculty, Computer Engineering Department.

Maity, K. P., & Bagal, D. K. (2015). Effect of process parameters on cut quality of stainless steel of plasma arc cutting using hybrid approach. *International Journal of Advanced Manufacturing Technology, 78*(1-4), 161–175. doi:10.1007/s00170-014-6552-6

Montgomery, D. C. (2001). *Design and Analysis of Experiments*. New York: Wiley.

Ozek, C., Caydas, U., & Unal, E. (2012). A fuzzy model for predicting surface roughness in plasma arc cutting of AISI 4140 steel. *Materials and Manufacturing Processes, 27*(1), 95–102. doi:10.1080/10426914.2011.551952

Radovanovic, M., & Madic, M. (2011). Modeling the plasma arc cutting using ANN. *Nonconventional Technologies Review, 4*, 43–48.

Ramakrishnan, S., Shrinet, V., Polivka, F. B., Kearney, T. N., & Koltun, P. (2000). Influence of gas composition on plasma arc cutting of steel. *Journal of Physics. D, Applied Physics, 33*(18), 2288–2299. doi:10.1088/0022-3727/33/18/313

Ross, P. J. (1996). *Taguchi Techniques for Quality Engineering* (2nd ed.). New York: McGraw Hill.

Roy, R. K. (1990). *A primer on the Taguchi method Dearborn*. Michigan Society of Manufacturing Engineers.

Salonitis, K., & Vatousianos, S. (2012). Experimental investigation of the plasma arc cutting process. *Procedia CIRP, 3*, 287–292. doi:10.1016/j.procir.2012.07.050

Taguchi, G. (1990). *Introduction to Quality Engineering*. Tokyo: Asian Productivity Organization.

Teulet, Ph., Girard, L., Razafinimanana, M., Gleizes, A., Bertrand, Ph., Camy-Peyret, F., & Richard, F. et al. (2006). Experimental study of an oxygen plasma cutting torch: II. Arc material interaction, energy transfer and anode attachment. *Journal of Physics. D, Applied Physics*, *39*(8), 1557–1573. doi:10.1088/0022-3727/39/8/015

Vasilev, K. V. (2003). Plasma-arc cutting – a promising method of thermal cutting. *Welding International*, *17*(2), 147–151. doi:10.1533/wint.2003.3096

Wang, J., Zhu, Z., He, C., & Yang, F. (2011). Effect of dual swirling plasma arc cutting parameters on kerf characteristics. *International Journal of Material*, *4*, 39–43.

Xu, W. J., Fang, J. C., & Lu, Y. S. (2002). Study on ceramic cutting by plasma arc. *Journal of Materials Processing Technology*, *129*(1-3), 152–156. doi:10.1016/S0924-0136(02)00600-3

Yang, L. J. (2001). Plasma surface hardening of ASSAB 760 steel specimens with Taguchi optimization of the processing parameters. *Journal of Materials Processing Technology*, *113*(1-3), 521–526. doi:10.1016/S0924-0136(01)00583-0

Yang, L. J. (2007). The effect of specimen thickness on the hardness of plasma surface hardened ASSAB 760 steel specimens. *Journal of Materials Processing Technology*, *185*(1-3), 113–119. doi:10.1016/j.jmatprotec.2006.03.117

Chapter 6
Optimizing the Electrical Discharge Drilling Process for High Aspect Micro Hole Drilling in Die Steel

Kamal Kumar
PEC University of Technology, India

ABSTRACT

Electric discharge drilling (EDD) is a thermo erosion process used to produce holes in high strength materials for various applications such as fuel injector, medical devices, turbine blades cooling channels etc. In this chapter, high aspect micro holes are drilled in die steel (of thickness 15 mm) using tubular electrodes of diameter 500μm. Using Taguchi' design of experiment method, four process parameters namely electrode material, discharge current (Ip), pulse on time (Ton) and pulse-off time (Toff) are investigated and optimized for two performance characteristics namely drilling rate (DR) and electrode wear rate (EWR). DR and EWR are opposite in nature, i.e. DR is higher the better type of characteristics while EWR is lower the better type of characteristics. Using Grey relational analysis (GRA) along with Taguchi method, both the characteristics are optimized simultaneously. Through GRA, grey relational grade has been computed as a performance index for predicting the optimal parameters setting for multi machining characteristics.

1. INTRODUCTION

Micro-holes are now becoming integral part of several components especially micro devices such as micro nozzles, micro-electro mechanical system (MEMS), inkjet printer nozzles, medical equipments, and aerospace applications such as drilling in turbine blades for cooling purpose etc. (Yu et al. 2009, D'Usro 2014). Drilling of micro holes with tight geometric tolerance relies on the micro technology used. Currently to produce micro geometries on different kind of materials, there are different manufacturing methods including micro-electrical discharge machining (EDM), electron beam machining

DOI: 10.4018/978-1-5225-2440-3.ch006

(EBM), laser machining, lithography, chemical etching, electro-chemical machining (ECM) and micro ultrasonic machining (MUSM) etc (Vaezi et al., 2013).

The micro-EDM is one of the most important technologies for fabricating micro 3D shapes such as micro holes, micro dies, micro shafts on variety of materials (Cho et al., 2008). Micro-EDM is highly preferred over other processes because of its comparatively better drilling rate, ability to machine hard and composite materials, burr free surface etc. EDM is a non contact machining process involving electrical thermal energy by repetitive sparks between tool electrode and work material (Kuppan et al., 2008).

Electrical discharge drilling (EDD) is a modified electrical discharge machining (EDM) process to produce high aspect macro to micro holes with tight tolerance in hard and composite conductive materials. EDD employs rotating tubular electrode of small diameter ranging from 0.2 mm to 3mm for drilling as represented in Figure 1. In case of tubular electrodes, the dielectric fluid is pumped through the hollow cavity of the electrode at very high pressure under the work material and tool electrode. The combination of very high dielectric pressure, rotation of electrode and servo controlled feed of electrode magnifies the material removal rate in EDD as compared to micro EDM. In EDM, there are many factors affecting the process performance; these can be related either to the process parameters such as voltage, peak current, pulse duration, spark gap and flushing conditions or to the system such as type of dielectric fluid, tool properties, chemical and physical properties of material (D'Urso and Merla, 2014).

In this chapter, Taguchi's design of experimental (Ross, 1996; Roy, 2001) approach is presented to determine the influence of electrode material, discharge current (Ip), Pulse-on time (Ton) and Pulse-off time (Toff) on performance characteristics of micro electrical discharge drilling. Drilling rate (DR) and electrode wear rate (EWR) are the performance characteristics while EDD in die steel. For drilling micro holes, copper and brass tubular electrodes of diameter 500μm are utilised. Taguchi method is designed to optimize the single performance characteristics only. The grey relational analysis (GRA) has been

Figure 1. Schematic diagram of EDD (modified I nYilmaz et al. 2010)

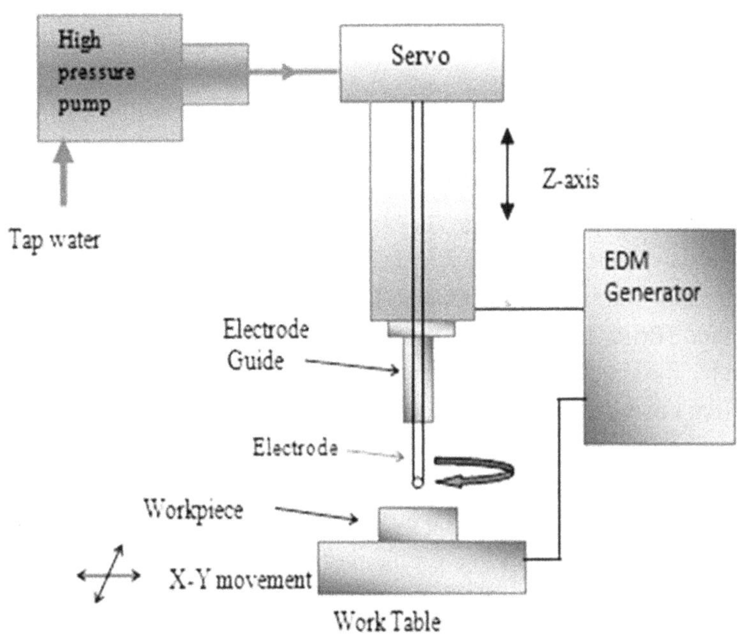

successfully utilised for optimizing the multiple performance characteristics in various engineering fields (Chiang et al., 2009; Vijayan et al., 2010; Jangra et al., 2011). By GRA, a grey relational grade is obtained to evaluate the multi performance characteristics. Therefore, using GRA along with Taguchi method, process parameters are optimized for DR and EW.

2. MICRO-DRILLING AND LITERATURE REVIEW

Conventional twist drilling process is not capable to produce high aspect macro or micro holes in hard and composite materials. Electrical discharge machining (EDM) is most common for drilling small holes in hard and high heat resisting materials such as nickel and titanium alloys (Kuppan et al. 2008, Wansheng et al., 2002). Yan et al. (2002) combined the EDM and ultrasonic machining to drill micro holes with diameter of about 150 µm and depth of 500 µm in borosilicate glass. Similar setup of EDM combined with Ultrasonic machining was utilised by Wanshen et al. (2002) to drill micro holes of diameter less than 0.02 mm and depth/diameter ratio of more than 15 in titanium alloy. Uno et al. (1999) demonstrated the capability of EDM to drill high aspect holes in silicon material. Kuppan et al. (2008) studied the EDM parameters such as peak current, pulse on time, duty factor, and electrode speed in deep hole drilling of Inconel 718 using copper tube of 3 mm diameter. Romoli et al. (2014) experimentally compared the inner surface in micro drilling of spray holes with ultrashort pulsed laser and EDM process. The mean square surface roughness derived from the scanned maps ranges between 220 and 560 nm for micro EDM, and between 50 and 100 nm for fs-pulses laser drilling.

Plaza et al., (2014) Compared the drilling performance of CuW and graphite electrodes of diameter 300µm for Ti6AI4V work material. CuW electrode melts up to 50% in drilling a depth of 5.5mm, while graphite electrode melts upto 400% in drilling the same depth of 0.5 mm. Liu et al., (2005) fabricated the micro holes in nickel alloy using micro EDM. Discharge current was the most important parameter to achieve the optimum results. Khanna et al., (2015) studied the influence of process variables namely pulse-on time, pulse-off time and flushing pressure on material removal rate and tool wear rate in drill EDM process. Using Grey relational analysis, material removal rate and tool wear were optimized simultaneously.

Electrode wear is an important issue in EDM as it affects the sparking conditions during process that makes the electrode taper and hence geometrical accuracy of the hole is impaired as the electrode protrudes into a work piece (Yuangang et al., 2009). In micro EDM, it is quite difficult and tedious to repeatedly achieve the desired 3D cavities in the same processing owing to quick changes of electrode shape. Therefore, materials having good electrical and thermal conductivity with a higher melting point are preferred to be used for fabricating electrodes (DT Pham et al., 2004).

The Taguchi method (Ross, 1996; Roy, 2001) is a systematic application of design and analysis of experiments to optimize the single performance characteristics only. Optimization of Multi performance characteristics is difficult using Taguchi method (Phadke, 1989). In recent years, grey relational analysis has become a powerful tool to analyze the processes with multiple performance characteristics. Tarang et al. (2002) utilized the grey-based Taguchi method to optimize the process parameters of submerged arc welding in hard facing, considering multiple weld qualities. Lin and Lin (2002) combined orthogonal array and grey relational analysis to solve the problem of multi characteristics optimization in electrical discharge machining process. Pan et al. (2007) combined Taguchi method with grey relational analysis to investigate the optimal design of cutting parameters for Nd: YAG laser welding titanium alloy plates.

Li and Tsai (2009) optimized the multiple performance characteristics in laser cutting of specially shaped electronic printed circuit board (PCB) carrier substrate of advanced integrated circuit (IC) using grey relational analysis. Jangra et al. (2012) optimized the multi machining characteristics in wire EDM of WC- 5.3% Co composite using Taguchi and grey relational analysis coupled with entropy method.

In EDD process, there are several controlling factors such as discharge current, pulse on time, pulse off time, type of dielectric fluid, fluid pressure, electrode rotational speed etc that need to be optimized to obtain burr free and geometrically accurate holes in hard and high heat resisting work materials. Previous research work reveals that the grey relational analysis is a better approach for optimization of multiple response characteristics in different fields.

3. HOLE DRILLING USING EDD

3.1. EDD Setup and Work Material

In this chapter, experimental results are presented for hole drilling that are performed on 3-axis electric discharge drilling machine (ED-25), where x-y movement of work table is manually controlled while electrode feed (z-axis) is servo controlled. Figure 2 depicts the experimental setup used for the experimentation. The machine tool have variable parameters in the following range; discharge current (Ip);1-9 amp., pulse-on time (Ton); 1-9μs, pulse-off time (Toff); 1-9μs and electrode rotational speed;130-160 rpm. Distilled water as a dielectric fluid is supplied at a pressure of 6 MPa through hollow electrode. In present machine tool, tubular electrode of maximum length 400mm and diameter ranging between 0.3 mm to 3 mm can be used.

In this experimentation, tubular electrodes of Copper and Brass with diameter 0.5 mm are used for drilling throughout holes in rectangular die steel plate having thickness of 15 mm. The composition of die steel constitutes Cr; 12.5%, C; 1.5%, Mn; 0.6%, Si; 0.6%, Co; 1%, Mo; 0.9%, V; 1.10%, Ni; 0.3% and remaining Fe. The density and hardness of die steel are 7.7 g/cm^3 and 680 HV respectively.

3.2. Performance Characteristics

In present experimentation, two performance characteristics namely drilling rate and electrode wear rate are measured. DR is expressed in mm/min., which is measured by depth of hole drilled in a given time. Electrode wear is expressed in mm^3/min. which is measured by the formula given in equation 1.

$$EWR = \frac{\pi}{4} \times (D^2 - d^2) \times h \tag{1}$$

where:

D = Electrode external diameter,
d = Electrode internal diameter,
h = Length of electrode wear during hole drilling.

Figure 2. Experimental setup for EDD

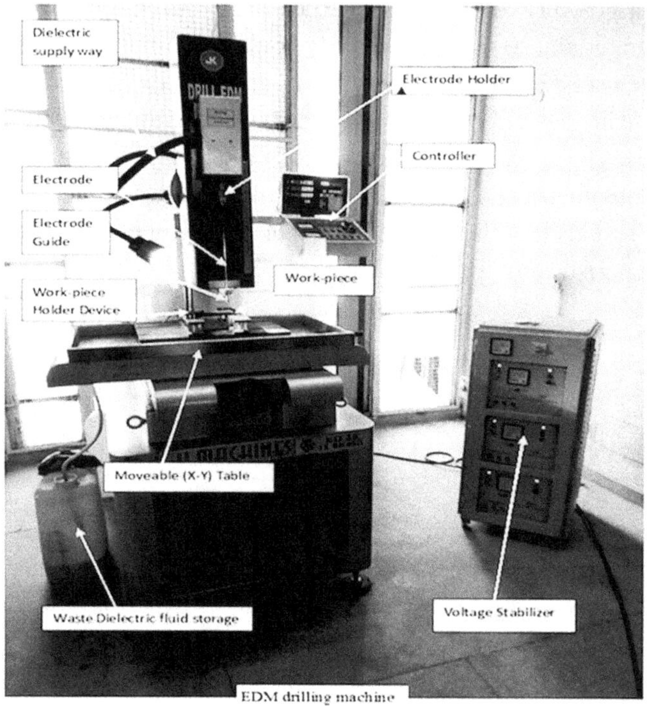

Internal and external diameters of the electrode are fixed. Length of electrode wear is calculated using a reference point on work surface and scale given on z-axis. The electrode wear is calculated by measuring the difference between length of electrode before and after drilling the throughout hole.

3.3. Selection of Process Variables and Experimental Design

Screening experiments were performed to select important process parameters and their effective range for conducting experimentation using design of experiment approach. Electrode material (Cu and Br) shows noticeable influence on performance characteristics. Low electrode rotation speed of 130 rpm resulted in high sparking and erosion; while at 160 rpm, minimum sparking was observed that makes the drilling process difficult. Therefore, in present work, a constant rotational speed of 145 rpm is taken for electrode. Based on screening experimentation, four process variables namely electrode material, Ip, Ton and Toff are selected. In present work, the effective variable range for Ip, Ton, Toff is found as 3-9 Amp, 2-8 μs, 3-7μs respectively for the electrode of diameter 500 μm. Table 1 shows the selected process parameters and their values.

For electrode material, two levels are selected (i.e. Cu and Br) while Ip, Ton and Toff are kept at three levels. Therefore, a mixed orthogonal array (L_{18}) is used for experimentation as listed in Table 2.

Table 1. Process parameters and their levels

Process Parameter	Symbol	Levels		
		Level 1	Level 2	Level 3
Electrode material	A	Copper (Cu)	Brass (Br)	--
Discharge Current (Ip), Amp.	B	3	6	9
Pulse on time (Ton), μs	C	2	5	8
Pulse-off time (Toff), μs	D	3	5	7

Table 2. Experimental design and results

Exp. No	Electrode Material	Ip (Amp)	Ton (μs)	Toff (μs)	DR (mm/min)	EWR (mm^3/min)
1	Cu	3	2	3	3.40	0.7907
2	Cu	3	5	5	7.23	1.2089
3	Cu	3	8	7	8.70	0.8940
4	Cu	6	2	3	4.80	1.9740
5	Cu	6	5	5	9.67	7.9960
6	Cu	6	8	7	11.30	9.5057
7	Cu	9	2	5	1.40	2.4670
8	Cu	9	5	7	1.90	12.1037
9	Cu	9	8	3	1.15	13.5600
10	Br	3	2	7	1.50	0.7900
11	Br	3	5	3	25.00	2.2163
12	Br	3	8	5	28.10	2.6247
13	Br	6	2	5	6.30	3.2890
14	Br	6	5	7	31.40	15.6900
15	Br	6	8	3	45.00	22.5700
16	Br	9	5	7	2.40	4.2430
17	Br	9	8	3	14.60	23.5600
18	Br	9	8	5	10.50	28.5600

4. RESULTS AND DISCUSSIONS

Using L_{18} orthogonal array mentioned in Table 2, holes are drilled and performance characteristics are calculated as listed in Table 2. Figure 3 shows the actual holes drilled in die steel. Using Taguchi's methodology, process parameters are analysed for two performance characteristics.

Figure 3. Drilled holes in die steel plate

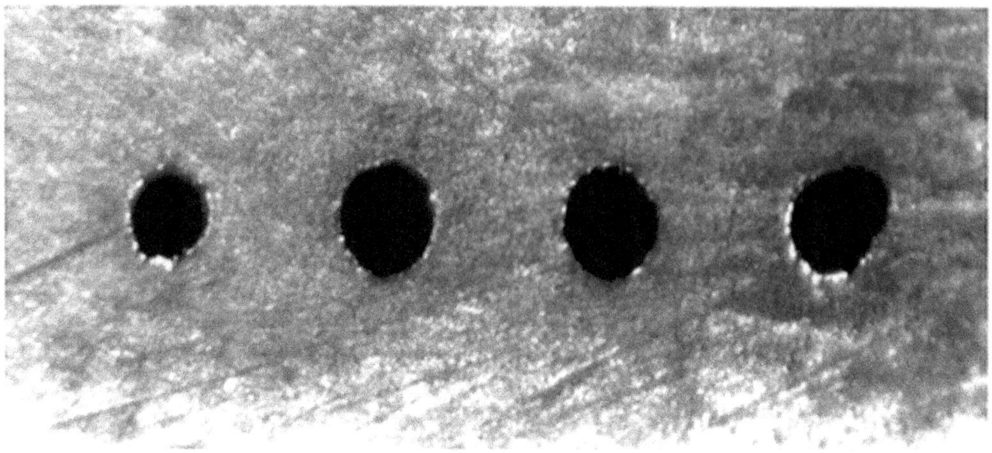

4.1. Drilling Rate

The response table using Taguchi method has been employed here to calculate the effect of each level of process parameter on drilling rate. It can be done by sorting the mean values of machining characteristics corresponding to levels of the process parameter in each column of the orthogonal array, and taking an average on those with same level. For example, the average effect of parameters A (electrode material) and B (Ip) at level 1 for DR can be calculated as follows:

$$A_1 = (3.40 + 7.23 + 8.70 + 4.80 + 9.67 + 11.30 + 1.4 + 1.9 + 1.15) / 9 = 5.506$$

$$B_1 = (3.40 + 7.23 + 8.70 + 1.5 + 25 + 28.10) / 6 = 12.32$$

Using this methodology, response tables for DR is obtained as listed in Table 3. The influence of each parameter is shown by the response graph for DR in Figure 4.

Figure 4 shows the influence of process parameters on DR. DR varies strongly with the variation in electrode materials, discharge current (Ip) and pulse-on time (Ton) while pulse-off time (Toff) produce little influence. DR shows higher drilling rate for brass electrode as compared to copper electrode. It employs that Brass electrode produces more sparking as compared to copper electrode. DR Increase with discharge current, Ip from 3 to 6 Amp but it further decreases at Ip of 9 Amp. Increasing Ip from

Table 3. Response table for mean DR

Level	EM (A)	Ip (B)	Ton (C)	Toff (D)
1	5.506	12.322	3.300	15.658
2	18.311	18.078	14.967	10.533
3	---	5.325	17.458	9.533

Figure 4. Influence of process parameters on DR

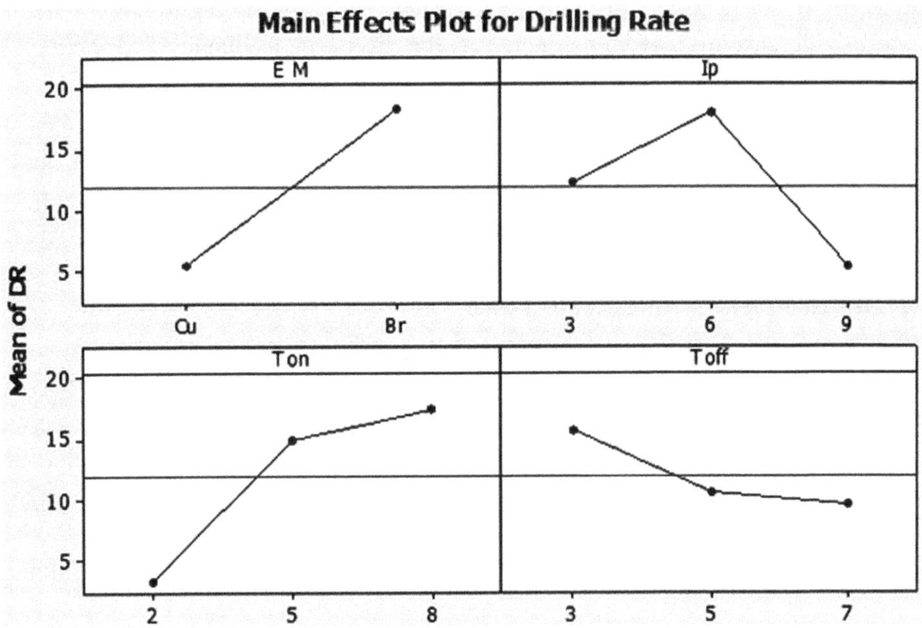

3 to 6 Amp; results into more ionisation in spark gap and melting of material which results into higher drilling rate. But at Ip; 9 Amp, the discharge energy is so high that electrode melts at rapid rate, which results into unstable machining and causes the lower drilling rate. Figure 5a and 5b shows the microscopic images of drilled hole for experiment no 5 and 8 respectively. Figure 5 clearly indicates the influence of discharge current on geometrical accuracy of drilled hole. At high discharge current, hole expansion takes place due to strong secondary discharges.

Figure 5. Hole geometry at parameters: (a) lp; 6Amp, Ton; 5μs, Toff; 5μs (b) lp; 9Amp, Ton; 5μs, Toff; 7μs

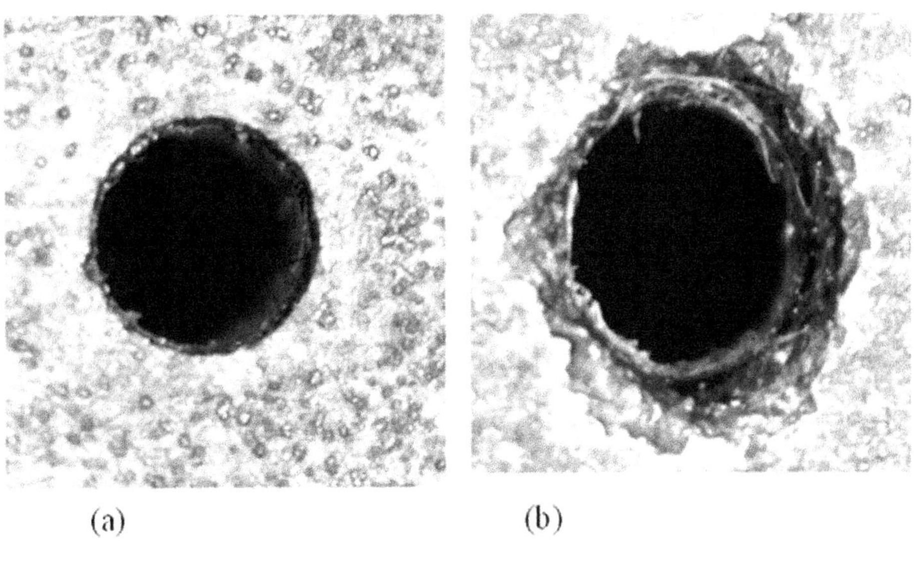

(a) (b)

Increase in pulse on time (Ton) increases the discharge frequency and hence the material erosion is more. The dielectric flushing at high pressure of 6 bar helps to flush out the melted debris of electrode and work material. Increase in Toff value decreases the discharge frequency and increase the overall drilling time.

In order to evaluate the significance or influence of process parameters for DR, analysis of variance (ANOVA) has been performed using statistical software, MINITTAB 15. Table 4 lists the F-value and p-value for four machining parameters. Table 4 reveals that electrode material (A), discharge current (B) and Ton (C) are significantly affecting the drilling rate under 95% confidence level (p-value ≤ 0.05).

4.1.1. Optimal Parameters for Drilling Rate

In Taguchi method, the basic method converts the objective parameters to signal-to-noise (S/N) ratio which is treated as the quality characteristics evaluation index. The least variation and the optimal design are obtained by means of the S/N ratio. Higher the S/N ratio, the more stable the achievable quality. Depending on the required objective characteristics, there are three types of S/N ratio- the lower-the-better, the higher-the-better and the nominal-the-better.

The S/N ratio for drilling rate is "larger-the-better" type characteristic that can be expressed as:

$$\eta_{ij} = -10\log\left\{\frac{1}{n}\sum_{i=1}^{n}\frac{1}{y_{ij}^{2}}\right\} \tag{2}$$

where y_{ij} is the *i*th experiment at the *j*th test and *n* is the total no. of tests.

Using equation 2, value of S/N ratio correspond to each experimental value is calculated. Similar to response graph 4, response graph for S/N ratio of DR is plotted as given in Figure 6.

The optimum parameters setting is correspond to the parameters having highest S/N ratio. For drilling rate, the highest S/N ratio is correspond to second level of electrode material; Brass, second level of discharge current (Ip; 6 Amp), second level of pulse-on time (Ton; 5μs) and first level of pulse-off time (Toff; 3μs). Therefore, the optimal setting of process parameters which yield maximum DR is $A_2B_2C_2D_1$.

Table 4. ANOVA for mean DR

Source	DF	Seq. SS	Adj. SS	Adj. MS	F	P
E.M. (A)	1	737.92	737.92	737.92	12.46	0.005
Ip (B)	2	489.48	489.48	244.74	4.13	0.049
Ton (C)	2	685.56	685.56	342.78	5.79	0.021
Toff (D)	2	129.56	129.56	64.78	1.09	0.372
Error	10	592.26	592.26	59.23		
Total	17	2634.78				

Figure 6. S/N ratio plots for drilling rate

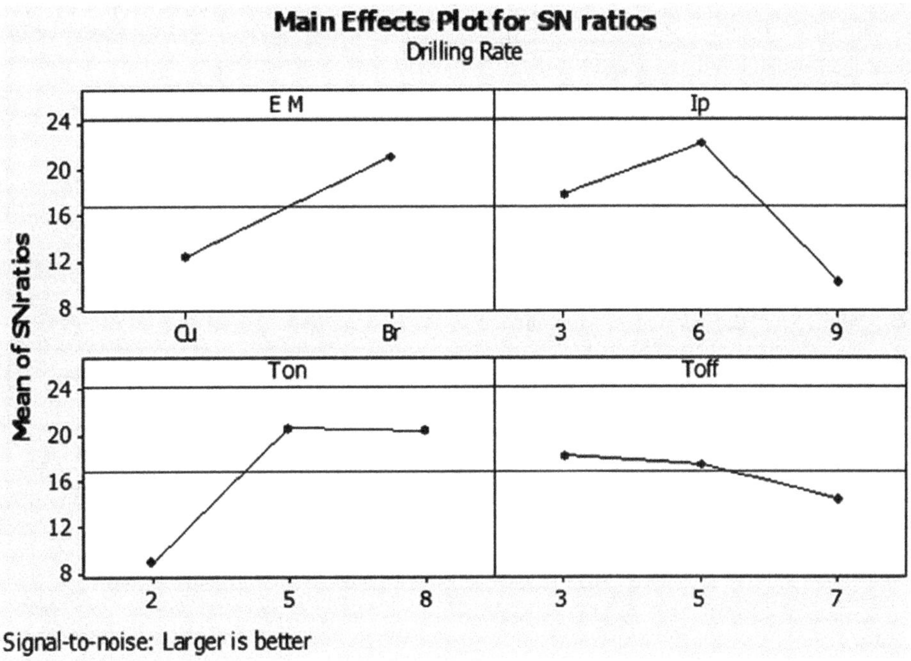

4.2. Electrode Wear Rate

Using the Taguchi methodology as discussed in section 4.1 to obtain the response table for DR, the response table for electrode wear rate is obtained as listed in Table 5. To illustrate the influence of process parameters on EWR, this response table is converted into response plot as shown in Figure 7.

Figure 7 shows that the electrode wear rate is higher in brass electrode as compared to copper electrode. This is due to the fact that the thermal conductivity of brass is lower as compared to copper, therefore, more heating and melting occurs on brass electrode. As a results electrode wear per unit time is more in brass electrode. High discharge current (Ip) and pulse on time (Ton), results into larger ionization of the spark gap, as a result more heating and melting occurs. This increases the EWR with increase in discharge current and pulse-on time. Figure 8 compares the two copper electrodes used in experiment 3 and 6. Figure 8 indicates that the electrode erodes not only at the bottom face but secondary discharges also take place. At higher discharge current, wear conditions are more severe for the electrode. EWR decreases with increase in pulse-off time. Increasing Toff, increases the duration between two sparks and

Table 5. Response table for mean EWR

Level	El. M (A)	Ip (B)	Ton (C)	Toff (D)
1	5.611	1.421	2.259	10.778
2	11.505	10.171	10.462	7.691
3	---	14.082	12.952	7.204

Figure 7. Influence of process parameters on EWR

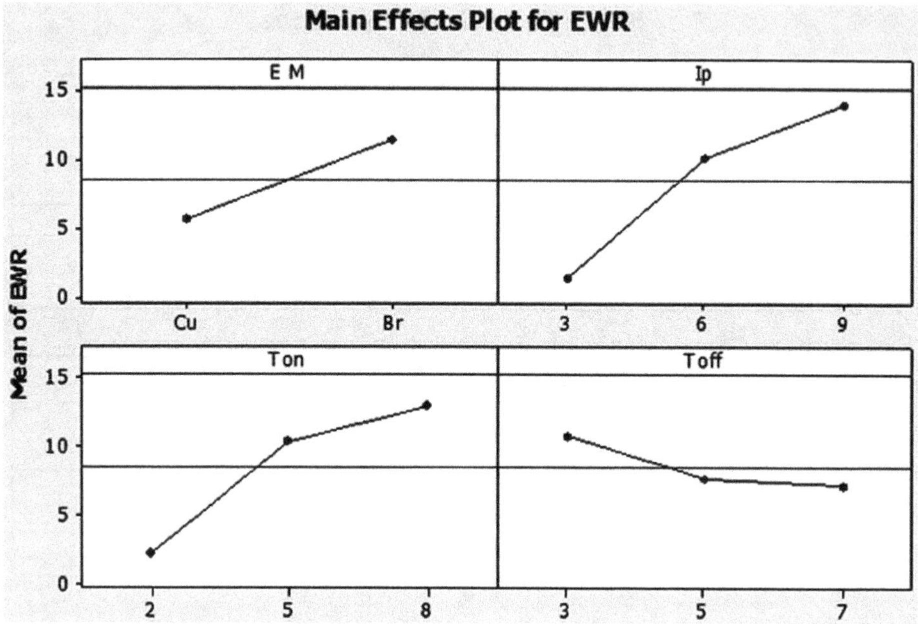

Figure 8. Electrode wear (a) Ip: 3Amp, Ton: 8μs Toff: 7μs (b) 6 Amp, Ton: 8μs Toff: 7μs

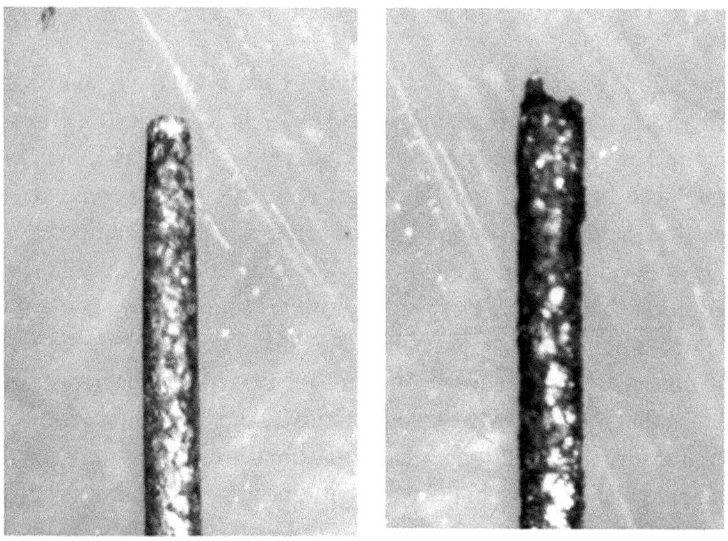

(a) (b)

this short time duration is utilised for flushing out the melted debris and cooling of the work material and electrode. Therefore, EWR decreases with increasing Toff.

Using analysis of variance (ANOVA), the significance or influences of process parameters for EWR are evaluated as listed in Table 6. From ANOVA Table 6, it is found that the electrode material (A), discharge current (B) and Ton (C) are significantly affecting the electrode wear rate under 95% confidence level (p-value ≤ 0.05).

4.2.1. Optimal Parameters for EWR

The S/N ratio for electrode wear rate is "smaller-the-better" type characteristic that can be expressed using equation 3 as given:

$$\eta_{ij} = -10\log\left\{\frac{1}{n}\sum_{i=1}^{n} y_{ij}^2\right\} \tag{3}$$

where y_{ij} is the ith experiment at the jth test and n is the total no. of tests.

Using equation 4, value of S/N ratio correspond to each experimental value is calculated. Response graph for S/N ratio of EWR is plotted as given in Figure 9.

For EWR, the highest S/N ratio is correspond to first level of electrode material (Copper), first level of discharge current (Ip; 3 Amp), first level of pulse-on time (Ton; 2μs) and third level of pulse-off time (Toff; 7μs). Therefore, the optimal setting of process parameters which yield minimum EWR is $A_1B_1C_1D_3$.

5. OPTIMIZATION OF MULTIPLE PERFORMANCE CHARACTERISTICS

DR is "larger the better" and EWR is 'smaller the better" type of performance characteristics. To optimize DR and EWR simultaneously, grey relational analysis (GRA) can be utilized along with Taguchi method. The grey theory proposed by Deng (1982, 1989) has been proven to be useful for dealing with the problems with poor, insufficient and uncertain information. For multi characteristics optimization, GRA involves following steps (Lu et al., 2009; Chiang et al., 2009; Jangra et al. 2011):

Table 6. ANOVA for mean EWR

Source	DF	Seq SS	Adj SS	Adj MS	F	P
E.M (A)	1	156.31	156.31	156.31	5.89	0.036
Ip (B)	2	504.35	504.35	252.18	9.50	0.005
Ton (C)	2	375.69	375.69	187.85	7.08	0.012
Toff (D)	2	45.09	45.09	22.54	0.85	0.456
Error	10	265.37	265.37	26.54		
Total	17	1346.82				

Figure 9. S/N ratio plots for EWR

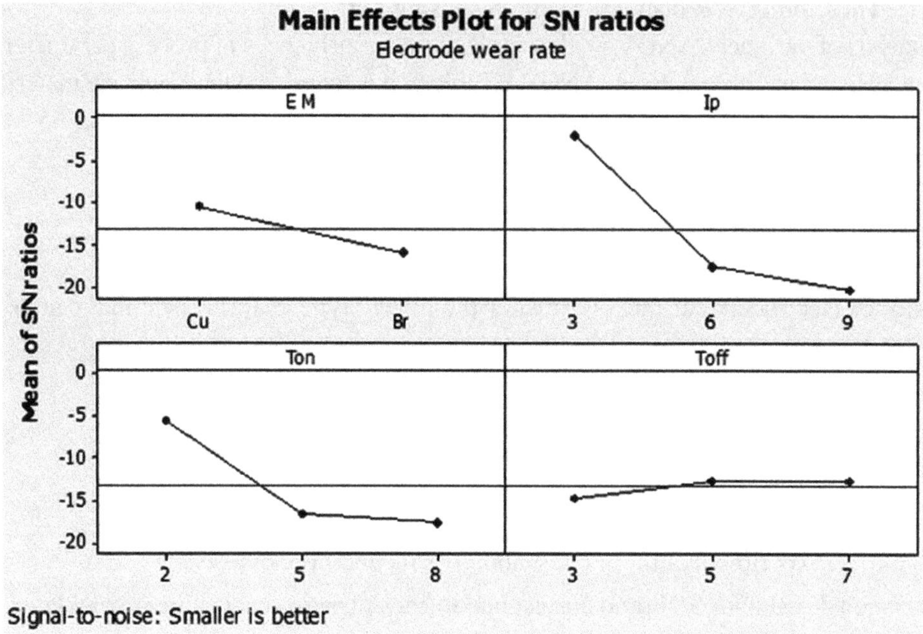

- Conduct the experiments at different settings of parameters based on the orthogonal array.
- Normalize the mean values.
- Perform the grey relational generating and calculate the grey relational coefficient.
- Calculate the grey relational grade by using the weighing factor for the performance characteristics.
- Analyse the experimental results using the grey relational grade and statistical analysis of variance (ANOVA).
- Select the optimal levels of process parameters.

5.1. Data Pre-Processing

Data pre-processing is the first step in GRA. Data pre-processing involves transforming an original sequence into a comparable sequence. A series of various units must be transformed to dimensionless quantities. Experimental results are thus normalized in a range of 0-1. Usually, each series is normalized by dividing the data in the original series by their average.

Let the original reference sequence and sequence for comparison be represented as $x_o(k)$ and $x_i(k)$, i=1, 2, . . ., m; k =1, 2, . . ., n, respectively, where m is the 18 number of experiment to be considered, and n is the 2 number of observation data. Data pre-processing converts the original sequence to a comparable sequence. Several methodologies of pre-processing data can be used in Grey relation analysis, depending on the characteristics of the original sequence

Drilling rate (DR) is a "larger-the-better" type of characteristic, therefore original sequence is normalized using equation 4 as given below:

$$x_i^*(k) = \frac{x_i^o(k) - \min x_i^o(k)}{\max x_i^o(k) - \min x_i^o(k)} \tag{4}$$

Electrode wear rate (EWR) is "smaller-the-better" type of characteristic; therefore, the original sequence is normalized using equation 5 as below:

$$x_i^*(k) = \frac{\max x_i^o(k) - x_i^o(k)}{\max x_i^o(k) - \min x_i^o(k)} \tag{5}$$

The notations $x_i^o(k)$ and $x_i^*(k)$ refers original data and comparability sequence, respectively. Using equation 4 and 5, data for DR and EWR is normalized as shown in Table 7.

5.2. Grey Relational Coefficient

The grey relational coefficient is calculated to express the relationship between the best (reference) and actual values. The grey relational coefficient is expressed using equation as given:

$$\gamma_{o,i}(k) = \frac{\Delta_{\min} + \varsigma.\Delta_{\max}}{\Delta_{o,i}(k) + \varsigma.\Delta_{\max}} \tag{6}$$

$$0 < \gamma_{o,i}(k) \leq 1,$$

where;

$\Delta_{o,i}(k)$= deviation sequence of reference sequence $x_o^*(k)$ and comparability sequence $x_i^*(k)$ i.e:

$$\Delta_{o,i}(k) = \left| x_o^*(k) - x_i^*(k) \right|$$

is the absolute value of the difference between $x_o^*(k)$ and $x_i^*(k)$,

$$\Delta_{\min} = \min.\min.\Delta_{o,i}(k)$$

$$\Delta_{\max} = \max.\max.\Delta_{o,i}(k)$$

ζ is the distinguishing coefficients $\zeta \in \left|0,1\right|$. ζ is taken as 0.5 in this study. The purpose of defining this coefficient is to show the relational degree between the reference sequences $x_o^*(k)$ and the compa-

Table 7. The sequence after data pre-processing

Experiment No.	DR	EWR
Reference Sequence $(x_o^*(k))$ Comparability sequence $(x_i^*(k))$	1.0000	1.0000
1	0.051311	0.999975
2	0.138655	0.984915
3	0.172178	0.996255
4	0.083238	0.957364
5	0.194299	0.740511
6	0.231471	0.686147
7	0.005701	0.939611
8	0.017104	0.592593
9	0	0.540151
10	0.007982	1
11	0.5439	0.948639
12	0.614595	0.933932
13	0.117446	0.910011
14	0.689852	0.46345
15	1	0.2157
16	0.028506	0.875657
17	0.306727	0.18005
18	0.213227	0

rability 18 sequences $x_i^*(k)$. Where $i = 1, 2, 3 \ldots m$ and $k = 1, 2, \ldots n$ with $m = 18$ and $n = 2$ in this study.

Using Table 7, the deviation sequence Δ_{01} for data for DR and EWR can be calculated as follows:

$$\Delta_{o1}(1) = |1.0000 - 0.051311| = 0.948689$$

$$\Delta_{01}(2) = |1.0000 - 0.999975| = 0.000025$$

Table 8 list the deviation sequence for 18 sequences.

Using Table 8 and equation (6), the grey relational coefficient can be calculated as follows:

Table 8. Deviation sequence

Deviation Sequences	DR $\Delta_{01}(1)$	EWR $\Delta_{01}(2)$
$i=1$	0.948689	0.000025
$i=2$	0.861345	0.015085
$i=3$	0.827822	0.003745
$i=4$	0.916762	0.042636
$i=5$	0.805701	0.259489
$i=6$	0.768529	0.313853
$i=7$	0.994299	0.060389
$i=8$	0.982896	0.407407
$i=9$	1.000000	0.459849
$i=10$	0.992018	0.000000
$i=11$	0.456100	0.051361
$i=12$	0.385405	0.066068
$i=13$	0.882554	0.089989
$i=14$	0.310148	0.536550
$i=15$	0.000000	0.784300
$i=16$	0.971494	0.124343
$i=17$	0.693273	0.819950
$i=18$	0.786773	1.000000

$$\gamma(x_o^*(1).x_1^*(1)) = \frac{0.0000 + 0.5\times 1.0000}{0.948689 + 0.5\times 1.0000} = 0.3451$$

$$\gamma(x_o^*(2).x_1^*(2)) = \frac{0.0000 + 0.5\times 1.0000}{0.000025 + 0.5\times 1.0000} = 0.99995$$

Similar calculations are performed for all 18 sequences as listed in Table 9.

5.3. Grey Relational Grade

The grey relational grade is a weighting-sum of the grey relational coefficients. The overall evaluation of multiple performance characteristics is based on the grey relational grade. It is defined as follows:

Table 9. Grey relational coefficient and grey relational grade

Sr. No.	Grey Relational Coefficients		Grey Relational Grade
	$\gamma_1(DR)$	$\gamma_2(EWR)$	$\Gamma_{o,i} = \frac{\gamma_1 + \gamma_2}{2}$
1	0.3451	0.9999	0.6725
2	0.3673	0.9707	0.6690
3	0.3766	0.9926	0.6846
4	0.3529	0.9214	0.6372
5	0.3829	0.6583	0.5206
6	0.3942	0.6144	0.5043
7	0.3346	0.8922	0.6134
8	0.3372	0.5510	0.4441
9	0.3333	0.5209	0.4271
10	0.3351	1.0000	0.6676
11	0.5230	0.9068	0.7149
12	0.5647	0.8833	0.7240
13	0.3616	0.8475	0.6046
14	0.6172	0.4824	0.5498
15	1.0000	0.3893	0.6947
16	0.3398	0.8008	0.5703
17	0.4190	0.3788	0.3989
18	0.3886	0.3333	0.3610

$$\Gamma_{o,i} = \sum_{k=1}^{n} w_k \gamma_{o,i}(k),\ i = 1,2,...m \quad (7)$$

w_k represent the weight of the kth machining characteristics, and $\sum_{k=1}^{n} w_k = 1$.

In present work, equal weights are assigned to both the characteristics. Table 9 shows the grey relational grade (GRG) for 18 sequences.

5.4. Optimization of Process Parameters

Using grey relational analysis, optimization of the multiple performance characteristics is converted into optimization of single grey relational grade. To separate out the influence of each level of process

Table 10. ANOVA for grey relational grade

Source	DF	Seq SS	Adj SS	Adj MS	F	P
E.M. (A)	1	0.000707	0.000707	0.000707	0.12	0.734
Ip (B)	2	0.144859	0.144859	0.072430	12.53	0.002
Ton (C)	2	0.020324	0.020324	0.010162	1.76	0.222
Toff (D)	2	0.001307	0.001307	0.000654	0.11	0.894
Error	10	0.057823	0.057823	0.005782		
Total	17	0.225021				

Table 11. Experimental values of DR and EWR at optimized parameters setting

Methodology	Performance Characteristics	Optimal Parameters	Experimental Value
Taguchi Method	DR	$A_2B_2C_2D_1$	45.00 mm/min
	EWR	$A_1B_1C_1D_3$	0.70 mm^3/min.
Taguchi + Grey relational Analysis	DR	$A_2B_1C_1D_1$	5.2 mm/min
	EWR		1.6 mm^3/min

parameters on the grey relational grade, response graph as shown in Figure 10 is constructed for grey relational grade using the Taguchi methodology as discussed in section 4.1.

Basically, the larger the grey relational grade, the better is the multiple performance characteristics. The parameters combination of A_2 (Brass) B_1(3 amp.) C_1 (2μs) D_1 (3μs) gives highest value of the grey relational grade for multiple characteristics. Analysis of variance (ANOVA) is performed on GR grade values to analyse the significant parameters as listed in Table 10. From ANOVA tables only one process parameter namely discharge current (B) is found most significant affecting the GRG values under 95% confidence level (because $p \leq 0.05$). It implies that irrespective of the electrode material, controlling of discharge current is most essential factor that helps to maintain a optimum drilling rate for minimum electrode wear rate. Table 11 shows the summary of results optimized using Taguchi and Grey relational analysis.

6. CONCLUSION

In this chapter, the experimental results are presented for electrical discharge drilling process to drill micro holes of diameter 500 μm in die steel plate of thickness 15mm. Four process variables are selected for investigation using Taguchi's design of experiment method. The following main conclusions are drawn:-

Figure 10. Influence of parameters on grey relational grade values

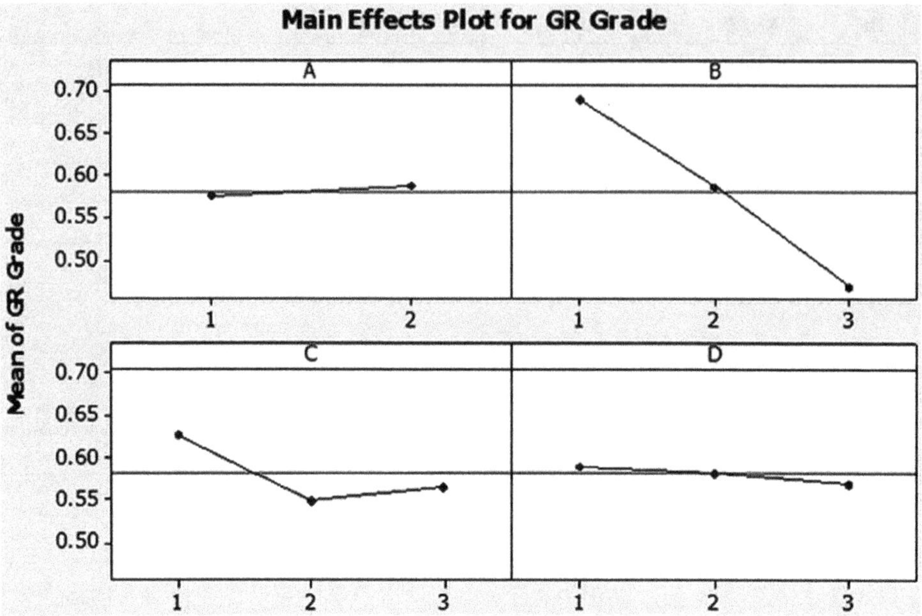

1. Copper and Brass are selected as the electrode materials, while three levels of Ip, Ton and Toff are selected for investigation. Two performance characteristics namely drilling rate (DR) and electrode wear rate (EWR) are measured in drilling process.
2. Brass electrode shows better drilling rate as compared to copper electrode but the electrode wear is more in case of brass electrode. Therefore, a trade off is required between the process parameters to chose one electrode material for better drilling rate and lower electrode wear.
3. Increasing Ip and Ton, DR and EWR increases which is due to the higher spark frequency which increases the melting of work material and electrode. Increasing the values of Toff, DR and EWR decreases.
4. Using Taguchi method, the optimized setting of parameters is obtained for individual characteristics. The optimized parameters setting for DR is second level of electrode material (Brass), second level of discharge current (Ip; 6 Amp), second level of pulse-on time (Ton; 5μs) and first level of pulse-off time (Toff; 3μs). The minimum EWR is obtained correspond to first level of electrode material (E. M.; Copper), first level of discharge current (Ip; 3 Amp), first level of pulse-on time (Ton; 2μs) and third level of pulse-off time (Toff; 7μs).
5. In order to optimize DR and EWR simultaneously, grey relational analysis (GRA) is utilized along with Taguchi method. The parameters combination of Electrode material; Brass, Ip; 3 amp, Ton; 2μs, Toff; 3μs gives the optimum values for the multiple characteristics.

In future work, electrical discharge drilling process must be optimized for improving the cylindricity, circularity and aspect ratios of holes in high heat resisting alloys and composite materials.

ACKNOWLEDGMENT

The research grant for this work is sponsored by PEC University of Technology, Chandigarh, India under the RIPA project entitled "Parameter Optimization For Small Hole Drilling in Hard and Composite Materials using Drill EDM" undertaken by Dr. Kamal Kumar.

REFERENCES

Chiang, Y. M., & Hsieh, H. H. (2009). The use of the Taguchi method with grey relational analysis to optimize the thin film sputtering process with multiple quality characteristics in color filter manufacturing. *Computers & Industrial Engineering*, *56*(2), 648–661. doi:10.1016/j.cie.2007.12.020

Cho, P. J., Kim, B. H., Choi, D. K., & Chu, C. N. (2008). Characteristics of MEDM for micro-holes with respect to circuit elements. *Journal of Mechanical Science and Technology*, *22*(2), 300–307. doi:10.1007/s12206-007-1039-0

Deng, J. (1982). Control problems of grey system. *Systems & Control Letters*, *5*, 288–294.

Deng, J. (1989). Introduction to grey system. *Journal of Grey System*, *1*(1), 1–24.

DUrso, G., & Merla, C. (2014). Workpiece and electrode influence on micro-EDM drilling performance. *Precision Engineering*, *38*(4), 903–914. doi:10.1016/j.precisioneng.2014.05.007

Jangra, K., Grover, S., & Aggarwal, A. (2011). Simultaneous optimization of material removal rate and surface roughness for WEDM of WC-Co composite using grey relational analysis along with Taguchi method. *International Journal of Industrial Engineering Computations*, *2*(3), 479–490. doi:10.5267/j.ijiec.2011.04.005

Jangra, K., Grover, S., & Aggarwal, A. (2012). Optimization of multi machining characteristics in WEDM of WC-5.3%Co composite using integrated approach of Taguchi, GRA and entropy method. *Frontiers of Mechanical Engineering*, *7*(3), 288–299. doi:10.1007/s11465-012-0333-4

Khanna, R., Kumar, A., Garg, M. P., Singh, A., & Sharma, N. (2015). Multiple performance characteristics optimization for Al 7075 on electric discharge drilling by Taguchi grey relational theory. *Journal of Industrial Engineering International*, *11*(4), 459–472. doi:10.1007/s40092-015-0112-z

Kuppan, P., Rajadurai, A., & Narayanan, S. (2008). Influence of EDM Process parameters in deep hole drilling of Inconel 718. *International Journal of Advanced Manufacturing Technology*, *38*(1-2), 74–84. doi:10.1007/s00170-007-1084-y

Leao, F. N., & Pashby, I. R. (2005). Optimisation of EDM fast hole-drilling through Evaluation of dielectric and electrode materials. *18th International Congress of Mechanical Engineering*.

Li, C. H., & Tsai, M. J. (2009). Multi-objective optimization of laser cutting for flash memory modules with special shapes using grey relational analysis. *Optics & Laser Technology*, *41*(5), 634–642. doi:10.1016/j.optlastec.2008.09.009

Lin, J. L., & Lin, C. L. (2002). The use of the orthogonal array with grey relational analysis to optimize the electrical discharge machining process with multiple performance characteristics. *International Journal of Machine Tools & Manufacture*, *42*(2), 237–244. doi:10.1016/S0890-6955(01)00107-9

Liu, H. S., Yan, B. H., Huang, F. Y., & Qui, K. H. (2005). A study on the characterization of high nickel alloy micro-holes using micro-EDM and their applications. *Journal of Materials Processing Technology*, *169*(3), 418–426. doi:10.1016/j.jmatprotec.2005.04.084

Lu, H. S., Chang, C. K., Hwang, N. C., & Chung, C. T. (2009). Grey relational analysis coupled with principle component analysis for optimization design of the cutting parameters in high-speed end milling. *Journal of Materials Processing Technology*, *209*(8), 3808–3817. doi:10.1016/j.jmatprotec.2008.08.030

Pan, L. K., Wang, C. C., Wei, S. L., & Sher, H. F. (2007). Optimizing multiple quality characteristics via Taguchi method based grey relational analysis. *Journal of Materials Processing Technology*, *182*(1-3), 107–116. doi:10.1016/j.jmatprotec.2006.07.015

Phadke, S. M. (1989). *Quality Engineering Using Robust Design*. Englewood Cliffs, NJ: Prentice Hall.

Pham, D. T., Ivanov, A., Bigot, S., Popov, K., & Dimov, S. (2007). An investigation of tube and rod electrode wear in micro EDM drilling. *International Journal of Advanced Manufacturing Technology*, *33*(1–2), 103–109. doi:10.1007/s00170-006-0639-7

Plaza, S., Sanchez, J. A., Perez, E., Gil, R., Izquierdo, B., Ortega, N., & Pombo, I. (2014). Experimental study on micro EDM-drilling of Ti6Al4V using helical electrode. *Precision Engineering*, *38*(4), 821–827. doi:10.1016/j.precisioneng.2014.04.010

Romoli, L., Rashed, C. A., & Fiaschi, M. (2014). Experimental characterization of the inner surface in micro-drilling of spray holes: A comparison between ultra short pulsed laser and EDM. *Optics & Laser Technology*, *56*, 35–42. doi:10.1016/j.optlastec.2013.07.010

Ross, P. J. (1996). *Taguchi Techniques for Quality Engineering*. New York: McGraw-Hill.

Roy, R. K. (2001). *Design of Experiments using the Taguchi approach: 16 steps to Product and Process Improvement*. New York: Wiley-Interscience.

Tarng, Y. S., Juang, S. C., & Chang, C. H. (2002). The use of grey based Taguchi methods to determine submerged arc welding process parameters in hard facing. *Journal of Materials Processing Technology*, *128*(1-3), 1–6. doi:10.1016/S0924-0136(01)01261-4

Uno, Y., & Okada, A. et al.. (1999). High efficiency fine boring of monocrystalline silicon ingot by electrical discharge machining. *Precision Engineering*, *23*(2), 126–133. doi:10.1016/S0141-6359(98)00029-4

Vaezi, M., Seitz, H., & Yang, S. (2013). A review on 3D micro-additive manufacturing technologies. *International Journal of Advanced Manufacturing Technology*, *67*(5-8), 1721–1754. doi:10.1007/s00170-012-4605-2

Vijayan, S., Raju, R., & Rao, S. R. K. (2010). Multi objective Optimization of Friction Stir Welding Process Parameters on Aluminium Alloy AA5083 using Taguchi Based Grey Relational Analysis. *Materials and Manufacturing Processes*, *25*(11), 1206–1212. doi:10.1080/10426910903536782

Wansheng, Z., Zhenlong, W., Shichun, D., Guanxin, C., & Hongyu, W. (2002). Ultrasonic and electric discharge machining to deep and small hole on titanium alloy. *Journal of Materials Processing Technology, 120*(1-3), 101–106. doi:10.1016/S0924-0136(01)01149-9

Yan, B. H., Tsai, H. C., & Lin, Y. C. (1997). Study on EDM characteristics of cemented carbides. *Proceedings of the 14th National Conference on Mechanical Engineering*, 157 - 164.

Yan, B. H., Wang, A. C., Huang, C. Y., & Huang, F. Y. (2002). Study of precision micro-holes in borosilicate glass using micro EDM combined with micro ultrasonic vibration machining. *International Journal of Machine Tools & Manufacture*, *42*(10), 115–112. doi:10.1016/S0890-6955(02)00061-5

Yilmaz, O., & Okka, M. A. (2010). Effect of single and multi-channel electrodes application on EDM fast hole drilling performance. *International Journal of Advanced Manufacturing Technology*, *51*(1-4), 185–194. doi:10.1007/s00170-010-2625-3

Yu, Z. Y., Zhang, Y., Luan, L. J., Zhao, F., & Gua, D. (2009). High aspect ratio micro-hole drilling aided with ultrasonic vibration and planetary movement of electrode by micro-EDM. *CIRP Annals-Manufacturing Technology*, *58*(1), 213–216. doi:10.1016/j.cirp.2009.03.111

Yuangang, W., Fuling, Z., & Jin, W. (2009). Wear resist Electrodes for Micro-EDM. *Chinese Journal of Aeronautics*, *22*(3), 339–342. doi:10.1016/S1000-9361(08)60108-9

Chapter 7
Optimization of Laser Transmission Welding Parameters Using Chicken Swarm Optimization Algorithm:
Chicken Swarm Algorithm Optimization of Laser Transmission Welding

Bappa Acherjee
Birla Institute of Technology, Mesra, India

Arunanshu S. Kuar
Jadavpur University, India

Debanjan Maity
Indian Institute of Technology, Kharagpur, India

Souren Mitra
Jadavpur University, India

Dipten Misra
Jadavpur University, India

ABSTRACT

The ever-increasing demand of high quality joints with improved cycle times leads the industries, those involve welding of plastics, to use laser for welding plastics. Laser transmission welding is the latest development in the field of welding of plastics. In present research work, experimental investigation, parametric modeling and optimization of laser transmission welding of polycarbonate is carried out. A systematic experimental plan is executed using central composite design of response surface methodology (RSM). Mathematical models are developed using RSM for establishing an interrelationship between the process parameters and the responses of interest. Chicken swarm optimization (CSO) algorithm is employed with the developed RSM models to set the optimal process parameters combinations to achieved desired weld qualities. Here, the objectives are set as to increase the weld strength and decrease the weld width independently, as well as simultaneously. The interdependent parametric trends are also studied and discussed.

DOI: 10.4018/978-1-5225-2440-3.ch007

1. INTRODUCTION

Now-a-days, the applications of plastics can be found almost everywhere, from a simple product to the extremely hi-tech. As the use of plastic components become more widespread, joining techniques play an important role in their processing. Since making the complex plastic components in one piece is not always practical and cost effective, various joining techniques have been developed including plastic adhesive joining, mechanical fastening and welding. The general advantages of welding techniques over other joining techniques are the fast and easy processing, tightness of the joint and high strength. The most widely used plastic welding procedures are hot gas or hot tool welding, vibration welding, and ultrasonic welding. In the case of hot gas and hot tool welding the residues and contaminations may result in undefined weld strength of weld zone. Again, the fluff development in the ultrasonic welding is still a problem (Herfurth et al, 1999; Bonten and Tüchert, 2002).

In this context, more favorable presupposition is offered by heating of infra-red radiation, because the heat is transmitted without any contact to the plastic parts. Infra-red radiation can be transmitted by infra-red lamps and also by infra-red lasers. Laser can be used in two general ways: (i) irradiating the surfaces to be joined directly or (ii) passing through one transparent material and directly heating only the second material, precisely at the mating surface. This later process is known as laser transmission welding (LTW) process. This is a clean and flexible process for joining similar or dissimilar plastics (Baylis, 2002). In this process the heat is generated at the joint interface, the area where melting is required to join the mating surfaces, thus, reducing the energy input to a minimum (Hansch et al., 1998). Figure 1 shows the operational procedure of laser transmission welding process.

Figure 1. The working principle of laser transmission welding

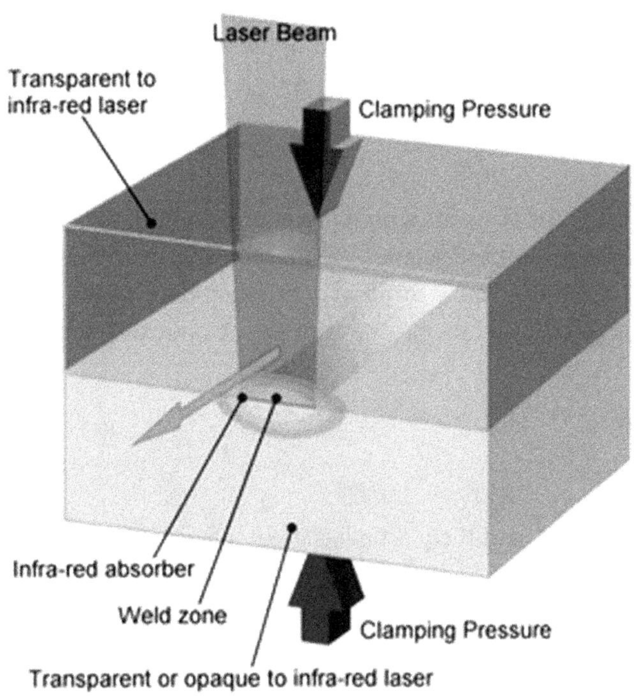

Laser transmission welding process is gaining its popularity because it is simple, faster and reliable than its mechanical counterparts, and the use of LTW results in higher productivity and lower rejection rate. Moreover, it has several other advantageous features like contactless yielding of joining energy, absence of process induced vibrations, inducing minimum thermal and mechanical stress to the parts, and no particle formation. This novel technology permits both reliable high quality joining of mechanically and electronically highly sensitive micro components and hermetic sealing of macro components (Coelho et al., 2000; Bachmann and Russek, 2002).

In present research, laser transmission welding of polycarbonate is studied through experimental investigation, mathematical modeling and process optimization. First, an overview of laser transmission welding process is presented which includes the process fundamentals, process variants and application potential. Then the experimental investigations have been carried out and the mathematical models are developed using response surface methodology (RSM). Finally, the developed RSM models are used with chicken swarm optimization (CSO) algorithm to find feasible solutions to obtain optimal process parameters and also to predict parametric trends.

2. LASER TRANSMISSION WELDING (LTW)

Plastics are generally transparent or translucent to the laser radiation in their natural states. Plastic are made opaque or laser absorbing by compounding it with appropriate additives. If a small quantity of absorbing additive is used in plastic, then the radiation energy is absorbed over a broad layer of that material. This phenomenon is termed as volumetric absorption. In this case, absorbed light energy, converted to heat, that is equivalent to total internal heat generation in the plastic. The ability of absorption of radiant energy in absorbing plastic is determined by the Beer-Lambert's law, which states that the intensity of a beam of monochromatic radiation in an absorbing medium decreases exponentially with penetration distance, and is expressed as:

$$I(z) = I_{(z=0)} e^{-kz} \tag{1}$$

where, I is radiation intensity, z is distance within the material and K is the total extinction coefficient caused by the laser beam absorption and scatter. For the case of amorphous polymers, the extinction is determined by the absorption only. The absorption coefficient depends on the quality and the color of the plastic material. It is defined as the reciprocal value of the optical penetration depth dp (Bonten and Tüchert, 2002).

$$K_a = dp^{-1} \tag{2}$$

Thus, the absorption coefficient of the plastic has great influence on the laser transmission welding process.

Plastics containing sufficient amount of laser absorbing additives, absorb the incident beam energy in a very thin layer of that material. This phenomenon is known as surface absorption. In this case, the absorbed light energy is converted into heat at the surface itself, similar to metal, and the laser beam may be considered to be equivalent to a surface heat flux.

The volumetric heat generated or the surface heat flux deposited in this way is transferred to the surrounding materials by thermal conduction. Thus, the heat generated at the interface transferred to the deeper layers of absorbing bottom plastic part and also to the plastic part that is transparent to the laser beam. Some part of that heat is also transferred to the surroundings through convection and radiation. When heated to a temperature above the melting point, or melting range, a thin layer of plastic in both the parts starts melting. Clamping pressure ensures contact between the parts to be joined and also an increase in molten material flow at the weld zone. Molecular diffusion occurs and after solidification a firm joint is formed.

2.1. Basic Requirements

The two plastic parts to be welded together must have different optical properties, that is, top plastic part should be transparent or translucent and the bottom plastic parts should be laser absorbing. In some cases, where both the plastic parts are transparent, a laser absorbing third material is to be placed at the area of heat generation between the joining surfaces. Sufficient contact between the mating parts is needed to allow for the heat to be conducted from the absorbing material to the transparent material. In case of dissimilar material welding, both the materials should be chemically compatible and their melting temperatures are comparable (Baylis, 2002).

2.2. Types of Laser Welding Systems

Three types of lasers, used for welding of polymer based materials, are CO_2 laser (10.6 μm), Nd:YAG laser (1.06 μm) and diode laser (808 ± 10, 830 ± 10 or 940 ± 10 nm). Their wavelengths of operation are the most critical feature for transmission welding application. Since most plastics absorb strongly at 10.6 μm, the CO_2 laser is impractical for most applications. An exception may be very thin materials. In contrast, both the Nd:YAG laser and diode laser operate in the near-infrared where plastics have minimal intrinsic absorption, which allows welding of plastic parts having millimeter thickness, successfully.

2.3. Joint Configurations

Lap joint is the basic joint arrangement for laser transmission welding of plastics. A lap joint is comprised of a transmitting polymer placed on top of an absorbing polymer. Other joint configurations involve variations of the lap joint and variations of the butt joint. The butt joint can be more difficult because the weld joint will be some distance from the end of the sample and it can be difficult to apply pressure. Regardless of what joint type is chosen there are two criteria, which must be met. The first criterion is the laser light must reach to the joint interface, and the second criterion is to ensure that there should be suitable provision to apply the weld pressure to the joint as shown in Figure 2.

2.4. Process Variants

2.4.1. Contour Welding

In contour welding the focused laser beam is moved over the materials surface, following the weld pool geometry. The laser beam is either moved with a robot or the work-piece is moved with an X-Y table

Figure 2. Laser transmission welding with lap and butt joint configurations

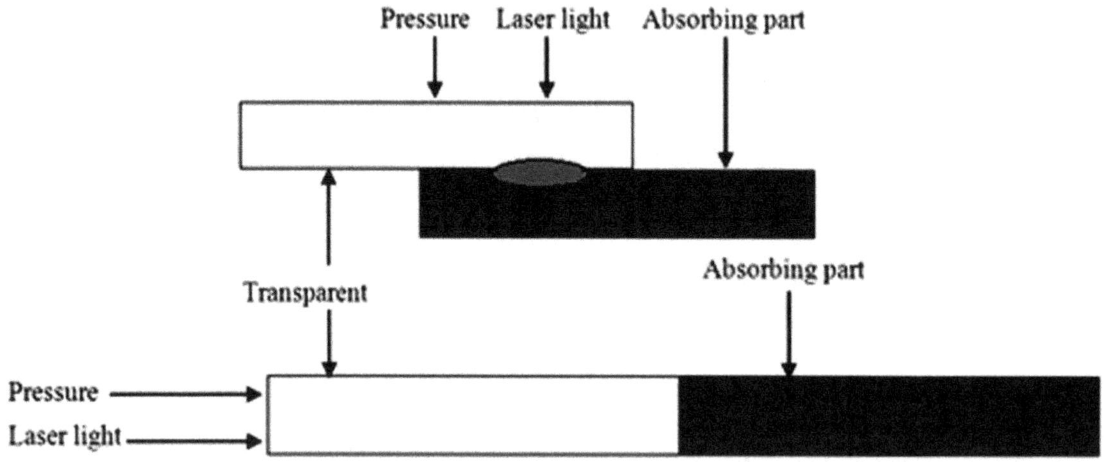

(Jansson et al., 2003). This method provides a simple, very flexible, easy controllable and cost efficient method for laser polymer welding.

2.4.2. Simultaneous Welding

The compact and modular set up of high power diode lasers permits simultaneous laser welding. The beam generated by each module consists of straight light and homogeneous intensity distribution has to be generated by appropriate beam forming (Russek et al., 2003). To obtain water- or air-tight welded parts, the challenge consists in positioning single power diode laser modules appropriate to the weld contour and avoids inhomogeneity in the entire intensity distribution. This means to control perfectly the position and intensity of several individual beams. Modules are arranged in a way to avoid beam recovering and non-irradiated areas in the entire seam contour (Abed et al., 2004).

2.4.3. Quasi-Simultaneous Welding

In quasi-simultaneous laser welding the laser head and the work piece are held in fixed position. The laser beam scanned several times around or across the weld path very fast with a galvo mirror system. Because of the low thermal conductivity of polymers, the welding path heats up gradually and approximately evenly, such that a quasi-simultaneous melting of the entire welding track occurs (Jansson et al., 2003). In quasi-simultaneous welding, short welding times can be achieved with different parameter combinations. The principal process parameters are laser power, scanning speed and the number of laser scans. Scanning speed and number of scans are parameters, which determine the total welding time (Jansson et al., 2005).

2.4.4. Mask Welding

Another way to circumvent the limitation in simultaneous welding regarding the seam geometry is to put a mask between the laser and the work piece. This mask is con-formal with the seam, so those places, which shall not be exposed to the laser radiation, are protected. The lasers may be assembled, so all open areas in the mask are filled simultaneously or a line scan can be performed. The process efficiency however is reduced by the fact, that power is blocked from the mask and thus, not used for the process (Bachmann and Russek, 2002).

2.5. Applications

If properly performed and if the materials combination is suited for laser welding, the laser transmission welding generates a strong and cosmetically perfect, almost invisible, seam with little thermal load to the product. Furthermore, laser welding of polymers is flexible, gentle and environmentally friendly (Bachmann and Russek, 2002; Kagan et al., 2002). Laser welding of thermoplastic parts is at the beginning stage for wide industrial application. The types of industries that would use laser welding of plastics include: medical device manufacturing, packaging, automotive, electrical and electronic device manufacturing, light manufacturing, information and communication technologies, petrochemical, chemical, house hold goods, and any other type of industry that involves the joining of polymers. Applications that have been advertised to date include: Mercedes remote door key, synthetic textile for watertight clothing, automotive dashboard components, automotive auxiliary components, and biomedical sensors.

3. RESEARCH METHODOLOGIES ADOPTED

3.1. Response Surface Method (RSM)

Response surface method is a statistical technique which is used to develop a mathematical relationship between the input parameters and the response parameters of a process or a system (Montgomery, 2001). The relationship between the independently controllable input parameters (ξ_1, ξ_2, ..., ξ_k) and the response parameter (y) can be formulated as:

$$y = f\left(\xi_1, \xi_2, ..., \xi_k\right) + \varepsilon \tag{3}$$

where, ε is the small experimental error, which is termed as noise and f is the response function. The response function can further be modified by assuming that the ε is normally distributed with mean zero and variance σ^2. Then the modified expression can be written as:

$$E(y) = \eta = E\left[f\left(\xi_1, \xi_2, ..., \xi_k\right)\right] + E(\varepsilon) = f\left(\xi_1, \xi_2, ..., \xi_k\right) \tag{4}$$

As the input parameters are of different degrees and dimensions, and they may have different measuring units, thus, in RSM the input parameters are converted into dimensionless coded variables x_1, x_2, ..., x_k, having mean zero and the same standard deviation. Then the response function equation can be written in terms of coded variables as:

$$\eta = f(x_1, x_2, ..., x_k) \tag{5}$$

As the pattern of response function *f* is unknown, experimenter must approximate it. In fact, successful use of RSM is critically dependent upon the experimenter's ability to develop a suitable approximation for *f*. Usually, a low-order polynomial in some relatively small region of the independent variable space is appropriate. In many cases, either a first-order or a second-order model is used.

A first order RSM model is generally expressed as:

$$\eta = \beta_0 + \beta_1 x_1 + \beta_2 x_2 + ... + \beta_k x_k \tag{6}$$

and, the second-order model is:

$$\eta = \beta_0 + \sum_{j=1}^{k} \beta_j x_j + \sum_{j=1}^{k} \beta_{jj} x_j^{\,2} + \sum_{i<}\sum_{j=2}^{k} \beta_{ij} x_i x_j \tag{7}$$

In some infrequent situations, approximating polynomials of order greater than two are used.

3.2. Chicken Swarm Optimization (CSO) Algorithm

Chicken swarm optimization is a very recent and almost unexplored algorithm in the field of swarm intelligence based meta-heuristic algorithm. The algorithm was proposed by Meng et al. (2014). The algorithm is based upon the hierarchical order of chicken swarm and behavior of chicken swarms. Like many swarm intelligence algorithms, the CSO is a kind of stochastic optimization algorithm, using an iterative approach to solve the objective problem. Through the observation of the individual and the entire flock of chickens, the researchers found that chicken has mature cognitive ability, communication skill and learning ability, and in the flock there exists an almost strict hierarchical order. Some idealized rules are implemented to simplify the algorithm; those are (Meng et al., 2014):

1. Several groups exist in a chicken flock, which includes a dominant rooster, a couple of hens, and chicks.
2. Based on the fitness values of the chickens, chicken swarm have been divided into several groups (roosters, hens and chicks). Again in a group of swarm, rooster is the chicken with best fitness value, whereas, chicks are having the worst fitness values. Remaining are the hens which choose a group randomly to live in. The mother-child relationship among the hens and chicks is also established randomly.
3. However, in a selected group, the hierarchal order, relationship preeminence and mother-child relationship will remain unchanged. These statuses only update every several (G) time steps.
4. The chickens of same group go after the roosters for searching foods while the roosters may prevent the chickens from eating their own food.

Assuming that the R_N is the number of roosters, H_N is the number of hens, C_N is the number of chicks, and M_N is the number of mother hens. The performance of the CSO depends appreciably on the flock structure, which is again, depends on the sizes of R_N, H_N, C_N and M_N. It is supposed that there are N numbers of chickens in the entire flock, and r_p is the proportion of the roosters and h_p is the proportions of hens in the entire flock. Among the hens present in the flock, m_p is the proportion of the mother hens in the entire flock. The correlation can be represented as (Chen et al., 2015):

$$R_N = N \times r_p \tag{8}$$

$$H_N = N \times h_p \tag{9}$$

$$C_N = N - (R_N + H_N) \tag{10}$$

$$M_N = H_N \times m_p \tag{11}$$

3.2.1. Identification of the Chicken

At any time step *t*, the positions of all *N* numbers of chickens are identified by $x_{i,j}^{t}$, as they are searching for food in *D* dimensional space, where $i \in [1,2,....N]$ and $j \in [1,2,....D]$.

3.2.2. Movement of Chickens

3.2.2.1. Rooster Model

The model is based on the fact that the roosters having higher fitness values can search their food in a larger area than others those have lesser fitness values. This can be mathematically represented as (Meng et al., 2014):

$$x_{i,j}^{t+1} = x_{i,j}^{t} \times (1 + Randn(0, \sigma^2)) \tag{12}$$

where,

$$\sigma = \begin{cases} 1, & if\ f_i \le f_k \\ \exp\left(\dfrac{f_k - f_i}{|f_i| + \varepsilon}\right), & otherwise, k \in [1, N], k \ne i \end{cases} \tag{13}$$

where $Randn(0, \sigma^2)$ is a Gaussian distribution with mean = 0 and standard deviation = σ, *k* is the rooster's index which is randomly selected from rooster's group, ε is the smallest constant that is used during computation to avoid any zero division error, and *f* is the fitness value of the corresponding *x* (Meng et al., 2014).

3.2.2.2. Hen Model

As compared to rooster group and chick groups, the hen groups are more complex and diverse in nature. Hens can choose to follow their group mate roosters in the hunt for food or they may hunt food by themselves. Hens may snatch the good foods collected by other chickens. The dominant hens have the competitive advantage of food searching than the submissive ones. (Meng et al., 2014). These facts can be mathematically described as:

$$x_{i,j}^{t+1} = x_{i,j}^{t} + s1 \times Rand \times (x_{r1,j}^{t} - x_{i,j}^{t}) + s2 \times Rand \times (x_{r2,j}^{t} - x_{i,j}^{t}) \tag{14}$$

$$s1 = \exp\left(\frac{f_i - f_{r1}}{\left|f_i\right| + \varepsilon}\right) \tag{15}$$

$$s2 = \exp(f_{r2} - f_i) \tag{16}$$

where $Rand$ is a uniform random number over [0, 1], $r1 \in [1, 2, ...N]$ is termed as an index of the group mate rooster of i_{th} hen, and $r2 \in [1, 2, ...N]$ is termed as an index of the chicken which may be randomly chosen rooster or hen, from the swarm $(r1 \neq r2)$ (Meng et al., 2014). It is understandable that $f_i > f_{r1}$, $f_i > f_{r2}$, thus $s2<1<s1$. If $s1 = 0$, then i_{th} hen would go to search for the food in the path followed by other chickens. If i_{th} hen possesses a smaller fitness value, then the nearer $s1$ approximates to 1 which minimizes the positional gap between the i_{th} hen and her head rooster. Hence, the more dominant hen would have better chance to eat the food than the submissive ones.

3.2.2.3. Chick Model

The chicks go after their mother for searching food, which can be expressed as (Chen et al., 2015):

$$x_{i,j}^{t+1} = x_{i,j}^{t} + FL \times \left(x_{m,j}^{t} - x_{i,j}^{t}\right) \tag{17}$$

where, $x_{m,j}^{t}$ is the position of i_{th} chick's mother where $m \in [1, N]$. FL is a parameter which signifies that the chick would go behind its mother to hunt for food (Meng et al., 2014), where $FL \in (0, 2)$.

4. MATERIALS AND PROCESSES

Natural transparent polycarbonate plaques and black (by adding 0.1% wt. carbon black) polycarbonate plaques are used as work piece materials to carry out the experimental investigations. All the work piece materials are injection moulded into 4 mm thick and 80 mm × 35 mm rectangular plaques.

The experimental set up used for performing the experimental works is presented in Figure 3. The set up consists of a diode laser system (Coherent® fiber array packaged diode laser) as a laser source, a CNC work station for moving the work piece and a hydraulic pressure equipment for applying weld pressure. The laser beam is delivered to the work station by using a fiber optic cable of 800 micron diameter. The output lens assembly attached to the delivery end of the fiber optic cable has a focal length of 13 mm. The maximum optical power of the laser beam delivered by the system is 30 W and the wavelength of the beam is 809.40 nm. The CNC table can move along three perpendicular axes which allow the movement of the work piece along X, Y and Z axes. Thus a contour welding operation can be performed using this experimental set up. The CNC table is interfaced with computer and can be controlled remotely. The stand-off distance, the distance between the top surface of the work piece and the optical lens assembly, can be varied by altering the position of CNC table along Z direction. The hydraulic pressure system is designed and developed to apply weld pressure during the experiments. The hydraulic system comprises a manually operated hydraulic pump and a hydraulic jack which is calibrated with respect to the overlapping area of the plaques by the help load cells. A welding fixture is also fabricated to hold the work pieces in position to ensure the fixed overlapping area during each and every experimental run.

Figure 3. Photograph of experimental setup
A = fiber optic transport cable, B = CNC z-axis motor driven carriage, C = laser optics, D = slot for passing laser, E = welding fixture, F = clamp pressure system, G = CNC x-y table, H = hose connected with hydraulic pump

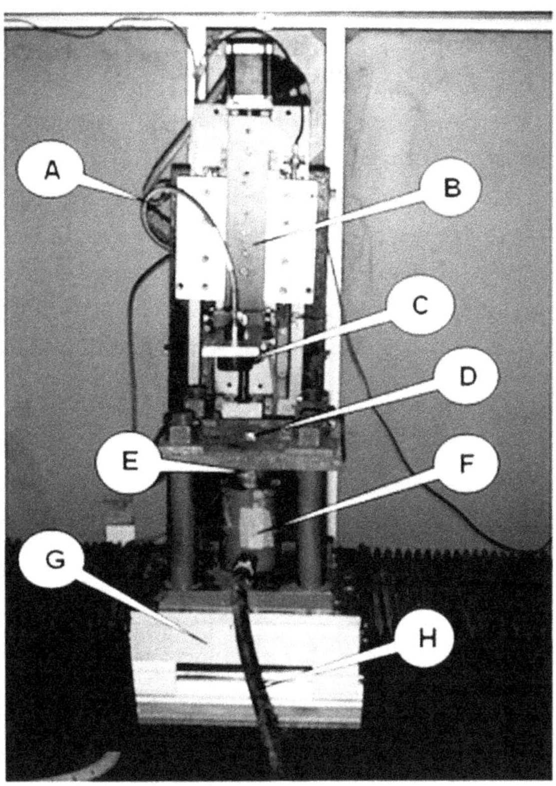

4.1. Developing the Design Matrix

Laser power (*A*), welding speed (*B*), stand-off distance (*C*) and clamp pressure (*D*) are selected as process parameters, which can be controlled and measured independently (Acherjee et al., 2012).

The quality of the weld is also depends on the thermo-physical properties of the material and the air gap between two mating parts. To minimize the detrimental effect of air gap, highly finished work materials are used and pressure is applied using specially designed hydraulic to hold the two materials in good contact. The ranges of parameters are decided based on the trial experiments. Process variables with their notation, units and levels are presented in Table 1.

The experimental matrix is designed based on CCRD (central composite rotatable design) of RSM using Design-Expert® statistical software.

4.2. Experimentation

The work piece materials are placed on the welding fixture by placing the black polycarbonate plaque at bottom and the natural polycarbonate plaque at the top in lap joint configuration. The overlapping are is maintained as 20 mm × 35 mm, for each and every experimental runs. The welding fixture is then placed on the top of the piston of the hydraulic jack, and raised the piston by using the hydraulic pump till the top polycarbonate plaque touches upper plate of the jack and starts pressing it. The pump is continuously operated till the desired pressure is obtained. The upper plate of the jack contains a slot to pass the laser. The polycarbonate plaques assembly is then exposed to the laser radiation and irradiated at the center of the overlapping portion. Contour welding is done by traversing the work piece materials by the help of the CNC table. Stand-off distance is also varied by varying the distance between the optical lens assembly and the CNC table by operating the motor driven carriage. The experiments are carried out in random order following the experimental matrix to avoid any systematic error. The widest heat affected area observed during this study is well below the length of overlap of the plaques. Hence, it is not possible for any 'melt-down' to occur. Meltdown is the decrease in total thickness of the samples after they have been welded, caused by melted material being driven out from between the plaques during welding (Prabhakaran et al., 2006). In contrast, meltdown can occur if the entire absorbing surface is heated above the melting point, as can be the case in small T-welded samples.

Table 1. Process variables with their notation, units and levels

Process Variables	Notations	Units	Levels				
			-2	-1	0	1	2
Laser power	*A*	W	10.0	12.5	15.0	17.5	20
Welding speed	*B*	mm/s	4	7	10	13	16
Stand-off distance	*C*	mm	25	28	31	34	37
Clamp pressure	*D*	MPa	1.2	2.4	3.6	4.8	6

4.3. Evaluation of Weld Qualities

The weld strength of each of the welded specimens are tested on a universal testing machine (Make: Instron®, Model: 8801) of 100 kN load capacity. Some modification of welded specimens is made to minimize the bending moment at weld line during the shear pull testing. During the shear testing the cross head speed is maintained at a constant value equal to 0.5mm/min. The breaking load is measured and the weld strength is determined by the breaking load per unit length of the weld seam (N/mm). For each experimental run, three samples are prepared and tested, and average of those three is taken as the weld strength result.

Weld seam widths are measured using an optical measuring microscope (Make: Olympus®, Model: STM 6) of sub-micron precision. For each specimen, weld width measurements are taken at three different positions along the weld line, and average of those three is taken as the weld width result. Table 2 presents the experimental design matrix and the measured response values.

5. DEVELOPMENT OF MATHEMATICAL MODELS USING RSM

Mathematical models are developed using RSM to predict the outputs for a combination of input parameters. The developed models are tested for their adequacies to obtain the best fit models by using sequential F-test, *lack-of-fit* test and the ANOVA (analysis-of-variance) techniques.

5.1. Analysis of Weld Strength

It is observed from the sequential F-test that the best fit model for the weld strength is a quadratic model, where the additional terms are significant and the model is not aliased. Backward elimination method of ANOVA is employed to remove the insignificant model terms to improve the adequacy of the model. Among the linear model terms, laser power (A), welding speed (B), and stand-off distance are significant terms, but to support hierarchy, the linear term of clamp pressure (D) is added to the model. Among the two factor interaction terms, the interaction of power and stand-off-distance (AC), and the interaction of welding speed and stand-off distance (BC) are the significant model terms. Among the quadratic model terms, the quadratic terms of the welding speed (B^2), stand-off distance (C^2), and clamp pressure (D^2) are the significant model terms. The model adequacy measures, namely R^2, *adjusted* R^2 and *predicted* R^2 are found to be equal to 0.97, 0.95, and 0.91, respectively, which are very close to unity. The very close agreement of R^2 values indicates that the developed model is adequate. The *adequate precision ratio* (the signal to noise ratio) is calculated and found equal to 28.38, is greater than the desired value of 4, which indicates the adequate model discrimination (Acherjee et al., 2015). The *lack-of-fit* F-value is not significant relative to the *pure error*, and this is desirable.

The developed mathematical model of weld strength (Y_{ws}) in terms of coded factors, is given below:

$$\begin{aligned} Yws\,(N\,/\,m) &= 51.21 + 7.05A - 6.33B + 6.87C + 0.03D \\ &+2.28AC - 2.04BC - 1.22B^2 - 1.16C^2 - 1.24D^2 \end{aligned} \tag{18}$$

Table 2. Experimental design matrix and measured response values

Exp. No.	Welding Parameters				Weld Strength (N/mm)	Weld Width (mm)
	A	*B*	*C*	*D*		
1	-1	-1	-1	-1	44.29	2.69
2	+1	-1	-1	-1	50.57	2.89
3	-1	+1	-1	-1	31.86	2.24
4	+1	+1	-1	-1	43.43	2.37
5	-1	-1	+1	-1	51.14	3.37
6	+1	-1	+1	-1	74.57	3.71
7	-1	+1	+1	-1	37.14	3.04
8	+1	+1	+1	-1	55.14	2.87
9	-1	-1	-1	+1	38.43	2.79
10	+1	-1	-1	+1	50.86	3.04
11	-1	+1	-1	+1	32.57	2.14
12	+1	+1	-1	+1	42.71	2.35
13	-1	-1	+1	+1	56.57	3.29
14	+1	-1	+1	+1	72.57	3.74
15	-1	+1	+1	+1	38.43	2.59
16	+1	+1	+1	+1	57.86	2.95
17	-2	0	0	0	37.43	2.53
18	+2	0	0	0	63.43	2.90
19	0	-2	0	0	57.29	3.80
20	0	+2	0	0	31.21	2.30
21	0	0	-2	0	30.46	2.40
22	0	0	+2	0	58.54	3.72
23	0	0	0	-2	44.43	2.81
24	0	0	0	+2	43.86	2.92
25	0	0	0	0	52.71	2.86
26	0	0	0	0	50.62	2.79
27	0	0	0	0	48.14	2.88
28	0	0	0	0	54.15	2.81
29	0	0	0	0	47.73	2.76
30	0	0	0	0	51.29	2.88

5.2. Analysis of Weld Width

For the weld width, the sequential F-test predicts that the quadratic model is the best fit model, where the additional terms are significant and the model is not aliased. Backward elimination method of ANOVA is employed to remove the insignificant model terms to improve the adequacy of the model. Among the linear model terms, laser power (A), welding speed (B), and stand-off distance are significant terms, but to support hierarchy, the linear term of clamp pressure (*D*) is added to the model. Among the two factor interaction terms, the interaction of power and welding speed (*AB*), power and clamp pressure (*AD*), and welding speed and clamp pressure (*BD*) are the significant model terms. Among the quadratic model terms, the quadratic terms of the power (A^2), welding speed (B^2), stand-off distance (C^2) are the significant model terms. The model adequacy measures, namely R^2, *adjusted* R^2 and *predicted* R^2 are found to be equal to 0.98, 0.96, and 0.93, respectively, which are very close to unity. The very close agreement of R^2 values indicates that the developed model is adequate. The *adequate precision ratio* (the signal to noise ratio) is calculated and found equal to 32.50, is greater than the desired value of 4, which indicates the adequate model discrimination (Acherjee et al., 2015). The *lack-of-fit F*-value is not significant relative to the *pure error*, and this is desirable.

The developed mathematical model of weld strength (Y_{ww}) in terms of coded factors, is given below:

$$\begin{aligned} Yww\ (mm) &= 2.83 + 0.10A - 0.33B + 0.32C - 2.92\text{E}(-003)D \\ &-0.044AB + 0.048AD - 0.043BD - 0.036A^2 + 0.048B^2 + 0.05C^2 \end{aligned} \tag{19}$$

6. OPTIMIZATION OF PROCESS PARAMETERS USING CSO

Single objective optimization and multi objective optimization of laser transmission welding process parameters are carried out by employing the chicken swarm optimization algorithm, using the developed RSM models. The objectives are set to maximize the weld strength and to minimize the weld width, independently during the single objective optimization, and simultaneously during the multi objective optimization. MATLAB® platform is used to develop the code of chicken swarm optimization algorithm and to execute the optimization process. The various algorithm specific parameters used are (a) *M* (maximum number of iterations) = 500, (b) *N* (population size) = 20, (c) *G* (how often the chicken swarm can be updated) = 10, (d) r % (population size of roosters) = 0.15, (e) h % (population size of hens) = 0.7, and (f) m % (population size of mother hens) = 0.5.

6.1. Single Objective Optimization

At first, single objective optimization is considered, where, weld strength and weld width are optimized, separately. Weld strength is a higher the better criteria, and thus, the desired objective is to maximize the value of weld strength. On the other hand, the weld width is the lower the better criteria, and thus, the lower value of weld width is a desired optimization criterion. The developed empirical RSM models (Eqs. (18) & (19)) are incorporated as the objective functions in chicken swarm optimization algorithm. The optimum value of weld strength is determined as 99.47 N/mm, whereas, the optimum value of weld width that can be obtained within the selected design space is 1.38 mm. Table 3 furnishes the results of single objective optimization using CSO. The optimum parameter setting and the values of responses obtained by the RSM based optimization technique (using optimization module of Design-Expert® statistical software) is also presented in the same table, which shows that the CSO predicts superior results than the conventional RSM based optimization techniques.

Table 3. Results of single objective optimization

Response	Nature of Optimization	Optimization Technique	Optimal Value	Optimal Parameter Setting			
				A	*B*	*C*	*D*
Weld strength	Maximization	CSO	99.47	20.00	4.00	37.00	3.62
		RSM	86.78	19.64	4.10	34.66	2.09
Weld width	Minimization	CSO	1.38	10.00	16.00	25.00	6.00
		RSM	1.84	11.38	14.82	25.96	4.09

Figure 4(a) shows the CSO algorithm convergence diagram for weld strength which shows that the CSO has very fast convergence capability toward the global optima. Figure 4(b) shows the CSO algorithm convergence diagram for weld width. The computational time for optimization of weld strength, weld width using CSO and the average computational time of CSO are recorded as 1.10 seconds, 1.15 seconds, and 1.125 seconds, respectively. Figure 4 also compares the convergence diagrams of CSO with two other metaheuristic algorithms namely harmony search algorithm (HS) and biogeography based optimization algorithm (BBO), to show its superiority over these two widely used and well established metaheuristics algorithms, in terms of convergence speeds.

6.2. Multi-Objective Optimization

In this section, weld strength and weld width, both, are optimized simultaneously, using chicken swarm optimization algorithm. Weld strength and weld width are of conflicting nature, and thus, a compromise is to be done between these two objectives to achieve the desired overall objective. For multi-objective optimization, the following objective function is developed:

$$\min Z = \frac{w_1 \times Yww}{Yww_{\min}} - \frac{w_2 \times Yws}{Yws_{\max}} \qquad (20)$$

Figure 4. Algorithm convergence diagrams relating to (a) weld strength and (b) weld width

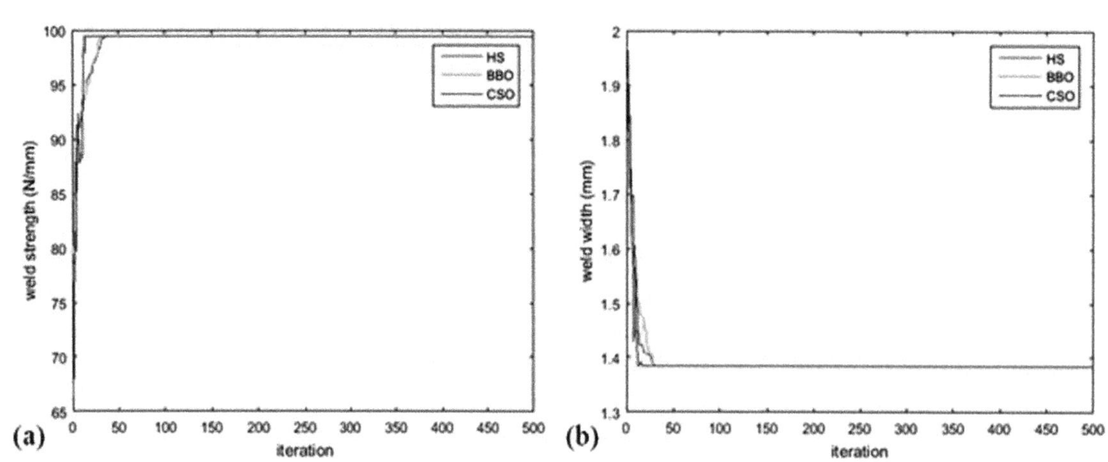

where, w_1 and w_2 are the weights (relative importance) assigned to Y_{ww} and Y_{ws}, respectively (such that $w_1+w_2 = 1$), and the min and max values in the denominator of the expression (Eq. 20) are those obtained from the results of single objective optimization. The multi objective function Z is minimized in all the cases. The results of multi objective optimization and the optimal settings of process parameters are furnished in Table 4. It is found from the results furnished in the Table 4 that CSO algorithm provides better results for weld width but gives inferior results for weld strength when compared to the results of conventional RSM based optimization technique for Case 1. However, CSO algorithm provides better results for weld strength and gives inferior results for weld width when compared to results of RSM for Case 2. The variation of the results is due to the compromise of two conflicting responses to give an overall best result. Increasing the value of one response would lead to decrease of another response. Figure 5 shows the comparison of algorithm convergence diagram for multi objective optimization. It can be clearly observed that CSO provides better and fast convergence towards optima compared to other metaheuristic algorithms like harmony search (HS) and biogeography based optimization (BBO). The computational times recorded for the multi response optimization for case 1 and case 2 are 1.04 seconds and 1.08 seconds, respectively.

Table 4. Results of multi objective optimization according to the selected criteria

Conditions	Optimization Technique	Z_{min}	WS	WW	*A*	*B*	*C*	*D*
Case 1: $w_1 = w_2 = 0.5$	CSO	0.43	13.57	1.38	10.00	16.00	25.00	6.00
	RSM	-	61.60	2.64	20.00	12.27	31.73	2.83
Case 2: $w_1 = 0.1$, $w_2 = 0.9$	CSO	-0.56	99.06	4.65	20.00	4.00	37.00	2.92
	RSM	-	74.57	3.18	20.00	10.02	33.95	2.65

Figure 5. Algorithm convergence diagram for multi objective optimization problem relating to (a) case 1, and (b) case 2

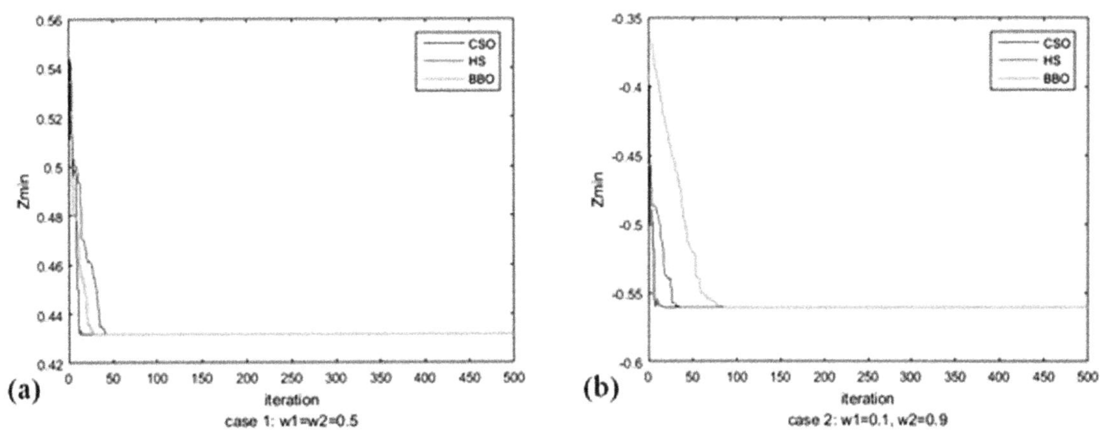

7. ANALYSIS OF PARAMETRIC EFFECTS

Figure 6 shows the variations of weld strength with laser transmission welding process parameters. During the prediction of parametric trends by using CSO, no parameter is held constant, and thus, the CSO can predict the overall parametric trends of responses, accurately. It is seen from Figure 6(a) that weld strength has a positive trend with laser power. This is due to the fact that the heat input to the weld zone increases with laser power which melts more base materials and form a larger weld pool, which in turn increases the strength of the weld. It is observed from Figure 6(b) that welding strength follows a negative trend with welding speed. This is because the increase of welding speed reduces the time of interaction between the laser beam and the work piece materials, which causing the low heat input at the weld zone, resulting in narrow and weak joint. It is evident from Figure 6(c) that weld strength has a positive tendency with the stand-off distance. Increasing the stand-off distance increases the beam spot area, which irradiates a wider zone at weld interface, increasing the weld area that can resist larger applied load, thus increases the strength of the weld. The result furnished in Figure 6(d) shows that clamp pressure does not have significant impact on weld strength. However, clamp pressure ensures intimate contact between two mating parts and influence the molten material flow inside weld pool, which is required for a good bond formation.

Figure 6. Scatter diagrams of weld strength for (a) laser power, (b) welding speed, (c) stand-off distance, and (d) clamp pressure

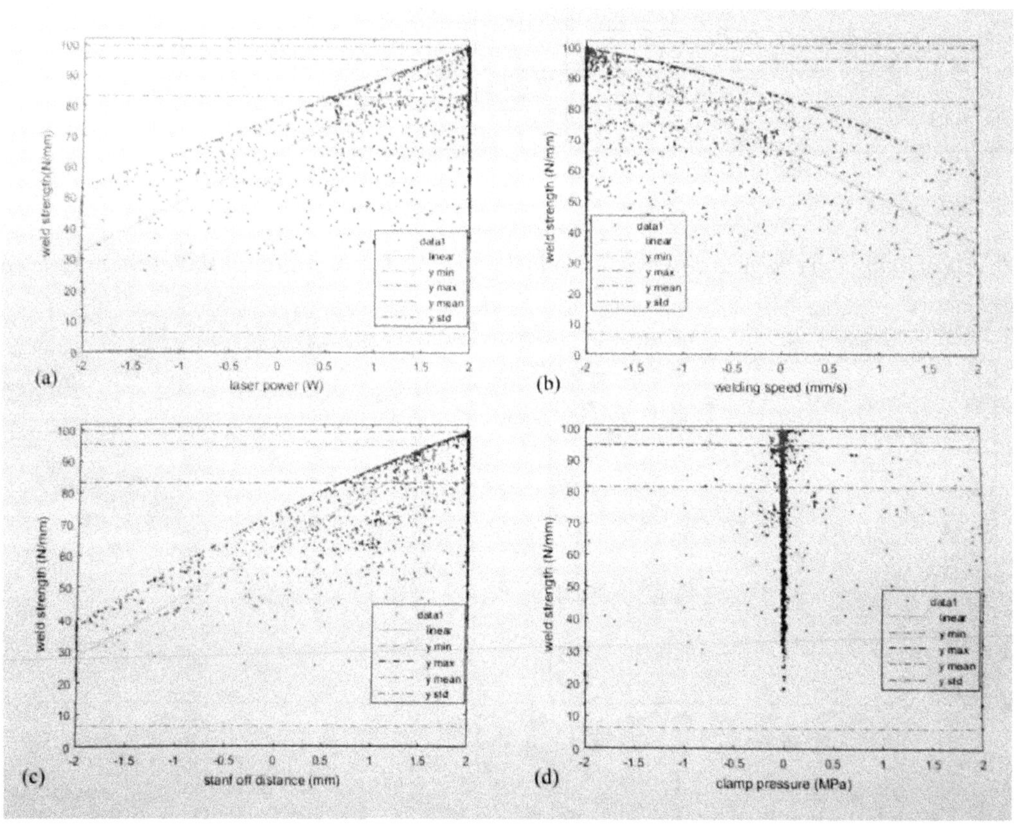

The parametric trends of weld width are presented in Figure 7. It is seen from Figure 7(a) that weld width has a positive trend with laser power. This is due to fact that the increase in laser power increases the heat input, thus melts more base materials and form a wider weld seam. It is observed from Figure 7(b) that weld width follows a negative trend with welding speed. This is because the increase of welding speed reduces the irradiation time, thus less heat is delivered to the weld zone which reduces the volume of melt pool, resulting in narrow weld seam. It is evident from Figure 7(c) that the stand-off distance has a positive effect on weld width. Beam diameter increases with stand-off distance, and thus increases the laser spot on the interface, which melts wider zone to form wider weld seam. It is seen from Figure 7(d) that clamp does not have prominent effect on weld width.

8. CONCLUSION

In present work, experimental investigations, mathematical modeling and process optimization of laser transmission welding of polycarbonates is performed using RSM and CSO. Following conclusions are drawn for this study within the considered factors limits:

Figure 7. Scatter diagrams of weld width for (a) laser power, (b) welding speed, (c) stand-off distance, and (d) clamp pressure

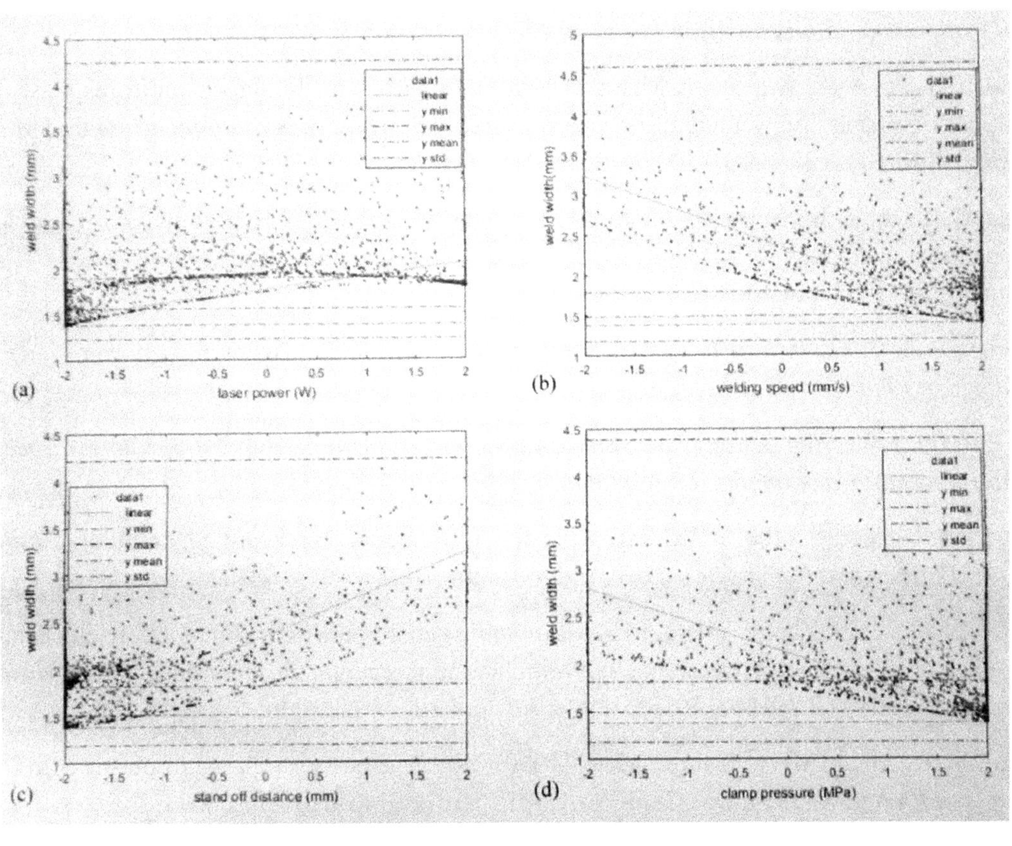

1. The developed RSM models are adequate and they can be used for prediction of responses for a set of process parameters, with significant accuracy (minimum 95% confidence level).
2. Weld strength and weld width, both, show increasing trends with laser power and stand-off distance; whereas, both the responses show decreasing trends with the increase of welding speed.
3. Clamp pressure has insignificant effect on the weld strength and the weld width, both.
4. Chicken swarm optimization algorithm is implemented successfully for prediction of optimal combination of process parameters to fulfill single objective independently, as well as, multiple objectives simultaneously.
5. Chicken swarm optimization algorithm posses a fast convergence rate and take very low computational time.
6. Chicken swarm optimization algorithm is also capable of predicting the parametric trends of responses.
7. The optimal results obtained using chicken swarm optimization algorithm are compared with the optimum results computed using RSM, and found that the CSO gives better results, as it can predict the global optima.

REFERENCES

Abed, S., Knapp, W., Traub, M., Hoffmann, D., Poprawe, R., & Loosen, P. (2004). Development of simultaneous laser welding process applied to thermoplastic polymers. *Proceedings of the 23rd International Congress on Applications of Lasers & Electro-Optics*, 97.

Acherjee, B., Kuar, A. S., Mitra, S., & Misra, D. (2015). Empirical modeling and multi-response optimization of laser transmission welding of polycarbonate to ABS. *Lasers in Manufacturing and Materials Processing*, *2*(3), 103–123. doi:10.1007/s40516-015-0009-0

Acherjee, B., Kuar, A. S., Mitra, S., Misra, D., & Acharyya, S. (2012). Experimental investigation on laser transmission welding of PMMA to ABS via response surface modeling. *Optics & Laser Technology*, *44*(5), 1372–1383. doi:10.1016/j.optlastec.2011.12.029

Bachmann, F. G., & Russek, U. A. (2002). Laser welding of polymers using high power diode lasers. *Proceedings of Photonics West Conference*.

Baylis, B. (2002). Welding thermoplastic elastomers to polypropylene with a diode laser. *Proceedings of the 21st International Congress on Applications of Lasers & Electro-Optics*, 94.

Bonten, C., & Tüchert, C. (2002). Welding of plastics-Introduction into heating by radiation. *Journal of Reinforced Plastics and Composites*, *21*(8), 699–710. doi:10.1177/073168402128988436

Chen, Y. L., He, P. L., & Zhang, Y. H. (2015). Combining penalty function with modified chicken swarm optimization for constrained optimization. *International Conference on Information Sciences, Machinery, Materials and Energy (ICISMME 2015)*, 1899-1907. doi:10.2991/icismme-15.2015.386

Coelho, J. M. P., Abreu, M. A., & Pires, M. C. (2000). High-speed laser welding of plastic films. *Optics and Lasers in Engineering*, *34*(4-6), 385–395. doi:10.1016/S0143-8166(00)00071-3

Hansch, D., Haff, D., Putz, H., Treusch, H. G., Gillner, A., & Poprawe, R. (1998). Welding of plastics with diode laser. *Proceedings of the 17th International Congress on Applications of Lasers & Electro-Optics*, 85, 131-140.

Herfurth, H., Ehlers, B., Heinemann, S., & Haensch, D. (1999). New approaches in plastic welding with diode lasers. *Proceedings of the 18th International Congress on Applications of Lasers & Electro-Optics*, *87b*, 48-56.

Jansson, A., Kouvo, S., & Kujanpää, V. (2005). Quasi-simultaneous laser welding of polymers - the process and applications for mass-production. *Proceedings of the 24th International Congress on Applications of Lasers & Electro-Optics*, 98, 381-387.

Jansson, A., Kouvo, S., Salminen, A., & Kujanpää, V. (2003). The effect of parameters on laser transmission welding of polymers. *Proceedings of the 22nd International Congress on Applications of Lasers & Electro-Optics*, 95, 124-133.

Kagan, V. A., Bray, R. G., & Kuhn, W. P. (2002). Laser transmission welding of semi-crystalline thermoplastics: part I: Optical characterization of nylon based plastics. *Journal of Reinforced Plastics and Composites*, *21*(12), 1101–1122. doi:10.1177/073168402128987699

Meng, X., Liu, Y., Gao, X., & Zhang, H. (2014). A new bio-inspired algorithm: Chicken swarm optimization. *Advances in Swarm Intelligence. Lecture Notes in Computer Science*, *8794*, 86–94. doi:10.1007/978-3-319-11857-4_10

Montgomery, D. C. (2001). *Design and Analysis of Experiments* (5th ed.). New York: John Wiley & Sons, Inc.

Prabhakaran, R., Kontopoulou, M., Zak, G., Bates, P. J., & Baylis, B. K. (2006). Contour laser – Laser-transmission welding of glass reinforced nylon 6. *Journal of Thermoplastic Composite Materials*, *19*(4), 427–439. doi:10.1177/0892705706062200

Russek, U.-A., Staub, H., Palmen, A., & Kind, H. (2003). Simultaneous laser beam welding of thermoplastics – innovations and challenges. *Proceedings of the 22nd International Congress on Applications of Lasers & Electro-Optics*, 95, 99-107.

Chapter 8
Surface Response Methodology Approach for Multi-Objective Optimization During Electrochemical Grinding of Al2O3/Al Interpenetrating Phase Composite

Goutam Kumar Bose
Haldia Institute of Technology, India

Pritam Pain
Haldia Institute of Technology, India

ABSTRACT

Now a day the advances in the material science lead to the development of advanced engineering materials like super alloys. The current research work focus on the selection of significant machining parameters initially depending on single objective and then multi objective responses, while machining alumina-aluminum interpenetrating phase composites during electrochemical grinding. Control parameters such electrolyte concentration (C), voltage (V), depth of cut (D) and electrolyte flow rate (F) have been considered for experimentation. Initially single objective optimal parametric setting is generated from Taguchi Methodology and Regression analysis. Further it is optimize using Response Surface Methodology. The contradictory responses like higher material removal rate (MRR), lower surface roughness (Ra), lower overcut (OC) and lower cutting force (Fc) are ensured by using Overlaid contour plots and Desirability functions. These soft computing techniques corroborates well during the parametric optimization of electrochemical grinding process.

DOI: 10.4018/978-1-5225-2440-3.ch008

1. INTRODUCTION

Rapid globalization has drastically changed the dynamics of the manufacturing industries across the globe which necessitates the use of alternative and concurrently superior materials for several applications commensurable with the deliberated use. In order to persevere in this competitive business environment industries are enduring several modifications by reorienting the traditional manufacturing process to non-traditional manufacturing process. In order to survive with the changes application of these evolving manufacturing techniques along with the contemporary materials are becoming popular gradually. Electrochemical Grinding (ECG) process is an amalgamation of electrochemical dissolution and mechanical abrasion which is one such very widely used non-conventional machining processes for machining complicated shapes in high strength electro-conductive materials, difficult-to-machine alloys, fragile, hardened and also heat sensitive parts (El-Hoffy, 2005). In ECG electrolyte is pumped in a small gap between the cathode grinding wheel and anode work piece. Direct current is passed through the system which helps in the dissolution of electrolyte solution. The major parameters influencing ECG process are Electrolyte Concentration (C), Voltage (V), Depth of cut (D) Flow rate (F), physical properties of the work piece, type of abrasive in the wheel, grain size of the abrasive particle, speed of the grinding wheel etc. (Reddy et al., 1998). As the material removal process is done by electrochemical dissolution of work material, the machined surfaces are free from residual stress and burr, thereby eliminating rejection caused by stress and cracks (Kaldos,1974). In this machining process there is no need of time consuming secondary operations.

Some of the past research works is presented here. Qu et al. (2015) prolong the wheel life in ECG while machining Inconel 718, by replacing the electrodeposited diamond wheel by the brazed diamond wheel. Senthilkumar et al. (2011) has studied on the process optimization of the machining parameter in electrochemical grinding for contradictory result by employing non-dominated sorting genetic algorithm-II (NSGA-II) for multi-objective optimization. Geva et al. (1976) has experimented on electrochemical surface parameters in ECG process and proposed a suitable process mechanism based on the output results. Molla and Manna (2013) has optimized the machining parameter of $Al/(Al_2O_3+ZrO_2)$ while machining in electrochemical grinding, and they find out that the roughness (R_a and R_t) are lesser at moderated grinding wheel speed. Bhandari and Shukla (2015) have studied on the optimization procedure of the control parameter of electrochemical grinding by using Particle Swarm Optimization Technique. Goswami et al. (2008) have investigate the effect of the process parameters while machining Al_2O_3/Al Interpenetrating Phase Composite material using ECG process and beside that they find out the effect of different kind of electrolytes and degree of reduction in this course and finally they compare the conventional grinding with electrochemical grinding of this material. Roy et al. (2007) experimentally studied using statistical procedure was made to evaluate the effect of the voltage on the surface finish of P-20 material. Puri and Banerjee (2013) comprehensively evaluated the influence of voltage and cutting speed on ECG while machining composite carbide inserts material. Levinger and Malkin (1979) have investigated on ECG of WC-Co cemented carbides and find out that electrochemical dissolution into the cobalt phase occurs initially at much faster rate than into the carbide phase and also their experimental result explains the interaction between the electrochemical and mechanical process. Kaczmarek and Zachwieja (1966) have investigated on the MRR by ECG of cutting tool materials and find out that MRR is depend on electrochemical solution as well as the grainsize of the grinding wheel. Atkinson (2007) has experimentally exhibits how surface hardness changes after machining AISI 01 tool steel in different regimes.

The present research work highlights on machining of Al_2O_3/Al interpenetrating phase composite (IPC) employing electrochemical grinding for finding out the effect of machining parameters namely Electrolyte Concentration (C), Voltage (V), Depth of cut (D) and Flow rate (F) keeping other control parameters unchanged on performances response viz. Material Removal Rate (MRR), Surface Roughness (Ra), Overcut (OC) and Cutting Force (Fc).

2. PLANNING FOR EXPERIMENTATION

In the current research study electro chemical grinding of Al_2O_3/Al interpenetrating phase composites material (Al_2O_3-68.9% and Al-31.1%) synthesized at 1150°C/24 hour has been considered. The mechanical properties of the Al_2O_3/Al IPC are shown in Table 1.

The experiment is executed in a developed (microprocessor controlled) surface grinder that utilizes a grinding wheel of 150 mm diameter and 12.27 mm width having metal bonding and diamond abrasive particles. The width of the layer comprising the diamond is 3 mm thick where the abrasive grit size is 100. A DC motor rotating at a speed of 3500 rpm drives the grinding wheel mounted at the tip of the spindle. While machining, current is recorded by using an ammeter connected to the electrical circuit. Figure 1 exhibits the machining setup with feed drive mechanism attachment.

The work includes study of the characteristic properties of the ECG process with various process parametric combinations likewise flow rate (lit/sec), electrolytic concentration (gm/lit), depth of cut (mm) and voltage (volts), on MRR (gm/min), Ra (μm) Overcut (mm) and cutting force (kgf). Table 2 shows the different levels and their units of process control parameters during machining operation. However, there are other factors like feed rate (0.4 mm/sec), type of power (continuous), rpm of the grinding wheel etc. which may affect the measured performance, are kept unaltered throughout the experimentation. Sodium chloride solution is used as electrolyte. The work pieces considered for experimentation are flat rectangular plates of dimension 23 mm x 15 mm x 5 mm thick.

During machining, the feed drive mechanism of the worktable is regulated by a microprocessor controller unit. The feed drive mechanism consists of a stepper motor (Type: STM – 1100), comprising of torque 10 kg-cm having angular resolution of 1.8° / step with voltage phase: 24 VDC and current phase: 0.67 Ampere. The machining time is recorded with the help of the stopwatch reading. Before and after

Table 1. Properties of the Al_2O_3/Al IPC work material

Properties	Values (Synthesis Condition is 11500/24 Hour)
Grain size (micron)	Al2O3 – 4.27, Al - 1.42
Fracture origin: Principal	Separation at Al_2O_3/Al grain boundary
Bulk density (gm/cm^3)	3.54
Micro hardness No.(VHN)	364
Compressive Strength (MPa)	576
Bend Strength (MPa)	458 ± 15
Elastic modulus (GPa)	67
Conductivity (105ohm-1cm-1at RT)	0.4

Figure 1. The Electrochemical grinding setup

Table 2. Control parameters and their levels

Operating Parameters	Symbol	Units	Levels		
			L1	L2	L3
Electrolyte Concentration	C	gm/lit	20	25	30
Supply Voltage	V	volts	15	20	25
Depth of Cut	D	mm	0.04	0.08	0.12
Electrolyte Flow Rate	F	lit/sec	0.1	0.2	0.3

each machining run the work pieces are cleaned properly and the respective weights are measured with the help of an electronic weighing machine (Make: Mettler Toledo; Model no-AG285) of accuracy 0.0002 mg. Perthometer-M1, of Mahr Gmbh make is used to measure the surface roughness. For each combination three trials are performed and average is computed for analysis. A well-designed experimental configuration is used to reduce the total number of experiments. Initially the experiment is conducted with the L_9 orthogonal array based on the Taguchi methodology to find out the most significant parameter and the parametric combination affecting the responses shown in Table 3. Finally, the experimentation is designed for L_{81} orthogonal array in order to conduct regression analysis and subsequently response surface methodology analysis shown in Table 4.

Table 3. Parametric settings and responses for L_9 experimental run

Exp. No.	Control Parameters				Responses			
	C gm/lit	V volts	D mm	F lit/sec	MRR gm/min	Ra μm	OC mm	Fc kgf
1	20	15	0.04	0.1	0.07	0.17	0.01	0.67
2	20	20	0.08	0.2	0.14	0.21	0.01	1.74
3	20	25	0.12	0.3	0.05	0.84	0.05	2.19
4	25	15	0.08	0.3	0.13	0.20	0.05	1.15
5	25	20	0.12	0.1	0.15	0.52	0.02	2.56
6	25	25	0.04	0.2	0.08	0.92	0.00	3.86
7	30	15	0.12	0.2	0.37	0.21	0.11	0.69
8	30	20	0.04	0.3	0.52	0.16	0.04	0.55
9	30	25	0.08	0.1	0.09	0.40	0.39	0.30

Table 4. Parametric settings and responses for L_{81} experimental run

Exp. No.	Control Parameters				Responses			
	C (gm/lit)	V (volts)	D (mm)	F (lit/sec)	MRR (gm/ min)	Ra (μm)	Overcut (mm)	Force (kgf)
1	20	20	0.08	0.1	0.17	0.76	0.00	0.71
2	30	20	0.08	0.3	0.67	0.49	0.28	0.09
3	30	20	0.08	0.1	0.30	0.19	0.22	0.43
4	25	25	0.12	0.1	0.12	1.26	0.06	1.19
5	25	20	0.08	0.2	0.25	0.62	0.14	0.31
6	25	20	0.04	0.3	0.57	0.50	0.23	0.03
7	25	20	0.12	0.3	0.30	1.03	0.11	0.12
8	20	15	0.12	0.1	0.24	0.49	0.04	0.24
9	30	20	0.04	0.3	0.80	0.22	0.34	0.17
10	20	25	0.08	0.2	0.04	1.43	0.07	1.00
11	20	20	0.12	0.3	0.07	1.31	0.00	0.26
12	20	25	0.08	0.1	0.23	1.28	0.11	1.26
13	30	20	0.08	0.2	0.49	0.34	0.25	0.17
14	25	25	0.04	0.2	0.32	0.88	0.10	0.78
15	30	15	0.08	0.2	0.55	0.19	0.35	0.38
16	25	25	0.04	0.3	0.51	1.03	0.13	0.52
17	20	15	0.08	0.3	0.26	0.52	0.17	0.36
18	20	25	0.08	0.3	0.14	1.57	0.04	0.73
19	20	20	0.08	0.2	0.02	0.90	0.03	0.45
20	20	25	0.12	0.3	0.01	1.84	0.10	0.81
21	25	25	0.04	0.1	0.14	0.74	0.06	1.04

continued on following page

Table 4. Continued

Exp. No.	Control Parameters				Responses			
	C (gm/lit)	V (volts)	D (mm)	F (lit/sec)	MRR (gm/ min)	Ra (μm)	Overcut (mm)	Force (kgf)
22	25	20	0.12	0.2	0.12	0.88	0.08	0.38
23	30	15	0.12	0.1	0.23	0.07	0.26	0.04
24	30	25	0.08	0.1	0.24	0.72	0.11	0.97
25	25	15	0.08	0.3	0.50	0.24	0.28	0.50
26	30	20	0.04	0.1	0.43	0.07	0.28	0.35
27	25	25	0.08	0.1	0.01	1.00	0.00	1.12
28	30	20	0.12	0.3	0.54	0.75	0.22	0.02
29	30	15	0.04	0.1	0.49	0.60	0.38	0.19
30	20	25	0.04	0.3	0.27	1.31	0.02	0.66
31	20	15	0.04	0.2	0.21	0.11	0.20	0.17
32	30	25	0.12	0.3	0.48	1.28	0.11	0.53
33	20	15	0.04	0.3	0.39	0.26	0.23	0.43
34	25	15	0.04	0.2	0.44	0.17	0.31	0.31
35	30	25	0.04	0.2	0.56	0.60	0.20	0.64
36	20	15	0.12	0.3	0.13	0.78	0.11	0.28
37	20	25	0.04	0.2	0.09	1.17	0.01	0.92
38	30	25	0.04	0.1	0.37	0.46	0.17	0.90
39	25	15	0.08	0.1	0.13	0.05	0.21	0.02
40	25	15	0.04	0.3	0.63	0.02	0.34	0.57
41	25	15	0.12	0.1	0.00	0.21	0.15	0.10
42	20	15	0.12	0.2	0.05	0.64	0.08	0.02
43	30	15	0.04	0.2	0.68	0.45	0.42	0.45
44	20	15	0.08	0.2	0.08	0.37	0.14	0.10
45	30	15	0.04	0.3	0.86	0.30	0.45	0.71
46	20	15	0.08	0.1	0.11	0.23	0.11	0.16
47	25	20	0.12	0.1	0.06	0.74	0.05	0.64
48	25	20	0.08	0.3	0.44	0.77	0.17	0.05
49	25	25	0.12	0.3	0.24	1.56	0.01	0.67
50	25	25	0.12	0.2	0.06	1.41	0.03	0.93
51	25	15	0.04	0.1	0.26	0.32	0.27	0.05
52	30	25	0.08	0.2	0.43	0.87	0.14	0.71
53	20	25	0.12	0.2	0.17	1.69	0.14	1.07
54	30	15	0.12	0.3	0.60	0.22	0.33	0.56
55	20	20	0.04	0.1	0.03	0.49	0.06	0.64
56	30	20	0.12	0.2	0.35	0.60	0.19	0.24
57	30	25	0.12	0.1	0.11	0.98	0.05	1.05

continued on following page

Table 4. Continued

Exp. No.	Control Parameters				Responses			
	C (gm/lit)	V (volts)	D (mm)	F (lit/sec)	MRR (gm/ min)	Ra (μm)	Overcut (mm)	Force (kgf)
58	30	25	0.04	0.3	0.74	0.75	0.24	0.38
59	30	20	0.12	0.1	0.17	0.46	0.15	0.50
60	20	25	0.12	0.1	0.36	1.55	0.17	1.33
61	20	20	0.08	0.3	0.20	1.05	0.07	0.19
62	25	20	0.04	0.1	0.20	0.21	0.17	0.50
63	30	15	0.08	0.1	0.36	0.33	0.32	0.12
64	25	15	0.08	0.2	0.31	0.09	0.25	0.24
65	30	15	0.12	0.2	0.41	0.08	0.29	0.30
66	30	25	0.08	0.3	0.61	1.01	0.18	0.45
67	20	20	0.12	0.2	0.11	1.16	0.03	0.52
68	25	25	0.08	0.3	0.38	1.29	0.07	0.59
69	25	25	0.08	0.2	0.19	1.15	0.03	0.85
70	25	15	0.12	0.3	0.36	0.50	0.22	0.42
71	20	20	0.04	0.3	0.33	0.78	0.13	0.11
72	30	20	0.04	0.2	0.62	0.08	0.31	0.09
73	30	15	0.08	0.3	0.73	0.04	0.39	0.64
74	25	20	0.04	0.2	0.38	0.36	0.20	0.23
75	30	25	0.12	0.2	0.29	1.13	0.08	0.79
76	20	15	0.04	0.1	0.03	0.03	0.17	0.09
77	25	15	0.12	0.2	0.18	0.36	0.19	0.16
78	25	20	0.08	0.1	0.07	0.47	0.11	0.57
79	20	20	0.04	0.2	0.15	0.64	0.09	0.38
80	20	25	0.04	0.1	0.09	1.02	0.05	1.18
81	20	20	0.12	0.1	0.30	1.02	0.06	0.78

3. THE TAGUCHI METHODOLOGY

Taguchi methodology is regarded as the foremost proponent of robust parameter design, which is applied for product or process design that emphases on reducing variation and/or sensitivity to noise. Taguchi designs render an effective and efficient technique for designing products that operate consistently and optimally under various circumstances. The Taguchi method calculates a response table for each characteristic i.e., mean, signal-to-noise ratio, standard deviation and slope (Reddy et al., 1999). Response tables can indicate which factor has the biggest impact on the response and which level of the factor is related to higher or lower response characteristic values. Analysis of Variance (ANOVA) is a useful process for evaluating error variance and determining the relative importance of various process variables (Park, 1996). The experimental outcomes are explored to study the role of different process variables on various responses by using S/N ratio graphs and ANOVA.

Signal-to-Noise (S/N) Ratio

S/N ratio determines the contribution of different process variables on several outputs. The aim is to find out an optimal set of control factor settings that accomplish robustness against (insensitivity to) noise factors. Depending upon the nature of objective(s), it can be categorized as follows-

Lager Is Better

The signal-to-noise (S/N) ratio is deliberate for each factor level sequence. The formulation for the larger-is-better S/N ratio using base 10 log is:

$$S / N = -10log\left(\sum \frac{(1 / y^2)}{n}\right) \tag{1}$$

where

Y= responses for the given factor level combination,
n= number of response in the factor level combination.

Smaller Is Better

The Signal-to-noise (S/N) ratio is deliberate for every factor level sequence. The formulation for the smaller-is-better S/N ratio using base 10 log is:

$$S / N = -10log\left(\sum \frac{y^2}{n}\right) \tag{2}$$

where

Y= responses for the given factor level combination,
n= number of responses in the factor level combination.

Result Analysis Using the Taguchi Methodology

The experimental responses from Table 3 are explored to study by using S/N ratio and ANOVA. The statistical analysis is carried out by the software MINITAB 17 environment.

Analysis to Maximize MRR

Table 5 shows the ANOVA results for MRR

It is evident from Table 5 that the F values of Electrolyte Concentration (C) assume a value 3.0598 with a yield of 50.49% in case of MRR. This implies that this variable has significant effects on MRR

Table 5. Analysis of variance for MRR

Source	DF	Adj SS	Adj MS	F-Value	%-Value
C	2	0.10483	0.052415	3.0598	50.49
V	2	0.05865	0.029325	1.7119	28.25
D	2	0.01699	0.008495	0.4959	8.18
F	2	0.02714	0.013570	0.7922	13.07
Error	0				
Total	8	0.20761			
Pooled Error	6	0.102780	0.017130	6.0598	

with respect to the other three parameters. The signal-to-noise (S/N) ratio analysis for MRR is carried out on the basis of larger is better option. The S/N ratio plot for MRR is shown in Figure 2.

The S/N ratio graph shows maximum MRR at the following parametric combination: C 30 gm/lit; V 20 volts; D 0.12 mm; F 0.2 lit/sec. The S/N ratio graph of MRR continuously increase with increase in electrolyte concentration as because higher concentration accelerates the ionization of Aliminum into liquid medium. In case of change in voltage, MRR initially increase with increase in voltage to a certain point and beyond which MRR decrease with increase in voltage, because initially increase in voltage increase the decomposition rate of the working material and attain an optimum value, beyond which increase in voltage adversely effect on MRR. While depth of cut increase, initially MRR reduce by the cogging effect between work surface and grinding wheel, but further increase in the depth of cut

Figure 2. S/N ratio plot for MRR

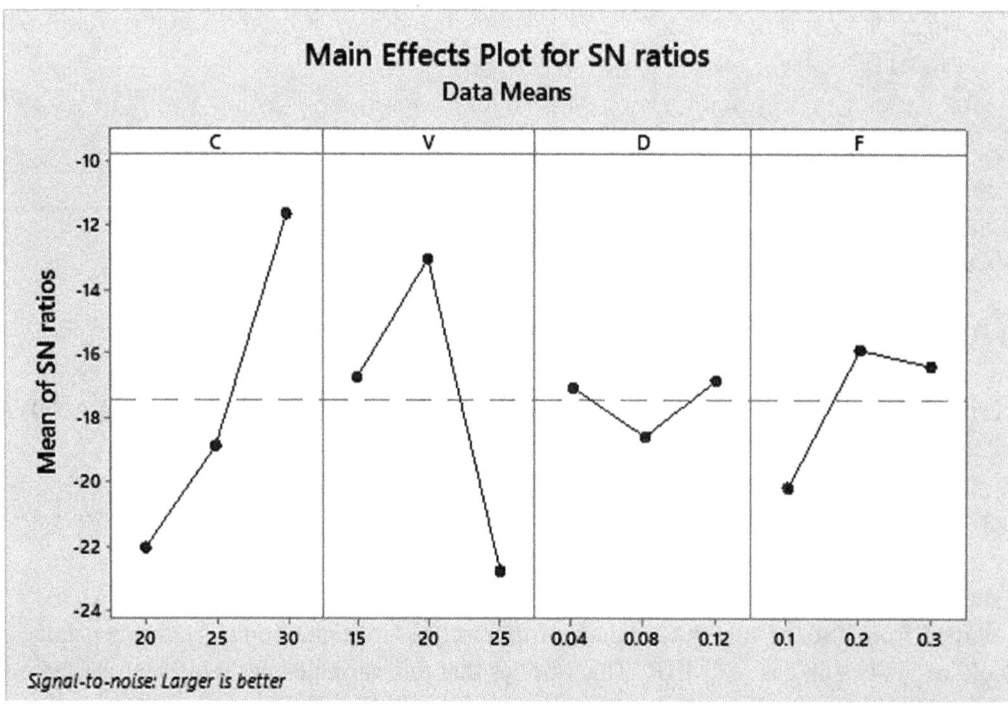

causes mechanical abrasion which increase the MRR. It is evident from the plot that increase in flow rate increase the MRR but at very high flow rate MRR reduce as because at high flow of electrolyte, the contact time of the liquid to working surface deduce.

Analysis to Minimize Ra

Table 6 shows the ANOVA results for Ra

It is evident from the table that the F values of Voltage (V) assume a value 9.0030 with a yield of 67.20% in case of Ra. This implies that this variable has significant effects on Ra with respect to the other three parameters. The signal-to-noise (S/N) ratio analysis for Ra is carried out on the basis of smaller is better option. The S/N ratio plot for Ra is shown in Figure 3.

Table 6. Analysis of variance for Ra

Source	DF	Adj SS	Adj MS	F-Value	%-Value
C	2	0.12623	0.063115	2.3953	17.88
V	2	0.47446	0.237230	9.0030	67.20
D	2	0.09416	0.047080	1.7867	13.34
F	2	0.01124	0.005620	0.2133	1.59
Error	0				
Total	8	0.70609			
Pooled Error	4	0.105400	0.026350	13.3983	

Figure 3. S/N ratio plot for Ra

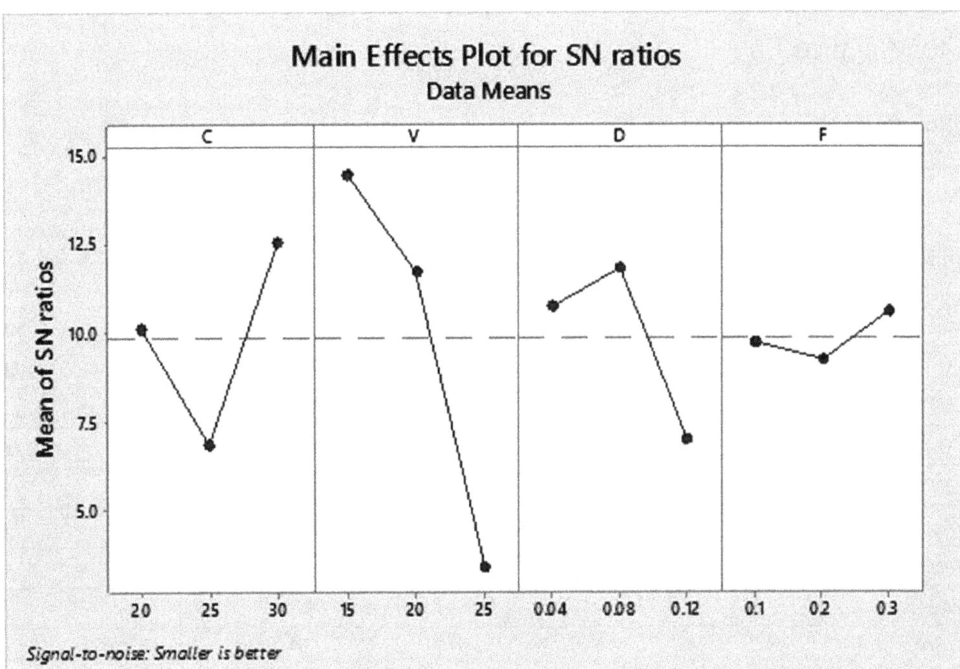

The S/N ratio graph shows minimum Surface roughness (Ra) at the following parametric settings: C 30 gm/lit; V 15 volts; D 0.08 mm and F 0.3 lit/sec. The S/N ratio plot for Ra shows that at the lowest level of concentration, electrochemical dissolution is minimum and machining is initially controlled by mechanical abrasion. With the surge in the concentration the dissolution rate increases which causes the Al_2O_3 texture to reveal resulting in poor surface finish, further increase in concentration reduce surface roughness. Increase in the voltage increase surface roughness considerably. This is due to the fact that higher machining voltage attended by amplified current density causes extra ionization of aluminum into the liquid medium, exposing Al_2O_3 texture, which makes the work surface rough. Increase in the depth of cut, the surface roughness initially decreases steadily. Further increase of the depth of cut roughness increases. Surface roughness initially increase with increase in flow rate of electrolyte and further increase of flow rate decrease surface roughness.

Analysis to Minimize OC

Table 7 shows the ANOVA results for OC

From the ANOVA table it is apparent that none of the parameters gives a clear indication of being most significant parameter. But from the F- value of electrolyte concentration (C) influences OC more than the rest of the parameters and having 40.59% of contribution. The Signal to noise (S/N) ratio analysis for OC is represented on the basis of smaller is better. The S/N ratio plot for OC is shown in Figure 4.

Overcut is one of the most general drawbacks in electrochemical grinding and it is higher at higher concentration due to extra chemical etching along grain pull out at wheel width end. The S/N ratio plot illustrates minimum overcut at the following parametric combination: i.e. C 25 gm/lit; V 20 volts; D 0.04 mm; and F 0.2 lit/sec. Increase in voltages initially reduce OC but further increase in voltage, due to sparking and partial etching overcut increases to a certain point. Increase in depth of cut increase OC significantly to a certain point, beyond which increase in depth of cut reduce overcut. Electrolyte flow initially reduce OC but further increase in electrolyte flow increase OC.

Analysis to Minimize Fc

Table 8 shows the ANOVA results for Fc

Table 7. Analysis of variance for OC

Source	DF	Adj SS	Adj MS	F-Value	%-Value
C	2	0.04857	0.024285	2.0497	40.59
V	2	0.02481	0.012405	1.0470	20.73
D	2	0.02762	0.013810	1.1656	23.08
F	2	0.01866	0.009330	0.7875	15.59
Error	0				
Total	8	0.11966			
Pooled Error	6	0.071090	0.011848	5.0497	

Figure 4. S/N ratio plot for OC

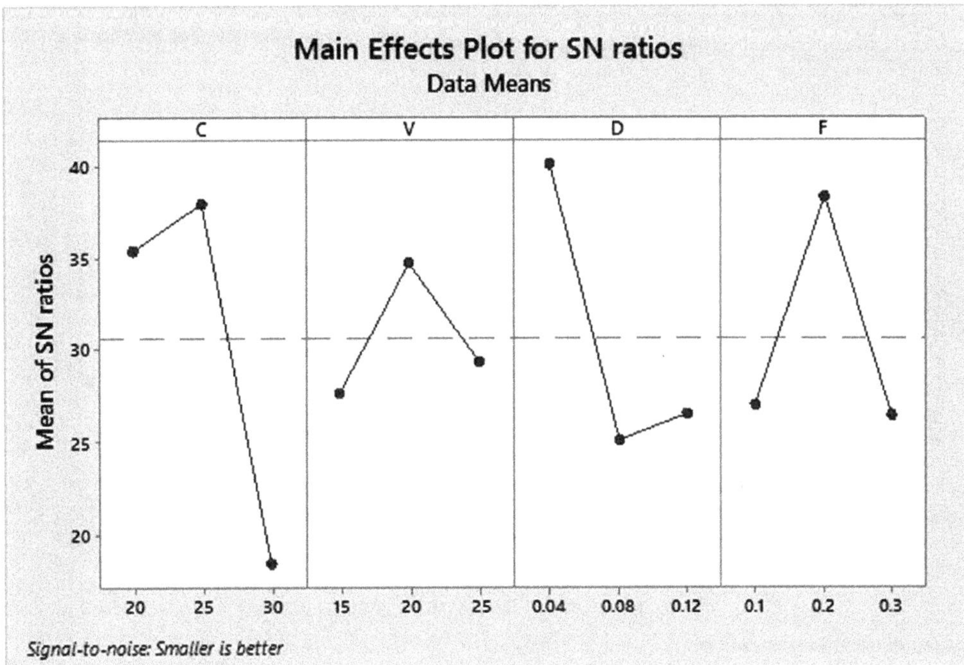

Table 8. Analysis of Variance for Fc

Source	DF	Adj SS	Adj MS	F-Value	%-Value
C	2	6.0606	3.0303	3.6573	54.9365
V	2	2.4968	1.2484	1.5067	22.6323
D	2	0.9738	0.4869	0.5876	8.8270
F	2	1.5008	0.7504	0.9057	13.6041
Error	0				
Total	8	11.0320			
Pooled Error	6	4.9714	0.8286	6.6573	

It is clear from the table that in case of Fc, Electrolyte concentration (C) alone is the most significant contributor having F value of 3.6573 and having % contribution of 54.9365. Other parameters behave insignificantly for the response. The signal-to-noise (S/N) ratio analysis for Fc is carried out on the basis of smaller is better option. The S/N ratio plot for Ra is shown in Figure 5.

The S/N ratio plot shows minimum cutting force at the following parametric combination: C 30 gm/lit; V 15 volts; D 0.08 mm; and F 0.1 lit/sec electrolyte flow rate. It is evident from the S/N ratio plot that with the surge in the concentration the cutting force primarily increases and then decreases sharply. With the upsurge in voltage the current density becomes high resulting in more dissolution of the conductive. Aluminium particles generates a layer over the work surface. This results in increase in the cutting force initially. It is observed that with rise in the depth of cut value the cutting force initially decreases and

Figure 5. S/N ratio plot for Fc

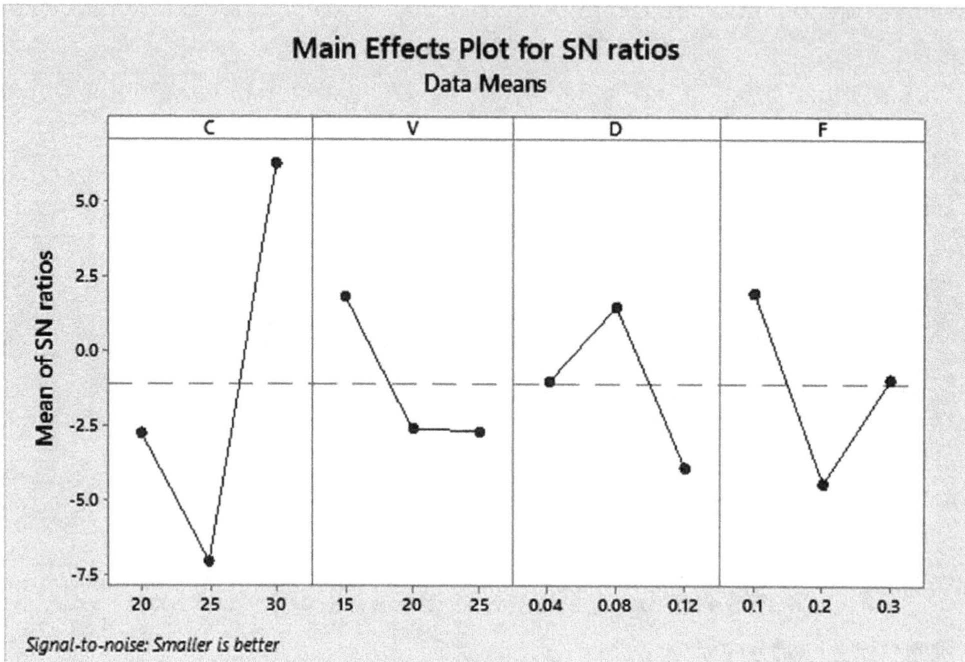

then increases. As the depth of cut surges to its maximum the space between the conductive grinding wheel and the work surface becomes little, as a result the tangential force increases. It is detected that with the increase in the electrolyte flow rate the cutting force initially rises to its maximum value but as the flow rate reaches its maximum more and more Aluminium gets softened and the work surface also gets washed resulting in less cutting force.

4. REGRESSION ANALYSIS

Developing a regression model that contains only a subset of the available regression involves two confliction objectives such as (Montogomery et al., 2003):

- Regression model has to include as many regressors as possible in order to make the resultant equation useful for predictive purpose.
- Regression model has to maintain as few regressors as possible due to the cost involved in obtaining and maintaining information on a large scale.

In this present experimentation backward elimination method is applied on L_{81} OA. Three graph generated from the regression calculation, namely histogram of residual, normal probably plot and residual versus fits plot.

Histogram of residuals is an exploratory tool to show general features of the data which incorporates typical values, dispersion, shape and also abnormal values in the data.

Normal Probability Plot is used to evaluate the fit of a distribution for the given data. It also estimates percentiles, and compare different sample distributions.

Residuals versus fits plot indicate a random pattern of residuals on both sides of 0. Fitted line plots graphically represent the fitted values for all x-values in the region of significance. These plots are a convenient way to compare fitted values to actual data values in order to assess model fit.

Analysis for MRR

The empirical model for MRR from L_{81} OA is shown in equation 3 below:

The estimated regression equation for MRR is:

$$MRR = -1.06129 + 0.045717C + 0.0245V + 6.7833D - 0.945F - 0.00111CV$$
$$-0.3075CD + 0.169CF + 0.07375VD - 0.0505VF - 13.3125DF \quad (3)$$

The regression analysis result is shown in Table 9. R-square as well as R-square (adjusted) assumes a value of 90.13% and 88.73% respectively, that implies the model is poised to explain 90.13% variability with process variable C, V, D and F. From the T values of the process variables, it can be concluded that C is the most significant process variables followed by D, V and F.

Figure 6, Figure 7 and Figure 8 represent the histogram of residual, normal probably plot and residual versus fits plot for MRR respectively. From Figure 6 Histogram for MRR shows a symmetric distribution. From Figure 7 the normal probability plot illustrates each value with respect to the proportion of values in the sample that are smaller than or equal to it, along a fitted distribution middle line. The points in this plot form a straight line for all the responses, since the residuals are normally distributed. As in the Residuals versus fits plot of MRR i.e Figure 8 has a recognizable pattern, so these model has room for improvement.

Table 9. Estimated regression coefficient for MRR

Term	Coef	SE Coef	T-Value	P-Value	VIF
Constant	-1.059	0.330	-3.20	0.002	
C	0.0456	0.0118	3.87	0.000	37.00
V	0.0244	0.0138	1.78	0.080	50.50
D	6.79	2.00	3.39	0.001	68.50
F	-0.950	0.801	-1.19	0.239	68.50
C*V	-0.001104	0.000474	-2.33	0.023	62.50
C*D	-0.3078	0.0592	-5.19	0.000	44.50
C*F	0.1692	0.0237	7.14	0.000	44.50
V*D	0.0735	0.0592	1.24	0.219	31.00
V*F	-0.0506	0.0237	-2.14	0.036	31.00
D*F	-13.28	2.96	-4.48	0.000	13.00

S=0.0710982; R-sq=90.13%; R-sq(adj)= 88.73%; R-sq(pred)= 85.96%

Figure 6. Histogram for MRR

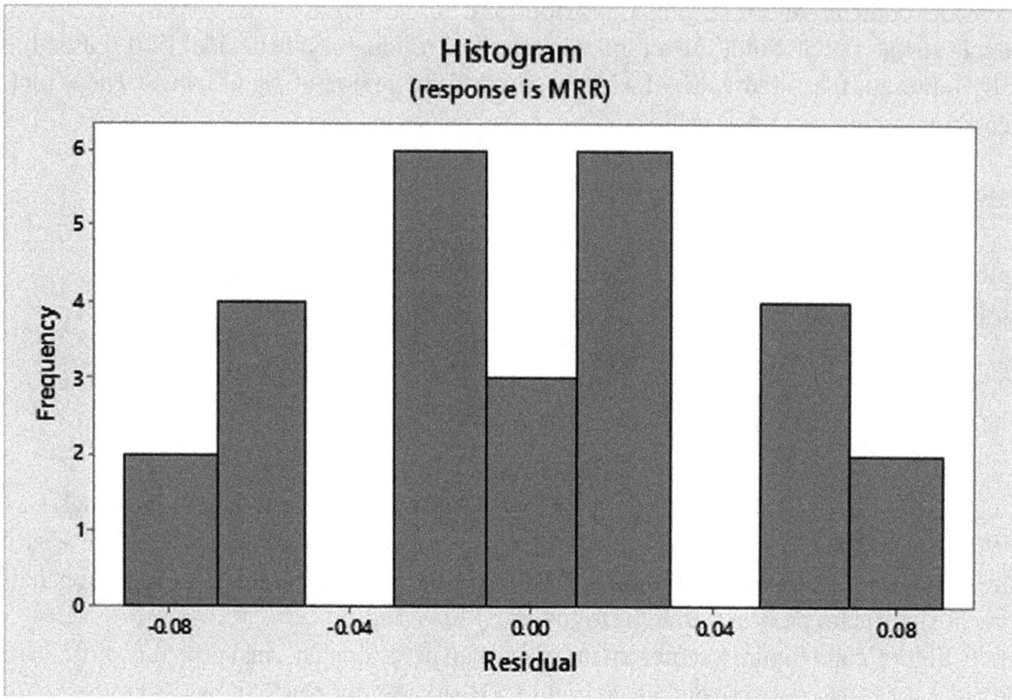

Figure 7. Normal Probability plot for MRR

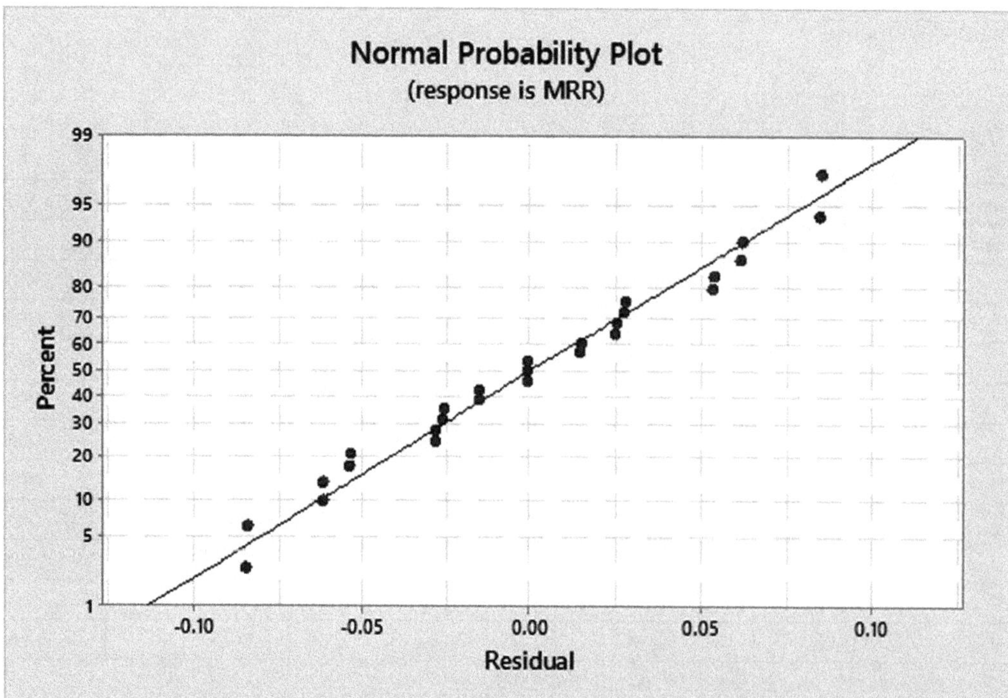

Figure 8. Residuals versus fits plot for MRR

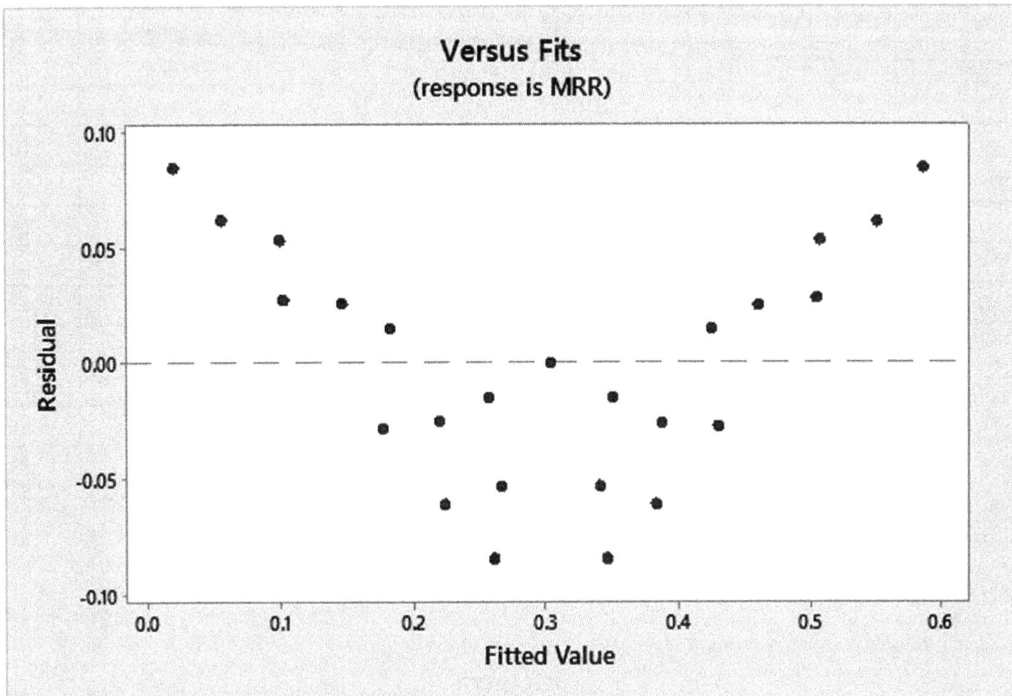

Analysis for Ra

The empirical model for Ra from L_{81} OA is shown in equation 4 below:

The estimated regression equation for Ra is:

$$Ra = -2.47969 + 0.089983C + 0.1312V + 2.1021D - 0.00083F - 0.00433CV \\ -0.365CD - 0.077CF + 0.50625VD + 0.1145VF + 8.625DF \quad (4)$$

The regression analysis result is shown in Table 10. R-square as well as R-square (adjusted) assumes a value of 93.78% and 92.89% respectively, that implies the model is poised to explain 93.78% variability with process variable C, V, D and F. From the T values of the process variables, it can be concluded that V is the most significant process variables followed by C, D and F.

Figure 9, Figure 10 and Figure 11 represent the histogram of residual, normal probably plot and residual versus fits plot for Ra respectively. From figure 9 Histogram of Ra shows a flat distribution. From Figure 10 the normal probability plot illustrates each value with respect to the proportion of values in the sample that are smaller than or equal to it, along a fitted distribution line. As the residuals are normally distributed the points in this plot form a straight line for all the responses. In Figure 11 Residuals versus fits plot of Ra has a recognizable pattern, so these model has room for improvement.

Table 10. Estimated regression coefficient for Ra

Term	Coef	SE Coef	T-Value	P-Value	VIF
Constant	-2.477	0.575	-4.31	0.000	
C	0.0899	0.0205	4.39	0.000	37.00
V	0.1311	0.0239	5.48	0.000	50.50
D	2.12	3.48	0.61	0.545	68.50
F	-0.01	1.39	-0.01	0.993	68.50
C*V	-0.004329	0.000825	-5.25	0.000	62.50
C*D	-0.365	0.103	-3.54	0.001	44.50
C*F	-0.0767	0.0412	-1.86	0.067	44.50
V*D	0.506	0.103	4.90	0.000	31.00
V*F	0.1148	0.0412	2.78	0.007	31.00
D*F	8.56	5.15	1.66	0.101	13.00

S=0.123687; R-sq=93.78%; R-sq(adj=92.89%; R-sq(pred)=90.95%

Figure 9. Histogram for Ra

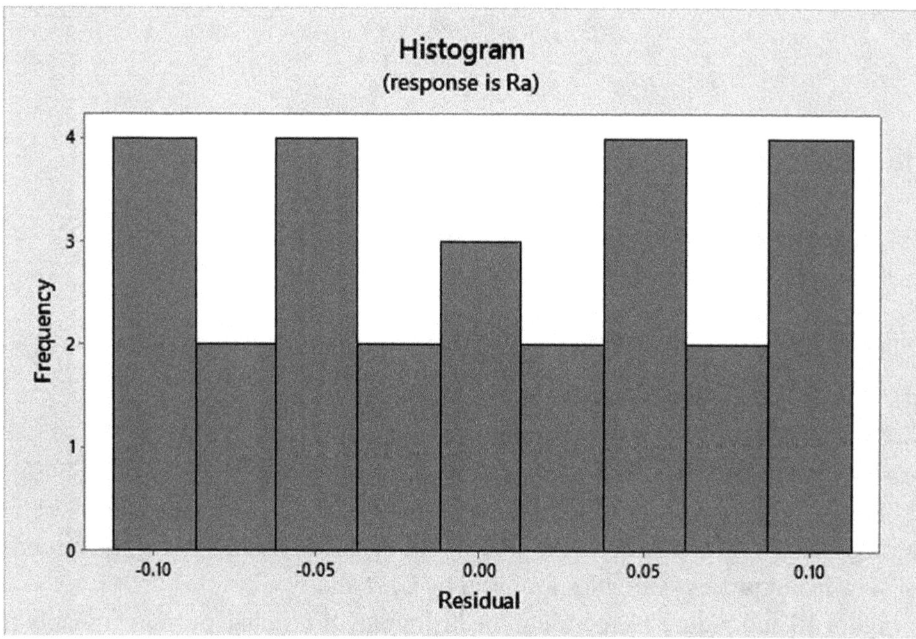

Analysis for OC

The empirical model for OC from L_{81} OA is shown in equation 5 below:

The estimated regression equation for OC is:

$$OC = -0.70444 + 0.050767C + 0.018533V - 0.0229D + 0.23917F - 0.00153CV \\ -0.12125CD + 0.027CF + 0.12VD - 0.027VF - 1.9375DF \quad (5)$$

Figure 10. Normal probability plot for Ra

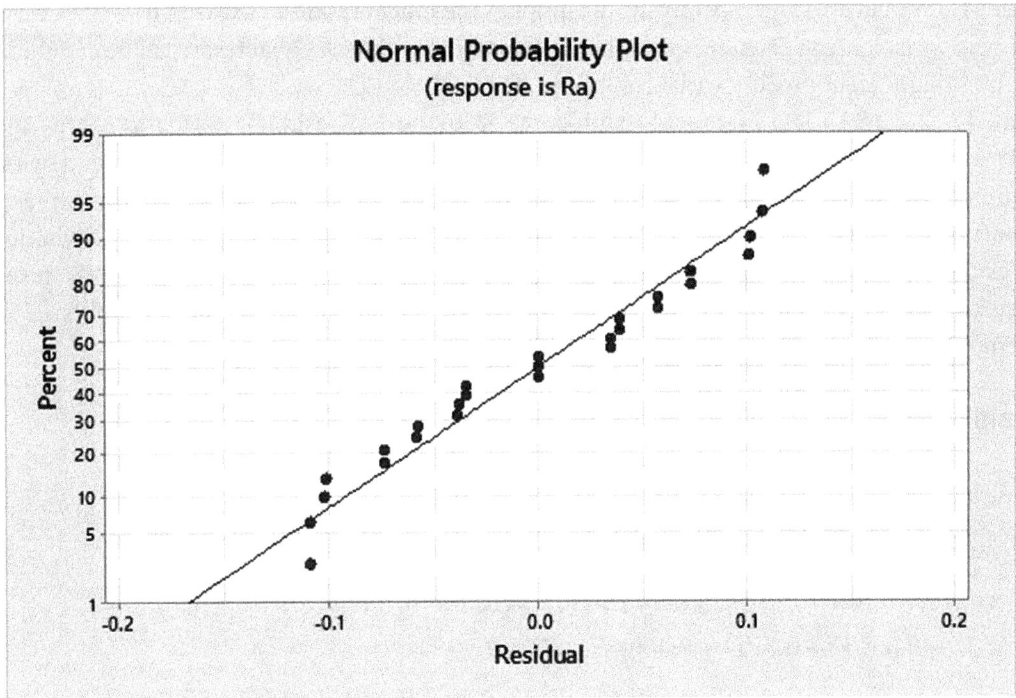

Figure 11. Residuals vs. fits plot for Ra

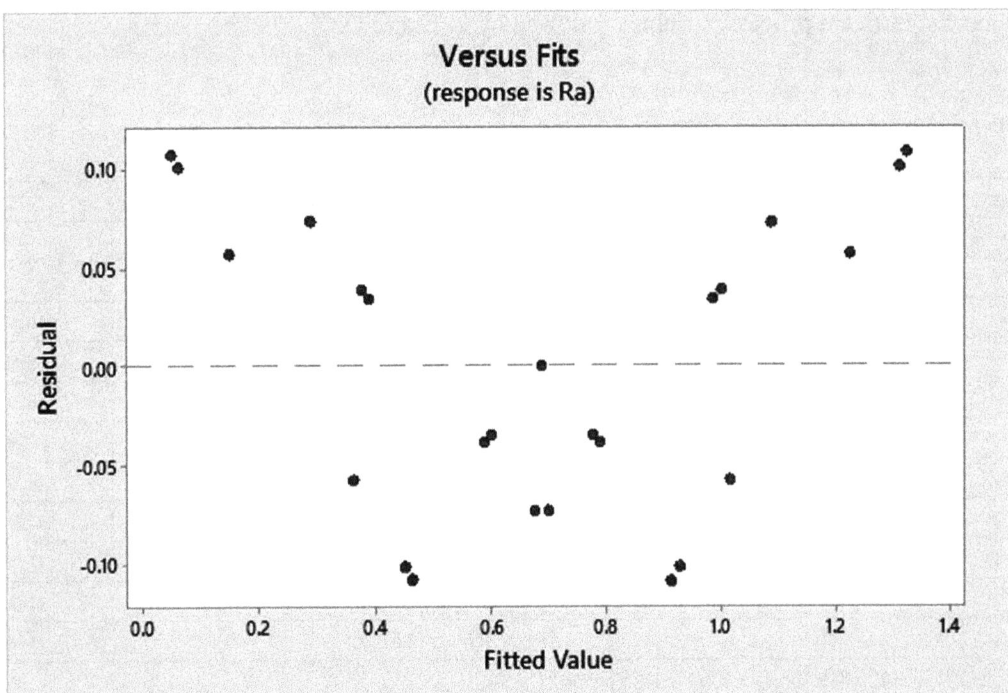

The regression analysis result is shown in Table 11. R-square as well as R-square (adjusted) assumes a value of 91.35% and 90.12% respectively, that implies the model is poised to explain 91.35% variability with process variable C, V, D and F. From the T values of the process variables, it can be concluded that C is the most significant process variables followed by V, F and D.

Figure 12, Figure 13 and Figure 14 exhibits the histogram of residual, normal probably plot and residual versus fits plot for OC respectively. From Figure 12 Histogram of OC also shows a symmetric distribution as MRR. From Figure 13 the normal probability plot illustrates each value with respect to the proportion of values in the sample that are smaller than or equal to it, along a fitted distribution line. As the residuals are normally distributed the points in this plot form a straight line for all the responses. In Figure 14 Residuals versus fits plot of OC has a recognizable pattern, so these model has room for improvement.

Analysis for Fc

The empirical model for Fc from L_{81} OA is shown in equation 6 below:

The estimated regression equation for Fc is:

$$\begin{aligned} Fc &= -3.6511 + 0.068C + 0.23683V - 1.917D + 6.358F - 0.0045CV \\ &-0.0875CD + 0.09CF + 0.3VD - 0.455VF - 6.25DF \end{aligned} \tag{6}$$

The regression analysis result is shown in Table 12. R-square as well as R-square (adjusted) assumes a value of 81.43% and 78.77% respectively, that implies the model is poised to explain 81.43% variability with process variable C, V, D and F. From the T values of the process variables, it can be concluded that V is the most significant process variables followed by F, C and D.

Table 11. Estimated regression coefficient for OC

Term	Coef	SE Coef	T-Value	P-Value	VIF
Constant	-0.703	0.161	-4.36	0.000	
C	0.05079	0.00575	8.83	0.000	37.00
V	0.01834	0.00672	2.73	0.008	50.50
D	-0.016	0.978	-0.02	0.987	68.50
F	0.236	0.391	0.60	0.549	68.50
C*V	-0.001528	0.000232	-6.60	0.000	62.50
C*D	-0.1218	0.0289	-4.21	0.000	44.50
C*F	0.0268	0.0116	2.31	0.024	44.50
V*D	0.1207	0.0289	4.17	0.000	31.00
V*F	-0.0264	0.0116	-2.28	0.025	31.00
D*F	-1.97	1.45	-1.36	0.177	13.00

S=0.0347292; R-sq=91.35%, R-sq(adj)= 90.12%; R-sq(pred)=87.62%

Figure 12. Histogram for OC

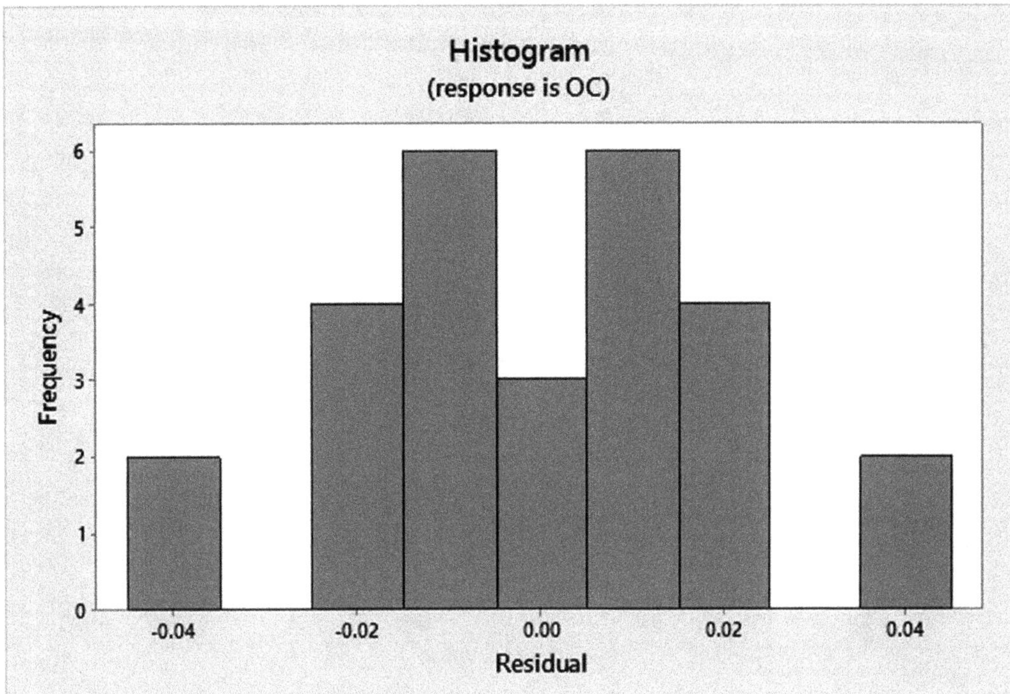

Figure 13. Normal probability plot for OC

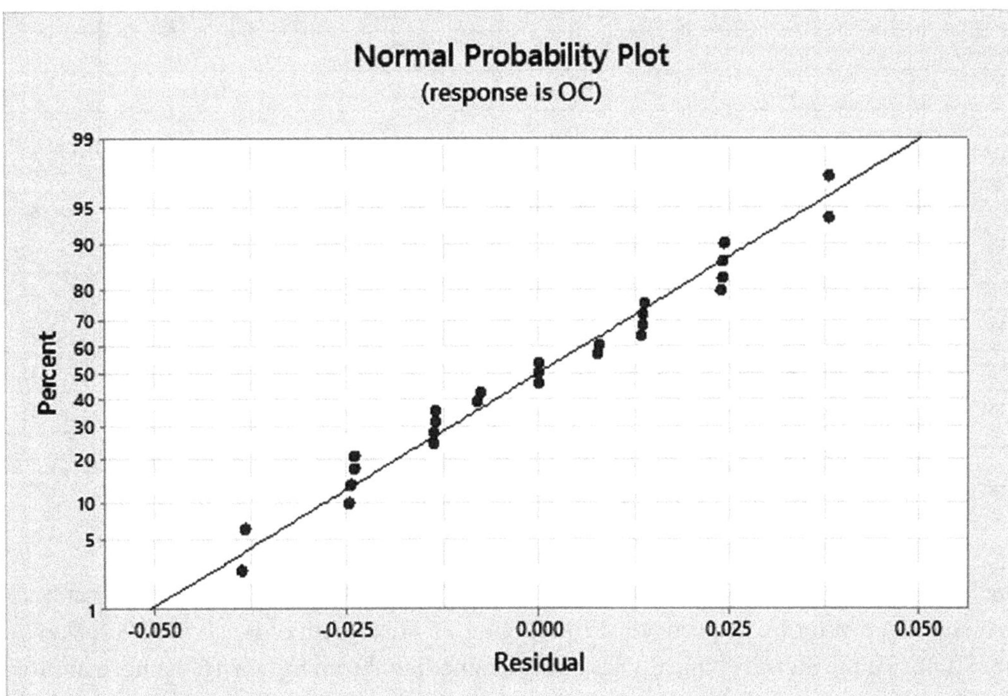

Figure 14. Residuals vs. fits plot for OC

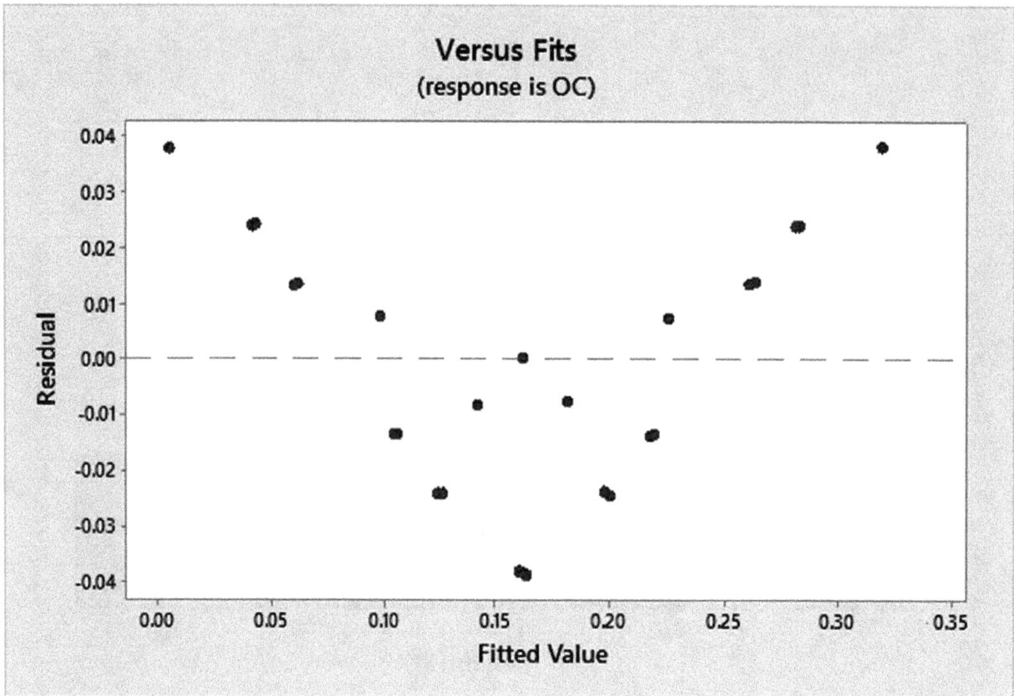

Table 12. Estimated regression coefficient for Fc

Term	Coef	SE Coef	T-Value	P-Value	VIF
Constant	-3.650	0.736	-4.96	0.000	
C	0.0687	0.0262	2.62	0.011	37.00
V	0.2368	0.0306	7.73	0.000	50.50
D	-2.25	4.46	-0.51	0.615	68.50
F	6.38	1.78	3.58	0.001	68.50
C*V	-0.00453	0.00106	-4.29	0.000	62.50
C*D	-0.081	0.132	-0.61	0.544	44.50
C*F	0.0867	0.0528	1.64	0.105	44.50
V*D	0.306	0.132	2.32	0.024	31.00
V*F	-0.4533	0.0528	-8.59	0.000	31.00
D*F	-5.83	6.60	-0.88	0.380	13.00

S=0.158373; R-sq=81.43%; R-sq(adj)= 78.77%; R-sq(pred)=75.51%

Figure 15, Figure 16 and Figure 17 represent the histogram of residual, normal probably plot and residual versus fits plot for Fc respectively. From Figure 15 Histogram of Fc shows a left skewed characteristic. So, for Fc the mean is typically less than the median. From Figure 16 the normal probability plot illustrates each value with respect to the proportion of values in the sample that are smaller than or equal to it, along a fitted distribution line. As the residuals are normally distributed the points in this plot

Figure 15. Histogram for Fc

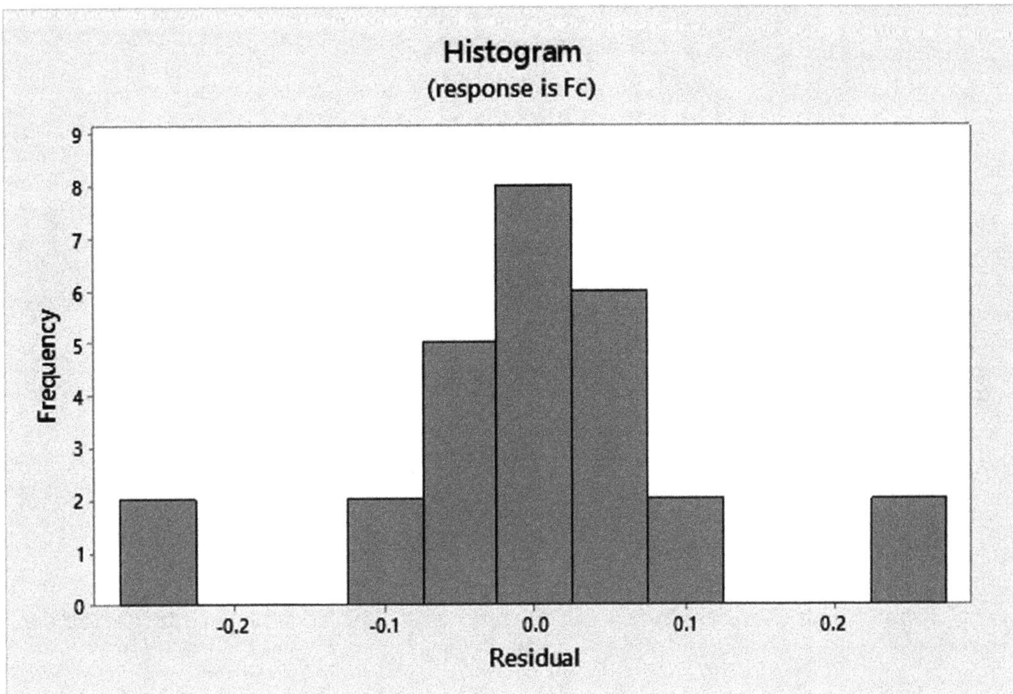

Figure 16. Normal probability plot for Fc

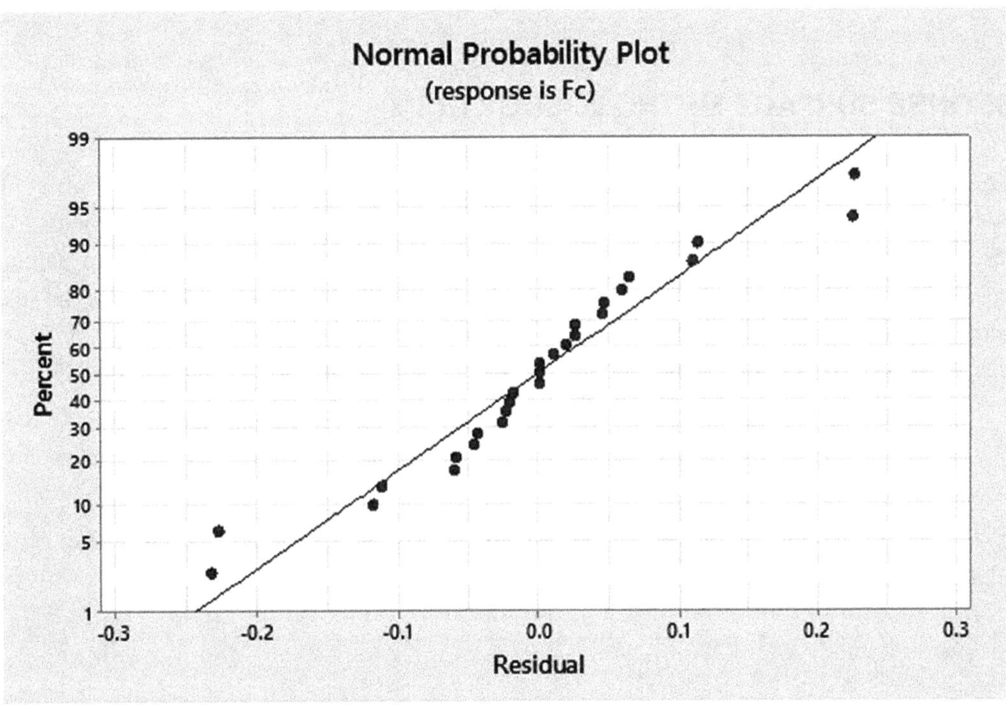

Figure 17. Residuals vs. fits plot for Fc

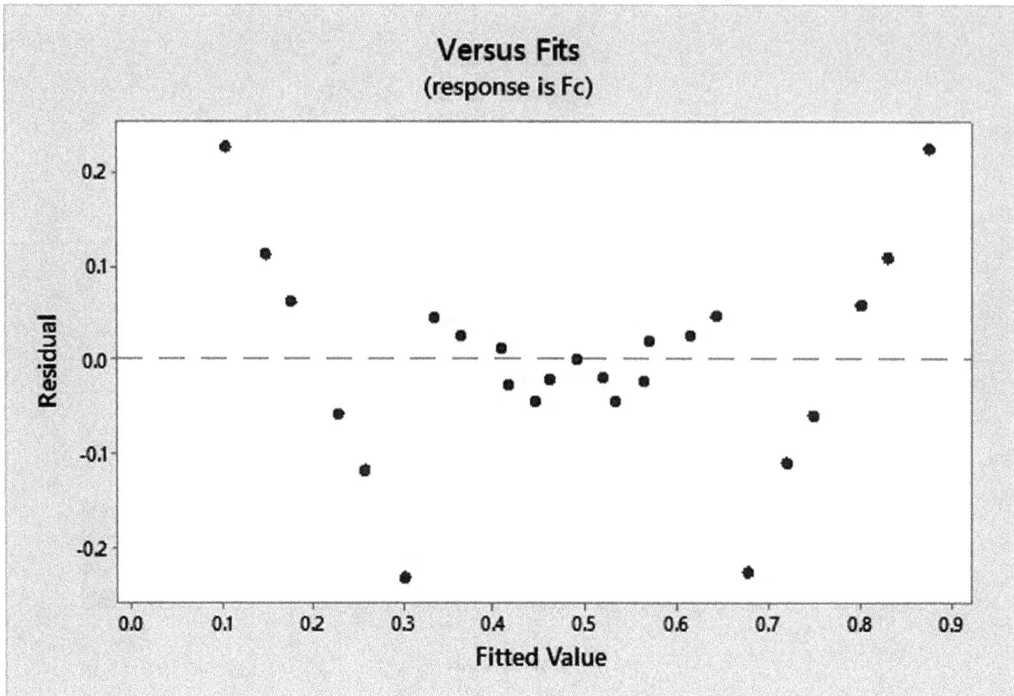

form a straight line for all the responses. In Figure 17 Residuals versus fits plot of Fc has a recognizable pattern, so these model has room for improvement.

5. RESPONSE SURFACE METHODOLOGY (RSM)

Response surface methodology (RSM) examines the relation between several control variables and one or more responses. The principal idea of RSM is to apply a sequence of designed experiments to obtain responses. The independent variables are governed by the experimenter, in a designed research, while the process response is an observed output during experimentation. Figure 18 illustrates the inscribed type response surface design which has 5 level of factors and over all accuracy is good.

The response surface (output) can be related with the number of controllable variables $X_1, X_2 \ldots\ldots X_k$ as

$$y = f(X_1, X_2 \ldots\ldots X_k) + \varepsilon \tag{7}$$

A second order mathematical model is developed to form an input-output relationship efficiently that takes the generic form:

$$y = \beta_0 + \sum_{i=1}^{k} \beta_i x_i + \sum_{i=1}^{k} \beta_{ii} x_i^2 + \sum_{i=1}^{k} \beta_{ij} x_i x_j + \varepsilon \tag{8}$$

Figure 18. Inscribed type response surface design

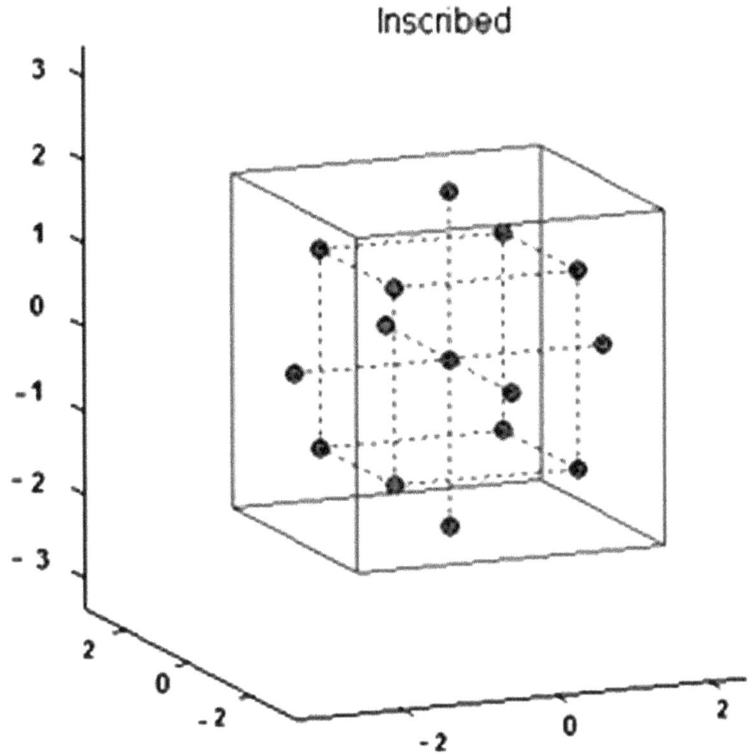

The calculated response for the model is

$$\hat{y} = \hat{\beta}_0 + \sum_{i=1}^{k} \hat{\beta}_i x_i + \sum_{i=1}^{k} \hat{\beta}_{ii} x_i^2 + \sum_{i=1}^{k} \hat{\beta}_{ij} x_i x \tag{9}$$

In the present research work, Box-Behenken Design is used which is based on 2^k (k = 4) factorials with incomplete designs and found to be very efficient. The process variables combinations and the corresponding responses are presented in Table 4.

Analysis of the Test Result for MRR

The response surface plots of MRR with respect to C, V, D and F are shown in Figure 19. In these plots, MRR is taken in the z-axis and at a time two control parameters are varied in x-axis and y-axis respectively, while other two parameters are kept constant. It is evident from the graph that the highest levels of the control parameters yield maximum response. Since the response is dependent on the variables therefore they cannot have any stationary point which is evident from the surface plot.

Figure 19. Surface plot for MRR

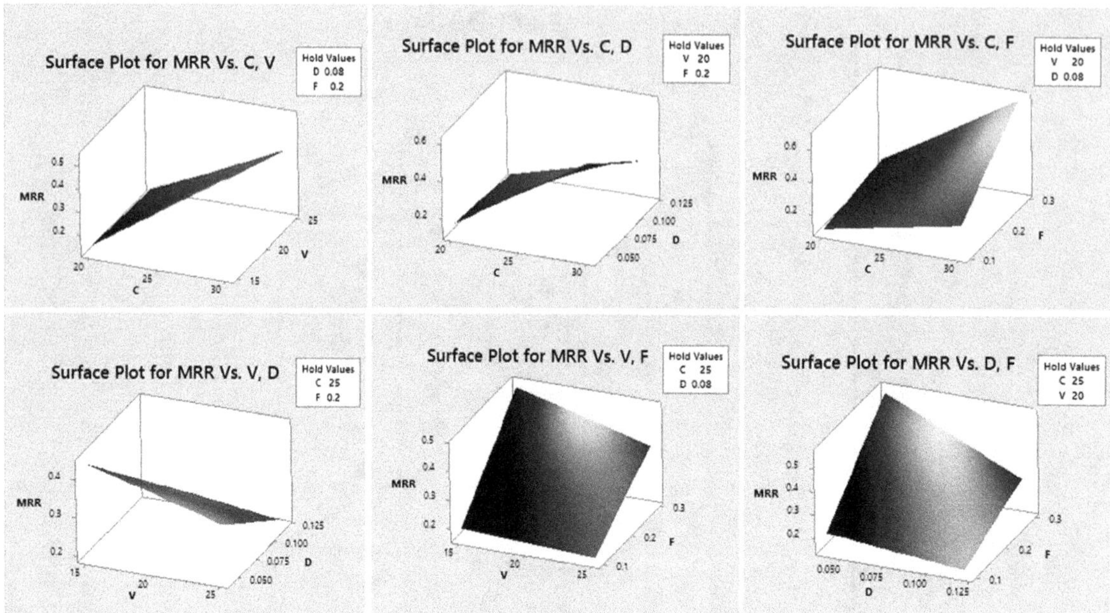

Analysis of the Test Result for Ra

The response surface plots of Ra with respect to C, V, D and F are shown in Figure 20. In these plots the variation of Ra is taken in the z-axis and at a time two control parameters are considered in x-axis and y-axis respectively, while other two parameters are kept constant. The graph depicts that the highest levels of the control parameters generate maximum response. Since the response is reliant on the variables therefore they cannot have any stationary point which is evident from the surface plot.

Analysis of the Test Result for OC

The response surface plots of OC with respect to C, V, D and F are shown in Figure 21. In these plots, the variation of OC is shown in the z-axis and two control parameters are taken in x-axis and y-axis respectively, while other two parameters are kept constant. The graph manifest that the highest levels of the control parameters produce maximum response. Subsequently the response is supported by the variables therefore they cannot have any stationary point which is evident from the surface plot (Lee and Park, 2015).

Analysis of the Test Result for Fc

The response surface plots of Fc with respect to C, V, D and F are shown in Figure 22. In these plots, Fc is considered in the z-axis and at a time two control parameters are varied in x-axis and y-axis respectively, while remaining two parameters are kept constant. The graph illustrates that the highest levels of the control parameters generate maximum response. Since the response is supported by the variables therefore they cannot have any stationary point which is evident from the surface plot.

Figure 20. Surface plot for Ra

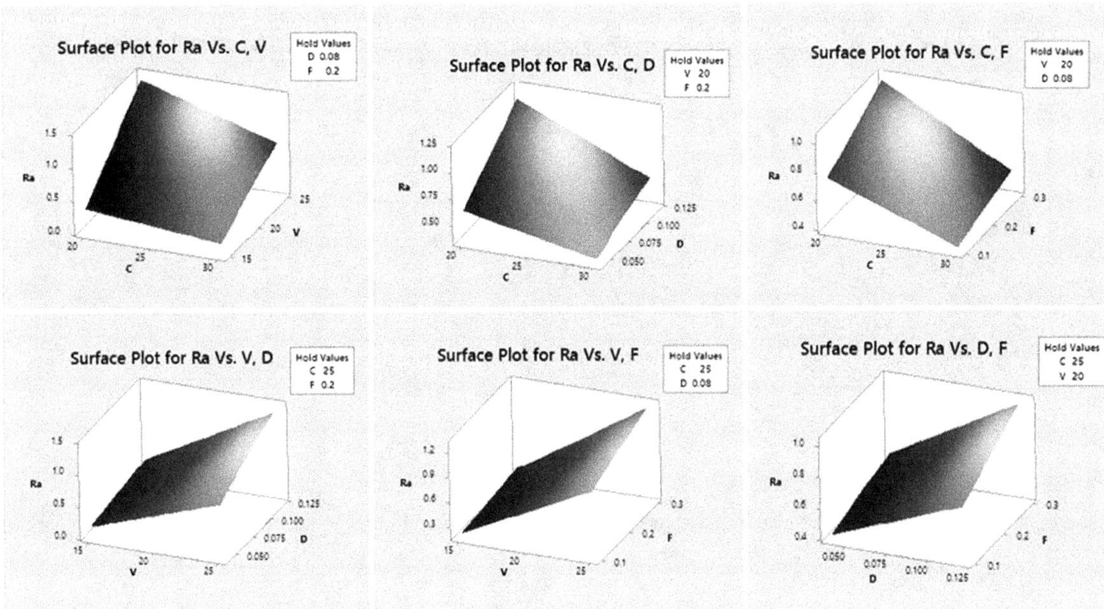

Figure 21. Surface plot for OC

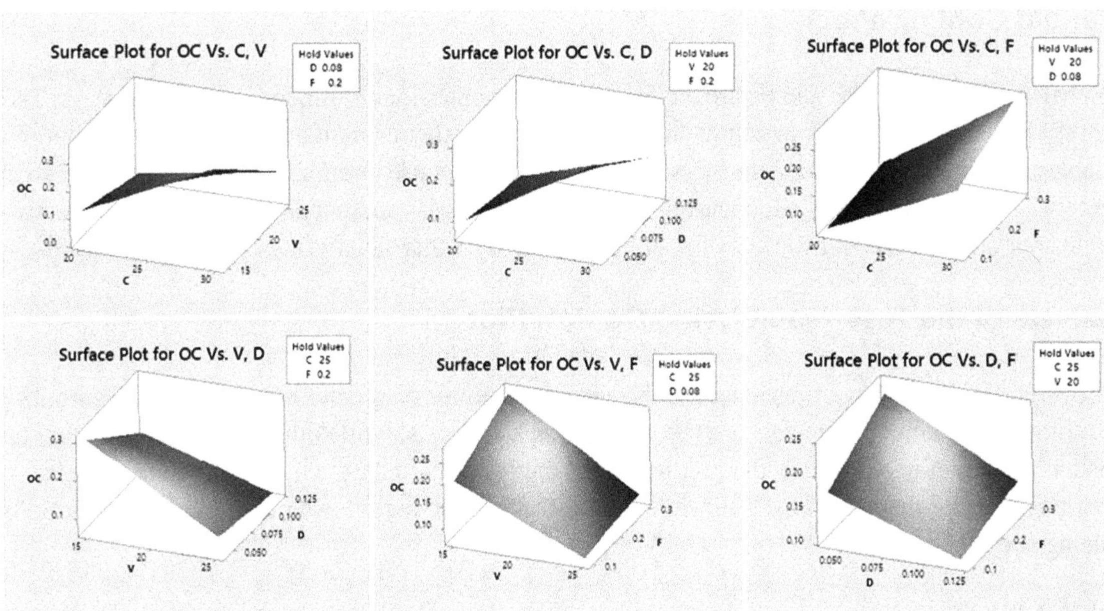

Figure 22. Surface plot for Fc

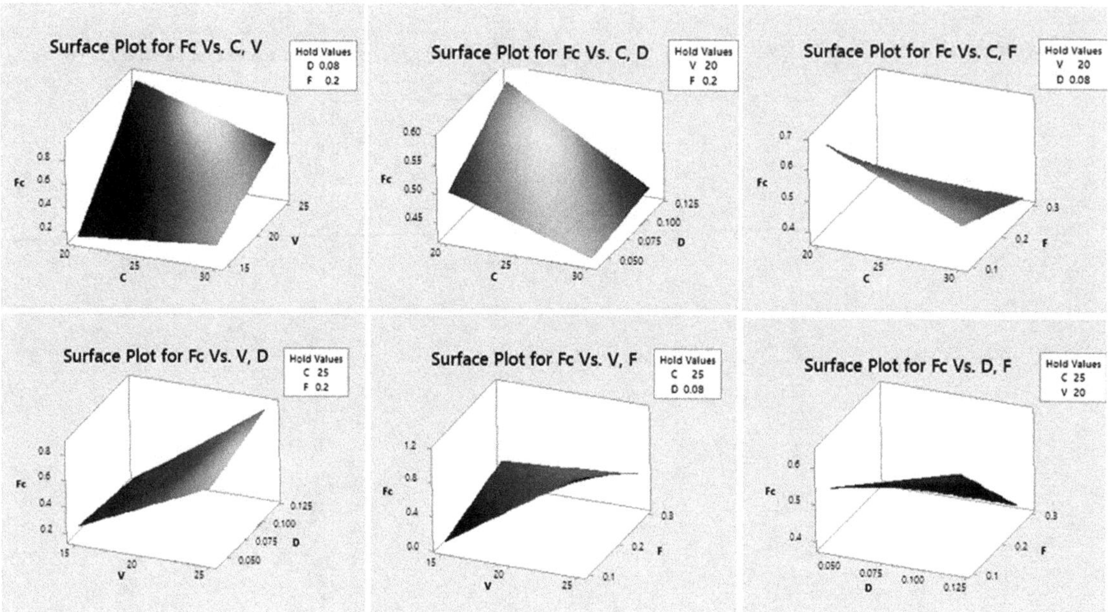

6. MULTI-RESPONSE OPTIMIZATION

Overlaid Contour Plots

High MRR and low Ra, OC and Fc are the four major attributes for optimizing the ECG process. These four responses are conflicting in nature and hence achieving them simultaneously by a set of optimum variables combination is difficult. In this section the multi response optimization is conceded out so that four conflicting goals are fulfilled concurrently. We resort to overlay contour plots which are comparatively simple approach to review the levels of operating parameters that satisfy four constrained objectives.

Analysis of the Test Result Using Control Plot

The overlaid control plots of responses with respect to control parameters are shown in Figure 23. In these plots two control parameters are taken in x-axis and y-axis, while other two parameters are kept constant. The continuous line in the plot represent the lower boundary while dotted line represent the upper boundary of the responses. The white area is the feasible region, where all the responses are optimizing simultaneously.

Desirability Functions

Response Optimizer helps to help recognize the factor settings which optimize a single response or a set of responses. For multiple responses, the necessities for all the responses in the set must be fulfilled (Cho and Park, 2006; Jeong & Kim, 2009). The general approach is to first convert each response into an individual desirability function that varies over the range (0 to 1) (Salmasnia, 2013). Here the goal,

Figure 23. Control plots of responses

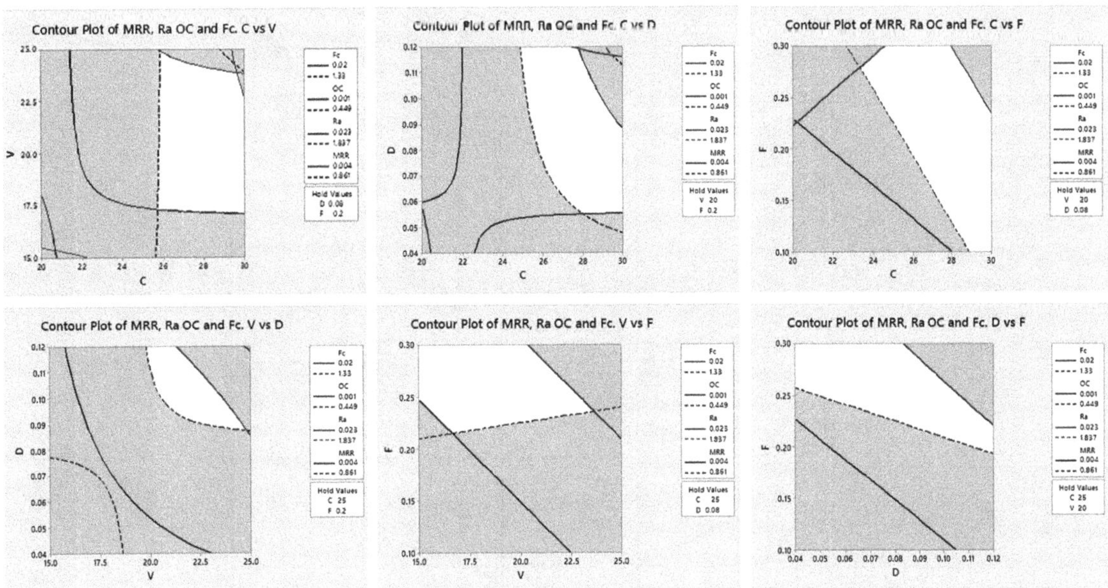

lower, target, upper, and weight characterize the desirability function for every response. The importance (Import) parameters decide how the desirability functions are combined into a single composite desirability. The overall desirability and composite desirability can be computed as the geometric mean of the individual desirability.

Analysis of the Test Result Using Desirability Functions

The response optimization is shown in Table 13.

Figure 24 represents the optimization plot of the responses (MRR, Ra, OC and Fc) with the process variables. It shows how the factors affect the predicted responses and allows to modify the factor settings interactively. The desirability of MRR becomes 100% having the predicted response at = 0.6724

Table 13. Desirability function result

Parameters	Goal	Lower	Upper	Weight
MRR	Maximum	20	30	1
Ra	Minimum	15	25	1
OC	Minimum	0.04	0.12	1
Fc	Minimum	0.10	0.30	1
Predicted Responses				**Global Solution**
MRR = 0.6724, desirability = 1.0000 (100%) Ra =0.8030, desirability = 1.0000 (100%) OC = 0.1719, desirability = 0.83022 (83.022%) Fc = 0.305, desirability = 0.88433 (88.433%) Composite Desirability = 0.9257 (92.57%)				C=29.9980 V=25 D=0.0602 F=0.30

Figure 24. Plot showing responses (MRR, Ra, OC and Fc) vs. process variables

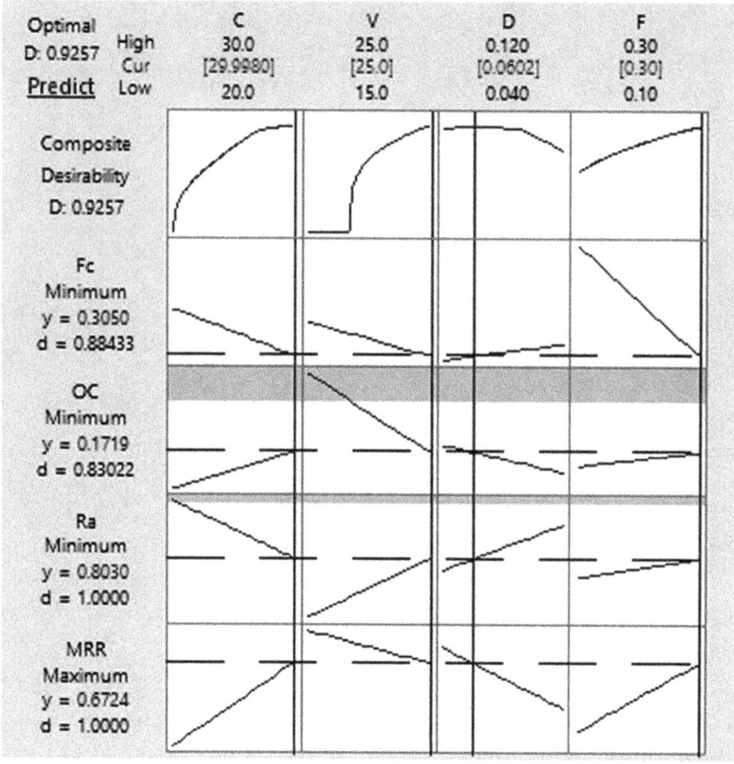

gm/min, similarly the desirability of Ra becomes 100% having the predicted response at 0.8030 μm, the desirability of OC becomes 83.022% having the predicted response 0.1719 mm, and the desirability of Fc becomes 88.433% having the predicted response 0.305 kgf. Hence composite desirability is 92.57% having the paramedic combination of C = 29.9980 gm/lit; V = 25 volts; D = 0.0602 mm; F = 0.30 lit/sec for multi objective optimal solution for higher MRR, and lower Ra, OC, and Fc.

7. CONCLUSION

The present paper attempts to validate and optimize the machining variables while machining of Alumina-Aluminium IPC by ECG. While analysing the individual responses applying the Taguchi Methodology it is found that the MRR, OC and Fc is mostly influence by electrolyte concentration (C) but, Ra is mostly guided by the voltage (V). Also from the S/N ration analysis the following parametric combinations are found: Maximum MRR is obtained at C 30 gm/lit, V 20 volts, D 0.12 mm and F 0.2 lit/sec. Minimum Ra is obtained at C 30 gm/lit, V 15 volts, D 0.08 mm and F 0.3 lit/sec. Lower OC is obtained at C 25 gm/lit, V 20 volts, D 0.04 mm and F 0.2 lit/sec. Minimum Fc is obtained at C 30 gm/lit, V 15 volts, D 0.08 mm and F 0.1 lit/sec.

The regression models are generated to be creditable to express input-output relationship with a very high degree of predictability. It is found that the MRR and OC is mostly influence by electrolyte concentration (C) but, Ra is typically guided by the voltage (V). Histogram for MRR shows a symmetric distribution. In case of Ra this graph shows a flat distribution. For OC histogram also shows a symmetric distribution. But in case of Fc, histogram shows a left skewed characteristic. So, for Fc the mean is typically less than the median. The points in the normal probability plot form a straight line for all the responses, since the residuals are normally distributed. For each residual versus response plots have recognizable pattern, so these model has room for improvement. The optimization is further carried out using RSM that is promised to tender near optimal solution with modest effort. Here the multi objective optimization is done using overlaid control plot. The overlaid contour plot is a decent visual utility to recognize the feasible region in regard to a set of input variables. The feasible region is obtained for MRR is 0.861 gm/min to 0.004 gm/min, Ra is 1.837 μm to 0.023 μm, OC is 0.449 mm to 0.001 mm and Fc is 1.33 kgf to 0.2 kgf. The interpretations drawn from the regression analysis is accentuated with the desirability functions. In order to achieve high MRR, and lower Ra, OC, and Fc the possible parametric combination is C = 30 gm/lit; V = 25 volts; D = 0.06 mm; F = 0.30 lit/sec.

Therefore, this experimental analysis for considering the optimal ECG parametric combination while machining of Al_2O_3/Al interpenetrating phase composites material can act as important and an operative guideline for manufacturing of products of similar material. This set of inputs can be used for further optimization of other functions like machining cost and therefore can act as a back bone of adaptive control strategies.

REFERENCES

Atkinson, J. (2007). Work piece Surface Hardness as an Indicator of Process Regime in Peripheral Electrochemical Grinding. In *Proceedings of the 35th International MATADOR Conference* (pp. 89-94). London: Springer London. doi:10.1007/978-1-84628-988-0_20

Bhandari, S., & Shukla, N. (2015). Parametric Optimization of Electrochemical Grinding Operation by Particle Swarm Optimization Technique. *SSRG International Journal of Mechanical Engineering, 2*(4), 1-6.

Cho, Y. H., & Park, S. H. (2006). Optimization in Multi Response Model with Modification Desirability Function. *Asian Journal on Quality, 7*(3), 46–57. doi:10.1108/15982688200600026

El-Hoffy, H. (2007). *Fundamentals of Machining Process.* CRC Press Taylor & Francis.

Geva, M., Lenz, E., & Nadiv, S. (1976). Peripheral electrochemical grinding of sintered carbides -Effect on surface finish. Technion — Israel Institute of Technology, 38(2), 325-339.

Goswami, R. N., Mitra, S., & Sarkar, S. (2009). Experimental investigation on electrochemical grinding (ECG) of alumina-aluminum interpenetrating phase composite. *International Journal of Advanced Manufacturing Technology, 40*(7), 729–741. doi:10.1007/s00170-008-1382-z

Jeong, I. J., & Kim, K. J. (2009). An interactive desirability function method to multi-response optimization. *European Journal of Operational Research, 195*(2), 412–426. doi:10.1016/j.ejor.2008.02.018

Kaczmarek, J., & Zachwieja, T. (1966). Investigation on the material removal rate by electrochemical grinding of cutting tool materials in dependence on the properties of the grinding wheel. *International Journal of Machining Tool Design and Research, 6*(1), 1–13. doi:10.1016/0020-7357(66)90002-3

Kaldos, E. (1974). A study of electrochemical grinding of small diameter specimens made of tungsten carbide. *Periodica Polytechnica*, *8*, 21–32.

Lee, P., & Park, K. S. (2015). An Interactive Approach to Multiple Response Optimization. *Journal of the Korean Operations Research and Management Science Society*, *40*(3), 49–61. doi:10.7737/JKORMS.2015.40.3.049

Levinger, R., & Malkin, S. (1979). Electrochemical Grinding of WC-Co Cemented Carbides. *Journal of Engineering for Industry*, *101*(3), 285. doi:10.1115/1.3439509

Molla, K. Z. & Manna, A. (2013). Optimization of Electrochemical Grinding Parameters for Effective Finishing of Hybrid Al/(Al2O3+ZrO2) MMC. *International Journal of Surface Engineering and Interdisciplinary Material Science*, *1*(2), 35-45.

Montogomery, D. C., Peck, E. A., & Vining, G. G. (2003). *Introduction to linear regression analysis* (3rd ed.). New York: Wiley.

Park, S. H. (1996). *Robust Design and Analysis for Quality Engineering*. Chapman & Hall.

Puri, A. B., & Banerjee, S. (2013). Multiple-response optimization of electrochemical grinding characteristics through response surface methodology. *International Journal of Advanced Manufacturing Technology*, *64*(5), 715–725.

Qu, N. S., Zhang, Q. L., Fang, X. L., Ye, E. K., & Zhu, D. (2015). Experimental Investigation on Electrochemical Grinding of Inconel 718.*15th Machining Innovations Conference for aerospace Industry*, *35*,16-19. doi:10.1016/j.procir.2015.08.055

Reddy, H. P., Ramamoorthy, K. B., & Nair, K. P. (1998). Estimation of Optimum Grinding Conditions to improve the quality of the Ground Surface by Applying Taguchi Techniques.*International Conference on CAD/CAM, Robotics & Factories of Future, India*, 473-479.

Reddy, H. P., Ramamoorthy, K. B., & Nair, K. P. (1999). Investigation on the influence of grinding parameters on the Microstrain of Medium Carbon Steel.*ASME Mechanics and Materials Conference*.

Roy, S., Bhattacharrya, A., & Banerjee, S. (2007). Analysis of effect of voltage on surface texture in electrochemical grinding by autocorrelation function. *Tribology International*, *40*(9), 1387–1393. doi:10.1016/j.triboint.2007.03.008

Salmasnia, A., Bashiri, M., & Salehi, M. (2013). *A robust interactive approach to optimize correlated multiple responses*. Academic Press.

Senthilkumar, C., Ganesan, G., & Karthikeyan, R. (2011). *Parametric optimization of electrochemical machining of Al/15%SiCp composite using NSGA-II*. Academic Press.

Section 2

Chapter 9
Investigations on Machinability Characteristics of Hardened AISI H13 Steel With Multilayer Coated Carbide Tool Using Statistical Techniques

R. Suresh
M. S. Ramaiah University of Applied Sciences, India

Ajith G. Joshi
Canara Engineering College, India

ABSTRACT

Hard turning with multilayer coated carbide tool has several benefits over grinding process such as, reduction of processing cost and increased productivity. The objective was to establish a correlation between cutting parameters with cutting force, tool wear and surface roughness on workpiece. In the present study, machinability of AISI H13 steel with TiC/TiCN/Al2O3 coated carbide tool using statistical techniques. An attempt has been made to analyze the effects of process parameters on machinability aspects using design of experiments. Response surface plots are generated for the study of interaction effects of cutting conditions on machinability factors. The obtained results revealed that, the optimal combination of low feed rate and low depth of cut with high cutting speed is beneficial for reducing machining force. The cutting tool wears increases almost linearly with increase in cutting speed and feed rate. The combination of low feed rate and high cutting speed is necessary for minimizing the surface roughness.

1. INTRODUCTION

1.1. Hard Turning

Hard machining is a term referred to machining of hardened alloys possessing hardness generally greater than 45 HRc. Generally the alloys which are having hardness in the range of 45-65 HRc are considered in many engineering applications such as automotive, power plants, etc. They are gaining prime signifi-

DOI: 10.4018/978-1-5225-2440-3.ch009

cance due to their properties such as excellent tribological characteristics, better fatigue strength and so on (Suresh et al., 2013). Heat treated ferrous alloys, usually owes to the hardness in this range when they are in order to obtain desired mechanical properties combined with surface integrity, dimensional accuracy and shape. Stellites, Inconel, high chrome coating, carburized and nitride iron are other candidates of hard materials (Mohandas et al., 2012).

The challenges are still being machining of such hardened alloys. Hard turning is being considered as one of the significant machining technology for machining hardened alloys. The present developments in manufacturing industries are towards replacing grinding of hardened steels with hard turning due to capabilities such as reduced costly equipments, long setup time and manufacturing cycles. In specific, finish hard turning is the evolving technique towards replacement of grinding and electrical discharge machining for manufacturing components which requires greater tolerances and surface finish such as bearings, transmission shafts, axles, etc. The major advantages of hard turning over grinding are complex contour shape, the comparable surface finish and dry machining can be achieved. While, hard turning is a technology driven process that requires performance characteristics of machine tool such as rigidity, good work clamping devices, tooling systems and so on. Hence hard turning cannot be accomplished on all types of machine tools that limit the application of hard turning. However, technology is being developed and imparted in present manufacturing CNC lathes with greater rigidity required for hard turning.

The fundamental issues related to hard turning such as rapid tool wear and surface integrity still exists, which needs attention to make it feasible process. Since, only very hard and advanced tools are used in the process that adds to the cost of tools subsequently large down time due to frequent tool change. Hard and brittle white layer occurrence is another major issue that causes component to get premature failure or affects the service performance. The white layer formation occurs because of tensile residual stress over the machined surface and subsurface. Thus, white layer formation favors premature failure of components due to reduced fatigue strength at points such as subsurface voids, cracks or inclusions. While the compressive residual stress arrests cracks formation and growth at such subsurface defects. Normally hard turning is employed at high cutting speeds. In such condition, both material as well as tool experiences extreme conditions of high strain rate of order 10^5 and high temperature around 1200° C (Ashley, 1995, Aouici & Yallese, 2013). Hence cutting tool and optimized process parameter selection are two major problems which are focused in the current chapter.

1.2. Literature Survey

Aouici et al. (2013) studied the influence of cutting speed, feed, workpiece hardness and depth of cut on cutting force components and surface roughness during machining of hardened AISI H11steel using CBN tools. The relationship between the variables and the technological parameters was determined through the response surface methodology (RSM), using a quadratic regression model. Caydas et al. (2010) carried out machinability evaluation in dry turning of hardened AISI 4340 steel with different cutting tools using statistical techniques. A detailed statistical investigation was made by using analysis of variance, Tukey—Kramer comparison, and correlation tests. The best combination of cutting parameters for determining optimal conditions and the influence of each factor on the response were revealed.

Sahoo and Sahoo (2013) studied the performance of coated and uncoated carbide inserts on surface roughness, flank wear, cutting forces and chip morphology during hard turning of AISI 4340 steel using response surface methodology and grey-based Taguchi method. It was foundthat the economic feasibility of utilizing of multilayer coated carbide inserts was better than uncoated carbide inserts and also model adequacy was checked using correlation coefficients. Suresh et al. (2012) presented the mathematical models for modeling and analyzing surface roughness, tool wear, cutting force, machine power during machining of AISI4340 steel with cemented carbide tools. Empirical models have been developed using analysis of variance (ANOVA). In order to analyze the response of the system, experiments were carried out for various cutting speeds, depths of cut and feed rates. To increase the confidence limit and reliability of the experimental data, full factorial experimental design has been carried out.

Bouacha et al. (2010) examined the effect of dry cutting on hardened AISI 52100 bearing steel using CBN cutting tools. The relationship between cutting parameters (cutting speed, feed rate and depth of cut) and machining output variables (surface roughness, cutting forces) through the response surface methodology (RSM) were analyzed and modeled. The combined effects of the cutting parameters on machining output variables are investigated while employing the analysis of variance (ANOVA). The quadratic model of RSM associated with response optimization technique and composite desirability was used to find optimum values of machining parameters with respect to objectives (surface roughness and cutting force values).

Gaitonde et al. (2009) analyze the effects of depth of cut and machining time on machinability aspects such as machining force, power, specific cutting force, surface roughness and tool wear using second order model and artificial neural network models during turning of high chromium AISI D2 cold work tool steel with ceramic inserts. The experiments were planned as per full factorial design (FFD). From the parametric analysis, it is revealed that, the wiper insert performs better with reference to surface roughness and tool wear.

Waleed et al. (2016) investigated the effect of process parameters on surface roughness and tool flank wear in hard turning of AISI 4340 steel (69 HRC) with CBN cutting tool using parametric design optimization techniques. The model have been developed using signal to noise (S/N) ratio analysis, analysis of variance (ANOVA), and multiple regression analysis were carried out on the experimental datasets to assert the dominance of each machining variable in dictating the machined surface roughness and tool wear to optimize the machining parameters.

Batish et al. (2014) studied the effect of different process parameters on machining forces, surface roughness, dimensional deviation and material removal rate during hard turning of EN31, SAE8620 and EN9 tool steels using genetic algorithm to determine the optimum combination of input parameters. Chandrasekaran et al. (2010) reviewed number of soft computing tools viz. neural networks, fuzzy sets, genetic algorithms, simulated annealing, ant colony optimization, and particle swarm optimization, all of which can conveniently be applied to the machining process depending on the complexity of the variable involved.

From the above literature it can be revealed that the hard-turning process, cutting conditions determine the cutting time and cost of production, which ultimately affect the quality of the final product. Optimization of the machining process not only improved the overall machining economics, but also the product quality to a great extent.

2. MATERIALS AND EXPERIMENTATION

2.1. Materials

The major part of industrial applications for manufacturing of dies, moulds, machine tool and automotive components includes work hardened tool steels, high speed steels, bearing steels, die steels and case hardened steels. Normally the hardness of such steels ranges more than 45 HRc with high indentation resistance, high abrasiveness, low ductility and high value of hardness to modulus of elasticity ratio (Nakayama et al., 1998). In current study, the AISI H13 hot work tool steel is used as workpiece material. AISI H13 die steel is chromium-molybdenum-vanadium alloyed tool steel. The AISI H13 steel is widely used in the production of forging dies, hot extrusion dies, hot punches, jigs and fixtures, backup and support tools. The composition and properties of AISI H13 steel is illustrated in Table 1 and 2 respectively. The chemical composition of the workpiece was determined through an optical emission vacuum spectrometer.

For machining test run, workpiece materials of 100 mm diameter and 400 mm length (L) of 10 pieces were selected from a single bar (L/D ratio is 1:4). The L/D ratio at the end of experimental work was approximately 8 in order to meet the requirement of ISO 3685, that assure the required stiffness of the chuck/workpiece/cutting tool system. Each workpiece was pre-machined and center drilled on one side. It was necessary in order to support the workpiece from both sides while turning on CNC lathe and in turn, reducing the vibration of workpiece materials and minimizing any impact forces on the cutting tool. The workpiece material was heat treated at 1025 °C followed by oil quenching and later tempered in the range 395 °C for 2 hrs to achieve better combination of toughness level and high hardness of 54 HRc.

2.2. Cutting Tool Materials and Geometry

The characteristics of hardened steel material have to be considered, subsequently careful selection of cutting tool material is necessary to achieve better machinability. The complexity lies with properties such as high hardness to modulus ratio, high thermal conductivity, high abrasive wear resistance and high thermal, physical and chemical stability (Tonshoff et al., 2000). Generally advanced tool materials such as coated carbide tools, ceramic, cubic boron nitride (CBN) and polycrystalline CBN tools are widely utilized. In current study, the coated carbide inserts of ISO geometry CNMG 120408 were used for machining. The multilayer coated carbide tool with CVD (chemical vapor deposition) coating of TiN/TiCN/Al_2O_3 on cemented carbide substrate was chosen. The inserts were mounted on PCLNL2525 M12 (ISO) tool holder that leads to geometry as follows: rake angle = -6°; clearance angle = -6°; included angle = 80°; approach angle = 95°; nose radius = 0.8 mm.

Table 1. Showing the chemical composition of H13 tool steel

Fe	**C**	**Si**	**Mn**	**P**	**S**	**Cr**	**Ni**	**Mo**	**V**	**Hardness (HRc)**
90.06	0.4	1.05	0.35	0.03	0.003	5.3	0.3	1.4	1	54

Table 2. Illustrates the properties of AISI H13 steel

Sl. No.	Properties	Values
1	Density (kg/m^3)	7800
2	Poisson's Ratio	0.3
3	Ultimate tensile strength (MPa)	1990
4	Yield tensile strength (MPa)	1650
5	Modulus of Elasticity (GPa)	210
6	Thermal Expansion (10^{-6}/°C)	10.4
7	Thermal Conductivity (W/m-K)	28.6
8	Specific Heat (J/m^3-°C)	3124025

2.3. Experimentation Details

Generally minimum quality lubrication and dry cutting conditions are utilized and also favorable for machining of hard turning process. In current study, dry cutting condition was adapted for machining. It is employed on CNC lathe of 22kW spindle power and maximum 5000 rpm spindle speed. Axial and run out was evaluated on machine, which was in the acceptance error limit. During test run, cutting force was recorded using Kistler Model 9263A piezo-electric dynamometer coupled with suitable load amplifier and data acquisition software. They allow for continuous monitor and simultaneous graphical visualization of the evolution of cutting force. In each run, fresh cutting tool was used for machining time of 4.0 minutes and experiments were repeated twice at each cutting condition in order to keep minimum experimental error. After each trial the tools were cleaned with HCl solution and acetone, later it is observed through optical microscope to measure the tool wear. The surface roughness was measured through stylus method using Mitutoyo made Talysurf surface roughness tester. The averages of five readings were considered for determining arithmetic surface roughness (Ra).

2.4. Plan of Experiments

The experimental data are used to set up the empirical relation or to express in terms of plots, which facilitates to enlarge the range of applications. Since, majority of data available are irregular and exhibits non-linear relationship, hence complexity exists in capture and systematic arrangement of data, and subsequently it is time consuming and man power consumable. Thus response surface method (RSM) was used to establish the model to correlate the process parameters with machining characteristics such as cutting force, tool wear and surface roughness. RSM adopts both mathematical and statistical techniques. It can be utilized for modeling and analysis of problems in which a response of interest is influenced by several variables. RSM attempts to analyze the influence of independent variable on a specific response. The aim to establish empirical model is to correlate the cutting parameters with machinability characteristics. The mathematical model commonly used for the machining response Y is represented as quadratic model as follows:

$$y = a_0 + \sum_{i=1}^{n} a_i x_{iu} + \sum_{i<j}^{n} a_{ij} x_{iu} x_{ju} + \sum_{i=1}^{n} a_{ii} x_{iu}^2 \tag{1}$$

where $y = f(y\text{-}\varepsilon)$ is expected response on higher order polynomial and x_i's are cutting parameters and a's are regression coefficients associated with corresponding terms, can be calculated by linear multiple regression analysis.

In the present study, Box-Behnken design based response surface method of design of experiments was used for acquiring data in a systematic manner. Based on the pilot experiments done before the start of experimental tests, constraints imposed by the machine and based on the objective of the present study, the parameters and their levels are chosen and is illustrated in Table 3.

3. RESULTS AND DISCUSSION

Table 4 reveals the results obtained in accordance to Box-Behnken design for fitting the response surface with 15 experimental rows corresponding to trial in random order. The combination of parameters considered in the study are shown in column 1-3, while the responses considered in the study viz. cutting force, tool wear and surface roughness are shown in the column 4, 5 and 6 respectively. The aim of investigation was to establish the mathematical model to predict the responses within range of parameters considered in study. The quadratic terms are neglected in the models, since the coefficients indicating corresponding terms were found to be relatively very less significant. Analysis of variance (ANOVA) method was used to find the significant parameters and their effect on responses. The confidence interval considered for ANOVA was 95% (α=0.05). Response surfaces are developed based on the response equations obtained.

3.1. Effect of Machining Parameters on Cutting Force

Figure 1 shows the response surface developed for cutting force based on response surface equation, Eq. (2). Table 5 illustrates the ANOVA results for cutting force. It indicates that, feed rate has the highest significance followed by depth of cut, besides cutting speed was found to be least significant parameter. The interactions were found to be less significant.

It is observed that, cutting force decreases with the increase in cutting speed, the cutting forces is also influenced by the hardness and shear strength of the materials. In hard turning process, the material hardness is higher initially hence greater cutting force is obtained. The increase in cutting speed tends to decrease in cutting force. This is attributed to the fact that, increase in cutting speed results in the

Table 3. Illustrates the parameters and their levels

Parameters	Level 1 (-1)	Level-2 (1)
Cutting speed (m/min	150	250
Feed rate (mm/rev)	0.06	0.18
Depth of cut (mm)	0.3	0.6

Table 4. Results obtained based on Box-Behnken design based experimental layout

Cutting Speed (m/min)	Feed Rate (mm/rev)	Depth of Cut (mm)	Cutting Force (N)	Tool Wear (mm)	Surface Roughness (μm)
150	0.06	0.45	488	0.060	0.66
250	0.06	0.45	350	0.120	0.45
150	0.18	0.45	734	0.090	2.10
250	0.18	0.45	650	0.280	1.40
150	0.12	0.30	480	0.066	1.00
250	0.12	0.30	436	0.170	0.70
150	0.12	0.60	520	0.082	1.12
250	0.12	0.60	486	0.190	0.85
200	0.06	0.30	385	0.080	0.46
200	0.18	0.30	554	0.150	1.50
200	0.06	0.60	513	0.110	0.60
200	0.18	0.60	865	0.180	1.80
200	0.12	0.45	580	0.120	0.96
200	0.12	0.45	580	0.120	0.96
200	0.12	0.45	580	0.120	0.96

Figure 1.

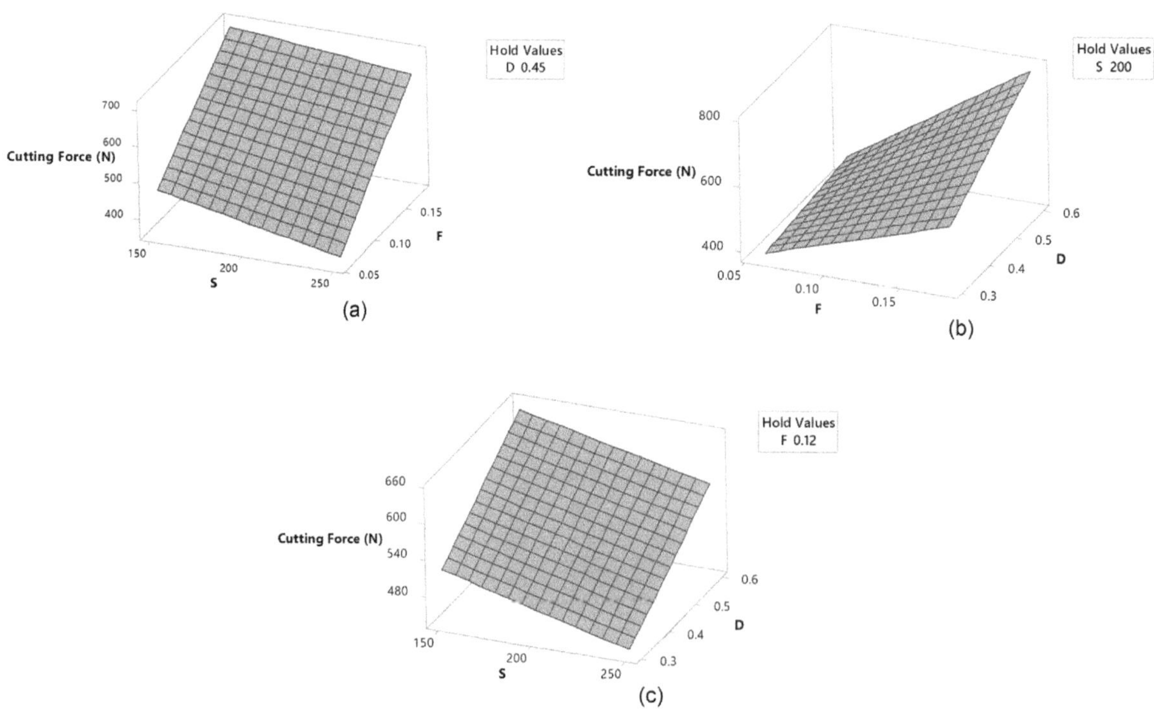

Table 5. ANOVA for cutting force

Parameter	Sum of Squares (SS)	DOF	Variance	F-Value	P	Remarks
Feed Rate (F)	142311	1	142311	26.29	0.001	Significant
Depth of Cut (D)	34980	1	34980	6.46	0.035	Significant
Cutting Speed (S)	11250	1	11250	2.08	0.187	Insignificant
F X D	8372	1	8372	1.55	0.249	Insignificant
F X S	729	1	729	0.13	0.723	Insignificant
S X D	25	1	25	--	0.947	Insignificant
Error	43299	8	5412			
Total	240967	14				

increase of temperature in cutting zone owing to reduction in the shear strength of materials (Lima et al., 2007). In addition, increased cutting speed leads to increase in shear angle that results in chip thickness reduction, consequently machining forces also get reduces.

Regression equation for cutting force:

$$Cutting\ Force\left(N\right) = 644 - 1.44S - 965F - 236D + 4.5S * F + 0.33S * D + 5083F * D \quad (2)$$

R-Sq.= 92.03%, R-Sq. (Adj.) = 88.55%

The increase in feed rate increases the cutting force. In this case, irrespective of material hardness, increase in feed rate results in the increased shear plane. It perhaps due to the fact that, increased feed rate increase the volume of material to be cut or material chip thickness, besides establishing a dynamic effect. Thus, corresponding normal contact stress at the tool-chip interface and tool-chip contact area increases (Suresh et al., 2012). Hence, increasing trend of cutting force was observed.

The increase in depth of cut leads to increase in cutting force. Similarly to feed rate, increase in depth of cut leads to increased tool work contact length (Suresh et. al., 2012). Subsequently, induces the higher volume of deformed and cut material resulting in increased chip thickness. Thus, requires greater cutting force for the removal of material. Hence, increase in cutting force was observed.

3.2. Effect of Machining Parameters on Tool Wear

Regression Equation for tool wear

$$Tool\ wear\left(mm\right) = 0.0517 - 0.000205S - 1.479F + 0.053D$$
$$+0.01083S * F + 0.000133S * D + 0.000F * D \quad (3)$$

R-Sq. = 97.61%, R-Sq. (Adj.) = 95.82%

Table 6. ANOVA for tool wear

Parameter	Sum of Squares (SS)	DOF	Variance	F-Value	P	Remarks
Cutting Speed (S)	0.026681	1	0.026681	191.19	0.00	Significant
Feed Rate (F)	0.013613	1	0.013613	97.55	0.00	Significant
Depth of Cut (D)	0.001152	1	0.001152	8.26	0.021	Significant
F X S	0.004225	1	0.004225	30.28	0.001	Significant
F X D	0.000004	1	0.000004	0.03	0.870	Insignificant
S X D	---	1	---	---	1.000	Insignificant
Error	0.001116	8	0.000140			
Total						

Figure 2 shows the response surface developed for tool wear based on Eq. (3). Table 6 illustrates the ANOVA results for tool wear. It depicts that, cutting speed has the highest significance followed by feed rate and their interactions were found to be statistically significant. Besides, depth of cut was found to be least significant parameter and other interactions were found to be insignificant. Tool material, tool geometry, workpiece material and machining conditions are influencing factors on tool wear. Generally observed tool wear mechanisms in hard turning are abrasion, adhesion and diffusion, while each wear mechanism is affected by the factors workpiece, machining conditions and tool geometry and material.

Figure 2.

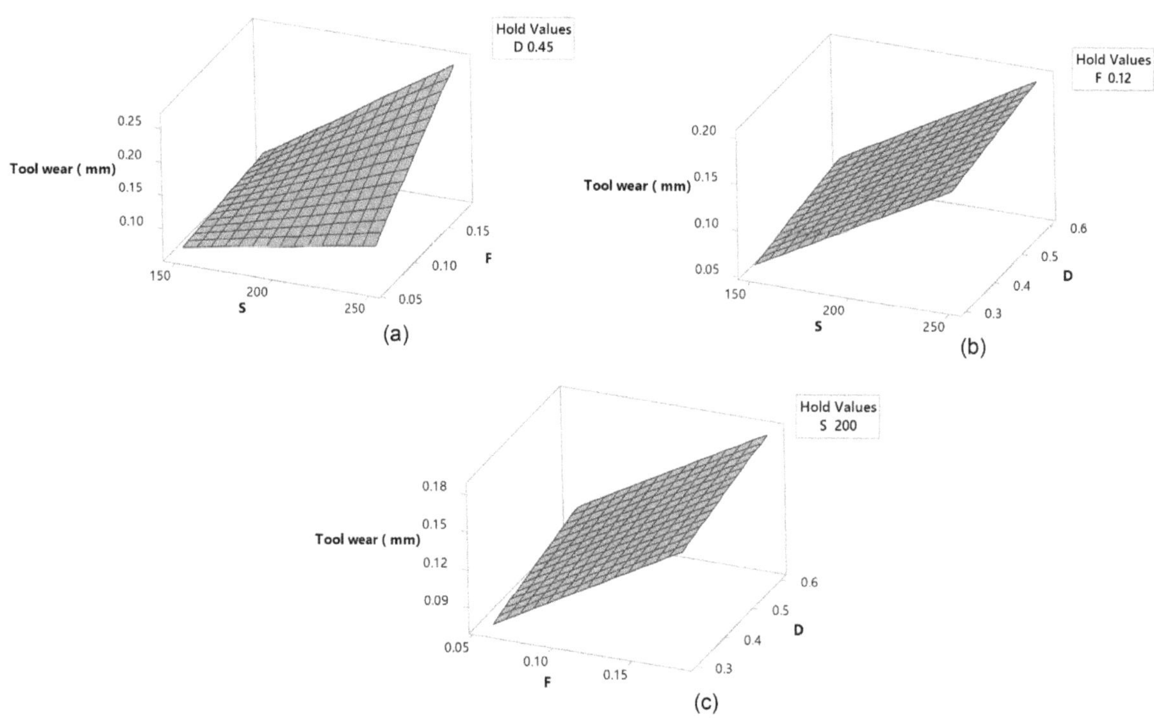

From the Figure 2, it can be observed that increase in cutting speed increased the tool wear. As discussed earlier, the temperature generated at the machining zone increase at higher cutting speed compared than at its lower level. The increase in temperature also may be because of increased rubbing action/interaction between tool/workpiece within the short interval of time exists. The increased temperature is even more than the thermal stability of the cutting tool materials, thus softens material. Hence high tool wear was exhibited.

It was also found that, increase in feed rate increases the tool wear linearly. It perhaps, increase in feed rate leads to increase in cutting temperature and stress generated on rake and flank faces close to nose area that has probably caused the yield strength of the tool to reduce (Lee and Mathew, 2007). In addition, increase in feed rate also causes the thermal softening of tool material due to elevated temperature at higher feed rate (Ashtakov (2007).

It is observed that depth of cut does not show much effect on the tool wear. Similarly, depth of cut has also increased the tool wear; however it was high at larger depth of cut and feed rate interaction values. The increase in tool flank wear has occurred probably due to abrasion at rake and flank face of cutting tool. Further, increase in depth of cut may be lead to the increased specific contact stress at the tool-chip interface and average contact temperature, thereby increasing tool wear was observed.

3.3. Effect of Machining Parameters on Surface Roughness

Regression Equation for surface roughness:

$$\begin{aligned} &Surface\ Roughness\left(\mu m\right) = -0.30 + 0.00075S + 15.81F - 0.14D \\ &-0.0408S * F + 0.00100S * D + 4.44F * D \end{aligned} \quad (4)$$

R-Sq. = 95.06%, R-Sq. = 91.35%

Figure 3 shows the response surface developed for surface roughness based on Eq. (4). Table 7 illustrates the ANOVA results for surface roughness. It depicts that; feed rate has the highest significance.

Table 7. ANOVA for surface roughness

Parameter	Sum of Squares (SS)	DOF	Variance	F-Value	P	Remarks
Feed Rate (F)	2.67961	1	2.67961	133.73	0.000	Significant
Cutting Speed (S)	0.27380	1	0.27380	13.66	0.006	Significant
Depth of Cut (D)	0.06301	1	0.06301	3.14	0.114	Insignificant
F X S	0.06002	1	0.06002	3.00	0.122	Insignificant
F X D	0.00640	1	0.00640	0.32	0.918	Insignificant
S X D	0.00022	1	0.00022	0.01	0.587	Insignificant
Error	0.16030	8	0.02004			
Total						

Figure 3.

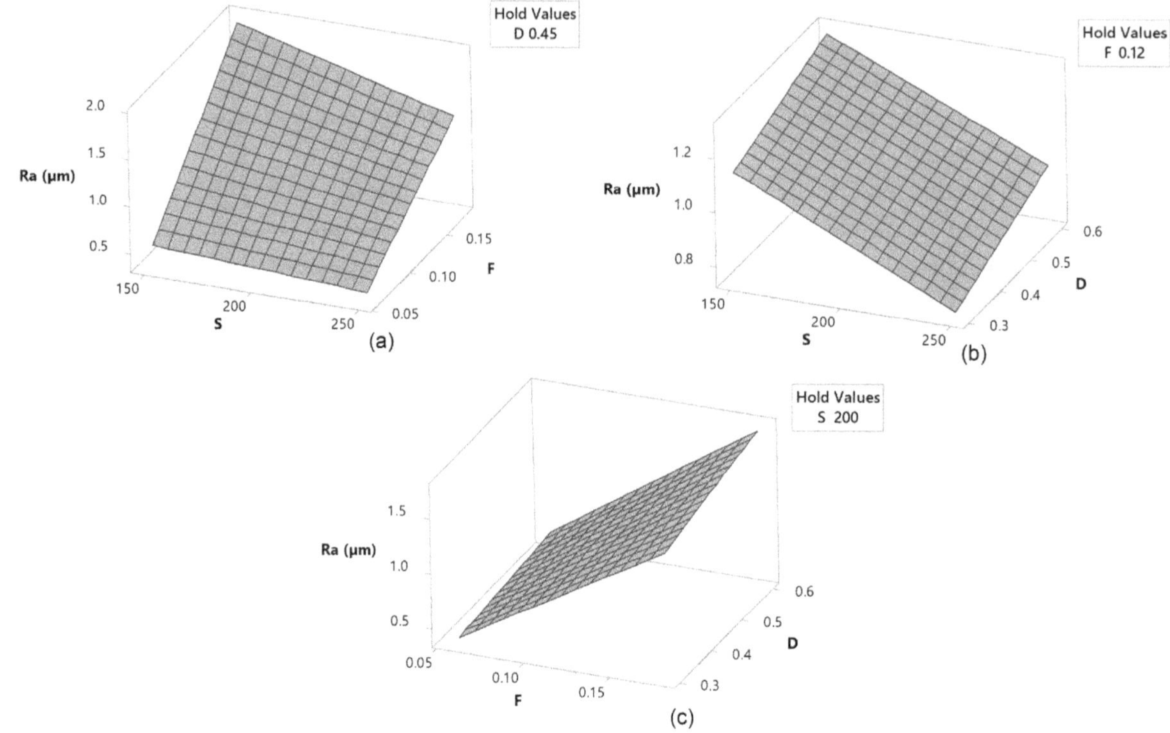

While the cutting speed was found to be less significant on surface roughness. The depth of cut and interactions of parameters were statistically found to be insignificant.

The surface roughness evaluation is a important aspect and requires greater attention. It is a method of measuring irregularities over the machined surface. It is influenced by many factors such as nose radius, work hardness, cutting tool material, machine tool rigidity, cutting conditions as well as other factors like environmental conditions and so on. From Figure 3, it can be noticed that, surface roughness found to be increased with increase in cutting speed. The probable reason might be less dissipation of heat to the work material as it is swept away in the chip during machining of hardened steel at higher cutting speeds. In conventional turning process, usual explanation is based on built-up-edge (BUE). While, in hard turning process BUE formation is minimum or completely eliminated due to high hardness. According to Chen (2000), the properties of metals are influenced by deformation velocity, thus higher velocity leads to less significant plastic behavior. In present study also, less lateral plastic flow of material has probably occurred. Hence increased cutting speed, that is beyond BUE region lead in low surface roughness.

It is also observed that, surface roughness increases with increase in feed rate. The increase in feed rate increases the cutting force that leads to vibration. In addition, amount of heat generated increases with increase in feed, since tool-work interaction as well as shear plane increases due to growth of material to cut. It can also be noticed that, surface roughness is susceptible to variation in feed rate at low cutting speed compared to higher cutting speed. The surface roughness was found to be minimum at low feed rate and high cutting speed.

The surface roughness was found to increase with the increase in depth of cut. The increased depth of cut probably has lead to increased tool work contact length. Subsequently, increases cutting temperature and higher surface roughness. However, the surface roughness was less susceptible with variation in depth of cut.

3.4. White Layer Formation

Figure 4 (a) and (b) shows the typical micrographic images of machined surface specimens were examined under optical microscope in the etched conditions. From Figure 4 (a), it is observed that the white layer does not exist at the cutting conditions of cutting speed (Vc) of 150 m/min, feed rate (f) of 0.12 mm/rev and depth of cut (d) of 0.30 mm. This is due to the fact that, the material will require higher temperature and longer time to undergo phase transformation. Figure 4 (b) illustrates the micrographic image of machined surface at Cutting speed (Vc) of 250 m/min, feed rate (f) of 0.06 mm/rev and depth of cut (d) of 0.30 mm. It is observed that the deformed layer just beneath (3 ~ 4 μm) the white layer might be the consequence of either mechanically or thermally dominated deformation or both depending on the cutting conditions. It can be revealed that the higher cutting speed with lower feed rate and depth of cut leads to increased depth of white layer formation on the machined surface. This type of subsurface deformation is attributed to the heavy plastic deformation in the machined surface. This indicates that

Figure 4.

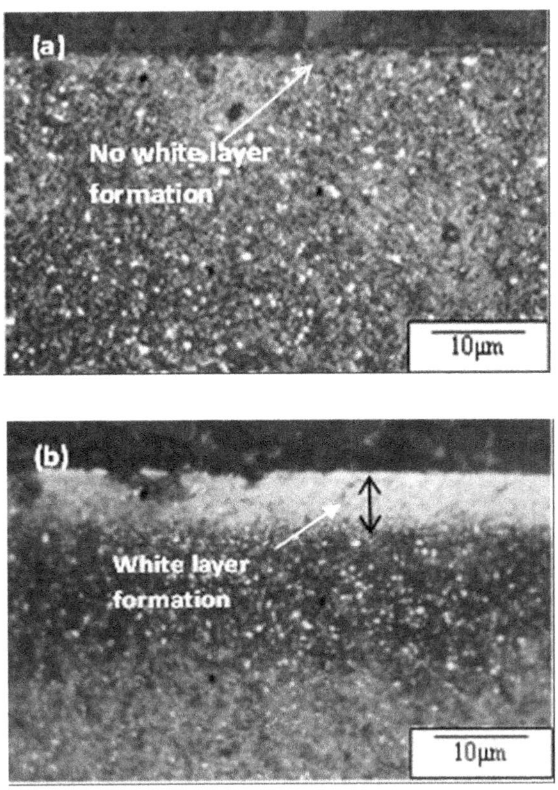

the austenization temperature has been exceeded within this layer leading to the generation of white layer caused by rapid cooling has been reported (Rech and Moisan, 2003). It can be deducing that martensite in steel transformed into austenite to form the white layer due to higher temperature conditions.

4. CONCLUSION

The following were major conclusions drawn from the present study.

1. The cutting force increases with increasing feed rate and depth of cut, while decreased with the increased in cutting speed. This is probably due to increase in shear angle and reduced chip thickness.
2. The tool wear increases with increase in cutting parameters and was found to be more susceptible with the variation in feed rate compared to depth of cut and cutting speed.
3. The surface roughness was also found to be increased with increase in feed rate and depth of cut, while decreased with increase in cutting speed. However, it was more susceptible with variation in cutting speed and feed rate.
4. Mathematical models based on response surface method are presented to correlate the cutting parameters with response parameters within range of parameters considered in the study.
5. At cutting speed (Vc) of 250 m/min, feed rate (f) of 0.06 mm/rev and depth of cut (d) of 0.30 mm, the deformed layer just beneath (3 ~ 4 μm) the white layer was noticed and might be the consequence of either mechanically or thermally dominated deformation or both depending on the cutting conditions

REFERENCES

Aouici, H., & Yallese, M. A. (2013). Experimental Investigation of Cutting Parameters influence on Surface Roughness and Cutting forces in Hard Turning of X38CrMoV5-1 with CBN Tool. *Sadhana*, *384*, 29–445.

Ashley, S. (1995). High speed machining goes mainstream. *Mechanical Engineering (New York, N.Y.)*, *117*, 56–61.

Ashtakov, V. P. (2007). Effects of the cutting feed, depth of cut, and workpiece (bore) diameter on the tool wear rate. *International Journal of Advanced Manufacturing Technology*, *34*(7-8), 631–640. doi:10.1007/s00170-006-0635-y

Batish, A., Bhattacharya, A., Kaur, M., & Cheema, M. S. (2014). Hard turning: Parametric optimization using genetic algorithm for rough/finish machining and study of surface morphology. *Journal of Mechanical Science and Technology*, *8*(5), 1629–1640. doi:10.1007/s12206-014-0308-y

Bouacha, K., Yallese, M. A., Mabrouki, T., & Rigal, J. F. (2010). Statistical analysis ofsurface roughness and cutting forces using response surface methodology in hard turning of AISI 52100 bearing steel with CBN tool. *International Journal of Refractory Metals & Hard Materials*, *28*(3), 349–361. doi:10.1016/j.ijrmhm.2009.11.011

Caydas. (2010). Machinability Evaluation in Hard Turning of AISI 4340 Steel with Different Cutting Tools using Statistical Techniques. *Journal of Engineering Manufacture*, *224*, 1043–1055.

Chandrasekaran, M., Muralidhar, M., Murali Krishna, C., & Dixit, U. S. (2010). Application of soft computing techniques in machining performance prediction and optimization: A literature review. *International Journal of Advanced Manufacturing Technology*, *46*(5–8), 445–464. doi:10.1007/s00170-009-2104-x

Chen, W. (2000). Cutting forces and surface finish when machining medium hardness steel using CBN tools. *International Journal of Machine Tools & Manufacture*, *40*(3), 455–466. doi:10.1016/S0890-6955(99)00011-5

Gaitonde, V. N., Karnik, S. R., Figueira, L., & Davim, J. P. (2009). Machinability investigations in hard turning of AISI D2 cold work tool steel with conventional and wiper ceramic inserts. *International Journal of Refractory Metals & Hard Materials*, *27*(4), 754–763. doi:10.1016/j.ijrmhm.2008.12.007

Lee, T.H., & Mathew, P. (2007). Experimental and Theoretical Investigation of AISI D2 hardened steel machining with varying nose radius CBN tools. *International Journal of Machining and Machinability of Materials*, *2*, 254-269.

Lima, J. G., Avila, R. F., & Abrao, A. M. (2007). Turning of hardened AISI 4340 steel using coated carbide inserts. *IMechE Part B: J. Engineering Manufacture*, *221*(8), 1359–1366. doi:10.1243/09544054JEM847

Mohandas, K. N., Ramesh, C. S., Prasad, K. E., Balashanmugam, N., & Shashikumar, P. V. (2012). Study of the flank wear during hard turning of hard chrome plated surfaces. *Int. Conf. on materials processing and characterization.*

Nakayama, K., Arai, M., & Kanda, T. (1998). Machining characteristics of hard materials, CIRP. *Annals of CIRP*, *37*(1), 89–92. doi:10.1016/S0007-8506(07)61592-3

Rashid, W. B., Goel, S., Davim, P. J., Shrikrishna, N., & Joshi, S. N. (2016). Parametric design optimization of hard turning of AISI 4340 steel (69 HRC*). International Journal of Advanced Manufacturing Technology*, *82*(1), 451–462. doi:10.1007/s00170-015-7337-2

Rech, J., & Moisan, A. (2003). Surface integrity in finish hard turning of case-hardened steels. *International Journal of Machine Tools & Manufacture*, *43*(5), 543–550. doi:10.1016/S0890-6955(02)00141-4

Sahoo, A. K., & Sahoo, B. (2013). Performance Studies of Multilayer Hard Surface Coatings (TiN/TiCN/Al2O3/TiN) of Indexable Carbide Inserts in Hard Machining: Part-II (RSM, greyrelational and techno economical approach). *Measurement*, *46*(8), 2868–2884. doi:10.1016/j.measurement.2012.09.023

Suresh, R., Basavarajappa, S., Gaitonde, V. N., Samuel, G. L., & Davim, J. P. (2013). State-of-the-art research in machinability of hardened steels, *Proc. IMechE Part B: J. Engineering Manufacture*, *227*(2), 191–209. doi:10.1177/0954405412464589

Suresh, R., Basavarajappa, S., & Samuel, G. L. (2012). Some studies on hard turning of AISI4340 steel using multilayer coated carbide tool. *Measurement*, *45*(7), 1872–1884. doi:10.1016/j.measurement.2012.03.024

Tonshoff, H. K., Arendt, C., & Ben, A. R. (2000). Cutting hardened steels. *Annals of CIRP*, *49*(2), 215–220. doi:10.1016/S0007-8506(07)63455-6

Chapter 10
Fault Detection Through Vibration Signal Analysis of Rolling Element Bearing in Time Domain

Pankaj Gupta
Maulana Azad National Institute of Technology, Bhopal, India

M. K. Pradhan
Maulana Azad National Institute of Technology, Bhopal, India

ABSTRACT

Mechanical failure prevention and condition monitoring have been one of the concerns of mechanical engineers in recent years due to the personal safety, cost of failure, reliability and downtime issues of equipment. Rotating machines are one of the most important actuators in the industrial applications as well as in every day applications. Rolling element bearings are very critical components of rotating machines and the presence of defects in the bearing may lead to failure of machines. Hence, early identification of such defects along with the severity of damage under operating condition of the bearing may avoid malfunctioning and breakdown of machines. Defective bearings are source of vibration and these vibration signals can be used to assess the faulty bearings. This chapter presents the brief review of recent trends in research on bearing defects, sources of vibration and vibration measurement techniques in time domain, frequency domain and time-frequency domain. Detailed explanation of defect detection through scalar indicators in time domain.

INTRODUCTION

Different types of machinery are present in the industries. The majority of these machines have rotating components. Machines must perform as per their designed specification and installed capacity to obtain high production rates. Thus, machines must be in proper condition to ensure that they run without significant downtime. Machine condition monitoring deals with the maintenance aspect of these machines

DOI: 10.4018/978-1-5225-2440-3.ch010

based on their past and present condition. Maintenance reduces the risk due to machine failure in a cost effective manner. In order to know the condition of machine sensors are installed around the machine and relevant information is collected, analyzed and decisions are made to take corrective actions so that machine can run as per its design objective. In a plant without proper protocol, in place the end result could be an eventual loss of the plant. Generally, prevention of failure can be identified as the process of fault detection, diagnosis, and prognosis:

- Fault detection is the process in which the measured system data and system status information is observed and it is compared with a normal range of observed attributes to determine whether some measurements fall outside the range representing the healthy condition of the system.
- In Diagnosis condition of failing components is determined, and failure causes is identified.
- In Prognosis impending component failures or abnormal system states are predicted before they actually occur, and estimates remaining useful life of machine.

During the operation machine gives out information or signal in the form of noise vibration, temperature, lubrication oil condition, and the like. These signals are acquired from the machine by installing transducer to measure the mechanical parameter of the machine.

PRINCIPLES OF MAINTENANCE

A machine that has been designed and manufactured to perform a certain function, is expected to do so when installed in a plant for its designed life span. However, for reasons beyond ones control, such a machine may fail to do so because of reasons like faulty design of the machine, inferior material and workmanship, incorrect installation and wrong operational procedure, among many others. The machine thus has to be maintained to avoid such failures. All over the world, three different types of maintenance techniques for machines have been adopted by the plant operators, known as the reactive maintenance, preventive maintenance, and predictive maintenance. The benefits of the maintenance are as follows:

- It eliminates unnecessary maintenance,
- Reduce lost production caused by failure,
- Reduce repair part inventory,
- Improve product quality,
- Increase production life of plants,
- Increase production capacity,
- Reduce overall maintenance cost.

Reactive Maintenance

Reactive maintenance is also known as breakdown maintenance is an expensive maintenance method. In this type we need to react because machine is failed completely. The reactive maintenance method result in inefficiencies with significant loss due to unplanned downtime, overtime cost and damaged machinery. Attributes of reactive maintenance programe are as follows:

1. Expenses involved are high,
2. Spare parts inventory costs are high,
3. Machine downtime is high.

Preventive Maintenance

In preventive maintenance, the maintenance on a particular machine is done in a regular periodic manner at a fixed frequency. This type of maintenance is also known as periodic maintenance. The guiding principle behind this maintenance technique is that the machine is up to performing its desired function at any time, since it has been well maintained. Even though this method assists inaccurate diagnosis of some faults, unnecessary disassembling often induces secondary damage and high downtime costs. Also, preventive maintenance might miss the fault that are beginning to occur, which may lead to serious in near future.

Predictive Maintenance

In predictive maintenance, maintenance is done on a machine depending on its need. The decision to conduct maintenance or not to conduct maintenance depends upon the machines past and present condition. Preventive maintenance is a need-based maintenance that depends on the condition of the machine. This type of maintenance is also known as condition-based maintenance. It has been recognized as the most effective maintenance program because it can

- Minimize the cost of unnecessary and ineffective maintenance.
- Improve personal and environmental safety.
- Reduce the quantity of unexpected and unplanned breakdown to maximize production availability.

In order to get above benefits this method should be able to accurately characterize the machinery condition and provide the information of imminent failure. However, in order to know a machines condition, additional instrumentation is required on the machine to measure its "health" parameters. This instrumentation includes transducers, signal conditioners, data acquisition units, computer-based signal analysis systems, and software-driven diagnostic routines. The typical process of predictive maintenance includes three: data acquisition; feature extraction; and classification of these features to diagnose the type and cause of fault as shown in the Figure 1.

There are many kinds of measurement used for monitoring, and which method is optimal for a given situation depends on the type of machinery, accessibility, and the purpose of monitoring. The methods commonly used for the monitoring are, vibration, acoustic, wear debris and oil analysis thermography,, motor current signature analysis, and non-destructive test technique like ultrasonic, radiography, eddy current testing and acoustic emission. Among these techniques vibration is the most widely used technique. In 70% cases, vibrations are used as monitoring technique for machines. Feature extraction is a very important step in the performance of predictive maintenance. By analyzing the collected measurements, the PdM must be able to reduce the amount of data, identify and select the fault indicating features for classification and diagnosis. If the features are not well chosen, the PdM may give frequent

Figure 1. Processes of predictive maintenance

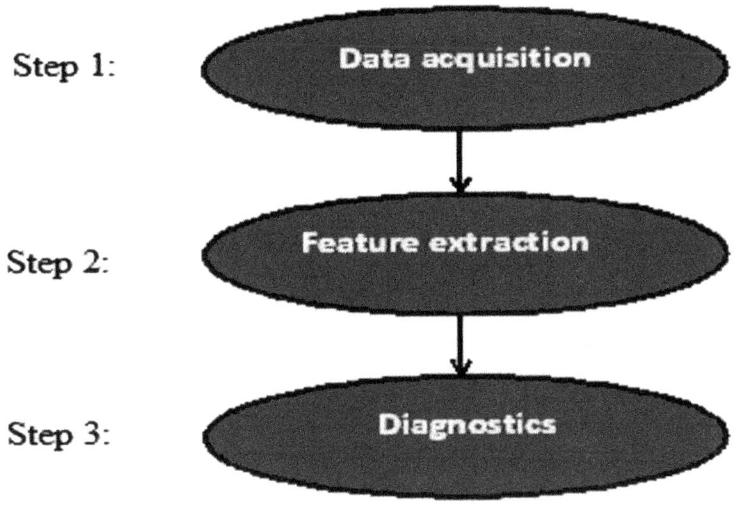

false alarms and be sensitive to legitimate failure condition. The ultimate aim of PdM is to determine the machinery condition and diagnose the actual problem. At last, to recognize what needs to be done is classification of problem.

FAULT PROGNOSIS

Faults in machines can be diagnosed by analyzing the signals obtained from transducers installed around the machines. The fault diagnosis depends on the characteristics of the signals, and incipient faults can also be detected with efficient algorithms. Once the faults have been detected and diagnosed, the next question is how long the machine will last in the present condition or what is the Remaining Useful Life (RUL) of the machine under observation. Many deterministic and stochastic approaches are available for predicting the RUL of a machine.

FUTURE NEEDS

New techniques for fault diagnosis in machines are always evolving with the advancement of technology. There has been improvement in all aspects of condition monitoring, such as transducers and instrumentation, digital signal processing, and decision making, although with the complexities of machines increasing day by day and their size reducing, their maintenance is also a challenge. The use of wireless systems for data transfer from the measurement location to the analysis station is preferred because it eliminates the need for long cables from the measurement location to the analysis station, though the current available wireless transfer speeds, limit high- speed data acquisition of dynamic vibration signals. Improvements are being made throughout the world by researchers and developers.

IMPORTANCE OF BEARING DEFECT DETECTION

The bearing is a very essential part of all rotating machines whose function is to support the machines and to permit the rotational motion of the shaft with respect to a fixed structure. More than 90% of machines use rolling element bearings and their failure may lead to catastrophic failure of the machine. Thus, they are considered as the most critical components in industrial applications. The presence of bearing defect causes significant increase in vibration level (Sunnersjö,1978; Tandon & Choudhury, 1999). Many techniques have been used for defect detection in bearings using vibration and acoustic measurement like vibration measurement in the time domain, frequency domain, time-frequency domain, shock pulse method, acoustic emission technique. Many researchers have worked on vibration signal analysis techniques and they have published numbers of research papers. Several researchers have published review papers based on fault diagnosis techniques. PY (1983) researched on vibration and wear debris analysis. In 1984, Mathew and Alfredson (1984) have presented a brief review of vibration monitoring techniques in time and frequency domains and their results on rolling element bearings. Tondon and Nakra (1992) also present a paper on vibration and acoustic methods, such as vibration measurements in time and frequency domains. Therefore, the aim of this study is to present the recent research on vibration measurement technique for detection of fault in rolling element bearings.

Several researchers explained the mechanism of vibration in bearings (Choudhury and Tandon, 1998; Sunnersjo, 1978; Tallian and Gustafsson, 1965; Wardle and Poon, 1983). Even a healthy bearing generate vibration, but the presence of defect increases vibration levels significantly. A large number of factors, the most common of which are fatigue, wear, plastic deformation, corrosion, brineiling, poor lubrication, faulty installation and incorrect design, can cause premature bearing failures. Bearing defects are classified in two categories as distributed defects and localized defects.

1. **Distributed Defects:** Surface roughness, waviness, misaligned races and off-size rolling elements are considered as the distributed defects (Choudhury & Tandon, 1998; Meyer, 1980). The main causes of distributed defects are manufacturing error, abrasive wear, and improper installation (Sunnersjo, 1978). In distributed defects, the contact force between rolling elements and raceways varies which results in vibration. Vibration response due to distributed defect is mainly used for the quality inspection and condition monitoring of bearings.
2. **Localized Defects:** This category of defects includes pits, cracks, spalls that may develop over the rolling surfaces. Of these, spalling is the dominant mode of failure. Actually, fatigue crack begins below the surface and propagates towards the surface until the material fails and leaves localized defects. Bentley (1989) in his paper showed that 90% percent of the total bearing faults involves damage of the inner ring, outer ring and rolling elements due to localized defects.

Sources of Vibration

The components of rolling contact bearings i.e. Inner raceway, outer raceway, rolling elements, cage interact through a combination of rolling and sliding to produce complex vibration signatures. The factors on which level of vibration depends are, energy of impact, the point at which vibration is measured and the bearing construction. The various sources of vibrations are as follows:

1. **Variable Compliance:** Even a geometrically perfect bearing generate vibration under radially acting load, but it does not indicate poor quality of bearing. Such type of vibration is called variable compliance. To support external loads, finite no of bearings are used whose position continuously varies with time with respect to the line of action of the load resulting in vibration. Number of rolling element bearing supporting the external load and running clearance causes vibration due to varying compliance. More is the number of loaded rolling element, lesser is the vibration. Variable compliance increases with running clearance in radially loaded or misaligned bearings, as it determines the extent of loaded region. Variable compliance vibration can be reduced to insignificant levels by using ball bearings with accurate level of pre-axial loads.
2. **Geometrical Imperfection:** Due to the nature of the manufacturing process, geometrical imperfections are always present in the bearing components. When an axial load is acting on the bearing and rotating with moderate speed then form and surface finish of critical rolling surface are the largest sources of vibration. Thus, surface roughness and waviness are the critical controlling components during manufacturing process because it may not produce significant vibration but may affect bearing life. Geometrical imperfections are measured in terms of wavelength as compared to the Hertzian width of contact of the rolling element raceway. Roughness is surface features of the wavelength of the order of the contact width, whereas waviness is features with longer wavelengths.
3. **Surface Roughness:** High level of surface roughness compared with lubricating film thickness generated between the rolling element raceway contacts, will cause significant vibrations because asperities break through the film and interact with the opposing surface resulting in vibration. Random small impulses in the vibration excite the natural frequency of bearing and support structure and level of vibration energy increases. Vibration produced due to surface roughness, occurs at frequencies above sixty times the rotational speed of the bearing. Thus, the high frequency part of the spectrum usually appears as a series of resonances.
4. **Waviness:** Surface features of longer wavelength are waviness. The peak curvatures are small compared with that of the Hertzian contacts and the rolling element, which follows surface contours, has continuous rolling motion. Surface geometry and vibration level bears a complicated relationship. The factors on which it depends are the bearing and contact geometry as well as conditions of load and speed. Vibrations produced due to waviness may occur at frequencies up to about three hundred times the rotational speed, but is usually predominant at frequencies below sixty times rotational speed.
5. **Discrete Defects:** These defects in bearing may be very small but can have a major impact on vibration-critical equipment and may lessen bearing life also. These defects may be present in the form of indentation, scratch, pits, abrasive particles embedded in lubricants. Rolling surfaces may damage due to contamination, mounting, assembly, poor maintenance, operation etc. Bearing manufacturers have adopted simple vibration measurements on the finished product to detect such defects but these tend to be limited by the type and size of bearing.
6. **Raceway Defect:** Any discrete defect on inner raceway generates pulses of high energy at a ball pass frequency to the inner raceway. As inner race rotates, the defect enters and leaves the load zone causing variation in contact force between rolling element and raceway. The highest amplitude of pulses occurs in the load zone and decrease as the defect leaves the load zone. The resulting signal is amplitude modulated at inner ring rotational frequency. If a discrete defect is present in the outer race, the resulting vibration occurs at frequency equal to ball pass frequency to the outer

race. Outer race is stationary thus, the high-energy pulses generated have the same amplitude and there will be a single discrete peak in the frequency domain.

7. **Rolling Element Defect:** Frequency generated by defects in rolling element can be twice the ball spin frequency and the fundamental train frequency. Two times the rolling element spin frequency is produced when the defect interacts both raceways, but sometimes, frequency is not this high because when the defect interacts with the mating surface ball may not be in the load zone and energy is lost as the signal passes through other structural boundaries as it strikes the inner raceway. In the presence of more than one defective rolling element frequency produced is the sum of all the ball spin frequencies.
8. **Cage Defect:** Rotation of cage is 0.4 times the speed of inner ring and has low mass and hence unless there is defect from manufacturing process it is generally not visible. During cage failure, signals obtained are in the form of random bursts of vibration as the ball slides and cage wears and deforms and wide band of frequencies occur.

Bearings are an important machine element that supports the rotor system. They are designed to provide less friction at the supports and carry the loads. In machineries, the most common type of bearings used is the antifriction bearing or the rolling element bearing. Rolling element bearings have four very important components, two circular races i.e. inner race and outer race between which the rolling elements rollers, tapered rollers, or spherical balls are held. The cage or the separator is the fourth important element in a rolling element bearing. The cage ensures that no two rolling elements come in contact with each other and maintain equal. Figure 2 shows the various components of ball bearings.

Vibration Response Due to Localized Defect

Due to interaction between local defect on the element and its mating surface abrupt change in contact stresses produces pulses of short duration .These pulses produce vibration which can be monitored

Figure 2.

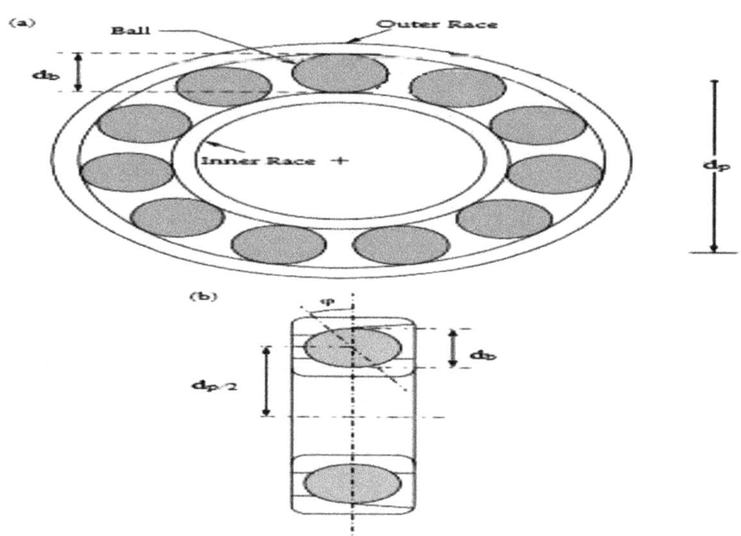

(Sunnersjö, 1978). The vibration signal generated by the faulty bearing can be analyzed in time domain or frequency domain or time-frequency domain.

Time Domain Approach

It is the simplest technique for detecting faults in bearings. Time domain analysis scalar indicators are used to determine the bearing condition. In time domain analysis, signal is analyzed through temporal vibrational signal data and based on its value bearing condition can be estimated. In the time domain approach usually the overall root-mean-square (RMS) level and crest factor, i.e., the ratio of peak value to RMS value of acceleration is measured. The resultant RMS values are compared with recommended values to estimate the bearing condition (Tandon, 1994). This method has been applied with limited success for the detection of localized defects (Miyachi & Seki, 1986; Tandon & Nakra, 1993). Dyer and Stewart (1978) proposed, kurtosis, a fourth moment, normalized with respect to the fourth power of standard deviation. The kurtosis is given by the expression:

$$Ku = \frac{1}{l}\sum_{k=1}^{l}\frac{\left(u_{k-\mu}\right)^{4}}{\sigma^{4}} \tag{1}$$

where u_k is the instantaneous amplitude μ is the mean and is the standard deviation of the data n is the sample length.

For healthy bearing, value of kurtosis is 3 and a value greater than 3 indicates defected bearings. However, the disadvantage of kurtosis is that when the damage is well advanced its value comes down to the level of an undamaged. This is because the value is sensitive to pulse response generated due to defects in the vibration signal. These pulse responses are dependent on nature of damage therefore, when the damage progresses vibrational signals become more and more random and its value reduces to a normal bearing level. Oack and Lapora (2005) also discussed such behavior in their paper. Kurtosis has not become popular in the industries. Gustafsson and Tallian (1965) proposed a method of defect detection based on the number of peaks crossing a pre-set voltage level. Researchers have also tried to develop some new indicators. Sassi et al. (2006) proposed two new indicators TALAF and THIKAT by combining conventional descriptors. Paliwal et al. (2014) proposed a new indicator CRIS which was obtained by the combination of RMS value, Crest factor, Impulse factor and standard deviation. Karacay and Akturk (2009) reviewed that the peak-to-peak value, RMS, shape factor, Crest factor and kurtosis do not give any information about the location of defect e.g. inner race, outer race, cage or the roller. Some researchers have also proposed band pass filter technique based on time domain in which impulse loading excites structural resonance in the high frequency zone that can be detected by a transducer. Shock pulse method (Barthel, 1978) is based on band pass filtering techniques. This method is widely used in industries (Butler, 1973; Alfredson & Mathew, 1985) but cannot effectively detect defects at low speed (Smith, 1982; Ray, 1980). Dwyer (1983) initially introduced the concept of spectral kurtosis and Antoni & Randall (2006) implemented it in the field of fault diagnosis of rolling element bearing. Spectral Kurtosis (SK) is a statistical tool that is helpful in identifying the presence of series of transients and its position in the frequency domain. Antoni & Randall (2006) in his paper listed various properties of SK for the first time.

Frequency Domain Approach

Frequency domain or spectral analysis is the most widely used approach for fault diagnosis in bearings. Frequency-domain techniques convert time-domain vibration signals into discrete frequency components using a fast Fourier transform (FFT). Analysis of raw vibration signal in frequency domain can be done through Discrete Fourier transform (DFT) and Fast Fourier transform (FFT). DFT is a straight mathematical procedure but is inefficient while fast Fourier transform (FFT) analyzers diagnosis are an easier and efficient way of obtaining narrow band spectra. In a frequency spectrum plot, the X-axis represents frequency and the Y-axis represents amplitude of displacement, velocity, or acceleration. Frequency domain technique is capable of detecting certain frequency components of interest very easily which is an advantage over time domain technique. James Taylor (1994) well described the order of appearing and disappearing of peaks in the spectrum. Some researchers (Tandon and Nakra, 1993; Reif and Lai 1989) explained high frequency range of the spectrum to be an effective method of predicting the condition of rolling element bearings. Tandon and Nakra (1992) predicted the values of these resonance frequencies to be higher than 5kH. Fault signal can be identified based on bearing fundamental frequencies which depends on the kinematic consideration i.e. bearing geometry and rotor speed. For a bearing with stationary outer race, these frequencies have been explained below.

1. **Inner/Outer Race Ball Pass Frequency (BPFI/BPFO):** The inner and outer races of a bearing are connected to other component. The inner race is usually attached to the shaft and has the same rotational speed as the shaft and the outer race is stationary. The inner and outer race defect frequencies are referred to as ball pass frequencies

$$BPFI = \frac{n}{2} f_r \left(1 + \frac{d_b}{d_p \operatorname{Cos} \varphi}\right) \tag{2}$$

$$BPFO = \frac{n}{2} f_r \left(1 - \frac{d_b}{d_p \operatorname{Cos} \varphi}\right) \tag{3}$$

2. **Ball Pass Spin Frequency (BSF):** The ball defect frequency is usually referred to as the ball spin frequency. Ball spin frequency is the frequency at which the roller (ball element) makes contact with one of the races.

$$BSF = \frac{n}{2} f_r \left(1 - \frac{d_b^2}{d_p^2 \operatorname{Cos}^2 \varphi}\right) \tag{4}$$

3. **Fundamental Train Frequency (FTF):** Due to rotation of rolling element about its own axis its velocity is equal and opposite at the point of the contact with the inner and outer races. Therefore, the axes of the rolling elements and the cage holding the rolling elements must move with a veloc-

ity which is mean of the velocities of the inner and outer races so that the races do not slide. The rotational frequency train frequency of cage or train frequency can be calculated by considering races diameter, pitch angle and rolling element diameter.

$$FTF = f_r \left(1 - \frac{d_b}{d_p \operatorname{Cos} \varphi} \right) \tag{5}$$

where, n = number of balls in rolling element bearing, d_b = ball diameter, d_p = bearing pitch diameter, ϕ = bearing contact angle, f_r = rotor frequency.

We assume that there is no sliding and contact between the rolling elements and the races are only pure rolling. However, there is always some sliding in reality. These defect frequencies lie in the low frequency range and are usually less than 500 Hz. If skidding and sliding is considered between rolling element and races, actual defect frequency is slightly different from that calculated from the above equations. Several researchers (Miyachi and Seki, 1986; Tandon and Nakra, 1993; Martins and Gerges, 1984) have reported success in bearing defect detection by identifying these rotational frequencies. In some work (Osuagwu and Thomas 1982; Chaturvedi and Thomas 1982) it is also reported that significant peaks are difficult to be detected when signal is noisy and defect is small. In such situations, envelope detector (ED) or high frequency resonance technique (HFRT) has often been used with frequency domain analysis to identify bearing faults (Broderick et al., 1972). Another important bearing fault detection technique is Power cepstrum. It is defined as the inverse Fourier transform of the logarithmic power spectrum. Bearing frequency. In general, practice, the vibration signal is bandpass filtered around one of the resonant frequencies excited by defect striking within the bearing, which eliminates most of the unwanted vibration signal. The bandpass filtered signal is then demodulated by an envelope detector where signal gets rectified and smoothed by low pass filter which in turn eliminate the carrier or bandpass filtered resonant frequen cy. The spectrum of the envelope signal in low frequency range obtains the characteristic frequencies associated with the faulty bearing. The efficiency of this technique has been evaluated by several researchers (McFadden and Smith, 1984; Prashad et al., 1985; Miyachi and Seki, 1986). Ho and Randall (2000) suggested that self-adaptive Noise Cancellation (SANC) technique gives better result in conjunction to an envelope detection method. FFT is effective when signals are non-stationary and signal to noise ratio is low (Burgess, 1988).

Time-Frequency Domain

In rotating machines, bearing vibration signal is a combination of periodic components, dominated by the machine rotation, with signals of a random nature, dominated by a possible bearing fault or imperfection. This phenomenon is periodic in nature, has time invariance and non-stationary. This has led to the development of time-frequency analysis methods. The advantage of time-frequency domain techniques over FFT is that it has capability to handle both, stationary and non-stationary vibration signals. A number of time-frequency analysis techniques, like the Short Time Fourier Transform (STFT), Wigner-Ville Distribution (WVD), and Wavelet Transform (WT), are the techniques in time-frequency domain. STFT uses window function which involves division of the initial signal into segments with short time window and application of the Fourier transform to all time segments to identify the frequencies present in that segment. Wigner-Ville Distribution is a Fourier transform of a product of the signal with its complex conjugate. Thus, it resembles the computation of the power spectral density. Wavelet transform (WT) is the most popular and widely used time-frequency domain technique for bearing defect detection. A waveform of limited duration that has an average value of zero. In waveletanalysis, signal is broken up into shifted and scaled versions of the original wavelet. For the diagnosis of fault in machines, the main features of wavelets are timefrequency analysis of signals, fault feature extraction, detection of singularity for signals, extracting weak signals, compression of vibration signals and the system identification (Burgess, 1988). Because wavelet is flexible and implemented effectively in computation, it i s a very effective tool for machine condition monitoring and fault diagnosis (Peng and Chu, 2004) and bearing fault detection in particular (Prabhakar et al., 2002; Shi et al., 2004; Nikolaou and Antoniadis, 2002). Wavelet transform are continuous and discrete wavelet type. Continuous Wavelet Transform is product of signal with translate and dilate of wavelet. CWT uses short windows at high frequencies and long windows at low frequencies. It is given by the expression

$$CWT(S,\tau) = \frac{1}{\sqrt{|s|}} \int_{-\infty}^{\infty} x(t)\varphi^{*}\left(\frac{t-\tau}{s}\right) dt \tag{6}$$

where x (t) denote signal, S denote scale index which is a reciprocal of frequency and τ is a time shifting parameter, ψ (t) is wavelet function. Discrete Wavelet Transform (DWT) uses a small number of scales to analyze signal and number of translation vary at each scale. The discrete wavelet transform is a discretization of the CWT through sampling specific wavelet coefficients. DWT analyses signals at different frequency bands with different resolution. FFT is ineffective when signals to be extracted are weak and hence wavelet transform is used. Discrete Wavelet Transform was applied to predict spalling defect in bearings (Mori et al., 1996). DWT can be used to detect single and multipoint defect in races of ball bearings (Prabhakar et al., 2002). Reliable condition monitoring is based on denoising of vibrational signals so that weak signals are easily detected. Wavelet decomposition based and wavelet filter based are the two denoising based approaches used by Qiu et al. (2006) and he compared the two approaches and concluded that wavelet filter based approach is better for detection of weak signal. To enhance feature extraction, Wang and Gao (2003) applied wavelet transforms and fast Fourier transform in conjunction to improve feature extraction. Junsheng et al. (2007) proposed scale wavelet power spectrum comparison and auto-correlation analysis of time-wavelet power. Kumar and Singh (2013) reports the use of Symlet wavelet for signal decomposition to extract the fault size on the outer race of taper roller bearing.

It is clear that with the advent of vibration based fault detection techniques a lot of work has been done to identify defects in the components of rolling element bearing. Though several techniques have been proposed for analyzing the vibration signal in time domain, frequency domain and in time-frequency domain, researchers are still attempting improvement in signal processing techniques.

EXPERIMENTAL SETUP

Experiments were conducted using a 2hp Reliance Electric motor, and acceleration data was measured at locations nearto and remote from the motor bearings. Vibration signals from test rig-2 of bearing data center at Case Western Reserve University, USA.

As shown in Figures 3 and 4, the test stand consists of a 2 hp motor (left), a torque transducer/encoder (center), a dynamometer (right), and control electronics (not shown).

Size and Location of Defect in 6205 SKF Bearings

Motor bearings were seeded with faults using electro-discharge machining (EDM). Faults ranging from 0.007 inches in diameter to 0.028 inches in diameter were introduced separately at the inner raceway, rolling element (i.e. ball) and outer raceway. Faulted bearings were reinstalled into the test motor and vibration data was recorded for motor loads of 0 to 3 horsepower (motor speeds of 1797 to 1720 RPM). The defect specification is shown in Table 1.

Figure 3. Bearing test-rig

Figure 4. Schematic view of the test rig

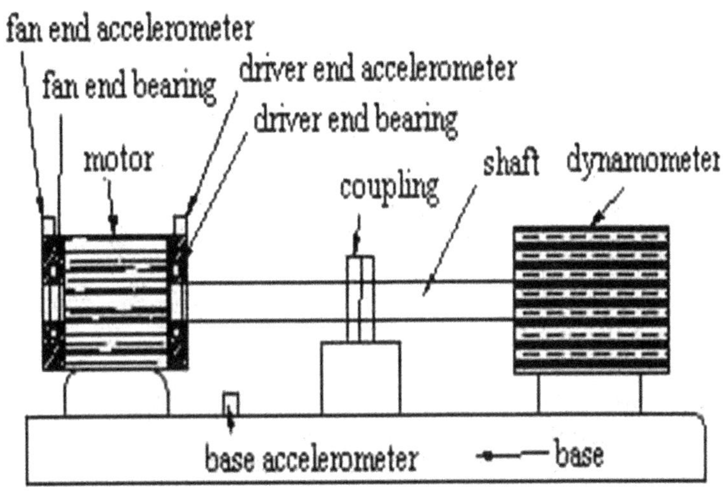

Table 1. Defect size and depth of inner race

Bearing	Fault Location	Fault Size	Fault Depth	Manufacturer
Drive End	Inner Raceway	.007	.011	SKF
Drive End	Inner Raceway	.014	.011	SKF
Drive End	Inner Raceway	.021	.011	SKF
Drive End	Inner Raceway	.028	.050	NTN

The photographic view of spall defect in bearing is shown in Figure 5 and Figure 6.

During experiments, an accelerometer was attached to the motor supporting base plate as well. Vibration signals were collected using a 16 channel DAT recorder, and were post processed in a Matlab environment. Digital data was collected at 12,000 samples per second, for drive end bearing faults. Speed and horsepower data were collected using the torque transducer/encoder and were recorded by hand.

Outer raceway faults are stationary faults, therefore placement of the fault relative to the load zone of the bearing has a direct impact on the vibration response of the motor/bearing system. In order to quantify this effect, experiments were conducted for both fan and drive end bearings with outer raceway faults located at 3 o'clock (directly in the load zone), at 6 o'clock (orthogonal to the load zone), and at 12 o'clock

TEST BEARING SPECIFICATION AND DIMENSIONS

The test bearing used is 6205-2RS JEM SKF, deep groove ball bearing at the drive end. Fault is induced at the inner race through electro-discharge machining. If a defect is present in the bearing it results in shock as the defect strikes the mating surface. The defect frequencies of the bearing depends on the

Figure 5. Spall defect in inner race

Figure 6. Spall defect in ball

speed, bearing geometry and dimensions. The bearing specification and defect frequency of bearing is shown in the Table 2 and Table 3 respectively

The experimental data obtained from the Case Western Reserve Bearing Data Centre is available in matlab file. The data has been processed in matlab to generate signals for different defect sizes at 1730 and 1750 RPM for inner race as well as ball defect and all the parameters in time domain has been calculated and tabulated.

FAULT ASSESSMENT USING SCALAR INDICATOR FOR ROLLER ELEMENT BEARING

When a failed rolling element bearing is removed from the machine it mostly displays cracks and flaking. Early identification of such defects along with the severity of damage under operating condition of bearing may avoid malfunctioning and breakdown of machines. Assessing defect severity from the initial stage of degradation can be done through processing of experimental signals considering faulty bearing with different defect sizes. Time domain processing of vibration signal is a simple technique to get information of bearing condition. Designing of a new scalar indicator may provide a way to constantly monitor

Table 2. 6205-2RS JEM SKF, deep groove ball bearing Size: (inches) drive end bearing

Inside Diameter	Outside Diameter	Thickness	Ball Diameter	Pitch Diameter
0.9843	2.0472	0.590	0.3126	1.537

Table 3. Defect frequencies (multiple of running speed in Hz)

Inner Ring	Outer Ring	Cage Train	Rolling Element
5.4152	3.5848	0.39828	4.7135

the aggravation in inner race, rolling element and outer race defect. The aim of this Chapter is to assess increase in bearing defect severity by applying a new scalar indicator on time domain vibration signal.

Monitoring Machine Health Using Scalar Indicator

For the supervision and condition monitoring of machine certain no of scalar indicators must be chosen. An indicator is used for the identifying the defects and tracking their evolution; it can also be used for the pausing or stopping of the installation.

Vibrations are induced in machine due to dynamic loads generated by moving parts. It is widely used for the condition monitoring of machine. Wide list of surveillance indicators can be defined that are more or less sensitive to severity of fault. Surveillance indicators are classified under two categories.

- Scalar indicators, which follow the evolution of parameter linked to the amplitude of vibration in time domain.
- Spectral indicators that follow the evolution in frequency and in amplitude of each of its components.

A universal indicator which can identify the defect at its intial stage is not present. An ideal indicator would give predefined alarm threshold whose value is independent of the nature of the defect, the machine, and its operating conditions.

A scalar indicator obtained from the time domain gives a scalar number which may not be essentially significant but as it evolves with time its values indicates the level of aggravation of a defect. The evolution in time of a scalar indicator is more important than its intrinsic value. Scalar indicators may provide information on the strategic decisions related to any immediate replacement of the failed bearing. Defining a scalar indicator in the time domain requires choosing:

- A parameter related to kinematics which gives the vibratory movement (acceleration, velocity, displacement) according to the frequency content of the signal.
- A parameter which represents amplitude of signal (Max-peak amplitude, RMS value, Crest factor, Kurtosis).
- A bandwidth over which the retained parameter will be evaluated.
- A duration of analysis.

Six most commonly used indicators are Peak, RMS, Kurtosis (Ku), Crest Factor (CF), Impulse Factor (IF), Shape Factor (SF).

Review of Scalar Parameters

The scalar indicators mentioned below can be used to analyze the data obtained in time domain for the defected bearings.

1. **Peak Value:** It is a magnitude of maximum value within the certain time range of vibration signal. It can be computed as $\max|u_K| 1 \leq k \leq l$

2. **Average Value:** It should be closer to zero for healthy bearing. Diversion of average value from zero indicates the severity of fault in the bearing. The value can be obtained through:

$$Average\ Value = \frac{1}{l}\sum_{k=1}^{l} u_k \tag{7}$$

'l' is the sample length and u_k is discrete values of signal

3. **Root Mean Square Value (RMS):** It gives is an indication of overall level of discrete signal and is a powerful tool to estimate the average power in system vibration. It can be very effective in detecting a major out of balance in rotating systems. RMS value also increases with the appearance of shock pulses (bearing related signal). The numerical value of RMS can be calculated by:

$$RMS = \sqrt{\frac{1}{l}\sum_{k=1}^{l} u_k^2} \tag{8}$$

RMS value of a signal is equal to the standard deviation for 0 mean signals

4. **Crest Factor (CF):** The crest factor is a measure of the impulsiveness of a noise or vibration signal. It is often used when dealing with shocks, impulsive noise and short events. The crest factor for a sine wave is 1.414and for a truly random noise is generally less than 3. It can be computed by identifying peak level and RMS of vibration signal. It is given by:

$$CF = \frac{Peak\ value}{RMS\ value} \tag{9}$$

The value of Crest Factor may reach between 2 and 6 in normal operations. A value above 6 is usually associated with machinery problems.

5. **Kurtosis (Ku):** It is a fourth moment, normalized with respect to the fourth power of standard deviation. It can be used as an indicator of major peaks in a signal. The theoretical value of kurtosis for a good surface finish is3. Its value can be obtained by calculating:

$$Ku = \frac{1}{l}\sum_{k=1}^{l} \frac{\left(u_{k-\mu}\right)^4}{\sigma^4} \tag{10}$$

'u_k' is the signal sampling value, μ is the average value of signal, σ is the standard deviation

6. **Impulse Factor (IF):** It can also be used to measure the presence of defect induced impulses in the vibration signal. Value of this indicator is given by:

$$IF = \frac{peak \sim value}{\frac{1}{l}\sum_{k=1}^{l}\left|u_k\right|} \tag{11}$$

7. **Shape Factor (SF):** Shape Factor is a dimensionless quantity that defines the shape of an object or a signal. A shape Factor equal to one usually represents an ideal case or maximum symmetry. Its value can be obtained through:
8. **Skewness:** Machined or ground surfaces in bearings show a random distribution of asperities that are commonly described with normal distribution function. Various statistical moments like Skewness can describe the shape of distribution curves and asses bearing surface damage level. Basically, Skewness measures the asymmetry of real random variable probability distribution and its computation formula can be given by:

$$Skewness = \frac{1}{l}\sum_{k=1}^{l}\frac{\left(u_{k-\mu}\right)^3}{\sigma^3}$$

where, μ and σ is the average and standard deviation respectively. For noise only, or at low SNR skewness is zero.

DEFECT SEVERITY ASSESSMENT THROUGH COMMONLY USED SCALAR PARAMETERS

Impulsive vibrations are generated due to shock within rolling bearing. If a defect is present on one surface it strikes mating surface and generate a pulse. Shock is produced which excites the resonances of the bearing as well as the overall mechanical system. Thus, the pulsation generated by defects in bearing excites vibration at specific defect frequencies as well as a high-frequency response in the overall machine structure. The scalar indicators obtained from time domain signals are physical parameters specially used for the recognition of the vibration origin in order to identify its nature and its degree of severity. The signals of bearings at moderate speed, with, discrete values representing velocity, have been considered for investigation. Experimental signals for SKF 6205 deep groove ball bearings running at 1730 and 1750 RPM for, respectively, healthy bearing and with different sizes of inner race and ball defect are shown in Figure 7. The defect sizes in both the bearings have been varied from 0.18 mm diameter to a large defect size of 0.72 mm diameter with constant depth of 0.28mm.

Measurement of RMS and Peak

Peak and RMS values are generally used to indicate the presence of defect severity. RMS signal is a simple and inexpensive measurement and is computed by the root mean square level of the time record. However RMS level does not indicate early detection of defect because the over all measurements do not significantly change unless the defect becomes severe. As an alternate to RMS, the peak level of signal is used. For a new machine a baseline "peak" level is defined and any deviation from this level indicates the change in machine condition.

Figure 7. Signals w.r.t. time for SKF 6205 bearing at 1730 RPM with defect sizes: (a) healthy bearing, (b) 0.18 mm, (c) 0.36 mm, (d) 0.54 mm, (e) 0.72 mm and at 1750 RPM, (f) healthy bearing, (g) 0.18 mm, (h) 0.36 mm, (i) 0.54 mm, (f) 0.72 mm inner race defect diameter

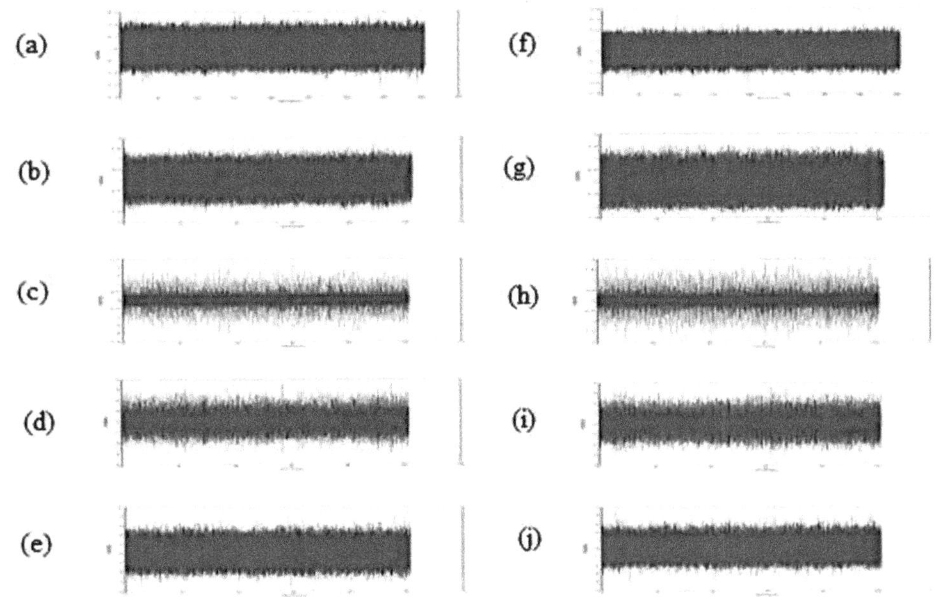

The values of six most commonly used scalar indicators i.e. peak, RMS, Kurtosis, Crest Factor, Impulse Factor, Shape Factor are extracted from the signals for different defect sizes for SKF 6205 bearing at 1730 and 1750 RPM for inner race and ball are given in the Table 4, 5, 6, and 7. These values are then are plotted with respect to defect sizes varying from 0.18 mm to 0.72 mm

Figures 8, 9, 10, and 11 show a typical evolution of peak and RMS values at 1730 RPM and 1750 RPM when the inner race and ball race of a bearing is deteriorating. It is observed that initially when a localized defect appears the small shock pulse increases the peak level but RMS is not influenced much. As the defect size grows significant impact results in increased RMS level along with peak level.

Table 4. Scalar parameters w.r.t aggravation of inner race defect in SKF6205 bearing at 1730 RPM

Indicator	Healthy Bearing	(0.18 mm)	(0.46 mm)	(0.54 mm)	(0.72mm)
Peak	0.2837	1.6715	2.1269	3.6150	4.4469
RMS	0.0659	0.4136	0.1808	0.4487	0.8231
Kurtosis	2.9572	5.2911	18.1140	8.4451	3.4169
Impulse Factor	5.4595	7.4575	18.4151	11.9329	6.7208
Crest Factor	4.4063	5.4928	11.7649	8.0560	5.2811
Shape Factor	1.2446	1.4992	1.5653	1.4812	1.2726

Table 5. Scalar parameters w.r.t aggravation of inner race defect in SKF6205 bearing at 1750 RPM

Indicator	Healthy Bearing	(0.18 mm)	(0.46 mm)	(0.54 mm)	(0.72mm)
Peak	0.4592	1.6396	1.8542	3.6231	4.5410
RMS	0.0643	0.2995	0.1631	0.4889	0.8413
Kurtosis	2.9251	5.5638	21.6862	8.0580	3.2863
Impulse Factor	6.9252	7.6735	18.7847	11.1192	6.8403
Crest Factor	5.5847	5.4743	11.4707	7.4113	5.4973
Shape Factor	1.2400	1.4017	1.6520	1.5003	1.2673

Table 6. Scalar parameters w.r.t aggravation of ball defect in SKF6205 bearing at 1730 RPM

Indicator	Healthy Bearing	(0.18 mm)	(0.46 mm)	(0.54 mm)	(0.72mm)
Peak	0.2837	0.6715	1.9892	0.5763	11.4639
RMS	0.0659	0.1536	0.1337	0.1180	2.1450
Kurtosis	2.9572	2.8897	14.8587	3.1056	3.8991
Impulse Factor	5.4595	5.4497	21.6375	6.1556	7.0120
Crest Factor	4.4063	4.4708	14.8786	4.8855	5.2980
Shape Factor	1.2446	1.2468	1.4543	1.2600	1.4255

Table 7. Scalar parameters w.r.t aggravation of ball defect in SKF6205 bearing at 1750 RPM

Indicator	Healthy Bearing	(0.18 mm)	(0.46 mm)	(0.54 mm)	(0.72mm)
Peak	0.4592	0.6046	1.8391	0.6455	10.4419
RMS	0.0643	0.1473	0.1435	0.1073	2.1457
Kurtosis	2.9251	2.8314	9.7522	3.4006	3.7723
Impulse Factor	6.9252	5.1085	18.2236	7.6284	6.4060
Crest Factor	5.5847	4.1058	12.8198	6.0177	4.8664
Shape Factor	1.2400	1.2442	1.4215	1.2676	1.4164

Measurement of Kurtosis, Crest Factor, Impulse Factor, and Shape Factor

Crest Factor and Kurtosis are the most suitable scalar indicators used to characterize the vibrations. The Crest factor and Kurtosis are sensitive to the spikiness of the vibration signals and less dependent on the vibration level, and they can provide an early indication of significant changes in vibration signals. The Impulse and Shape factors are functions of the redressed signal average.

Monitoring Crest factor of acceleration time waveform is a simple technique because it does not require elaborate signal processing and interpretation is straightforward. For a healthy bearing crest factor is below 4 for random waveform. As fault increases waveform becomes impulsive with high peak levels, while RMS value doesn't change significantly thus, Crest factor increases above 4. With further increase in fault RMS level increases significantly and the crest factor decreases.

Figure 8. Plot of peak and RMS vs. defect size at 1730 RPM for inner race defect

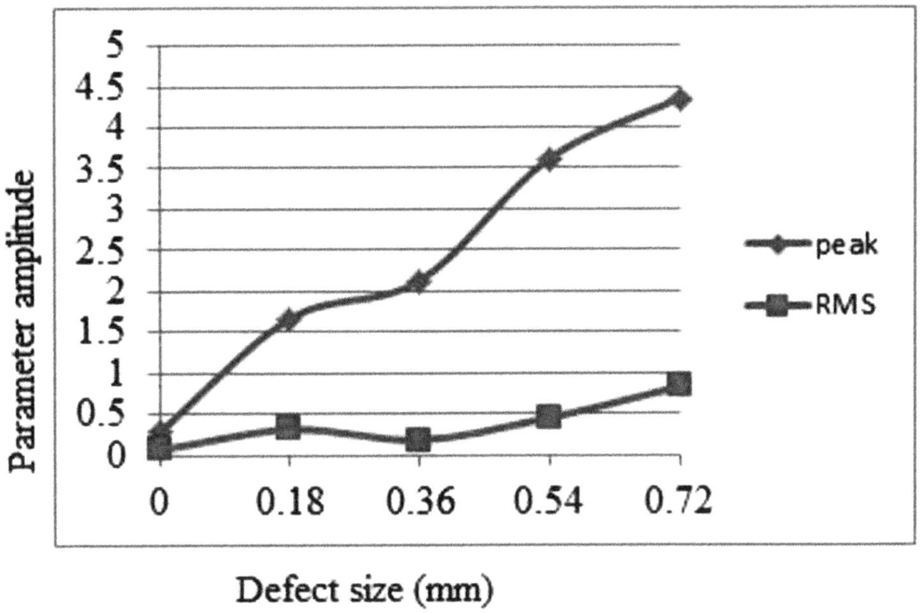

Figure 9. Plot of peak and RMS vs. defect size at 1750 RPM for inner race defect

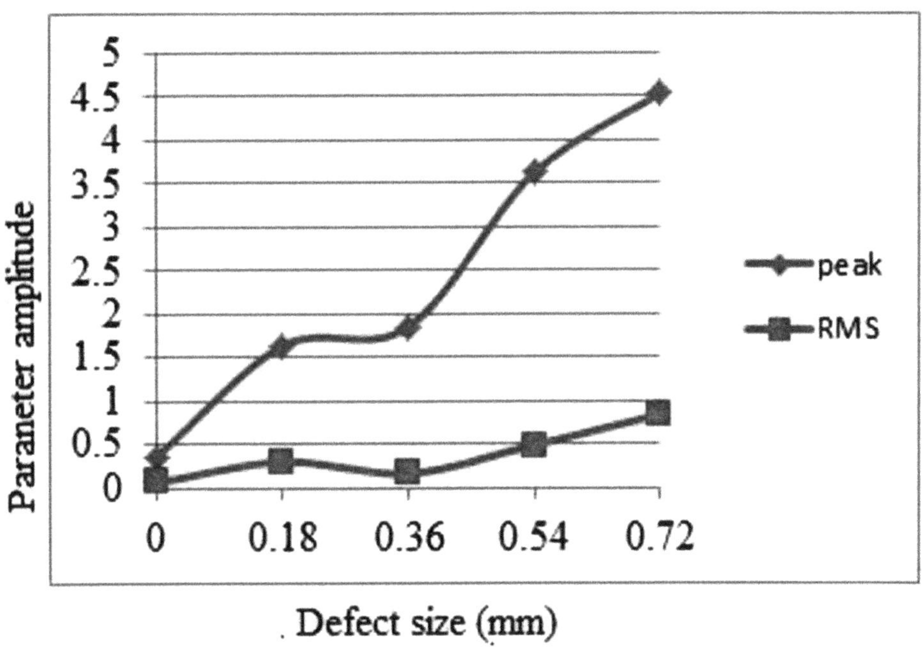

Figure 12, 13, 14 and 15 describes the evolution of the Kurtosis, Impulse Factor, Crest factor and Shape Factor at 1730 RPM and 1750 RPM with the defect size when the inner race of a bearing is deteriorating. It can be noticed that the levels of the indicators increases at the beginning of the deterioration when the RMS level is constant and then a decrease is seen as the bearing further deteriorates. As the size of

Figure 10. Plot of peak and RMS vs. defect size at 1730 RPM for ball defect

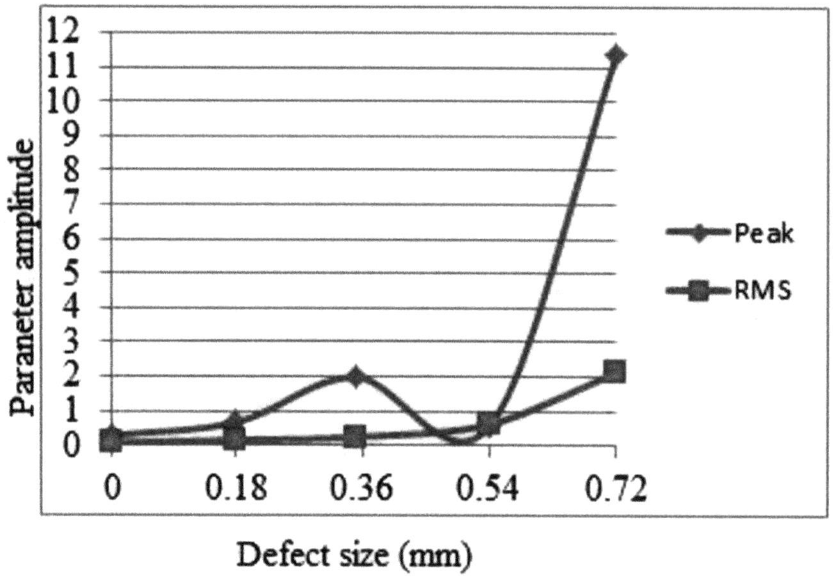

Figure 11. Plot of peak and RMS vs. defect size at 1750 RPM for ball defect

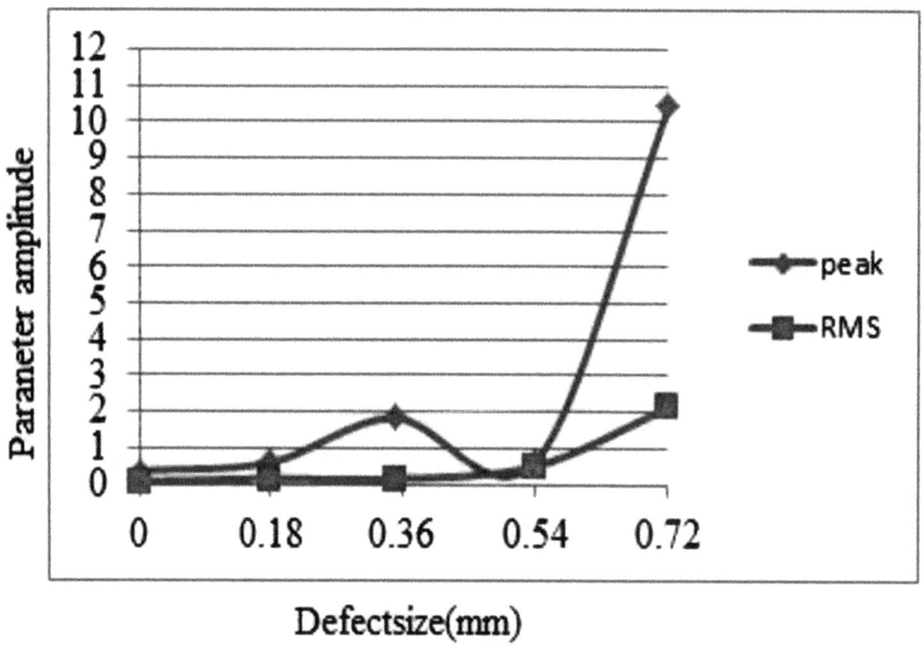

defect increase these values are reverting back to the undamaged bearing level as the RMS values increase dramatically. The Kurtosis value is close to 3 for healthy bearing (corresponding to a pure random signal), and increases until it reaches a value close to 18. As the defect further increases signal becomes more and more random and the Kurtosis, impulse factor, crest factor decreases down to a level of undamaged

Figure 12. Plot of scalar indicators vs. defect size at 1730 RPM for inner race defect

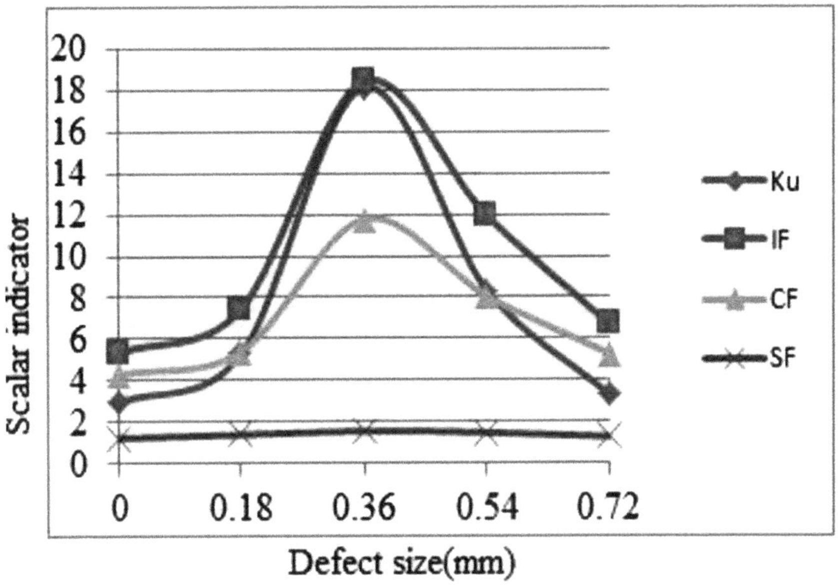

Figure 13. Plot of scalar indicators vs. defect size at 1750 RPM for inner race defect

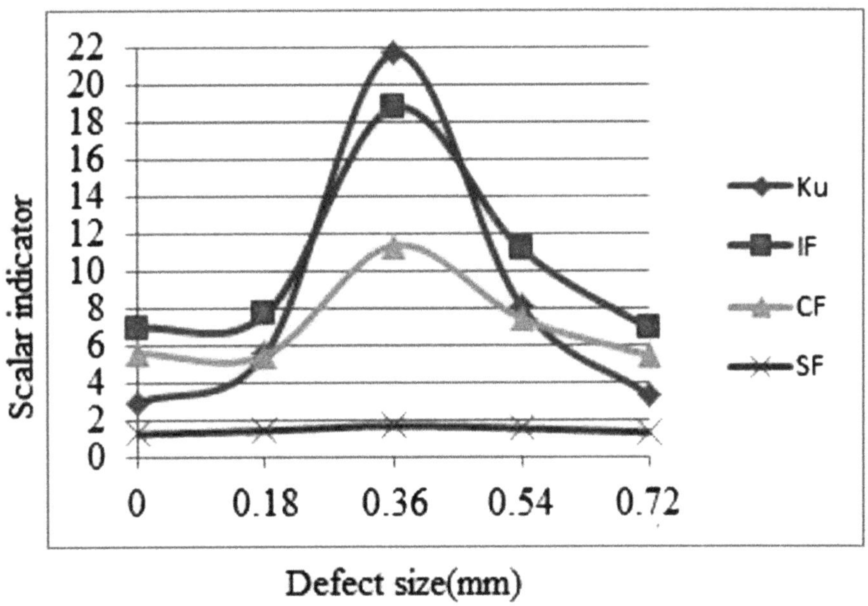

one this makes the surveillance of bearing condition difficult and may deliver misleading information especially when the defect becomes severe. When all the scalar indicators are compared, it is observed that Kurtosis is the most sensitive indicator at the beginning of the deterioration whereas impulse factor becomes most sensitive at the end of the bearing life. The shape factor can be seen to be least sensitive

Figure 14. Plot of scalar indicators vs. defect size at 1730 RPM for ball defect

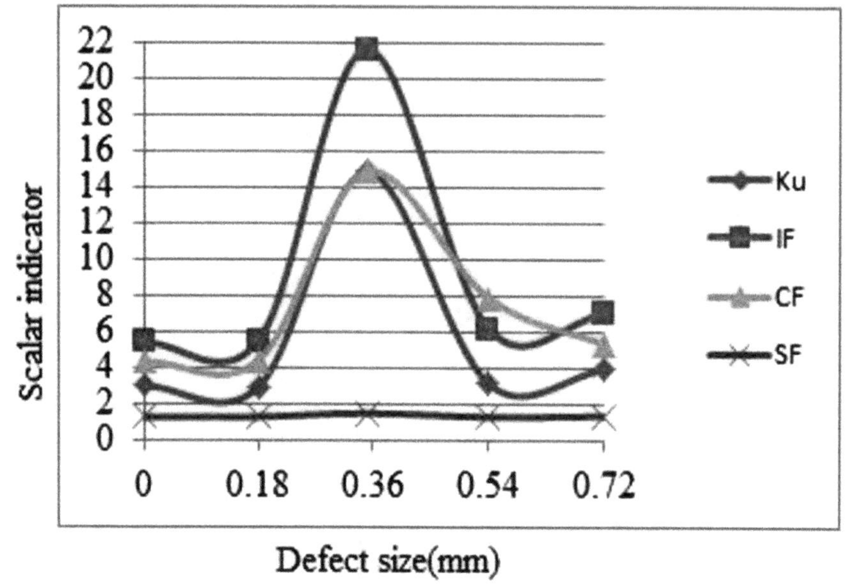

Figure 15. Plot of scalar indicators vs. defect size at 1750 RPM for ball defect

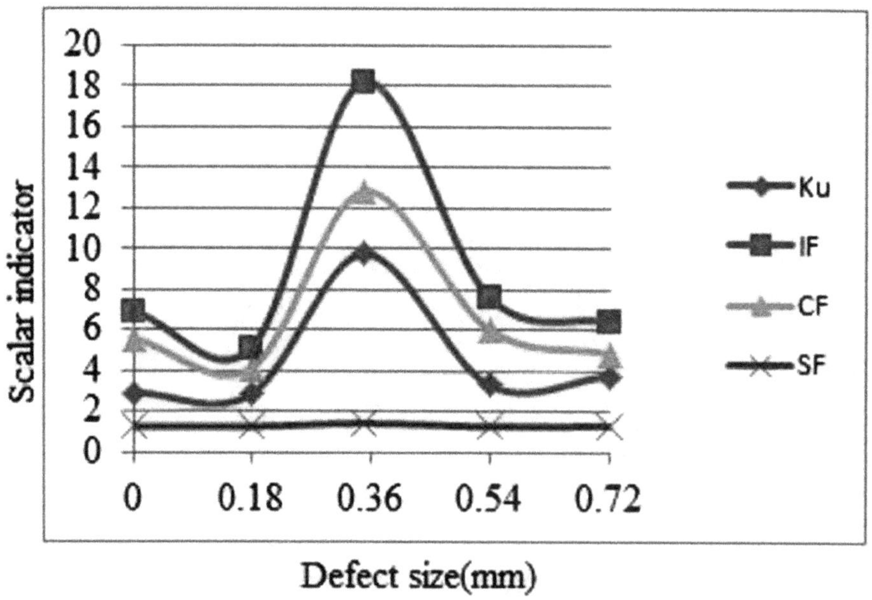

and cannot be used for the detection of defects. A trend chart of monitoring process uses Kurtosis, Crest factor, Impulse factor until the bearing vibration indicators reach their maximum value. When the slope becomes negative, it means bearing is approaching the end of its life. When the indicators are weak, trend should be checked whether the bearing is at its first stage of degradation or at end of its life.

CONCLUSION

The Kurtosis and, to some degree, the Impulse Factor and the Crest factor are three scalar indicators for detection of fault during the beginning of degradation. However, the values of these indicators becomes decreasing monotonous functions of the deterioration after the defect reaches a maximum. Furthermore, if a trend analysis is not conducted, they are difficult to use as surveillance indicators without associating it to the evolution of the RMS value of the amplitude of the signal. They reveal fault propagation, but do not predict when the fault will become excessive this may lead to false alarm to the machine operator.

REFERENCES

Alfredson, R. J., & Mathew, J. (1985). Frequency domain methods for monitoring the condition of rolling element bearings. Transactions of the Institution of Engineers, Australia. *Mechanical Engineering (New York, N.Y.)*, *10*(2), 108–112.

Antoni, J., & Randall, R. B. (2006). The spectral kurtosis: Application to the vibratory surveillance and diagnostics of rotating machines. *Mechanical Systems and Signal Processing*, *20*(2), 308–331. doi:10.1016/j.ymssp.2004.09.002

Barthel, K. (1978, February). The Shock Pulse Method for Determining the Condition of Anti-Friction Bearings.*Proceedings of the Machinery Vibration Monitoring Analysis Seminar.*

Bently, D. (1989). *Predictive maintenance through the monitoring and diagnostics of rolling element bearings*. Bently Nevada Co. *Applications Note*, *ANO44*, 2–8.

Broderick, J. J., Burchill, R. F., & Clark, H. L. (1972). *Design and fabrication of prototype system for early warning of impending bearing failure*. Academic Press.

Burgess, P. F. (1988). Antifriction bearing fault detection using envelope detection. *Transactions of the Institution of Professional Engineers New Zealand: Electrical/Mechanical/Chemical Engineering Section, 15*(2), 77.

Butler, D. E. (1973). The shock-pulse method for the detection of damaged rolling bearings. *Non-Destructive Testing*, *6*(2), 92–95. doi:10.1016/0029-1021(73)90116-3

Chaturvedi, G. K., & Thomas, D. W. (1982). Bearing fault detection using adaptive noise cancelling. *Journal of Mechanical Design*, *104*(2), 280–289. doi:10.1115/1.3256337

Choudhury, A., & Tandon, N. (1998). A theoretical model to predict vibration response of rolling bearings to distributed defects under radial load. *Journal of Vibration and Acoustics*, *120*(1), 214–220. doi:10.1115/1.2893808

Dwyer, R. (1983, April). Detection of non-Gaussian signals by frequency domain kurtosis estimation. In *Acoustics, Speech, and Signal Processing, IEEE International Conference on ICASSP'83* (Vol. 8, pp. 607-610). IEEE. doi:10.1109/ICASSP.1983.1172264

Dyer, D., & Stewart, R. M. (1978). Detection of rolling element bearing damage by statistical vibration analysis. *Journal of Mechanical Design*, *100*(2), 229–235. doi:10.1115/1.3453905

Ho, D., & Randall, R. B. (2000). Optimisation of bearing diagnostic techniques using simulated and actual bearing fault signals. *Mechanical Systems and Signal Processing*, *14*(5), 763–788. doi:10.1006/mssp.2000.1304

Junsheng, C., Dejie, Y., & Yu, Y. (2007). Application of an impulse response wavelet to fault diagnosis of rolling bearings. *Mechanical Systems and Signal Processing*, *21*(2), 920–929. doi:10.1016/j.ymssp.2005.09.014

Karacay, T., & Akturk, N. (2009). Experimental diagnostics of ball bearings using statistical and spectral methods. *Tribology International*, *42*(6), 836–843. doi:10.1016/j.triboint.2008.11.003

Kumar, R., & Singh, M. (2013). Outer race defect width measurement in taper roller bearing using discrete wavelet transform of vibration signal. *Measurement*, *46*(1), 537–545. doi:10.1016/j.measurement.2012.08.012

Martins, L. G., & Gerges, S. N. Y. (1984). Comparison of Signal Analyses for Detecting Incipient Bearing Damage. *Condition Monitoring, 84*, 191-204.

Mathew, J., & Alfredson, R. J. (1984). The condition monitoring of rolling element bearings using vibration analysis. *Journal of Vibration, Acoustics, Stress, and Reliability in Design, 106*(3), 447-453.

McFadden, P. D., & Smith, J. D. (1984). Vibration monitoring of rolling element bearings by the high-frequency resonance technique—a review. *Tribology International*, *17*(1), 3–10. doi:10.1016/0301-679X(84)90076-8

Meyer, L. D., Ahlgren, F. F., & Weichbrodt, B. (1980). An analytic model for ball bearing vibrations to predict vibration response to distributed defects. *Journal of Mechanical Design*, *102*(2), 205–210. doi:10.1115/1.3254731

Miyachi, T., & Seki, K. (1986, September). An Investigation of the Early Detection of Defects in Ball Bearings Using Vibration Monitoring, Practical Limit of Detectability and Growth Speed of Defects. *Proceedings of the International Conference of Rotordynamics*, 403-408.

Mohanty, A. R. (2014). *Machinery Condition Monitoring: Principles and Practices*. CRC Press.

Mori, K., Kasashima, N., Yoshioka, T., & Ueno, Y. (1996). Prediction of spalling on a ball bearing by applying the discrete wavelet transform to vibration signals. *Wear*, *195*(1), 162–168. doi:10.1016/0043-1648(95)06817-1

Nikolaou, N. G., & Antoniadis, I. A. (2002). Demodulation of vibration signals generated by defects in rolling element bearings using complex shifted Morlet wavelets. *Mechanical Systems and Signal Processing*, *16*(4), 677–694. doi:10.1006/mssp.2001.1459

Ocak, H., & Loparo, K. A. (2005). HMM-based fault detection and diagnosis scheme for rolling element bearings. *Journal of Vibration and Acoustics*, *127*(4), 299–306. doi:10.1115/1.1924636

Osuagwu, C. C., & Thomas, D. W. (1982). Effect of inter-modulation and quasi-periodic instability in the diagnosis of rolling element incipient defect. *Journal of Mechanical Design*, *104*(2), 296–302. doi:10.1115/1.3256339

Paliwal, D., Choudhury, A., & Tingarikar, G. (2014). Wavelet and scalar indicator based fault assessment approach for rolling element bearings. *Procedia Materials Science*, *5*, 2347–2355. doi:10.1016/j.mspro.2014.07.478

Peng, Z. K., & Chu, F. L. (2004). Application of the wavelet transform in machine condition monitoring and fault diagnostics: A review with bibliography. *Mechanical Systems and Signal Processing*, *18*(2), 199–221. doi:10.1016/S0888-3270(03)00075-X

Prabhakar, S., Mohanty, A. R., & Sekhar, A. S. (2002). Application of discrete wavelet transform for detection of ball bearing race faults. *Tribology International*, *35*(12), 793–800. doi:10.1016/S0301-679X(02)00063-4

Prashad, H., Ghosh, M., & Biswas, S. (1985). Diagnostic monitoring of rolling-element bearings by high-frequency resonance technique. *ASLE Transactions, 28*(4), 439-448.

Py, L. K. (1983). A review of rolling element bearing health monitoring.*Proceedings of Machinery Vibration Monitoring and Analysis Meeting*, 145.

Qiu, H., Lee, J., Lin, J., & Yu, G. (2006). Wavelet filter-based weak signature detection method and its application on rolling element bearing prognostics. *Journal of Sound and Vibration*, *289*(4), 1066–1090. doi:10.1016/j.jsv.2005.03.007

Ray, A. (1980, September). Monitoring rolling contact bearings under adverse conditions.*2nd International Conference of Vibrations in rotating Machinery, Institution of Mechanical Engineers*, 1-4.

Reif, Z., & Lai, M. S. (1989). *Detection of developing bearing failures by means of vibration*. New York: The American Society of Mechanical Engineers.

Sassi, S., Badri, B., & Thomas, M. (2006, October). "TALAF" and "THIKAT" as Innovative Time-Domain Indicators for Tracking Ball Bearings. *Proceedings of the 14th Seminar on Machinery Vibration*, 24-27.

Shi, D. F., Wang, W. J., & Qu, L. S. (2004). Defect detection for bearings using envelope spectra of wavelet transform. *Journal of Vibration and Acoustics*, *126*(4), 567–573. doi:10.1115/1.1804995

Smith, J. D. (1982). Vibration monitoring of bearings at low speeds. *Tribology International*, *15*(3), 139–144. doi:10.1016/0301-679X(82)90130-X

Sunnersjö, C. S. (1978). Varying compliance vibrations of rolling bearings. *Journal of Sound and Vibration*, *58*(3), 363–373. doi:10.1016/S0022-460X(78)80044-3

Tallian, T. E., & Gustafsson, O. G. (1965). Progress in rolling bearing vibration research and control. *ASLE Transactions, 8*(3), 195-207.

Tandon, N. (1994). A comparison of some vibration parameters for the condition monitoring of rolling element bearings. *Measurement*, *12*(3), 285–289. doi:10.1016/0263-2241(94)90033-7

Tandon, N., & Choudhury, A. (1999). A review of vibration and acoustic measurement methods for the detection of defects in rolling element bearings. *Tribology International*, *32*(8), 469–480. doi:10.1016/S0301-679X(99)00077-8

Tandon, N., & Nakra, B. C. (1992). Vibration and acoustic monitoring techniques for the detection of defects in rolling element bearings—a review. *The Shock and Vibration Digest, 24*(3), 3-11.

Tandon, N., & Nakra, B. C. (1993). Detection of defects in rolling element bearings by vibration monitoring. *Indian Journal of Mechanical Engineering Division, 73*, 271–282.

Taylor, J. I. (1994). *The vibration analysis handbook.* Vibration Consultants.

Wang, C., & Gao, R. X. (2003). Wavelet transform with spectral post-processing for enhanced feature extraction . *IEEE Transactions on Instrumentation and Measurement, 52*(4), 1296–1301. doi:10.1109/TIM.2003.816807

Wardle, F. P., & Poon, S. Y. (1983). Rolling bearing noise-cause and cure. *Chartered Mechanical Engineer, 30*, 36-40.

Chapter 11
Impact of Human Resources on Quality After Just-in-Time Implementation

Teresita Molina
Universidad Autónoma de Ciudad Juárez, Mexico

Valeria Martínez Loya
Universidad Autónoma de Ciudad Juárez, Mexico

Jorge Luis García-Alcaraz
Universidad Autónoma de Ciudad Juárez, Mexico

Nadia Sofia Tanino
Texas A&M University, USA

Diego Tlapa
Universidad Autónoma de Baja California, Mexico

ABSTRACT

Nowadays maquiladoras require improving processes and reducing costs in order to stay in the market and be competitive. Just in Time (JIT), the lean manufacturing tool, works specifically on materials handling, but critical success factors (CSFs) for its implementation are unknown. For this reason, in this chapter the results of a survey applied to 372 maquiladora workers of Ciudad Juárez (México) are reported. Six hypotheses are proposed and validated through a structural equations model executed in WarpPLS 5.0® software, where four latent variables related with Management Commitment, Human Resources, Suppliers and Quality benefits in a JIT environment are associated. Direct effects, indirect effects and total effects of each one of the variables are analyzed. The results show there is enough statistical evidence to state that there is a direct and positive impact of Human Resources on the obtained Quality in a JIT environment application.

INTRODUCTION

The organization of this study presents a background of literature review, the research problem, the hypotheses according to the proposed model, the methodology and the results as is presented in the structural equation model evaluated, conclusions and the suggested future research. The objective of the study is the validation of six hypotheses through a structural equations model.

DOI: 10.4018/978-1-5225-2440-3.ch011

BACKGROUND

Facing the utmost challenges with stiff competition globally and locally, firms are forced to search alternatives to reduce costs, quality improvement and demand needs expectation of their customer (Nadarajan et al., 2013).

Lean Production has been adopted by companies in various service and industrial sectors in recent decades and these companies have subsequently moved forward in its implementation. In many cases this has enabled them to improve their results and competitiveness (Martínez-Jurado et al., 2014), Womack et al. (1991) first coined the term "Lean Production" in their seminal book, The Machine that Changed the World; however, the origin of lean thinking is generally attributed to Toyota, whose production system was originally referred to as JIT, but is now commonly called the Toyota Production System (TPS) (Fullerton et al., 2014). The TPS provided the basis for what was to become popularly known as JIT.

Initially JIT was applied to the manufacturing industry. This successfully helped to accomplish better product quality, preventive maintenance and employee motivation, and it also improved workers' involvement and commitment to their organization (Low Sui and Gao, 2011). The use of JIT practices can greatly reduce the number of adjustments and the defect rate, which mean JIT practices can improve product quality and reduce costs (Zhao et al., 2014).

Manufacturers can obtain improvements in quality and delivery by motivating their suppliers to adopt JIT production and JIT delivery (Jayaram et al., 2010). In the enterprise using JIT production system, the implementation and operation of flexible production plan must be accompanied by flexible allocation of human resources and incentive policies (Li and Yang, 2010).

The study of Flynn et al. (1997) illustrated that the JIT practices can affect the quality performance from three aspects: firstly, JIT practices will reduce waste and the stock, so the enterprise will improve their quality from more than one aspect; secondly, the question that reduce stock revealed will be reflected in the quality performance; thirdly, JIT practices will also have an impact on the butch, which will affect the quality performance because of timely feedback and thus reduce the variability of process (Zhao et al., 2014).

In JIT, long-term business relationships are maintained and information is exchanged closely with a small number of suppliers who have been selected according to such criteria as high quality, low-costs and punctual delivery. This provides benefits than can be enjoyed by both, suppliers and assemblers in the way of profits generated through more moderate transaction costs and through a learning effect brought about by the accumulation of production experience (Kaneko and Nojiri, 2008).

Although lean management is widely regarded as a business strategy and implementation of lean techniques improves business competitiveness and organizational performance, few researchers have concentrated on the validation of its positive link with business performance (Büyüközkan et al., 2015).

Research Problem

The literature review mentions that JIT practice, if implemented at management will have an impact on human resources practices and suppliers' deliveries; these three areas working together with JIT practices will achieve an overall improvement in quality. Literature also mentions that even though many authors recognize the benefits of JIT implementation on the improvement of quality many companies fail to achieve a full implementation (Bergenwall et al., 2012; Tillema and van der Steen, 2015).

The maquiladora industry in Ciudad Juárez counts with 326 businesses that import raw materials and export finished goods on high quantity levels, creating a necessity to improve manufacturing processes and optimize strategies to achieve quality objectives and to reduce costs on a day to day basis, in order to stay in the competitive market. This optimization requires a fast identification of CSFs that leads maquiladoras to an effective and efficient JIT implementation.

Currently, the maquiladora industry in Ciudad Juárez does not have a validated instrument or mechanism whereby the relation between analyzed variables can be determined; this mechanism would allow the industry to know the impact of *Human Resources* on *Quality*.

This study was performed with the objective to identify the CSFs for JIT implementation, and to analyze the interrelations among four latent variables (*Management Commitment*, *Human Resources*, *Suppliers*, and *Quality*) in maquiladora industry.

HYPOTHESES

According to the literature review, the four latent variables are related as follow.

Management Commitment and Human Resources in JIT

Companies must always put people on the first place. Respect for each employee, and help them to establish self-esteem and courage, which is half of a success. We should think about human resource management from a strategic perspective. If companies want to implement JIT successfully, they must give full play to the role of human resource policy (Li and Yang, 2010). One of the most important aspects of JIT is the capability to leverage the human capital of the whole enterprise by developing problem solving skills for the whole workforce in order to have all the employees contribute to the day by day improvement of the enterprise, to ensure the whole management functions as problem solvers at their level and coaches for the levels below them in order to foster those improvement skills. The individuals working in the company interact to create an evolving system that improves continuously, even without strong top down directions, and in a more relevant way, since the focus will be on delivering added value to the customer, while removing waste along the way (Masai et al., 2015).

Additionally, Hiltrop (1992) emphasized the need to invest in long-term training programs when firms implement JIT. Wafa and Yasin (1998) stressed, among other aspects, that obtaining good results depended on having well-trained workers. Power and Sohal (2000) found in a study that JIT companies needed to actively promote the development of a multi-skilled and flexible workforce by means of training and effective employee-development programs. Workers also need training to engage in improvement activities and problem-solving techniques. Training is also essential for tasks related to maintenance of tools and machinery. In short, training can advance employees' capabilities of accepting new skills and using new knowledge (Bonavia and Marin-Garcia, 2011).

Some of the JIT objectives are educating employees and having employee involvement and commitment (Jirarat et al., 2011; Low and Show, 2008; Savino and Mazza, 2015).

Table 1 shows Management Commitment attributes and Table 2 shows Human Resources attributes and authors who endorse these attributes.

Table 1. Management commitment JIT attributes

Item	Items or Attributes	Authors Who Endorse
1	Communication and constant coordination between different departments of the plant	(Bonavia and Marin-Garcia, 2011; Chen and Tan, 2011; Masai et al., 2015; Savino and Mazza, 2015)
2	Supervisors strive to promote teamwork, encouraging operators to cooperate and express their views	(Bonavia and Marin-Garcia, 2011; Chen and Tan, 2011; Low and Show, 2008; Low Sui and Gao, 2011; Masai et al., 2015)
3	Managers, engineers and operators constantly interact with each other	(Bonavia and Marin-Garcia, 2011; Chen and Tan, 2011; El Dabee et al., 2013; Masai et al., 2015; Savino and Mazza, 2015)
5	Senior management has a culture that encourages compliance with the deadlines	(Chen and Tan, 2011; Danese et al., 2012; Li and Yang, 2010)

Table 2. Human resources JIT attributes

Item	Items or Attributes	Authors Who Endorse
6	Employees are hired for their ability to solve problems and teamwork	(Bonavia and Marin-Garcia, 2011; Chen and Tan, 2011; Jirarat et al., 2011; Li and Yang, 2010; Masai et al., 2015; Savino and Mazza, 2015)
7	There are teams to solve production problems and encourage employee participation	(Bonavia and Marin-Garcia, 2011; El Dabee et al., 2013; Jirarat et al., 2011; Kaneko and Nojiri, 2008; Li and Yang, 2010; Masai et al., 2015; Savino and Mazza, 2015)
9	Many problems with machinery and equipment are solved due to the suggestions of workers	(Bonavia and Marin-Garcia, 2011; Kaneko and Nojiri, 2008; Villa and Taurino, 2013)
10	Operators are responsible for inspecting their own work on their machine	(Low and Show, 2008)
11	Both managers and operators have authority to stop production if detected inconsistencies	(Summers and Scherpereel, 2008)

The general idea is that it should be a management commitment focused to train and involve the employees to obtain the expected JIT benefits. Hypothesis H1 is formulated under this logic:

H_1: *Management Commitment* has a direct and positive impact on *Human Resources* efficiency in a JIT environment.

Management Commitment and Suppliers in JIT

The JIT principles include top management commitment, long term working relationships with suppliers and continuous improvement (Pheng et al., 2011). Several studies extended the analysis of JIT practices beyond the company's boundaries and considered JIT production and JIT supply as strictly interconnected practices. To be effective JIT production requires a faster and more intense transmission of information and a greater degree of co-ordination with suppliers (Danese et al., 2012). It is a good rule of JIT production system managed to have a reputable supplier deliveries and possibly localized deposits in the vicinity of enterprise customer. Such an arrangement makes safer to agree on periodic deliveries, lower procurement costs and transport, smaller batches with each delivery, allowing reduced inventories (Villa and Taurino, 2013). The procurement of parts and materials is a very important issue in the successful and effective implementation of JIT; thus, supplier selection and performance evaluation in long-term relationships have became more critical in JIT production environments (Aksoy and Öztürk, 2011). JIT low manufacturing costs including low capital costs result from low level of inventory due to close relationship with suppliers and customers, improved communication and cooperation among manufacturing processes and functions (Mia and Winata, 2014).

Refer to Table 1 to see the JIT Management Commitment attributes and Table 3 shows Suppliers JIT attributes and authors who endorse the attributes.

To carry out a team work, management should put special attention to suppliers and certify their work system to accomplish the specifications required for quality and on-time delivery; so the hypothesis H2 is proposed:

H_2: *Management Commitment* has a direct and positive impact on *Supplier* JIT practices.

Management Commitment and Quality in JIT

One of the points of Ohno (1988) gives as the definition of JIT is Total Quality Control (TQC) that is implemented on the principle of full participation by all personnel (Kaneko and Nojiri, 2008). JIT aims to achieve zero defects which are to eliminate errors. For this aim to be cost effective, the error should be identified and corrected at the source where the work is performed. The responsibility for quality will be given to the worker, with the goal to make it right (perfect) the first time, all the time (Low and Show, 2008). Additionally, Paul et al. (2004) found that effective implementation of TQM is enhanced by JIT implementation. Eker and Pala (2008) found a strong, positive relationship between JIT practices and TQM practices. TQM requires a focus on the customer's definition of quality, continuous improvement of the production processes, and the use of statistical quality control techniques (Zelbst et al., 2010). When successfully implemented, the JIT concept will reap the following benefits: improved quality levels, reduced scrap and rework rates, improved customer service levels (Pheng et al., 2011). Also, the goals of JIT, TQM and TPM are similar, which are continuous improvement and waste reduction. Therefore, in practice, manufacturing plants are likely to combine the implementation of JIT, TQM, and TPM (Jirarat et al., 2011).

Refer to Table 1 to see JIT Management Commitment attributes and Table 4 shows JIT Quality attributes and authors who endorse these attributes.

Table 3. Suppliers JIT attributes

Item	Items or Attributes	Authors Who Endorse
1	Suppliers are integrated into the company through a pull system	(Aksoy and Öztürk, 2011; Danese et al., 2012; El Dabee et al., 2013; Low and Show, 2008; Mia and Winata, 2014)
2	Deliveries are received daily from most suppliers	(Low and Show, 2008; Villa and Taurino, 2013)
3	The company has long-term contracts with its suppliers	(Aksoy and Öztürk, 2011; Kaneko and Nojiri, 2008; Pheng et al., 2011)
4	The suppliers of the company are certified	(Aksoy and Öztürk, 2011; Kaneko and Nojiri, 2008; Matsuo, 2015; Pheng et al., 2011)

Table 4. Quality JIT attributes

Item	Items or Attributes	Authors Who Endorse
1	Statistical control is used to control and reduce process variation	(Zelbst et al., 2010)
2	Quality initiatives are customer-oriented	(Pheng et al., 2011; Savino and Mazza, 2015; Zelbst et al., 2010)
3	They have quality metrics	(Jirarat et al., 2011; Low and Show, 2008; Pheng et al., 2011)
4	It has implemented a program of total productive maintenance (TPM)	(Jirarat et al., 2011)
5	They have been implemented principles and tools of Total Quality Management (TQM)	(Aksoy and Öztürk, 2011; Amasaka, 2014; Chen and Tan, 2011; El Dabee et al., 2013; Holl et al., 2013; Inman et al., 2011; Jayaram et al., 2010; Jirarat et al., 2011; Kaneko and Nojiri, 2008; Low and Show, 2008; Villa and Taurino, 2013; Zelbst et al., 2010)

Consequently, management requires work in a continuing basis to achieve the quality objectives through JIT quality practices; so the hypothesis H3 is generated:

H_3: *Management Commitment* has a direct and positive impact in *Quality* indicators obtained in a JIT environment process production

Human Resources and Suppliers in JIT

There are some critical principles for successful implementation of a JIT system such as people involvement, training and education, supplier relations, waste elimination, kanban or pull system, uninterrupted work flow, and total quality control (El Dabee et al., 2013). JIT deliveries from supplier or JIT supply include practices such as vendor-kanban for raw materials and outsourced components and pull systems for inbound logistics (Danese et al., 2012). Besides the involvement of employees, maintaining a good relationship with suppliers also play a vital role to ensure continuous material flow into the factory in the right quantities at the right time (Low and Show, 2008).

Refer to Table 2 to see Human Resources JIT attributes and Table 3 to see Suppliers JIT attributes that show authors who endorse the attributes.

The human resources participation and suggestions have consequences in suppliers, therefore hypothesis H4 is proposed:

H_4: *Human Resources* have a direct and positive impact on *Suppliers* practices in a JIT environment

Human Resources and Quality in JIT

People can contribute to variations in system performance due to fatigue, improper training, and lack of motivation. In order to alleviate these causes, JIT implement a decentralized approach of problem solving by empowering workers to make decisions and also adopt the "Respect for people" slogan as one of its tenets. Its employees are trained on the job, taught to treat any problem that arises as a learning opportunity, and encouraged to develop suggestions for improvement (Jayaram et al., 2010). JIT emphasizes quality control. Quality is even more important than throughput. Workers are trained in quality control, and when problems are found, workers have authority to stop production (Summers and Scherpereel, 2008). JIT is associated with "kaizen", or continuous improvement, and "quality circles" or groups of employees involved in continuous monitoring of production, by themselves implemented (Villa and Taurino, 2013).

Refer to Table 2 to see *Human Resources* JIT Attributes and Table 4 to see *Quality* JIT attributes that show authors who endorse the attributes.

Human resources are associated with quality due to people constantly working and taking decisions to achieve production objectives, in this way hypothesis H5 is proposed:

H_5: Human *Resources* have a direct and positive impact in *Quality* indicators obtained in a JIT environment process production

Suppliers and Quality in JIT

Manufacturers practicing JIT require suppliers that punctually supply materials and outsourced parts – in the appropriate quantity and with consistent quality. Because reliable suppliers enable manufacturers to reduce inventory costs and improve product quality, it is understandable that manufacturers are increasingly concerned about supplier selection (Aksoy and Öztürk, 2011). The JIT close supplier relationship is attributed to the high quality of product and the competitive cost due to the removal of wasteful general specifications. The use of customized components also provides an opportunity for the suppliers to improve both the component design and its production processes (Matsuo, 2015). In JIT especially, long term business relationships are maintained and information is exchanged closely with a small number of suppliers who have been selected according to such criteria as high quality, low-costs and punctual delivery (Kaneko and Nojiri, 2008). Golhar and Stamm (1991) reviewed 211 articles published from 1970s to 1980s defined JIT practice as four types of activities: (1) elimination of waste; (2) employee involvement in decision making; supplier participation; and (4) total quality control (Chen and Tan, 2011).

Refer to Table 3 to see Suppliers JIT Attributes and Table 4 to see Quality JIT attributes that show authors who endorse the attributes.

Suppliers' deliveries are highly important to maintain quality JIT standards, for that reason hypothesis H6 is established:

H_6: Suppliers practices have a direct and positive impact in *Quality* indicators obtained in a JIT environment process production.

This study was performed with the objective to identify the CSFs for JIT implementation, and to analyze the interrelations among four latent variables (Management Commitment, Human Resources, Suppliers, and Quality) in maquiladora industry. The proposed model is shown in Figure 1.

METHODOLOGY

The methodology followed to prove the hypotheses discussed before, which is divided into eight stages.

Figure 1. Proposed model

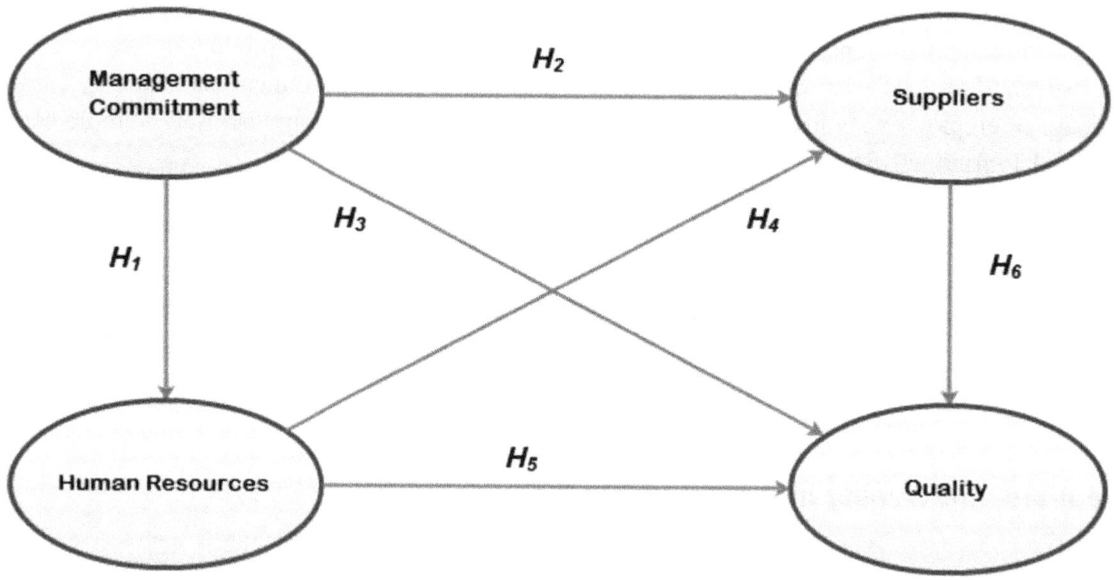

Review of Available Knowledge

A literature review was made to verify the available knowledge of JIT, defining the currently prevailing aspects concerning the implementation of JIT. There were several databases consulted such as Science Direct, Ebscohost, Ingenta, among others. Several keywords like "JIT" human resources", 'suppliers", "JIT-Quality", among others were used for the consultation.

The objective of this review is to determine the observed variable items that can measure in a better way the latent variables that have been identified. This literature review identified the items that are found in Table 1, Table 2, Table 3 and Table 4 for Management Commitment, Human Resources, Suppliers and Quality, respectively.

Preparation of the Questionnaire

With the revision of the previous literature, each and every one of the identified items were integrated in a preliminary questionnaire, which represents a rational validity based on the state of the art of the subject matter hereof. However, once complemented this activity, a development and validation by judges to the questionnaire is performed, since all items are drawn from cases applied in other countries, with different types of businesses and working conditions. Therefore it is required an adjustment to the proper context in which it is applied.

In the opinion of experts it is determined whether the items have relevance, coherence, sufficiency and clarity to be included in the questionnaire. Regarding the relevance it is judged whether it is important that the item is included. Coherence determines whether the item has logical relation to the dimension you are measuring. Sufficiency determines whether the item is sufficient for measuring the dimension to which they belong. Finally, clarity determines if the item is properly written and understandable.

At the conclusion of this assessment, the final format was approved by the judges involved. The questionnaire consists of 56 items divided into 6 categories or dimensions: 5 items in relation to the Management Commitment, 11 items related to the integration of Human Resources, 15 items related to Production Techniques, 5 items related to Suppliers, 5 items related to Quality and 15 items related to the Benefits of the implementation of JIT.

In addition, at the final part of the questionnaire a section corresponding to the demographics was added. In this section there are four demographic questions: the years of the respondent in the position, where there is five options: 0 to 1, 1 to 2, 2 to 5, 5 to 10, and over 10 years; sector to which the respondent belongs, with six options: aeronautics, electrical, automotive, electronics, medical, and metal forming; gender of the respondent with two options: male and female; and hierarchical position in the company, with five options: manager, engineer, supervisor, operator and technician.

Application of the Questionnaire

The final questionnaire was administered to people in charge of the departments of logistics and supply chains of manufacturing companies in Ciudad Juárez, whom should have at least one year performing the role. As a principle of inclusion the companies had to have at least 2 years of formally implementing the JIT philosophy. The questionnaire must be answered on a Likert scale, where each item can be classified on a scale of 1 to 5, where 1 corresponds to "never", 2 to "hardly ever", 3 to "regularly", 4 to "almost always", and 5 "always"(Inman et al., 2011).

The questionnaire is aimed at people responsible for the areas of logistics and supply chains. To answer the questionnaire, personal interview was used. The period of application was from November 2015 to January 2016. An appointment is set to answer the questionnaire with the heads of companies. If for some reason the respondent cannot meet on the first occasion, it proceeds to scheduling a second interview within 15 days. In case of three failed attempts the case is abandoned.

Capture of Information

Once the questionnaires were applied in the above described time period, it proceeds to capture the information obtained, for which the software SPSS 21® is used. A database is generated, where the rows represent each of the cases or collected questionnaires, while the columns represent the observed variables in each of the latent variables. Demographic variables are defined according to the type of information contained (string or numeric), while in the 56 items it is used a numerical scale.

Debugging the Database

Because there are times when some respondents do not answer certain items, these values are considered to be lost. They are replaced by the median of that item of the survey where it was answered. As the scale of measurement is ordinal the average cannot be used. However, if some of the surveys or cases have more than 10% of missing values, those cases are discarded and removed from the analysis (Hair et al., 2013).

Similarly, it is possible that in polls the uncommitted respondent will give the same answer to all the questions. In Excel standard deviation is estimated for each survey (except demographic information). If this value is less than 0.15 then the survey or questionnaire is omitted from the database.

Also, it is possible to make errors in the capture of data where there could be extreme values. The values must be between 1 and 5, so you may not have values outside that range. To identify these extreme values, it is necessary to standardize all variables observed and assessed by subjective judgments. Values above 4 in absolute value are considered outliers. A box and whisker plot is used to identify them. The outliers are replaced by the median in given variable, the same way as in Excel (Tabachnick and Fidell, 2013).

Descriptive Analysis of the Sample

With a refined data base, we proceed to carry out the descriptive analysis of the sample in order to generate charts and summaries to allow its characterization. These analyses permit to know information about respondents, such as the years they have in the position to measure the reliability of the data provided. One of the ways to analyze the information is with crosstabs.

Descriptive Analysis of the Items

To determine, in an univariate way, the performance of the items in each of the latent variables, a descriptive analysis is performed. As a measure of central tendency it has been used the median or 50th percentile, since the data has been obtained on an ordinal scale and the mean is not a representative parameter. In addition, although there has been a process of purging data search and elimination of outliers, the median is less affected if any of them is not detected properly. High median values indicate that the activity is almost always done or that the benefit is almost always obtained, on the contrary, low values in the median indicate that the activity is not performed or that the benefits are not obtained (Green Jr et al., 2014).

Similarly, to measure the dispersion of data the interquartile range (IR) is used. The IR is obtained by the difference between the first and third quartiles, which represents 50% the data analyzed, easily identified by a box plot and whiskers. A high value of the interquartile range indicates that respondents do not have a consensus on the true value of the activity or benefit analyzed but low values indicate that there is a consensus (Tastle and Wierman, 2007). In this research, items with IR above two are considered an anomaly and a bad translation.

Structural Equation Model

To validate the hypothesis of the model proposed in Figure 1, the methodology of structural equation model (SEM) has been used. The SEM methodology is widely used in determining causal relationships between latent variables, especially in the area of logistics. In this research WarpPLS 5 software was used for running the model, it is integrated based on partial least squares routines. In this case the model has been raised with latent variables reflective type, since there is much correlation between the observed variables or items that compose it (Ned Kock, 2010).

However, before determining whether the model is statistically stable, each of the latent variables has been validated as follows variance (N. Kock, 2013):

1. **Predictive Validity:** It is measured by R-squared and adjusted R-squared indices in a parametric way and with the Q-squared in a nonparametric way. It is recommended that the R-squared and Q-squared values are similar. These indices require values above 0.2.
2. **Convergent Validity:** It is measured by the average variance extracted (AVE) and values above 0.5 are required, indicating that each latent variable must extract at least 50%.
3. **Internal Validity:** It is estimated by Cronbach's alpha index and composite reliability index, which must have values greater than 0.7.
4. **Collinearity:** It is measured by variance inflation factors, which must be less than 5 and preferably lower than 3.3.

Once the latent variables have exceeded the validation process, they were integrated to the SEM with the amount of items that belonged to them, where they have used the following efficiency indices of the model:

1. **Average Path Coefficient (APC):** Which is associated with a p value and the null hypothesis being tested is that the average beta values is zero, versus the alternative hypothesis that is different from zero (García et al., 2014).
2. **Average R-Squared (ARS) and Average Adjusted R-Squared (AARS):** These indices measure the predictive validity of the model and like the APC, is associated with a p-value testing the null hypothesis of ARS and AARS that are zero, versus the alternative hypothesis that is different from zero (Ketkar and Vaidya, 2012).
3. **Average Block VIF (AVIF) and Average Full Collinearity VIF (AFVIF):** They are used to measure rates of overall collinearity of the model. These indices are associated with a value, which must be less than 5, but ideally lower than 3.3 (Ned Kock, 2015) .
4. **Tenenhaus GoF (GoF):** It is considered as an index of goodness of fit, it is recommended to be greater than 0.36 to be considered acceptable.

In this model it was estimated three types of effects, which are defined below (Kock, 2013):

- Direct effects to validate the relationship between the latent variables or established hypothesis. The value of dependence between the latent variables appears as a value of beta (β) and is a standardized value indicating the intensity of exchange between the dependent and independent variables. To determine the statistical significance of this effect, the beta is associated with a p value.
- Indirect effects that occur between latent variables and occurring through mediating variables, so that more than two segments are used. Also the value of beta is associated with a p-value to determine statistical significance.
- Total Effects, which represent the sum of the direct and indirect effects. They are also associated with a p-value to determine statistical significance.

RESULTS

The results of this research have been divided into several sections in order to be more explanatory, as illustrated below.

Descriptive Analysis of the Sample

A total of 375 questionnaires from the maquiladora industry of Cd. Juárez were obtained, but only 372 results are analyzed, as 3 were eliminated by having minor deviations of 0.15. Demographic information of the 372 respondents is described below:

Table 5 shows the sector and gender of respondents, where you can observe that the questionnaires were answered by 105 women and 245 men. You can also observe that the most surveyed sectors were the automotive and metal forming; however, it is important to note that only a total of 350 cases are reported, due to 22 respondents not answering the sector in which they work.

Table 6 shows the hierarchical positions of respondents and level of experience they have. This can be summarized as follows: 0-1 years 25.6%, 1-2 years 20.9%, 2-5 years 22.2%, 5-10 years 18.3%, and 13% over 10 years. Hierarchical position in the company: 6% managers, 16.1% supervisor, technical 27.2%, operator 23.5%, and operator engineer 27.2%.

Descriptive Analysis of the Items

Table 7 shows the descriptive analysis of the items, where the name of the latent variable is in bold and the name of the items that compose the second, third, and fourth columns shown in the first column is

Table 5. Sector and gender of the respondent from IBM SPSS

Sector	Gender of the Respondent		Total
	Feminine	Male	
Automotive	41	115	156
Metal forming	35	47	82
Electrical	10	45	55
Electronics	14	29	43
Medical	3	9	12
Aeronautics	2	0	2
Total	105	245	350

Table 6. Hierarchical position in the company and years in the position from IBM SPSS

Hierarchical Position in the Company	Years in the Position				Total
	1-2	2-5	5-10	More than 10	
Manager	2	1	9	7	19
Supervisor	18	16	13	9	56
Technical	52	13	20	8	93
Operator	57	15	6	5	83
Engineer	25	32	16	15	88
Total	154	77	64	44	339

Table 7. Descriptive analysis of the items

Management Commitment	Q1	Q2	Q3	IR
Communication and constant coordination between different departments of the plant	3.3	4.20	4.88	1.58
Senior management has a culture that encourages compliance with the deadlines	3.23	4.11	4.81	1.58
Supervisors strive to promote teamwork, encouraging operators to cooperate and express their views	3.12	3.94	4.71	1.59
Managers, engineers and operators constantly interact with each other	2.91	3.71	4.52	1.61
Human Resources	**Q1**	**Q2**	**Q3**	**IR**
Both managers and operators have authority to stop production if detected inconsistencies	2.55	3.70	4.64	2.09
There are teams to solve production problems and encourage employee participation	2.79	3.67	4.50	1.71
Many problems with machinery and equipment are solved due to the suggestions of workers	2.47	3.43	4.31	1.84
Employees are hired for their ability to solve problems and teamwork	2.39	3.39	4.29	1.90
Suppliers	**Q1**	**Q2**	**Q3**	**IR**
The suppliers of the company are certified	3.24	4.12	4.83	1.59
The company has long-term contracts with its suppliers	3.12	3.93	4.69	1.57
Deliveries are received daily from most suppliers	3.04	3.84	4.63	1.59
Suppliers are integrated into the company through a pull system	2.7	3.66	4.53	1.83
Quality	**Q1**	**Q2**	**Q3**	**IR**
Quality initiatives are customer-oriented	3.36	4.18	4.84	1.48
They have quality metrics	3.34	4.18	4.86	1.52
They have been implemented principles and tools of Total Quality Management (TQM)	3.07	4.01	4.77	1.70
It has implemented a program of total productive maintenance (TPM)	3.03	3.98	4.75	1.72
Statistical control is used to control and reduce process variation	3.09	3.97	4.71	1.62

illustrated containing the first quartile (Q1), second quartile (Q2), third quartile (Q3), and IR respectively. Note that the items are sorted in descending order according to the second quartile or median.

In the case of *Management commitment*, it is observed that the most important item refers to *The constant communication and coordination between different departments of the plant*, which is not surprising, since in similar studies, communication has been key to the achievement business objectives as it relates to the ability to guide all human resources on a single goal, in this case, the implementation of JIT. Similarly and of great importance it is that management and people responsible of decision-making processes fosters a culture focused on meeting times and dates set in the programs.

Regarding human resources, it is noted that the item with the highest median means that *Both managers and operators have authority to stop production if detected inconsistencies* and it refers to the ability of the company to grant empower to their workers, which can be reached only after a good training program in which they are taught to solve problems quickly and accurately.

In relation to the latent variable associated with suppliers, it is observed that the most important item is the one that refers to *The suppliers of the company are certified*, which guarantees that the raw material entering the production process meets specifications of required quality. This may result that *The company has long-term contracts with its suppliers*, which indicate that they are fully integrated into the supply chain from the manufacturer.

Finally, there is the latent variable quality, where the item with the highest median was *Quality initiatives are customer-oriented*, because if you do not have this approach, these initiatives often have no justification. However, it should be remembered that there must *have quality metrics* to quickly determine deviations to the program and commitments.

Structural Equation Model

Before running the model, composed of four latent variables, one should consider whether they have sufficient validity required to be integrated. Table 8 illustrated below display these indices, where first the R- squared and adjusted R-squared values measure parametric validity, which are higher than 0.2. This leads to the conclusion that this requirement is fulfilled. Likewise, it is observed that the Q-squared values are greater than 0.2 and similar to those that have square R-squared, therefore it is concluded that there is also a predictive validity from a nonparametric view (Ned Kock, 2015).

With regard to internal validity, composite reliability index and Cronbach's alpha are higher than 0.7 for the four latent variables analyzed, so this requirements are met adequately. Also, in relation to collinearity of the latent variables, it is noted that for all the obtained value is less than 3.3, so it is concluded that there is no such problem.

Model Fit Indices

Since all latent variables have passed the validity, they were integrated into the model presented in Figure 1. The results are seen in Figure 2, where the values for the beta parameters are presented and the p values as well, which measures the statistical significance of the parameter. Thus, according to the p values associated with beta parameters it is concluded that all outcomes are statistically significant.

However, before proceeding to interpret the model it is necessary to know the efficiency indices of it. In Table 9 the indices are illustrated. In this case, it is observed that the APC, ARS and AARS indices are associated with a p value less than 0.001, indicating that evidence is sufficient to state that it has predictive validation of the model and that on average the value of betas is different from zero.

Similarly, in relation to AVIF and AFVIF indices are used to measure the collinearity in the model, where it shows that they are less than 3.3, so it is concluded that the model does not have that kind of problem. Finally, it is noted that the Tenenhaus index is greater than 0.36, concluding that the model is sufficiently explained.

Table 8. Latent variable coefficients

Index	Management Commitment	Human Resources	Suppliers	Quality
R-squared		0.230	0.314	0.529
Adj. R-squared		0.228	0.310	0.525
Composite reliability	0.872	0.828	0.871	0.927
Cronbach's alpha	0.804	0.723	0.802	0.901
Avg.var.extrac.(AVE)	0.631	0.548	0.628	0.718
Full collin.VIF	1.646	1.432	1.877	2.093
Q-squared		0.231	0.315	0.528

Figure 2. Structural equation model evaluated

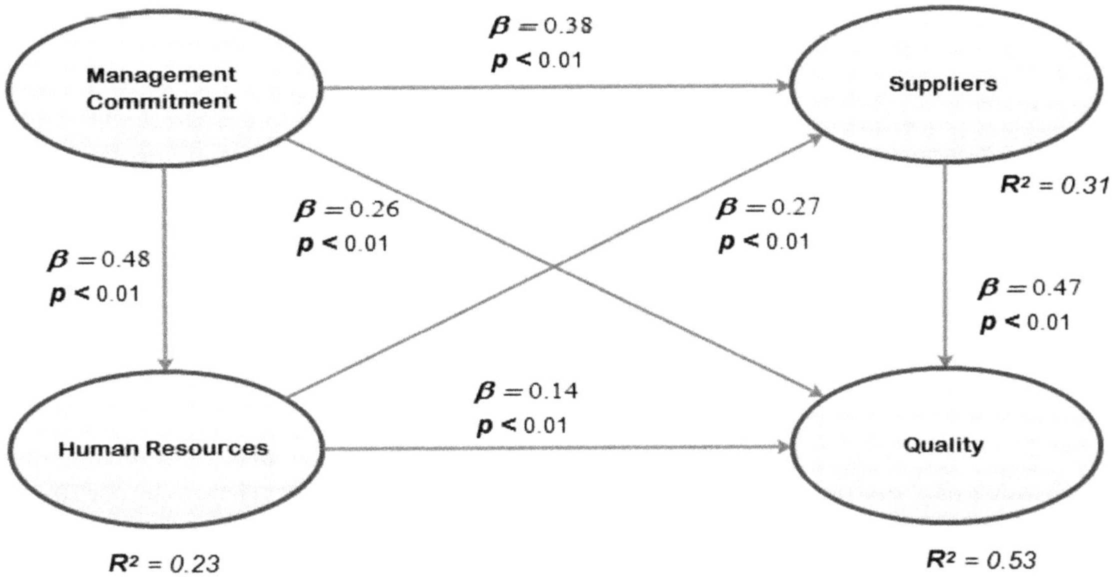

Table 9. Model fit and quality indices

Index	Acronym	Value
Average path coefficient	APC	0.334, P<0.001
Average R-squared	ARS	0.357, P<0.001
Average adjusted R-squared	AARS	0.354, P<0.001
Average block VIF	AVIF	1.386
Average full collinearity VIF	AFVIF	1.762
Tenenhaus GoF	GoF	0.475

Direct Effects

Based on the values observed in Figure 2 of the evaluated model, it can be concluded in relation to the hypotheses stated. In Table 10, a summary is shown of the related variables, the value of direct effect, the p value, and finally the decision established in relation to these hypotheses.

Thus, based on the above, the conclusions based on the assumptions are as follows:

H_1: There is enough statistical evidence to state that *Management Commitment* has a direct positive impact on *Human Resources* efficiency in a JIT environment, because when former changes by one standard deviation, the latter consequently directly changes in 0.48 units. In other words, *Management Commitment* is required for a good *Human Resources* practices.

H_2: There is enough statistical evidence to state that *Management Commitment* has a direct positive impact on *Suppliers* efficiency in a JIT environment, because when former changes by one standard deviation, the latter consequently directly changes in 0.38 units. Moreover, there is also an indirect

Table 10. Direct effects

From	To	Direct Effect	P value	Decision
Management Commitment	Human Resources	0.479	< 0.001	Accept
Management Commitment	Suppliers	0.380	< 0.001	Accept
Management Commitment	Quality	0.261	< 0.001	Accept
Human Resources	Suppliers	0.273	< 0.001	Accept
Human Resources	Quality	0.141	0.003	Accept
Suppliers	Quality	0.468	< 0.001	Accept

impact of 0.131, which makes a total effect change. In other words, *Management Commitment* is required for a good *Suppliers* practices.

H_3: There is enough statistical evidence to state that *Management Commitment* has a direct positive impact on *Quality* efficiency in a JIT environment, because when former changes by one standard deviation, the latter consequently directly changes in 0.26 units. Moreover, there is also an indirect impact of 0.306 through Human Resources and Suppliers variables, which makes a total effect change. In other words, *Management Commitment* is required for a good *Quality* practices.

H_4: There is enough statistical evidence to state that *Human Resources* has a direct positive impact on *Suppliers* efficiency in a JIT environment, because when former changes by one standard deviation, the latter consequently directly changes in 0.27 units. In other words, *Human Resources* is required for a good Suppliers practices.

H_5: There is enough statistical evidence to state that *Human Resources* has a direct positive impact on *Quality* efficiency in a JIT environment, because when former changes by one standard deviation, the latter consequently directly changes in 0.14 units. Moreover, there is also an indirect impact of 0.128, which makes a total effect change. In other words, *Human Resources* is required for a good *Quality* practices.

H_6: There is enough statistical evidence to state that *Suppliers* has a direct positive impact on *Quality* efficiency in a JIT environment, because when former changes by one standard deviation, the latter consequently directly changes in 0.47 units. In other words, *Suppliers* is required for a good *Quality* practices.

Indirect Effects

Indirect effects are those that occur between latent variables by means of a third mediating variable. Although there are many effects of this type only the total sum of these is made. In this case, in Table 11 the total indirect effects are presented where the dependent variables, the independent variables, the effect size, and p-value are shown. It is seen that all are statistically significant, since p values are less than 0.05, the maximum allowed level of significance (N. Kock, 2013).

According to its magnitude, it is important to note the relationship between the two latent variables, Management Commitment and Quality. Being 0.245 the largest quantity, indicating that whenever the first latent variable increases its standard deviation in one unit, the second does in 0.245 units. A similar interpretation can be made for other relationships and indirect effects between latent variables.

Table 11. Indirect effects

From	To	Indirect Effect	P Value
Management Commitment	Suppliers	0.131	< 0.001
Management Commitment	Quality	0.245	< 0.001
Management Commitment	Quality	0.061	0.020
Human Resources	Quality	0.128	< 0.001

Total Effects

The sum of direct and indirect effects results in total effects. In this case Table 12 illustrates these effects; where the independent variable and the dependent variable illustrate the magnitude of the effect and the associated p-value for the hypothesis test of statistical significance. They are in descending order according to the magnitude of the effect. According to the magnitude of p values it can be concluded that all full effects are statistically significant, as they are less than 0.05, maximum allowable significance.

It can be seen that the effect more important according to their magnitude is associated with the relationship between Management Commitment and Quality, where every time the first latent variable increases its standard deviation by one unit, the second does in 0.567 units. The second ratio according to the magnitude of the effect is between Management Commitment and Suppliers, where every time the first latent variable increases its standard deviation by one unit, the second does in 0.511 units. Similar interpretations can be made for relations between the other latent variables. It is important to note that the total effects of Management Commitment on the other latent variables is the one with the highest effects, since in this model is shown to serve as the initial latent variable and therefore it affects all other latent variables.

CONCLUSION

To obtain the benefits of JIT technique it is required the sum of many CSFs involved. In this study only four CSFs are analyzed, which are Management Commitment, Human Resources, Suppliers, and Quality. Based on the results obtained the following industrial implications are made:

Table 12. Total effects

From	To	Total Effects	P value
Management Commitment	Quality	0.567	< 0.001
Management Commitment	Suppliers	0.511	< 0.001
Management Commitment	Human Resources	0.479	< 0.001
Suppliers	Quality	0.468	< 0.001
Human Resources	Suppliers	0.273	< 0.001
Human Resources	Quality	0.268	< 0.001

- Management commitment is the generator of the implementation of JIT as it requires high hierarchical levels of the company to have the necessary knowledge and convincement of its benefits for transmission to all areas of the company, which will be leaders of the implementation processes.
- Human Resources, which is the participation of people of productive areas, is another trigger for the implementation of JIT as it is the people who work daily in the processes which causes a change in production methods; however, they too must be fully convinced of the benefits that JIT will bring them.
- Similarly, the active participation of suppliers in order to have the JIT methodology deliveries is required, since they are the beginning of the supply chain. If the delivery times of raw material by the supplier start badly, surely it will end badly at the time of delivery of finished product.
- The joint participation of the three elements, Management Commitment, Human Resources, and Suppliers, make an impact on the quality benefits expected to implement JIT.

FUTURE RESEARCH

It is suggested to continue applying the method of SEM to analyze the relationships of the CSFs and that maquiladoras may count with this information, to know in which aspects they need to focus for an implementation of JIT and achieve the expected benefits to its optimum using the resources of the company. In addition, in this research there were only studied four CSF's, however, many others have been discovered and analyzed in other environments and contexts. Therefore, the industry established in Ciudad Juarez justifies it due to its volume and economic impact.

REFERENCES

Aksoy, A., & Öztürk, N. (2011). Supplier selection and performance evaluation in just-in-time production environments. *Expert Systems with Applications*, *38*(5), 6351–6359. doi:10.1016/j.eswa.2010.11.104

Amasaka, K. (2014). New JIT, New Management Technology Principle: Surpassing JIT. *Procedia Technology*, *16*, 1135–1145. doi:10.1016/j.protcy.2014.10.128

Bergenwall, A. L., Chen, C., & White, R. E. (2012). TPSs process design in American automotive plants and its effects on the triple bottom line and sustainability. *International Journal of Production Economics*, *140*(1), 374–384. doi:10.1016/j.ijpe.2012.04.016

Bonavia, T., & Marin-Garcia, J. A. (2011). Integrating human resource management into lean production and their impact on organizational performance. *International Journal of Manpower*, *32*(8), 923–938. doi:10.1108/01437721111181679

Büyüközkan, G., Kayakutlu, G., & Karakadılar, İ. S. (2015). Assessment of lean manufacturing effect on business performance using Bayesian Belief Networks. *Expert Systems with Applications*, *42*(19), 6539–6551. doi:10.1016/j.eswa.2015.04.016

Chen, Z. X., & Tan, K. H. (2011). The perceived impact of JIT implementation on operations performance: Evidence from Chinese firms. *Journal of Advances in Management Research*, *8*(2), 213–235. doi:10.1108/09727981111175957

Danese, P., Romano, P., & Bortolotti, T. (2012). JIT production, JIT supply and performance: Investigating the moderating effects. *Industrial Management & Data Systems*, *112*(3), 441–465. doi:10.1108/02635571211210068

El Dabee, F., Marian, R., & Amer, Y. (2013). An optimisation model for a simultaneous cost-risk reduction in just-in-time systems. *Australian Journal of Multi-Disciplinary Engineering*, *9*(2), 139–147. doi:10.7158/N13-GC03.2013.9.2

Fullerton, R. R., Kennedy, F. A., & Widener, S. K. (2014). Lean manufacturing and firm performance: The incremental contribution of lean management accounting practices. *Journal of Operations Management*, *32*(7–8), 414–428. doi:10.1016/j.jom.2014.09.002

García, J. L., Maldonado, A. A., Iniesta, A., Robles, G., & Hernández, G. (2014). A systematic review/survey for JIT implementation: Mexican maquiladoras as case study. *Computers in Industry*, *65*(4), 761–773. doi:10.1016/j.compind.2014.02.013

Green Jr, K. W., Inman, R. A., Birou, L. M., and Whitten, D. (2014). Total JIT (T-JIT) and its impact on supply chain competency and organizational performance. *International Journal of Production Economics, 147, Part A*, 125-135. doi:10.1016/j.ijpe.2013.08.026

Hair, J. F., Hult, G. T. M., Ringle, C., & Sarstedt, M. (2013). *A Primer on Partial Least Squares Structural Equation Modeling (PLS-SEM)*. SAGE Publications.

Holl, A., Pardo, R., & Rama, R. (2013). Spatial patterns of adoption of just-in-time manufacturing * Spatial patterns of adoption of just-in-time manufacturing. *Papers in Regional Science*, *92*(1), 51–67. doi:10.1111/j.1435-5957.2011.00406.x

Inman, R. A., Sale, R. S., Green, K. W. Jr, & Whitten, D. (2011). Agile manufacturing: Relation to JIT, operational performance and firm performance. *Journal of Operations Management*, *29*(4), 343–355. doi:10.1016/j.jom.2010.06.001

Jayaram, J., Das, A., & Nicolae, M. (2010). Looking beyond the obvious: Unraveling the Toyota production system. *International Journal of Production Economics*, *128*(1), 280–291. doi:10.1016/j.ijpe.2010.07.024

Jirarat, T., Ketlada, K., & Nuttapon, S. (2011). Relationship model and supporting activities of JIT, TQM and TPM. *Songklanakarin Journal of Science & Technology*, *33*(1), 101–106.

Kaneko, J., & Nojiri, W. (2008). The logistics of Just-in-Time between parts suppliers and car assemblers in Japan. *Journal of Transport Geography*, *16*(3), 155–173. doi:10.1016/j.jtrangeo.2007.06.001

Ketkar, M., & Vaidya, O. S. (2012). Study of Emerging Issues in Supply Risk Management in India. *Procedia: Social and Behavioral Sciences*, *37*(0), 57–66. doi:10.1016/j.sbspro.2012.03.275

Kock, N. (2010). Using WarpPLS in E-collaboration Studies: An Overview of Five Main Analysis Steps. *International Journal of e-Collaboration*, *6*(4), 1–11. doi:10.4018/jec.2010100101

Kock, N. (2013). Using WarpPLS in e-collaboration studies: What if I have only one group and one condition. *International Journal of e-Collaboration*, *9*(3), 12. doi:10.4018/jec.2013070101

Kock, N. (2015). *WarpPLS 5.0 User Manual*. USA: ScriptWarp System.

Li, J., & Yang, Q. (2010). *The Analysis of Human Resource Allocation and Incentive Based on JIT System*. Paper presented at the Information Science and Management Engineering (ISME), 2010 International Conference of.

Low, S. P., & Show, M. Y. (2008). Facilities design incorporating just-in-time principles for ramp-up light factories in Singapore. *Facilities*, *26*(7/8), 321–342. doi:10.1108/02632770810877967

Low Sui, P., & Gao, S. (2011). The Application of the Just-in-Time Philosophy in the Chinese Construction Industry. *Journal of Construction in Developing Countries*, *16*(1), 91–111.

Martínez-Jurado, P. J., Moyano-Fuentes, J., & Jerez-Gómez, P. (2014). Human resource management in Lean Production adoption and implementation processes: Success factors in the aeronautics industry. *BRQ Business Research Quarterly*, *17*(1), 47–68. doi:10.1016/j.cede.2013.06.004

Masai, P., Parrend, P., & Zanni-Merk, C. (2015). Towards a Formal Model of the Lean Enterprise. *Procedia Computer Science*, *60*, 226–235. doi:10.1016/j.procs.2015.08.122

Matsuo, H. (2015). Implications of the Tohoku earthquake for Toyota's coordination mechanism: Supply chain disruption of automotive semiconductors. *International Journal of Production Economics*, *161*, 217–227. doi:10.1016/j.ijpe.2014.07.010

Mia, L., & Winata, L. (2014). Manufacturing strategy and organisational performance: The role of competition and MAS information. *Journal of Accounting & Organizational Change*, *10*(1), 83–115. doi:10.1108/JAOC-10-2011-0048

Nadarajan, S., Chandren, S., & Mohamed Elias, E. (2013). Hiccups in just–in–time practices for electrical & electronic manufacturing. *International Journal of Academic Research*, *5*(5), 269–271. doi:10.7813/2075-4124.2013/5-5/B.41

Pheng, L. S., Arain, F. M., & Fang, J. W. Y. (2011). Applying just-in-time principles in the delivery and management of airport terminal buildings. *Built Environment Project and Asset Management*, *1*(1), 104–121. doi:10.1108/20441241111143812

Savino, M. M., & Mazza, A. (2015). Kanban-driven parts feeding within a semi-automated O-shaped assembly line: A case study in the automotive industry. *Assembly Automation*, *35*(1), 3–15. doi:10.1108/AA-07-2014-068

Summers, G. J., & Scherpereel, C. M. (2008). Decision making in product development: Are you outside-in or inside-out? *Management Decision*, *46*(9), 1299–1312. doi:10.1108/00251740810911957

Tabachnick, B., & Fidell, L. (2013). *Using Multivariate Statistics*. Pearson.

Tastle, W. J., & Wierman, M. J. (2007). Consensus and dissention: A measure of ordinal dispersion. *International Journal of Approximate Reasoning*, *45*(3), 531–545. doi:10.1016/j.ijar.2006.06.024

Tillema, S., & van der Steen, M. (2015). Co-existing concepts of management control: The containment of tensions due to the implementation of lean production. *Management Accounting Research*, *27*, 67–83. doi:10.1016/j.mar.2015.01.002

Villa, A., & Taurino, T. (2013). From JIT to Seru, for a Production as Lean as Possible. *Procedia Engineering*, *63*, 956–965. doi:10.1016/j.proeng.2013.08.172

Zelbst, P. J., Kenneth, W., Green, J., Abshire, R. D., & Sower, V. E. (2010). Relationships among market orientation, JIT, TQM, and agility. *Industrial Management & Data Systems*, *110*(5), 637–658. doi:10.1108/02635571011044704

Zhao, W., Yu, Q.-q., Li, H.-s., & Tian, Y.-z. (2014). *Study on the relationship between JIT practices and operational performance based on the cost leading strategy.* Paper presented at the Management Science & Engineering (ICMSE), 2014 International Conference on.

KEY TERMS AND DEFINITIONS

Critical Success Factors (CSF): Practices or conditions that determine that a project can be executed and produce the expected results.

Direct Effects: Result that a variable produces in the following variable related by SEM.

Human Resources: Activities productive area where people participate.

Implementation: Implement procedures or activities of a methodology.

Indirect Effects: The result that a variable produced by a mediating variable in another variable.

Management Commitment: Activities fostered by the active participation of members of the managerial levels in a company.

Maquiladora: Company where to manufacture or assemble products or transnational companies. The final products after processing are exported.

Process: Stage where certain tasks or work to transform or produce a final product are made.

Total Effects: The sum of the direct effects and indirect effects of one latent variable on a latter variable.

Chapter 12
Modeling of Polypropylene Modified Bitumen Mix Design Results Using Regression Analysis

Kaval Chhabra
VIT University, India

Divesh Agrawal
VIT University, India

Saladi S. V. Subbarao
VIT University, India

ABSTRACT

This study investigates the effects of mixing Polypropylene waste plastics in the bituminous mix for the design of Flexible Pavement. Since, obtaining Marshall Test results from the bituminous mix is time-consuming, so if the practitioners measure the values of stability and flow by mechanical testing, other computations can be done by applying simple mathematical calculations. So, this study carried out stability and flow tests on different specimens made with varying bitumen and polypropylene plastic content. From the initial test results, the optimum bitumen and plastic contents are found. Further, the test results obtained from Marshall Test are modelled by identifying various input variables, which are various physical properties of the mix such as plastic content, bitumen content, air voids and unit weight. The regression modeling framework is adopted in this study to predict the Marshall stability and flow value. Since the developed models have yielded good results, these can be effectively used in parameter estimation, and thus aids the future researchers.

DOI: 10.4018/978-1-5225-2440-3.ch012

INTRODUCTION

The rapid growth in industrialization coupled with population growth has led to enormous rise in production of various types of wastes worldwide. In a recent study concerning the top 20 countries that have dumped the most plastic waste into the oceans, India (at 12th) is one of the worst performers. It has dumped up to 0.24 million tons of plastic waste into the ocean every year, thereby polluting the natural resources. One recent study (2014-15) reported that approximately 8 kg of waste plastics dumped by the each individual in India. These plastic wastes are needed to be recycled, if not, it gets disposed by the available waste disposal techniques like Land filling and Incineration. Though these techniques have been widely in use but these too have certain impacts on the environment which are discussed as under.

Land filling refers to a process of disposal wherein the waste materials are buried in a specific area that is situated far away from the city. This process has certain impacts on the environment like affecting water recharge, reducing soil microbial activity, and also it leads to drainage and water line clogging. Such clogging in turn leads to production of gases such as methane, which affects Greenhouse effect. Also, land availability for filling is a major point of concern. Further, Incineration refers to a process of burning the wastes above 700°C. Incineration of polymers like polypropylene, polystyrene, etc. evolves gases like CO_2, CO etc. which in turn causes global warming, pollution. If polyvinyl chloride mixes with the waste, it results in the production of Chlorine, HCl and sometimes Dioxin which is a carcinogenic and a poisonous gas. In fact, this process requires scientific monitoring and it can accentuate pollution problems, if strict standards aren't being enforced.

Since both landfilling and incineration have certain negative impact on the environment. Under these circumstances, an alternate use for the waste plastics is the need of the hour. Several investigations in India and many other countries have revealed that properties of bituminous mixes can be enhanced to meet the present requirements of pavement with the inculcation of certain additives in bituminous mixes. These additives are termed as "Bitumen Modifiers" and the bitumen that is premixed with these modifiers is termed as "Modified Bitumen". Modified bitumen is expected to provide higher strength and durability depending upon the type of additives used and modification process involved. The time period of next renewal is expected to be extended by 50 percent in case of surfacing with modified bitumen in comparison to normal period specified for conventional bituminous mix. Certain Performance analysis studies undertaken under the protection of Central Road Research Institute, New Delhi; Highway Research Station, Chennai; Rubber Board, Kottayam (Kerala); Ministry of Road Transport and Highways, New Delhi; Kerala Public Works Department and Gujarat Engineering Research Institute, Vadodara have revealed that the usage of Modified Bitumen in construction as well as maintenance of bituminous roads is quite cost effective, when life-cycle cost is taken under consideration.

Plastic waste when mixed with hot bitumen, it melts to form an oily coat over the aggregate and the mixture is laid over the road surface as similar as a normal tar road. The plastic waste can be effectively used as a binder material in flexible pavements. They may be used to enhance strength, enhance fatigue resistance to deformation and thus makes the pavements more durable. Plastic coated bitumen can result in higher stiffness than that of conventional bitumen, but without undergoing a large increase in flexibility.

PAST RESEARCH ON MODIFIED BITUMEN

Since last few decades, researchers are focussing more on usage of modified bitumen in pavement construction. In this process, modifying the properties of bitumen by using waste plastics is gaining importance due to its nature of enhancing the strength and economical mix. Kumar et al. (2003) reported that there occurs comparatively plastic bitumen blends less than that of neat bitumen, as observed during thin film oven test, which in turn controls the premature failure of a flexible pavement due to cracking. There is significant improvement in the indirect tensile strength and a considerable reduction in the rutting characteristic of the modified mixes prepared as compared to conventional ones. It has been reported that the Marshall Stability value of modified bituminous mix modified with 5 per cent of waste plastics increased by about 23%. Vasudevan et al. (2006) reported that even though Plastics-Modified-Bitumen (PMB) improves the quality of the roads and associated properties, the process (dry or wet process of mixing) of using these plastics for blending decides the enhanced strength of the bonding. Dry process is far better process than wet process. They also reported that polymer will not be leaching out after laying the road using waste plastic coated aggregate bitumen mix. Rajmane et al (2007) conducted a study on effective utilization of waste plastic in construction of flexible pavement for improving their performance and have found out that Plastic roads would be a boon for India's hot and extremely humid climate, where temperatures frequently cross 50°C and torrential rains create havoc, leaving most of the roads with big potholes.

Vasudevan (2007) has emphasized upon the use of dry process of mixing over wet process because bitumen binding strength in the Wet process is less compared to Dry Process. Reena Gupta et al (2011) worked upon a novel approach to improve road quality by utilizing plastic waste in road construction and have found out that no toxic gas is produced as the maximum temperature to coat the aggregates is 150-160 degree C in technology related to dry process of plastic waste utilization. Jalal J. Jafar (2015) under his study for Utilisation of Waste Plastic in Bituminous Mix for Improved Performance of Roads has proposed a mechanism to explain the effect of the chemical additives on the increase in the stiffness of the bituminous mixture which attributed to an increase in the bonding forces between the aggregates and the bitumen. Further, SerkanTapkın et al (2009) conducted a study on Prediction of Marshall test results for polypropylene modified dense bituminous mixtures using neural networks and have found that the addition of polypropylene fibres results in the improved Marshall Stabilities and Marshall Quotient values.

Though the above studies reported substantial literature on utilization of waste plastics for modifying the properties of bitumen in order to enhance the strength of the bitumen mix, still there is a scope for investigating the changes in properties of bitumen after adding the polypropylene plastics.

PROVEN CASE STUDIES ON UTILIZATION OF MODIFIED BITUMEN IN INDIA

Several live projects have been reported in India for effective utilization of modified bitumen in flexible pavement design. Some of the studies are reported in this section.

1. Laboratory studies have been carried out at the Centre for Transportation Engineering of Bangalore University, in which the waste plastic was used as an additive to heated bitumen in different proportions (ranges from zero to 12% by weight of bitumen). The results from these laboratory investigations reported that, the addition of waste plastic of about 8.8% by weight of bitumen, helps in substantially enhancing the strength, stability, fatigue life and many other desirable properties of bituminous mix, even during adverse water-logging conditions. The addition of 8.0% by weight

of waste plastic for the preparation of modified bitumen resulted in a saving of 0.4% bitumen by weight of the mix or about 9.6% bitumen per cubic meter of bituminous mix.

2. In the state of Tamil Nadu, length of roads around 1000 m in various stretches have been constructed using waste plastic as an additive to bituminous mix under the scheme of "1000 km Plastic Tar Road", and it was found that, the performance characteristics of all the road stretches are satisfactory.
3. The performance characteristics of the road stretches constructed using waste plastic in Karnataka are also found to be satisfactory.

The construction of roads using Waste Plastics in the above states was based on the guidelines developed by Bangalore University and College of Engineering, Madurai. However, standard specifications are not available on the use of waste Plastic in Bituminous road Construction. So in this regard, IRC was specially requested by NRRDA for the preparation of such Guidelines for enabling the construction of these Rural Roads under PMGSY using Waste Plastic. In order to facilitate the development of Guidelines for this, an Expert Group was appointed by NRRDA for the preparation of interim guidelines for the usage of plastics which would be sent to IRC for further approval.

It is quite evident that the above studies reported on utilizing the waste plastics in flexible pavement design, but most of these studies followed the laborious experiments for calculating the Marshall Stability and flow value. Hence, by keeping this in mind, this study concentrated on developing a model for the estimation of Marshall Stability, so that the model results serves as an aid for future researches in this area.

GOALS AND OBJECTIVES

Motivation behind the study is too efficiently utilise the waste plastic in a constructive and a cost effective manner so that it can be beneficial to the society. Main objective of the current project work are:

1. To identify the optimum proportion of waste plastic that is to be added in the bituminous mix for getting the desired strength.
2. To compare the experimental results with the conventional pavement details and perform the analysis.
3. To prepare a statistical model to analyse the parameter dependencies and generate a relation between dependent and independent variables.
4. To assess the extent with which various physical properties of constituents affecting the stability and flow.

METHODOLOGY

Plastic Aggregate Bitumen Interaction Model

The shredded plastics when sprayed over the hot aggregate melt and spread over the aggregate in the form of a thin coating at the aggregate surface. When the aggregate is heated around 160°C, the coated plastics remains in the softened state. Over these coated aggregates, hot bitumen (heated at 160°C) is added. The added bitumen then spreads over the aggregate. At this temperature both the coated plastics and bitumen are in the liquid state and so are capable of easy diffusion at the interphase. This process is further aided by the increase in contact area. These observations may be reported as follows- Waste polymers i.e. polypropylene are hydrocarbons with long chains and the bitumen is a complex mixture of asphaltenes and maltenes that are also long-chain hydro carbons. When bitumen gets mixed with plastic coated aggregate a percentage of bitumen diffuses through the plastic layer and binds with aggregates. The plastic layer has already strongly bonded with aggregates. During this phase, a three-dimensional internally cross linked network structure develops between bitumen constitutes and polymer molecules. Therefore, the bonding becomes stronger and removal of bonded bitumen becomes hard.

Materials Used in the Study

1. **Aggregate:** Stones passing through 12.5mm sieve and retained on 10mm sieve; Sand.
2. **Bitumen:** VG30 Bitumen Grade.
3. **Waste Plastic:** Polypropylene (PP) waste plastic in the shredded form (2.36 to 4.75mm).

Process Description

As indicated in the flow chart, the primary step is the material procurement wherein the different constituents like aggregates, bitumen and waste plastic are collected in the desired form and quantity. Aggregates composed of stones that are retained on 10 mm sieve but passed through 12.5 mm sieve;

Figure 1. Shows the detailed methodology which was followed in this study

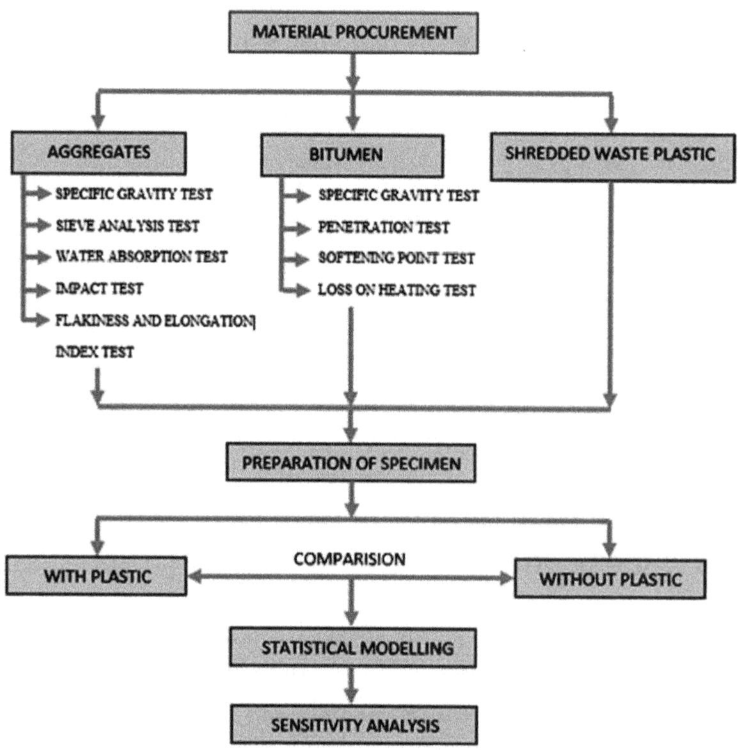

and sand. The grade of bitumen used in this study is VG30 grade bitumen (i.e. 50-70 penetration grade bitumen). Polypropylene waste plastics have been used in shredded form (2.36mm to 4.75mm). Shredding of these plastics has been done in a plastic factory in Vellore. The choice of Polypropylene types of waste plastics was based on the availability, suitability and its efficient properties. Not all types of waste plastics can be used in flexible pavement design. Plastics like Polyvinyl chloride can't be used for this purpose due to the release of harmful gases on its applications.

The next step is to find out various physical properties of the constituents, using laboratory testing. These physical properties of constituents are tested in order to assure that the selected constituents are suitable pavement design. Following this, Marshall Specimens have been prepared varying the percentages of plastic and bitumen and then the results obtained from the modified bitumen have been compared with that of conventional bitumen. The main objective of this test was to obtain the Optimum contents of Bitumen and Plastic in order to get the desired strength. Post analysis have been done using Regression modeling that yielded a relationship between parameters and this relationship was then used to perform sensitivity analysis.

Processes for Manufacturing Bituminous Mix Using Waste Plastic

There are two different processes namely, dry process and wet process used for design of bituminous mix flexible pavement. However, this study uses the dry Process of Mixing due to following reasons:

1. Advantage of Dry Process Over Wet Process:
    a. Harmful gases are not evolved in dry process as no burning of the plastic waste is done.
    b. Dry process is simple while the wet process is complex.
    c. Ease in handling in case of dry process as the dealing is done with the normal dry shredded waste plastic unlike the wet process wherein the dealing is done with the hot molten waste plastic.

Figure 2. Shows the steps involved in dry process of mixing

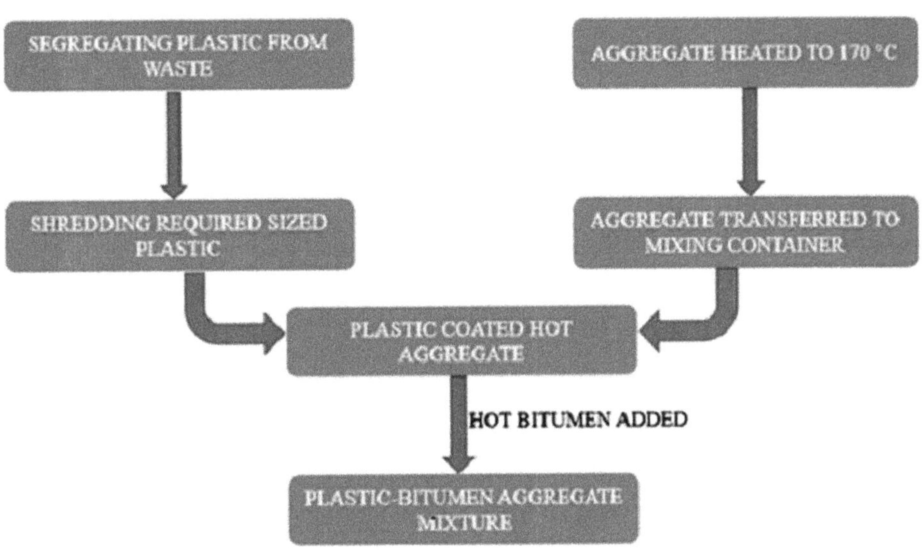

d. Unlike wet process, there is no risk of uneven mixing of molten plastic and bitumen and thus there will not be any loose binding of aggregates.

EXPERIMENTATION

Tests on Aggregates

1. **Specific Gravity:** It is the ratio of the weight of a given volume of aggregate to the weight of an equal volume of water. It is considered to be a measure of strength or quality of the material. Stones having low specific gravity are generally weaker than those with higher specific gravity values. The specific gravity of coarse aggregates used is found out to be 2.5g/cc. The specific gravity of aggregates normally used in road construction ranges from about 2.5 to 3 with an average of about 2.68.
2. **Sieve Analysis Test:** A sieve analysis (or gradation test) is a laboratory procedure used to assess the particle size distribution (also called gradation) of a granular material. The size distribution is of critical importance as it is an indicative to the way the material performs when in use.
3. **Water Absorption Test:** Aggregate absorption is the increase in mass due to water in the pores present in the material. Aggregate absorption is a useful quality measure because:
    a. High values of water absorption can indicate non-durable aggregate.
    b. Absorption also indicates the amount of asphalt binder the aggregate will absorb.
    c. Water absorption is found out to be 0.9%. According to IRC, the maximum allowable limit of water absorption for any aggregate used in highway construction should not exceed 2% by weight of aggregates.
4. **Impact Test:** Aggregates should be tested for their resistance against shock or impact loads caused due to the movement of vehicles at high speeds. Aggregate impact strength is found out to be 11.64%. According to IRC, the maximum allowable aggregate impact is 30%.
5. **Flakiness and Elongation Test:** Flakiness index is the percentage by weight of flaky aggregates whose least thickness is less than 0.6 times the average size of the aggregate. Elongation index is the percentage by weight of aggregates whose longest dimension is more than 1.8 times the average size of aggregate. The combined flakiness and elongation index is found out to be 24%. According to IRC, the maximum allowable combined flakiness and elongation indices is 30%.

Tests on Bitumen

1. **Specific Gravity:** It is the ratio of mass of a given volume of bitumen to the mass of an equal volume of water, both taken at a specified temperature. The specific gravity of pure bitumen ranges from 0.97 to 1.02. The specific gravity of bitumen used is found out to be 0.99 g/cc. According to Indian standards (BIS), minimum specific gravity of paving bitumen at 27°C shall be 0.99.
2. **Penetration Test:** Penetration value is a measure of hardness or consistency of bituminous material. It is the vertical distance penetrated or traversed by the point of a standard needle into the bituminous material under specific conditions of load, temperature and time. This distance is measured in one tenths of a millimetre. This test is an indicative of consistency of bitumen. Higher penetration values implies softer grade bitumen. Penetration value of bitumen used is found out to be 52.

3. **Softening Point Test:** The Softening Point of bitumen is the temperature at which the substance attains particular degree of softening. As per IS: 334-1982, it is the temperature in °C at which a standard ball passes through a sample of bitumen in a mould and falls through a height of 2.5 cm, when heated under glycerine or water at specified conditions of test. The binder should possess sufficient fluidity before its application in road uses. The softening point is an indicative of the temperature up to which a bituminous binder should be heated for various road-use applications. The softening point of bitumen used is found out to be 35°C.
4. **Loss on Heating Test:** The loss in weight (exclusive of water) of bituminous materials when heated is called loss on heating of bituminous materials. The loss on heating is observed to be 0.4%. According to IRC, the maximum allowable value of matter loss by heating should not exceed 1%.

Tests Results

From the tests conducted, it can be obtained that the given materials meet the desired specifications and hence can be successfully used in further study.

Marshall Test

This test is used to measure the resistance to plastic deformation of a compacted cylindrical specimen of bituminous mixture when the specimen is diametrically loaded at a deformation rate of 5 cm per minute. The Marshall stability of a mix is defined as the maximum load carrying capacity of the specimen at a standard test temperature of 60°C. The flow value is defined as the deformation that the test specimen undergoes during loading up to the maximum load. Flow is measured in 0.25 mm units. In this test, Marshall Specimens are prepared varying the bitumen and the plastic content in order to find

Table 1. The test results for bitumen tests

S. No.	Tests	Results	Permissible Limits	IS Code
1	Specific gravity test	0.99 g/cc	0.97 g/cc to 1.02 g/cc	IS 1202
2	Penetration test	52	50 to 70 (VG30)	IS 1203
3	Softening point test	35°C	40°C to 50 °C	IS 1205
4	Loss on heating test	0.4%	≤ 1%	IS 1212

Table 2. The test results for aggregate tests

S. No.	Tests	Results	Permissible Limits	IS Code
1	Specific gravity test	2.5 g/cc	2.5 to 3.0	IS 2386 Part II
2	Sieve analysis test	Uniformly graded	-	IS 2386 Part I
3	Water absorption test	0.9%	≤ 2%	IS 2386 Part III
4	Impact test	11.64%	≤ 30%	IS 2386 Part IV
5	Flakiness and elongation index test	24%	≤ 30%	IS 2386 Part I

the optimum bitumen content and the optimum plastic content. All the samples are prepared as per the guidelines given in ASTM D1559. Cylindrical moulds (10cm diameter and 7.5cm in height) consisting of a base plate and collar extension have been used.

Steps of Design

1. Determine the proportion of coarse and fine aggregates to be used as per the desired grading.
2. Determine the specific gravity of each of the constituents: aggregate constitutes, bitumen and the waste plastic.
3. Prepare the trial mixes by varying the bitumen and the plastic content.
4. Measure the specific gravity of compacted specimens.
5. Perform the stability test on specimens.
6. Compute percentage air voids, voids filled with bitumen and voids in mineral aggregates for each specimen.
7. Record the optimum bitumen content and the optimum plastic content from the data obtained.
8. Evaluate the design and perform analysis.

Formulae Used in the Study

$$G_m = \frac{W_a}{W_a - W_w} \tag{1}$$

where,

G_m = Bulk or actual specific gravity (g/cc)
W_a = Weight of compacted specimen in air (Kg)
W_w = Weight of compacted specimen in water (Kg)

$$G_t = \frac{W_1 + W_2 + W_3 + W_b}{\frac{W_1}{G_1} + \frac{W_2}{G_2} + \frac{W_3}{G_3} + \frac{W_b}{G_b}} \tag{2}$$

where,

G_t = Theoretical specific gravity (g/cc)
W_1 = Weight of coarse aggregate (Kg)
W_2 = Weight of fine aggregate (Kg)
W_3 = Weight of filler (Kg)
W_b = Weight of bitumen in total mix (Kg)

G_1 = Specific gravity of coarse aggregate (g/cc)
G_2 = Specific gravity of fine aggregate (g/cc)
G_3 = Specific gravity of filler (g/cc)
G_b = Specific gravity of bitumen (g/cc)

$$V_v = \frac{G_t - G_m}{G_t} * 100 \tag{3}$$

where,

V_v = Percent air voids in the mix (%)

$$V_b = \frac{\frac{W_b}{G_b}}{\frac{\sum W}{G_m}} * 100 \tag{4}$$

where,

V_b = Percent bitumen content in the mix (%)

$$VMA = V_v + V_b \tag{5}$$

where,

VMA = Volume of voids in the mineral aggregates (%)

$$VFB = \frac{V_b}{VMA} * 100 \tag{6}$$

where,

VFB = Voids filled with bitumen (%)

Tables 3-6 show the Marshall Test results.

Table 3. Marshall test results for specimens without plastic

Bitumen Content (%)	Air Voids (%)	VFB (%)	VMA (%)	Unit Weight (Kg/m3)	Stability (Kg)	Flow (mm)
3	8.988	53.258	19.229	2378	1120	2.73
4.5	4.886	72.701	17.898	2442	1210	2.92
6	3.196	82.746	18.524	2446	780	4.94
7.5	2.882	86.291	21.023	2416	670	7.37

Table 4. Marshall test result for specimens modified with 3% of polypropylene plastics

Bitumen Content (%)	Air Voids (%)	VFB (%)	VMA (%)	Unit Weight (Kg/m3)	Stability (Kg)	Flow (mm)
3	9.226	52.899	19.588	2356	1210	2.94
4.5	5.567	66.152	18.319	2416	1430	3.5
6	3.298	82.094	18.418	2450	940	4.82
7.5	2.996	85.688	20.933	2424	730	7.2

Table 5. Marshall test result for specimens modified with 4.5% of polypropylene plastics

Bitumen Content (%)	Air Voids (%)	VFB (%)	VMA (%)	Unit Weight (Kg/m3)	Stability (Kg)	Flow (mm)
3	9.896	50.564	20.018	2354	1240	3.89
4.5	7.015	64.546	19.786	2391	**1450**	3.57
6	4.219	78.795	19.896	2421	950	4.02
7.5	3.159	85.201	21.346	2412	790	4.91

Table 6. Marshall test result for specimens modified with 6% of polypropylene plastics

Bitumen Content (%)	Air Voids (%)	VFB (%)	VMA (%)	Unit Weight (Kg/m3)	Stability (Kg)	Flow (mm)
3	10.998	46.987	21.016	2329	1230	3.27
4.5	7.777	60.244	20.746	2355	**1370**	3.58
6	5.886	71.821	20.888	2384	890	4.04
7.5	4.998	77.177	21.899	2378	650	5.5

Result Analysis and Discussion

Based on the Marshall Test Results, variation of unit weight, air voids, VFB, VMA, stability and flow for each of the plastic content, is plotted against Bitumen content and the analysis is performed as under.

It is quite evident from the Figure 3 and Figure 7 that at the lower bitumen contents, unit weight as well as the stability are less but with the increase in bitumen content, these values attains a peak and

Figure 3. Variation in unit weight with bitumen content

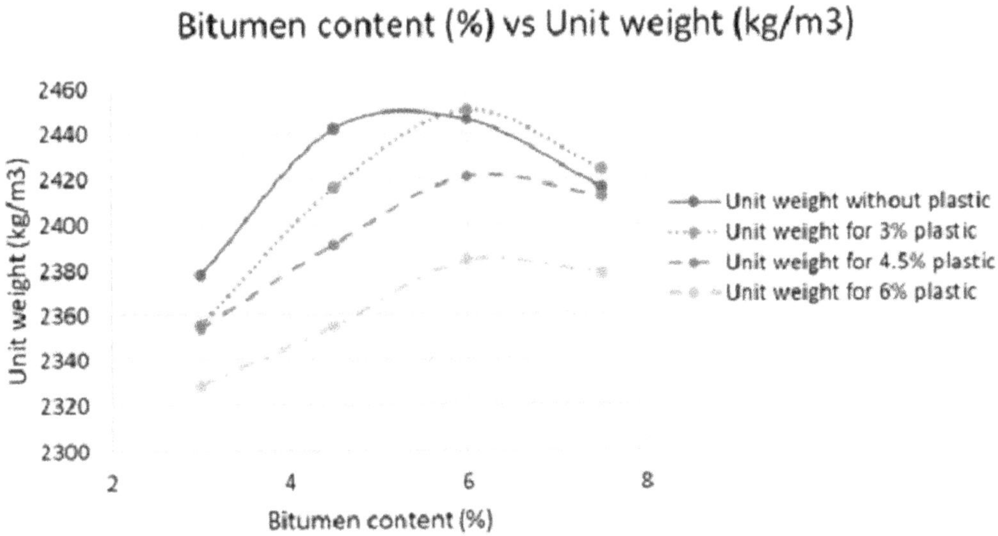

afterwards reflects a decreasing trend. At lower bitumen contents, specimens fails at lower loads because the quantity of bitumen mixed with aggregates is not sufficient enough to develop efficient bonding between aggregates to withstand applied loads. On the other hand, specimens with excess quantity of bitumen also fail at smaller loads due to a thick film of bitumen at the interface of aggregates which prevents the effective transmission of loads onto the aggregates. Hence, only at a particular quantity of bitumen (optimum content), maximum values of unit weight as well as Marshall Stability are observed.

From the Figure 4 and Figure 5, it can be observed that as the bitumen content increases, Voids filled with bitumen also increases which in turn leads to decrease in air voids present in the specimen as more and more bitumen tends to fill empty voids in the specimen.

As the bitumen content increases, a slight variation is observed in the values of Voids in mineral aggregates, as reflected in Figure 6. This variation attributes to the adjustment of mineral aggregates with the increased bitumen content.

Flow is the measure of the deformation caused due to applied loads. From the Figure 8, it is observed that the flow increases with the increase in bitumen content because of the presence of a thick bitumen film around aggregates that prevents strong aggregate interlocking.

From the Figure 7, as the maximum stability occurs at 4.5% bitumen content, therefore the Optimum Bitumen Content is 4.5%. Then a curve is plotted between stability and plastic content at the optimum bitumen content and it is observed from the Figure 9 that the maximum stability value occurs at 4.5% plastic content, so the optimum plastic content is 4.5%.

MODELLING OF MARSHALL STABILITY AND FLOW

Regression Analysis

Regression analysis is the technique employed here to model Marshall Stability and flow. Regression analysis is a statistical tool in order to assess the relationship between variables. Regression analysis is

Figure 4. Variation in VFB with bitumen content

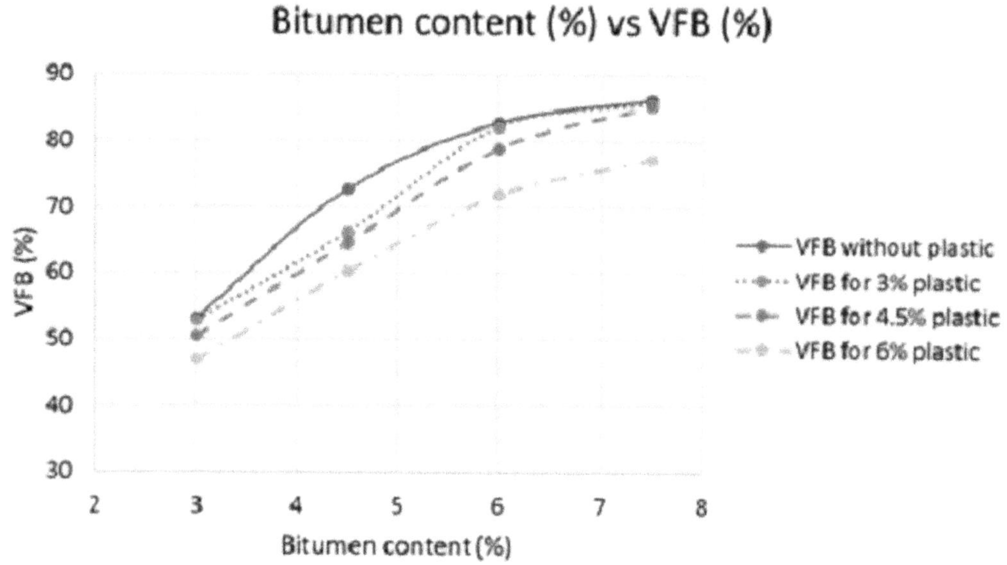

Figure 5. Variation in air voids with bitumen content

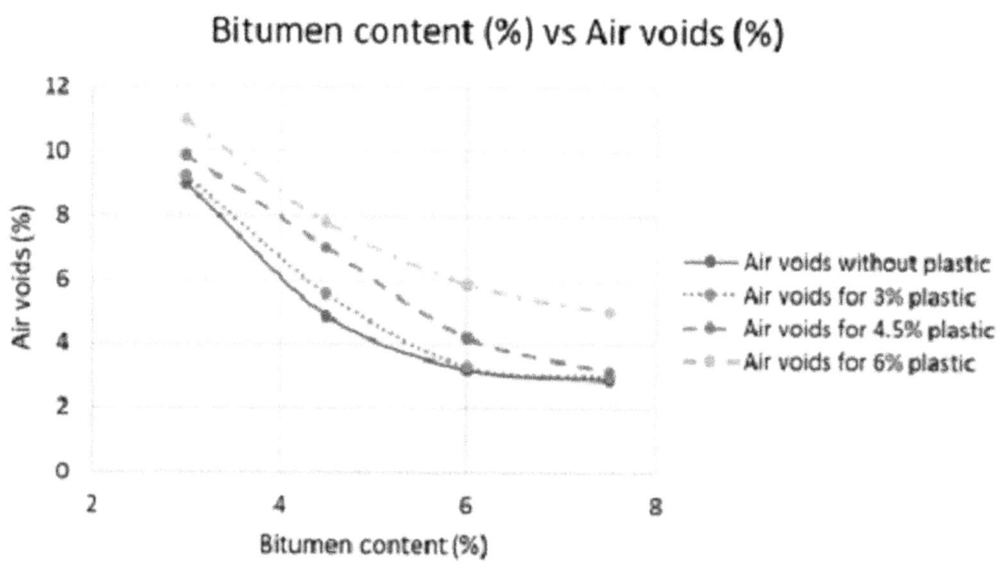

widely used for prediction and forecasting purposes. Usually, it is used to ascertain the causal effect of one variable upon another, as for example, the effect of increase in price of a commodity upon its demand, or the effect of change in money supply upon inflation rates. To explore such issues, the data on the underlying variables of interest is assembled and regression is employed to predict the quantitative effect of the causal variables upon the variable that those variable influence. This analysis is also used to assess the 'statistical significance' of the estimated relationships, i.e. the degree of confidence that the true relationship is close to the estimated relationship.

Figure 6. Variation in VMA with bitumen content

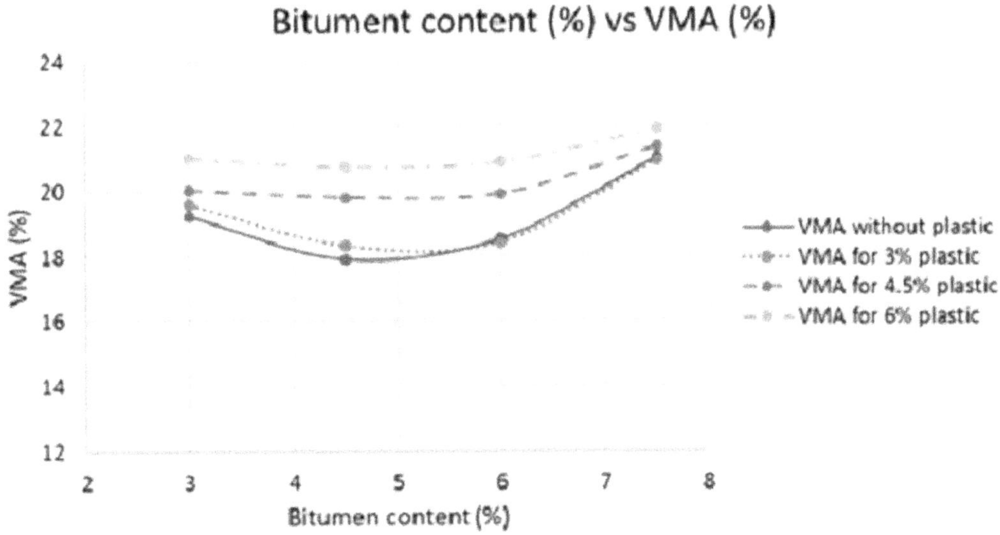

Figure 7. Variation in stability with bitumen content

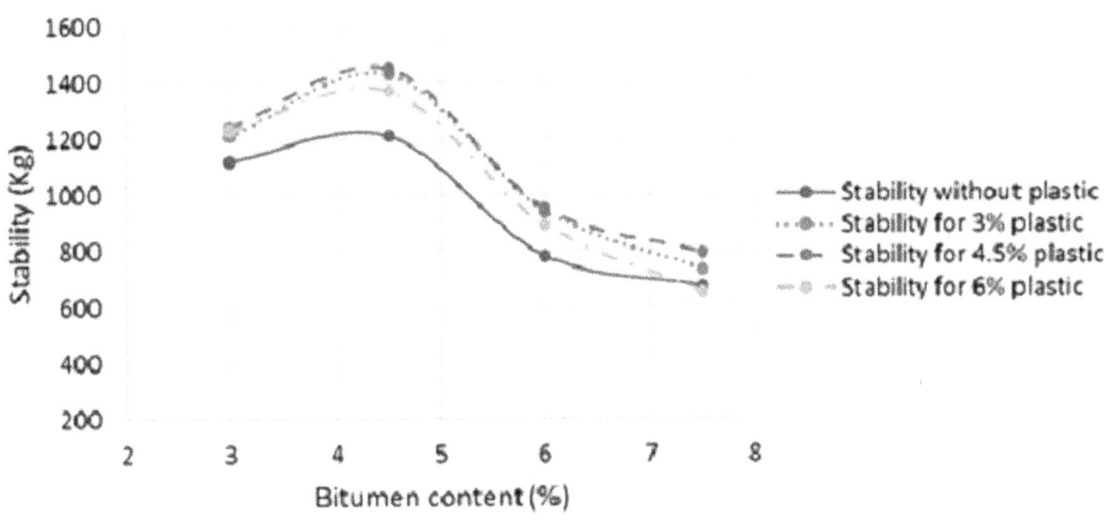

This study inculcated the use of 'Multiple Linear Regression model' since there are multiple parameters involved. Multiple regression analysis is a powerful tool used for predicting the unknown value of a variable from the known value of two or more variables, also termed as 'predictors'.

More precisely, multiple regression analysis helps to predict the value of Y for given values of X_1, X_2, …, X_n.

Figure 8. Variation in flow with bitumen content

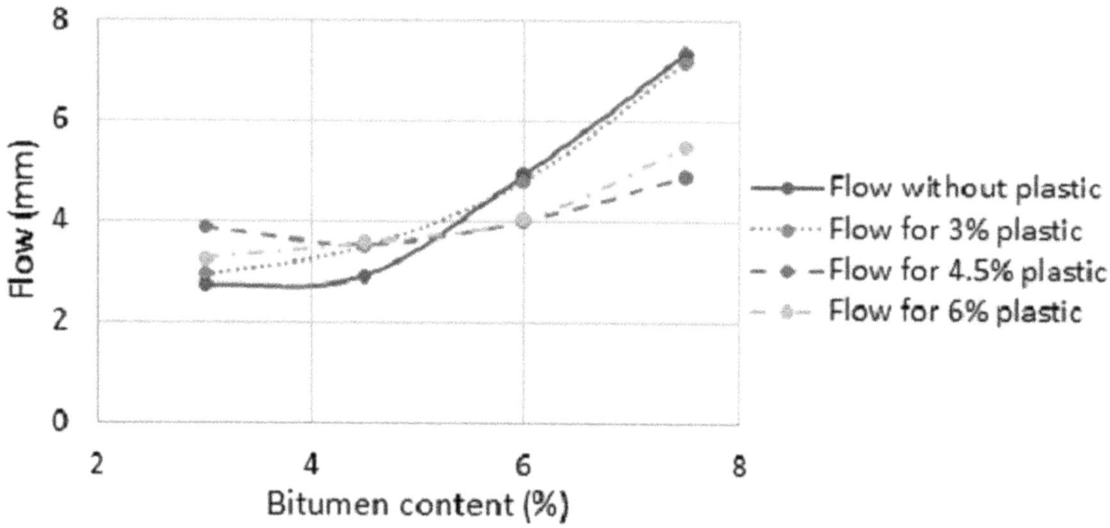

Figure 9. Variation in stability with plastic content

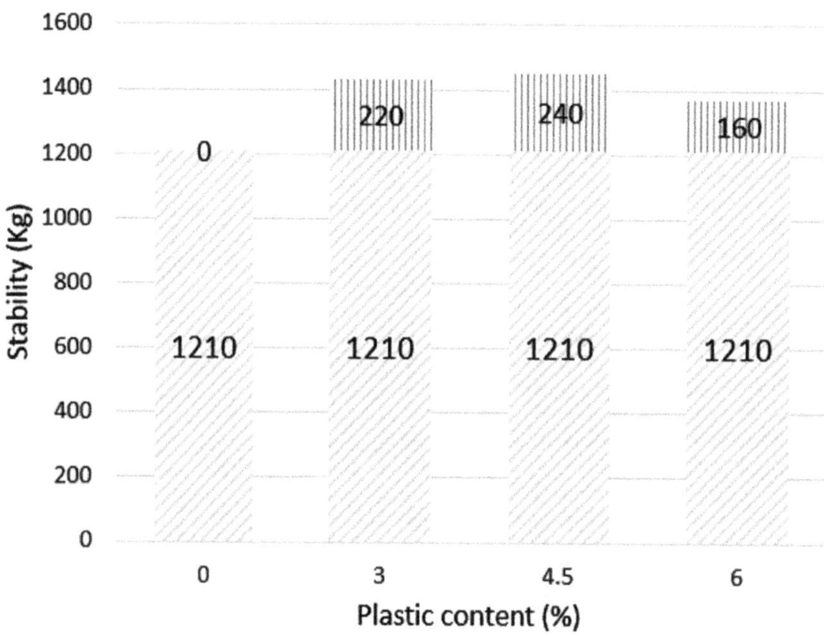

Dependent and Independent Variables

In multiple regression model, there is one dependent variable and two or more independent variables. A dependent variable is one whose value is to be predicted and the independent variables are those whose values are used for prediction. These independent variables are also referred to as 'exploratory' variables.

In general, a multiple regression equation of Y on $X_1, X_2, \ldots, X_n$ is of the form:

$$Y = b_0 + b_1X_1 + b_2X_2 + ... + b_nX_n \tag{7}$$

where,

Y = Dependent variable
X_i = Independent variables (where, i = 1, 2,….n)

Interpretation of Regression Coefficients

Here, b_0 is the intercept and b_1, b_2,….b_n are analogous to the slope in linear regression equation and are also termed as 'regression coefficients'. These could be indicated in similar way as slope. Thus, if b_i = 10, then it implies that Y will increase by 10 units if X_i increased by 1 unit. The appropriateness of the multiple regression model as a whole can be tested by using the t-test in the ANOVA (Analysis of Variance) table. A significant 't value' indicates a linear relationship between Y and at least one of the X's.

Test for Model Significance

In order to assess prediction ability of the developed regression equation, a good practice is to look upon 'Coefficient of Determination' (R^2 value). R^2 lies between 0 and 1. The closer is the value of R^2 to 1, the better is the developed model and its prediction.

In order to assess, if the independent variables chosen individually influences the dependent variable significantly or statistically, to test the null hypothesis that the relevant regression coefficient is zero; a t-test is required.

If the t-test of a regression coefficient is significant, it indicates that the concerned independent influences Y significantly while controlling for other independent variables.

Assumptions for Multiple Linear Regression

1. Multiple regression technique does not test if the data are linear. On the contrary, it proceeds by the assumption that the relationship between the Y and each of the X_i's is linear. Hence, it is a good practice to always look at the scatter plots of (Y, X_i), i= 1, 2,…,n. If any of the plots suggests non linearity, one must use a suitable transformation to attain linearity.
2. Another important assumption is the absence of multicollinearity i.e. the independent variables are not related among themselves. However, this can be tested by the computation of the correlation coefficient between each pair of independent variables.
3. Other assumptions include those of normality and homoscedasticity (i.e. the variance around the regression line is same for all values of the predictor variable (X_i)).
4. Multiple regression analysis is generally useful when one is interested in predicting a continuous dependent variable from a number of independent variables. In the case, the dependent variable is dichotomous, then the logistic regression should be used.

For this study, the dependence of Marshall Stability and Flow upon different variables is assessed. Herein, parameters are chosen based on the correlations between them i.e. there should be significant correlation between dependent and independent variables.

The multiple regression equation for stability is:

$$Stability = 20.31579 + 0.042882 * \left(Plastic\ \ Content\right) - 0.40602 * \left(Bitumen\ \ Content\right)$$
$$-0.25557 * \left(Air\ \ voids\right) - 0.00659 * \left(Unit\ \ weight\right)$$

The multiple regression equation for flow is:

$$Flow\ = -4.701225554 - 0.208519892 * \left(Plastic\ \ Content\right)$$
$$+1.344323532 * \left(Bitumen\ \ Content\right) + 0.476473045 * \left(Air\ \ voids\right)$$

Figure 10. The table of correlations between the parameters involved

		Plastic_content	Bitumen_content	Air_voids	VFB	VMA	Unit_weight	Stability	Flow
Plastic_content	Pearson Correlation	1	.000	.326	-.248	.625**	-.579*	.150	-.140
	Sig. (2-tailed)		1.000	.218	.354	.010	.019	.578	.604
	N	16	16	16	16	16	16	16	16
Bitumen_content	Pearson Correlation	.000	1	-.889**	.936**	.402	.577*	-.820**	.836**
	Sig. (2-tailed)	1.000		.000	.000	.123	.019	.000	.000
	N	16	16	16	16	16	16	16	16
Air_voids	Pearson Correlation	.326	-.889**	1	-.989**	.049	-.879**	.673**	-.694**
	Sig. (2-tailed)	.218	.000		.000	.857	.000	.004	.003
	N	16	16	16	16	16	16	16	16
VFB	Pearson Correlation	-.248	.936**	-.989**	1	.076	.819**	-.748**	.751**
	Sig. (2-tailed)	.354	.000	.000		.781	.000	.001	.001
	N	16	16	16	16	16	16	16	16
VMA	Pearson Correlation	.625**	.402	.049	.076	1	-.498*	-.421	.449
	Sig. (2-tailed)	.010	.123	.857	.781		.050	.104	.081
	N	16	16	16	16	16	16	16	16
Unit_weight	Pearson Correlation	-.579*	.577*	-.879**	.819**	-.498*	1	-.405	.398
	Sig. (2-tailed)	.019	.019	.000	.000	.050		.120	.126
	N	16	16	16	16	16	16	16	16
Stability	Pearson Correlation	.150	-.820**	.673**	-.748**	-.421	-.405	1	-.785**
	Sig. (2-tailed)	.578	.000	.004	.001	.104	.120		.000
	N	16	16	16	16	16	16	16	16
Flow	Pearson Correlation	-.140	.836**	-.694**	.751**	.449	.398	-.785**	1
	Sig. (2-tailed)	.604	.000	.003	.001	.081	.126	.000	
	N	16	16	16	16	16	16	16	16

**. Correlation is significant at the 0.01 level (2-tailed).

*. Correlation is significant at the 0.05 level (2-tailed).

Table 7. Gives the regression model for stability

Variables	Coefficients	Standard Error	t Stat
Constant	20.31579	7.381433	2.752283
Plastic Content	0.042882	0.008386	5.113287
Bitumen Content	-0.40602	0.063546	-6.38942
Air voids	-0.25557	0.067728	-3.77347
Unit weight	-0.00659	0.002793	-2.36032

Note: For a variable to be significant, the corresponding absolute t stat value should be greater than 2.131 i.e. the t stat value corresponding to 95% Confidence level and 15 degrees of freedom.

Figure 11. The regression plot for stability

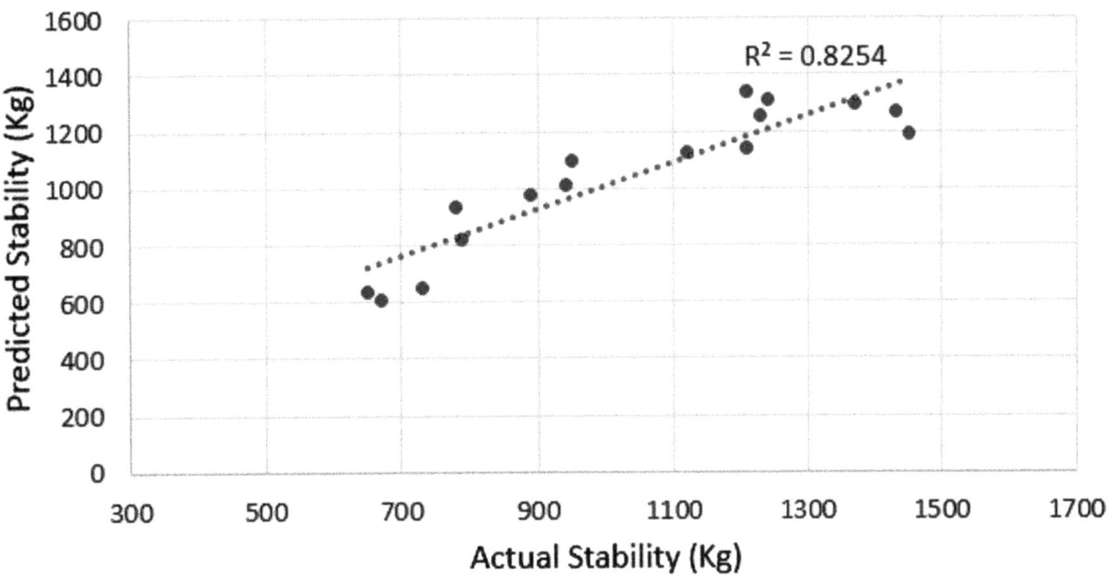

Table 8. The regression model for flow

Variables	Coefficients	Standard Error	t Stat
Constant	-4.701225554	1.320386109	-3.560493043
Plastic Content	-0.208519892	0.044535712	-4.682082834
Bitumen Content	1.344323532	0.157989793	8.508926473
Air voids	0.476473045	0.106305775	4.482099337

Note: For a variable to be significant, the corresponding absolute t stat value should be greater than 2.131 i.e. the t stat value corresponding to 95% Confidence level and 15 degrees of freedom.

Figure 12. Shows the regression plot for flow

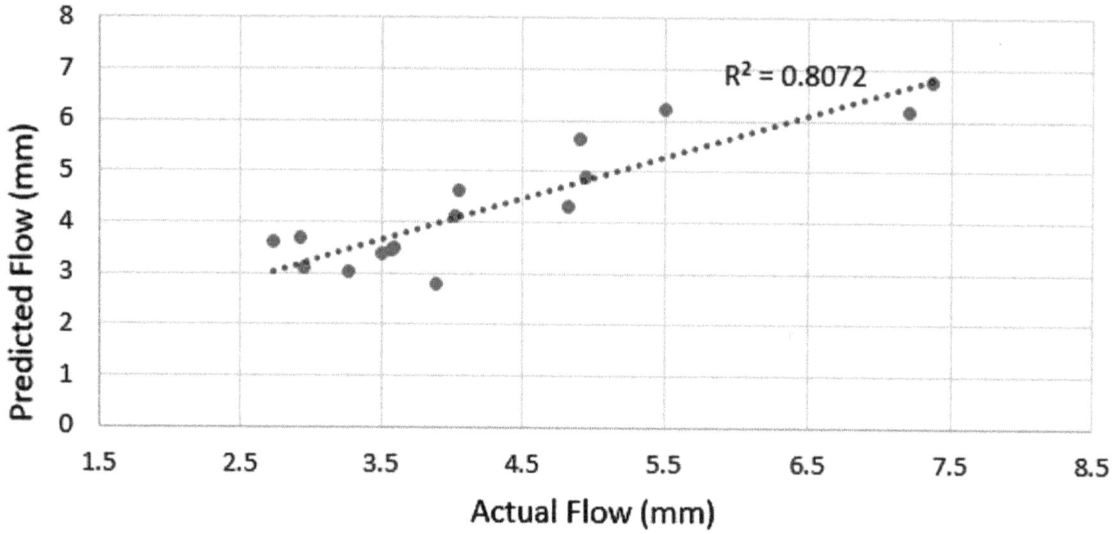

CONCLUSION AND FUTURE SCOPE OF WORK

Conclusion

This study explored the effective utilization of waste plastic (polypropylene) as a binder replacement in the flexible pavement design as the obtained results in case of plastic additive reflects higher strength values as compared to the conventional bituminous mixes. A considerable improvement in other pavement properties is also observed. The use of waste plastics not only helps in enhancing pavement properties, it also aids in curbing the waste disposal. The key findings of the study are the optimum bitumen content of 4.5% and the optimum plastic content of 4.5%. The modelling approach adopted in the study has produced quite satisfactory results. This regression modelling also shows that Bitumen Content is the most significant parameter affecting Marshall Stability and Flow. Since the Coefficient of determination values of both the models are close to 1, which means the predicted results are quite closer to the actual results, so the developed models can be efficiently used for prediction. These models can aid future researchers to obtain the Marshall Stability and Flow values directly from the physical properties of materials which will save a lot of time spent in conducting cumbersome and laborious tests.

Future Research Directions

The findings and the approach from this study can be extended in exploring the efficient utilization of other types of waste plastics as well as the other types of wastes from different industries. Besides, due to availability of less data, the study is limited to Regression modelling, however with the increase in the number of data collection, further advanced modelling can also be adopted to model Marshall Stability and flow which may validate the results more accurately.

REFERENCES

Afroz Sultana, S. K. (2012). Utilization of Waste Plastic as a Strength Modifier in Surface Course of Flexible and Rigid Pavements. *International Journal of Engineering Research and Applications, 2*(4), 1185-1191.

Al-Hadidy, A. I., & Yi-qiu, T. (2009). Effect of polyethylene on life of flexible pavements. *Construction & Building Materials, 23*(3), 1456–1464. doi:10.1016/j.conbuildmat.2008.07.004

Chavan. (2013). Use of plastic waste in flexible pavements. *International Journal of Application and Innovation in Engineering and Management, 2*(4), 540-552.

Dhodapkar, A. N. (2008). Use of waste plastics in road construction. Indian Highways Technical Paper Journal, 31-32.

Gupta, R. (2011). A novel approach to improve road quality by utilizing plastic waste in road construction. *Journal of Environmental Research and Development, 5*(4), 1036–1042.

Habib, N. Z. (2011). Effect of Mixing Process on Polypropylene Modified Bituminous Concrete Mix Properties. *International Journal of Civil, Environmental, Structural, Construction and Architectural Engineering Vol, 5*(:10), 477–482.

Jafar. (2015). Utilisation of Waste plastic in bituminous mix for improved performance of roads. *KSCE Journal of Civil Engineering*, 1-7.

Menaria, Y., & Sankhla, R. (2015). Use of Waste Plastic in Flexible Pavements-Green Roads. *Open Journal of Civil Engineering, 5*(03), 299–311. doi:10.4236/ojce.2015.53030

Ozgan, E. (2011). Artificial neural network based modelling of the Marshall Stability of asphalt concrete. *Expert Systems with Applications, 38*(5), 6025–6030. doi:10.1016/j.eswa.2010.11.018

Rajmane, P. B. (2007). Effective utilization of waste plastic in construction of flexible pavement for improving their performance. *IOSR Journal of Mechanical and Civil Engineering*, 27-30.

Swami, V. (2012). Use of waste plastic in construction of bituminous roads. *International Journal of Engineering Science and Technology, 4*(5), 2351–235.

Tapkın, S., Çevik, A., & Uşar, Ü. (2010). Prediction of Marshall test results for polypropylene modified dense bituminous mixtures using neural networks. *Expert Systems with Applications, 37*(6), 4660–4670. doi:10.1016/j.eswa.2009.12.042

Vasudevan, R. (2007). Utilization of Waste Polymers for Flexible Pavement and Easy Disposal of Waste Polymers. *International Conference on Sustainable Solid Waste Management*, 105-111.

Yilmaz, M. (2011). Effects of using asphaltite as filler on mechanical properties of hot mix asphalt. *Construction & Building Materials*, 1–8.

Chapter 13
Prediction of Biosorption Capacity Using Artificial Neural Network Modeling and Genetic Algorithm:
Prediction of Biosorption Capacity

Prakash Chandra Mishra
Fakir Mohan University, India

Anil Kumar Giri
Fakir Mohan University, India

ABSTRACT

Artificial neural network model is applied for the prediction of the biosorption capacity of living cells of Bacillus cereus for the removal of chromium (VI) ions from aqueous solution. The maximum biosorption capacity of living cells of Bacillus cereus for chromium (VI) was found to be 89.24% at pH 7.5, equilibrium time of 60 min, biomass dosage of 6 g/L, and temperature of 30 ± 2 °C. The biosorption data of chromium (VI) ions collected from laboratory scale experimental set up is used to train a back propagation (BP) learning algorithm having 4-7-1 architecture. The model uses tangent sigmoid transfer function at input to hidden layer whereas a linear transfer function is used at output layer. The data is divided into training (75%) and testing (25%) sets. Comparison between the model results and experimental data gives a high degree of correlation R2 = 0.984 indicating that the model is able to predict the sorption efficiency with reasonable accuracy. Bacillus cereus biomass is characterized using AFM and FTIR.

1. INTRODUCTION

ANNs are categorized as an artificial intelligence modeling technique due to their ability to recognize patterns and relationships in historical data and subsequently make inferences concerning new data. ANNs can be used for two broad categories of problems: data classification and parameter prediction.

DOI: 10.4018/978-1-5225-2440-3.ch013

(1) Data classification problems, the ANN uses a specified algorithm to analyze data cases or patterns for similarities and then separates them into a pre-defined number of classes. (2) Parameter prediction problems, the ANN learns to predict accurately the value of an output parameter when information is given with sufficient input parameter (Yetilmezsoy and Demirel, 2008). The main applications of the ANN technique in the water treatment industry are in the development of process models and model-based process-control and automation tools.

1.1. Types of ANN Models

ANN models of drinking water treatment process can assume two distinct forms: (1) process models and (2) inverse process models. (1) The process model predicts the value of one or more process outputs, given the values of the process input parameters. An example of this type of model is the prediction of clarifier effluent turbidity using influent water parameters and operational parameters. (2) The inverse process model predicts the value of one or more process inputs, given the values of the remaining process inputs and process output(s). This type of model is often used to predict the value of an operational parameter required to reach a target effluent quality (Dimopoulos et al., 1999, Fernandes and Lona, 2005 and Gob et al., 1999).

1.2. Key Components of ANN Models

There are several key components to ANN models that are collectively referred to as the ANN architecture. Processing units or neurons perform primitive operations such as scaling data, summing weights, and amplifying or threshold sums. Neurons are organized into layers with each layer performing a specific function. The input layer serves as an interface between the input parameter data and the ANN model. Most models also contain one or two hidden layers, although more are possible (Aleboyeh et al., 2008; Belanche et al., 2008). These layers perform most of the iterative calculations within the network. The output layer serves as the interface between the ANN model and the end-user, transforming model information into an ANN-predicted value of the output parameter(s). Each neuron is connected to every neuron in adjacent layers by weights; links that represent the 'strength' of connection between neurons. Each ANN model has a propagation rule that defines how the weights connected to a neuron are combined to produce a net input. The propagation rule is generally a simple summation of the weights. As discussed, the input layer serves as an interface between input parameter data and the model. In this layer, a scaling function is used to scale data from their numeric range into a range that the network deals with efficiently, typically 0 to 1. The hidden and output layers contain an activation function that defines how the net input received by a neuron is combined with its current state of activation to produce a new state of activation. The most common activation function used in process modeling is the sigmoidal activation function. Finally, each network has a learning rule that defines how the weights are modified in order to minimize prediction error.

1.3. ANN Modeling Requirements

The key requirement of the ANN modeling approach is the availability of relevant data to describe the process being modeled. Data must be available in a useable electronic format for each of the process input and output parameters. The data used in model development must be representative of plant operations, spanning the range of operating conditions that may be encountered during both routine operations and process upset conditions. Model development requires the use of appropriate ANN software. As previously indicated, software packages that use only proprietary algorithms should be avoided in favour of those that have a high degree of user input in model-building. With respect to hardware requirements, recent advances in computing technology have made even the most modest new home computing system capable of performing the modeling calculations. To ensure optimal performance however, a 500 MHz processor and 128 M of RAM should be considered to be the minimum system requirements (Chu, 2003 and Despagne and Massart, 1998).

1.4. Artificial Neural Network Modeling for Biosorption Capacity from Water

The ANN modeling technique holds several advantages over mechanistic modeling that make it particularly suitable to process modeling in the drinking water treatment industry. ANN models can handle non-linear relationships, and can provide predictions of output parameters in real time in response to simultaneous and independent fluctuations of the values of model input parameters. Data patterns where the value of one or more of the model inputs are missing can be incorporated into model building if necessary, although model prediction will improve if only complete data patterns are used. Similarly, when completed models are applied to new data patterns where values are missing, the value of model outputs can still be predicted. Yetilmezsoy and Demirel, (2008) observed that artificial neural network (ANN) approach for modeling of Pb(II) adsorption from aqueous solution by *Antep pistachio* (*Pistacia vera* L.) shells. Salari *et al.* (2005) observed that application of artificial neural networks for modeling of the treatment of wastewater contaminated with methyl tert-butyl ether (MTBE) by UV/H_2O_2 process. Strik *et al.* (2005) have used to prediction of trace compounds in biogas from anaerobic digestion using the MATLAB neural network toolbox. Many user-friendly ANN software packages exist, offering the user a myriad of modeling options and allowing the user to customize the modeling process to suit his or her knowledge of modeling heuristics. Hamed *et al.* (2004) proposed that to prediction of wastewater treatment plant performance using artificial neural networks. Aber *et al.* (2009) proposed that to removal of Cr(VI) from polluted solutions by electro-coagulation: modeling of experimental results using artificial neural network.

2. GENETIC ALGORITHM USING ARTIFICIAL NEURAL NETWORK MODELING

In recent years, considerable advancement in artificial intelligence techniques has taken place to predict the responses in complex and difficult situations using MATLAB 7.0 is used for simulation purpose. Such techniques can enhance predicting capability of the model when mathematical or statistical methods are difficult to formulate and fails to predict with desired accuracy. Therefore, artificial neural network (ANN) has been attempted in this work for prediction purpose because ANN has the capacity to map inputs and outputs efficiently in complex situations.

2.1. Back Propagation Neural Network Architecture (BPNNA)

A back propagation neural network (BPNN) architecture consisting of three layers such as input layer, hidden layer and output layer is considered. Functioning of neural network proceed in two stages viz., learning or training and testing or inferences (Yetilmezsoy and Demirel, 2008; Yetilmezsoy and Saral, 2007). The network architecture is represented as *l-m-n* where *l* neurons are present at input layer (equal to the number of inputs in the models), *m* neurons at the hidden layer (optimized through experimentation), and *n* neurons at the output layer depending on number of outputs desired from the model. Input layer receives information from the external sources and passes this information to the network for processing. Hidden layer receives information from the input layer, does all the information processing, output layer receives processed information from the network, and sends the results out to an external receptor. The input signals are modified by interconnection weight known as weight factor (W_{ji}), which represents the interconnection of ith node of the first layer to j^{th} node of the second layer. The sum of modified signals (total activation) is then modified by a sigmoid transfer function *(f)*. Similarly, outputs signal of hidden layer are modified by interconnection weight (W_{ji}) of kth node of output layer to j^{th} node of hidden layer. The sum of the modified signal is then modified by sigmoid transfer *(f)* function and output is collected at output layer.

Let $I_p = (I_{p1}, I_{p2}, \ldots\ldots\ldots I_{pl})$, $P = 1, 2 \ldots. N$ is the Pth pattern among N input patterns. Where W_{ji} and W_{kj} are connection weights between i^{th} input neuron to j^{th} hidden neuron, and j^{th} hidden neuron to k^{th} output neuron, respectively.

Output from a neuron in the input layer is,

$$O_{pi} = I_{pi}, i = 1,2,\ldots,l$$

Output from a neuron in the output layer is,

$$O_{pk} = f\left(\sum_{j=0}^{m} W_{kj} O_{Pj}\right), k = 1,2\ldots n$$

Sigmoid transfer function *(f)* is a bounded, monotonic, non-decreasing, S-shaped function that provides a graded nonlinear response. The logistic sigmoidal function is given as

$$f(x) = \frac{1}{1 + e^{-x}}$$

Sigmoid transfer function is recommended for obtaining activation signals from input and hidden layer neurons whereas linear transfer function is used to get the same at output layer neurons. Metal concentration, biosorbent dosage, different temperature and time are used as inputs to the ANN model. Sorption efficiency (%) is desired from the network as output. One hundred seventy one experimental data are used to develop the ANN model (Yetilmezsoy, 2006). The complete experimental are divided into two sets - training (75% of data) and test sets (25% of data). All data are normalized in the 0.1-0.9 range to avoid the scaling effect of parameter values. Therefore, all of the data (x_i) are converted to normalize values (x_{norm}) as follows.

$$X_{norm} = 0.8 \times \left(\frac{X_i - X_{min}}{X_{max} - X_{min}}\right) + 0.1$$

where X_i is i^{th} input or output variable X.

X_{min} and X_{max} are minimum and maximum value of variable X.

2.2. Learning or Training in Back Propagation Neural Network

Initially, the connection weights are generated randomly in the range of -1 to +1. Batch mode type of supervised learning has been adopted in the present case where interconnection weights are adjusted using delta rule algorithm after sending the entire training sample to the network. During training, the predicted output is compared with the desired output, and the mean square error is calculated. If the mean square error is more than a prescribed limiting value, it is back propagated from output to input, and weights are further modified till the error or number of iterations is within a prescribed limit. Mean square error, E_p for pattern p is defined as

$$E_p = \sum_{i=1}^{n} \frac{1}{2}(D_{pi} - O_{pi})^2$$

where, D_{pi} is the target output, and O_{pi} is the computed output for the ith pattern.

Weight change at any time t, is given by

$$\Delta W(t) = -\eta E_p(t) + \alpha \times \Delta W(t-1)$$

η = *learning rate i.e* $0 < \eta < 1$

α = momentum coefficient i.e $0 < \alpha < 1$

2.3. Testing of Back Propagation Neural Network

Entire experimental data set is divided into training set and testing set. The mean square error is monitored during the training phase. The error usually decreases during the initial phase of training. However, when the network begins to over fit the data, the error on the training set will typically begin to rise. When the training error starts increasing for a specified number of iterations, the training is stopped and the weights at the minimum value of the training error are returned. The testing data is then fed to the trained network to check the percentage variation of predicted output in comparison to the actual one (Shetty and Chellam, 2003).

3. PREDICTION OF CHROMIUM (VI) IONS BY BIOSORPTION PROCESS USING ARTIFICIAL NEURAL NETWORK (ANN) MODELING

Waste water treatment using biosorption process is a complex process because a large number of variables influence the removal efficiency. Modeling of such processes is difficult when standard procedure like simple linear multivariate correlation is used. Artificial neural network (ANN), being capable of mapping inputs and outputs efficiently, can recognize and reproduce the cause and effect relationships through training in a multiple input-output systems. Like any experimental investigation, study on biosorption by living cells demand substantial amount of time, energy and materials. Hence, there is a need for a prediction tool to supplement to the experiments. Over the last two decades, ANNs have been used by many researchers for a variety of engineering applications. ANNs are a family of massively parallel architectures that solve difficult problems via the cooperation of highly interconnected but simple computing elements (or artificial neurons) arranged in layers. ANN represents a powerful tool for the identification of the relevant parameters and their interactions especially when relationships are very complex and highly non-linear. Prediction of biosorption of heavy metals from aqueous solution has been attempted in the past by many researchers using ANN to a reasonably good degree of success.

3.1. Effect of pH on the Sorption Efficiency

The pH is one of the important factors affecting the sorption of metal/metalloid ions in the solution. Percentage removal of chromium (VI) at different pH is studied in batch experiments using 0.3 g of biosorbent in 50 ml synthetic solution, at ambient temperature (30 ±2°C) and contact time of 60 min for initial chromium (VI) concentration of 1 mg/L, 5 mg/L and 10 mg/L. The result indicates that the sorption capacity of biomass for chromium (VI) depends on the pH of the solution. Present investigation of the sorption efficiency was increased from 45.33% to 89.24% for 1 mg/L, 42.64% to 85.13% for 5 mg/L and 38.23% to 81.54% for 10 mg/L with the increase of pH from 3.5 to 7.5 than decreases with further increase in pH up to 9.5. Cr (VI) exists as $HCrO_4^-$, $Cr_2O_7^{2-}$ in solution at optimum sorption pH has a tendency to bind the protonated active sites of the biosorbent. The dominant form of hexavalent chromium at acidic pH is $HCrO_{4-}$ which arises from the hydrolysis reaction of the dichromate ion ($Cr_2O_7^{2-}$). But as pH of the solution increases, bacterial cell wall becomes more and more negatively charged due to functional groups, which repulse the negatively charged chromate ions thereby affecting Cr (VI) sorption on the bacterial surface, resulting in reduced sorption efficiency. At low pH (2.0-6.5) the surface of biosorbent is highly protonated and as a result, a strong attraction exists between oxyanion and positively charged surface of the biosorbent. The further decrease in metal uptake with increase in pH (7.5 - 9.5) may be due to the fact that at higher pH, the substrate may be negatively charged by adsorbing hydroxyl ions on the surface or by ionization of very weak acidic functional groups of the sorbent, or both. A repulsive force may develop between the negatively charged surface and the anions (Nourbakhsh *et al.*, 2002).

3.2. Effect of Biosorbent Dosage on the Sorption Efficiency

Biosorbent dose is an important parameter which determines the capacity of sorbent for an initial concentration of the sorbate. The effect of sorbent dose on the sorption of chromium (VI) is studied at ambient temperature (30±2°C) and contact time of 60 min for chromium(VI) concentration of 1 mg/L, 5 mg/L and 10 mg/L. Experimental results generally showed that as the sorbent mass increased from 1g/L to 6 g/L, the percentage removal of chromium (VI) also increased from 78.40% to 89.24%, 74.45% to 85.13% and 71.12% to 81.54% for 0.05-0.5 g/50 mL of living cells of *B. cereus* biomass and initial chromium(VI) concentration of 10 mg/L, 5 mg/L and 1 mg/L. It is observed that after dosage of 0.3g/ 50 mL, there is no significant change in percentage of removal of chromium (VI). It may be due to the higher dosage could produce a 'screen effect' on the cell wall, protecting the binding sites, thus in lower Cr(VI) uptake (Mashitah *et al.*, 1999). It can be seen that the ANN model satisfactorily predicts the trend of the experimental data.

3.3. Effect of Contact Time on the Sorption Efficiency

Biosorption of chromium (VI) at different contact time is studied for initial chromium (VI) concentration of 1 mg/L, 5 mg/L and 10 mg/L at pH 7.5 keeping all other parameters constant. It is clear that more than 90% removal takes place within 30 min and equilibrium is reached after 60 min. The change in the rate of removal might be due to the fact that initially all sorbent sites are vacant and also the solute concentration gradient is high. At higher concentrations, metals need to diffuse to the sorbent surface by intraparticle diffusion and greatly hydrolyzed ions will diffuse at a slower rate. Few researchers have reported that the initial metals concentration provides a driving force to overcome mass transfer resistance of chromium ions between the aqueous and solid phase. This indicates the possible monolayer formation of chromium (VI) ions on the outer surface. It can be seen that the ANN model satisfactorily predicts the trend of the experimental data.

3.4. Effect of Initial Concentration on the Sorption Efficiency

The biosorption of chromium(VI) concentration using optimum adsorbent dose (0.3 g/50 mL) at ambient temperature (30±2° C) and contact time of 60 min. The initial chromium (VI) concentration is increased from 1mg/L to 10 mg/L and the corresponding removal gradually decreases from 89.24% to 79.22%. It is clear that more than 80% sorption of chromium(VI) took place in first 30 min and equilibrium is established after 60 min. At higher concentrations, metals need to diffuse to the biosorbent surface by intraparticle diffusion and greatly hydrolyzed ions will diffuse at a slower rate. It can be seen that the ANN model satisfactorily predicts the trend of the experimental data.

3.5. Effect of Temperature of Sorption Efficiency

The effect of temperature on the sorption of chromium(VI) with initial concentration 1mg/L, 5 mg/L and 10 mg/L is studied using optimum adsorbent dose (0.3 g /50 mL). The results are represented as percentage removal of chromium (VI) versus temperature. The percentage removal of chromium (VI) with initial concentration 1 mg/L, increased from 79.24% to 92.24%, the percentage removal of chromium(VI) of initial concentration 5 mg/L, increased from 78.36% to 88.32% and the percentage removal of chromium(VI) of initial concentration 1mg/L, increased from77.22% to 85.56% for 15 - 40 ° C temperature. It is seen that increase in temperature the percentage removal increased slowly and reached almost 84% for arsenic(III) and 92% for chromium(VI). The increase in biosorption capacity with the increase in temperature indicates that the biosorption process is endothermic in nature.

3.6. Prediction of Biosorption Efficiency Using ANN Model

The prediction of removal efficiency of Cr(VI) ions from aqueous system using *Bacillus cereus* biomass is a complex proposition and hence, a neural network approach is adopted. One hundred seventy one experimental data is divided into training and testing sets. Seventy five percent of data (131) is used as training set whereas twenty five percent of data (40) are used for testing for chromium (VI). Figure 1 shows a BP algorithm with three-layer architecture (a single hidden layer) with a tangent sigmoid transfer function (*tansig*) at input and hidden layer and a linear transfer function (*purelin*) at output layer is used and run on MATLAB 7.0 using a Pentium IV PC (Chu, 2003).The distribution of output of training data is shown in Figure 2 for chromium (VI). Initially, the training data set is presented to the network (Chang *et al.*, 2013).

The number of nodes in the hidden layer (H) is decided by the relation below:

Figure 1.

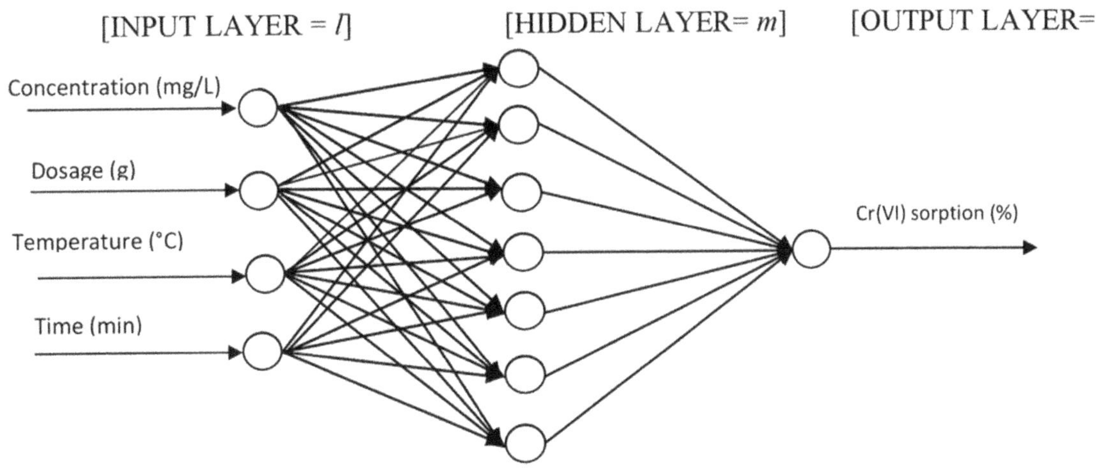

Figure 2.

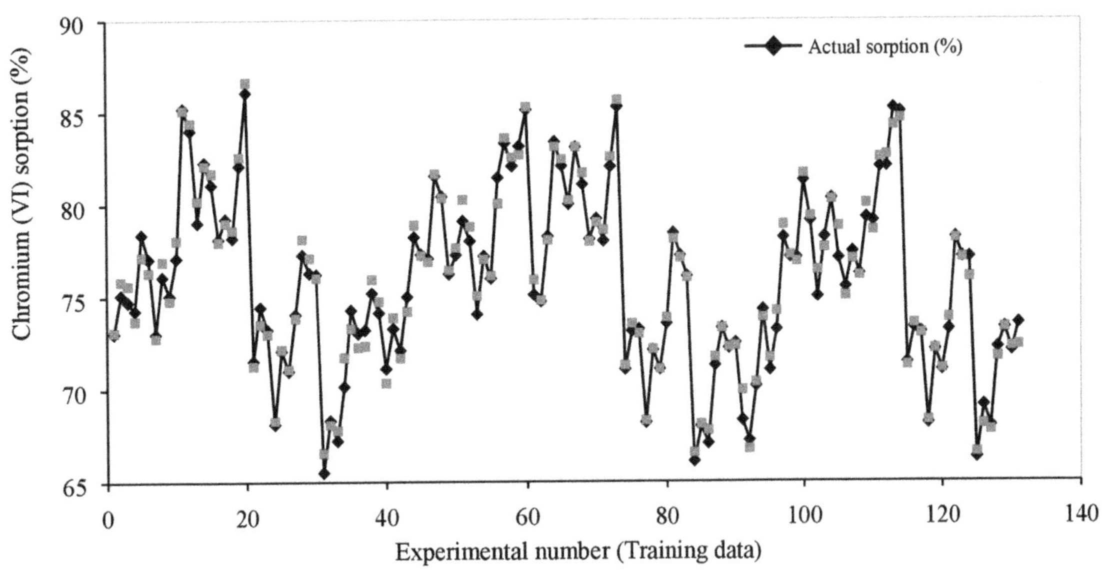

$$H = 2\sqrt{(I+1)}$$

where I is the number of nodes at the input layer.

The network is tested with different number of neurons to find the optimal number of neurons at the hidden layer by observing the mean squared error. Seven neurons are selected in the hidden layer when mean square error starts decreasing. Learning and momentum parameters are set at 0.20 during the training phase chromium (VI). During training phase, the output vector is computed by a forward pass in which the input is propagated forward through the network to compute the output value of each unit. The output vector is then compared with the desired vector which resulted into error signal for each output unit. In order to minimize the error, appropriate adjustments were made for each of the weights of the network. After several such iterations, the network was trained to give the desired output for a given input vector. The network is trained till minimum root mean square error is observed. A root mean square error of 0.68 is observed at epoch number 23, 835, 76 for chromium (VI). Training was stopped at this point and weights have been frozen for network to undergo testing phase. A high degree of correlation between actual and predicted sorption efficiency (%) is observed as shown in Figure 3 for chromium (VI). Coefficient of determination (R^2) of 0.984 for chromium (VI) obtained for training data set. When the network is well trained, testing of the network with testing data set is carried out (Debnath *et al.*, 2016; Podder and Majumder, 2016).

The prediction ability of the developed network model for responses of experimental data not forming part of the training set. During testing phase, output of the data is not presented to the network. A high degree of correlation (R^2=0.9773) between actual and predicted chromium(VI) sorption efficiency (%) is observed for testing data set. The mean absolute relative percentage error for training data is obtained as 0.749 whereas mean absolute relative percentage error for testing data is found to be 0.673. It can

Figure 3.

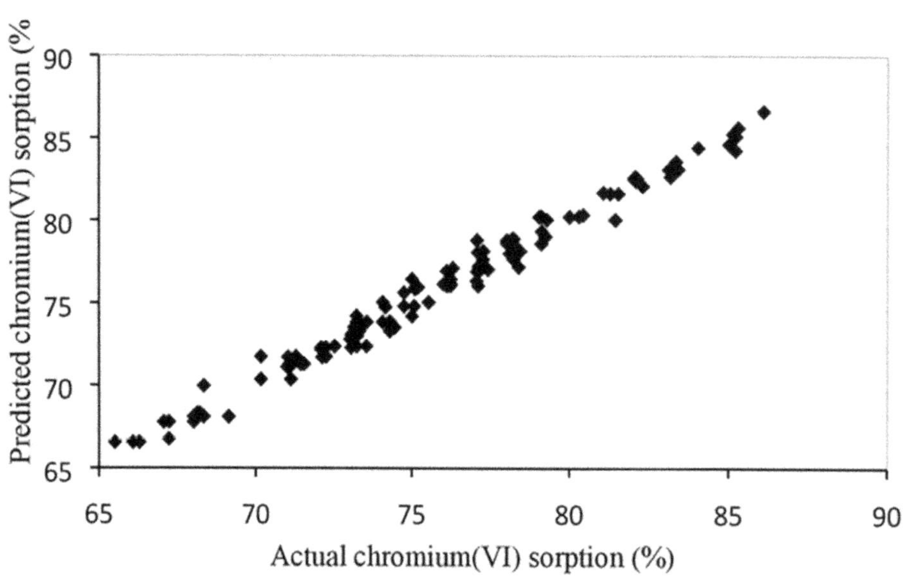

be observed that the residuals are distributed uniformly below and above zero line. Therefore, it can be assumed that the errors are normally distributed and the model can be used for prediction purpose with reasonable accuracy. The development of the proposed ANN model is an effort towards the growing interest in applying ANN modeling (see Table 1).

The technique to the area of biosorption of pollutants from water bodies have tested several ANN approaches to discover the optimum learning algorithm for modeling Zn(II) adsorption from leachate using a new biosorbent considering input parameters as initial pH, adsorbent dosage, contact time, and temperature. They have also suggested a design of experiment approach for finding out significant parameters. However, the present study is focused on modelling Cr(VI) biosorption using a new biosorbent. Both the studies conclude that ANN approach is quite efficient in modelling complex biological phenomenon (Sinha *et al.*, 2013; Chowdhury and Saha, 2013).

4. CHARACTERIZATION OF *Bacillus cereus* BIOMASS BEFORE AND AFTER BIOSORPTION OF CHROMIUM (VI) IONS

4.1. Atomic Force Microscopy (AFM) Analysis

Atomic force microscopy (AFM) is one of the for most tools for imaging, measuring, and manipulating matter at the nanoscale and was developed by Gred Binning and Heinrich Rohrer in the early 1980s at IBM research-zurich. The present investigation for surface morphology of *Bacillus cereus* biomass without and with sorption of chromium (VI) during biosorption process is observed with the help of Atomic Force Microscopy (Digital Instruments, Santa Barbara, CA, USA). *Bacillus cereus* bacteria without chromium (VI) ions exposure in the control blank are rod-like in shape with a smooth surface (the dimension of these cells is about 4.0 μm long and 1.0 μm wide, on average. After exposure of chromium (VI) ions can be clearly seen that the biomass shape has changed into a spindle-like structure. The morphological changes of the sample can be attributed to the interactions between chromium (VI) ions and the surface of *Bacillus cereus* cells. These agree with the results of FTIR spectra analysis.

4.2. FTIR Analysis

The FTIR spectra of the *Bacillus cereus* biomass without chromium (VI) ions loaded displays a number of absorption peaks, indicating the complex nature of the bacterial biomass and the spectra is presented in Figure 4. The spectra of *Bacillus cereus* biomass loaded with chromium (VI) is presented in Figure 4. The FTIR spectra of the *Bacillus cereus* biomass with and without chromium (VI) ions loaded displays a number of absorption peaks, indicating the complex nature of the bacterial biomass. The spectra of sorbent exhibits a broad absorption band at 3,448.94 cm^{-1} due to bonded –OH stretching vibration which is shifted to 3,450.21 cm^{-1} may be due to complexation of –OH groups with chromium (VI) (Gardea-Torresdey *et al.*, 2000; Roy *et al.*, 2014). The next absorption peaks at 1655.83 cm^{-1} has been shifted to 1,648.97 cm^{-1} possibly due to the complexation of amide group (N-H stretching and C=O stretching vibration) with chromium (VI) (Nourbakhsh *et al.*, 2002; Park *et al.*, 2004). Another shift is observed from 1,407.37 cm^{-1} to 1,412.83 cm^{-1}, corresponding to the complexation of nitrogen with chromium (VI) of the N-H group. The peaks at 1,074.06 cm^{-1} may be attributed to C-N stretching vibrations of amino groups is shifted to higher frequency and appeared at 1,118.07 cm^{-1} due the interaction of nitrogen from

Table 1. Comparison of experimental and predicted output for chromium (VI) (testing data)

Exp. No.	Concentration (mg/L)	Biosorbent (g/L)	Temp (°C)	Time (min)	Actual Sorption (%)	Predicted Sorption (%)	Residual (%)	(%) Error	Absolute Error
1	10	6	40	30	85.11	84.9976	0.1124	0.132064	0.132064
2	10	6	20	30	85.07	84.7944	0.2756	0.323968	0.323968
3	10	6	30	30	89.24	88.3401	0.8999	1.0084	1.0084
4	10	6	40	30	88.24	89.1302	-0.8902	-1.0088	1.0088
5	5	2	20	30	83.21	84.4803	-1.2703	-1.52661	1.52661
6	5	2	30	30	87.04	87.0811	-0.0411	-0.04722	0.04722
7	5	2	40	30	86.21	86.0977	0.1123	0.13026	0.13026
8	5	4	20	30	80.05	80.489	-0.4391	-0.54841	0.5484
9	1	4	30	30	81.55	81.1587	0.3913	0.479828	0.479828
10	1	4	40	60	80.51	80.2694	0.2406	0.298844	0.298844
11	1	6	20	60	79.48	80.1682	-0.6882	-0.8658	0.8658
12	1	6	40	60	78.45	79.5125	-1.0625	-1.37227	1.37227
13	5	4	20	60	77.426	77.9345	-0.5085	-0.65676	0.65676
14	5	4	30	60	76.395	77.3455	-0.9505	-1.24491	1.24491
15	5	4	40	60	75.364	75.7895	-0.4255	-0.56459	0.564592
16	5	2	20	60	74.333	73.3659	0.9671	1.301037	1.301037
17	10	2	30	60	73.302	73.2625	0.0395	0.053887	0.053886
18	10	2	40	60	72.271	72.621	-0.35	-0.48429	0.484229
19	10	4	20	60	71.24	71.8066	-0.5666	-0.79534	0.795343
20	1	4	30	30	70.209	69.4716	0.7374	1.050292	1.050292
21	1	6	40	30	69.178	70.3689	-1.1909	-1.7215	1.72152
22	10	2	20	30	84.21	84.9976	-0.7876	-0.93528	0.93528
23	10	2	30	30	87.44	86.7944	0.6456	0.738334	0.738334
24	10	2	40	30	85.11	86.3401	-1.2301	-1.44531	1.44531
25	5	4	20	30	81.05	81.1302	-0.0802	-0.09895	0.09895
26	5	4	30	30	82.35	82.4803	-0.1303	-0.15823	0.158227
27	5	4	40	30	82.419	81.0811	1.3379	1.62329	1.62329
28	5	4	20	30	80.488	80.0977	0.3903	0.484912	0.484912
29	1	2	30	30	81.457	81.489	-0.032	-0.03928	0.039284
30	1	2	40	30	79.426	79.1587	0.2673	0.336539	0.336539
31	1	6	20	30	82.395	82.2694	0.1256	0.152436	0.152436
32	10	6	30	90	74.164	74.1682	-0.00042	-0.00057	0.00057
33	10	6	40	90	75.333	76.5125	-1.1795	-1.56571	1.56571
34	10	6	20	90	73.402	73.9345	-0.5352	0.725457	0.725457
35	10	6	30	90	73.271	73.3455	-0.0745	-0.10168	0.101677
36	10	6	40	90	74.24	74.7895	-0.5495	-0.74017	0.74017
37	1	4	20	90	69.209	70.3659	-1.1569	-1.6716	1.67161
38	1	4	30	90	70.178	70.2625	-0.0845	-0.12041	0.12041
39	1	4	40	90	73.41	73.621	-0.211	-0.28748	0.28748
40	1	4	20	90	72.86	72.8066	0.0534	0.073291	0.073291

Figure 4.

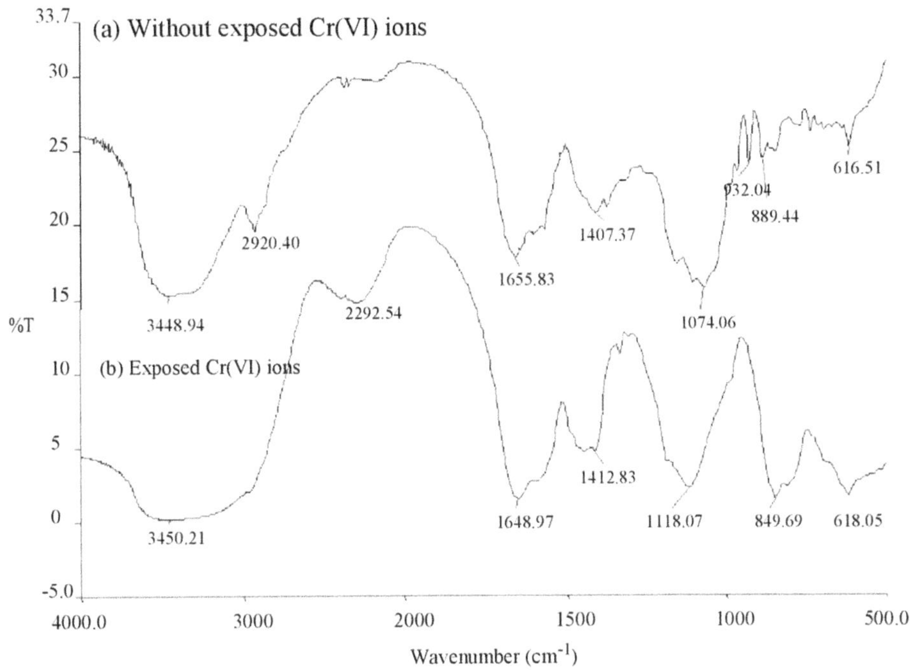

amino group with chromium(VI) (Mangaiyarkari *et al.*, 2011). The other weak absorption peak is shifted from 889.44 cm^{-1} to 849.69 cm^{-1}, corresponding to the O-C-O scissoring vibration of polysaccharide (Seki *et al.*, 2005; Podder and Majumder, 2016 (a)). The above changes in the spectra may be attributed to the interaction of chromium (VI) with the hydroxyl, amide and amino groups present on the *Bacillus cereus* biomass.

5. CONCLUSION

Artificial neural network (ANN) techniques can enhance predicting capability of the model when mathematical or statistical methods are difficult to formulate and fails to predict with desired accuracy. In the present work, estimation of sorption efficiency using mathematical and analytical tools is involved because the physical phenomenon for removal of chromium (VI) by living cells of *Bacillus cereus* is complex one. The network is found to be working satisfactorily as seven neurons are selected in the hidden layer when mean square error starts decreasing. Learning and momentum parameters are set at 0.20 during the training phase. The network is trained till minimum root mean square error is observed. A root mean square error of 0.68 is observed at epoch number 23, 835, 76. The network is found to be working satisfactorily as absolute relative percentage error of 0.567 during training phase. Comparison between the model results and experimental data gives a high degree of correlation $R^2 = 0.984$ indicating that the model is able to predict the sorption efficiency with reasonable accuracy.

REFERENCES

Aleboyeh, A., Kasiri, M. B., Olya, M. E., & Aleboyeh, H. (2008). Prediction of azo dye decolorization by UV/H2O2 using artificial neural networks. *Dyes Pigm.*, *77*(2), 288–294. doi:10.1016/j.dyepig.2007.05.014

Belanche, L., Valdes, J. J., Comas, J., Roda, I. R., & Poch, M. (2000). Prediction of the bulking phenomenon in wastewater treatment plants. *Artificial Intelligence in Engineering*, *14*(4), 307–317. doi:10.1016/S0954-1810(00)00012-1

Chang, F. J., Chen, P. A., Liu, C. W., Liao, V. H. C., & Lio, C. M. (2013). Regional estimation of groundwater arsenic concentrations through systematical dynamic-neural modeling. *Journal of Hydrology (Amsterdam)*, *499*, 265–274. doi:10.1016/j.jhydrol.2013.07.008

Chowdhury, S., & Saha, P. D. (2013). Artificial neural network (ANN) modeling of adsorption of methylene blue by NaOH-modified rice husk in a fixed-bed column system. *Environmental Science and Pollution Research International*, *20*(2), 1050–1058. doi:10.1007/s11356-012-0912-2 PMID:22562342

Chu, K. H. (2003). Prediction of Two-Metal Biosorption Equilibria Using a Neural Network. *European Journal of Mineral Processing and Environmental Protection*, *3*, 119–127.

Debnath, A., Majumder, M., Pal, M., Das, N. S., Chattopadhyay, K. K., & Saha, B. (2016). Enhanced Adsorption of Hexavalent Chromium onto Magnetic Calcium Ferrite Nanoparticles: Kinetic, Isotherm, and Neural Network Modeling. *Journal of Dispersion Science and Technology.*, *37*(12), 1806–1818. doi:10.1080/01932691.2016.1141100

Despagne, F., & Massart, D. (1998). Neural networks in multivariate calibration. *Analyst (London)*, *123*(11), 157R–178R. doi:10.1039/a805562i PMID:10396805

Dimopoulos, I., Chronopoulos-Serelia, J. A., & Lek, S. (1999). Neural network models to study relationships between lead concentration in grasses and permanenturban descriptors in Athens city (Greece). *Ecological Modelling*, *120*(2-3), 157–165. doi:10.1016/S0304-3800(99)00099-X

Fernandes, F. A. N., & Lona, L. M. F. (2005). Neural networks applications in polymerization processes. *Brazilian Journal of Chemical Engineering*, *22*(3), 401–418. doi:10.1590/S0104-66322005000300009

Gardea-Torresdey, J.L., Tiemann, K.J., & Armendariz, V. (2000). Characterization of Cr(VI) binding and reduction to Cr(III) by the agricultural byproducts of *Avena monida* (Oat) biomass. *J. Hazard. Mater., B 80*, 175-188.

Gob, S., Oliveros, E., Bossmann, S. H., Braun, A. M., Guardani, R., & Nascimento, C. A. O. (1999). Modeling the kinetics of a photochemical water treatment process by means of artificial neural networks. *Chemical Engineering and Processing*, *38*(4-6), 373–382. doi:10.1016/S0255-2701(99)00028-8

Hagan, T. M., Demuth, H. B., & Beale, M. (1996). *Neural Network Design*. PWS Publishing Company.

Hamed, M. M., Khalafallah, M. G., & Hassanien, E. A. (2004). Prediction of wastewater treatment plant performance using artificial neural networks. *Environmental Modelling & Software*, *19*(10), 919–928. doi:10.1016/j.envsoft.2003.10.005

Lek, S., & Guegan, J. F. (1999). Artificial neural networks as a tool in ecological modeling:an introduction. *Ecological Modelling, 120*(2-3), 65–73. doi:10.1016/S0304-3800(99)00092-7

Mary Mangaiyarkari, M. S., Vincent, S., Janarthanan, S., Subba Rao, T., & Tata, B. V. R. (2011). Bioreduction of Cr (VI) by alkaliphilic *Bacillus subtilis* and interaction of the membrane groups. *Saudi J. of Biol. Sci., 18*(2), 157–167. doi:10.1016/j.sjbs.2010.12.003 PMID:23961119

Mashitah, M. D., Zulfadhly, Z., & Bhatia, S. (1999). Binding mechanism of heavy metals biosorption by Pycnoporus sanguineus. J. Artificial Cells. *Blood Substitutes and Immobilization Biotechnology, 27*(5/6), 441–445. doi:10.3109/10731199909117717

Nourbakhsh, M. N. (2002). Biosorption of Cr6+, Pb2+ and Cu2+ ions in industrial wtewater on Bacillus sp. *Chemical Engineering Journal, 85*(2-3), 351–355. doi:10.1016/S1385-8947(01)00227-3

Nourbakhsh, M. N. (2002). Biosorption of Cr^{6+}, Pb^{2+} and Cu^{2+} ions in industrial wtewater on *Bacillus* sp. *Chemical Engineering Journal, 85*(2-3), 351–355. doi:10.1016/S1385-8947(01)00227-3

Park, Y. S., Chon, T. S., Kwak, I. S., & Lek, S. (2004). Hierarchical community classification and assessment of aquatic ecosystems using artificial neuralnetworks. *The Science of the Total Environment, 327*(1-3), 105–122. doi:10.1016/j.scitotenv.2004.01.014 PMID:15172575

Podder, M. S., & Majumder, C. B. (2016). *Corynebacterium glutamicum* MTCC 2745 immobilized on granular activated carbon/$MnFe_2O_4$ composite: A novel biosorbent for removal of As(III) and As(V) ions. *Spectrochimica Acta. Part A: Molecular and Biomolecular Spectroscopy, 168*, 159–179. doi:10.1016/j.saa.2016.05.056 PMID:27289352

Podder, M. S., & Majumder, C. B. (2016). Predictive approach for simultaneous biosorption and bioaccumulation of arsenic by *Corynebacterium glutamicum* MTCC 2745 biofilm supported on NL/$MnFe_2O_4$ composite. *Journal of Water Process Engineering., 11*, 8–31. doi:10.1016/j.jwpe.2016.03.010

Roy, P., Mondal, N. K., Das, K. (2014). Modeling of the adsorptive removal of arsenic: A statistical approach. *Journal of Environmental Chemical Engineering, 2*(1), 585-597.

Salari, D., Daneshvar, N., Aghazadeh, F., & Khataee, A. R. (2005). Application of artificial neural networks for modeling of the treatment of wastewater contaminated with methyl tert-butyl ether (MTBE) by UV/H2O2 process. *J. Hazard. Mater. B, 125*(1-3), 205–210. doi:10.1016/j.jhazmat.2005.05.030 PMID:15996818

Seki, H., Suzuki, A., & Maruyama, H. (2005). Biosorption of chromium (VI) and arsenic (V) onto methylated yet biomass. *Journal of Colloid and Interface Science, 281*(2), 261–266. doi:10.1016/j.jcis.2004.08.167 PMID:15571680

Shetty, G. R., & Chellam, S. (2003). Predicting membrane fouling during municipal drinkingwater nanofiltration using artificial neural Networks. *Journal of Membrane Science, 217*(1-2), 69–86. doi:10.1016/S0376-7388(03)00075-9

Sinha, K., Chowdhury, S., Saha, P. D., & Datta, S. (2013). Modeling of microwave-assisted extraction of natural dye from seeds of *Bixa orellana* (Annatto) using response surface methodology (RSM) and artificial neural network (ANN). *Industrial Crops and Products, 41*, 165–171. doi:10.1016/j.indcrop.2012.04.004

Strik, D. P. B. T. B., Domnanovich, A. M., Zani, L., Braun, R., & Holubar, P. (2005). Prediction of trace compounds in biogas from anaerobic digestion using the MATLAB Neural Network Toolbox. *Environmental Modelling & Software, 20*(6), 803–810. doi:10.1016/j.envsoft.2004.09.006

WHO. (1993). *Guidelines for drinking-water quality*. Geneva, Switzerland: World Health Organization.

Yetilmezsoy, K. (2006). Determination of optimum body diameter of air cyclones using a new empirical model and a neural network approach. *Environmental Engineering Science, 23*(4), 680–690. doi:10.1089/ees.2006.23.680

Yetilmezsoy, K., & Demirel, S. (2008). Artificial neural network (ANN) approach for modeling of Pb(II) adsorption from aqueous solution by Antep pistachio (Pistacia vera L.) shells. *Journal of Hazardous Materials, 153*(3), 1288–1300. doi:10.1016/j.jhazmat.2007.09.092 PMID:17980484

Yetilmezsoy, K., & Saral, A. (2007). Stochastic modeling approaches based on neural network and linear–nonlinear regression techniques for the determination of single droplet collection efficiency of countercurrent spray towers. *Environmental Modeling and Assessment, 12*(1), 13–26. doi:10.1007/s10666-006-9048-4

Section 3

Chapter 14
Out-of-Autoclave (OOA) Manufacturing Technologies for Composite Sandwich Structures

Laraib Alam Khan
CESAT, Pakistan

Wajid Ali Khan
University of Hafr Al Batin, Saudi Arabia

S. Ahmed
IIU, Pakistan

ABSTRACT

This chapter gives an overview of cost effective processing techniques, employed for manufacturing of sandwich structures, primarily for high performance applications. The scope of this chapter is to review the processes which may produce the autoclave cured quality parts, without the use of autoclave. The chapter starts with an overview of autoclave processing, highlighting the different features, which depicts the capacity of technique to produce highly repeatable parts. Studies related to manufacturing of sandwich structures in autoclave, with and without adhesive film, are reviewed and discussed. Process briefs and recent developments of different out of autoclave (OOA) techniques for manufacturing of sandwich structures are then discussed which includes, OOA prepregs systems, Quickstep processing, Vacuum Infusion Process, compression molding, double belt press and filament winding. A brief overview of manufacturing processes for thermoplastic based sandwich structures is also provided at the end.

1. INTRODUCTION

A sandwich consists of three major components i.e. two rigid, strong and relatively thin faces which are separated by weaker, light and thick core materials. The American Society for Testing and Materials (ASTM) defines a sandwich structure as follows:

DOI: 10.4018/978-1-5225-2440-3.ch014

A special sandwich is a special form of a laminated composites comprising of a combination of different materials that are bonded to each other so as to utilize the properties of each separate component to the structural advantage of the whole assembly.

Sandwich structures are used in almost every industrial sector ranging from building to aerospace applications. High performance sandwich structures, especially for aerospace applications, are usually manufactured using prepregs, followed by cure in autoclave. This requires high manufacturing temperature, pressure, heavy tooling and a cycle time of several hours. As the aerospace timescale requirements are shortening, an increase in production rates is required and this is difficult to achieve with traditional autoclave manufacturing. Also, high capital expenditure, infrastructure requirements and time to commission has made autoclave processing, increasingly undesirable.

There are several techniques, which have been established as out of autoclave (OOA) approaches for construction of sandwich structures, however, minor modifications in each technique may be required, because of different types of face-sheets and core materials used in sandwich construction. These core materials may be wood, foam or honeycomb. Several configurations are available in each category of core materials. Face-sheets are usually constructed using glass, carbon and Kevlar fiber, depending upon the application areas and are usually provided in prepregs form for high performance applications. These face-sheets materials, however, also being used in dry form in different OOA approaches.

Voids are among the most common manufacturing induced defects in the composites. The voids can be formed in the composite structures due to many reasons; some of them are (1) due to dissolved volatiles within premix resins, (2) bubbles evolved from mixing process itself, (3) air pockets trapped during lay-up (Boey, 1990), (4) due to moisture absorbed during storage, (5) due to resin rich areas in the cured part (Costa, Almeida, & Rezende, 2001), (6) due to volatile by-products during the cure reaction of the polymeric matrix, (7) due to use of high viscosity resin combined with closely packed fiber that are not completely wetted by resin, (8) due to fabrication mishaps such as leaking vacuum bag or a poor vacuum source (Costa, Almeida, & Rezende, 2005; Guo, Liu, Zhang, & Du, 2009). Air pockets trapped during lay-up of sandwich structure is reported as the major source of porosity (Kratz & Hubert, 2013), which not only weakened the skin performance but also affects skin to core adhesion properties.

Extensive research has been conducted on the development of new resin systems to match the existing processing techniques; however, the development of new processes for manufacturing of high performance sandwich structures are limited. This chapter will discuss the recent developments in materials and manufacturing techniques of composites sandwich structures, their approaches to reduce the porosity and enhanced skin-core bonding strength and modification (if any) required for different core materials. The chapter started with brief overview of topic and comments on chapter structure. Section 2 covered the autoclave processing, advantages and limitations of autoclaves with adequate number of references for further study. Section 3 listed down the OOA prepregs systems, which have gained popularity over the recent years in high performance applications. Brief comments are also provided on the issues associated with fabrication of sandwich structures with OOA prepregs systems. Section 4 explained the Quickstep processing, one of the emerging OOA approach for leading applications ranging from aerospace to sports goods. Few success stories are provided, depicting the suitability of the approach for different applications. Section 5, overviewed the liquid molding techniques like vacuum assisted resin infusion (VARI) for sandwich construction. Section 6 highlighted the commercial aspect of compression molding technique for fabrication of sandwich structures while section 7 briefly described different techniques like wet lay-up, double belt press and filament winding for construction of sandwich structures. Section 8 is the last section with brief comments on manufacturing techniques for thermoplastic sandwich structures. In the end, references along with additional reading section are provided for interested readers.

2. AUTOCLAVE MOLDING

2.1. Process Brief

The autoclave is a device that can generate a controlled pressure and temperature environment. In this process, the dry fiber/wet resin or prepregs are laid-up to produce a laminate of the required shape and dimensions, either manually or by using lay-up machines. The laminate then vacuum bagged and placed in the autoclave. Since, the pressure in the bag and autoclave are different, a compacting force is exerted on the laminate. This compacting force combined with the flexibility of the bag exerts pressure normal to the laminate surface resulting in uniform pressure distribution over the laminate surface. This uniform pressure distribution reduces the risk of void formation in the laminate. The autoclave is usually pressurized with nitrogen in order to avoid any fire hazards, which can be imposed by exothermic reaction of resin materials.

The laminate lay-up step is a very labor intensive activity, requiring many man hours and hence high labor cost. Tape lay-up machines have been developed, but their use remains marginal and dedicated for only aerospace part manufacturing (Seferis, Hillermeier & Buehler, 2000).

An important step is de-bulking. In this step, after laying-up of four layers, entrapped air is squeezed out either by using vacuum bagging or de-bulk table. Debulking step is essential because, once the laminate lay-up process is completed, it is almost impossible to remove entrapped air. However, this additional step usually leads to extended lay-up time and hence this tool dead time also plays a significant role in increasing operational cost (Seferis et al., 2000).

The laminates are then vacuum bagged and moved to autoclave for curing. In case of three dimensional complex structures, the vacuum bags usually have pleats in order to avoid the bridging over the edges of laminate. Bridging may produce low pressure which can lead to porosity and finally the bag failure. Incorporation of these pleats greatly increases the work time required and is highly dependent on operator's skills (Seferis et al., 2000).

The key processing variables in the autoclave process are temperature, pressure, dwell time, heating-up rate, vacuum and cooling rate. The usual procedure for autoclave processing is to first increase the pressure followed by increasing the temperature. Once the pressure reached its desired limit, it kept constant throughout the process. As the temperature rises, the resin viscosity drops thus facilitating the wetting out of the fiber. The resin viscosity then increases due to polymerization reaction and the wetting process continues till the gelation of resin occurs. The high pressure application assists in proper consolidation and void removal in the product during the wetting process. Once the gelation occurs, the consolidation process stops. Sometimes, voids remain in the product due to resin starvation. In order to avoid resin starvation, a complete chemo-rheology study of the resin is required.

2.2. Selection of Tooling Materials

The selection of tooling material is an important step in autoclaving. It usually depends on the required thermal expansion and durability behavior of the material. Composite tools are less durable but their thermal expansion characteristics can match with the laminate material. Metal tools are reliable but are quite expensive and their coefficient of thermal expansion can mismatch with that of laminate material. Due to these reasons very costly metal alloys 'Invar' are being used for high performance composites.

2.3. Adhesive-Less Prepregs and Prepregs With Adhesive Film

There are usually three methods are used, to bond the skin panel with the core i.e. Co-bonding, Co-curing and secondary bonding. In the co-bonding process, the adhesive film is used to bond the skin with the core materials. The adhesive film contributes significantly in the overall increase of structural weight, especially for thin skinned panels. Pre-cured skin requirement in secondary bonding, especially for contour profiles are difficult to achieve and thus practically infeasible. In the co-curing, the bonding of skin with the core is carried out by prepregs only. So, when the prepregs cured, this not only makes the face-sheet rigid, but simultaneously forming fillets with the wall of honeycomb (Yuan, Li, Zhang & Gu, 2008). Co-curing is considered as an efficient way of bonding, thus significantly reducing the manufacturing cost and time. It is reported that Co-cure method is the widely used method and can produce panels with significantly reduced porosities or de-bonding, appropriate fillet size and regular shape (Yuan et al., 2008).

Rion et al studied the ways of strong bonding between face-sheet and core with particular emphasis on fillet formation between the honeycomb cell walls and the skin. They developed a method for controlled flow of adhesive, forming appropriate fillet size, based on weight of adhesives. The shape of the adhesive fillet was also studied in detail and noticeable contact angle hysteresis, due to high viscosity and rough surface, was observed (Rion, Leterrier & Månson, 2008).

The aircraft primary structures, based on sandwich scheme, are primarily manufactured using prepregs and adhesive film, in between core and skins. Gandy analyzed the feasibility of adhesive-less prepregs (without adhesive film) based honeycomb sandwich structure (followed by autoclave cure) for manufacturing of secondary aircraft structures. The associated published literature was reviewed and the gap was scientifically assessed for application of said materials on aircraft primary structures. The gap was filled by detailed literature survey of recent developments in this aspect and experimental characterization was carried out for this particular study (Gandy, 2012). For this study, physical and mechanical responses of Cytec's 977-2 Carbon/epoxy prepregs and AF191 from 3M (used in primary aircraft structures) was studied and compared with Advanced Composites Group (ACG)'s MTM 45-1, without adhesive film. Manufacturing and testing conditions for panels manufactured from these materials, were kept identical, in order to ensure the transparency of comparison. The adhesive-less MTM 45-1 cured panels exhibited better physical and mechanical properties, except in Short beam strength, showing their suitability for aircraft's primary structures (Gandy, 2012).

2.4. Studies Related to Autoclave Curing

Several studies have been carried out related to autoclave processing regarding the effect of several processing parameters on the physical and mechanical properties (Boey & Lye, 1992; Fernlund, Rahman, Courdji, Bresslauer, Poursartip, Willden & Nelson, 2002; Liu, Zhang, Wang & Wu, 2006; Olivier, Cottu & Ferret, 1995; Purslow & Childs, 1986; Tarsha-Kurdi & Olivier, 2002; White & Kim, 1996), simulation and modeling of cure cycle for optimum properties (Fernlund et al., 2002; Frank-Susich, Laananen & Ruffner, 1993; Gopal, Adali & Verijenko, 2000; Joshi, Liu & Lam, 1999; LeGrand & Bellenger, 1998;

Pillai, Beris & Dhurjati, 2002), online cure monitoring (Maistros & Partridge, 1998; Saliba, Hofmann & Smolinski, 1994) and processing of thick parts (Joshi et al., 1999; Kim, Kim & Lee, 2004; Martinez, 1991; Oh & Lee, 2002; Young, 1995).

3. OUT-OF-AUTOCLAVE PREPREGS SYSTEMS

Conventional prepregs are formulated for autoclave curing and require refrigeration for storage, relatively shorter shelf life and require high temperature and pressure during curing for optimum properties. As the global usage of prepregs are growing and due to increase in demand of cost-effective production of composites components even in high-end applications like aerospace, manufacturers of prepregs in the recent years, introduced prepregs which do not require autoclave and commonly known as out-of-autoclave (OOA) prepregs. These prepregs can be vacuum bagged and cure in oven to produce autoclave quality parts with low porosity.

The initial cost of these prepregs is higher than materials used in traditional wet lay-up and spray-up processes, however, lower fabrication costs, faster productions and other benefits offered by these prepregs can reduce and even eliminate the cost difference (Stewart).

Among the leading prepreg manufacturer are Hexcel Corporation, Cytec Industries, Gurit AG, Toray industries, Advanced Composite Group (ACG) and Tencate advanced composites. A new material Hex-Ply M56 is introduced by Hexcel recently which is an OOA prepreg for aerospace composites structures (Ryder, 2009). The material is available with woven carbon, UD carbon tape and woven glass.

Cycom® 5320 is Cytec's newest toughened epoxy resin prepreg system. Specifically designed for out-of-autoclave (OOA) manufacturing, the system provides performance similar to autoclave with the benefits of lower processing costs and manufacturing flexibility. The material can be used to design large primary structures without the autoclave costs and size limitations (Herring, Mardel & Fox, 2010).

The MTM 44-1, MTM 45-1, MTM 46 and MVR 444 range from ACG are claimed to have excellent hot/wet conditions with the chemorheology of the resins being specifically designed for low pressure out-of-autoclave processing. ACG's dual cure prepreg MTM44-1 has been qualified by Airbus industries for structure applications in tape and fabric formats (Stewart).

3.1. An Optimization Study of Sandwich Structure Developed From OOA Prepregs

The key to OOA Prepregs is that they are specially formulated to reduce porosity in the structure by removing the air entrapped during the lay-up stage. So, when they used in sandwich construction, most of the air entrapped in the cells of core materials, causing the delamination and porosity at the skins and bond-line. Kratz et al studied this aspect and tried to resolve by different bagging arrangement for vacuum only process. The in-plane and through the thickness permeability of the prepregs were determined to evaluate the bagging configurations for optimum performance. It was found that bagging arrangement with one ply of non-perforated release film and edge breathing around the perimeter of assembly, produced the lowest skin porosity. The resin formulation for OOA prepregs are such that any resin loss resulted in dry spot, pinholes and voids in the structure. The edge breather around the perimeter, allows the air entrapped in the prepregs and core to escape via vacuum effect (Kratz & Hubert, 2011, 2013). It is also reported that excessive de-bulking of OOA sandwich structure may reduce the skin compaction pressure, again resulting in void formation and growth during cure. The optimal internal core pressure was experimentally determined and found to be between 35-55 kPa. A complete chemo-rheology study of resin was carried out for this optimization study (Kratz & Hubert, 2011, 2013).

4. QUICKSTEP PROCESSING

4.1. Process Brief

The technique is developed by Quickstep Technologies Pty Ltd in Perth, Western Australia and was patented in 1996 (Graham, 1997). Figure 1 highlights the principles involved in the Quickstep process.

The Quickstep functions by rapidly applying heat to the laminate placed between two floating rigid or semi-rigid mould by passing a heat transfer fluid (HTF) at high velocity over the part. The tool and the mould are separated from HTF by flexible membranes. Quickstep provides a much faster cure cycle than the autoclave process due to the fact that, increased heat capacity and thermal conductivity of the heat transfer fluid allows faster and more efficient cure than nitrogen or air filled autoclave.

Typical heating and cooling rates of 10 K min^{-1} have been made possible with the Quickstep setup. Due to high heating and cooling rates during the processing of composites, the reaction exotherms are not manifest as a significant temperature rise within the lay-up, because the circulating fluid removed any excess heat generated, during the process.

4.2. Comparison of Autoclave and Quickstep Processing

In contrast to autoclave processing, the Quickstep process is a low pressure process where approximately 10 kPa pressure is applied during processing as compared to 700 kPa external pressure typically applied

Figure 1. Schematic of quickstep process

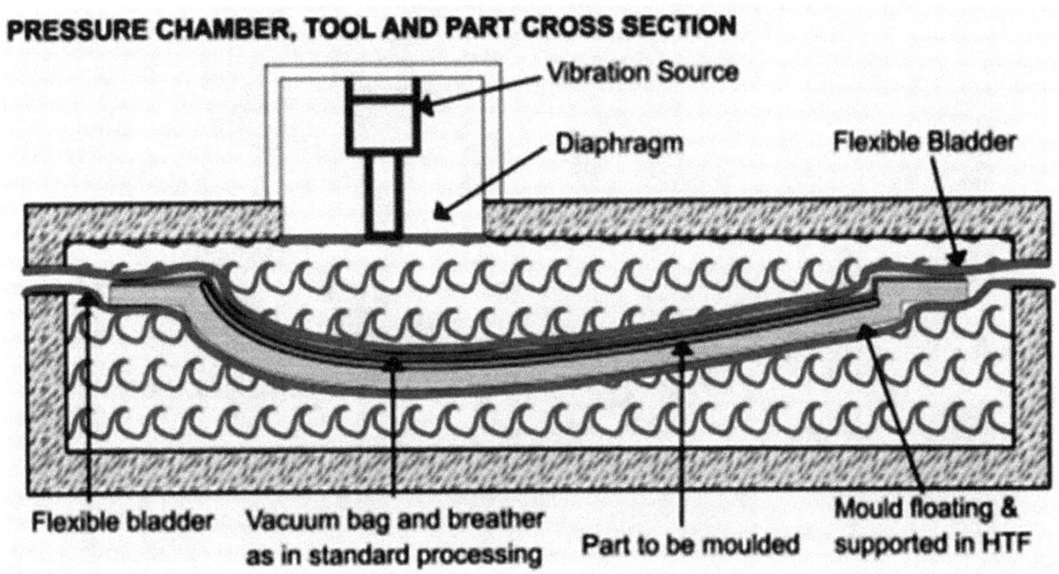

during autoclave processing. As external pressure of autoclave is used to collapse high pressure voids that may arise from volatile solvents, other avenues of void removal are utilized in Quickstep processing, which will be explained later in this chapter.

Several advantages of Quickstep over autoclave processing have been reported since the invention of Quickstep technology. The parts manufactured from Quickstep have found to be comparable or sometimes superior to autoclave curing in terms of strength, stiffness, surface finish and appearance while achieving aerospace grade void content of less than 2%. The other important parameter is low pressure employed for Quickstep processing which results in better safety, light moulds and hence in cost savings. The rapid processing reduces the production cost and hence these features enable Quickstep technology for mass production.

4.3. Quickstep Plant

The Quickstep plant can be explained with reference to Figure 2. The three fluid tanks; high, medium and low temperature tanks are shown in Figure 2. In order to release excess pressure developed in the tanks and to avoid any accident, two safety release valves named as FS 101 and FS 201 are installed on the top of high and medium temperature tanks. In addition to these valves, a complete safety system is also provided. If any set-point breach occurs, it will inform the operator automatically by security alarm and if the set-point breach is severe, system will shut down automatically.

The HTF in the tanks are stored under nitrogen atmosphere to prevent the degradation of the hot fluid. The nitrogen supply to all three tanks.

As Quickstep plant is a PLC based fully automated system, this computerized system provides information of almost every parameter of the plant before, during and after the operation. The parameters include

Figure 2. Quickstep plant showing three fluid tanks, safety valves, and heat exchangers

fluid level and fluid temperature in each tank, vacuum and mold pressure, the actual and programmed temperature of tool and mold, the temperatures and pressures at inlet and outlet valves.

This information also helps in running the desired cure cycle for processing of composites. Typical cure cycle employed for autoclave processing is as follows:

1. Heat from Room Temperature to 130 °C at 2 K min^{-1}.
2. Dwell at 130 °C for 60 minutes.
3. Heat from 130 °C to 180 °C at 2 K min^{-1}.
4. Dwell at 180 °C for 120 minutes.
5. Cool from 180 °C to room temperature at 2 K min^{-1}.

If this cure cycle for autoclave processing will be employed in Quickstep, the hot, intermediate and cold tanks will be maintained at 185-190 °C, 130-140 °C and 20-30 °C, respectively. For the first step, the fluid from the hot tank will circulate through the mould bed to quickly reach 130°C. Then, the fluid from intermediate tank will circulate through the mould bed to maintain this dwell temperature for 60 minutes. The hot fluid then will flow for the 3rd and 4th steps of cure cycle to reach and maintain the temperature of 180°C for 2 hours. For the final step of the cure cycle, the cold tank will be used to cool down the mould to room temperature.

The floating mould beds consisted of two fluid trays, which provided the HTF a path to flow through the mould beds in order to heat or cool the composite laminates. When the fluid enters in the tray, it passes through a series of small orifices into the mould. At the outlet, the suction provided by the displacement pump ensures the continuous circulation of the fluid through the plant.

The bagged composite lay-up are placed in between the mold beds and two thermocouples are attached to the part for the real time monitoring of the temperature and the cure cycle was based on these thermocouple readings. A slightly different vacuum bag arrangement is usually employed for the Quickstep process. A larger bagging film is used to allow the vacuum valve and connector to be put outside the mold beds.

4.4. Quickstep Processing of Sandwich Structures

A comparative study was conducted on a foam core sandwich structure, manufactured from autoclave and Quickstep cure. The skin was manufactured from Toray G83C/T600S Carbon UD tape, while the core material and film adhesive used was 12 mm thick ROHACELL 51 WF- HT and Cytec FM 94, respectively. The autoclave curing and mechanical testing of all samples were carried out by Degussa (Germany), while Quickstep processing was done at Deakin University, Australia. The results from climbing drum peel test showed the enhanced specific peel torque of Quickstep cured panel, as compared to autoclave cured one, while simultaneously reducing the manufacturing time and cost. A series of tests were then conducted on different core materials of ROHACELL and persistent result is observed. It is also reported that Quickstep cured panels exhibited more homogenous adhesive film distribution over the surface than its counterpart.

A comparative study was also conducted on a honeycomb core sandwich structure, manufactured from autoclave and Quickstep cure. The results from climbing drum peel test showed the 12% improved specific peel torque of Quickstep cured panel, as compared to autoclave cured one, while simultaneously reducing the manufacturing time and cost. It is also reported that Quickstep cured panels exhibited enhanced skin to core bonding than its counterpart which is shown in Figure 3.

5. RECENT DEVELOPMENTS IN LIQUID MOLDING TECHNIQUES FOR SANDWICH STRUCTURES

5.1. Introduction

There are number of composite manufacturing processes available, and they are identified by the facility (used to provide the cure cycle) or the type of tool/mold used in the process (Akovali, 2001; Davé & Loos, 2000). Wet lay-up or hand lay-up is one of the oldest techniques for composites manufacturing and is still widely used in the marine industry or where prototype part manufacturing is required (Miracle, Donaldson & Vander Voort, 2001). The major advantage of this technique is lower production cost and simplicity. Any fiber type material can be selected with any fiber orientation; however, the major disadvantages are that the technique are that it is labor intensive, inconsistent in the quality of part and leads to a low fiber volume fraction and high porosity. It is reported that fiber volume fraction up to 30% is possible with wet lay-up process and with vacuum bag compaction, this may increase up to 40% (Beckwith & Hyland, 1998). Fiber is stronger constituent of Composites and for aerial vehicles, fiber volume fraction of at least 60% is required. Another issue with wet lay-up process is exposure of operators to the hazardous chemicals. Due to the regulations on acceptable limits of volatile organic compounds (VOC's), there is always a need of manufacturing process, with the aim of reducing, or eliminating, exposure of operators to hazardous VOC's (Petrucci, Nisini, Ghelli, Santulli, Puglia, Sarasini, Minak & Kenny, 2012; Yuhazri & Sihombing, 2010).

Figure 3. Honeycomb surface: upper (autoclave-cured surface), lower (quickstep-cured surface)

An extension of wet lay-up process, vacuum bagging is employed mainly for these purposes: (1) It removes entrapped air between the laminas. (2) Compaction enables efficient force transmission among fiber bundles. (3) It prevents changing of fiber orientation during cure. (4) It reduces humidity. (5) It optimizes the fiber to resin ratio. (6) It allows resin to flow and absorbed without hydraulic lock. These advantages have for years, enabled aerospace and racing industries to maximize the physical and mechanical properties of advanced composites materials (Akovali, 2001; Davé & Loos, 2000; Miracle et al., 2001). Despite of all these advantages, the technique failed to produce consistent and high fiber volume fraction parts. Also, the issue of VOC's could not be resolved with this approach.

With slight modification of vacuum bagging (VB) process, the issues mentioned above may be resolved. Resin infusion (RI) is a widely used molding process for the manufacture of large composite structures (FENG & WANG, 2010; Schubel, 2010). Its popularity is due to the low cost of the tooling and the environmental safety (the process eliminates more than 90% of the VOC's, released by unsaturated polyester resins (UPE)). In addition, low operator involvement increases the repeatability of the process compared to wet lay-up and vacuum bagging process and the components are of relatively high fiber content, up to 60% by volume.

5.2. Vacuum Infusion of Sandwich Structures: A Success Story

In order to reduce the mass of the commercial vehicle (urban buses and delivery trucks), the rear of hybrid bus is redesigned and manufactured using composite sandwich structure, as a part of European collaborative project, HCV (Hybrid Commercial Vehicle) (Neves, Fernandes, Lantz, Carldén & Andersson, 2014). This part of project was joint venture of Volvo bus Co. (Sweden) and IDMEC, Portugal. The overall objective of this project is to develop the light weight structures, which will operate using second generation hybrid electric technology. For this purpose, an eighteen (18) ton passenger bus and a six (6) ton distribution vehicle will be developed, with light weight body and advanced hybrid technology. It is reported that the 'benefit' of weight reduction in terms of fuel consumption is higher for buses than any other medium of transportation (Neves et al., 2014).

First of all, numerical and modal analysis of a benchmark model, with proprietary load cases (both provided by Volvo) were carried out for subsequent comparative study of replacing Aluminum alloy superstructure with composites sandwich structures, which were to be produced by vacuum Infusion process. While designing the largest sandwich panel, several challenges were faced, like (1) vibration of plates may come out in resonance frequency range, due to excitation imposed by engine, exhaust and rear axle; (2) to get access to the engine from interior for maintenance, a cut out patch had to be provided at critical locations of the panel, resulting in the significant reduction of mechanical properties; (3) the placement of rear door imposed serious difficulties. An optimum design, consisted of four independent sandwich panels is provided to cater all these challenges.

Composite material properties, required in the design phase were evaluated by coupon testing. For this purpose, a plate of dimension (522 mm x 142 mm) was produced, the lay-up was validated by microscopy and average cured ply thickness was determined. The lay-up sequence of the sandwich panel is provided in the paper.

Instead of using barrier film and infusing the upper and lower skins separately, another method was used. The holes were drilled at regular pattern in the perforated foam to provide vertical channel for the resin, to flow from upper skin to lower skin during infusion. Although, the drilled holes account for only 0.8% of the total core volume, this resulted in enhanced core stiffness in the transverse direction. Since, the resin flowed from upper skin to lower skin, this accounted for variable ply thicknesses due to varying fiber volume fraction.

The composites sandwich panels as an alternative of metallic frames for the rear side of the bus were successfully modeled, fabricated and validated. Although, the weight of fabricated panels were not provided, the modeled data suggested the weight of main rear panel is 62.3 Kg and the floor panel together with two seat bases is 25.8 Kg (Neves et al., 2014). This resulted in effective weight reduction of 308 Kg in the rear part of the bus structure. It is also reported that the development of prototype bus has been completed and are ready for production phase. With a completely finished structure, a trial run of 10 Km has already made.

6. COMPRESSION MOLDING

6.1. Process Brief

Compression molding is a closed mold process which is used for processing of both thermosetting and thermoplastic materials. The molding compound is placed in an open and heated mold and matched metal mold cavity is used to close the mold. Metallic molds are used in this process and pressure is applied through a variety of methods including the tightening of screws, through press and other means to obtain high volume fraction of composites. Heat is also applied during the curing stage in some applications. The part size that can be obtained ranges from 1mm thickness to over 50mm thickness.

The range of materials which can be used in the process includes almost all the composite materials including fiberglass, aramid and all grades of carbon fiber. Both thermoset and thermoplastic matrices can be used such as epoxies and cyanate esters; and PEEK and PEI.

Figure 4. Compression molding process

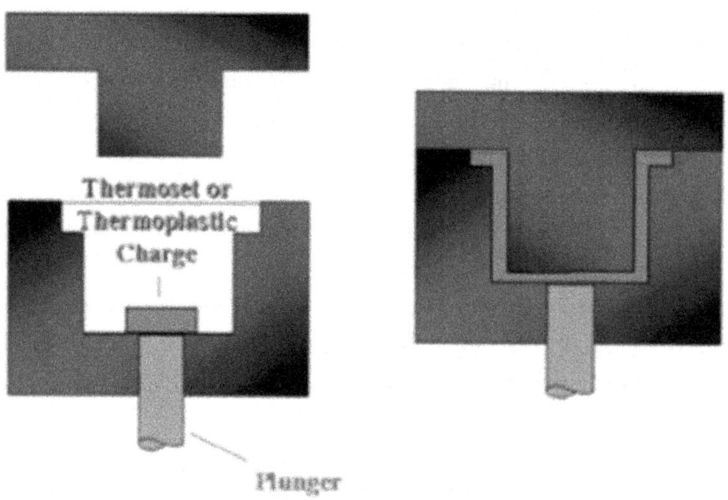

Figure 5. Compression molding of sandwich structures

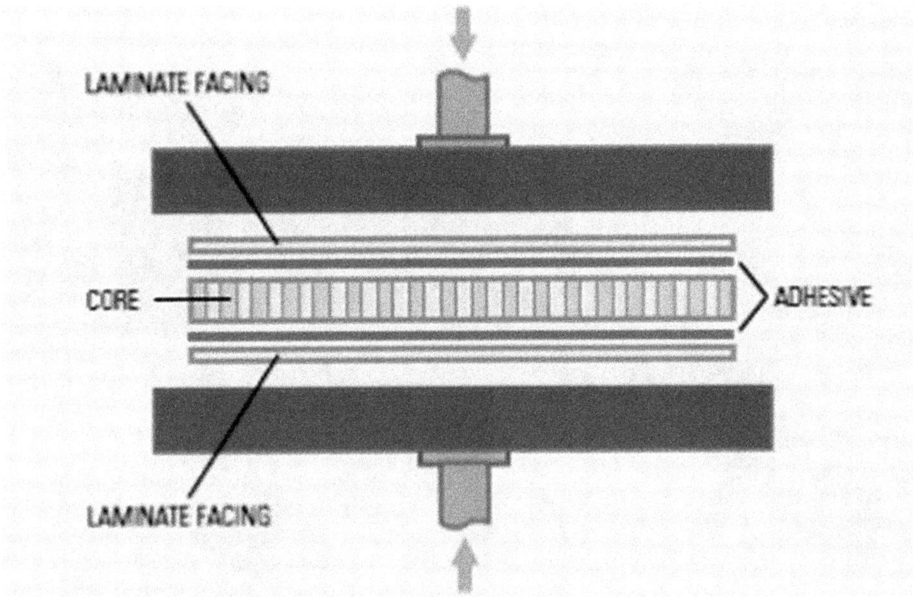

6.2. Geometry Dependency

The compression molding process can be used to parts with variable cross sections and complex geometries. Ribs and holes can easily be incorporated in the process to not only improve the structural properties but also to facilitate the assembly. The process can also be used to manufacture sandwiched composites. Some examples of compression molded composite parts are shown below.

Compression molded parts have found applications in aircraft and space sectors. These thermoplastic parts have a great potential for the aircraft industry, as these materials have certain advantages over thermosetting composites such as FST (Fire, smoke and toxicity) and durability. Some of the new variations of compression molding technique are discussed here. Thermal Press Curing (TPC) is a type of compression molding in which a formed thermoset composite laminate is clamped between a heated mold and rubber-coated mold with the required pressure and temperature. The curing mold is normally designed to have thin walls so that it can be heated up and cooled down quickly. Rubber forming is another popular type of compression molding process for sandwich composites which can be used for sandwich construction as it provides relatively uniform pressure for consolidation of flat or 2-D parts. The TPC process has the added advantage of having uniform pressure and temperature distributions across the entire 3-D thermoset composite part surface. The TPC process have been claimed to be offering significant reductions in energy consumption, cost, and complexity along with significant lead time. Some new process concepts of multi-cavity molding were presented in a conference by the Institute of Plastics Processing (IKV) in Aachen, Germany.

Figure 6. Example of a large compression molded part used on the V-22 Osprey helicopter (60 in length x 8 inch base, and 12 inches height 152x20x31 cm)
Image credit: TenCate.

Figure 7. Compression molded parts
Image credit: Tencate

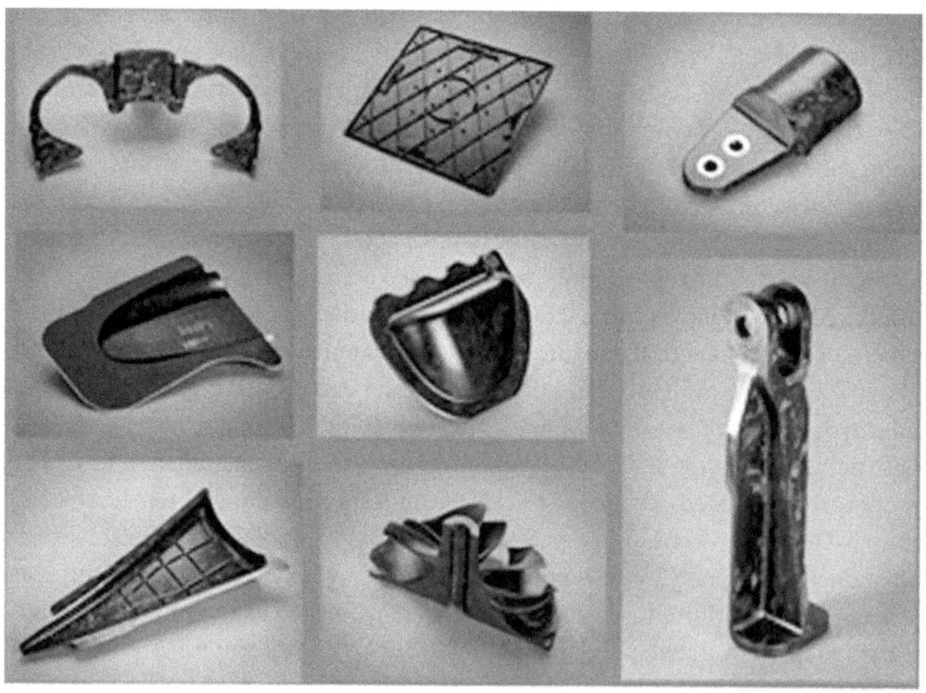

6.2.1. Injection Transfer

The injection transfer molding aims to produce better quality and uniformity in multi-cavity molding of small thermoplastic parts. This process is a combination of thermoset transfer molding and injection-compression with hot-runner molding. Heat transfer between the hot and cold halves is reduced by an insulating plate in the stationary half.

6.2.2. Melding Multiple Processes

Processes such as sandwich molding and injection-compression molding, are used to obtain ultra-thin wall parts and glass filled hollow pipes. The combination of these techniques can result in more optimum results. IKV has evaluated process combinations of sandwich molding with injection-compression and fluid-assist molding. IKV then devised a test utilizing a sandwich of 30% long-glass reinforced PP skins over an unfilled PP core.

6.3. Advantages of the Compression Molding Process

Following are the main advantages of compression molding as compared to other manufacturing processes.

- **Part Complexity:** The process has the ability to produce as complex shapes as can be obtained by injection molding process.
- **Good Flexibility:** In part design is possible as the inserts and rib stiffeners can be added. Metallic or other material inserts can be placed into the mold at the time of forming. These attachment hard points eliminate common secondary processes compared to other composite manufacturing methods.
- **Twisting and Shrinkage:** In product is reduced therefore dimensional accuracy is good. As it is a closed mold process, each part produced from this process is a dimensionally duplicate to the previous part.
- Raw material wastage is minimum. Lower finished part cost with the ability to make high volumes of parts. The process has the ability to produce finish parts with little or no secondary processes, thus reducing the cost of production.
- Production rate is high as the mold cycle time is in few minutes. The part is cured at high temperature and pressure resulting in quick curing.
- Residual stresses are absent or negligible in the molded component.

6.4. Disadvantages of Compression Molding Process

- The process generally needs high capital investment. As it is a closed mold process and part thickness and dimensions are controlled by the clearance between the two parts of the mold, the molds are generally precision machined to obtain the desired geometry of part.
- Another drawback is directly related to above point as due to high capital investment, the process is generally suitable for only large batch production.
- Like hand layup, the process is labor intensive. Automation can be achieved to a certain degree but that means higher initial costs.

- The secondary processing (trimming, machining) cannot be eliminated altogether.
- Uneven parting lines can be seen if the two parts of the mold are not properly matched.
- There is limitation on mold depth.

7. REVIEW OF OTHER PROCESSES

Out-of-autoclave (OOA) processing has emerged as an alternate processing technique to autoclave processing since the early 1990s. Many OOA processing techniques have since been developed which include automated-tape-placement (ATP), filament-winding, vacuum-assisted resin-transfer-molding (VARTM) and resin-transfer-mold (RTM).

The first problem in the design and production of composite sandwich structures is the selection of the most suitable manufacturing process for the production of a single part. The criteria for the selection of process are production rate, cost, strength, and size/shape features of the part with added complication of the intrinsic anisotropy and non-homogeneity of composite sandwich structures. The peculiar features of sandwich structures make the validation of model difficult, so the most popular method today for the manufacturing of sandwich structures is still either manual lay up or vacuum bag/autoclave molding.

7.1. Hand Lay-Up/Wet Layup

Hand Layup is the simplest, cheapest and oldest manufacturing process applied for the production of composite parts. It is a manual process and does not require highly skilled man power. The parts are produced in open molds or flat surfaces depending upon the design requirements. The mold is designed to match the design shape of the final structure with some clearances for fabric application. The process requirements are tooling, raw materials, and vacuum or pressure along with optional heating. It is still one of the most popular manufacturing methods for construction of sandwich composites due to its simplicity and economy.

Mold design is the first step of any manufacturing technique and hand layup is no exception.

As it is an open mold process, only one surface smooth surface is available. This process still lacks the sophisticated control over part thickness, void fraction, fiber content, and surface quality on the rear face, so this process is not recommended for a high stressed part, where dimensional accuracy might be critical. Furthermore, high fiber volume fraction parts cannot be manufactured using this process.

Hand layup has remained the dominant production technique for making the composites parts including sandwich composites, although other processes are gaining greater significance. The process is preferred for small batch processing of complex shapes and structures that cannot be obtained by press or other techniques.

Out of Autoclave processing of sandwich structures have certain benefits such as low processing cost but it also has certain disadvantages as well. The major challenge is to attain the high volume fraction with low void content, as without the additional pressure available in the autoclave the volume fraction drops from 60% to 55% and void content goes up from 1% to 5%. Higher void content is undesirable as these reduce the compression strength and inter-laminar shear strength. Another important effect is reduction in skin–core adhesion and the compaction of the prepregs plies within the skins. For the processing of sandwich structures for aerospace applications, the pressure can be tailored throughout the curing cycle through small wall perforations allowing the air to circulate between the cells but it is not possible for all applications, especially for the boat industry, where exposure to water can result in the filling of the core and failure of the structures.

The use of sandwich structures composites have increased exponentially over the years, thus there is a need to maintain and repair these sandwich structure in aircraft components to restore the original design strength without any compromise on overall weight and structural integrity, and the wet layup is the most popular method for such repairs today.

7.2. Double Belt Press

Double belt press is a variation of compression molding and there are mainly two types of this process. The first type is called isochoric system, in which the pressing gap between the two plates remains constant, and pressure is varied to maintain the gap; the second type is called isobaric system and pressure applied in this process remains constant, while the height of pressing gap varies, depending upon the property of the product. The feeding, heating, cooling and pressure can be varied to cater for a variety of applications. This process is used for composites parts especially of automotive and aviation industries where surface pressure is required.

For the isobaric systems, hydrostatic principle is used, as hydraulic oil is pressed directly against the backside of the belts, the oil can be heated or cooled as per process requirements. The process is used to make sandwich structures for building industry.

Sandvik is the premier manufacturer of the double belt systems, and provides the solution for not only sandwich composite manufacturing but also sheet molding, sheet casting and film casting. Romakowski is providing all components for the production of sandwich composites including the de-coiling and roll forming sections and has produced panels with a length of 24m.

Propex has designed a new thermoplastic material called Curv® that bridges the gap between pure plastics and conventional fiber-reinforced materials. It is a unique composite and have the same material as reinforcement and matrix. The product properties demand a very sophisticated manufacturing process

Figure 8. Isochoric double belt press for low pressure applications
Image credit: Sandvik.

to ensure the high quality of this material and isobaric double belt press has emerged the most suitable manufacturing process for the product. The double-belt press has the potential of high process capacity due to the continuous production method.

Some of the advantages of the process are listed below.

7.2.1. Advantages of Double Belt Press

- Maximum efficiency can be achieved by applying heat treatment or cooling to both sides of the product.
- Practically all common sources can be used for heating and cooling.
- Ambient air is not in contact with the product.
- Vapors and fumes are reduced and can be excluded.

7.3. Filament Winding

Filament winding is a well-known process for structural composites having a predominately cylindrical or conical shape. In this process a band of continuous resin impregnated rovings or monofilaments is wound around a rotating mandrel and then cured. The filament winding produces high speed and precise production of composites with many layers. The process is generally used for axis-symmetric structures, but prismatic shapes and more complex parts can be wound on machines equipped with the appropriate number of degrees of freedom.

The process has been used for sandwich composites where the winding is done in two distinct phases. In the first phase inner face is wound and the winding is stopped, then a flexible or pre-foamed core is applied onto the inner face. The winding is then resumed to apply on the outer face. This technique has

Figure 9. Isobaric double press with fixed roller
Image credit: Sandvik.

found many applications in the industry such as the self-supporting inner body of a railway passenger-car which is fabricated in three sections. The body sections are mounted through the roof and slid in to position.

7.3.1. Winding Methods

Wet winding is a process in which the fibers are passed through a resin bath and wound onto a rotating mandrel and in prepregs winding the pre-impregnated fiber tows are placed on the rotating mandrel.

7.3.2. Winding Patterns

- **Helical Winding:** In helical winding, mandrel rotates at a constant speed while the fiber feed carriage transverses back and forth at a speed regulated to generate the desired helical angles.
- **Hoop Winding:** Hoop winding is a high angle helical winding that approaches an angle of 90°. Each full rotation of the mandrel advances the band delivery by one full bandwidth.
- **Polar Winding:** In this process the fiber passes tangentially to the polar opening at one end of the chamber, reverses direction, and passes tangentially to the opposite side of the polar opening at the other end.

7.3.3. Recent Advances in Filament Winding Technology

Numerically controlled filament winding machines are replacing the old manual machines providing higher degrees of freedom for placing the fibers at required position for meeting the complex design configurations.

A software CADWIND is developed by MATERIAL S.A of Brussels, Belgium to generate winding patterns in a user friendly environment. It also calculates the required laminate for axis symmetric and non-axis symmetric mandrel geometries and stores the laminate structure as an interface for finite element method programs.

7.3.4. Advantages of Filament Winding Process

The filament winding is known as a process for cheap production of composites. Some of the advantages of the process are:

- The process generally uses relatively expensive reinforcement with inexpensive matrix materials. The production speed is relatively high.
- Filament winding is an accurate and repeatable process. The degree of repeatability amongst the layers and parts is better than tape placement process.
- It is a continuous process and the reinforcement fibers can easily oriented in the load directions.
- The process produces high volume fraction and tooling cost is lower than other processes.

Figure 10. Some of the examples of filament wound product

7.3.5. Disadvantages

- Mandrel is needed, which can be complex and adds cost. Also the component shape must be such that the mandrel could be removed. Different mandrel materials, because of differing thermal expansion and differing composite materials and laminate lay-up percentages of hoops versus helical plies, will demonstrate varying amounts of difficulty in removal of the part from the mandrel.
- It is difficult to wind reverse curvature and fiber path cannot be changed easily.
- The surface finish is poor and some secondary operations such as machining are required.

8. TECHNIQUES FOR THERMOPLASTIC-BASED SANDWICH STRUCTURES

There are two different types of reinforced thermoplastic composites with opposite property and cost spectrum. On the one side the low property and low cost engineering plastics are modified with various fillers for automotive applications. Automotive industry has led the development of this type of materials as the industry strives to reduce weight and noise and improve safety by developing stiffer and lightweight thermoplastics that offer tailoring of properties. There is a trend toward reinforcement with increasingly longer glass fibers and new products such as natural fibers and self-reinforced polypropylene (PP). Thermoplastic sandwich composites have high impact strength and produce attractive surface finish, are recyclable, can bond to other thermoplastics but get soft with the application of heat.

Some of the processing techniques which are used to manufacture thermoplastic sandwich composites are listed here.

Thermoforming is a very old process in its purest sense as heat and pressure are used to transform a sheet into the required shape. The sheet is preheated using either conduction via contact heating panels or rods, or convection using ovens or radiant heating achieved with infrared heaters. The sheet is then held at a specific temperature and cooled for final shape and the part is trimmed off. The process has many variations such as vacuum forming or pressure forming or matched die stamping.

Foam injection molding is a technique which is specifically used for manufacturing of sandwich construction of composites. The mold cavity is only partially filled and the remaining volume is filled by the generated foam pressure. This also reduces the internal mold pressure thus lower clamping pressures are possible. The foam which is used as the core material has a pressure which acts homogeneously and in all areas of the cavity, so that there is no need for a holding pressure phase. It results in lower internal stress and less distortion thus improving the dimensional accuracy of the molded part.

Continuous fiber composites are used as face sheets and its combination with FIM process results in cost-effective, automatable manufacture of sandwich structures. The bonding of the face sheets is done in a single processing step by introducing the core material with subsequent foaming.

Thermoplastic sandwich structures are conventionally based on foamed cores components, bonded through the use of adhesives. The high performance thermoplastics such as PEEK, PEI, PPS are used as matrices in applications where structural performance is the requirement such as the aerospace composite due to their high mechanical and thermal properties (continuous working temperatures between 150°C and 230°C). There are other engineering thermoplastics such as PEN which have comparable mechanical properties and good thermal behaviour (continuous working temperatures between 125°C and 150°C) at a fraction of their cost.

Figure 11. Schematic diagram of the processing steps for manufacturing thermoplastic sandwich structures with continuous-fiber-reinforced face sheets
Image credit: Kunststoffe international.

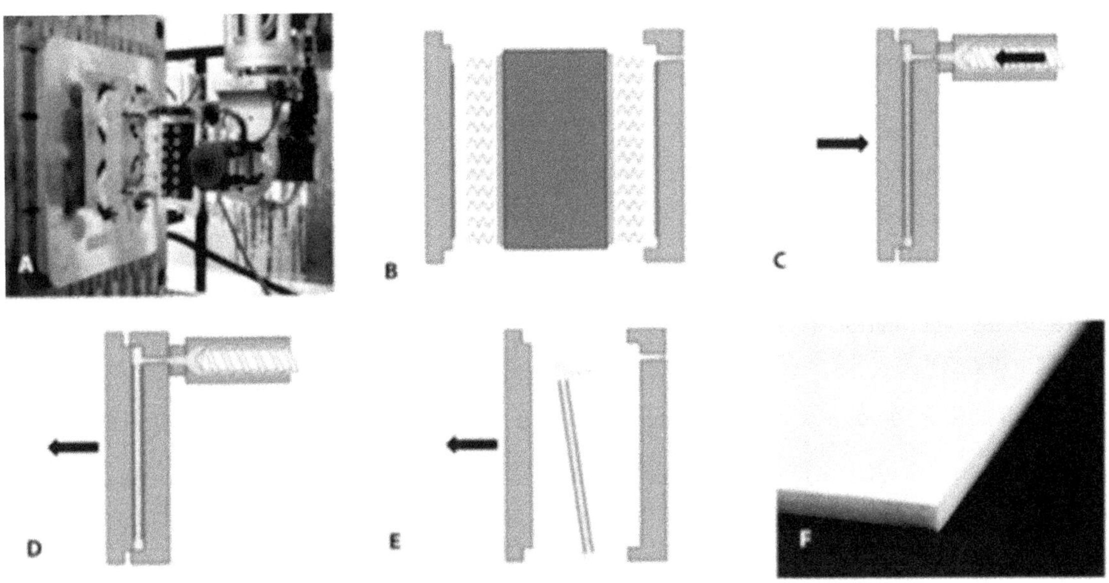

In order to produce lightweight structures, the matrix materials used to prepare the sandwich should also have good foaming properties. Polymeric foams are typically employed for a lot of applications such as packaging, thermal and acoustic insulation, and in the impact energy absorption.

The use of reinforced thermoplastics in composites certain benefits, some of which are listed here:

- The manufacturing cycles are short (typically 2 minutes or less).
- Better mechanical properties such as toughness and impact resistance.
- The option of post forming such as corrections and forming in multiple steps. The materials can be recycled during the manufacturing and after service life.
- The shelf life is unlimited.

SUMMARY

Several techniques for manufacturing of composite sandwich structures with and without autoclaves are presented. Effect of adhesive film to skin to core bonding is discussed and a case study is provided, depicting the efficacy of adhesive-less prepregs (without adhesive film) for the manufacturing of aircraft primary structures. OOA prepregs manufacturers list is provided and again with the support of case study, an optimum solution is discussed for manufacturing of sandwich structures with OOA prepregs. Quickstep is a relatively new technique for OOA processing of monolithic and sandwich structures. The principle of this technique is highlighted with comments on its advantages over autoclave cure. Comparative studies on foam and honeycomb based sandwich panels manufactured through autoclave and Quickstep are discussed. It was observed that Quickstep cured panels showed enhanced de-bonding resistance than autoclave cured ones. A success story is provided, in which composite sandwich components manufactured through vacuum Infusion process, successfully replaced the metallic frames in leading vehicles, a project funded by leading European automotive manufacturers. The manufacturing of sandwich panels using Compression molding, wet-layup, filament winding and double belt process is briefly discussed. In the end, manufacturing techniques for processing of thermoplastic based sandwich structures is provided.

REFERENCES

Abliz, D., Duan, Y., Zhao, X., & Li, D. (2014). Low-energy electron beam cured tape placement for out-of-autoclave fabrication of advanced polymer composites. *Composites. Part A, Applied Science and Manufacturing*, *65*(0), 73–82. doi:10.1016/j.compositesa.2014.06.004

Akovali, G. (2001). Handbook of composite fabrication. Shropshire, UK: iSmithers Rapra Publishing.

Bajer, K., & Braun, U. (2014). Different aspects of the accelerated oxidation of polypropylene at increased pressure in an autoclave with regard to temperature, pretreatment and exposure media. *Polymer Testing*, *37*(0), 102–111. doi:10.1016/j.polymertesting.2014.05.006

Bartolozzi, G., Baldanzini, N., & Pierini, M. (2014). Equivalent properties for corrugated cores of sandwich structures: A general analytical method. *Composite Structures*, *108*(0), 736–746. doi:10.1016/j.compstruct.2013.10.012

Beckwith, S. W., & Hyland, C. R. (1998). Resin Transfer Molding: A decade of technology advances. *SAMPE Journal*, *34*(6), 7–19.

Ben Ammar, I., Karra, C., El Mahi, A., El Guerjouma, R., & Haddar, M. (2014). Mechanical behavior and acoustic emission technique for detecting damage in sandwich structures. *Applied Acoustics*, *86*(0), 106–117. doi:10.1016/j.apacoust.2014.04.016

Boey, F. Y. C. (1990). Reducing the void content and its variability in polymeric fibre reinforced composite test specimens using a vacuum injection moulding process. *Polymer Testing*, *9*(6), 363–377. doi:10.1016/0142-9418(90)90029-D

Boey, F. Y. C., & Lye, S. W. (1992). Void reduction in autoclave processing of thermoset composites: Part 1: High pressure effects on void reduction. *Composites*, *23*(4), 261–265. doi:10.1016/0010-4361(92)90186-X

Butukuri, R., Bheemreddy, V., Chandrashekhara, K., Berkel, T., & Rupel, K. (2012). Evaluation of skin-core adhesion bond of out-of-autoclave honeycomb sandwich structures. *Journal of Reinforced Plastics and Composites*, *31*(5), 331–339. doi:10.1177/0731684412437267

Centea, T., & Hubert, P. (2011). Measuring the impregnation of an out-of-autoclave prepreg by micro-CT. *Composites Science and Technology*, *71*(5), 593–599. doi:10.1016/j.compscitech.2010.12.009

Centea, T., & Hubert, P. (2012). Modelling the effect of material properties and process parameters on tow impregnation in out-of-autoclave prepregs. *Composites. Part A, Applied Science and Manufacturing*, *43*(9), 1505–1513. doi:10.1016/j.compositesa.2012.03.028

Costa, M. L., & Almeida, S. F. M. (2005). Critical Void Content for Polymer Composite Laminates. *AIAA Journal*, *43*(6), 1336–1341. doi:10.2514/1.5830

Costa, M. L., Almeida, S. F. M., & Rezende, M. C. (2001). The influence of porosity on the interlaminar shear strength of carbon/epoxy and carbon/bismaleimide fabric laminates. *Composites Science and Technology*, *61*(14), 2101–2108. doi:10.1016/S0266-3538(01)00157-9

Davé, R. S., & Loos, A. C. (Eds.). (2000). *Processing of composites*. Munich, Germany: Hanser Verlag. doi:10.3139/9783446401778

Feng, X.-B., & Wang, W. (2010). Materials and process research of 2 megawatt wind turbine blade. *Fiber Reinforced Plastics, 4*, 22.

Fernlund, G., Rahman, N., Courdji, R., Bresslauer, M., Poursartip, A., Willden, K., & Nelson, K. (2002). Experimental and numerical study of the effect of cure cycle, tool surface, geometry, and lay-up on the dimensional fidelity of autoclave-processed composite parts. *Composites. Part A, Applied Science and Manufacturing*, *33*(3), 341–351. doi:10.1016/S1359-835X(01)00123-3

Frank-Susich, D., Laananen, D. H., & Ruffner, D. (1993). Cure cycle simulation for thermoset composites. *Composites Manufacturing*, *4*(3), 139–146. doi:10.1016/0956-7143(93)90098-S

Gandy, H. T. N. (2012). *Adhesiveless honeycomb sandwich structure with carbon graphite prepreg for primary structural application: a comparative study to the use of adhesive film*. Wichita State University.

Garschke, C., Weimer, C., Parlevliet, P. P., & Fox, B. L. (2012). Out-of-autoclave cure cycle study of a resin film infusion process using in situ process monitoring. *Composites. Part A, Applied Science and Manufacturing*, *43*(6), 935–944. doi:10.1016/j.compositesa.2012.01.003

Gopal, A. K., Adali, S., & Verijenko, V. E. (2000). Optimal temperature profiles for minimum residual stress in the cure process of polymer composites. *Composite Structures*, *48*(1–3), 99–106. doi:10.1016/S0263-8223(99)00080-X

Graham, N. (1997). *Patent No. EP 0805746 A1*. Munich, Germany: E. P. Office.

Grunenfelder, L., Centea, T., Hubert, P., & Nutt, S. (2013). Effect of room-temperature out-time on tow impregnation in an out-of-autoclave prepreg. *Composites. Part A, Applied Science and Manufacturing*, *45*, 119–126. doi:10.1016/j.compositesa.2012.10.001

Grunenfelder, L. K., Centea, T., Hubert, P., & Nutt, S. R. (2013). Effect of room-temperature out-time on tow impregnation in an out-of-autoclave prepreg. *Composites. Part A, Applied Science and Manufacturing*, *45*(0), 119–126. doi:10.1016/j.compositesa.2012.10.001

Guo, Z.-S., Liu, L., & Zhang, B.-M. (2009). Critical Void Content for Thermoset Composite Laminates. *Journal of Composite Materials*, *43*(17), 1775–1790. doi:10.1177/0021998306065289

Herring, M., Mardel, J., & Fox, B. (2010). The effect of material selection and manufacturing process on the surface finish of carbon fibre composites. *Journal of Materials Processing Technology*, *210*(6), 926–940. doi:10.1016/j.jmatprotec.2010.02.005

Joosten, M. W., David, M., Kelly, D., & Thomson, R. (2014). Static and dynamic evaluation of a multifunctional crashworthy pi-joint/sandwich energy absorbing structure. *Composite Structures*, *118*(0), 423–431. doi:10.1016/j.compstruct.2014.07.023

Joshi, S. C., Liu, X. L., & Lam, Y. C. (1999). A numerical approach to the modeling of polymer curing in fibre-reinforced composites. *Composites Science and Technology*, *59*(7), 1003–1013. doi:10.1016/S0266-3538(98)00138-9

Kim, D., Centea, T., & Nutt, S. R. (2014a). In-situ cure monitoring of an out-of-autoclave prepreg: Effects of out-time on viscosity, gelation and vitrification. *Composites Science and Technology*, *102*(0), 132–138. doi:10.1016/j.compscitech.2014.07.027

Kim, D., Centea, T., & Nutt, S. R. (2014b). Out-time effects on cure kinetics and viscosity for an out-of-autoclave (OOA) prepreg: Modelling and monitoring. *Composites Science and Technology*, *100*(0), 63–69. doi:10.1016/j.compscitech.2014.05.027

Kim, J. W., Kim, H. G., & Lee, D. G. (2004). Compaction of thick carbon/phenolic fabric composites with autoclave method. *Composite Structures*, *66*(1-4), 467–477. doi:10.1016/j.compstruct.2004.04.070

Kratz, J., & Hubert, P. (2011). Processing out-of-autoclave honeycomb structures: Internal core pressure measurements. *Composites. Part A, Applied Science and Manufacturing*, *42*(8), 1060–1065. doi:10.1016/j.compositesa.2011.04.009

Kratz, J., & Hubert, P. (2013). Anisotropic air permeability in out-of-autoclave prepregs: Effect on honeycomb panel evacuation prior to cure. *Composites. Part A, Applied Science and Manufacturing*, *49*(0), 179–191. doi:10.1016/j.compositesa.2013.02.013

LeGrand, M., & Bellenger, V. (1998). The cure optimisation of carbon/epoxy prepregs. *Composites Science and Technology*, *58*(5), 639–644. doi:10.1016/S0266-3538(97)00136-X

Liu, L., Zhang, B.-M., Wang, D.-F., & Wu, Z.-J. (2006). Effects of cure cycles on void content and mechanical properties of composite laminates. *Composite Structures*, *73*(3), 303–309. doi:10.1016/j.compstruct.2005.02.001

Maistros, G. M., & Partridge, I. K. (1998). Monitoring autoclave cure in commercial carbon fibre/epoxy composites. *Composites. Part B, Engineering*, *29*(3), 245–250. doi:10.1016/S1359-8368(97)00020-6

Martinez, G. M. (1991). Fast cures for thick laminated organic matrix composites. *Chemical Engineering Science*, *46*(2), 439–450. doi:10.1016/0009-2509(91)80005-J

Miracle, D. B., Donaldson, S. L., & Vander Voort, G. F. (2001). ASM handbook: Vol. 21. *Asm International Materials Park*. Academic Press.

Mortimer, S., Smith, M. J., & Olk, U. E. (2010). *Product development for out-of-autoclave (OOA) manufacture of aerospace structures*. Paper presented at the SAMPE conference.

Mustapha, S., & Ye, L. (2014). Leaky and non-leaky behaviours of guided waves in CF/EP sandwich structures. *Wave Motion*, *51*(6), 905–918. doi:10.1016/j.wavemoti.2014.03.004

Neves, P., Fernandes, A., Lantz, D., Carldén, L., & Andersson, K. (2014). *Bus BiW Weight Optimization Using Composite Sandwich Panels*. Paper presented at the Transport Research Arena (TRA) 5th Conference: Transport Solutions from Research to Deployment.

Oh, J. H., & Lee, D. G. (2002). Cure Cycle for Thick Glass/Epoxy Composite Laminates. *Journal of Composite Materials*, *36*(1), 19–45. doi:10.1177/0021998302036001300

Olivier, P., Cottu, J. P., & Ferret, B. (1995). Effects of cure cycle pressure and voids on some mechanical properties of carbon/epoxy laminates. *Composites*, *26*(7), 509–515. doi:10.1016/0010-4361(95)96808-J

Osei-Antwi, M., de Castro, J., Vassilopoulos, A. P., & Keller, T. (2014). Fracture in complex balsa cores of fiber-reinforced polymer sandwich structures. *Construction & Building Materials*, *71*(0), 194–201. doi:10.1016/j.conbuildmat.2014.08.029

Peng, T., Shi, H., Liang, X., Luo, F., & Wu, X. (n.d.). Experimental investigation on sandwich structure ring-type ultrasonic motor. *Ultrasonics*.

Petrucci, R., Nisini, E., Ghelli, D., Santulli, C., Puglia, D., & Sarasini, F. (2012). *Mechanical and impact characterisation of hybrid composite laminates based on flax, hemp, basalt and glass fibers produced by vacuum infusion.* Paper presented at the ECCM15 Conference, Venezia, Italy.

Pillai, V. K., Beris, A. N., & Dhurjati, P. S. (2002). Implementation of Model-Based Optimal Temperature Profiles for Autoclave Curing of Composites Using a Knowledge-Based System. *Industrial & Engineering Chemistry Research, 33*(10), 2443–2452. doi:10.1021/ie00034a029

Pimenta, S., & Robinson, P. (2014). Wavy-ply sandwich with composite skins and crushable core for ductility and energy absorption. *Composite Structures, 116*(0), 364–376. doi:10.1016/j.compstruct.2014.05.020

Purslow, D., & Childs, R. (1986). Autoclave moulding of carbon fibre-reinforced epoxies. *Composites, 17*(2), 127–136. doi:10.1016/0010-4361(86)90250-8

Rion, J., Leterrier, Y., & Månson, J. A. E. (2008). Prediction of the adhesive fillet size for skin to honeycomb core bonding in ultra-light sandwich structures. *Composites. Part A, Applied Science and Manufacturing, 39*(9), 1547–1555. doi:10.1016/j.compositesa.2008.05.022

Ryder, A. (2009). *The design and manufacturing of primary aircraft structures via out-of-autoclave techniques.* Paper presented at the ICMAC Conference, Belfast Waterfront, UK.

Saliba, T. E., Hofmann, D., & Smolinski, P. (1994). Development of an in situ Hall-effect sensor for on-line monitoring of thickness and compaction during composite curing. *Composites Science and Technology, 51*(1), 1–9. doi:10.1016/0266-3538(94)90151-1

Satasivam, S., & Bai, Y. (2014). Mechanical performance of bolted modular GFRP composite sandwich structures using standard and blind bolts. *Composite Structures, 117*(0), 59–70. doi:10.1016/j.compstruct.2014.06.011

Schlimbach, J., & Ogale, A. (2012). 14 - Out-of-autoclave curing process in polymer matrix composites. In S. G. Advani & K.-T. Hsiao (Eds.), *Manufacturing Techniques for Polymer Matrix Composites (PMCs)* (pp. 435–480). Woodhead Publishing. doi:10.1533/9780857096258.3.435

Schubel, P. (2010). Technical cost modelling for a generic 45-m wind turbine blade producedby vacuum infusion (VI). *Renewable Energy, 35*(1), 183–189. doi:10.1016/j.renene.2009.02.030

Seferis, J. C., Hillermeier, R. W., & Buehler, F. D. R. U. (2000). *2.20 Prepregging and Autoclaving of Thermoset Composites* (A. Kelly & C. Zweben, Eds.). Comprehensive Composite Materials.

Stark, W., Jaunich, M., & McHugh, J. (2013). Carbon-fibre epoxy prepreg (CFC) curing in an autoclave analogue process controlled by Dynamic Mechanical Analysis (DMA). *Polymer Testing, 32*(8), 1487–1494. doi:10.1016/j.polymertesting.2013.09.014

Stewart, R. (n.d.). New prepreg materials offer versatility, top performance. *Reinforced Plastics, 53*(5), 28-33.

Tarsha-Kurdi, K. E., & Olivier, P. (2002). Thermoviscoelastic analysis of residual curing stresses and the influence of autoclave pressure on these stresses in carbon/epoxy laminates. *Composites Science and Technology, 62*(4), 559–565. doi:10.1016/S0266-3538(01)00148-8

White, S. R., & Kim, Y. K. (1996). Staged curing of composite materials. *Composites. Part A, Applied Science and Manufacturing*, *27*(3), 219–227. doi:10.1016/1359-835X(95)00023-U

Young, W.-B. (1995). Compacting pressure and cure cycle for processing of thick composite laminates. *Composites Science and Technology*, *54*(3), 299–306. doi:10.1016/0266-3538(95)00067-4

Yuan, C., Li, M., Zhang, Z., & Gu, Y. (2008). Experimental investigation on the Co-cure processing of honeycomb structure with self-adhesive prepreg. *Applied Composite Materials*, *15*(1), 47–59. doi:10.1007/s10443-008-9056-4

Yuhazri, M., & Sihombing, H. (2010). A comparison process between vacuum infusion and hand lay-up method toward kenaf/polyester composite. *International Journal of Basic & Applied Sciences, 10.*

Chapter 15
Prediction of Hardness Distribution in Plasma Arc Surface Hardening Using Neural Network

Manoj Kumar
Mother Parwati Education Services, India

ABSTRACT

In this chapter, an attempt has been made to develop neural network models to predict the hardness distribution of hardened zone in plasma arc surface hardening process. The back propagation method with the Levenberg-Marquardt algorithm was used to train the neural network models. Hardness distributions were collected by the experimental setup in the laboratory and the associated data were used to train the neural network models. Furthermore, the prediction of neural network models were compared with those obtained from a statistical regression models. It is confirmed experimentally that the hardness distribution can be accurately predicted by the trained neural network models. The accuracy of hardness distribution prediction using neural network is superior to that using other statistical regression models.

1. INTRODUCTION

Plasma arc surface hardening is quite effective in achieving higher surface hardness, and keeping the modification of the surface down to a minimum. Owing to the high energy transfer efficiency of plasma arc (about 75%) (Yan and Zhu, 2008), the surface layer is heated above its austenite transformation temperature (A_{c3} temperature), but below the melting temperature, martensite structure can be produced in a short interaction time. Various types of steel such as cast iron, medium carbon steel and tool steel can be hardened by plasma arc to increase their hardness, wear resistance and corrosion resistance (Yan and Zhu, 2007; Bourithis *et al.*, 2012; Yan, 2003; Pan *et al.*, 2005). Plasma arc surface hardening is a promising technology in manufacturing, such as in the automobile and metal working industries (Krasposhin *et al.*, 2009). The rapidity, flexibility and lower cost of the method, can improve the competitiveness of these industries.

DOI: 10.4018/978-1-5225-2440-3.ch015

The method of surface hardening of steels and alloys with the heat source are widely used a concentrated energy flow such as laser beam, electron beam, or plasma arc (Ashby and Easterling, 2004). The surface hardening method is a modification of the surface structure of steel containing sufficient carbon to allow the transformation from austenite to martensite after the appropriate amount of heat is applied to the surface followed by rapid cooling of the heated layer by heat sink. The surface hardening process has been typically carried out by laser beam (Cheng and Lin, 2010; Chu *et al.,* 2012) and electron beam (Demuth and Beale, 2004; Ezugwu *et al.,* 2015). However, there is very little independent published information available on the use of plasma arc source for surface hardening (Hwang and Fung, 2014). As a heat source, a plasma arc possesses advantages over electron beams or lasers. Plasma arc does not require vacuum chamber as in the case of electron beam, or a complex and expensive optical-mechanical system for laser. In fact, the plasma arc is widely used for welding (Ion and Anisdahl, 2007), cutting (Kim *et al.,* 2013) and forming (Linnik *et al.,* 2015), as well as in surface modification of biomedical materials (Luo, 2013). Therefore, the plasma arc process is one of the most attractive methods. Plasma arc surface hardening is a promising technology in manufacturing, such as in the automobile and metal working industries. The rapidity, flexibility and lower cost of the method, can improve the competitiveness of these industries.

This technology began in 1980s, and from that time until now, the development of plasma arc surface hardening technology was more concentrated on the experimental studies (Cheng and Lin, 2010; Chu *et al.,* 2012; Demuth and Beale, 2004; Ezugwu *et al.,* 2015). Most of the studies have been developed by using trial and error or empirical methods. Classical experimental design methods are too complex and are not easy to use. A large number of experiments have to be carried out when the number of the process parameters increases. This also proves to be expensive and time consuming. The method of surface hardening is a modification of the surface structure of steel containing sufficient carbon to allow the transformation from austenite to martensite after the appropriate amount of heat is applied to the surface, followed by a rapid cooling of the heated layer by heat sink. The surface hardening process has been typically carried out by laser beam (Ruiz *et al.,* 2013; Selvan *et al.,* 2009) and electron beam (Hwang and Fung, 2014; Song *et al.,* 2013). However, there is very little independent published information available on the use of plasma arc source for surface hardening (Stich *et al.,* 2010). Plasma arc is a concentrated heat source and widely used in the welding (Luo, 2013), cutting (Wang *et al.,* 2013), and forming (Pan *et al.,* 2012), as well as in surface treatment of biomedical materials (Wang *et al.,* 2011). As a heat source, a plasma arc possesses advantages over electron beams or lasers. Plasma arc does not require vacuum chamber as in the case of electron beam, or a complex and expensive optical-mechanical system for laser. Therefore, the plasma arc surface hardening is one of the most attractive and effective methods in achieving higher surface hardness without any additional quenching mediums, so that the wear and corrosion resistance of the materials can be improved. Research efforts on surface hardening by plasma arc began in 1980s (Woo and Cho, 2008). Most of the studies have been concentrated on the single pass of plasma arc (Bourithis *et al.,* 2012; Linnik *et al.,* 2015; Kraposhin *et al.,* 2009; Nitkiewicz & Jeiorski, 2011; Samotugin, 2008; Yang, 2011). The plasma arc surface hardening with multiple passes to achieve a large hardened area has been reported (Pan *et al.,* 2005). Multiple passes in selective hardening are made continuously from one end of the surface to the other without intermittent stops. After making a pass, the hardened zone will cool down rapidly. At the same time, the heat is also conducted away raising the temperature of the substrate adjacent to it, including part of the area covered by the previous pass, i.e. the overlapped zone and the new area to be hardened. The rapidity of heat conduction depends on the thermal diffusivity of the material (Zhang *et al.,* 2009). The temperatures of the hardened zone of the previous pass and the new area to be hardened, depend on this factor as well as the size of the area to be hardened and the arc scanning velocity (Yang, 2011). For a small area, with a short length of travel, and high scanning velocity, the turnaround time, i.e. the time taken between the first and the second passes, is shorter. The temperature on the hardened zone of the previous pass may remain high while the new area to be hardened may remain at a lower temperature. On the other hand, for a larger area with a longer length of travel and a lower scanning velocity, the turnaround time is long. The temperatures on the hardened zone of the previous pass may be lower while the temperature on the new area to be hardened may be higher. Hence, different results would be expected from different processing conditions.

Surface hardening is well-known since decades to be an effective enhancement process to improve the performance and service life of a work-piece especially for carbon steels with carbon content below o. 6% or for low alloy steels. Surface hardening is commonly based on the use of high energy sources, as they offer a means of rapid heating and subsequent quenching below the melt, leading to fine microstructures and consequently to possible improvement of mechanical, corrosion or tribological properties. Laser surface hardening has drawn great attentions for its high energy heating source and considerable investigations were carried out in the use of laser for surface hardening of steel with martensitic transfornnation. However, only a few of investigations concern the use of the plasma transferred arc (PTA), although there are serious indications that its use may be quite attractive in industrial applications. However, the PTA surface processing is also characterized by extremely high temperature, excellent arc stability, which is due to the use of a constrictor nose. The nose restrains the column diameter of an electric arc established between a tungsten electrode and the work-piece in an inert gas atmosphere, usually argon. The heating efficiency of thermal plasma jet could be much higher and processing cost can be much lower than a laser beam, and the PTA hardening process is easy to operate and not related to surface reflectivity of the base material. These features impart the PTA an expectant heat source of a new surface modification technology. Usually, martensite could be obtained during the fast heating and cooling for the high energy of the plasma transferred arc hardening process. Bourithis *et al.,* (2012) evaluated the surface treatment of calmax tool steel. The geometric characteristics of melted and heat affected zones were studied as a function of working parameters. During the hardening process, the treated piece is usually working as the anode which generally moves relative to the torch cathode. Thus, treatment parameters, such as arc current, distance, plasma gas flow and moving speed between work-piece and cathode, will all affect the treatment results. It is essential to understand the effects of treatment parameters on the final properties of hardened specimens.

Hardness distribution in the vertical distance from the hardened surface can be applied to verify the hardened depth (Kim *et al.,* 2013). To predict hardness distribution, researchers used empirical phase-property relationships to express hardness as a function of microstructure and chemical composition (Ashby and Easterling, 2004; Ion and Anisdahl, 2007). However, little work has been reported concerning the relationship between process parameter and hardness distribution. The relationship between process parameters and hardness distribution in the plasma arc surface hardening process are not known completely. Hence, a new practical prediction method is desired. Neural networks have provided a means of successfully prediction studies in surface hardening processes (Woo and Cho, 2008; Stich *et al.*, 2010).

Artificial neural network is a non-linear algorithm which is used for modeling of process parameters, and this algorithm can learn and predict experimental knowledge because of ANNs' wide range of applicability and their capability to treat complicated and nonlinear problems. A common ANN model is established in a training process by using back propagation (BP) with the gradient descent algorithm.

However, to our best knowledge, little work has been done on the parameter optimization of the plasma hardening process using neural network. Therefore, in the present work, surface hardened by micro-plasma transferred arc process. Uniform experiment design was applied to arrange the experimental procedures with different plasma arc power, scanning speed, gas flow and hardening distance, considering the interactions of different parameters instead of one factor experiment. Based on the tested results, neural networks were applied to establish the nonlinear model of the relationship between the properties of hardening layers and the process parameters. Recently, neural networks have been widely utilized to tackle problems which cannot be satisfactorily handled by traditional analytical approaches. The advantages of neural networks include extreme computation, powerful memory and rapid learning as well as it can predict an output with accuracy even if the variable interactions are not completely understood. Neural networks have been applied successfully in a various fields of mathematics, engineering, medicine, economics, meteorology, psychology, neurology and many others (Kim *et al.*, 2013).

In this chapter, the relationship between the process parameters and hardness distribution during plasma arc surface hardening is established using neural network analyses. In addition, statistical regression models were used in this study to be compared with the proposed neural network models.

2. NEURAL NETWORK

Neural network simulates the behavior of human brain neurons. It is a parallel processing structure, which can be divided into several processing procedures, which can be trained simultaneously. A neural network model is constructed by using a set of data consisting of input and output variables. In the training process, the structure of the model adjusts itself to the data, and the final model can be used for prediction. One of the most important applications of neural networks is modeling a system with an unknown input–output relation (Zhang *et al.*, 2009). To date, many kinds of neural network architecture have been proposed. Neural network using the back-propagation training methodology is most prevalent in modeling and controlling applications, owing to its capability of learning system characteristics through non-linear mapping (Wang *et al.*, 2013). The operation of the neural network model can be divided into two main phases: forward computing and backward learning (Cheng and Lin, 2010).

2.1. Forward Computing

The input patterns applied to the neurons of the first layer are just a stimulus to the network. On the other hand, there is no computation in the input layer. As depicted in Figure 1, each neuron in the hidden layer determines a net input value based on all its input connections.

The net input is calculated by summing the input values multiplied by their corresponding weight. Once the net input is calculated, it is converted to an activation value. The weight on the connection from the ith neuron in the forward layer to the jth neuron is indicated as w_{ij}. The output value Y_j of neuron j is computed by the following equation:

$$net_j = \sum_{i=0}^{n} w_{ij} x_i \tag{1}$$

Figure 1. Architecture of an individual neuron for back propagation network

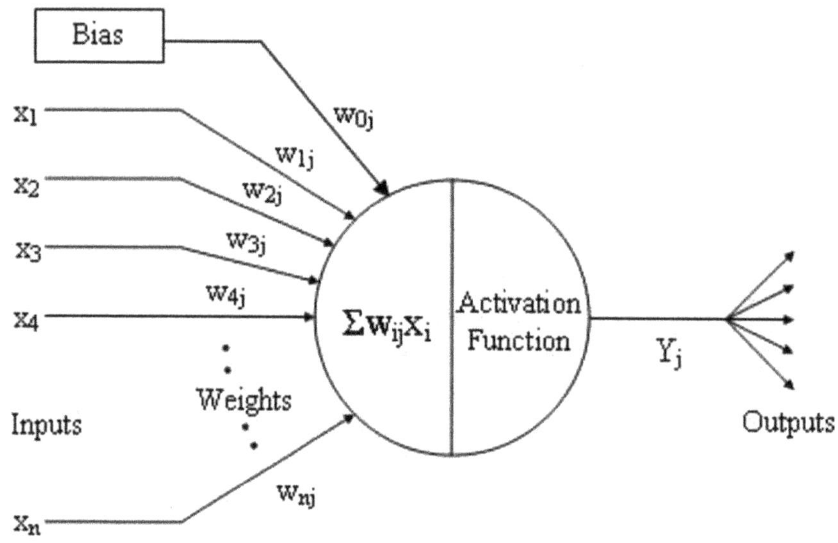

$$Y_j = f_{act}\left(net_j\right) \tag{2}$$

where net_j is the linear combination of each of the x_i values multiplied by w_{ij}, n is the number of inputs to the jth neuron, and f_{act} is the activation of neuron j. Sigmoid functions (S-shaped curves), such as logistic function and hyperbolic tangent function, are commonly adopted for the activation functions (Ezugwu *et al.*, 2015). The activation functions of logistic and hyperbolic tangent functions are, respectively.

$$Y_j = \frac{1}{1+\exp\left(-net_j\right)} \tag{3}$$

$$Y_j = \frac{1-\exp\left(-net_j\right)}{1+\exp\left(-net_j\right)} \tag{4}$$

2.2. Backward Learning

The generated output of the network is compared to the desired output, and an error is computed for each output neuron. The error vector E between desired values and the output value of the network is defined as

$$E = \sum_j E_j = \sum_j \frac{1}{2}\left(T_j - Y_j\right)^2 \tag{5}$$

where Y_j is the output value of the jth output neuron, T_j is the desired value of the jth output neuron. Errors are then transmitted backward from the output layer to each neuron in the forward layer. The process repeats layer by layer. Connection weights are updated by each neuron to cause the network to converge. The network was trained with Levenberg-Marquardt algorithm. This training algorithm was chosen due to its high accuracy in similar function approximation. The adjustment of weights and biases are done according to transfer function

$$\Delta w_{ij} = -\left(J^T J + \mu I\right)^{-1} J^T E \tag{6}$$

where J is Jacobian matrix of derivation of each error, μ is a scalar and E is error function.

3. EXPERIMENTAL WORK

Plasma arc surface hardening process was performed using a plasma arc machine with torch diameter of 1.6 mm, which integrated with a six degree-of-freedom articulated robot. The negative terminal of the power supply is connected to the cathode located inside the plasma torch and the work-piece is connected to the positive polarity of the power supply. Argon gas was used at 6 bar as plasma and shielding gas, to minimize oxidation. The nozzle work-piece standoff distance was kept constant at 13 mm. The selected currents of plasma arc were 30 A and 60 A. The scanning velocities of plasma arc were each set to 0.1, 0.2, 0.3, 0.4 and 0.5 m/s. ASSAB 618 and ASSAB DF3 steels were used in this study with carbon content of 0.37 wt.% and 0.90 wt.%, respectively.

Prior to the experiment, specimens of size 60 x 40 x 10 mm^3 were cut, ground and polished to 1000 grit silicon carbide paper in order to remove oxides and obtain a smooth surface. After the experiment, the hardened specimens were cut perpendicular to the scanning direction, polished, etched in 2% Nital and then used for hardness measurements. The hardness distribution over the depth of hardened zone was measured using the microhardness tester with a load of 200 g and an indentation time of 15 s. The hardness distributions of the hardened zone and base material were measured from a distance of 0.03 mm from the surface up to 0.25 mm (Line 1) as shown in Figure 2.

Figure 2. Schematic illustration of hardness distribution measurement

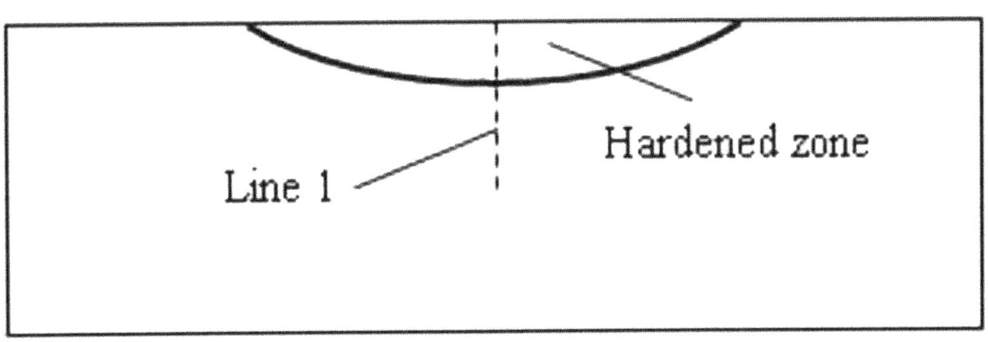

4. NETWORK TRAINING

In this study, the development and the training of the network is carried out using MATLAB Neural Network Toolbox (Demuth and Beale, 2004). The input/output dataset of the model is illustrated in Figure 3.

Here, there are four neurons in the input layer of network. The inputs are arc current, carbon content of material, scanning velocity and hardness location. The output layer has only one neuron, which gives the values of hardness. In this study, networks with one and two hidden layer(s) were used. With a learning rate of 0.1, the network was trained for 2000 iterations. The error between the desired and actual outputs is less than 0.001 at the end of the training process. The four variables data sets could not be trained by neural network in their original form due to the wide range of values among them. In order to become feasible input neurons, all the values in the input neurons had to be pre-processed by normalizing and transformed within the range of ±1, using the MATLAB subroutine *premnmx*. The normalized value (X_i) for each raw input/output dataset (d_i) was calculated as

$$X_i = \frac{2}{d_{max} - d_{min}}\left(d_i - d_{min}\right) - 1 \qquad (7)$$

where d_{max} and d_{min} are the maximum and minimum values of the raw data.

To evaluate the performance of the model, a total of 192 data pairs of hardening conditions with hardness locations; and corresponding hardness values were used. Among them, 128 pairs were used for modeling and the total data set including the remaining 64 pairs was used to assess the performance of the models. The criterion used to judge the efficiency and the ability of the model to predict hardened zone performances was the percentage error or deviation (Δ) which is defined in equation (8). With this criterion, it would be much easier to see how the proposed model fit and how the predicted values are close to the actual ones.

$$\Delta = \frac{\left|\text{Predicted value} - \text{Actual value}\right|}{\text{Actual value}} \times 100\% \qquad (8)$$

Figure 3. Schematic model of neural network for prediction of hardness distribution

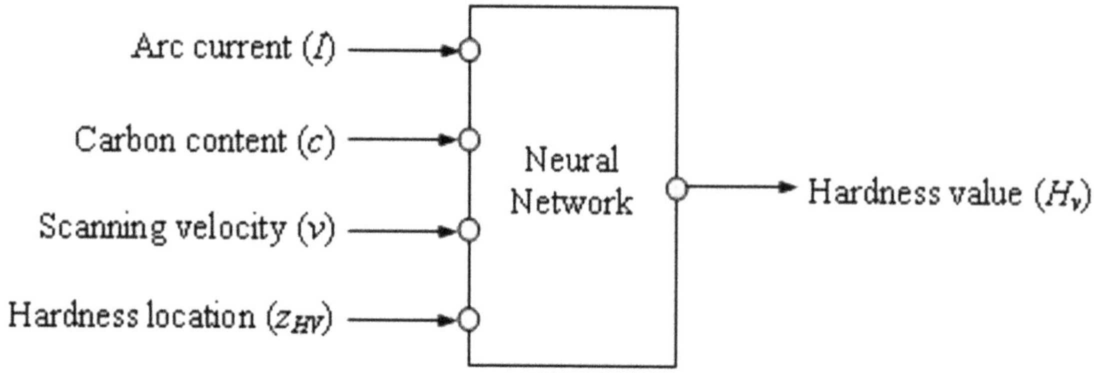

5. RESULTS AND DISCUSSION

In the first model, the network has one hidden layer and the number of neurons is examined between one and eight. The hyperbolic tangent function (*tansig*) was used in the hidden layer as an activation function. In addition to the previous model, a second model was considered. Here, the hidden layer comprised two layers with the number of neurons were between one and eight for each layer. The hyperbolic tangent function (*tansig*) and logistic function (*logsig*) were used for the first and second hidden layer, respectively. For both models, linear transfer function (*purelin*) was used in output layer. In order to reform the activation of network, one bias for each layer can be added. There is one bias for each layer of network except the output layer. After having finished the neural network training, neural network was tested using the different data from the trained data.

The performance of the neural network depends on the number of hidden layers and the number of neurons in the hidden layers. Therefore, many attempts have been carried out in choosing the optimal structure for the neural network by changing the number of hidden layers as well as the number of neurons in each of these hidden layers. To examine the effect of the different structure of the neural network models, the RMS errors were determined. Table 1 lists the results based on the RMS error of the training and testing sets. It is clear that the 4–8–1 structure had the lowest RMS error among all the structures with one hidden layer, and 4–8–7–1 was better than other structures with two hidden layers. Table 2 summarizes the percentage deviation of the above two structures.

In addition, two statistical models with the same input data were also employed to evaluate the results with neural network models. The linear regression and non-linear regression models are, respectively.

$$\begin{aligned} H_v &= 919 + 4.46\left(x_1\right) - 357\left(x_2\right) - 476\left(x_3\right) - 4174\left(x_4\right) \\ &+5.06\left(x_1x_2\right) - 28.6\left(x_1x_4\right) + 1670\left(x_3x_4\right) + 9410\left(x_4^{\,2}\right) \end{aligned} \tag{9}$$

$$H_v = \frac{12.46 - x_1^{\,0.522}}{x_2^{\,1.27} - x_3^{\,0.135} - x_4^{\,0.56}} e^{0.274 ln x_1 ln x_2} \tag{10}$$

Table 1. Summary of neural network with different structures and their training and testing RMS error

Number of Hidden Layer	Structure	Training RMS Error	Testing RMS Error
1	4 – 1 – 1	0.19264	0.16722
	4 – 2 – 1	0.17074	0.16357
	4 – 3 – 1	0.11960	0.10920
	4 – 4 – 1	0.11629	0.10268
	4 – 5 – 1	0.10308	0.09023
	4 – 6 – 1	0.09640	0.08395
	4 – 7 – 1	0.08257	0.03556
	4 – 8 – 1	**0.07585**	**0.03162**

continued on following page

Table 1. Continued

Number of Hidden Layer	Structure	Training RMS Error	Testing RMS Error
2	4 – 1 – 1 – 1	0.19246	0.16194
	4 – 2 – 1 – 1	0.14295	0.12475
	4 – 2 – 2 – 1	0.14087	0.10820
	4 – 3 – 1 – 1	0.11173	0.08992
	4 – 3 – 2 – 1	0.10802	0.06721
	4 – 3 – 3 – 1	0.06898	0.03702
	4 – 4 – 1 – 1	0.09612	0.07376
	4 – 4 – 2 – 1	0.07956	0.04223
	4 – 4 – 3 – 1	0.06864	0.03570
	4 – 4 – 4 – 1	0.05836	0.03259
	4 – 5 – 1 – 1	0.09057	0.07634
	4 – 5 – 2 – 1	0.07492	0.03654
	4 – 5 – 3 – 1	0.07608	0.04148
	4 – 5 – 4 – 1	0.06859	0.03935
	4 – 5 – 5 – 1	0.05280	0.03161
	4 – 6 – 1 – 1	0.07092	0.05806
	4 – 6 – 2 – 1	0.06961	0.03552
	4 – 6 – 3 – 1	0.06880	0.04027
	4 – 6 – 4 – 1	0.04933	0.03160
	4 – 6 – 5 – 1	0.04574	0.03159
	4 – 6 – 6 – 1	0.04051	0.03138
	4 – 7 – 1 – 1	0.07230	0.04681
	4 – 7 – 2 – 1	0.04904	0.03489
	4 – 7 – 3 – 1	0.03628	0.03142
	4 – 7 – 4 – 1	0.03987	0.03162
	4 – 7 – 5 – 1	0.03411	0.03162
	4 – 7 – 6 – 1	0.03175	0.03136
	4 – 7 – 7 – 1	0.03162	0.03097
	4 – 8 – 1 – 1	0.06607	0.03311
	4 – 8 – 2 – 1	0.04963	0.03241
	4 – 8 – 3 – 1	0.03635	0.03147
	4 – 8 – 4 – 1	0.03139	0.03125
	4 – 8 – 5 – 1	0.03161	0.03116
	4 – 8 – 6 – 1	0.03161	0.03000
	4 – 8 – 7 – 1	**0.03043**	**0.02382**
	4 – 8 – 8 – 1	0.03122	0.03052

Table 2. Summary of 4–8–1 and 4–8–7–1 neural network models

Result	4–8–1	4–8–7–1
Training cycle	2000	2000
Percentage deviation of the training data	3.98%	1.68%
Percentage deviation of the testing data	1.58%	1.02%

where H_v is the hardness value in [HV], x_1 is arc current [A], x_2 is the specimen carbon content [wt.%], x_3 is the scanning velocity [m/s], and x_4 is hardness location in thickness direction [mm]. In order to guarantee the reliability of the regression analysis, the regression model coefficients were determined using a backward elimination procedure in which insignificant terms were eliminated based on significance level of ≤ 0.05. Table 3 summarizes the percentage deviation of the training and testing data for each regression model. It was apparent that the linear model had a higher R^2 value and a lower percentage deviation between the predict values and the actual values.

Figure 4 shows the prediction error for the neural network models and regression models. Compared with the experimental results, the maximum error obtained by neural network model with 4–8–7–1 structure is not more than 10%, showing better accuracy than those of linear regression, non-linear regression and neural network 4–8–1 model with the maximum error of 75.3%, 62.3%, and 24.9%, respectively. To ensure the accuracy of the developed regression and neural network models for prediction of hardness distribution, all the measured and predicted results using the developed models were compared and represented in Figure 5. In the neural network, it can be seen that the distribution of data points for model with 4–8–1 structure is similar and close to the 'A = T' line. However, the predicted values obtained using the model with 4–8–7–1 structure is more accurate with accuracy of 99.7%. The prediction accuracy of other models was 98.5%, 87.1% and 78.1% for the 4–8–1 one hidden layer neural network model, linear regression model and non-linear regression model, respectively. In all of them, the 4–8–7–1 two hidden layer neural network model was found to be the best in terms of predictive ability.

The experimental results have been graphically compared with the testing results obtained from neural network models and regression models as shown in Figure 6. The values with the neural network models prediction were able to follow the trend better, as compared those with regression models prediction. Since the neural network models and the statistical models are both generated by back propagation and regression, respectively, a brief comparison is made between them. It is apparent that the neural network

Table 3. Summary of statistical regression models

Result	Linear Regression Model	Non-Linear Regression Model
R^2	0.8710	0.7814
Adjusted R^2	0.8698	0.7797
Percentage deviation of the training data	13.72%	14.98%
Percentage deviation of the testing data	14.76%	18.40%

Figure 4. Prediction error of hardness distribution using (a) neural network 4-8-1, (b) neural network 4-8-7-1, (c) linear regression and (d) non-linear regression model

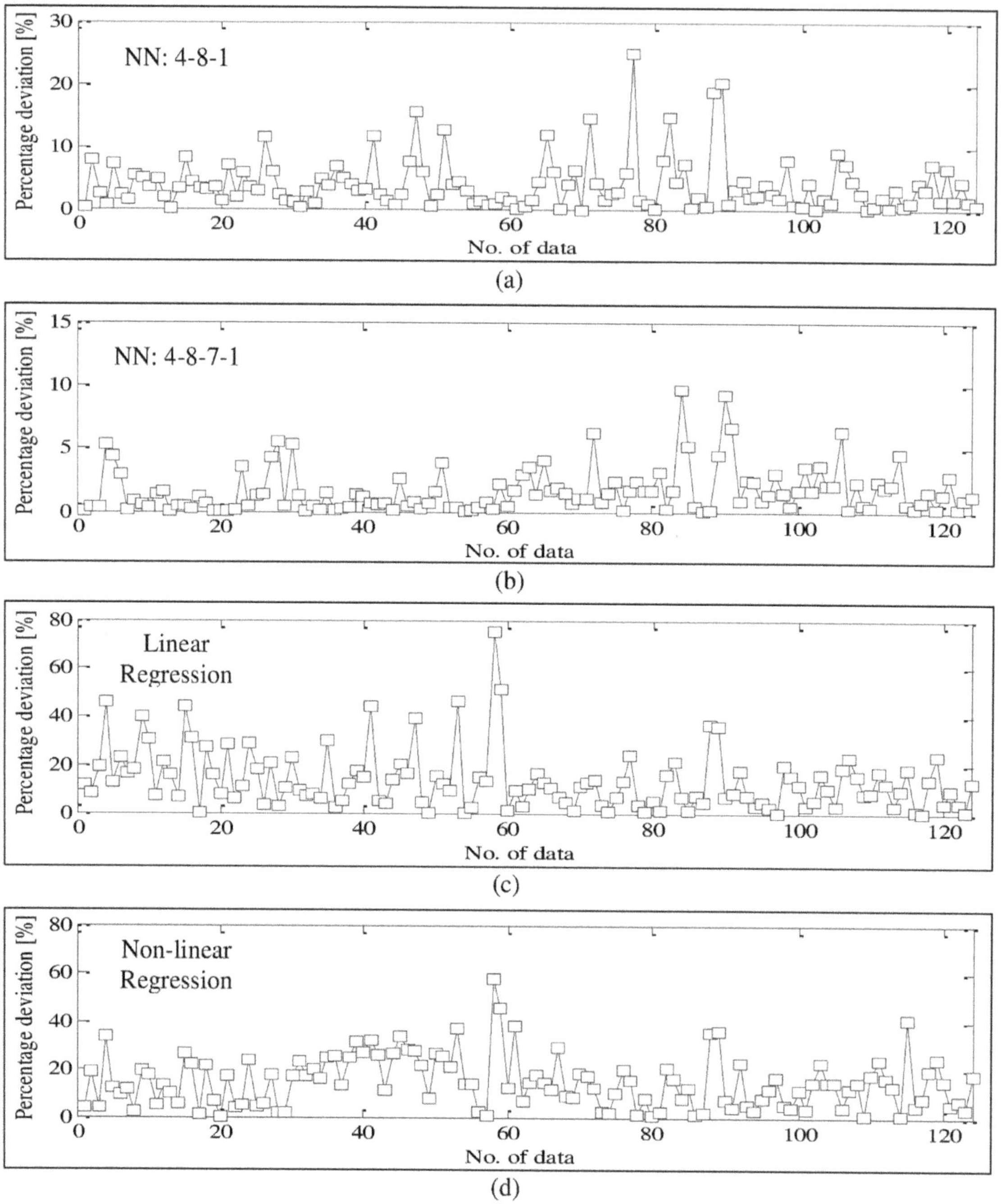

Figure 5. Comparison of measured and predicted results using (a) neural network 4-8-1, (b) neural network 4-8-7-1, (c) linear regression and (d) non-linear regression model

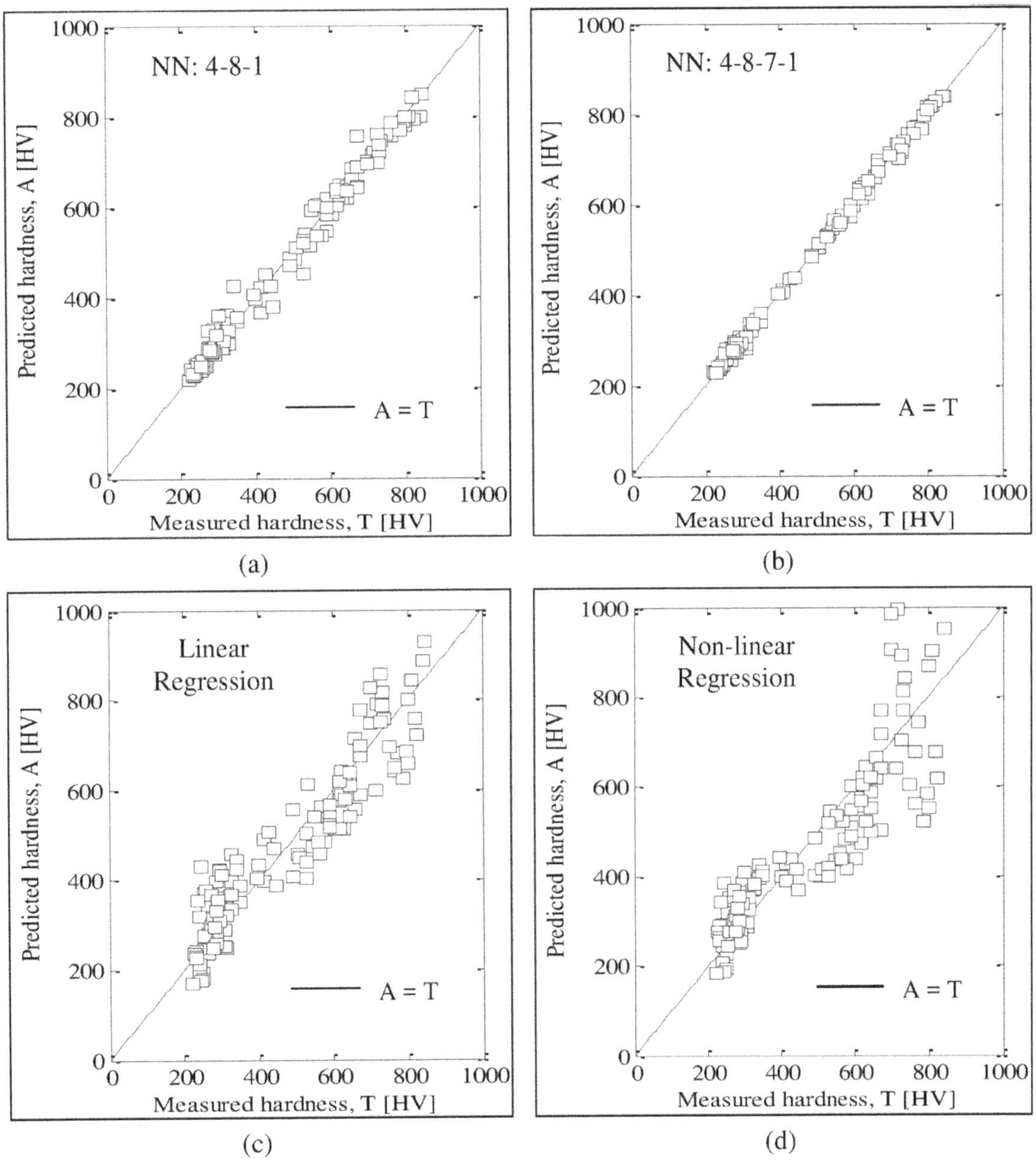

models generally have better predictive ability of both training and testing relative to their percentage deviation. The comparison chart confirms this finding as shown in Table 4. The predicted hardness values of the neural network models are much closer to the actual hardness values than those of the regression models. This is thought to be because the neural network expresses the non-linear relationship of the hardness values which is formed through the input variables better than the statistical regression.

Figure 6. Prediction models trend with experimental data order (a) neural network 4-8-1, (b) neural network 4-8-7-1, linear regression and (d) non-linear regression model

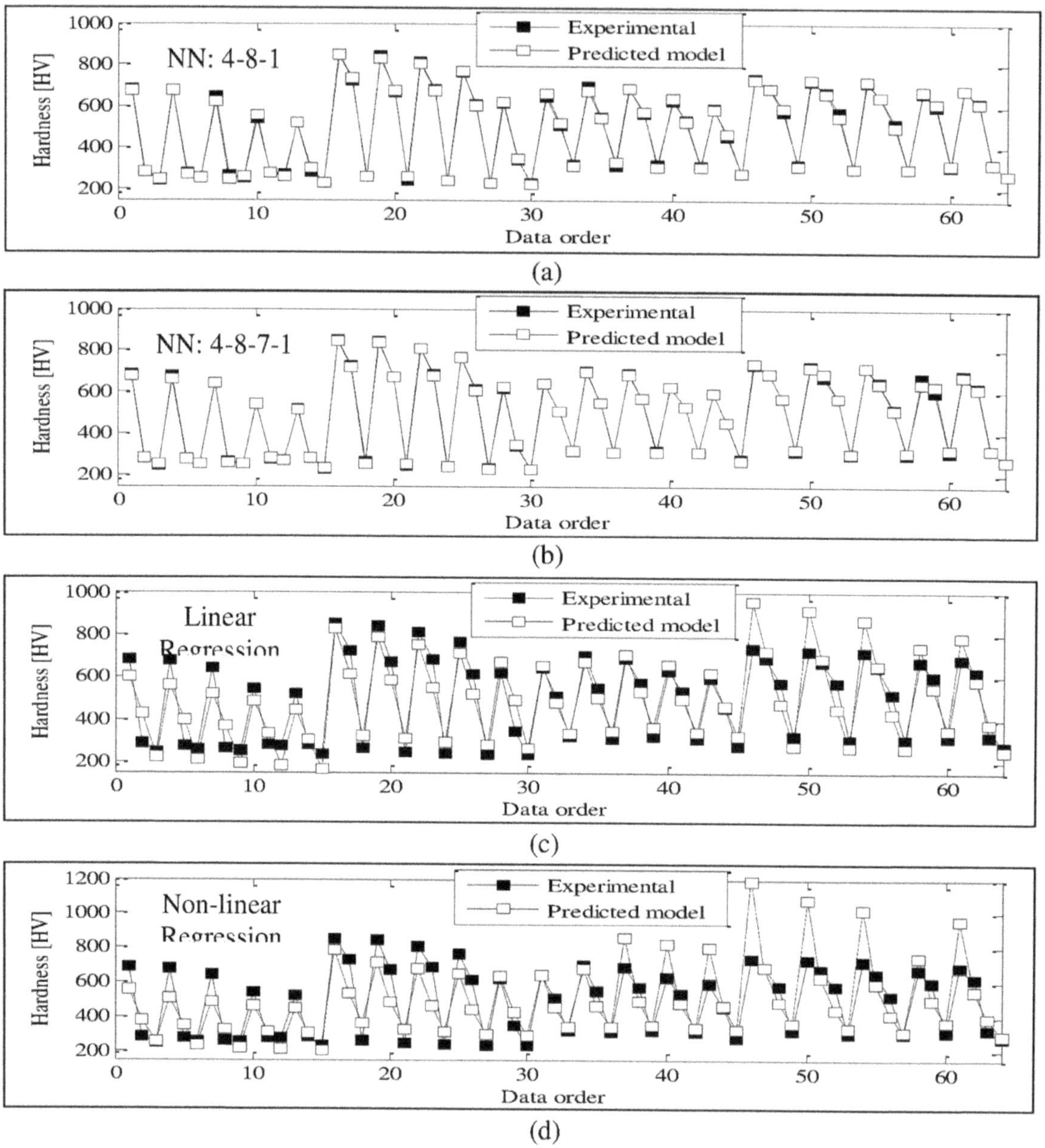

6. CONCLUSION

Two models of back propagation neural network with one and two hidden layers for predictions of hardness distribution in plasma arc surface hardening process have been established, and compared with two models of statistical regression i.e. linear and non-linear. The neural network models have better predictive ability compared with the regression models in predicting hardness distribution. The predicted

Table 4. Comparison between the regression modes and neural network model

Result	Regression		Neural Network	
	Linear	Non-Linear	1 Hidden Layer (4–8–1)	2 Hidden Layers (4–8–7–1)
Training Data				
Max error	184.48	287.74	85.70	30.03
Min error	0.46	0.53	0.01	0.04
Δ	13.72%	14.98%	3.98%	1.68%
Testing Data				
Max error	222.83	456.40	22.18	31.37
Min error	3.30	1.09	0.02	0.08
Δ	14.76%	18.40%	1.58%	1.02%

hardness values of the neural network models are much closer to the actual hardness values than those of the regression models.

In the neural network models, after training with 2000 cycles, the one hidden layer 4–8–1 structure model could achieve an accuracy of 98.5% precision. In addition, a two hidden layer 4–8–7–1 structure produces an accuracy of prediction of 99.7%. Thus, the two hidden layer was more accurate and effective in predicting the hardness distribution than the one hidden layer neural network model. Neural network is a powerful tool, easy-to-use in complex problems. Neural network can be used reliably, successfully and very accurately for the prediction of hardness distribution in surface hardening process.

REFERENCES

Ashby, M. F., & Easterling, K. E. (2004). The transformations hardening of steel surface by laser beams. *Acta Metallurgica, 32*(11), 1935–1948. doi:10.1016/0001-6160(84)90175-5

Bourithis, E., Tazedakis, A., & Papadimitriou, G. (2012). A study on the surface treatment of Calmax tool steel by a plasma transferred arc (PTA) process. *Journal of Materials Processing Technology, 128*(1), 169–177.

Cheng, P. J., & Lin, S. C. (2010). Using neural networks to predict bending angle of sheet metal formed by laser. *International Journal of Machine Tools & Manufacture, 40*(1), 1185–1197.

Chu, P. K., Chen, J. Y., Wang, L. P., & Huang, N. (2012). Plasma surface modification of biomaterials. *Materials Science and Engineering, 36*(1), 143–156.

Demuth, H., & Beale, M. (2004). *Neural Netwotk Toolbox User's Guide. Version 4.0.* The Mathworks Inc.

Ezugwu, E. O., Fadare, D. A., Bonney, J., Da Silva, R. B., & Sales, W. F. (2015). Modeling the correlation between cutting and process parameters in high-speed machining of Inconel alloy using an artificial neural network. *International Journal of Machine Tools & Manufacture, 45*(1), 1375–1385.

Hwang, J. R., & Fung, C. P. (2014). Effect of electron beam surface hardening on fatigue crack growth rate in AISI 4340 steel. *Surface and Coatings Technology*, *80*(1), 278–291.

Ion, J. C., & Anisdahl, L. M. (2007). A PC based system for produce development in laser transformation hardening. *Journal of Materials Processing Technology*, *65*(1), 261–267.

Kim, I. S., Son, J. K., & Yarlagadda, P. K. D. V. (2013). A study on the quality improvement of robotic GMA welding process. *Robotics and Computer-integrated Manufacturing*, *19*(1), 567–578.

Kraposhin, V. S., Bobrov, A. V., & Gaponenko, O. S. (2009). Surface hardening of 9KhF steel by heating with a plasma gun. *Journal of Materials Processing Technology*, *11*(1), 13–17.

Linnik, V. A., Onegina, A. K., Andrev, A. I., Aldarkin, K. K., Sinaiskii, V. M., & Grigorenko, L. P. (2015). Surface hardening of steel by plasma hardening. *Journal of Materials Processing Technology*, *5*(4), 2–23.

Luo, W. (2013). The corrosion resistance of 0Cr19Ni9 stainless steel arc welding joints with and without arc surface melting. *Materials Science and Engineering A*, *345*(1), 7–19.

Nitkiewicz, Z., & Jeiorski, L. (2011). Plasma heat treatment of steel: Microstructures, properties and applications. *Metal Science and Engineering*, *140*(1), 474–493.

Pan, C. X., Chen, Y. W., & Male, A. T. (2012). Microstructural development in plasma jet forming of sheet steels. *Materials Science and Technology*, *18*(1), 1151–1168.

Pan, W. X., Meng, X., Li, G., Fei, Q. X., & Wu, C. K. (2005). Feasibility of laminar plasma jet hardening of cast iron surface. *Surface and Coatings Technology*, *197*(1), 345–350. doi:10.1016/j.surfcoat.2004.06.043

Ruiz, J., Lopez, V., & Fernandez, B. J. (2013). Effect of surface laser treatment on the microstructure and wear behaviour of grey iron. *Materials & Design*, *17*(5/6), 273–287.

Samotugin, S. S. (2008). Plasma treatment of tool steels. *Welding International*, *12*(3), 228–235.

Selvan, J. S., Subramaniam, K., & Nath, A. K. (2009). Effect of laser surface hardening on En18 36–(AISI 5135) steel. *Journal of Materials Processing Technology*, *91*(1), 29–43.

Song, R. G., Zhang, K., & Chen, G. N. (2013). Electron beam surface treatment. Part I: Surface hardening of AISI D3 tool steel. *Journal of Materials Processing Technology*, *69*(1), 513–527.

Stich, T. J., Spoerre, J. K., & Velasco, T. (2010). The application of artificial neural networks to monitoring and control of an induction hardening process. *Journal of Information Technology*, *16*(1), 1–11.

Wang, J., Kusumoto, K., & Nezu, K. (2011). Plasma arc cutting torch tracking control. *Technology of Welding and Joining*, *6*(3), 154–156. doi:10.1179/136217101101538695

Wang, J. Y., Kusumoto, K., & Nezu, K. (2013). Modelling and prediction of cut shape for plasma arc cutting based on artificial neural network. *Science and Technology of Welding and Joining*, *4*(4), 195–200. doi:10.1179/136217199101537770

Woo, H. G., & Cho, H. S. (2008). Estimation of hardened layer dimensions in laser surface hardening processes with variations of coating thickness. *Surface and Coatings Technology*, *102*(1), 205–217.

Yan, M. (2003). Micro-beam plasma arc surface processing for ferrous and non-ferrous metals. *Journal of Materials Science*, *38*(1), 3219–3222. doi:10.1023/A:1025169517526

Yan, M., & Zhu, W. Z. (2007). Surface treatment of 45 steel by plasma arc melting. *Surface and Coatings Technology*, *91*(1), 183–191.

Yan, M., & Zhu, W. Z. (2008). A new approach of surface treatment: Micro-beam plasma arc remelting. *Materials Letters*, *34*(1), 222–225.

Yang, L. J. (2011). Plasma arc surface hardening of ASSAB 760 steel specimens with taguchi optimization of the processing parameters. *Journal of Materials Processing Technology*, *113*(1), 521–526.

Zhang, Y. F., Lu, L., & Yap, S. M. (2009). Prediction of the amount of PCA for mechanical milling. *Journal of Materials Processing Technology*, *89*(1), 260–265.

Chapter 16
Efficiency Analysis of Genetic Algorithm and Genetic Programming in Data Mining and Image Processing

Ayan Chatterjee
Sarboday Public Academy, India

Mahendra Rong
Bangabasi Evening College, India

ABSTRACT

Today, in the age of artificial intelligence and machine learning, Data mining and Image processing are two important platforms. GA and GP are value based and program based randomized searching tools respectively and these two are very much useful in the fields' data mining and image processing for handling different issues. In this chapter, a review is made on ability of GA and GP in some applications of these two fields. Here, the selected subfields of data mining are market analysis, fraud detection, risk management, sports analysis, protein interaction, classification of data, drug discovery and feature construction. The similar in image processing are enhancement and segmentation of images, face recognition, photo mosaic generation, data embedding, image pattern classification, object detection and Graphics Processor Unit (GPU) development. The efficiencies of GA and GP in these particular applications are analyzed with corresponding parameters, comparing with other non-GA and non-GP approaches of the corresponding subfields.

INTRODUCTION

In present decade, optimization is an essential part in our everyday life. In any type of efficiency analysis, this is an important in each platform. Optimization procedures are majorly divided into two parts- hard computing and soft computing. Soft computing is more realistic than hard computing in every situation. Genetic Algorithm (GA) is a useful optimization approach of soft computing.

DOI: 10.4018/978-1-5225-2440-3.ch016

Genetic Algorithm (GA) was invented by John Holland in mid-sixties in university of Michigan. But idea of natural selection and natural genetics were developed by David Goldbarg. This is probabilistic and global search depending on principles of evolution. It was built up on the basis of three different hereditary properties of animals- selection, cross-over and mutation. Actually, the concept of creation of children from previous generation is used to obtain global optimum value in this useful as well as interesting soft computing approach.

According to differentiation of the operators' selection, cross-over and mutation, different GA schemes are developed. According to number of objectives, there are two type genetic algorithms are made up like each type optimization approach- single objective genetic algorithms and multi objective genetic algorithms. Binary coded GA, Real coded GA, Messy GA etc are some single objective GA approaches and weighted sum, NSGA and NSGA-II etc are some multi objective GA approaches. In different platforms, different GA shows their optimization efficiencies in different ways. Among them, computer science, biology, economics, mathematics, pharmacology etc. are major application fields of genetic algorithm. In computer science, some efficient fields of GA are artificial intelligence, network security, data mining, image processing, pattern recognition etc.

Genetic Programming (GP) is an extended version of GA. In GA, population is created with a set of values and in GP; population is created with a set of sub programs, more specifically with a set of instructions. The concept of GP was first proposed by Lawrence J. Fogel in 1964. But at that time, he only introduced about this idea and simply implemented it in finite state automata. It was broadly used in 1990 with machine learning and artificial intelligence. Some specific application fields of GP are expert system, horn clause, networking, automation theory, software engineering, data mining, image processing etc.

In this chapter, we mainly focus on the application of GA and GP in data mining and image processing. We choose these two fields of application, because- these two are in a broad range of particular research application as well as GA and GP both are used in these two particular cases. So, we can analyze the efficiency of GA and GP with proper experiments.

BACKGROUND

In this section, we will describe some typical natures of GA and GP, observed in different situations as well as we mention about some comparative analysis of GA and GP in some application fields. In 2014, Mujahid Tabassum and Kuruvilla Mathew proposed a genetic algorithm analysis towards optimization solutions. In that paper, they discussed about some non linear, large and discrete type attributes containing problems with high degree complexity solution. Actually they mainly focused on different parts of artificial intelligence which are solved using GA. In this paper, the particular application fields are evolvable hardware applications, robotics, engineering design, data encryption and computer gaming. The comparative analysis is developed among GA and traditional methods with the mentioned application fields in that paper. In 1997 and 2006, two papers were published on data mining using GA and GP respectively. In the paper, application of GA in Data mining (published in 1997), Robert E. Marmaelstein proposed a comparative analysis between KNN classifier and genetic algorithm with Mushroom data set clustering. In that particular clustering, it is observed that, GA takes 15 generations to obtain 0% training error, whereas KNN classifier takes 3 generations to get the same result with same population size, probability of mutation and probability of cross-over. In the paper, application of GP in data mining (published in 2006), Saleh Mesbah Elkaffas and Ahmed A. Toony proposed a GP scheme to maintain a database of balance sheets, income details and cash flow details for some American public companies. It also gives a good result for clustering the data in particular data bases. In other words, in these two papers especially for clustering purpose, genetic algorithm is used. Another important field is image processing. In this field, GA and GP give a sound result in image segmentation, image transformation, image enhancement and photo mosaic generation. In the paper, A Review of Genetic Algorithm application for Image Segmentation, Raj Kumar Mohanta and Binapani Sethi proposed a comparative analysis of efficiency of GA and artificial neural network (ANN) in image processing. In the paper of Taras Kowaliw, Wolfgang Banzhaf, Nawwaf Kharma, and Simon on Genetic Programming-based Image Transforms, a GP scheme is proposed to specify a finite set of transform of grayscale image pixels. In the paper, a survey of Genetic Algorithms applications for Image Enhancement and Segmentation, Mantas Paulians and Andrius Usinskas proposed a review on GA application in image pre processing. Considering the mentioned papers on data mining and image processing, it is observed that there is no comparative analysis between GA and GP in data clustering, image segmentation, image transformation and image enhancement. Only in the paper, comparison between GA and GP performance for photo mosaic generation, we get excellent comparative experimental results in photo mosaic. But in other cases, we didn't get any efficiency analysis between GA and GP. In this chapter, we will build up a an analysis of GA and GP in the fields of data mining (data clustering) and image processing (image segmentation, image transformation and image enhancement) with proper experimental results.

AN OVERVIEW OF DATA MINING

In now a day, Information technology (IT) is one of the most important as well as useful platform of the world. Information technology is built on the basis of artificially information handling. Data storing, data extraction and data manipulation are the main facts related to that. Data mining is associated with data extraction procedure from a large data set or mine of data. To extract data from a data set automatically, it should be observed that how data are stored in the container or data base. Generally, data are stored according to their features or attributes. The basic thing of data mining is data clustering. Clustering is the process of organizing data according to their similarity. Moreover, according to similarity of attributes, data are arranged as well as stored in the information system. For the automatic clustering process of data, various techniques are developed. Actually, target of each data clustering process is to make groups or clusters from a large number of data with a huge no. of attributes. Generally, data are stored in an information system with matrix representation, where a particular data X_i is represented as

$$X_i = \left(X_{i1}, X_{i2}, \ldots\ldots, X_{in}\right) \tag{1}$$

where n is the number of attributes corresponding to the data X_i.

Actually, clustering is unsupervised classification. The major difference of it with supervised classification is the characteristics of type of set, dimension of data type, learning process and goal. In clustering process, there is no specification initially about the groups. So, it is more useful for accessing the set of data group wise.

Data mining is used in different daily required application areas, such as- market analysis, fraud detection, customer retention, production control, science exploration, corporate analysis, risk management etc. Also, now it is explored to the areas of astrology, sports, web search etc.

In this chapter, we will concentrate on GA and GP based data mining applications (some of mentioned previous) and their efficiencies in the particular cases.

APPLICATION OF GA IN DATA MINING FIELDS

Data mining is an important field for using GA, because- data extraction is a large scale as well as non linear optimization problem. Here, we discuss some particular fields of data mining, in which GA is used efficiently.

Market Analysis Using GA

Market analysis is a particular field of data mining. Here, we concentrate one particular type of market analysis. This is optimization of stock market investment in data mining. This particular part gives an efficient result using GA.

Optimization of Investment in Stock Market

In stock market, there is a technical trading rule for buying and selling of shares. So, the investors in stock market need to take proper decision before buying and selling shares. So, there should be some particular parameters for this purpose. Some popular rules or parameters in favor of this are filter rules, support and resistance, moving average and abnormal return. We give a short idea about these parameters here:

Filters Rule

The rule was developed by Alexandra. This rule was made on the basis of price momentum. The convention of this rule is rising and falling prices tend to continue their corresponding behavior. According to this rule, investors always observe the rate of change of percentage among previous low and high quantities. Depending on this fact, hold days and delay days are fixed by investors.

Support and Resistance

This is also a tool of decision making for the investors in stock market. Actually, resistance is considered as ceiling because price level obstruct the stock market from upwards direction of moving price. In other hand, support is considered as floor because the price of an asset is prevented from downwards pushing. To take decision on a short term investment, this is an important technical tool.

Moving Average

In statistical language, moving average is a result for analyzing data points through the series of different subsets of a full data set. The working formula of moving average is,

$$MA = \frac{\sum_{i=1}^{N} D_i}{N} \tag{2}$$

where,

N = No. of periods in the moving average
D_i = Demand in i-th period

For stock market investors, it is an important material to take decision on span of investment periodically.

Abnormal Return

Abnormal return is a special case in this particular purpose. It is the difference between single stock amount and probable expected return over a finite period of time. This is applied in Capital Asset Pricing Model (CAPM) to take decision in managerial level. The working formula of CAPM is,

$$r = \text{Rf} + \text{beta}^{*}\left(\text{Rm} - \text{Rf}\right) + \text{abnormal rate of return} \tag{3}$$

where,

r = The security return
Rf = Risk free rate of return
beta = Security's price volatility relative to overall market
Rm = Market return

Modified Filter Rule With GA

Filter rule was developed on the basis of three important facts. In other words, according to this rule, stock market investment decision follows three major facts and these three are equally important. These are- hold day (h), delay day (d) and previous day (p). These are used to create population. The chromosomes corresponding to hold day (h) and delay day (d) are generated randomly between 1 to 30 and corresponding to previous day (p) is generated randomly between 30 and 100. h, p and d all are the integer values. The operators (selection, crossover & mutation) are taken fractional values between 0 and 1. After that, simple real coded GA is operated with a finite number of iterations obeying basic features of GA iteration wise with previous mentioned operators and chromosomal values.

Fraud Detection Using GA

Artificially fraud detection system is one of the most important as well as most popular applications of data mining. Here, we analyze importance of GA also. In this portion, at first we concentrate on credit card payment system fraud detection. One of the best fraud detection was developed by K. Ramakalyani and D. Umadevi in 2012. At the time of credit card transaction online, different type fraud difficulties occur in present decade. In most of the cases, the major problem is that an amount is withdrawn unintentionally at a sudden moment. So, automatic online verification is very much important here. Also, the details of hacking should be known. In this particular case, GA helps us very much. In the particular paper of K. Ramakalyani, 20 different attributes are taken for verification purpose. These are- customers Id, authentication type, current balance, average bank balance, times of overdraft, credit card age, deducted amount, location of credit card (CC) used, time of CC used, average daily overdraft, amount of transaction, credit card type, time of using credit card, card holder income, card holder age, card holder position, card holder profession, card holder marital status, average daily spending and card frequency. So, here multi objective GA is used. An initial data set is formed with the data corresponding to the mentioned attributes. After that, these are sorted using fitness function. Then two types of selection procedures are used here- elitism selection and tournament selection. Elitism selection is used to create new population and before crossover operation, tournament selection is operated. After that, single point crossover and uniform mutation are performed. Then we are able to extract the particular fraud from a large scale data set. Another theme of fraud detection using GA was formed by Satvik Das and his co researchers in 2013. The speciality of this scheme is particular search space. Here, search space is created with the particular matrix position and corresponding costs are used to form population. Selection and crossover operations are same with the previous. But here no mutation operation occurs. It also gives a good result to detect the fraud. But, due to absence of mutation, time complexity is less over the previous one.

Risk Analysis of Decision Making Using GA

In all organizations, especially in corporate world, proper decision making at instant moment is very much important thing. But, in most of the cases, it is observed that decision is required with insufficiency of information. To help upper level employees of an organization, generally a decision support system (DSS) is developed artificially to take proper decision. But due to lack of all information accurately, a risk factor may occur at the time of decision making. So, risk analysis at the time of important decision making must be required. So, we have to find risk rate at that time. A lot of different methods are already developed for analyzing risk factor through risk rate. But to maintain it artificially, another drawback arises. If risk rate indicates that the risk is low, then we can accept the particular decision easily. But in opposite side, we have to take the decision that how we can reduce the risk. To reduce risk factor, GA helps us very much. A useful technique was developed by Alireza Tamjidyamcholo and Rawaa Al-Dabbag in 2012 related to this. They proposed that GA can be used to reduce risk factor. The basic components of GA are to create initial population, GA variables and fitness function depending on variables. In this particular type problem, four facts were taken as decision variables and these are information asset value (VA), likelihood of vulnerability (LV), percentage of risk mitigated (MC) and uncertainty of current knowledge vulnerability (UV). The ranges of the variables are defined in the following:

VA: 1 to 100
LV: 0 to 1
MC: 0% to 100%
UV: 0% to 100%

The fitness evaluation function is defined in the following as

$$Risk\ rate = VA \times LV - \left(VA \times LV\right) \times MC + \left(VA \times LV\right) \times UV \tag{4}$$

At the time of creating initial population, the values of the variables help to make initial chromosomes. After that, selection, crossover and mutation are operated. Approximately, after 51 iterations we obtain the solution of the problem. Types of selection, crossover and mutation are dependent on particular cases.

Another GA based risk analysis approach was developed by A Sumithra and Dr. E Ramraj in the same year. Here, crucial human factor involvement is analyzed beside the artificially decision making by secure software. Eight factors of human nature are taken as decision variables. These are competency in the technology, experience in the technology, exposure of security principles, ability of working to a team, experience in the domain, awareness of organizational security policies, motivation and personal commitment. Depending on these facts, GA was used here and we can check the efficiency of a particular decision support system (DSS).

Decision Making in Sports Using GA

In sports and games, optimization is required in different purposes. Actually, game theory is totally developed on optimization background. But for team selection and sports scheduling purpose with a large amount of data, genetic algorithm gives a powerful result. Here, we will discuss a football team selection procedure using GA. This procedure or technique was proposed by Prof. Khusali Deulkar, Vasudha Sarda and Prasham Sakaria in 2015. In now a day, there are many parameters, associated with team selection of a particular game or a series of particular games. These attributes include the quality of players and budget of a particular game. Quality of a player is generally measured through previous performance and experience of them. But high quality players demand a high amount of money. So, to maintain these two facts in football leagues, GA helps us very much. Actually there are two objectives in this problem. These are maximization of quality of players and minimization of costs of the games. The selected variables in this purpose are developed with related parameters of innovative and collaborative performances. In this purpose, Prof. Deulkar and his co researchers have developed and used INSGA II. This is improved version of NSGA II, developed by Kalyanmoy Deb. This particular team selection problem is NP hard in nature. To handle this and to reduce computational complexity of non dominated sorting process, they used INSGA II with round robin scheduling process. Requirement of no. of offenders and defenders of a particular team, budget of the team and demand of the players are used as variables. Actually, the beauty of INSGA II is dual consideration of interior and exterior population. Otherwise, tournament selection, simulated binary crossover and polynomial mutation are same with NSGA II. At the time of creation of initial population, both of interior and external set are developed and then NSGA II is carried on. For this, team selection process gives much better result using INSGA II over using only NSGA II.

Another sports optimization problem was developed in 2003. This was developed on sports scheduling with a large no. of constraints using genetic algorithms. This method was proposed by Andrew Brian Goldberg. This type problem is also NP hard just like previous type. This is a time scheduling and pair selection problem. In a league, the formation of groups initially and selection of two oppositions in the group level games are most important factors. But, here we should always maintain that all groups are more or less same standard. The standardization of the teams is parameterized through the previous performance of the teams and corresponding players. In GA part of decision making, initial population was created with different possible combination of the teams and fitness evaluation function was made to realize the strength of each group. The variables were set according to previous record. Then selection, crossover and mutation will occur. After a finite number of iterations, the final scheduling can be obtained. Since this is a large scale problem with a large no. of variables, then GA gives an excellent result. Here, simple weighted sum approach of multi objective GA was used for scheduling purpose. Other heuristic searches couldn't give better result, because of its NP hardness and use of qualitative parameters over quantitative.

APPLICATION OF GP IN DATA MINING FIELDS

In some particular fields or some different applications of data mining, genetic programming gives a good result over genetic algorithm. Actually, in some cases, both of clustering and feature selection are required on a large amount of data set. In this type of some cases, GP handles better optimization over GA. Some related specific fields with GP application are discussed in the following with proper specification.

Protein Interaction Using GP

Protein interaction process of biochemistry using GP was developed by M.L. Raymer, W.F. Punch, E.D. Goodman and L.A. Kuhn. In their proposed approach, the objective was development of artificially pattern recognition. Actually at the time of development of this problem using GA over K- means clustering, there was a restriction initially. The initial drawback was at the time of selection process; only linear combinations of features are selected always. But other features may be more prioritized in some particular cases. In this case, GP gives a better solution. It takes both of non linear and linear relationships of the features. So, if the selection process randomly occurs at the time of using GP, then the process gives also better result. But in GA, only linear combination of feature selection gives good result. In the protein binding problem of biochemistry, GP was applied with population size 100 and for four important features, four sub trees were taken. In this particular, mutation process was dropped. Only selection and cross over were used with probability of crossover (p_c) 0.9 or 90%. After 300 generations, the obtained clusters give better DB index and Dunn index than obtained clusters using GA.

Classification Task Using Cellular GP

Classification task or discovery knowledge in data base is an important part of data mining. Genetic programming (GP) is an important tool in this particular purpose. Moreover, cellular genetic programming model gives a fruit full result over other approaches in this particular. This model for the task of classification was developed by Alexandra Takac. Generally to extract a particular data from a large data set, we use SQL queries. In a system of information, a huge amount of data is stored of the organizations. But at the time of decision making, the particular data must be required. So, similar data should be stored in one field, another similar data set should be kept in another field and so on. Also, in now a day, most important thing is artificially decision making. In previous, we already observed that how GA is used in decision making. But here GP helps to take instant decision combining with cellular model of GA. In initial level, population is generated with the classification rules. These rules are represented with parse trees. After that three genetic operations- selection, crossover and mutation are performed. In this particular, selection process is Roulette wheel and that is same with GA selection. Simple crossover and mutation will occur among the nodes of parse tree. A fitness function is defined to evaluate the quality of solution iteration wise. As a result, after a finite number of iterations, new classification rules will be generated at the time of taking decision. So, more accurate result can be obtained over using only GA model.

Drug Recovery Using GP

Oral bioavailability is one of the most important implementation of genetic programming (GP) in the field of data mining. Drug discovery is a special division of oral bioavailability. The best GP application in this field was proposed by W.B. Langdon and S.J. Barrett in 2004. But before that many GP models were developed related to that. Before Langdon's proposal, GP was used in modeling experiments only in data mining area. Langdon's technique or Langdon's model of GP was used to combine different machine learning methods, such as- decision trees, naïve Bayes, artificial neural network etc. This method was used for human cryptochrome inhibition prediction and to find starting compounds in disease treatments drugs. Actually, both of feature selection and model building were done by using this specific GP scheme. The speciality of this GP scheme was development of single objective GP to multi objective nature. The basic requirements of GP are population, fitness function, selection, crossover and mutation. Here, initial population is generated with receiver's realizable operating characteristics. Fitness function is defined in the following:

$$Fitness = \frac{1}{2} True\ positive\ rate + \frac{1}{2} True\ negative\ rate \tag{5}$$

These two rates are individually single objective in nature with equal importance. Actually, this GP follows the weighted sum approach of multi objective optimization. So, the rules of selection, crossover and mutation are sequentially used maintain the architecture of single objective GP. After a finite number of iterations, we reach to a new structure of drug for a particular use. So, multi objective satisfaction is the positive side of using GP in drug discovery, specially obtained through this model.

Feature Construction Using GP

New feature construction is an interesting as well as powerful development of classification task using genetic programming. Actually, at the time of feature selection operation, mostly observe that due to inadequate of attribute corresponding to a set of data, information can't be extracted or decision tree can't be formed properly. To reduce this problem, Mohammed Ahmed Yahya Muharam proposed an excellent scheme in his PhD thesis in july, 2005. The main concept of his theme is reconstruction of feature space from the actual space using GP. So, it can be easily realized that actual feature space is the initial population for using GP. The next important thing for using GP is fitness evaluation function. In this particular case, four fitness functions are used. These are raw fitness, standardized fitness, adjusted fitness and normalized fitness. These four functions give four different indications. Raw fitness evaluates accuracy of classification or rate of error. Standardized fitness is used to select lower numerical value or better value iteration wise. Adjusted value or adjusted fitness is dependent on standardized fitness and it is defined by

$$a(i) = \frac{1}{1 + s(i)} \tag{6}$$

where $a(i)$ and $s(i)$ are adjusted fitness and standardized fitness of $i-th$ individual respectively. Normalized fitness is dependent on adjusted fitness in the following way:

$$n(i) = \frac{a(i)}{\sum_{k=1}^{M} a(k)} \tag{7}$$

Here $n(i)$ is the normalized fitness value of $i-th$ individual.

Depending on these fitness functions, reproduction will occur. After that, crossover and mutation procedure are simple like GA. But in this type problem, probabilities of crossover and mutation are very low. Termination will occur after getting the best score corresponding to best individual or maximum number of generations. As a result, the number of iterations or time complexity will be decreased due to the creation of new feature space from actual feature space for classification task.

AN OVERVIEW OF IMAGE PROCESSING

In the field of engineering science, image analysis is one of the most important things. More specifically, in electronics, computer science and IT fields, image handling and manipulation are required very much for various purposes. Image processing is the way of image handling, manipulation and making an image in different use. Especially, signal processing, computer graphics, computer vision etc. are particular application fields of image processing. Actually, image processing is not only processing of image files, but also processing of audio and video files.

Technically, an image is represented by a large set of image pixels. The different features of images are developed on different type pixel values, such as- RGB, entropy etc. Different type pixel values give different indications in different fields of the images. At the time of processing of audio and video files, audio and video frames are manipulated in different ways.

The most basic as well as important use of image processing is signal processing. Two major type image processing occur in this purpose. These are digital image processing and analog image processing. If the wave form of data is in analog form, then we can use analog image processing and if the wave form of data is in digital form, and then we use digital image processing. But digital image processing is more usable, because digital wave can't be affected very much with noise.

Now a day, signal processing and computer graphics are widely applicable. In present decade, the broad areas of image processing are television, photography, remote sensing, medical diagnosis, cyber security, industrial inspection, identification and so on. In this chapter, we choose those fields, in which GA and GP are used very much and give a fruit full result corresponding to that particular purpose. Also, in experimentation and result section, we will analyze the efficiencies of GA and GP in the particular fields of image processing.

APPLICATION OF GA IN IMAGE PROCESSING FIELDS

Genetic Algorithm (GA) is one of the most important techniques in different applications of image processing. Actually, manipulation of images, audios and videos depend on large quantity of data; more specifically on large number of different type image pixels. Therefore, to use different image features in different fields, GA should be a good tool. Generally to handle large scale optimization problem, GA gives an interesting as well as fruit full result. In this chapter, we consider four particular important applications of image processing, in which GA gives an excellent result. These are image enhancement and segmentation, face recognition, photo mosaic generation and steganography. Here, we discuss about these fields at first with a brief idea of each of these as well as analyze GA in the corresponding fields.

Image Enhancement and Segmentation Using GA

Image enhancement is basic requirement of us in different daily purposes and it is a part of image processing. Image segmentation is a related field of that. The main objective of image enhancement is to improve the quality of images. In many cases, we observe that a particular image can't be realized properly. In other words, the information within the image can't be extracted in correct way. To realize all the information properly, the particular image should be modified through enhancement. After enhancement, dynamic range of selected features will be increased. But there is no efficient method of image enhancement, through which every image quality can be improved always. There are a lot of ways for this purpose. All the methods are good in some particular purposes of image enhancement. The general image enhancement procedures are divided into four classes. These are point operation, spatial operation, transform operation and pseudo coloring. In point operation category, the techniques of enhancement are contrast stretching, window slicing, noise clipping and histogram modeling. Similar in spatial operation, the methods are median filtering, zooming, noise smoothing and LP, HP and BP filtering. Transform operation techniques are homomorphic filtering, linear filtering & root filtering and pseudo coloring methods are false coloring and pseudo coloring. Now, we will take short ideas about the methodologies of four image enhancement classes.

Point Operation

Point operation is a basic technique of image enhancement. Here, a pixel value is determined of the enhanced image depending on corresponding pixel value of input image using the following function:

$$S = M\left(r\right) \qquad (8)$$

i.e.

$$S = \begin{cases} 0 & if\, r < T \\ L-1 & if\, r > T \end{cases} \tag{9}$$

Here,

r : *pixel value of input image*
S : *Pixel value of output image*
T : *Defined fixed constant*

This transformation is called LUT transformation.

Spatial Operation

Spatial operation is another method of image enhancement. Here, each pixel value of input image is replaced by average of the neighborhood pixels using the following formula:

$$v(m,n) = \sum_{(k,l)\in w} \sum a_{k,l} y(m-k, n-l) \tag{10}$$

Here, $y(m,n)\, and\, v(m,n)$ are the images before and after enhancement respectively. Various methods of spatial operation are developed depending on the no. of neighborhood pixels of a particular pixel of input image.

Transform Operation

Transform operation is some developed method of image enhancement. In this methodology, the domain of input image is changed through a mathematical transformation. The most popular transforms, used in this purpose are Discrete Cosine Transform (DCT), Discrete Fourier Transform (DFT), and Fast Fourier Transform (FFT) etc. Therefore, the common format of transform domain is,

$$g(x,y) = f(x,y) * h(x,y) \tag{11}$$

Here,

f : *pixel value before enhancement*
h : *transforming matrix value*
g : *pixel value after enhancement*

Pseudo Coloring Operation

Pseudo coloring is assigned the gray pixels to some different color depending on particular criterion. It is generally used for proper visualization of human. The major technique of this operation is giving different false colors corresponding to different intensity levels of the images. As a result, the separate intense portion can be visualized properly as well as information of an image can be realized fully.

Image segmentation is partitioning of image to the meaningful structures. It is required in visualization, image analysis, object representation and more tasks of image processing. Obviously, the segmentation process is dependent on image pixels. But it is the fact that the pixels through which the particular object can be identified, that should be known to the user previously. There are many methods developed from many years ago like image enhancement. These methods are categorized into five phases. These are threshold based, edge based, region based, clustering based (image mining) and matching based segmentation. Among them, first two major image segmentation methods are used very much and these are described in the following with proper specifications.

Threshold Segmentation

Thresholding is one of the most basic operations of image segmentation. Here, pixels of an image are grouped or divided through intensity values. Global thresholding with an appropriate threshold T is defined in the following:

$$g\left(x,y\right)=\begin{cases}1, & if\, f\left(x,y\right)>T\\ 0, & if\, f\left(x,y\right)\leq T\end{cases} \tag{12}$$

So, all the pixel values will be converted in binary digits 0 and 1 using global thresholding. In similar way, variable thresholding and multiple thresholding are developed for different images with specific threshold values.

Edge Segmentation

Edge segmentation is also a very much applicable segmentation approach. Using this segmentation, a 3D object is segmented into a lot of 2D planes. Actually, the lines or edges of an image are obtained by keeping same positions as in the input image. Therefore, the structure as well as shape of particular image can be realized fully. It is very much applicable in the field of design analysis.

GA-Based Image Enhancement and Segmentation

For both of image enhancement and segmentation, GA is used in different ways. Among all of the GA based techniques for image enhancement and segmentation, we will discuss a particular scheme, which gives more fruit full result than other approaches. This particular scheme was developed by Miss. Komal. R. Hole, Prof. Nitin D. Shellokar and Prof. Vijay S. Gulhane. In their scheme, initial population for using GA is generated using the edges of a particular input image. To create this, the range of gray levels is 0 to 255. Obviously, this is a large scale optimization. For the initial phases of reproduction, i.e. for selection, a fitness function is required. Here, fitness function is sum of intensities of the edges of input image. Probability of crossover and mutation are dependent on particular image type as well as no. of generations is not specific always. But, in most of the cases it requires a less no. of generations to improve the quality of image comprising to others.

This technique is also applied for image segmentation purpose. Here, selected types of segmentations are based on grayscale, texture, motion and depth. The major categories of segmentation algorithms are dependent on edge, region and clustering based. So, at the time of creation of initial population, any one of the feature set can be taken corresponding to particular purpose. After that, fitness function is estimated by segmentation quality. Next genetic operators; selection, crossover and mutation criterion are same as used in image enhancement purpose.

Face Recognition Using GA

Face recognition is a part of image segmentation. More specifically, it is an important application of image segmentation. For verification and identification of a person's face, the face recognition system is required. Many methods are developed for this purpose. But GA gives a good result here. In this section, we will discuss a particular face recognition scheme, developed by Sarawat Anam and her co-researchers in 2009. Actually, they made a hybrid scheme of face recognition combining GA and neural network. The architecture of their proposed scheme is defined in the Figure 1.

Among all the phases, GA and BPN are used separately only in recognition part. For recognition part, GA gives an excellent result over other previous approaches of face recognition. Here, simple selection, crossover and mutation are applied. We will analyze its efficiency in experiment part.

Photo Mosaic Generation Using GA

The term photo mosaic comes from the term photographic mosaic. The general architecture of photo mosaic is that a picture is divided into small tiles with equally sized and each of them is replaced by a different photograph and after replacement, it matches with the target photo. Most of the cases, it is used

Figure 1. Architecture of face recognition system
Developed by Sarawat Anam and her co-researchers in 2009.

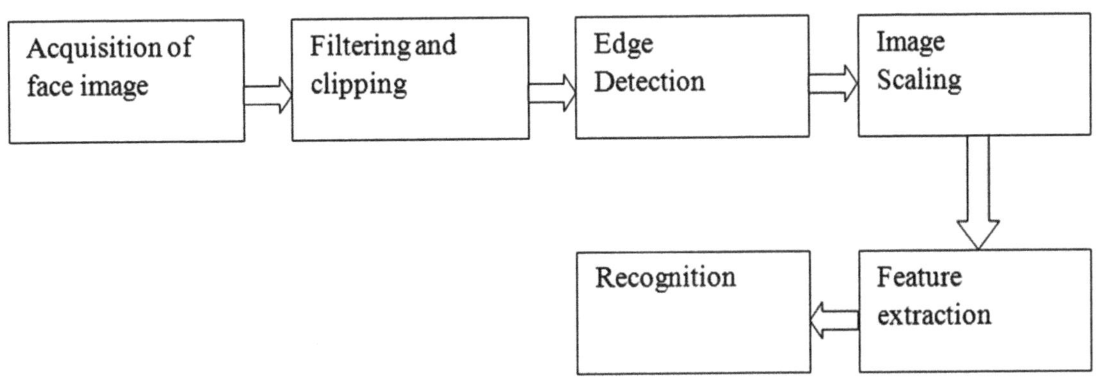

for authentication purpose. To do this, many techniques or methods are already developed previously. But, here randomized search GA gives a good result over other approaches. One of the most usable photo mosaic scheme was developed by Vic Ciesielski, Shahrul Badariah Mat Sah, Marsha Berry and Daryl D' Souza in 2008. Here, the initial population is generated by a canvas of m no. of pixels in each row and n no. of pixels in each column. So, there are m chromosomes in initial population and each takes n no. of genes. So, the no. of genes in the population is $m \times n$. The fitness evaluation function for this purpose is,

$$fitness = \sum_{i=1}^{K}\sum_{j=1}^{L}\left|target\left(i,j\right) - individual\left(i,j\right)\right| \tag{13}$$

Here, the symbols take the usual meanings. Iteration wise, our objective is to minimize this fitness function. Actually, to insert new tiles of another image corresponding to the position of actual image, mutation operation is required. In a single chromosome, at most 9 mutations are permeated in all iterations. According to their scheme, elitism rate is 0.05, crossover rate is 0.7 and mutation rate is 0.0001. Here, 2 point crossover is used in all iterations. The general tile size is 5×5 and target size is 120×100 pixels. As a result, a good photo mosaic is formed after a finite no. of iterations.

Steganography Using GA

Steganography is the way of secret communication through wireless network. Maintaining the basic features of information security, i.e. confidentiality, integrity and availability, steganography is built up. In this type of technology, the secret information are transferred to the intended receiver(s) from sender side using some cover medium, such as- image, audio or video with minimum distortion. In this particular purpose, image pixels are generally used for exchanging secret information among two or more intended parties. The diagramed architecture of steganography is shown in the Figure 2.

Figure 2. Diagramed architecture of steganography

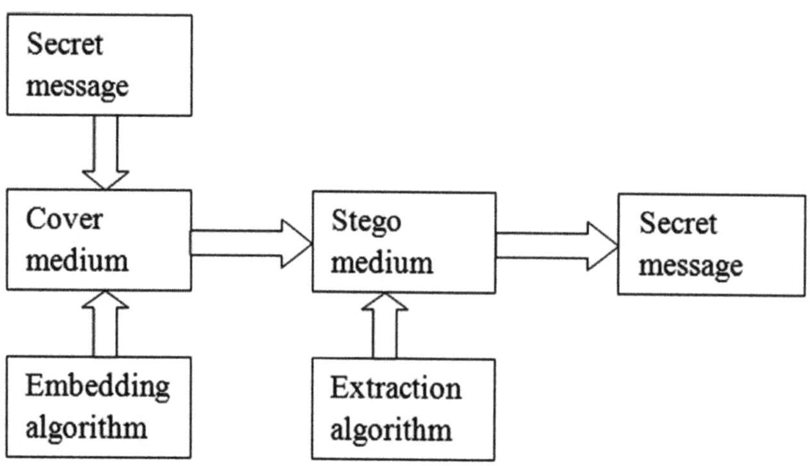

Generally, there are two types of steganography occur. These are spatial domain and transform domain. In spatial domain steganography, information is inserted directly to the image pixels. But in transform domain steganography, at first pixel values are transformed to another specific domain and secret information is inserted in the transformed domain. The specific mathematical transformations used in transform domain steganography are DCT, DWT, DFT, FFT etc. In the case of audio and video cover mediums, audio and video frames are used for inserting secret information. Many methods were developed previously in both of spatial domain and transform domain steganography. Actually, minimum distortion of cover medium is one of the most important factors. Here are two important parameters to measure the distortion and these are peak signal to noise ratio (PSNR) and mean squared error (MSE). The workings of MSE and PSNR are defined in the following:

$$MSE = \frac{1}{pq}\sum_{i=0}^{p-1}\sum_{j=0}^{q-1} I\left(i,j\right) - K\left(i,j\right) \tag{14}$$

$$PSNR = 20\log_{10}\frac{Max_i}{\sqrt{MSE}} \tag{15}$$

Here, I and K are original and stego image respectively. Therefore, our objective in steganography is to maximize PSNR and minimize MSE. In this type of problem, GA helps us very much. Here, population is created with image pixels. In this chapter, we will discuss two important image steganography schemes using GA, where GA gives a good PSNR as well as the methods is safe from different unintentional attacks.

Among them, the first one is spatial domain steganography using GA and another one is transformed domain (DCT based) using GA. The first one was developed by S. Talebi, M. Saleimanpour and H. Azadi Motlag. In that scheme, the population is created with an arbitrary collection of image pixels from an image. Each pixel is taken as chromosome. After that, simple roulette wheel selection, one point crossover and random mutation are operated iteration wise. In general way, it can be said that simple operations of real coded genetic algorithm are applied here. Insertion of secret information occurs through the mutation phase. As a result, we can get a good PSNR or good quality of stego image. So, it is very much protected from visual attack. But since it is a spatial domain steganography, then it's secrecy can be broken through pseudo random number generator (PRNG). To reduce this problem, we use transform domain steganography. In this chapter, we consider one of the finest GA based technique using DCT. This scheme was proposed by J. K. Mandal and Amrita Khamrui in 2013. Here, GA is applied on the DCT values of image pixels. The architecture of their scheme is shown in the Figure 3.

Input: Cover image and secret information.
Output: Stego image.

Here, we will discuss only authentication part, where GA is applied. Crossover operation is very much important of the scheme. After IZT, initial population is created with 32 bit mask. Then in row wise, two pixels are crossed over and LSB of two consecutive pixels are swapped. As a result, at the

Figure 3. Architecture of described transform domain steganography using GA

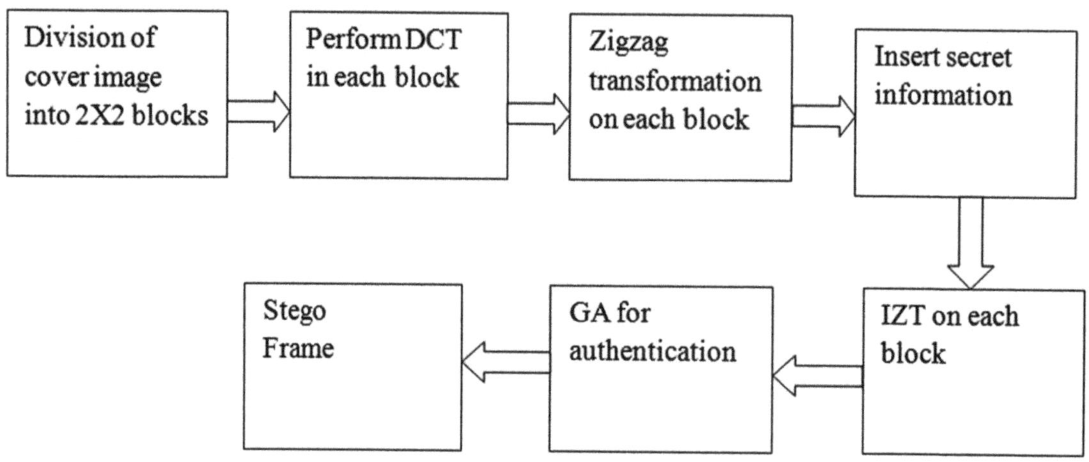

receiver end integrity of information can be checked easily. Also, due to use of transform domain, it is protected from unintentional attack PRNG.

APPLICATION OF GP IN IMAGE PROCESSING FIELDS

In the fields of image processing, GP gives some better solutions over GA. Actually there are some particular fields of image processing, where GP gives better results over GA. Here, we have selected four important applications of GP in image processing and these are development of classification of image pattern, detection of objects, reducing computational cost of execution of kernel on Graphics Processor Unit (GUI) and generation of photo mosaic. In these fields, GP gives well defined systematic procedures to handle these type problems. The particulars of these using GP are discussed in this section separately.

Development of Classification of Image Patterns Using GP

In image mining, image pattern is very much important thing. To develop image pattern artificially, various approaches are developed indifferent times. Among them, we will discuss a novel approach of properly image features arrangement using GP based transformation. The general architecture of this is extraction of a moment value from the GP transformed image helps very much for classification task. Actually this approach was implemented for detection of cell images dystrophy indicating inclusions. In other words, it was applied for classification task of medical image. General procedure of image classification or image recognition is selection of some particular features of the feature space to minimize dimensionality. So, the most important factor in this purpose is selection of the standard features in the feature space. At first, for breaking the image to numeric values, intensity $f(p)$ is taken corresponding to each pixel $(p_x, p_y) \in I$. Then using these, a moment value is generated. The most basic usable moments are geometric moments and central moments. The working formula of geometric moment is,

$$M_{mn} = \sum_{p \in I} p_x^{\ m} p_y^{\ n} f(p) \tag{16}$$

In similar way, the working formula of central moment is,

$$C_{mn} = \sum_{p \in I} (p_x - x_1)^m (p_y - y_1)^n f(p) \tag{17}$$

Here, (x_1, y_1) is the centroid and that is equal to (M_{01}, M_{10}). Depending on these moments and Shannon entropy function, the classification occurs. This is one of the oldest ways of classification. The next better technique was feature creation. An artificial feature is created from the original feature space and using this, we do not process of classification. After that, GP developed the feature selection operation. This technique was developed by Wolfgang Banzhaf, Taras Kowaliw, Simon Harding and Nawwaf Kharma in 2009. Actually their model was developed by combining parallel distributed GP and linear GP. Using the different level of intensities, a graph is formed with image features. This graph is formed in such a way that the graph has two different connections of each node to two previous nodes (x and y). Here, a particular function x/y is used and this is sound due to the condition that if the denominator is in neighborhood of 0, then the function returns the value 1. The inputs are taken through square neighborhood and the final node represented value is output. As a result, we can get a new form of classified image pattern, obtained from previous other techniques.

Detection of Objects Using GP

Quality improvement of the corrupted images is an important requirement of us as well as it is an important application of image processing field. Cartesian Genetic Programming (CGP) helps us very much in this purpose. Previously, we observed that GA gives a good result in image enhancement. But, GP gives a better output over GA. The criterion required for this purpose is performing the mapping of genotype- phenotype. At first, we start from the output node and then we will follow the connections from inputs to output. Next, CGP is used and the final node of CGP-IP is taken as the output of genotype. In the phenotype part, program input is the input image and after that, a forward parse is performed. A threshold operation occurs in the case of binary classification. This combining operation of phenotype and genotype brings an efficient output of actual object detection.

Application of GPU Using GP

In now a day, use of parallel processors in graphics card is a new attraction. This is used to increase evaluation speed of execution as well as to decrease computational cost. For this purpose, an extended version of GP, used in filtering of image, is used. This is also a special kind of filtering. This technique was developed by Simon Harding and Wolfgang Banzhaf. Actually, here filtering process is done iteration wise by using a GP function. Here, initial population is generated with a set of consecutive instructions of a processor. Then each individual of the population takes a counter that counts the no. of iterations of filtering. The value of counter lies between 1 and 5. Moreover, sharing of sub trees using GP is very much important here. As a result, to perform a particular operation in the processor, space requirement will be decreased as well as space complexity will be decremented corresponding to a problem with finite no. of inputs.

Photo Mosaic Generation Using GP

The problem definition and purposes of photo mosaic generation was already described in the GA application section. Here, we only concentrate on GP application part of photo mosaic generation. We will discuss the GP technique, developed by Shahrul Badariah and his research team in 2008. The fitness evaluation function is same as the function, used in GA application. The population is generated through collection of selected tiles from an image. Each tile is associated with an index. Then to create phenotype, first C tiles are collected for photo mosaic operation. After that a simple GP is carried out with some fixed rate of parameters. Here, elitism rate is 0.05, crossover rate is 0.70, mutation rate is 0.25, maximum and minimum depths are 7 and 2 respectively and it will occur approximately 1000 generations. Actually, the system is that, if the no. of terminals of a particular parse tree is less than C, then that parse tree will be cancelled and another parse tree will be taken for evaluation. As a result, better photo mosaic will be created over GA. Some experimental results corresponding to photo mosaic generation are shown in the experiment section.

EXPERIMENTS AND RESULTS

In this section, we will observe some experimenting results of applications of GA and GP in data mining and image processing. The particular GA and GP based applications with corresponding approaches are already described previously. Now, corresponding to each particular field of data mining and image processing, we will observe at least one experimenting result in this section.

Application of GA in Data Mining Fields

Here, we selected four GA based sub fields of data mining and these are stock market analysis, fraud detection, risk management and sports analysis.

Stock Market Analysis Using GA

Here, GA is applied to take decision artificially before investment in share market for share holders. The particular described approach was tested in some cases of stock exchange in Australia. After giving all present information in the decision support system, it gives buy and sells signals corresponding to particular ranges. Also, the main interesting thing is less computational cost. Time analysis says that GA takes approximately 10 seconds to get particular decision, whereas time requirement of greedy algorithm is 10 hours for obtaining the same decision.

Fraud Detection Using GA

We have discussed about two methods on detection of frauds through payment system of credit card using GA. In the proposed technique of K. Ramakalyani, fraud detecting system is developed on general 20 basic attributes (mentioned previous) and compared using GA among the information database and transaction data base. Here, an experiment is shown with the parameters card Id, current balance, average bank balance, CCAge, Loc, ODT, LocT, Auth, CU and AmtT. Survey result says that every type of fraud, based on usage frequency, usage location, overdraft, book balance and average daily spending are evaluated separately in correct way. Another technique, we have discussed previously, was developed under supervision of Satvik Das. This is the fraud detection at the time of online purchase. Here, the co-ordinates of search space are compared with actual stored data set. According to specified example in this paper, there was only one fraud in the block of the space and that was evaluated easily.

Risk Analysis Using GA

In data mining area, one of the most important risk analysis is the analysis of risk factor at the time of decision making by the decision support system. In the methodology of Alireza Tamjidyamcholo, the GA based application is very much interesting due to requirement of less no. of iterations. They took a sample, where a fitness function was constructed with 4 variables and no constraints. The correct decision was obtained only after 51 iterations. In other side, to analyze the crucial factor of human involvement in decision making purpose, the technique of A Sumithra is very important. Here, multi objective GA shows that human factor is dependent on three specific facts and these are competency, experience and motivation. These are prioritized with different percentages according to different situation.

Sports Analysis Using GA

For various purposes of decision making of the games, GA is used in different approach. In the proposed method of Vasudha Sarda, a GA based technique was developed for football team selection. Through the GUI, the inputs are taken with the information- no of forwards, no of midfielders, no of defenders and obviously the budget of a particular team. Their technique gives a formed team as output maintaining good optimality. In the general formation of the last technique of the game scheduling was developed by A. B. Goldberg and that was highly constraint problem. In their experimentation, the best, worst and random elite selection rates are 0.2, 0.4 and 0.02; the best, worst and random crossover probabilities are 0.2, 0.6 and 0.0 and similar sequences of mutation probabilities in the iterations are 0.1, 0.5 and 1.0. It is observed that, for a problem of population size 50, 1000 generations are required. Therefore, this methodology gives a better performance and requires low computational cost over previous approaches.

Application of GP in Data Mining Fields

In this chapter, we consider four important sub fields of data mining, namely protein interaction, classification task, drug discovery and feature construction. All these cases, GP gives a good result over some other approaches. The experimental results of the previously described techniques of the particular sub fields are analyzed in this section.

Protein Interaction Using GP

For protein interaction purpose, we consider the methodology of M.L. Raymer. This particular technique is implemented in water displacement testing and it was training and unbiased. The training set '1700 waters' is used as data set and both of GA and GP are implemented there. Using GA, 121 site waters are predicted out of 157, whereas 124 are predicted using GP. In other words, the accuracy in GA is 77.1% and the same in GP is 79%.

Classification Task Using Cellular GP

For classification task, the approach of Alexandra Takac is considered. This is tested in the project STATLOG and other some popular approaches are also tested here. Average train costs and test costs are evaluated separately. Performances of these schemes are shown in Table 1.

Here, GP is used with population size 400 and 200 epochs are used in this purpose.

Drug Recovery Using GP

In the field of drug discovery, the methodology of W.B. Langdon is chosen. It is implemented on two different set and these are human and rat data records. In the first data set, there are 481 human records and the number of randomly drawn compounds is 321. The second data set consists of 2013 rats' record and 1342 compounds are chosen randomly. In both of cases, GP gives good results over other approaches.

Feature Construction Using GP

In the case of GP analysis for feature construction, the proposed scheme of Mohammed Ahmed Yahya is selected. This is implemented in different data sets and error rates are evaluated. The types of errors are Information Gain (IG), Gini Index (GI), combination of IG and GI, Chi^2 and Bias check. From the testing results, it is observed that the performances of C5, CHAID, CART and ANN are efficient in the particular data sets and these are evaluated properly using GP.

Application of GA in Image Processing Fields

GA is very much useful in the fields of image processing like data mining. In this chapter, we have explained about four GA based sub fields and these are image enhancement and segmentation, face recognition, photo mosaic generation and steganography.

Table 1. Experimental result of German credit set

Algorithms	Average Training Cost	Average Test Cost
Alloc 80	0.597	0.584
KNN	0	0.694
Backprop	0.446	0.772
Castle	0.582	0.583
Cellular GP	0.472	0.539

Image Enhancement and Segmentation Using GA

To check efficiency of GA in image enhancement and segmentation, we have taken the mcthod of Miss. Komal. R. Hole. This technique is discussed in previous fully. This approach is implemented in an image of car for enhancement and an image of house for segmentation. The results of these two show the abilities of GA in the corresponding purposes.

Face Recognition Using GA

Face recognition is an important application of GA. Here, the selected scheme is developed by Sarawat Anam. This scheme is broadly used for face verification purpose. It is implemented in 115 different cases and number of successful recognized face is 95. So, in 82.61% cases faces are recognized successfully.

Photo Mosaic Generation Using GA

For photo mosaic generation, we have selected the GA based technique, provided by Shahrul Badariah Mat Sah. It is implemented in some particular cases with 1000 generations. The mean of five runs give a fruitful result. Elitism rate is fixed to 0.1 and five runs are carried out again. Then good photo mosaics are generated.

Steganography Using GA

In this part, we have taken two approaches- first one is GA based spatial domain approach and second one is transformed domain. First one is developed by S. Talebi and second one is developed by J.K. Mandal. The first one is implemented in six different images with a standard length secret message. The average PSNR value of the stego images is near about 53. The second one is implemented in 7 different stego images and the mean of PSNR of the stego images is approximately 44. So, from the PSNR results, it is realized that GA maintains good quality of stego images.

Application of GP in Image Processing fields

In this chapter, we have considered four popular sub fields of image processing, in which GP give good results. The selected sub fields are classification of image pattern, detection of objects, GPU development and photo mosaic generation.

Image Pattern Classification Using GP

Here, the methodology of Taras Kowaliw is taken in our consideration. In the experiment of the proposed scheme, the data set Cells DB is taken and it is 24 features set. The different classifiers 5-NN, 1-NN and decision tree are implemented to compare efficiency of the proposed scheme over others. The considered parameters are sensitivity specificity (SS), true positive rates for healthy cells (TRP (H)) and true positive rates for sick cells (TRP (S)). Among the classifiers, TEF is used in 1-NN and decision tree. SS, TRP (H) and TRP (S) of 5-NN without transformation are 0.580, 0.774 and 0.750 respectively. Similarly, these values of 1-NN with TEF are 0.776, 0.878 and 0.873 respectively. Also, decision tree with TEF give these results are 0.801, 0.912 and 0.872 sequentially. So, using GP transformation, 38% improvement is possible for classification over other approaches without transformation.

Object Detection Using GP

For object detection purpose of image processing, we have taken a GP based approach developed by Simon Harding, J. Schmidhuber and J. Leitner. Here, Cartesian Genetic Programming (CGP) is used. In general purpose, fitness function is developed as the difference between the target image and output image through evolved program pixel wise. For binary classification, fitness is calculated by *1-|MNC|*, where

$$MCC = \frac{TP \times TN - FP \times FN}{\sqrt{(TP + FP)(TP + FN)(TN + FP)(TN + FN)}} \tag{18}$$

The specifications of the symbols are

TP: Number of true positives
FP: Number of false positives
TN: Number of true negatives
FN: Number of false negatives

This method is implemented in a corrupted image with pepper and salt noise and evolved filtering gives a clear image reducing the noises.

GPU Performance Development Using GP

In this part, the scheme of Harding and Banzhaf is selected for our analysis. In the methodology analysis, it is observed that using GP, computational cost is reduced at the time of using graphics card. Different filters are used in this purpose in various ways. The obtained lowest error in Dialate filter is 0.57, Dialate 2 is 5.84, Emboss is 3.86, Erode is 0.56, Erode 2 is 5.70, Motion is 2.08, Neon is 1.32, Sobel is 8.41, Sobel 2 is 1.70 and unsharp is 5.85. So, in all the approaches, GP is able for reducing multiple programs to a single program. As a result, computational cost is reduced.

Photo Mosaic Generation Using GP

In this purpose, the scheme of S.B. Mat Sah is really very much interesting. In that paper, a comparative study occurs between GA and GP in the purpose of photo mosaic generation. In the analysis part of GA applications, the response of GA in this purpose is already analyzed. Here, the importance of GP in this particular purpose is checked. In some same type images, GA and GP are applied both for photo mosaic generation. The previous approach of this scheme generated tree with 1093 nodes, 364 tile joints and parse tree with 729 tiles. The general structure says that less number of nodes reduces the complexity of the problem. For this, this scheme was developed. This scheme consists of a tree with 730 nodes, 243 tile joint and 487 tiles in each parse tree. So, this GP based approach is applicable properly for photo mosaic generation.

FUTURE RESEARCH DIRECTION

Data mining and image processing, both are large fields of machine learning. So, majorly in computer science and IT, these are widely applicable. GA and GP are randomized searching approaches of soft computing, moreover of soft optimization. In this chapter, only four important type applications from each section are selected and one of the best GA or GP approach corresponding to each is reviewed and compared with other non GA or non GP approaches. But it can be extended in future with review of other particular applications using GA and GP. Also, more than one scheme of the particular subfields will be discussed with a comparative analysis among them in future. Only two major fields, data mining and image processing are considered here. But there are so many fields and corresponding subfields, which are the excellent applications of GA and GP, are not considered here. These will be taken in future also. Another important fact is that in this chapter the gradually improvement of GA and GP approaches in data mining and image processing are not described. This will be kept as future scope of our considered efficiency analysis.

CONCLUSION

In this comparative analysis, a critical review of GA and GP are explored in data mining and image processing. Here, efficiency of GA and GP are analyzed in the particular fields comparing with other approaches. Actually, GA and GP give fruitful result in large scale problems. Data mining is the process of handling of a huge number of data and similarly, image processing is the manipulation of large number of image pixels. So, for both of them, GA and GP are two suitable approaches. After analyzing all the facts, it is observed that GA is applied in a huge number of fields with proper efficiency. GP is also good optimization tool in these fields. In some times, GP gives better result over GA. But GP is not widely applicable in different application fields, comparing with GA. But it is concluded that these two are soft computing approaches and obviously, it is observed that the efficiencies of them are dependent on particular problems and situations.

REFERENCES

Anam, S., Islam, M. S., Kashem, M.A., Islam, M.N., Islam, M.R., & Islam, M.S. (2009). *Face Recognition Using Genetic Algorithm and Back Propagation Neural Network.* Paper presented at International Multi Conference of Engineers and Computer Scientists, IMECS 2009, Hong Kong, China

Chatterjee, A. (2015). A Comparative Analysis of Three Important Data Hiding Schemes in Steganography. *International Journal of Innovative Trends In Engineering*, *11*(2), 61–65.

Chatterjee, A. (2015). A Survey on Different Digital Media Based Steganography Techniques Using Genetic Algorithms. *International Journal of Innovative Research in Computer and Communication Engineering*, *3*(12), 12990–12998.

Elkaffas, S. M., & Toony, A. A. (2006). *Applications of Genetic Programming in Data Mining*. Paper presented at World Academy of Science, Engineering and Technology.

Goldberg, A. B. (2003). *Highly Constrained Sports Scheduling with Genetic Algorithm* (Bachelor's dissertation). Amherst College.

Han & Kamber. (2006). Data Mining: Concepts and Techniques (2nd ed.). Urbana, IL: Elsevier

Harding, S., & Banzhaf, W. (2008). *Genetic Programming on GPUs for Image Processing*. Paper Presented in First International Workshop on Parallel and Bioinspired Algorithms. doi:10.1504/IJHPSA.2008.024207

Harding,, S., & Leitner,, J., & Schmidhuber. (2013). Cartesian Genetic Programming for Image Processing. *Genetic and Evolutionary Computation: Genetic Programming Theory and Practice*, *10*, 1–17.

Hole, K. R., Gulhane, V. S., & Shellokar, N. D. (2013). Application of Genetic Algorithm for Image Enhancement and Segmentation. *International Journal of Advanced Research in Computer Engineering & Technology*, *2*(4), 1342–1346.

Khamrui, A., & Mandal, J.K. (2013). *A Genetic Algorithm Based Steganography using Discrete Cosine Transformation (GASDCT)*. Paper presented at International Conference on Computational Intelligence: Modeling Techniques and Applications (CIMTA 2013). doi:10.1016/j.protcy.2013.12.342

Kowaliw, T., Banzhaf, W., Kharma, N., & Harding, S. (2009). *Evolving novel image features using genetic programming based image transforms*. Paper presented at IEEE conference on Commerce and Enterprise on Evolutionary Computation. doi:10.1109/CEC.2009.4983255

Langdon, W. B., & Barrett, S. J. (2004). *Genetic Programming in Data Mining for Drug Discovery. In Evolutionary Computing in Data Mining* (pp. 211–234). Springer.

Lin, L., Cao, L., Wang, J., & Zhang, C. (2007). *The applications of Genetic algorithms in Stock Market Data Mining Optimisation*. Paper presented at NSW Australia on Capital Market CRC, Sydney, Australia.

Marmelstein, R. E. (1997). *Application of Genetic Algorithms to Data Mining*. Paper presented at Modern Artificial Intelligence and Cognitive Science Conference.

Mat Sah, S. B., Ciesielski, V., D'Souza, D., & Berry, M. (2008). Comparison between Genetic Algorithm and Genetic Programming for Photo mosaic Generation. In *LNCS 5361* (pp. 259–268). Springer.

Muharam, M. A. Y. (2005). *The Application of Genetic Programming for Feature Construction in Classification* (Unpublished doctoral dissertation). School of Computer Science, University of East Anglia.

Paulians,, M., & Usinskas, A. (2007). A Survey of Genetic Algorithms Applications for Image Enhancement and Segmentation. *Information Technology and Control*, *36*(3), 278–284.

Ramakalyani, K., & Umadevi, D. (2012). Fraud Detection of Credit card Payment System by Genetic Algorithm. *International Journal of Scientific & Engineering Research*, *3*(7), 1–6.

Raymer, M. L., Punch, W. F., Goodman, E. D., & Kuhn, L. A. (1996). *Genetic Programming for Improved Data Mining: application to the biochemistry of protein interactions*. Paper presented at 1st annual conference on genetic programming.

Sarda, V., Sakaria, P., & Deulkar, K. (2015). Football Team Selection Using Genetic Algorithm. *International Journal of Engineering and Technical Research*, *3*(2), 153–156.

Schoreels, C., Logan, B., & Garibaldi, J. M. (2014). *Agent based genetic Algorithm employing financial Technical analysis for making trading decisions using historical equity market data*. Paper presented at International Conference on Intelligent agent Technology.

Soleimanpour, M., Talebi, S., & Azadi Motlagh, H. (2013). A Novel Technique for Steganography Method Based on Improved Genetic Algorithm optimization in Spatial Domain. *Iranian Journal of Electrical & Electronics Engineering*, *9*(2), 67–75.

Stefanowski, J. (2009). *Data Mining- Clustering. Presented at SE Master Course*. Poznan, Poland: Poznan University of Technology.

Sumithra,, A., & Ramraj, E. (2012). A Genetic Algorithmic Approach to the Identification of the Crucial Human Factors Involved in the Development of Secure Software. *International Journal of Emerging Trends & Technology in Computer Science*, *1*(2), 256–258.

Takac, A. (2004). Cellular Genetic Programming Algorithm applied to classification task. Comenius University.

Tamjidyamcholo, A., & Al-Dabbagh, R. D. (2012). Genetic Algorithm Approach for Risk Reduction of Information Security. *International Journal of Cyber-Security and Digital Forensics*, *1*(1), 59–66.

Vats, S., Dubey, S.K., & Dubey, N.K. (2013). *Genetic Algorithms for credit card fraud detection*. Paper presented at International Conference on Education and Educational Technologies.

Zaiane, O. R. (1999). Introduction to Data mining. In *CMPUT690, Principles of Knowledge Discovery in Databases* (pp. 1–15). University of Alberta Publishing.

Zaki, & Meira, Jr. (2014). Data mining and Analysis. Fundamental Concepts and Algorithms. New York: Cambridge University Press.

KEY TERMS AND DEFINITIONS

Artificial Intelligence: Intelligence of machine or machinery tools to do particular task perfectly and quickly.

Crossover: Interchanging one (or more) portion(s) of chromosomes between a pair, selected arbitrarily from a population.

Genotype: Full hereditary information of organism, used in GA and GP.

Machine Learning: Study on capability of learning without explicating programs.

Mutation: Genetic operator for maintaining genetic diversity of population, i.e. changes on genes from one generation to another.

Phenotype: Actual observed properties of an organism, used in GA and GP.

Selection: Choosing individual chromosomes (or nodes) from a population for next generation of GA (or GP) using particular operator.

Chapter 17
Development of Hybrid Cellulose Bio Nanocomposite From Banana and Jute Fiber

Ayush Rathore
Maulana Azad National Institute of Technology, Bhopal, India

Mohan Kumar Pradhan
Maulana Azad National Institute of Technology, Bhopal, India

ABSTRACT

In the era of globalization and industrialization the concern is limited only in development, without taking the environment into consideration this leads to global warming and big ecological changes in recent year. The material like Synthetic materials used in many applications due to ease of fabrication and higher strength, but the major disadvantage with it is, its neither recyclable nor bio-degradable. Therefore, the researchers develop a new material and technique for the sustainable development. A lot of researches were carried out in the reinforcing potential in the polymer matrix composite, reinforcing can be of two kinds synthetic and natural fiber. Natural fiber is gaining importance in the last decade due to its ecofriendly nature and does not leave carbon foot print, for better utilization of banana and jute fiber for making value added products. Hence, in this work the objective is to develop a new class of hybrid nano-materials from natural fiber such as banana and jute fiber. This chapter sees an opportunity of enhancement of interface property.

OBJECTIVE

- To develop a new class of material using banana jute as a reinforcement with the epoxy as a base material in construction, automobile segment such as door panels, railing, window, frames, parts of engine such as insulators which can resist various stresses.
- To prepare nano-composite from banana and jute fibber in epoxy as a matrix, having good strength and other mechanical properties.

DOI: 10.4018/978-1-5225-2440-3.ch017

- To characterise the banana, jute, fiber based epoxy nano-composite, to analyse various properties such as surface morphology, mechanical and chemical properties.
- To analyse effect of various process parameter on strength of nano-cellulose.
- To analyze the change in material properties with different weight percent of banana and jute nano-particles. The variation in mechanical properties are studied and analyzed. Here, the tensile strength has calculated by universal testing machine, impact strength has calculated by pendulum impact tester and flexural strength has calculated by universal testing machine with flexural test arrangement of the specimen. Then the treated and untreated specimens are analyzed and compared through Scanning Electron Microscope to study about its adhesion between fiber and resin matrix and surface morphology by AFM.
- A new class of differentiation can be analyzed, materials is designed with fibers as a reinforcing, same fiber as a particle reinforcing to compare the results. Found that nano-cellulose can be used as an excellent reinforcing agent in biodegradable polymer systems due to its high surface area, unique morphology and mechanical strength. This may be suitable for automotive applications.

PROPERTY AND STRUCTURE OF BANANA FIBER

It is outstanding to notice that natural fibers like jute, coir, banana, sisal, etc. are richly available in developing countries like European country, Thailand and a few of the African countries, but are not optimally utilized. The different mechanical Property of natural fiber as shown in (Table 1) (Kalia, Kaith & Kaur, 2009). At present, these fibers are measured used for the assembly of yarns, ropes, mats and matting in addition as in creating fancy articles like decoration, table mats, purses and purses. Fibers like cotton, banana and pineapple are utilized in creating material additionally to being used within the paper trade. its a widely known in controvertible fact that banana is one in every of the earliest cultivated plants within the world. Banana-trees turn out usually thirty massive leaves (almost 3 m long and 35-65 cm wide) (Sakthivel and Ramesh, 2013). The micrograph of the longitudinal section and cross section of the banana fiber strands were taken the cross sectional space of the banana fiber was investigate d by victimization optical light beam instrumentality, and it absolutely was found to be zero, 3598 MM^2. The plot of stress vs. proportion of strain for banana fiber is more or less linear; with a stress value of around 570 MPa, once the sharp of the strain is there. A truck model Manaca was developed and tested by, victimization banana fiber and epoxy resin (Pickering, Efendy & Le, 2013). However, some special and significant panels were made from hybrid composites of glass/banana fibers. The vehicle went through a few years of road performance tests and provided wonderful results.

PROPERTY AND STRUCTURE OF JUTE FIBER

It is produced from plants of the genus (Cor-chorus), as shown in (Figure 1). It is one of the cheapest natural fibers and is presently available in the highest production in China, India, Bangladesh. Different researchers investigate the jute fiber properties with its dynamic and static properties and its chemical modification. Jute is decortified from there source of fiber and treated chemically and finally alkali treated and fiber are separated for better dispersion of the fiber in base material (Hossain, Islam, Vuurea, Verpoest, 2013). The treatment should be for short time to restore the properties of hemi-celluloses and

Figure 1. Jute fiber

for improved the crystalline properties a little longer treatment is required for better interface property. The tensile property of the jute fiber in the direction of load increases the elongation up to 50%.

Jute fiber reinforced with natural polymer shows a good adhesion properties with the thermoset polymers and creep and dynamic behavior. The various environment effect such as UV rays, degradable. The jute fiber are tested for durability, thermal properties, Eco-design of the specimen, magnetic properties (Chandramohan and Marimuthu, 2011). Jute fiber reinforced with synthetic polymer shows good property but the degradable issue is the may concern, and the strength obtained is higher than the matrix of natural rubber (Pentakota, Prasad, Ramji, Pusty, 2014). The dielectric property of the jute is also the major concern in the matrix interface phenomenon. The other fiber such as Kenaf fiber can replace the other natural fiber with low cost and high mechanical strength and higher modulus of elasticity than any other natural fiber. Lower cost alternative that is environmentally friendly (Rashed, Islam, Rizvi, 2006). Hybrid composites are prepared to the check the individual properties they impart in the matrix and shows the hybrid effect. The non-hybrid composite shows less strength as compared to the hybrid composite with increased moduli and tensile and interface property.

BANANA FIBER REINFORCED THERMOSET COMPOSITES

Composite of varied thermosetting matrices (Polyester, Phenol gas, organic compound gas and epoxy) bolstered with banana fibers were investigated by varied researchers as shown in (Figure 2) fiber reinforced with polyester matrix was widely investigated. The research shows that the impact of the fiber length and content on mechanical properties of the Composite (Madhukiran, Rao, Madhusudan, 2013). The investigation shows that the fiber length of 35-45 mm linear unit and 50% content has higher mechanical properties. Further, aging decreases the mechanical properties of the composite because of the affinity of the banana fibers towards wetness. The mechanical and water absorption behavior of banana fiber composite ready by organic compound transfer molding methodology shows that the most tensile, flexural and impact strength is achieved at 35% linear unit fiber length and 40%. It conjointly showed that the most diffusion, absorption, and permeability constant are achieved at 50%. The twisted type of

Table 1. Mechanical property of natural and synthetic

Fiber	Density	Length	Strain	Strength	Young's Modulus	Specific Tensile Strength	Specific Young's Modulus
Ramie	64	10	5	N.S	N.S	N.S	20
Flax	71.5	20.4	13	0.2	0.5	1.2	25
Hemp	1.5	52	1.6	1110	70	740	47
jute	1.3	110	1.7	800	55	610	39
Sisal	1.4	900	2.4	855	28	610	20
Cotton	1.6	50	3.0	800	13	530	8.4
Banana	1.35	6	6	914	32	600	42
Coir	1.2	140	30	220	6	180	5
Wool	1.3	32	35	315	5	242	3.8
E-glass	2.5	Continuous	2.5	3000	70	1400	29

Sumaila, Ambera, & Bawa, 2013.

Figure 2. Banana fiber

banana yarn was placed on the warp direction and interchange bundles of banana and yarns area unit weaved in filling direction. It indicates that the tensile strength was the most for the 2 superimposed composite, whereas flexural strength is that the maximum for the tri-layer and also the impact strength will increase with the quantity of layers. The storage modulus is most for the four layers woven composite, and any addition of layers shows the addition of peaks for the loss modulus of the hybrid woven composite (Pujari, Ramakrishna, Padal, n.d.). The impact of chemical treatment on the flexural, impact and water absorption properties of wove n banana polyester composite was analyzed. The result indicates that up to 20vol% and 25vol% the flexural and impact strength of the treated composite redoubled. Further, the addition of fibers ends up in degrading the upper-class to poor adhesion. The chemical modification of banana fiber exploitation silane because the coupling agent has showed that the material property increases to greater extent (Dedeepya, Raju, Kumar, 2012). It also show that the material constant activity can function a tool for predicting the fiber-matrix adhesion. The impact of the layering collection on the storage modulus, loss modulus and damping property of banana/sisal hybrid composite was studied.

CONSTITUENT OF NATURAL FIBER COMPOSITE

The different Chemical Composition of various natural fiber was shown in (Table 2).

1. **Cellulose:** It is most abundant in natural fiber which provides strength stiffness, and other structural properties. It is a compound of hemicelluloses and lignin. Cellulose is used in different textile industries, fiber reinforcement etc. The chemical structure of cellulose consists of three hydroxyl groups (OH). Two of them form hydrogen bonds within the cellulose macromolecules(intermolecular) while the rest of the group forms hydrogen bond with other cellulose molecules(intermolecular).
2. **Hemicellulose Hemi:** cellulose is only component that directly affects the properties of moisture absorption in the hybrid cellulose composites that changes the the shape or volume of the composite. It is also responsible thermal degradation. The impurities is usually removed by treating with chemicals.
3. **Lignin:** It is most responsible for ultraviolet degradation of fibers. This is also removed from chemical treatment.
4. **Pectin:** Pectin plays an important role in the natural fibers as a component that binds the fiber into bundles and also determines the luster and touch of the fiber. Pectins are present in plant tissue to different degrees, being found predominantly in fruit stem peel and gums unretted flax contains between 3% and 4% pectins.
5. **Wax:** Waxes are also important in terms of mechanical parameters. They exhibit light touch, low friction and thus the case of moving the fiber. In flax fiber, waxes are present mainly in the outer part of the stem, in epidermis, and in smaller amounts in fiber cells. Waxes have different types of alcohols substances that are soluble in water as well as in acids, such as phenolic, oleaginous and stearic acid.

INTRODUCTION

What Is Composite?

There are necessary two segment of composite material, in which one is known as matrix material and another one is well-known as reinforcing material. The reinforce material is surrounded over matrix material. The matrix material is discontinuous phase and continuous phase and reinforcing. The reinforcing phase is much stronger than matrix phase (Harikrishnan, Shivakumar, n.d.). In composite material matrix phase remove the stresses between reinforcing phase and also protect from mechanical and ecological break. The role of reinforcing material is to look up mechanical and thermal properties of composites.

Table 2. Chemical composition of banana and jute fiber

Composition	Cellulose	Hemicellulose	Lignin	Pectin	Wax	Water Soluble
Banana	64	10	5	N.S	N.S	N.S
Jute	71.5	20.4	13	0.2	0.5	1.2

Babu, Babu, & Gowd, 2013.

Composite are hybrid of two or extra material such as reinforced polymer, metal or ceramics. The reinforcement may be in the form of particles, fiber, lamellae, whisker and are included in a proper matrix, thereby given that a material that merges the most useful properties of the constituent. Generally the properties of composites are better to those of its individual constituents. Glass fiber reinforced plastics, High structural strength were developed in the early 1945s and the technology of reinforced polymers has progress considerably since then (Lokare, Gholap, Gayade, n.d.). In a typical glass fiber reinforced composite, the stiffness and strength are provided by the glass fiber while the temperature capability of the composites is administrate by the plastic matric. These composites are verdict increasing application in automobile air craft, industry.

Matrix Material

A multiplicity of matrix polymer, ceramics, carbon, intermetallic, cement etc. have been used for manufacture composites. The matrix material serves several functions which are imperative to the show of the hybrid composite material. These function also depend upon the type of support such as fibers, particulates, dispersions discontinuous or continuous fibers whispers, The matrix enables the composite to withstand tensile, impact, compression, flexural strength tensile masses. The polymer material for a matrix can be thermosetting or thermoplastic resins. Thermoset resins, in the occurrence of a catalyst reaction or cure (Ezema, Menon, Obayi, Omah, n.d.). Once cured these material can no Longer flow. Thermosetting resins such as, bismaldeimides and cyanates, epoxies can provide good strong point properties. Thermoplastic resins are in general solids at room temperature but Soften or liquefy when heated to high temperatures and become solid when chilled distinctive thermoplastic resins include polyethylene terephthalate (PET), polyamide, polyetherulfone, polyphenylene sulphide (PPS), polymide polyether ether ketone, and polyetherulfone (Pujari, Ramakrishna, Kumar, 2014).The specialist epoxy resins and activity agent designed to be used in high performance composites have glorious elevated temperature resistance, sensible mechanical properties, low water absorption and comparatively high glass transition temperature. therefore a correct combination of epoxy resin, modifying rosin, reinforcing fiber, activity agent and process procedure square measure needed to take full benefits of their properties. Since the comparatively low strength polymers is converted into high force composites by support this field of the composite industry is well recognized and is growing quicker into newer function.

INTERFACE ADHESION BETWEEN MATRIX AND REINFORCEMENT

The bonding between the natural fiber and the polymer matrix plays an important role in determining the properties of the composite. In simple terms it could be imagined that there is a zone in which the properties of the fibers transition to the properties of the polymer matrix and vice versa. This transition zone must function optimally so that each part does its job properly. Designing this interface is very important to achieve the best possible properties of the composite. But why is interface in NFC a concern. One of the properties of natural fibers is that they can attract moisture i.e. natural fibers are said to be strongly polar and hence hydroponic. This characteristic is mainly caused by the different biochemical components in the natural fiber itself. On the other hand the man-made polymer matrix ends strobe a polar and generally hydrophobic. So essentially it is a compatibility problem between natural and man-made materials. natural fibers themselves are composite materials consisting of different polymers and fibrous structures.

Bu t nature has resolved the interface issues elegantly. The natural fibers are able to optimally fulfill their functions within the plant. Over the years researchers have recognized and proposed strategies to deal with this compatibility issue in natural fiber composite. The key strategies are to modify the surfaces of the natural fiber or to supplement the polymer matrix properties with additives to enhance the adhesion between the two materials. physical treatment, such as corona and plasma treatment attempt to modify the fiber surface area so that the bonding with the polymer matrix is improved. It is also possible to follow a chemical route. The simplest method is to treat the natural fibers with a chemical such as an alkaline (e.g., 5% NaOH) solution for a certain amount of time. However, if not done correctly, such treatments can weaken the overall mechanical properties of the natural fibers because lignin, the natural binding material or matrix in natural fibers, could also be weakened. acetylating is another method to make the surface of the natural fiber hydrophobic. Below is a brief information regarding the raw materials used for the matrix and enforcement (Siddika, Mansura, Hassan, 2013).

EPOXY RESIN LY-556 Anhydride-cured, low-viscosity standard matrix system with extremely long pot life. The reactivity of the system is adjustable by variation of the accelerator content. The system is easy to process, has good fiber impregnation properties and exhibit s excellent mechanical, dynamic and thermal properties. It has an excellent chemical resistance, especially to acids at temperatures up to 80°C.

HARDENER PROPERTIES

The hardener properties are shown in (Table 3).

TYPES OF COMPOSITE MATERIALS

On the basis of matrix material composite materials can be classified into three groups. They are: Metal Matrix Composites (MMC) Ceramic Matrix Composites (CMC) Polymer Matrix Composites (PMC)

- **Ceramic Matrix Composite:** Ceramics square measure characterised by lightness, hardness, corrosion and reaction resistance and superior elevated temperature properties. Thus, several ceramic matrix composites may be used as temperatures on top of those of the compound and metal matrix composites. vital ceramic matrixes square measure oxides, carbides, nitrides, borides, glasses, glass-ceramics and silicates. The reinforcement used square measure set on, ZRO_2, AL_2O_3, SI_3N_4,

Table 3. Properties of epoxy resin

Property	Value
Aspect(Visual)	Clear,pale yellow liquid
Epoxy Content(ISO 3000)	5.30-5.45[eq/kg]
viscosity at 25°C(ISO 9317B)	10000-12000[mPas]
Density at 25°C[ISO 1675]	1.15-1.20[g/cmMM^3]
Flash Point(ISO 2719)	>200°C
Storage Temperature	2-40°C

Table 4. Properties of hardener

Property	Value
IUPAC	N-[2-(2-aminoethlyamino]ethane]
Molecular formula	$C_6H_18N_4$
Color	moderately viscous,yellowish Liquid
Density at 25°C	0.98g/cmMM^3
Flash Point	290°C

own and C within the type of fibers, whiskers or particulates (Baker, Chen, Lague, n.d.). These composites square measure processed by sintering, hot pressing, hot isotactic pressing, infiltration, reaction bonding and combustion synthesis. presently ceramic matrix composites square measure used as cutter inserts wear composites resistant, space vehicle tiles and aerospace elements. Different potential application includes engine elements, vanguard, and armor of military vehicles application in region and warm temperature corrosion resistant elements. Different potential application includes super conducting energy of magnet system storage.

- **Polymer Matrix Composite:** In crate of the strengthened plastics, the characteristics of the required ending like size, shape, operate and value verify the strategy by that the fundamental materials combined, moulded, cure and machined. distinguished moulding method for strengthened plastics includes hand lay-up, scatter up moulding, prepreg lay-up (bags of vacuum and auto-clave moulding), press moulding (SMC, BMC etc.), organic compound injection moulding and thread winding. to create helpful structure, the composite materials have to be compelled to be joined and machined. The manufacturing applications of strengthened plastics have open out a large spectrum of client goods, chemical plants, marine constructions and road transportation and part elements (Baker, Chen, Lague, n.d.). Mechanical engineering product embody cylinders, rolls, shafts, coupling, spindles, robot arms, covering, etc. the part trade used a large vary of product together with floor panels, elevators, skin panels flaps, covers, tanks, wings, struts and blades of rotor.

ARRANGEMENT OF FIBERS IN COMPOSITES

The natural fibers can be placed in the matrix in different forms and ways as shown (Figure 3). The vocabulary and even the theoretical concepts generally used have their roots in the man-made fiber reinforced composites industry. Some care is needed in using the terms to natural fibers. The orientation and the dispersion of the fibers in the base material play a crucial role in determining the mechanical properties of the composite (Khalifa & Chappar, 2014). In the present context, when reference is made to fibers, it means the performs or the derived products (semi-products) of fibrill. Unlike man-made synthetic filaments/fibers, natural fibers have a finite length. However bundles of natural fibers can be processed to form yarns, slivers, tows and roving are that have very long lengths. The performs of natural fibers can also be pre-impregnated or mixed (compounded) with a polymer. These pre-pregs or compounds can be purchased ready for further processing.

The (natural fiber) composites can be described in many ways. A common description is the form of the fiber part. They can be

1. **Flake Reinforcement:** Flakes could have positive effect compared to particulate reinforcement because the dimensional geometry; they could exhibit equal strength in all directions. Potentially it is very difficult to exploit flake reinforcement in NFC due to the costs of producing natural fiber flakes.
2. **Particulate Reinforcement:** It is assumed that particles, generally, do not to have long dimensions. They could have the same dimension in all directions. In many composites particles can be randomly oriented in the matrix. Particles do not contribute to improvement of strength but could

improve other properties. It may not be advisable to use natural fibers as particulate reinforcement, unless as a filler to reduce costs. This is because fibers naturally have a high length to diameter ratio that can be beneficial in composites.

3. **Fiber Reinforcement:** For the sake of completeness particulate and flake reinforcement are briefly described. Fiber reinforcement is the key perform typed is discussed in this section because it has the most diverse range of semi-products of natural fibers. The key characteristic of fibrous materials in composites is the length (L) to cross-sectional dimensions or diameter (D). The ratio L/D is called aspect ratio. Different natural fibers can have different aspect ratios (Khalifa, Chappar 2014). But within each fiber type itself the aspect ratio could vary. So the question is what is meant by fiber length, Fibers used in composites are commonly referred to as short fibers long fibers and continuous fibers. There is no internationally agreed nomenclature therefore these terms can be confusing and even misleading. For example if the natural fibers length in a polypropylene (PP) based compound, used for injection molding is used in long-fiber-technology(LFT) is considered long. Another example could be found in coir fiber non-woven(or random mats) production. In this perform fiber lengths lower than 100-120 mm would be considered short and 180-200 mm as long and may be even problematic for non-woven production. Therefore the product designer should really seek further clarification when the terms short and long fibers are used. Generally fiber reinforcement would improve most mechanical properties of the matrix. The degree to which the fibers materials improve the mechanical or engineering properties would depend on factors such as:
    a. The form of the fiber materials e.g. discontinuous or continuous.
    b. The orientation of the fibrous material in the matrix i.e. random or oriented.
    c. The aspect ratio (L/D), as in the matrix.

BACKGROUND AND MOTIVATION

While fabrication of the specimen this can be generalized as a product value as output. The most important benefit for the project is that it is biodegradable and recyclable and can replace the synthetic material which hazardous and non-recyclable. The study influences the thinker that there is a proper sequence of fabrication should be carried out to develop the product (Sanjay, Arpitha, Yogesha, 2015). The basic idea behind the project is development a variety of natural fiber materials. The objective of this project is to provide a glimpse into the world of Natural Fiber Composites especially for product designer. It is proposed as a beginning investigation of the broad area of Natural Fiber Composites (NFC). NFC is a actually multi-corrective field and therefore it is a challenge to cover in depth all the related areas that is useful for the project to execute the line of manufacturing of the natural fiber composite so it readily influences the others if the proper working is carried out in this field so it is greatly a long run task to accomplish it and still is going on. Why the focus on natural fiber composites Is the biggest question the reason behind is that NFC could put in towards amplified use of renewable materials such as natural fibers and accelerate developments towards Bio based product wealth way towards sustainable societies (Rout & Satapathy, 2012). In the recent past the polymer industry has increased its efforts to make polymers base on renewable eco-friendly materials. consequently composites could become more and more bio-based. The ultimate purpose must be to construct composite materials using renewable sources based on short carbon cycles. This can bring towards sustainability and reduce the belief on fossil/petro based

Figure 3. Types of reinforcement

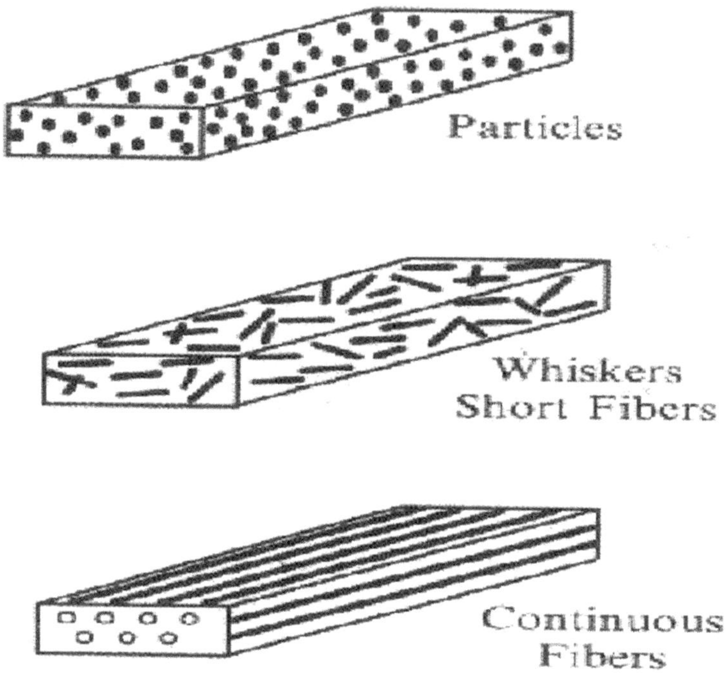

products. In the above para designer word is there Finally the question is why product designers as the target audience to introduce the subject of NFC. The respond to this is quite at once forward. Product architecture, be it industrial product designers, crafts persons or even artists, could capture the thoughts of probable users by infused aesthetic aspects into a product. So it is challenging to present a material like NFC to both groups as several issues should be reconciled to be appealing to both disciplines. It is not the intention to approach the subject from a hard core industrial design.

FABRICATION OF HYBRID COMPOSITE

Fabrication of composite is done by conventional method called hand lay-up method. In this process a layer of epoxy is added and then banana nanoparticles followed by jute fiber mat is placed one after the other in a mold of dimension 160 × 160 × 10. A sheet is solidify after few days and then taken out from the mold thus the material is fabricated and after that it is cut into the required dimension for the test. The composite to develop a new class of hybrid nano-materials from natural fiber such as banana and jute fiber which is very crucial task to accomplish here greater emphasis has been focused in preparation of natural fiber which is very laborious task (Jarek & Kubik, 2015). Ball milling is done in a tungsten jar in planetary ball milling machine. Banana and jute composites were treated chemically in sodium hydroxide to remove lignin and moisture content and further bleached in sodium hypochlorite solution for 24 hrs and then hydrolyzed in H_2SO_4 and filtered to get nanoparticles and mechanically in grinding machine to cut fiber into small pieces and after-wards in planetary ball milling machine for 30 hrs and then having jute fiber weight percentage sequences 5%, 10%, 15% whereas for banana is 20%, 30%, 40%.

PROCESS OF CHEMICAL TREATMENT

- The fibers are cut to the required size of 1 cm.
- Then the fibers are cleaned normally in clean running water and dried.
- A glass beaker is taken and 6% Na-OH is added and 80% of distilled water is added and a solution is made.
- After adequate drying of the fibers in normal shading for 2 to 3 hours the fibers are taken and soaked in the prepared Na-OH solution.
- Soaking is carried out for different time intervals depending upon the strength of fiber required.
- For our project the fibers are soaked in the solution for three hours.
- After the soaking process is complete the fibers are taken out and washed in running water and dried for another 2 hours.
- After that fibers are grinded.
- After this grinded fiber is put in tungsten jar of planetary ball milling machine with agate balls as shown in Figure 5.

MERITS OF CHEMICAL TREATMENT

1. It is done to increase the strength of the fiber while doing this moisture in the fiber is removed.
2. It also enhances the flexural strength of the material.
3. It removes all the impurities present in the fiber and thereby decreases the gas defect in the fiber.

PROCESS INVOLVED IN MAKING OF NATURAL FIBER SHEETS

Step wise follow up of the processes involved in the fabrication technique in the development of Banana Jute Epoxy composite can be over-viewed below:

1. Gathering the important materials i.e. Epoxy resin, Chemically treated Jute, Epoxy hardener, Banana, Acetone.
2. Preheating of the Epoxy resin to 80°C in the magnetic stirring for 1 hour at 400 RPM.
3. In plane Epoxy sample with banana fiber no coupling agent is needed.
4. Percentage variation of Banana and jute fiber, according to the sample is added in the Epoxy resin and magnetic stirred for another 1 hour at 80°C at 400 RPM.
5. After this the mixture is placed in ultrasonic bath for 30 min which is maintained at 80°C for uniform dispersion of banana particles as shown in Figure 4.
6. Then getting the mixture to cool down and attain the room temperature so that acetone can be added.
7. In banana and jute varied samples acetone is added as a coupling agent.
8. Once the mixture reaches to the room temperature, hardener (HY-951) is mixed in the ratio of 15:1, i.e. in 200 gm of epoxy 12.5 gm of hardener is added. As the reaction after adding the hardener is exothermic it emits heat and starts getting hard in next 10 to 15 min.

Figure 4. Ultrasonic bath

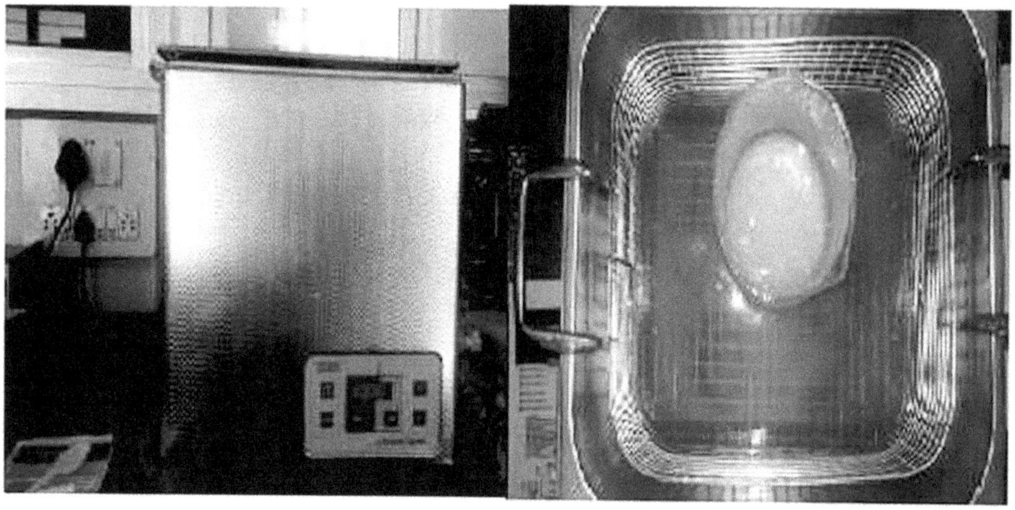

Figure 5. Ball milling machine

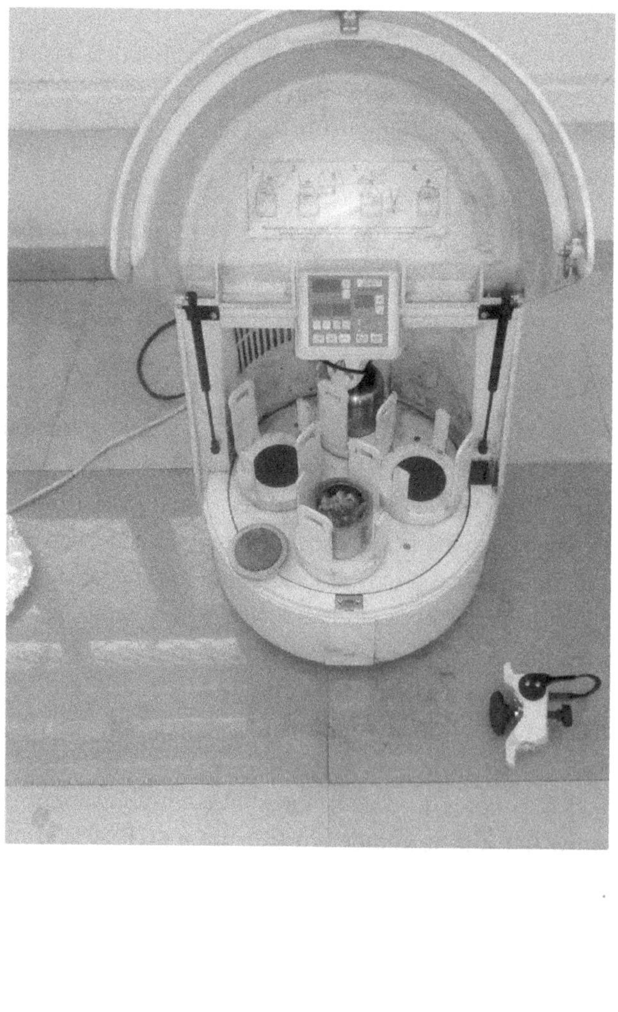

9. Dies are preheated and then a coat of mold relief agent is applied on the inner surfaces of the die to facilitate easy removal of a sample after curing.
10. Now the mixture along with the hardener is poured into the die, then the jute fiber is fixed in the die with the help of the hand roller and above that rest of the mixture is poured. And then when solidification starts the upper plate is closed and the lower and upper die are screwed.
11. Then both the dies are left for 10 to 12 hrs curing at room temperature and then the samples are removed from the die.
12. The sample plate is then cut in the desired dimensions according to ASTM standards with the fiber cutting machine.

Processing Technologies for Natural Fiber Reinforced Thermosets (Biswas & Satapathy, 2010)

The various technologies are used in advance fiber composite making in textile engineering as well but the natural fiber processing is being quite difficult due to its moisture absorption property of raw material. The various chemical treatment before use in production is gone through a series of process which is quite lengthy take time. The very efficient method in this is to make the sheets of natural fiber of the required size by compression mounding technique, or by putting the fiber by hand lay-up process in die. The different technology are processing is been carried out but extraction of raw fiber, its processing and treatment is laborious proce ss a proper line up is not there. The Product design of the fiber is very important task in natural fiber structural design. Th e various methods used are as follows

1. **Hand Layup Method:** Probably the oldest, simplest and traditional method to product designers to make natural fiber composite Products. The release agent is applied to the mold along with the gel coat to improve the surface finish and aesthetic effect. Reinforcement placed into the mold and then resin is applied by brushing, pouring Or spraying. The air particle is removed by rollers. The curing temperature at ambient conditions mostly without any ad-additional heat. Cure time of resin could be improved by additives like accelerators and catalysts.
2. **Spray Up Method:** Release agent is applied to the mold. complemented by a gel coat to improve the surface in his hand Aesthetic. NF reinforcement as roving is cut with in a nozzle before being sprayed long with the Resin into the mold. The entrapped air is re-moved by rollers. Additional reinforcement and resin added to desired properties. NF mats can be incorporated to improve properties of special care as. Curing at Ambient conditions, mostly without additional heat. Cure time of resin could be improved by additives Like accelerators and catalysts. increasingly over-taken by Resin Transfer Molding (RTM) methods.
3. **Vacuum Assisted Resin Infusion (VARI):** The easiest method to produce quality NFC products. Release agent is applied to the mold. complemented by a gel coat to improve the surface finish and aesthetic effect. Reinforcement placed into mold. Several layers of complimentary sheets are used to ensure proper wetting and low of resin. Finally covered by a vacuum bag. All edges are sealed to ensure no leakages and to maintain vacuum in the system. The air is removed by vacuums system prior to introducing resin. Resin is infused and transferred through the woven or non-woven NF layers. Curing at ambient conditions, mostly without additional heat. Curing can also be in special oven so autoclave. Cure time of resin could be improved By additives like accelerators and catalysts.

4. **Resin Transfer Molding:** A closed mold similar to VARI, release agent is applied to the mold complemented by a gel coat to improve the surface finish and aesthetic effect. Reinforcement placed into the mold. The air is removed By proper venting prior to resin pouring in the die. Resin is infused in woven and non woven natural fiber composites. The various pressure are 1 bar 2 bar 3 bar. Generally resin and hardener are kept separately and then mixed is nozzle before infusion. Curing in the oven or with heat source. Cure time of Resin could be improved by additives like accelerators and catalysts.
5. **Sheet Molding:** This is at waste process. SMC is manufactured in 1st stage and supplied with long fibers chopped and dispersed in the resin. Resin paste is deposited to and covered by carrier limos both sides. The polymers are often highly loaded with additive. There in paste is thickened to facilitate molding. The sheets can be cut and the charge is about 70% area of mold size. Part produced by compression in heated molds. The charge lows in the mold to in final shape.
6. **Compression Molding:** There are many variations to compression molding. The technique can be combined with other processes. Polymer pre-impregnate done-woven mats produced and supplied in rolls or sheets. The mats are cut to size and placed in heated molds and then consolidated and cured with heat and pressure. Compressed parts can be laminated to desired surface finish.
7. **Injection Molding:** There are many variations and advancements in this technique. Natural fibers are com-pounded with the polymer in an extruder and formed to granules for further process-ignite NF/polymer granules are fed to injection mold-in equipment, the injection screws haves specific length to diameter ratio. The Screw consists mostly of a convey-in section, a plasticizing sections and a pumping section. The plasticized material is injected into a closed mold held in a clamp mechanism. Mold held closed for A short period. Cooling period. Molds opened and in all parts are ejected. Properties of mix improved by additives.
8. **Direct Injection Molding:** Long Fiber Injection Molding There are many variation sand advancements in injection molding techniques. Natural fibers and the thermoplastic matrix in mixed directly in the extruder. The plasticized NF/polymer material is injected in to a closed mold held in a clamp mechanism. Molds open part is ejected. properties of mix improved by additives.
9. **Pultrusion:** A continuous process to produce linear profiles, rods, and etc. NF roving taken through bath to impregnate with resin. The impregnated fibers are cured in heat Eddie. The strands are then pulled through fibers. Cut to required profile length.
10. **Long Fiber Technology (LFT) and Direct Long Fiber Technology:** The ideal concept would be to have one system to deal with material processes and end-part production on site. LFT-D is a partial solution to compression molded parts. The plasticizing of the thermoplastic in an extruder and the mixing of long natural fibers (10-25) mm result in dough. The dough is to the transferred immediately to a mold in a press. The dough compressed to desired shape and part removed/ejected from mold.
11. **Filament Winding:** An open mold and auto-mated process with fiber yarn sort owes pass through areas in bath to be wetted and wound around a mandrel with a desired form. The mandrel is the mold. Depending on the desired product several winding patterns can be used. End caps can be separately attached for vessels etc. Curing on mandrel by heat. Cure time of resin could be improved by additives like accelerators and catalysts.
12. **Autoclave:** Autoclave is a molding method. Release agent is applied to the mold. Complemented by a gel-coat to improve the surface finish and aesthetic fact. Fabric spree-impregnated with res-

ins and partially cured are cut and placed into the molds. Placement is mostly done by hand as in vacuum bagging. The number of layers of prepares used depends on the desired part properties. Several layers off functional materials are used for optimal operations finally covered by a vacuum bag. All edges are sealed to ensure no leakages and to maintain vacuum in the system. The air and emissions removed by vacuum system. Curing at in the autoclave with pressure and temperature. Curing can also be in special ovens or autoclave. Cure time of resin could be improved by additives like accelerators and catalysts as shown in Figure 5.

a. Possibility of recycling and wastage produced during making of composite and molding.
b. The production of natural fibers can be started with a low capital investment and with a lower cost.
c. Strength can be increased with chemical treatment of fibers.
d. Emission of toxic substances are less upon treatment of fibers.

LIMITATION OF POLYMER MATRIX NATURAL COMPOSITE (JABBAR, MILITKY, KALE, RWAWIIRE, NAWAB, & BAHETI, 2016)

- Temperature limitations for processing.
- The properties such as strength, stiffness etc. could be different in different directions. Therefore composites can exhibit isotropic properties. The isotropic properties of composites could pose problems when there are multidirectional forces acting on the object. An isotropic material has uniform properties in all directions
- Properties could be dependent on the manufacturing process. However, by appropriate design the impact of these limitations can be reduced. For example the isotropic characteristics could be reduced by the fiber architecture or by stacking/layering.

CHARACTERISTICS OF POLYMER MATRIX-FIBER COMPOSITES

Very of literature tends to focus on the mechanical properties of composites. The idea is that composites should show (not limitative): In addition to mechanical properties the aesthetic properties such as look, feel and sound could be of importance in product design. The material canals obey apart of the products brand image (Vilay, Mariatti, Taib, & Todo, 2008). Increasingly the environmental aspects of composites are gaining attention. As composites made of bio based materials are becoming readily available.

- It has high specific strength.
- It has high specific stiffness.
- It has high impact resistance.
- It is light in weight.
- Good abrasion properties.
- Good corrosion properties.
- Good fatigue strength.
- Design flexibility in shapes form.

ADVANTAGES OF NATURAL FIBER

- Less in weight, high stiffness and strength.
- Good mechanical properties as compared to glass based composites.

DISADVANTAGE OF NATURAL FIBER (SCHELLBACH, MONTIERO, & DRELICH, 2016)

One big disadvantage of fiber in terms of their industrial application is the strong hydrophilicity of weak interface, thus reducing the mechanical performance of the composites. It is due to OH group found in cellulose. It has ability to make H-bond.

1. Lower strength especially tensile and impact strength.
2. Limitation on maximum processing temperature.
3. Price uncertainty because of random harvesting results and agricultural policies.

These disadvantages can be reduce to some amount due to surface treatment of fibres (Kitamura, Ito, Teramura, Zhang, Hamada, n.d.). Compatibility with the matrix material is increased by surface chemical modification and making the CNC or CNW compatible with epoxy by surface grafting with long aliphatic isocyanate chain done by These chemical treatments are useful and suitable for removing amorphous fractions.

CONCLUSION

The present chapter reports the utilization of cellulose nanoparticles, whiskers and MFC, as reinforcements in polymer matrices. This chapter is aimed toward providing information to stimulate any analysis during this space. it's been shown that in the last fifteen years an increasing range of researches within the space of cellulose nanocomposites are developed. the chance of surface chemical modification and Nano sized dimensions of cellulose nanoparticles are extensively utilized in a large type of application, e.g., packaging, adhesives and electronic show materials. several studies are examined, reviewed and highlighted during this chapter concerning the importance of the interface, the influence of varied kinds of surface modifications, differing types of matrices used for the composites, yet as fabrication strategies, and therefore the performance of composites. any analysis is needed to beat obstacles like wet absorption, inadequate toughness, and reduced semi-permanent stability for out of doors applications. in particular, completely different weathering conditions, like temperature, humidity, and ultraviolet light radiation all have an effect on the service lifetime of the merchandise. vital analysis is presently current round the world to deal with and overcome the obstacles mentioned higher than. This effort to develop bio-composite materials with improved performance for world applications is an in progress process.

REFERENCES

Babu, G. D., Babu, K. S., & Gowd, B. (2013). Optimization of machin ing parameters in drilling hemp fiber reinforced composites to maximi ze the tensile strength using design experiments. *Indian Journals of Engineering and Materials Science*, *20*, 385–390.

Baker, M., Chen, Y., Lague, C., Landry, ¨. H., Peng, Q., Zhong, W., & Wang, J. (2010). Hemp fibre decortications using a planetary ball mill. *Canadian Biosystems Engineering*, *52*(2), 2–7.

Biswas, S., & Satapathy, A. (2010). A comparative study on erosion characteristics of red mud filled bamboo–epoxy and glass–epoxy composites. *Materials & Design*, *31*(4), 1752–1767. doi:10.1016/j.matdes.2009.11.021

Chandramohan, D., & Marimuthu, K. (2011). Drilling of natural fibe r particle reinforced polymer composite material. *International Journal of Advanced Engineering Research and Studies*, *1*(1), 134–145.

Dedeepya, Raju, & Kumar. (2012). Effect of alkali ne treatment on mechanical and thermal properties oftypha angustifolia fiber reinforced composites. *International Journal of Mechanical and Industrial Engineering*.

Ezema, Menon, Obayi, & Omah. (2014). Effect of surface treatment and fiber orientation on the tensile and morphological properties of banana stem fiber reinforced natural rubber composite. *Journal of Minerals and Materials Characterization and Engineering*.

Harikrishnan & Shivakumar. (n.d.). *Mode i fracture toughness of jute/glass fibre hybrid composite–an experimental and numeric al study*. Academic Press.

Hossain, M. R., Islam, M. A., Van Vuurea, A., & Verpoest, I. (2013). Tensile behavior of environment friendly jute epoxy laminated composite. *Procedia Engineering*, *56*, 782–788. doi:10.1016/j.proeng.2013.03.196

Jabbar, A., & Militky, J. (2016). Modeling and analysis of the creep behavior of jute/green e poxy composites incorporated with chemically treated pulverized nano/micro jute fibers. *Industrial Crops and Products*, *84*, 230–240. doi:10.1016/j.indcrop.2015.12.052

Jarek, B., & Kubik, A. (2015). The examination of the glass fiber reinforced polymer composite rods in terms of the application for concrete reinforcement. *Procedia Engineering*, *108*, 394–401. doi:10.1016/j.proeng.2015.06.163

Kalia, S., Kaith, B., & Kaur, I. (2009). Pretreatments of natural fibers and their application as reinforcing material in polymer compositesa review. *Polymer Engineering and Science*, *49*(7), 1253–1272. doi:10.1002/pen.21328

Khalifa, M., & Chappar, S. S. (2014). Experimental investigati on of tensile properties of banana reinforced epoxy composites. *International Journal of Science and Research, 3*(6), 832–836.

Kitamura, T., Ito, K., Teramura, S., Zhang, Z., & Hamada, H. (2014). Application of multilayered paper processing to hybrid random natural fiber mat. *Energy Procedia*, *56*, 247–254. doi:10.1016/j.egypro.2014.07.155

Krishnan & Ramesh. (n.d.). *Synthesis and characterization of cellulose nanofibers from coconut coir fibers*. Academic Press.

Lokare, Gholap, & Gavade. (n.d.). *To develop and evaluate impact behaviour of natural fibre reinforced composite-review*. Academic Press.

Madhukiran, J., Rao, S., & Madhusudan, S. (2013). Fabrication and testing of natural fiber reinforced hybrid composites banana/pineapple. *International Journal of Modern Engineering Research*, *3*(4), 2239–2243.

Pentakota, S. N., Prasad, V., Ramji, K., & Pusty, G. (2014). Synthe sis of bio–degradable banana nanofibers. *IJITR*, *2*(1), 730–734.

Pickering, K. L., Efendy, M. A., & Le, T. M. (2016). A review of recent developments in natural fibre composites and their mecha nical performance. *Composites. Part A, Applied Science and Manufacturing*, *83*, 98–112. doi:10.1016/j.compositesa.2015.08.038

Pujari, Ramakrishna, & Padal. (n.d.). *Experimental investigation of mechanical properties of uni directional jute and banana fiber composites*. Academic Press.

Pujari, Ramakrishna, & Suresh Kumar. (2014). Comparison of jute and banana fiber composites: A review. *International Journal of Current Engineering and Technology*.

Ramesh, M., Atreya, T. S. A., Aswin, U., Eashwar, H., & Deepa, C. (2014). Processing and mechanical property evaluation of banana fiber reinforced polymer composites. *Procedia Engineering*, *97*, 563–572. doi:10.1016/j.proeng.2014.12.284

Rashed, H., Islam, M., & Rizvi, F. (2006). Effects of process para meters on tensile strength of jute fiber reinforced thermoplastic co mposites. *Journal of Naval Architecture and Marine Engineering*, *3*(1), 1–6.

Rout, A. K., & Satapathy, A. (2012). Study on mechanical and tribo-performance of rice-husk filled glass–epoxy hybrid composites. *Materials & Design*, *41*, 131–141. doi:10.1016/j.matdes.2012.05.002

Sakthivel, M., & Ramesh, S. (2013). Mechanical properties of nat ural fiber (banana, coir, sisal) polymer composites. Science Park, 1(1), 1–6.

Samat, S. N. A., Azahari, B., & Rashid, A. A. (2013). Effcct of ball milling time on the particle size and morphology of banana stem powder soaked in sodium hypochlorite solution. *Journal of Chemistry and Chemical Engineering*, *7*(3), 206.

Sanjay, M., Arpitha, G., & Yogesha, B. (2015). Study on mechanic al properties of natural-glass fibre reinforced polymer hybrid composites: A review. *Materials Today: Proceedings*, *2*(4), 2959–2967. doi:10.1016/j.matpr.2015.07.264

Schellbach, S. L., Monteiro, S. N., & Drelich, J. W. (2016). A novel method for contact angle measurements on natural fibers. *Materials Letters*, *164*, 599–604. doi:10.1016/j.matlet.2015.11.039

Siddika, S., Mansura, F., & Hasan, M. (2013). Physico-mechanic al properties of jute-coir fiber reinforced hybrid polypropylene composites. In Proceedings of World Academy of Science, Engineering and Technology. World Academy of Science, Engineering and Technology (WASET).

Sumaila, M., Amber, I., & Bawa, M. (2013). Effect of fiber length on the physical and mechanical properties of random oriented, nonwoven short banana (Musa balbisiana) fibre/epoxy composite. Asian Journal of Natural and Applied Sciences, 2(1), 39–49.

Vilay, V., Mariatti, M., Taib, R. M., & Todo, M. (2008). Effect of fiber surface treatment and fiber loading on the properties of bagasse fiber– reinforced unsaturated polyester composites. *Composites Science and Technology*, *68*(3), 631–638. doi:10.1016/j.compscitech.2007.10.005

Section 4

Chapter 18
Prediction of Temperature Evolution During Self-Pierced Riveting of Sheets

Deepak Mylavarapu
IIT Guwahati, India

Manas Das
IIT Guwahati, India

Ganesh Narayanan R
IIT Guwahati, India

ABSTRACT

Weight reduction of automotive components by tailoring materials is the state of the art. This basically has resulted in the development of advanced joining methods like clinching, friction stir welding, self-pierced riveting etc. to assemble similar or dissimilar materials, with significant change in sheet properties. In the present work, the main aim is to predict the temperature evolution during Self-Pierced Riveting (SPR) of sheets by Finite Element (FE) analyses. Load evolution is also predicted. Generally temperature estimation during SPR is not attempted. The influence of a few selected SPR parameters has been studied on the temperature and load evolution through FE simulations. The relationship between these parameters and the temperature and the load evolution are revealed. Later a neural network model is developed to predict the temperature rise during SPR. The same has been validated at 20 intermediate levels and the predictions are accurate. Thus a hybrid FEM-ANN model for SPR has been developed to predict the SPR outputs efficiently.

INTRODUCTION

The overall weight of an automobile has a great impact on fuel efficiency and vehicle emissions. According to Polmear (1995), fuel consumption can be reduced by 5.5% for every 10% reduction in weight. Moreover, for 1 kg weight reduction, 20 kg reduction in CO_2 emission can be achieved for a vehicle running for 1,70,000 km. This results in an increased need of light weight vehicle structure having

DOI: 10.4018/978-1-5225-2440-3.ch018

lightweight materials in automotive industries. Some of these light weight materials are difficult or impossible to weld with conventional spot welding technology. Hence, researchers have started developing new technologies suitable for joining lightweight materials such as friction stir welding (He et al., 2014), laser welding (He et al., 2012), mechanical clinching (Mucha, 2011), adhesive bonding (He, 2014), and different hybrid joining processes.

Self-Piercing Riveting (SPR) is a high speed mechanical fastening technique for joining similar and dissimilar sheet materials which does not require predrilling (He et al., 2015; Mori et al., 2014). Hence, its usage is increasing in automobile industry having light weight Al structure for most of the applications. In SPR process, a semi tubular rivet is pressed by a punch into two or more sheets of materials that are clamped between a die and a blank holder. The die shape causes the rivet to flare (plastic deformation) inside the bottom sheet to form a mechanical interlock. During SPR process, the upper and lower sheets are under severe plastic deformation and the upper sheet gets pierced.

The four different stages of SPR process with respect to load requirement (Figure 1) are discussed below.

1. **Clamping:** Initially the sheets are clamped against the die with the help of a blank holder and it prevents wrinkling of the sheets.
2. **Piercing:** A semi tubular rivet is pushed by a punch into the sheets. After some displacement the upper sheets gets pierced by the rivet tip.

Figure 1. Load-displacement behaviour of SPR of two sheets

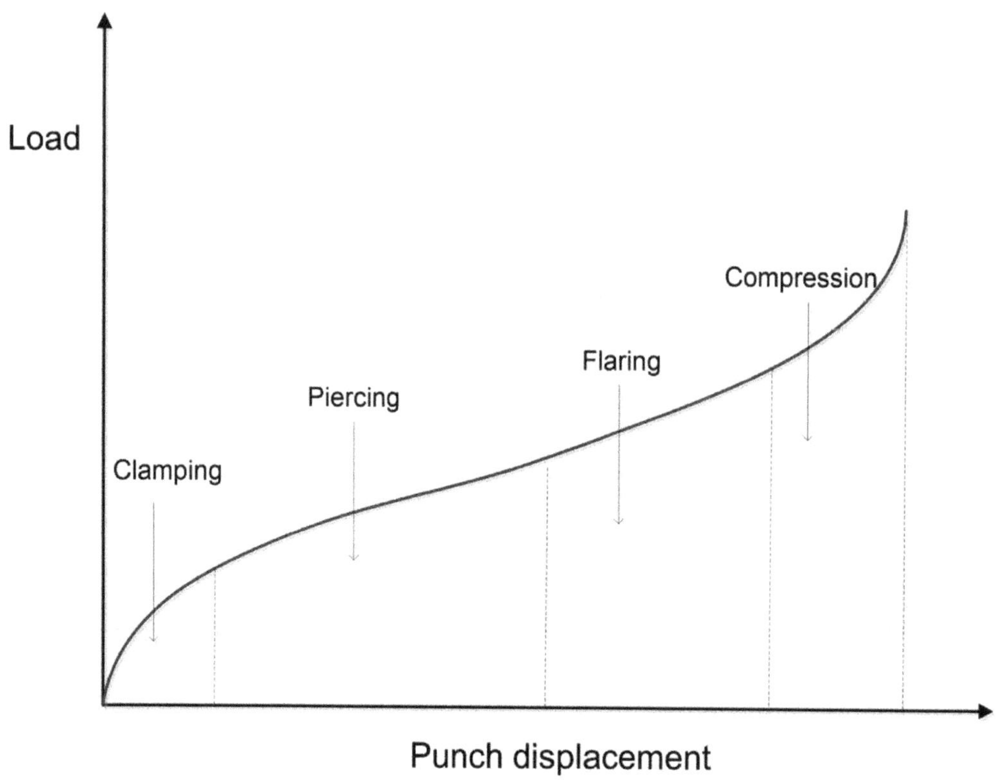

3. **Flaring:** The lower sheet undergoes deformation along with the upper sheet. Then the die is completely filled with the lower sheet. Due to the restriction by the die, the rivet flares in the lower sheet and mechanical interlock is formed.
4. **Compression:** Finally to bring the rivet head upper surface along the sheet, it is compressed till it reaches the required tolerance.

When metals undergo plastic deformation, significant amount of heat is generated, especially in case of localised deformation. If the deformation process is rapid, heat generation can lead to large temperature rise, since there is little time to conduct heat away from the deformation zone due to adiabatic condition. To predict the rise in temperature, one should know the material mechanics and coupling between plastic deformation and heat evolution. Part of the mechanical energy spent in mechanical deformation is converted into heat and remaining part is stored in the material.

For elastoplastic solids, a number of specializing assumptions are often made, including infinitesimal deformation, additive decomposition of strain into elastic and plastic part, a relation between stress and elastic strain identical to that of isotropic linear thermo-elasticity and linear Fourier heat conduction law. Considering above assumptions, the first law of thermodynamics or energy balance equation reduces to the following form under uniaxial stress condition as shown in Eq. (1) (Rosakis et al., 2000).

$$\rho c\dot{\theta} - k\theta_{xx} = \beta\sigma\dot{\varepsilon}^{p} - \alpha E\theta\dot{\varepsilon}^{e} + frictional\ heat \tag{1}$$

Here, θ is absolute temperature, σ, ε^{e} and ε^{p} are components of stress, elastic strain, and plastic strain, respectively as a function of coordinate x and time t. Subscripts, θ_{xx} indicates partial derivative with respect to the corresponding variable, and superscript $\dot{\varepsilon}^{p}$ denotes time derivative. The material constants ρ, c, k, α, E, and Å are mass density, specific heat, thermal conductivity, thermal expansion coefficient, Young's modulus, and Poisson's ratio, respectively. The detailed discussion of the theoretical models and assumptions involved in Eq. (1) is available in Rosakis et al. (2000). In most of the thermo-mechanical response studies, heat losses due to convention and radiation are neglected (Zehnder and Rosakis, 1993; Zehnder and Rosakis, 1991). Convection and radiation heat losses can also be eliminated by wrapping the specimen with fiberglass insulation.

Most of the plastic energy is converted into heat energy and it is defined as the ratio β. Zehnder (1991) and Zehnder et al. (1998) reported that the fraction of plastic energy converted into heat depends on plastic strain, yield strain, and strain hardening of the material and it decreases with the increase in strain hardening. However, for metals, Zehnder and Rosakis (1993; 1991), Kallivayalil and Zehnder (1994) and Weichert and Schonert (2011) considered the value of β as 0.9, while Wong and Kirby (1990) reported its value as 0.8-0.95 during low cycle fatigue. Several researchers reported the rise in temperature due to plasticity and localized plasticity. Eisenlohr et al. (2012) studied the stress drop along with localised straining and temperature burst in Fe having 3 wt.% Si at room temperature. They reported that a strong adiabatic heating occurred during the tensile testing in plastic region. Ravichandran et al. (2002) investigated the conversion of plastic work into heat during high strain rate deformation. Kolsky pressure bar experiment was used in their experiments for obtaining high strain rate on aluminium 2024-T3 alloy and measured temperature during experiment by using infrared thermography.

SPR is a rapid process and it takes place for a fraction of a second. Both sheets and rivet undergo deformation. All the forces exerted during joining of the sheets are concentrated at the interface between the sheet and the rivet. Hence, localised plasticity can be considered at this small region of the sheet. As discussed above, when plasticity-energy is converted into heat, it leads to a temperature burst in the material.

The temperature produced during SPR process is confined surrounding the rivet in the sheet. If thermocouple is used during riveting, it is damaged due to the forces produced during deformations of the sheets. Apart from the failure of the thermocouple, the positioning of the thermocouple at the required location is difficult. Hence, the best way to find temperature for SPR is by modelling and simulation.

The main objective of the present work is to predict and understand the influence of chosen process and material conditions on the temperature and load evolution during SPR. Since temperature evaluation by experiments is time intensive and uncertain, only modelling using Finite Element (FE) simulations are aimed. SPR process is simulated using a finite element based code ABAQUS® using Explicit-Temperature-Displacement module. The simulations helped in understanding the role of chosen material and friction conditions in the temperature and load evolution. The simulations using ABAQUS® are computationally expensive due to its longer computational time. Hence, it is difficult to simulate the process for each unknown parameter values. Hence, Artificial Neural Network (ANN) is utilized to optimize and predict the temperature evolution as a function of process and tool parameters within its range. Here developing a hybrid FEA-ANN model is the objective. The ANN model is trained with the help of data generated by FE simulations. The predictions from ANN and finite element simulations are compared and validated. The novelty of the present work lies in the prediction of temperature evolution during SPR, which is not generally attempted in the present literature. Only load and strain evolution, interlocking distance predictions are available.

METHODOLOGY

Finite Element Simulation Details

A 2D-axisymmetric model (Figure 2) is generated in ABAQUS®, using four-node 2D elements with four Gauss points, and hourglass control (assumed strain co-rotational stiffness form). The model contained six different parts namely rivet, bottom plate, top plate, punch, blank holder, and die. The punch, blank holder and die are assumed to be rigid, while the material of the rivet and sheets, which undergo plastic deformation, are modelled as elastic–plastic materials. The rivet geometry is taken such that its nominal diameter is 5 mm, and it is made of high-strength steel. The die geometry has a flat bottom surface. The blank holder and punch geometries are created in such a way that they have sufficient clearance between their surfaces. The punch is of 5 mm diameter with 2 mm as its corner radius. The blank holder is modelled as an annular ring of rectangular cross section with 10 mm inner radius and 20 mm outer radius.

The assembly of the SPR process (Figure 2) is done in such a way that the rivet touches the top sheet surface, while the punch is in contact with the rivet. Also the sheets are clamped between blank holder and die. All the contacts are created by auto-positioning rather than absolute translation of the part instances. For proper fixture of the sheets over the die, a step is given at 10 mm radius on the die surface which has same height as that of the die tip height.

Figure 2. SPR model for finite element analyses

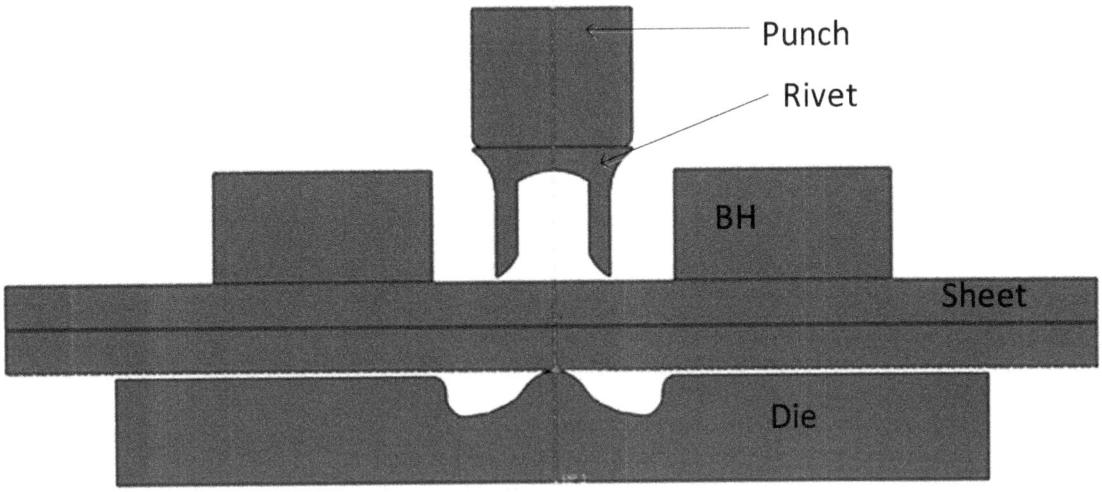

Material Properties

The sheets belong to a 5xxx grade and the stress-strain behaviour is shown in Figure 3. The rivet material is modelled proportional to the stress-strain curve of the sheet material with some factor (Figure 3). Hence, the material properties of sheet and rivet are considered elasto-plastic in nature. The die, punch and the blank holder are modelled as being made of tool steel which has very high Young's modulus, thus leading to nearly rigid property. The other material properties required to simulate the explicit temperature-displacement model are given in Table 1.

Boundary Conditions and Meshing Strategy

The contact was modelled using surface-to-surface penalty formulation available in ABAQUS®. The interactions are formed in SPR process between rivet and sheet, two sheets, blank holder and upper sheet, rivet and punch, die and lower sheet. The static coefficient of friction in all these interactions is set to 0.3. Later this has been varied within a practical range for understanding the temperature and load evolution during SPR. A displacement is then prescribed to the punch in order to push the rivet through the sheets until the joint is formed. The punch starts from the position where the rivet tip touches the upper sheet and the end position is given to be 6.5 mm below the initial state. The rate of punch motion is fixed in such a way that the total displacement occurs in a fixed time practically. Similar boundary condition is given to the blank holder to clamp the sheets in between die and blank holder.

In the present work, a four-noded plane strain quadrilateral element (CPS4RT) with average mesh size of 0.4 mm × 0.4 mm is considered having a reduced integration along with temperature degree of freedom and hourglass control with four Gauss points. Reduced integration corresponds to lowering of the order of integration as compared to that of full integration. For CPS4RT element, the order of integration corresponds to a single point per element and it is the centroid of the element. Now, this can

Figure 3. True stress-strain curve for Al 5xxx sheet and rivet

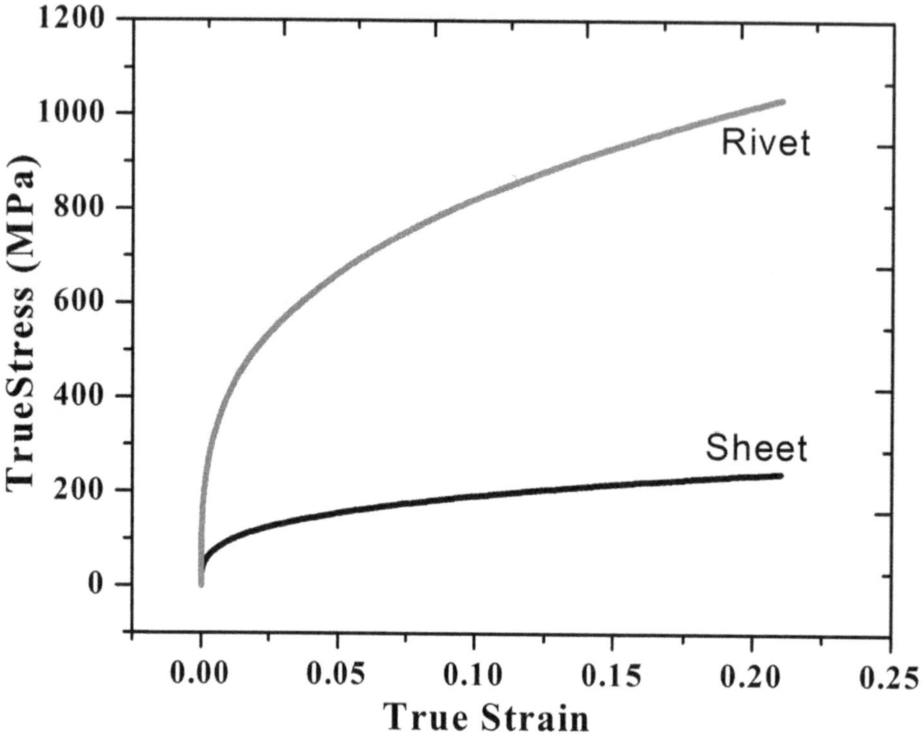

Table 1. Material properties required for FE simulation

Material	Properties					
	Density (kg/ m³)	**Specific Heat (J/Kg-K)**	**Thermal Conductivity (W/m-K)**	**Inelastic Heat Factor**	**Elasticity (GPa)**	**Poisson's Ratio**
Sheet Material (Al 5xxx)	2700	900	205	0.9	68	0.33
Rivet material	7830	490	50.2	0.9	188	0.3
Tool material	7830	490	50.2	0.9	210	0.3

lead to development of zero-energy modes, where zero-energy modes are those which can cause the element's integration point to experience no strain, even though there is deformation. In the absence of these zero-energy modes, reduced integration can increase FEM accuracy. Also hourglass control is used to control these zero-energy modes. The meshing of sheets and blank holder is done in structured form. However, the meshing of the rivet, die, and the punch is done in free meshing form (Figure 4). This causes the rivet to have varying element dimensions. Once the above modeling of SPR process is done, simulation is carried out and temperature and load evolution with displacement of punch has been predicted.

Figure 4. Meshing of the computational domain in SPR process

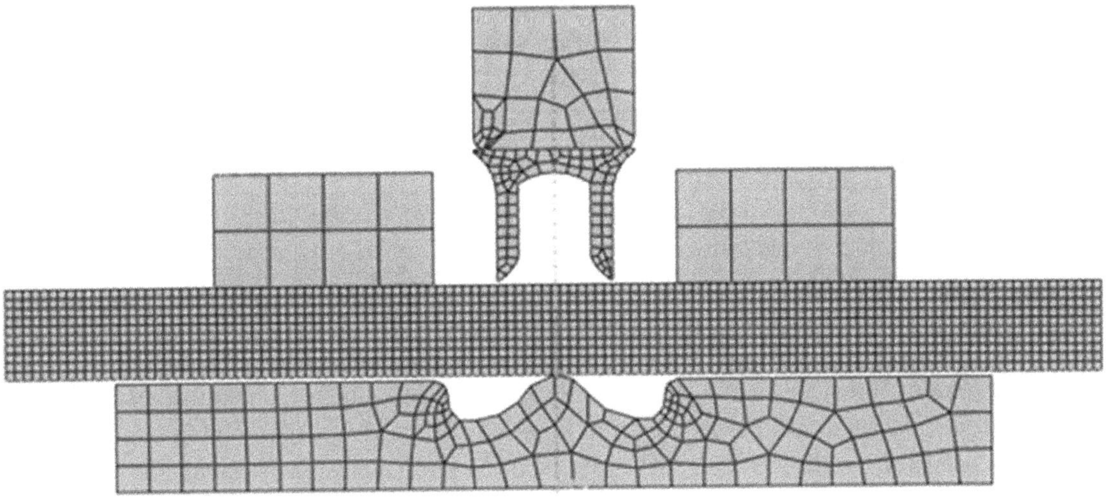

Parameters in SPR Process

The parameters which affect the SPR process are mentioned below. There are several sheet properties, tool geometries, process conditions that can be varied. But the scope of the present work is restricted to the parameters given in the Table 2. The range of each parameter was decided initially depending on the practical availability of the material and friction coefficient. Later equally spaced levels are decided for studying the influence of individual parameters through FE simulations (Table 2). The temperature and load evolution were predicted for each simulation case to understand the influence of these parameters. The data thus generated will be included during ANN modelling as well.

Artificial Neural Network Modelling

FE simulations were carried out for different parameters at different levels. The temperature-punch displacement was plotted for each combination. In the present work, ANN modelling was restricted to temperature evolution prediction and not for the load evolution. A total of 270 FE simulations were carried out for five parameters (Table 3) with several levels for each parameter. In order to predict the

Table 2. Parameters and levels chosen for FE simulations and analyses

Parameters (P)	Levels (L)				
	1	2	3	4	5
Coefficient of friction (μ)	0.1	0.2	0.3	0.35	0.4
Upper sheet K-value (U_K), MPa	310	350	390	430	460
Upper sheet n-value (U_n)	0.16	0.17	0.21	0.31	0.43
Lower sheet K-value (L_K), MPa	310	350	390	430	460
Lower sheet n-value (L_n)	0.16	0.17	0.21	0.31	0.43

Table 3. Parameters and levels used for ANN modelling

P	L																
	1	2	3	4	5	6	7	8	9	10	11	12	13	14	15	16	17
μ	0.1	0.12	0.14	0.16	0.18	0.2	0.22	0.24	0.26	0.28	0.30	0.32	0.38	0.4	0.43	-	-
U_K	100	125	150	175	200	225	250	275	300	310	330	350	370	390	410	430	460
U_n	0.1	0.14	0.16	0.17	0.21	0.23	0.25	0.27	0.29	0.31	0.33	0.34	0.37	0.39	0.4	0.43	0.47
L_K	100	125	150	175	200	225	250	275	300	310	330	350	370	390	410	430	460
L_n	0.1	0.14	0.16	0.17	0.21	0.23	0.25	0.27	0.29	0.31	0.33	0.34	0.37	0.39	0.4	0.43	0.47

U_K in MPa, L_K in MPa

temperature evolution (the whole curve) during SPR, the temperature-displacement curves were divided into segments at equal intervals. For example, in Figure 5, the plots for four combinations out of 270 combinations are shown with the segments made at different displacements. Here, for 6 mm punch displacement, 10 segments are considered. Hence, for every 0.6 mm a partition line is drawn. For one partition line, the curves intersect at four different points. These four temperatures are considered as output for a particular displacement at four different conditions. Similar exercise has been followed for all the segments resulting in large data set for temperature evolution. A sample table for four combina-

Figure 5. Temperature-punch displacement for four combinations of parameters

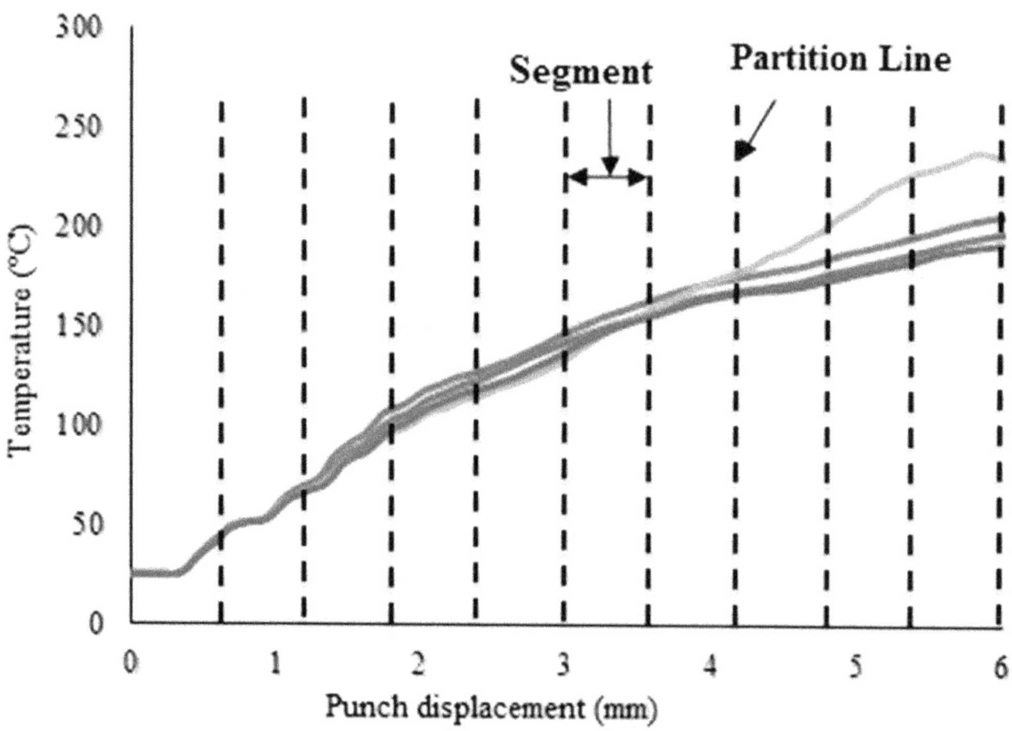

tions at one partition line is given in Table 4. The eleven different temperature values for a condition will be trained to generate an ANN model. In this way, from 270 simulations, five parameters with different levels as input and eleven temperature of each curve as output, the ANN model was developed.

The ANN model was trained by using Back-Propagation Algorithm using an in-house C++ code. The algorithm is based on the error-correction learning rule. In the current application, the objective is to use the network to learn mapping between input and output patterns. The components of the input pattern consist of the control variables of the SPR operation (Table 3), whereas the output pattern components represent the temperature rise during the process. The nodes in the hidden layer are necessary to implement the nonlinear mapping between the input and output patterns. Each neuron is connected to all the other neurons in the adjacent layer through the weighted connections. In the forward pass, the synaptic weights remain unaltered throughout the network and the function signals of the network are computed on a neuron-by-neuron basis (Chaturvedi, 2008). Specifically, the function signal appearing at the output of neuron, j, is computed as,

$$y_j(n) = \Phi(v_j(n)), \tag{2}$$

where v_j (n) is the net internal activity level of neuron, j, defined by:

$$v_j(n) = \sum_{i=0}^{p} w_{ji}(n) y_i(n), \tag{3}$$

where p is the total number of inputs applied to the neuron, i, and $w_{ij}(n)$ is the synaptic weight connecting neuron, i, to neuron, j, and $y_i(n)$ is the input signal of neuron, j or, equivalently, the function signal appearing at the output of neuron, i. If the neuron, i, is in the first hidden layer of the network, then the index, i, refers to the ith input terminal of the network, for which we write

$$y_i(n) = x_i(n), \tag{4}$$

Table 4. Sample input table for ANN code

Cases	Input 1	Input 2	Input 3	Input 4	Input 5	Output
	Upper Sheet K-Value (MPa)	Upper Sheet n-Value	Lower Sheet K-Value (MPa)	Lower Sheet n-Value	Coefficient of Friction	Temperature (°C)
Combination1	390	0.31	390	0.31	0.47	226.993
Combination2	390	0.22	390	0.31	0.1	196.067
Combination3	390	0.26	390	0.31	0.1	187.93
Combination4	390	0.31	390	0.31	0.17	183.478

where $x_i(n)$ is the ith element of the input pattern. If, on the other hand, neuron i, is in the output layer of the network, the index, i, refers to the ith output terminal of the network, for which we write

$$y_i(n) = z_i(n), \tag{5}$$

where $z_i(n)$ is the ith element of the output pattern (Chaturvedi, 2008). This output is compared to the desired response, $D_i(n)$, for obtaining the error signal for the ith output neuron. Thus at the end of forward propagation, error signal for each neuron as output will be delivered. The backward pass starts at the output layer by passing the error signals backward through the network, layer-by-layer, and recursively calculating the local gradients, δ, for each neuron. This recursive process permits the weights of the network to undergo changes in accordance with the delta rule. For a neuron located in the output layer, the δ, is simply equal to the error signal of that neuron multiplied by the first derivative of this non-linearity. In this way, the changes in the weights of all the connections feeding into the output layer are calculated.

Training and prediction is done separately for each output temperature (i.e., for each partition line). Thus all the eleven outputs are trained separately, generating eleven ANN architectures. The final ANN architecture contains five input neurons, one hidden layer with ten neurons, and one output neuron corresponding to temperature at each displacement. The final ANN architecture has been decided based on numerous trial exercises. For first eight displacement levels, tan-sigmoid transfer function has been used between input and hidden layer, while log-sigmoid between hidden layer and output layer. But for the last three displacement levels, tan-sigmoid transfer function has been used between input and hidden layer and hidden layer and output layer. The transfer functions used for ANN model are described briefly.

Log-sigmoidal transfer function: It generates the output lying in a range of (0.0, 1.0) and becomes equal to 0.5, corresponding to an input u= 0.0. The output 'O' of the transfer function can be expressed as the function of input 'u' as follows.

$$O = \frac{1}{1 + e^{-au}}, \tag{6}$$

where 'a' represents the coefficient of transfer function.

Tan-sigmoidal transfer function: This transfer function yields an output lying in the range of (-1.0, 1.0). The output and input relationship can be expressed as follows.

$$O = \frac{e^{au} - e^{-au}}{e^{au} + e^{-au}}, \tag{7}$$

where 'a' represents the coefficient of transfer function.

The coefficient of transfer function between the input and hidden neurons, coefficient of transfer function between the hidden neurons and output, learning constant, momentum constant are decided based on trial and error and accuracy of the network. The trained ANN has been validated with the FE simulation results at 20 intermediate levels or cases (Table 5) for checking the adequacy of the ANN model.

Table 5. Validation cases for checking the adequacy of ANN model

Validation Cases	Upper Sheet *K*-Value (MPa)	Upper Sheet *n*-Value	Lower Sheet *K*-Value (MPa)	Lower Sheet *n*-Value	Coefficient of Friction
Case 1	390	0.31	390	0.31	0.17
Case 2	390	0.31	390	0.31	0.25
Case 3	390	0.31	390	0.31	0.39
Case 4	**390**	**0.31**	**390**	**0.31**	**0.47**
Case 5	112	0.31	390	0.31	0.1
Case 6	167	0.31	390	0.31	0.1
Case 7	**237**	**0.31**	**390**	**0.31**	**0.1**
Case 8	350	0.31	390	0.31	0.1
Case 9	390	0.12	390	0.31	0.1
Case 10	390	0.26	390	0.31	0.1
Case 11	390	0.35	390	0.31	0.1
Case 12	**390**	**0.22**	**390**	**0.31**	**0.1**
Case 13	**390**	**0.31**	**112**	**0.31**	**0.1**
Case 14	390	0.31	167	0.31	0.1
Case 15	390	0.31	237	0.31	0.1
Case 16	390	0.31	350	0.31	0.1
Case 17	390	0.31	390	0.12	0.1
Case 18	390	0.31	390	0.26	0.1
Case 19	**390**	**0.31**	**390**	**0.35**	**0.1**
Case 20	390	0.31	390	0.22	0.1

RESULTS AND DISCUSSION

Influence of Chosen SPR Parameters on the Load-Displacement and Temperature-Displacement Behaviour

Influence of Coefficient of Friction

As the coefficient of friction between the sheet and the rivet increases, the temperature produced in SPR process is also increased (Figure 6a). Temperature produced is directly proportional to the heat dissipation by frictional resistance. This frictional resistance depends on the coefficient of friction (μ) and the normal force. So larger the friction coefficient, larger should be temperature rise. It is observed that up to 3 mm rivet displacement, the temperature rise due to conversion of plasticity energy is more and frictional temperature rise is negligible so the temperature profile almost coincides in all the cases. But after 3 mm, frictional temperature rise is significant and larger temperature rise is obtained for higher friction coefficient during SPR. There is about 40°C increase in temperature with increase in friction

Figure 6. (a) Temperature-displacement and (b) load-displacement behaviour for varying coefficient of friction

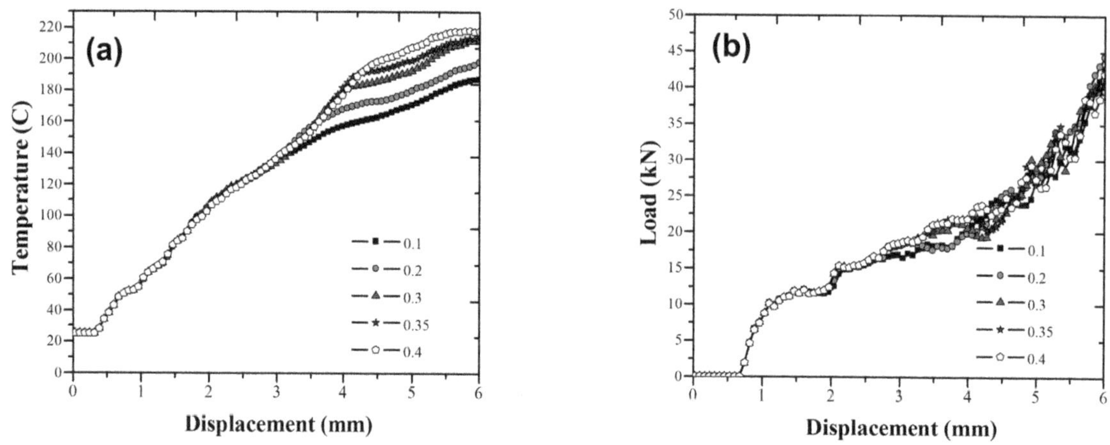

coefficient at the end of the SPR process. As shown in Figure 7b, friction coefficient has negligible effect on the load-displacement behaviour throughout the process. A typical SPR simulation at three different stages is shown in Figure 7.

Influence of Upper Sheet "*K*" Value

As the strength coefficient in strain hardening law of the upper sheets is increased, the load required for deformation during SPR has increased by about 10-20 kN (Figure 8). This is expected as it requires larger load to deform a stronger sheet. Similarly as the strength coefficient increases, the plasticity energy (i.e., stress × strain) at any point of time during the process increases, which leads to larger temperature rise in the sheet during the process (Figure 8). But the effect of *K*-value on the temperature evolution is much stronger than on the load evolution. There is about 60°C increase when the *K*-value is increased from 310 to 460 MPa.

Influence of Upper Sheet "*n*" Value

The effect of '*n*' value is similar to that of effect of '*K*' value. As the upper sheet '*n*' value increases, the temperature and load requirement decreases, which is expected (Figure 9 a, b). The effect on load can be understood from the effect of '*n*' value on the flow stress from any strain hardening law. Lower is the load requirement, lower is the plasticity energy conversion and hence lower will be the temperature rise.

Influence of Lower Sheet "*K*" Value

As the lower sheet strength coefficient increases, its stiffness increases, resulting in larger load to push the sheets inside the die. This consequently has raised the temperature evolution during SPR (Figure 10 a, b). So the effect remains same as that of upper sheet '*K*' value. But the effect is relatively less in the temperature evolution for changing lower sheet '*K*' value as compared to upper sheet value (Figure

Figure 7. FE simulation of SPR process, (a) initial stage, (b) after 3 mm punch displacement, (c) end of the process

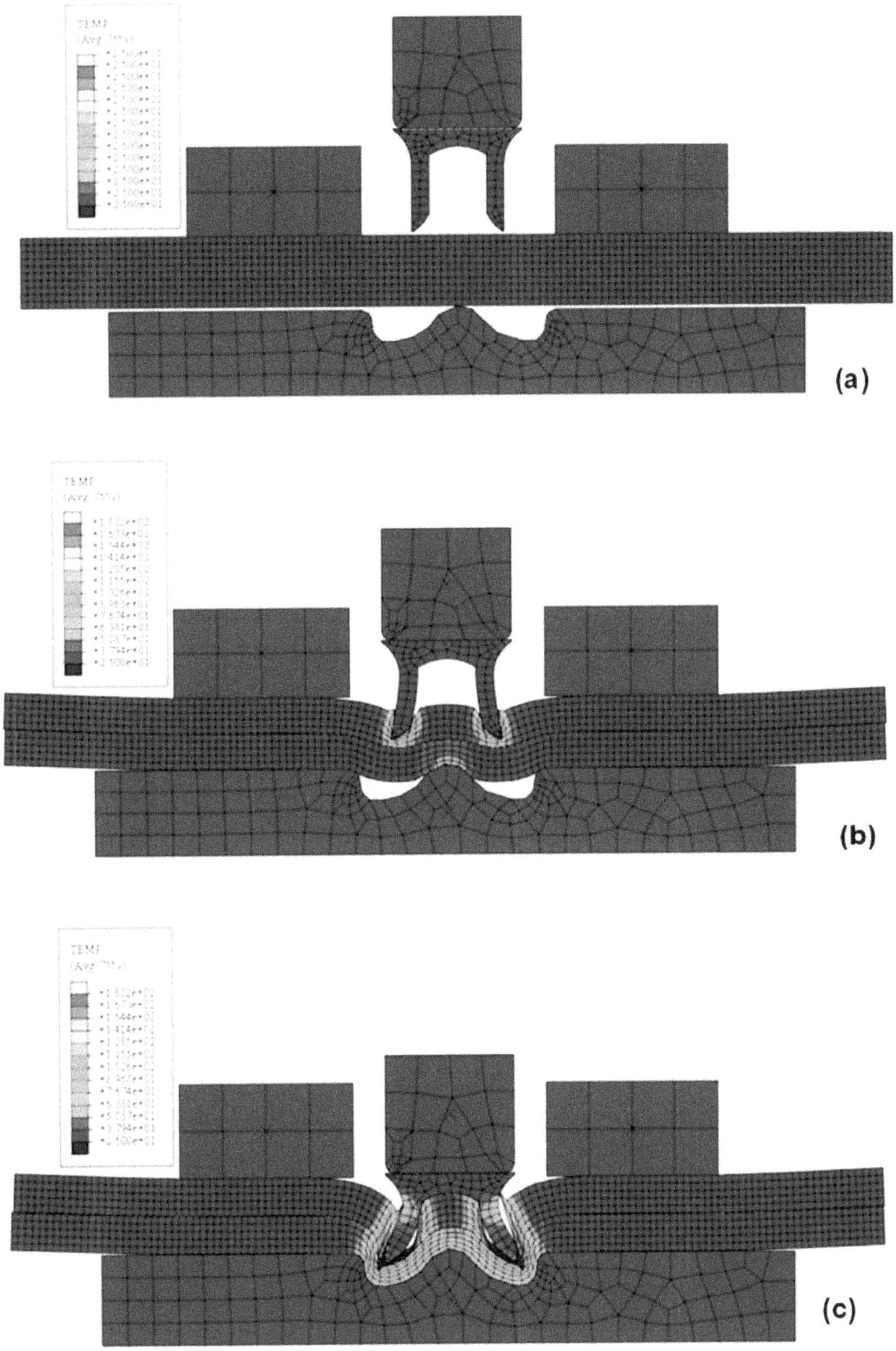

Figure 8. (a) Temperature-displacement and (b) load-displacement behaviour for varying upper sheet strength coefficient (K)

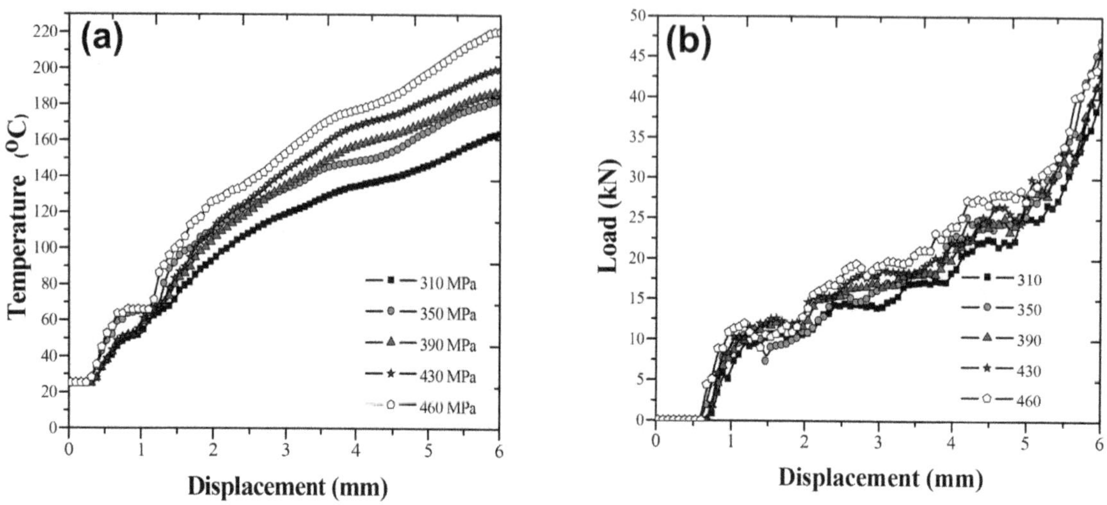

Figure 9. (a) Temperature-displacement and (b) load-displacement behaviour for varying upper sheet strain hardening exponent (n-value)

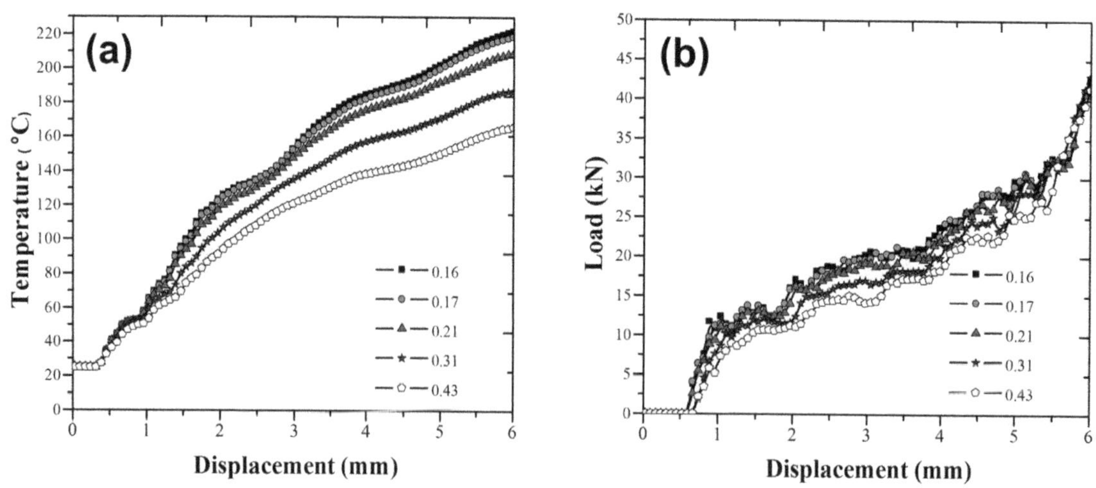

8a). The upper sheet undergoes larger deformation and it gets pierced by the rivet tip, when compared to the lower sheet. This has resulted in larger temperature difference when the upper sheet '*K*' value is changed as compared to lower sheet value. The load evolution is almost same in both the cases (Figure 8b and 10b).

Influence of Lower Sheet "*n*" Value

The effect is similar to the effect of upper sheet '*n*' value. The temperature and load requirement decreases with increase in '*n*' value of the lower sheet (Figure 11). The effect on load can be understood from the

Figure 10. (a) Temperature-displacement and (b) load-displacement behaviour for varying lower sheet strength coefficient (K-value)

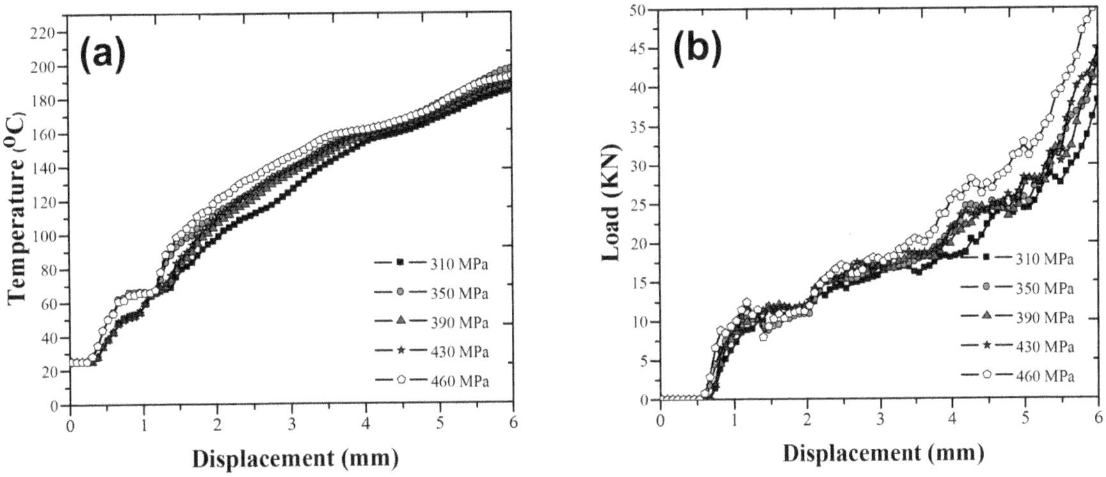

Figure 11. (a) Temperature-displacement and (b) load-displacement behaviour for varying lower sheet strain hardening exponent (n-value)

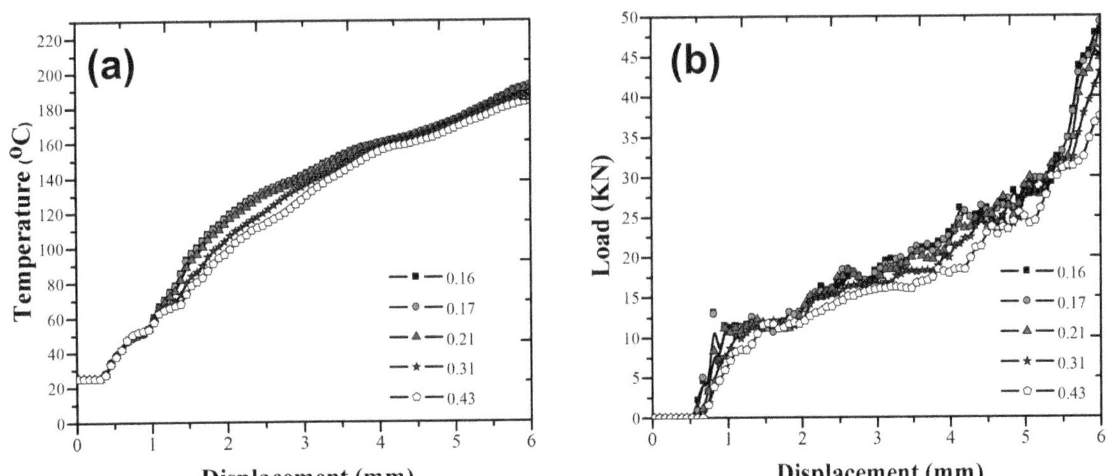

effect of '*n*' value on the flow stress from any strain hardening law. Lower is the load requirement, lower is the plasticity energy conversion and hence lower will be the temperature rise. Though this is the case, the effect is considerably less as compared to upper sheet '*n*' value (Figure 9 a, b). This is again due to the severe plastic deformation undergone by the upper sheet as compared to the lower sheet during SPR.

Just to summarize the effect of chosen SPR parameters, in order to have lower load requirement and lesser temperature rise during SPR, the preferred selection of parameter combination would be lubricants with lesser friction coefficient, the upper and lower sheets having lesser strength coefficient (*K*) and higher strain hardening exponent (*n*). There are many other SPR parameters like rivet dimensions and geometry including rivet height, rivet tip radius, rivet diameter, die tip height, die shape, and sheet properties including thickness, anisotropy etc. that affect the SPR joint quality. One should consider all these parameters for a comprehensive analyses and process design. Such an effort will certainly help shop floor engineers to design the process, material and tool selection successfully.

ANN Prediction

The trained ANN model has been validated for 20 intermediate cases (as given in Table 5) that were not part of the training set. Out of the 20 cases, validation for five different cases namely case 4, case 7, case 12, case 13, and case 19 (all in Table 5) are shown in Figure 12 (a-e). It is observed that the temperature

Figure 12. Validation of ANN predictions with FE simulation predictions for various parameter cases: (a) case 4, (b) case 7, (c) case 12, (d) case 13, (e) case 19 (all in Table 5), (f) ANN vs. FE simulation results for all the 20 cases

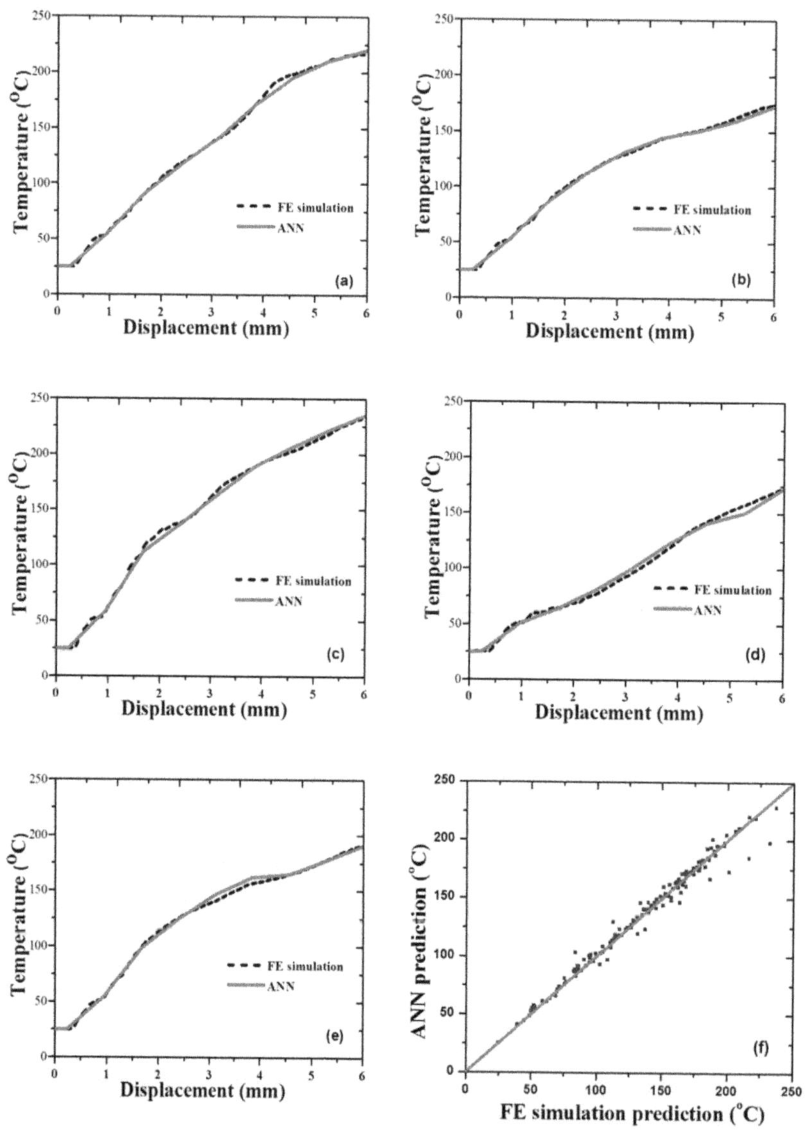

evolution predictions from ANN agree well with that of from FE simulations. Similar accuracy levels are observed in all other cases as well. Figure 12 (f) shows the accurate predictions for all the 20 cases of SPR. With the FEM-ANN hybrid model, one can design and optimize the process, tool and material conditions during SPR with the highest efficiency, which otherwise will be time and resource intensive.

CONCLUSION

The following conclusions are drawn from the present analyses.

1. Practical temperature measurement in SPR process is challenging and difficult task to perform. The temperature produced during the process is inside the sheet and the process takes place for a fraction of a second within a span of 10 mm. But from the present work, it can be understood that the temperature rise during SPR process can be predicted by following a robust FE simulation strategy for any combination of sheet materials, process conditions, and tool dimensions. Analyses of SPR process using temperature evolution is not attempted until now.
2. From the present analyses, it has been suggested to use lubricants with lower friction coefficient, the upper and lower sheets having lesser strength coefficient (K) and higher strain hardening exponent (n) to have lesser load requirement and temperature rise during SPR.
3. The ANN model is accurate enough to predict the temperature evolution during SPR. It has been also demonstrated that FEM-ANN hybrid model can be used as a 'virtual machine' to predict the effect of few SPR parameters on the temperature evolution during joint formation.

ACKNOWLEDGMENT

The authors are thankful to Ms. Anwesa Barman, a PhD research scholar in the Department of Mechanical Engineering, IIT Guwahati for helping them in ANN modelling.

REFERENCES

Chaturvedi, D. K. (2008). *Soft Computing: Techniques and its applications in electrical engineering*. Springer-Verlag Berlin and Heidelberg GmbH & Co. KG. doi:10.1007/978-3-540-77481-5

Eisenlohr, A., Gutierrez-Urrutia, I., & Raabe, D. (2012). Adiabatic temperature increase associated with deformation twinning and dislocation plasticity. *Acta Materialia*, *60*(9), 3994–4004. doi:10.1016/j.actamat.2012.03.008

He, X. (2012). Finite Element Analysis of Laser Welding: A State of Art Review. *Materials and Manufacturing Processes*, *27*(12), 1354–1365. doi:10.1080/10426914.2012.709345

He, X. (2014). Influence of boundary conditions on stress distributions in a single-lap adhesively bonded joint. *International Journal of Adhesion and Adhesives*, *53*, 34–43. doi:10.1016/j.ijadhadh.2014.01.009

He, X., Gu, F., & Ball, A. (2014). A review of numerical analysis of friction stir welding. *Progress in Materials Science*, *65*, 1–66. doi:10.1016/j.pmatsci.2014.03.003

He, X., Zhao, L., Deng, C., Xing, B., Gu, F., & Ball, A. (2015). Self-piercing riveting of similar and dissimilar metal sheets of aluminum alloy and copper alloy. *Materials & Design*, *65*, 923–933. doi:10.1016/j.matdes.2014.10.002

Kallivayalil, J. A., & Zehnder, A. T. (1994). Measurement of the temperature field induced by dynamic crack growth in Beta-C titanium. *International Journal of Fracture*, *66*(2), 99–120. doi:10.1007/BF00020077

Mori, K., Abe, Y., & Kato, T. (2014). Self-pierce riveting of multiple steel and aluminium alloy sheets. *Journal of Materials Processing Technology*, *214*(10), 2002–2008. doi:10.1016/j.jmatprotec.2013.09.007

Mucha, J. (2011). The analysis of lock forming mechanism in the clinching joint. *Materials & Design*, *32*(10), 4943–4954. doi:10.1016/j.matdes.2011.05.045

Polmcar, I. J. (1995). *Light Metals: Metallurgy of the light metals* (3rd ed.; Vol. 3). London: Halsted Pr.

Ravichandran, G., Rosakis, A. J., Hodowany, J., & Rosakis, P. (2002). AIP Conf. Proc.: Vol. 620. *On the Conversion of Plastic Work into Heat During High-Strain-Rate Deformation*. doi:10.1063/1.1483600

Rosakis, P., Rosakis, A. J., Ravichandran, G., & Hodowany, J. (2000). Thermodynamic internal variable model for the partition of plastic work into heat and stored energy in metals. *Journal of the Mechanics and Physics of Solids*, *48*(3), 581–607. doi:10.1016/S0022-5096(99)00048-4

Weichert, R., & Schonert, K. (2011). Heat Generation at the tip of a moving crack. *Journal of the Mechanics and Physics of Solids*, *26*(3), 151–161. doi:10.1016/0022-5096(78)90006-6

Wong, A. K., & Kirby, G. C. III. (1990). A hybrid numerical/experimental technique for determining the heat dissipated during low cycle fatigue. *Engineering Fracture Mechanics*, *37*(3), 493–504. doi:10.1016/0013-7944(90)90375-Q

Zehnder, A. T. (1991). A model for the heating due to plastic work. *Mechanics Research Communications*, *18*(1), 23–28. doi:10.1016/0093-6413(91)90023-P

Zehnder, A. T., Babinsky, E., & Palmer, T. (1998). Hybrid Method for Determining the Fraction of Plastic Work Converted to Heat. *Experimental Mechanics*, *38*(4), 295–302. doi:10.1007/BF02410392

Zehnder, A. T., & Rosakis, A. J. (1991). On the temperature distribution at the vicinity of dynamically propagating cracks in 4340 steel. *Journal of the Mechanics and Physics of Solids*, *39*(3), 385–415. doi:10.1016/0022-5096(91)90019-K

Zehnder, A. T., & Rosakis, A. J. (1993). Temperature Rise at the Tip of Dynamically Propagating Cracks: Measurements Using High-speed Infrared Detectors. Experimental Techniques in Fracture III. Society for Experimental Mechanics.

Chapter 19
An Insight on the Texture and Electrical Properties of Tomato Ketchup on a Temperature Scale

Indu Yadav
National Institute of Technology Rourkela, India

Suraj Kumar Nayak
National Institute of Technology Rourkela, India

Preeti Madhuri Pandey
National Institute of Technology Rourkela, India

Dibyajyoti Biswal
National Institute of Technology Rourkela, India

Arfat Anis
King Saud University, Saudi Arabia

Kunal Pal
National Institute of Technology Rourkela, India

ABSTRACT

The current study delineates the effect of incubation temperature on mechanical and electrical properties of ketchup. The analysis was done in the temperature range of 5°C and 25°C. Microstructure analysis of ketchup suggested the presence of cell wall components. Mechanical properties of ketchup were analyzed by spreadability and stress relaxation studies. Electrical properties of ketchup were analyzed by measuring the impedance in the frequency range of 100 Hz and 15 KHz. Spreadability study showed a higher firmness and cohesiveness in samples incubated at a lower temperature. The stress relaxation study suggested that inherent mechanical stability of the samples reached a critical lower limit at 15°C. Stability of the electrical equivalent circuit of samples was highest at 5°C. An increase in temperature drastically compromised the stability of electrical equivalent circuit of the samples. In gist, a variation in incubation temperature of the tomato ketchup can allow modulating not only the mechanical and viscoelastic properties, but also the electrical properties of ketchup.

DOI: 10.4018/978-1-5225-2440-3.ch019

INTRODUCTION

Tomato ketchup is one of the commonly used food products across the globe (Berta, Wiklund, Kotzé, & Stading). It may be produced either by cold or hot extraction process (Sánchez-Moreno, De Ancos, Plaza, Elez-Martínez, & Cano, 2008). The rheological properties of the tomato ketchup have been found to modulate the consumer acceptability. The thickening property of the starch and other hydrocolloids is dependent on temperature (Juszczak, Oczadły, & Gałkowska, 2013). Some of the commonly used hydrocolloids include guar gum, sodium alginate, pectin, carboxymethyl cellulose, xantham gum and gum acacia (Ma & Boye, 2015; Mert, 2012; Villay, et al., 2012). The use of these hydrocolloids reduces the chances of occurrence of serum separation (Mert, 2012). Usually, an increase in the temperature of hydrocolloids leads to a decrease in the thickening property of the hydrocolloids containing sols (Xu, Gong, Dong, & Li, 2015). Hence, it is quite expected that the rheological properties of the hydrocolloids containing tomato ketchup will also show altered rheological properties. The viscosity enhancing properties of the hydrocolloids is due to their ability to form physically crosslinked network structure (Kadam, Pochat-Bohatier, Sanchez, & El Ghzaoui, 2015). Hence, the addition of these hydrocolloids plays an important role in altering the rheological properties of ketchup (Foster, 2011). The physical properties of the hydrocolloid gels are altered by the alteration in the storage temperature (Albano, Franco, & Telis, 2014). Hence, the properties of the ketchup are also altered when stored at different temperatures. Since there is an alteration in the rheological properties of the ketchup with the variation in the storage temperature, it is highly likely that there will have a marked variation in the electrical properties of the ketchup. But a thorough literature survey indicated no study on the variation in the electrical properties of the ketchup and their relationship with the viscoelastic properties. Taking a note of the above, in this study, the alteration in the viscoelastic and electrical properties of the commercially available ketchup were studied in the temperature range of 5 °C to 25 °C. The viscoelastic properties of the ketchup were estimated from the large-scale stress relaxation study. The viscoelastic parameters were calculated using Kohlarusch (Enrione, et al., 2012) and Maxwell-Weichert (Fu, Chang, & Shiau, 2015) model of stress relaxation. The electrical properties were modeled using $(RQ)Q_1$ model (Biswal, et al., 2016), a commonly used electrical model for the analysis of soft materials. In addition to the above studies, the microstructure and the spreadability of the ketchup, incubated at different temperatures, were also studied.

BACKGROUND

Ketchup is an important tomato product prepared by hot or cold extraction of tomatoes. It can also be obtained directly from concentrates like tomato paste and tomato puree (Mert, 2012; Sahin & Ozdemir, 2004). The preparation of ketchup from tomatoes requires washing (for removal of dirt, stack and outside materials) and sorting of ripened tomatoes (for eliminating rotten, unripened and crushed tomatoes) followed by the mixing of peeled onion (1% of tomato weight) and garlic (0.1% of tomato weight). The mixture is then crushed (using a mixer), boiled (90 min), filtered (for removal of peel and seeds), concentrated and homogenized. Finally, modified corn starch, blended with flavorings (salt, sugar, grape vinegar and spice), is added to ketchup followed by pasteurization at 90 °C for 15 min and cooling to 25 °C (Yılmaz, Karaman, Cankurt, Kayacier, & Sagdic, 2011). However, Varela *et al.* (2003) have reported that the addition of corn starch can cause some quality problems like retrogradation (Varela, Gambaro, Giménez, Duran, & Lema, 2003). Therefore, various researchers have focused on the addition of gums into the ketchup because of their ability to behave as water binding and bodying agents (Gujral, Sharma, & Singh, 2002; Koocheki, Ghandi, Razavi, Mortazavi, & Vasiljevic, 2009; Sahin & Ozdemir, 2004; Şahin & Özdemir, 2007; Sidhu, Bawa, & Singh, 1997; Varela, et al., 2003). These studies have suggested that the incorporation of gums into the ketchup results in a higher consistency and a lower serum separation in the ketchup, although the use of non-tomato thickeners like gums into tomato ketchup is prohibited in certain countries and consumers also prefer to buy food products having minimal additives.

The consistency and flow behavior of ketchup during processing are regarded as important physical parameters. This is attributed to the fact that they are essential to obtain a constant and desired quality of ketchup (Barringer, Azam, Heskitt, & Sastry, 1998; Bottiglieri, et al., 1991; Jin Choi, et al., 2006). Therefore, various methods have been suggested to monitor the consistency and flow behavior of ketchup during processing. The consistency of ketchup is generally monitored by a Bostwick Consistometer. It provides a fast and cheap method for measuring the consistency of ketchup, which is regularly accomplished for samples taken during processing and for the final product. The correlation between the measurement of viscosity by shear rheology and the Bostwick measurement has been explored by various researchers (Barringer, et al., 1998; Bottiglieri, et al., 1991; Haley & Smith, 2003; McCarthy, Sacher, & Garvey, 2008). But, the rotational viscosity measurement becomes more expensive and the bostwick measurement is susceptible to a number of factors like skills of the user, cleanliness of instrument, the temperature of the ketchup and shape of guiding edge of the flow of the ketchup. In order to overcome these shortcomings, in-line measurement of ketchup viscosity has been proposed by various researchers, although at a higher cost. McCarthy *et al.* (2009) have reported the in-line measurement of viscosity using magnetic resonance (McCarthy & McCarthy, 2009). Similarly, Barringer *et al.* (1998), Haley *et al.* (2003) and Choi *et al.* (2006) have proposed the in-line measurement of viscosity using pressure difference, in-line absorption photometry and ultrasonic Doppler velocimetry, respectively (Barringer, et al., 1998; Haley & Smith, 2003; Jin Choi, et al., 2006). All these methods are characterized with shear flow based on the assumption that shear flow takes place in the pipe during processing. This assumption holds good only when the liquid sticks to the wall of the pipe and shearing flow occurs. However, it fails with the occurrence of any abrupt restriction in the cross-section of the pipe. In this case, the liquid accelerates because of the rapid change in cross-section of the pipe and the extensional flow becomes dominant (Cruz & Pinho, 2003; Debbaut & Crochet, 1988). Extensional flow is found to take place often during the processing of ketchup and squirting of ketchup out of a bottle.Therefore, Wikstrom *et al.* (1999) and Stading *et al.* (2001) have reported the offline extensional viscosity measurement of ketchup by hyperbolic contraction flow (Jin Choi, et al., 2006; Stading & Bohlin, 2001).

The main quality parameters of ketchup are color, flavor and flow properties. Numerous studies have reported the relationship between the quality parameters and the rheological attributes of ketchup (Anthon, Diaz, & Barrett, 2008; Bayod, Willers, & Tornberg, 2008; Koocheki, et al., 2009; PokorNý, 2009; Sahin & Ozdemir, 2004; Valencia, Sanchez, Ciruelos, & Gallegos, 2004; Varela, et al., 2003). The rheological attributes of ketchup are dependent on the quantity of enzymatic degradation, pectin, pectin/protein proportion, pulp amount and the process of homogenization. The rheological properties of ketchup, like viscoelastic properties and yield stress, are considered as important as they can divulge detailed information regarding the microstructure (Valencia, et al., 2004). These properties also play an important role during handling, processing and storage of ketchup (Sharoba, Senge, El-Mansy, Bahlol, & Blochwitz, 2005). The afore-mentioned studies, describing the relationship between the quality parameters and the rheological properties of ketchup, have suggested the use of some kind of hydrocolloids for enhancement of the rheological properties. Hydrocolloids are polysaccharides having water-solubility and high molecular weight, which perform a number of functions in food systems like improvement of viscosity, construction of gel structures, making of a film, regulation of crystallization, retardation of syneresis, improvement of texture, encasement of flavors and enhancement of the physical stability (Dickinson, 2003; Garti & Reichman, 1993; Glicksman, 1991; Sahin & Ozdemir, 2004). Another method of improving the rheological properties of ketchup is valve homogenization. In this method, the ketchup is passed through a microscopic orifice in the homogenizing valve. It results in high turbulence and shear along with compression, acceleration, pressure drop and impact. Therefore, breakdown and distribution of the tomato solids throughout the ketchup take place resulting in an improvement in the quality of ketchup.

The electrical impedance of any matter is a function of its structure, which may be used to differentiate whether the ketchup is having accumulated solid particles or uniform mixed particles (Dean, Ramanathan, Machado, & Sundararajan, 2008). Halter *et al.* (2007) have reported that impedance spectroscopy can be used to differentiate between normal and cancerous tissues (Halter, Hartov, Heaney, Paulsen, & Schned, 2007). Many researchers have used impedance spectroscopy for superior comprehension of the electric and dielectric characteristics of biological tissues as it divulges important information essential for an understanding of electrical stimulations and electroporation (Pethig, 1987; Pliquett & Prausnitz, 2000). The electrical impedance of a substance at a series of frequencies provides information about the change in the electrical behavior of that substance. Impedance spectroscopy provides information about the electric pulses like electroporation and electric stimulation (Pliquett & Prausnitz, 2000).

Viscosity is one of the important quality determining characteristics of tomato ketchup. Very less research has been reported on the non-linear rheological properties of the tomato ketchup. Mackley *et al.* (1994) have studied the linear and non-linear rheological behavior of the commercial ketchup by using controlled-strain rheometer (Mackley, Marshall, Smeulders, & Zhao, 1994). They have proposed the steady-state flow of ketchup by fitting the data obtained from the experiment with non-linear viscoelastic Wagner model. Based on the results, they have revealed the influence of processing variables like sieve orifice size and breaking temperature on the viscous flow and viscoelastic behavior of tomato pastes prepared from a type of tomato having a reduced amount of pectolytic enzyme called polygacturonase (PG). Valencia *et al.* (2003) have reported the characterization and modeling of non-linear rheological parameters of tomato paste samples prepared under diverse situations like different sieve orifice size and breaking temperature. This study included flow measurement, dynamic linear viscoelastic and stress relaxation studies performed on the afore-mentioned tomato paste products (Valencia, et al., 2003). Yoo *et al.* (1994) have reported the consequence of unimodal particle size and pulp content on rheological characteristics of tomato purees (Yoo & Rao, 1994). Their study suggests that tomato paste can be regarded as the scattering of solid particles (pulp) in the aqueous medium (serum) obtained from the concentration of tomato juice after removing seed and peels of the tomato (Hayes, Smith, & Morris, 1998) (Yoo & Rao, 1994).

MAIN FOCUS OF THE CHAPTER

In recent years, there has been an increased research on the analysis of the properties of the food products. Ketchup is one of the commonly studied food product. Though composition-dependent viscoelastic properties of ketchup samples have been reported by many researchers, no reports were found on the dependence of the viscoelastic properties on a temperature scale. In this paper, we have tried to analyze the viscoelastic properties of a commercially available ketchup sample on a temperature scale and correlate the results with the impedance spectroscopic analysis. Impedance spectroscopy is a non-destructive testing method unlike texture analysis. Also, the ease of conducting impedance spectroscopy is an added advantage. The present study may open a new dimension in analyzing the textural properties of food products.

MATERIALS

Tomato ketchup (Capital Food Pvt. Ltd, Nasik, India) was procured from the local market.

METHODS

Sample Preparation

5 g of the tomato ketchup was collected in 50 ml glass beaker and was subsequently incubated in a temperature-controlled cabinet for 1 h. The samples incubated at 5 °C, 10 °C, 15 °C, 20 °C, and 25 °C were regarded as F1, F2, F3, F4 and F5, respectively. The temperature equilibrated tomato ketchup samples were used for further analysis.

Microscopic Study

The microstructure of the ketchup samples was visualized using a bright-field microscope (DM 750, Leica Microsystems, Germany).

Texture Analysis

The mechanical properties of the ketchup samples were analyzed by spreadability and stress relaxation studies using static mechanical tester (TA-HD plus, Stable Microsystems, U.K.).

Electrical Study

The electrical property of the ketchup samples was measured in the frequency range of 100 Hz and 15 KHz. The study was performed using ELVIS II+ (National Instruments, U.S.A.). All the studies were performed in triplicate and the results are reported as mean ± standard deviation.

RESULTS AND DISCUSSION

Microscopy

The micrographs of the ketchup samples have been shown in Figure 1. The microstructure of the ketchup samples observed in our study was in concurrence with the microstructure of the ketchup reported by the other authors (Tan & Kerr, 2015). The samples did not show the presence of tomato skin and seeds.

Figure 1. Microstructure of the ketchup samples: (a) F1, (b) F2, (c) F3, (d) F4, and (e) F5

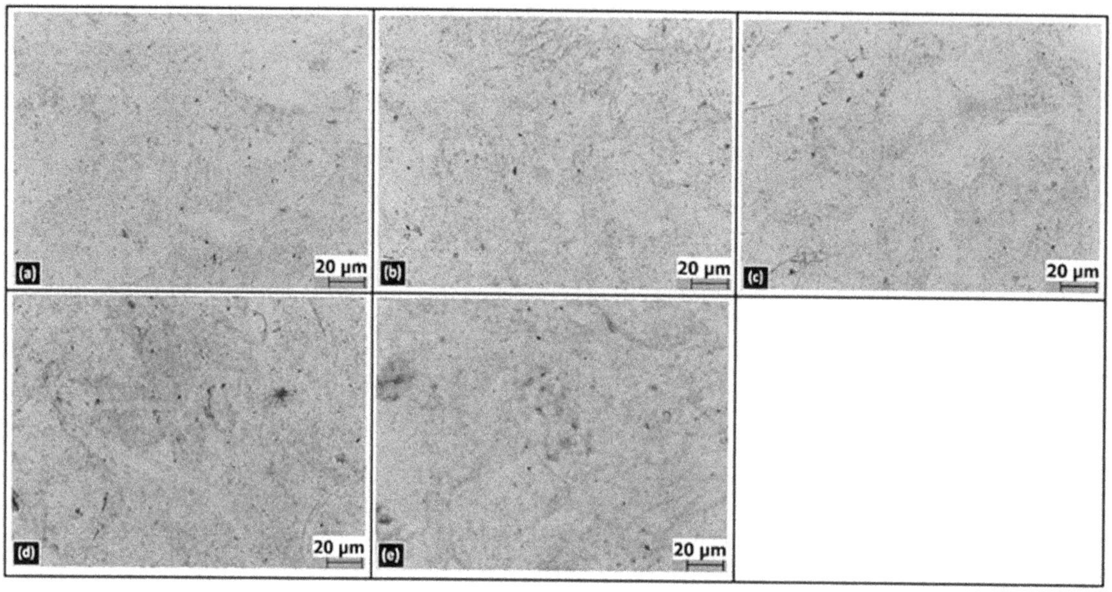

This is because the skin and seeds are removed during processing of the ketchup. The presence of fibrils was seen throughout the ketchup matrices. All the ketchup samples showed the presence of several fragments of the tomato cell wall. Apart from the cell wall fragments, randomly distributed cellular materials were also observed. Some intact cells from the mesocarp layer (mainly soft parenchymal cells) were also observed (Tan & Kerr, 2015). Clusters of cellular arrays were also observed. No significant changes in the microstructure of the ketchup samples were observed.

Texture Analysis

The initial mechanical property of the ketchup samples was measured by spreadability study (Table 1). The peak force (F_m) during the compression stage was highest in F1 and F2, which showed similar peak forces. Thereafter, an increase in the temperature to 15 °C (in F3) resulted in a drastic reduction in the peak force of the ketchup sample. Thereafter, when the incubation temperature was increased to 20 °C (in F4), there was a further reduction in the peak force. A further increase in the incubation temperature to 25 °C did not significantly alter the peak force attained. The peak force attained during the compression cycle is generally inversely related to spreadability (Hanim, Chin, & Yusof, 2015). From the results, it can be predicted that an increase in the incubation temperature of the tomato ketchup results in the decrease in the spreadability (Figure 2a). The area under the positive peak provides information about the energy required during spreading (Eroglu, et al., 2015). The energy required for spreading the ketchup samples was higher in ketchup samples incubated at a lower temperature. The spreading energies (proportional to the area under the positive peak; A_m) were significantly different from each other. Similar to the maximum force attained during the compression stage, the magnitude of the force sensed during the retraction of the probe (negative force; F_n) was also found in the same order. The maximum negative force provides information about the stickiness of the samples (Kaaki, Baghdadi, Najm, & Olabi, 2012). The results suggest that there was a decrease in the stickiness of the ketchup samples as the incubation

Table 1. Mechanical parameters of the ketchup samples

Test	Model	Parameters	Samples				
			F1	F2	F3	F4	F5
Spreadability	--	F_m	56.051 ± 0.511	56.166 ± 0.472	51.675 ± 0.643	48.09 9 ± 0.322	48.05 4 ± 0.624
		F_n	-33.825 ± 0.942	-32.728 ± 0.367	-32.013 ± 0.154	-28.821 ± 0.166	-28.436 ± 0.281
		A_p	177.764 ± 0.383	167.657 ± 0.365	163.775 ± 0.462	150.594 ± 0.399	147.438 ± 0.342
		A_n	-92.712 ± 0.664	-96.604 ± 0.992	-87.836 ± 0.270	-82.054 ± 0.821	-81.742 ± 0.201
Stress relaxation	--	F_o	5.422 ± 0.810	4.874 ± 0.010	3.602 ± 0.771	3.475 ± 0.572	3.208 ± 0.021
		F_r	2.112 ± 0.365	1.801 ± 0.011	1.444 ± 0.370	1.212 ± 0.101	1.101 ± 0.108
		%SR	61.040 ± 0.568	63.048 ± 0.517	59.911 ± 0.562	65.122 ± 0.745	65.679 ± 0.134
	Kohlrausch	σ_∞/σ_0	0.476 ± 0.028	0.453 ± 0.047	0.385 ± 0.037	0.428 ± 0.031	0.428 ± 0.021
		σ_1/σ_0	0.524 ± 0.028	0.546 ± 0.047	0.614 ± 0.037	0.571 ± 0.031	0.571 ± 0.021
		T	3.929 ± 0.434	3.481 ± 0.364	1.335 ± 0.392	1.331 ± 0.130	1.286 ± 0.296
		B	1.183 ± 1.744	1.399 ± 0.858	0.630 ± 0.261	0.449 ± 0.377	0.417 ± 0.412
		R^2	0.938	0.939	0.927	0.936	0.932
	Weichert	P_0	0.466 ± 0.022	0.329 ± 0.048	0.259 ± 0.115	0.273 ± 0.083	0.271 ± 0.122
		P_1	0.159 ± 0.026	0.161 ± 0.033	0.137 ± 0.031	0.161 ± 0.053	0.175 ± 0.005
		τ_1	14.628 ± 8.039	6.417 ± 2.548	7.040 ± 1.864	6.573 ± 2.303	7.335 ± 2.901
		P_2	0.355 ± 0.001	0.352 ± 0.002	0.444 ± 0.037	0.357 ± 0.007	0.357 ± 0.046
		τ_2	0.997 ± 0.513	0.735 ± 0.3681	0.676 ± 0.299	0.761 ± 0.314	0.745 ± 0.2122
		P_3	0.016 ± 0.011	0.163 ± 0.002	0.176 ± 0.124	0.213 ± 0.150	0.207 ± 0.145
		τ_3	169.650 ± 0.142	169.677 ± 0.122	169.676 ± 0.125	168.782 ± 0.633	168.783 ± 0.753
		R^2	0.996	0.995	0.998	0.995	0.995

temperature was increased. The area under the negative peak (A_n) provides information about the cohesive nature of the samples (Bi, et al., 2015). A higher area under the negative peak suggests higher cohesive force amongst the components of the samples. The analysis of the results suggested a decrease in the cohesiveness of the ketchup samples when the incubation temperature is increased. The above results may be explained by the presence of hydrocolloids, used as a thickening agent (Hong, Wu, Liu, & Gu, 2015). Usually, an increase in the temperature of the aqueous dispersion of the hydrocolloids decreases the viscosity of the dispersion. This may be explained by the decrease in the interpolymeric interactions among the hydrocolloid chains with the increase in the temperature.

Though spreadability study provides indirect information about the interactions amongst the components of the samples, the study fails to provide information about the viscoelastic property of the samples. Hence, to analyze the viscoelastic property of the ketchup samples, the stress relaxation study

Figure 2. Mechanical properties of the ketchup samples: (a) spreadability profiles, (b) stress relaxation profiles, (c) modeling of relaxation profiles using Kohlrausch model, and (d) modeling of relaxation profiles using Weichert model

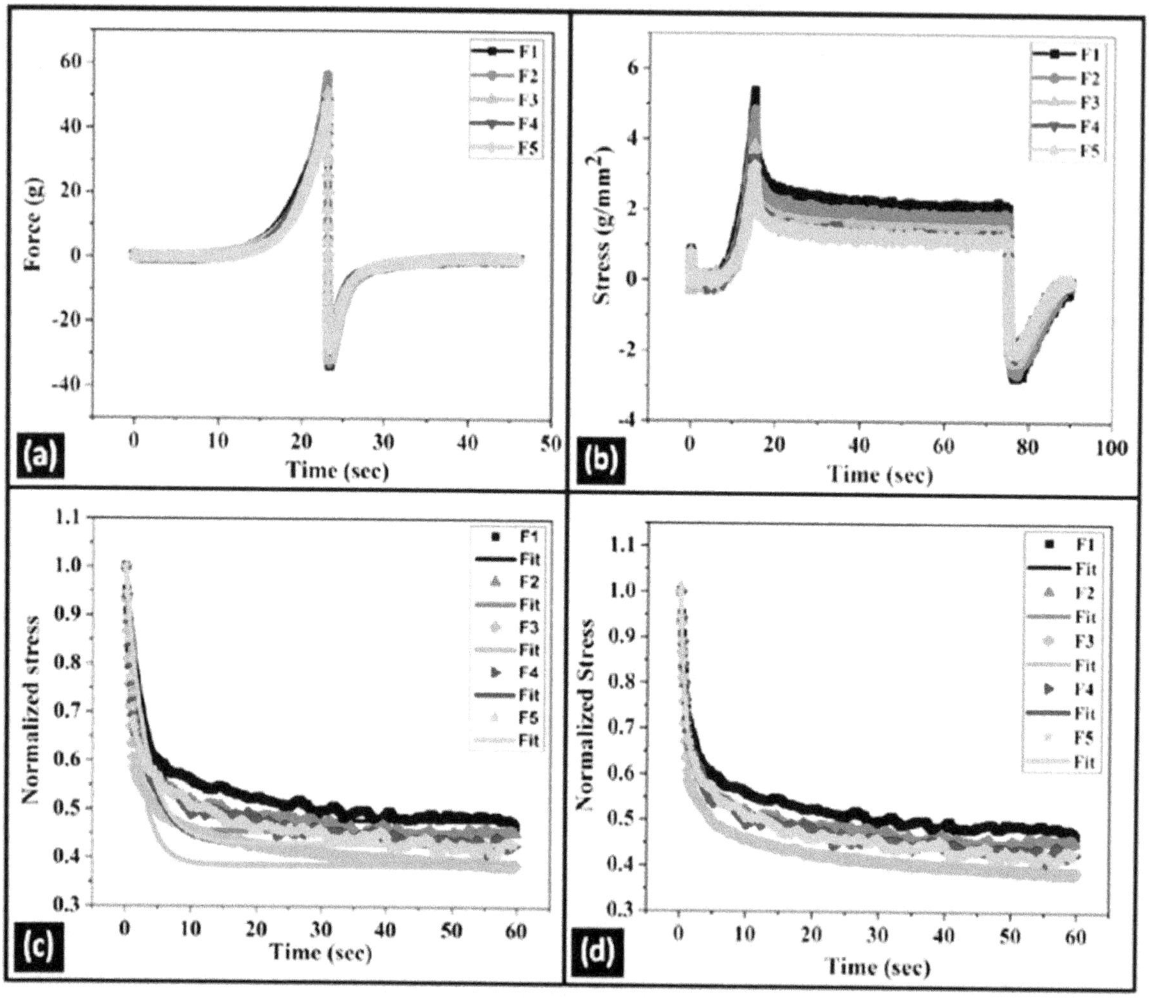

was performed (Figure 2b) (Myhan, Markowski, Daszkiewicz, Zapotoczny, & Sadowski, 2015). During the compression stage of stress relaxation study, the maximum force (F_0) was lower in samples incubated at a higher temperature. This result is in concurrence with the result (peak force) obtained from the spreadability study. The results suggest a reduction in the firmness of the ketchup samples with a corresponding increase in the incubation temperature. During the relaxation stage, the maximum force underwent an exponential decay to a residual force (F_r). The residual force was also in the same order as that of maximum force F_0. % stress relaxation (%SR; Equation 1) was calculated from the maximum and residual forces (Findley & Davis, 2013). An increase in the incubation temperature resulted in the corresponding increase in the stress relaxation values. Samples exhibiting a %SR value of '100' are categorized as viscous materials, whereas, %SR value of '0' suggest the sample under test is a true elastic material (Findley & Davis, 2013). The %SR values suggested that the ketchup samples behaved as viscoelastic materials, having a higher degree of fluidity (Augusto, Falguera, Cristianini, & Ibarz, 2011).

$$\%SR = \frac{W_2 - W_1}{W_1} \times 100 \quad (1)$$

where, F_0 is the maximum force attained during the compression stage, and F_r is the residual force at the end of the relaxation period.

The relaxation profiles of the ketchup samples were preliminarily analyzed by Kohlrausch model of viscoelasticity (Table 1, Equation 2, Figure 2c) (Maria, et al., 2014). The model fitted well with the relaxation profiles (acquired experimentally) ($R^2 > 0.90$). The limiting stress value (σ_∞/σ_0) was highest for F1. An increase in the temperature resulted in the corresponding decrease in the limiting stress values till F3. Thereafter, a further increase in incubation temperature in F4 (20 °C) and F5 (25 °C) resulted in the slight increase in the limiting stress values. The limiting stress values of F4 and F5 were equal. The alteration in the limiting stress values suggested that there has been an alteration in the interactions amongst the ketchup components at the molecular level (Maria, et al., 2014). In general, the limiting stress values are a marker of the retained elastic component of the samples after the relaxation process. From the results, it may be predicted that an increase in the temperature from 5 °C to 15 °C resulted in the decrease in the inherent elastic components of the samples. When the incubation temperature was increased to 20 °C and 25 °C, there was a slight increase in the inherent elastic components of the ketchup samples. This slight increase in the inherent elastic components of the samples may be due to the altered interaction among the components of the ketchup, which might be different than the interactions prevalent at lower temperature. The transient stress value (σ_1/σ_0) is a representative factor for the viscous component (Maria, et al., 2014). It is important to note that the addition of the limiting stress and the transient stress values equal to one. Hence, it was expected that the samples showing lower limiting stress values would show higher transient stress. The analysis of the results indicated that an increase in the incubation temperature from 5 °C to 15 °C correspondingly increased the viscous component of the samples. A further increase in the incubation temperature of the ketchup samples to 20 °C and 25 °C resulted in the slight decrease in the viscous component of the ketchup samples. Interestingly, the relaxation time (τ) of the ketchup samples was lowered as the incubation temperature of the ketchup samples was correspondingly increased. The relaxation time is a marker of rearrangement of the molecules of the samples when subjected to strain. The results suggested that an increase in the incubation temperature of the ketchup samples promoted molecular rearrangement. The Kohlrausch model of viscoelasticity is an over-simplistic model (Maria, et al., 2014). To extract detailed information on viscoelasticity, the relaxation profiles were fitted to Weichert model (Equation 3, Figure 2d) of viscoelasticity with three relaxation times (Maria, et al., 2014). The P_0 values (inherent elastic property) were in the similar order as that of the limiting stress values (calculated from the Kohlrausch model). The relaxation time of the viscoelastic materials during the relaxation process can be segregated into instantaneous relaxation time (τ_1), intermediate relaxation time (τ_2), and delayed relaxation time (τ_3) (Maria, et al., 2014). The instantaneous relaxation time was higher in the samples incubated at higher temperatures. This suggested an initial quick dissipation of the stress in the samples incubated at lower temperatures. The instantaneous relaxation values are a marker of molecular rearrangement when subjected to strain. Hence, it can be predicted that at lower temperatures, the molecular rearrangement was predominant and can be explained by the better inter-component interactions. Since the inter-component interactions were better in the samples incubated at lower temperatures (as predicted from the cohesive property of the samples), the dissipation of the induced stress was superior. The intermediate relaxation time of F1 was much higher than the other samples. The intermediate relaxation time provides information about the magnanimity of molecular interactions. The analysis of the results suggested much faster disruption of the molecular interactions in F2, F3, F4 and F5 as compared to F1. Delayed relaxation time provides information about the structural breakdown of the samples. A faster-delayed relaxation time is an indicator of quicker structural breakdown. The results suggested a quicker structural breakdown of F1 as compared to the other samples. It can be concluded from the above that though the disruption in the inter-component interactions in F1 occurred relatively slowly, the structural breakdown happened much quickly.

$$\frac{\sigma_\tau}{\sigma_0} = \frac{\sigma_\infty}{\sigma_0} + \frac{\sigma_1}{\sigma_0} \times e^{\left[-\left(\frac{t}{\tau}\right)\beta\right]} \quad (2)$$

where, τ is the relaxation time, β is the stretching parameter, σ_1/σ_0 is the transient stress, and σ_∞/σ_0 is the limiting stress.

$$P = P_0 + P_1.e^{\left(\frac{-t}{\tau_1}\right)} + P_2.e^{\left(\frac{-t}{\tau_2}\right)} + P_3.e^{\left(\frac{-t}{\tau_3}\right)} \quad (3)$$

where, P_0, P_1, P_2 and P_3 are the spring constants; and τ_1, τ_2 and τ_3 are time constants of the dashpots.

Electrical Study

The electrical properties of the samples were measured in the frequency range of 100 Hz to 15 KHz. A preliminary analysis of the impedance profiles showed that the samples which were incubated at lower temperatures had higher impedances. This can be explained by the increase in the local viscosity which results in the decrease in the mobility of the electrical carrier molecules (Mohanty & Mishra, 2013). The area under the positive peak of the spreadability profile also supports the increase in the local viscosity. The impedance of the ketchup samples reduced to residual impedance values on a frequency-dependent manner. The rate of decrease of the impedance to the residual impedance value was slower when the samples were incubated at a lower temperature. The decrease in the impedance values, as the frequency of the injecting current is increased, occurs due to the decrease in the polarization effect at the sample-electrode interface (Huang, et al., 2015). The nature of the impedance profiles suggested that the rate of decrease of polarization was slower in the samples incubated at lower temperatures. The impedance profile of the ketchup samples was modeled using (RQ)Q_1 electrical model (Equation 4). The element 'R' stands for resistive component of the circuit, whereas, the element 'Q' stands for constant phase element. A constant phase element is considered as the non-ideal capacitor associated with an inhomogeneous constant (n) (Mac Kenna, Calvert, & Morrin, 2015). The 'Q' in the first parenthesis provides information about the material capacitive element. On the other hand, 'Q_1' provide information about the sample-electrode interface capacitance. A good acceptance amongst the experimental and the predicted data was obtained (Table 2). The model parameters suggested that the resistive components of the ketchup samples were equal. A reduction in the capacitive value of the samples was observed with the reduction in the incubation temperature. This explains the higher polarization effect exhibited by the ketchup samples incubated at lower temperatures. Both the sample and the electrode-sample-

Table 2. Electrical parameters of the ketchup samples

Model	Parameters	Samples				
		F1	F2	F3	F4	F5
$(RQ)Q_1$ model	R (Ω)	1997.793	1998.499	2000.24	2000.43	2000.483
	Q (F)	9.66E-03	1.21E-02	1.90E-02	2.00E-02	6.37E-02
	Q_1 (F)	1.88E+00	1.75E-01	3.19E-01	6.99E-05	9.35E-03
	n	1.34E-01	1.27E-01	0.08	0.08	0.01
	n_1	0.13	0.12	0.23	0.98	0.27
	R^2	0.996	0.997	0.997	0.999	0.997

electrode interface capacitive components behaved as a non-ideal capacitor. This was evident from the inhomogeneous constants (n and n_1).

$$Z = \frac{R}{1+(jw)^{n}CR} + \frac{1}{1+(jw)^{n1}C1} \quad (4)$$

$$C = (jw)^{n-1}Q \quad (5)$$

$$C1 = (jw)^{n1-1}Q1 \quad (6)$$

Z is defined as the impedance of the ketchup, R is the resistance of the ketchup, C and C1 are the capacitance of the ketchup, w is the frequency component, n and n1 are the inhomogeneous constants. Q and Q1 are the constant phase elements

The impulse response, Nyquist and Bode plots were calculated using the $(RQ)Q_1$ model parameters. A MATLAB program was used for calculating the plots. Impulse response profile is the output of the equivalent electrical circuit when an impulse signal is used as the input (Assambo & Burke, 2012). The impulse response profile of a stable electrical circuit rapidly decays to a zero residual value. A non-zero residual value is a marker of an unstable system. In general, higher the residual value, lower is the stability of the electrical circuit. The impulse response profile (Figure 3c) of the equivalent electrical circuit of the ketchup samples showed a higher residual value as the incubation temperature was increased. This indicated a decrease in the electrical stability of the ketchup samples with the increase in the incubation temperature. This may be due to the higher reactance exhibited by the samples incubated at lower temperature, as was evident from the constant phase element values.

The Nyquist plot (Figure 4) of F1 showed the formation of a semicircular curve in the low-frequency region (Warren, Gately, O'Brien, & Gorkin, 2014). This suggested a good dielectric property exhibited by F1 and can be explained by the highest stability of the equivalent electrical circuit of F1, as was predicted from the impulse response profiles. As the incubating temperature was increased, though the ketchup samples showed initiation of the formation of a semicircular curve, the curve was converted to spike at

Figure 3. (a) Impedance profile of the ketchup samples, (b) fitting of the experimental data using (RQ) Q_1 model, and (c) impulse response of the equivalent electrical circuit of the samples

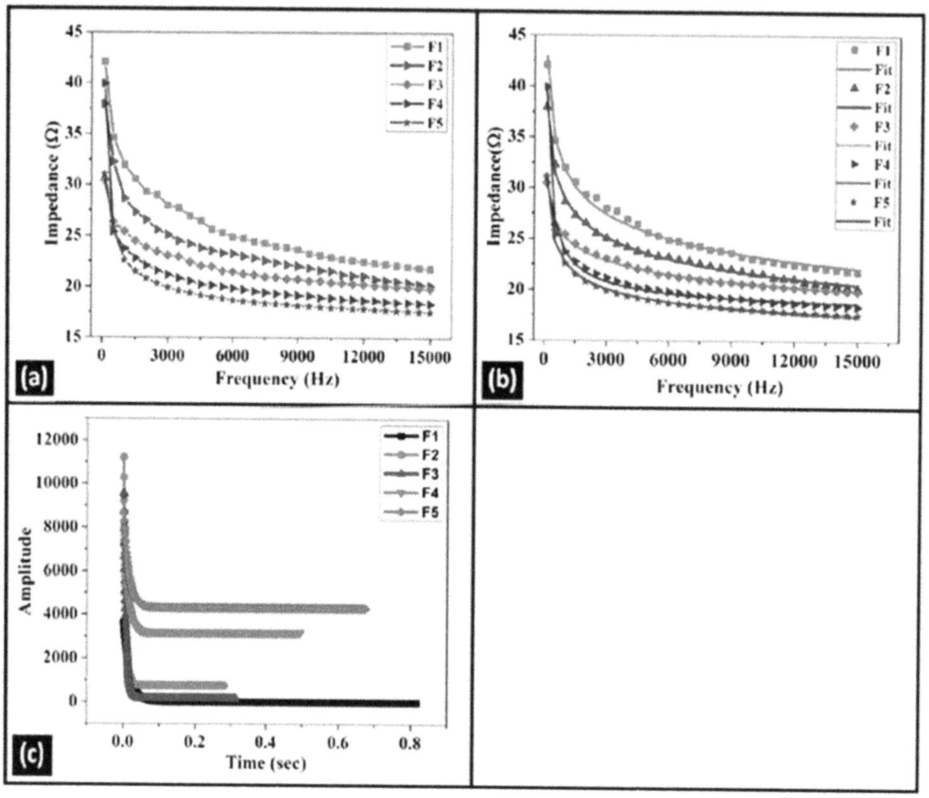

Figure 4. Nyquist plot of the ketchup samples: (a) F1, (b) F2, (c) F3, (d) F4, and (e) F5

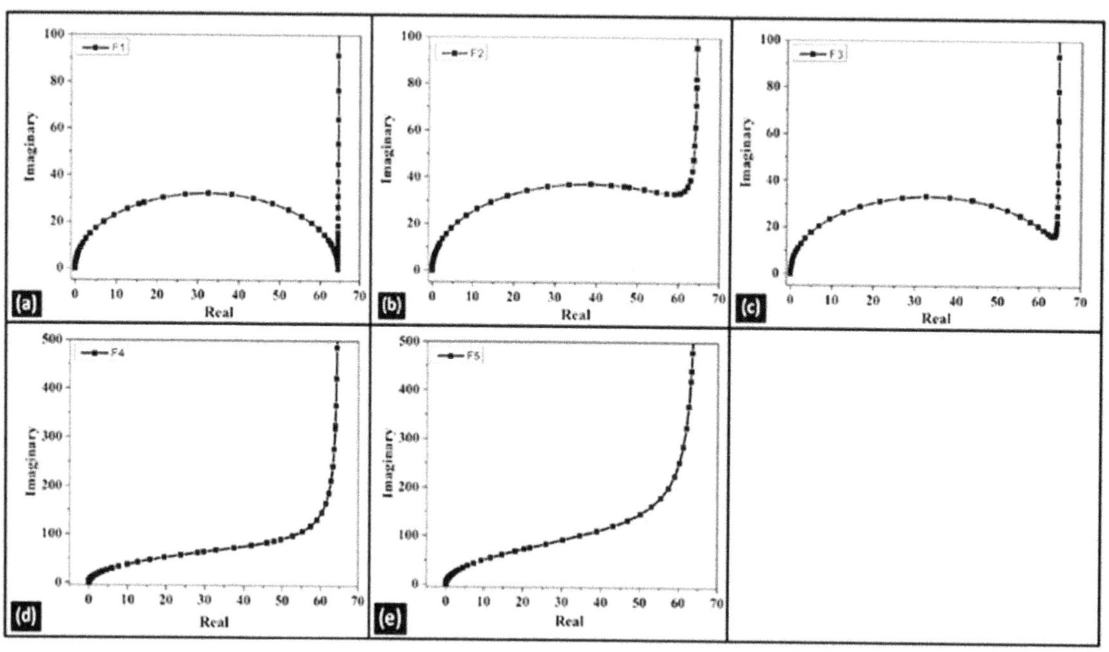

the high-frequency region. This suggested that the samples behaved as a poor dielectric medium when incubated at a high temperature and can be explained by the relatively unstable nature of the equivalent electrical circuit of the samples.

The magnitude response (Bode plot, Figure 5) of the samples showed three distinct regions (Warren, et al., 2014). A monotonously decreasing amplitude was observed in the low-frequency region, followed by a constant magnitude in the mid-frequency region and again a monotonously decreasing amplitude in the high-frequency region. Corresponding to the mid-frequency region the phase response of the samples showed a peak. The area under the phase response peak provides information about the energy required to mobilize the molecules within the samples. In general, a higher area under the curve of the phase response suggests that a higher amount of energy is required to induce the mobility of the molecules of the samples. From the phase plot, it can be observed that the area under the curve was highest in F1. An increase in the temperature resulted in the phase plot with a lower area under the curve. The analysis of the Bode plot also suggests that the stability of the samples increases with the decrease in the incubation temperature.

FUTURE RESEARCH DIRECTIONS

Though the results obtained from the current study are encouraging, calculation of tomato ketchup shear and extensional viscosity can be performed during the future studies for the better understanding of the ketchup quality. The rheological properties of ketchup should also be studied with the addition of natural fibers made from fresh tomato paste during the future studies.

Figure 5. Bode plot of the ketchup samples: (a) F1, (b) F2, (c) F3, (d) F4, and (e) F5

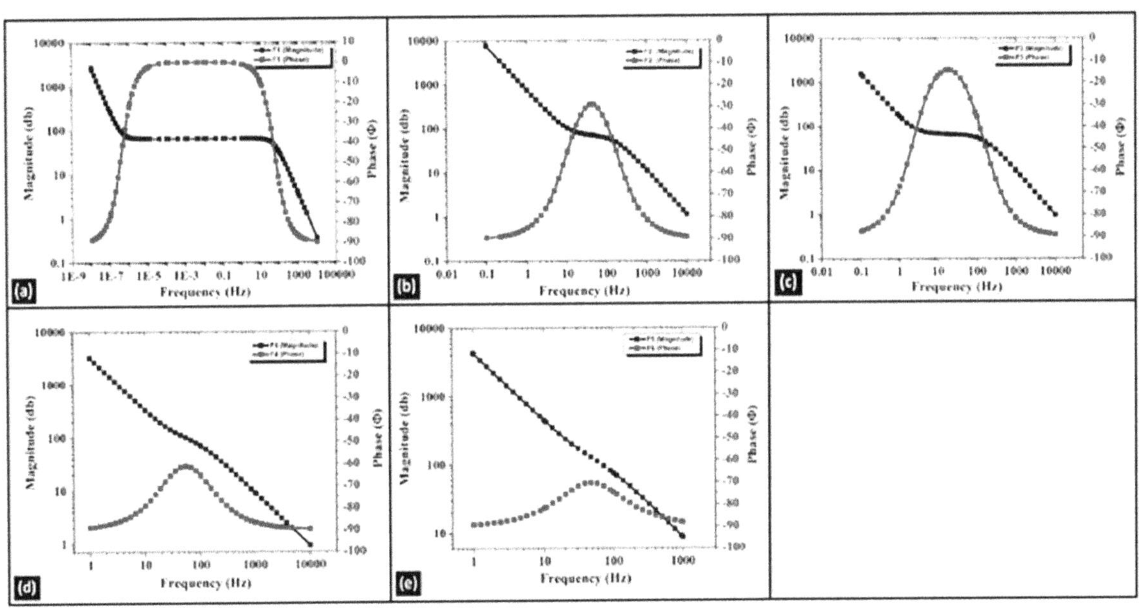

CONCLUSION

In this study, an attempt was made to understand the effect of temperature on the mechanical and the electrical properties of the ketchup samples. The temperature of the ketchup samples was varied in the temperature range of 5 °C and 25 °C. The microscopic study showed the presence of cellular components of the tomato, a common characteristic of the tomato ketchup. No significant changes amongst the microstructure of the ketchup samples were observed. An increase in the temperature of the incubation temperature reduced the firmness of the ketchup samples. This was due to the increase in the cohesive interactions amongst the components of the ketchup. An increase in the fluidic component of the samples was observed with the increase in the incubation temperature. An in-depth analysis of the viscoelastic nature of the samples showed that the inherent mechanical stability of the samples reached a critical lower limit when the samples were incubated at 15 °C. The impedance analysis showed that the bulk resistance was equal in all the samples. On the other hand, the capacitive element of the samples increased with the increase in the incubation temperature. Since capacitance is inversely related to the reactance (impedance due to the capacitor) offered by the capacitive element, an increase in the impedance of the samples with the decrease in the incubating temperature can be well explained by the capacitive dominant nature of the samples. The decrease in the rate of polarization was slower in the samples incubated at a lower temperature and can be associated with the nature of the capacitive element. This can also explain the electrical stability of the equivalent circuit due to the variation in the local viscosity of the samples. From the analysis of the ketchup samples, it was found that an alteration in the incubation temperature of the samples can significantly tailor the mechanical and the electrical properties.

ACKNOWLEDGMENT

The authors acknowledge the support provided by National Institute of Technology Rourkela for the completion of this study.

REFERENCES

Albano, K. M., Franco, C. M., & Telis, V. R. (2014). Rheological behavior of Peruvian carrot starch gels as affected by temperature and concentration. *Food Hydrocolloids*, *40*, 30–43. doi:10.1016/j.foodhyd.2014.02.003

Anthon, G. E., Diaz, J. V., & Barrett, D. M. (2008). Changes in pectins and product consistency during the concentration of tomato juice to paste. *Journal of Agricultural and Food Chemistry*, *56*(16), 7100–7105. doi:10.1021/jf8008525 PMID:18605731

Assambo, C., & Burke, M. J. (2012). *Low-Frequency Response and the Skin-Electrode Interface in Dry-Electrode Electrocardiography*. INTECH Open Access Publisher. doi:10.5772/22777

Augusto, P. E., Falguera, V., Cristianini, M., & Ibarz, A. (2011). Viscoelastic properties of tomato juice. *Procedia Food Science*, *1*, 589–593. doi:10.1016/j.profoo.2011.09.089

Barringer, S. A., Azam, A. T. M. S., Heskitt, B., & Sastry, S. (1998). On-line prediction of bostwick consistency from pressure differential in pipe flow for ketchup and related tomato products1. *Journal of Food Processing and Preservation*, *22*(3), 211–220. doi:10.1111/j.1745-4549.1998.tb00346.x

Bayod, E., Willers, E. P., & Tornberg, E. (2008). Rheological and structural characterization of tomato paste and its influence on the quality of ketchup. *LWT-Food Science and Technology*, *41*(7), 1289–1300. doi:10.1016/j.lwt.2007.08.011

Berta, M., Wiklund, J., Kotzé, R., & Stading, M. (n.d.). Correlation between in-line measurements of tomato ketchup shear viscosity and extensional viscosity. *Journal of Food Engineering*.

Bi, W., Li, X., Zhao, Y., Zhao, W., He, S., Ge, W., et al. (2015). Imitation cheese manufacture using Rapid Visco-Analyzer and its optimization. *International Journal of Food Properties*.

Biswal, D., Anupriya, B., Uvanesh, K., Anis, A., Banerjee, I., & Pal, K. (2016). Effect of mechanical and electrical behavior of gelatin hydrogels on drug release and cell proliferation. *Journal of the Mechanical Behavior of Biomedical Materials*, *53*, 174–186. doi:10.1016/j.jmbbm.2015.08.017 PMID:26327452

Bottiglieri, P., Sio, F., Fasanaro, G., Mojoli, G., Impembo, M., & Castaldo, D. (1991). Rheological characterization of ketchup. *Journal of Food Quality*, *14*(6), 497–512. doi:10.1111/j.1745-4557.1991.tb00089.x

Cruz, D., & Pinho, F. (2003). Turbulent pipe flow predictions with a low Reynolds number k–ε model for drag reducing fluids. *Journal of Non-Newtonian Fluid Mechanics*, *114*(2), 109–148. doi:10.1016/S0377-0257(03)00119-8

Dean, D., Ramanathan, T., Machado, D., & Sundararajan, R. (2008). Electrical impedance spectroscopy study of biological tissues. *Journal of Electrostatics*, *66*(3), 165–177. doi:10.1016/j.elstat.2007.11.005 PMID:19255614

Debbaut, B., & Crochet, M. (1988). Extensional effects in complex flows. *Journal of Non-Newtonian Fluid Mechanics*, *30*(2-3), 169–184. doi:10.1016/0377-0257(88)85023-7

Dickinson, E. (2003). Hydrocolloids at interfaces and the influence on the properties of dispersed systems. *Food Hydrocolloids*, *17*(1), 25–39. doi:10.1016/S0268-005X(01)00120-5

Enrione, J. I., Sáez, C., López, D., Skurtys, O., Acevedo, C., Osorio, F., & Hill, S. et al. (2012). Structural relaxation of salmon gelatin films in the glassy state. *Food and Bioprocess Technology*, *5*(6), 2446–2453. doi:10.1007/s11947-011-0618-3

Eroglu, A., Bayrambaş, K., Eroglu, Z., Toker, O. S., Yilmaz, M. T., & Karaman, S. (2015). Steady, dynamic, creep/recovery, and textural properties of yoghurt/molasses blends: Temperature sweep tests and applicability of Cox–Merz rule. *Food Science & Technology International*. PMID:25614154

Findley, W. N., & Davis, F. A. (2013). *Creep and relaxation of nonlinear viscoelastic materials*. Courier Corporation.

Foster, T. (2011). Natural structuring with cell wall materials. *Food Hydrocolloids*, *25*(8), 1828–1832. doi:10.1016/j.foodhyd.2011.05.016

Fu, J.-T., Chang, Y.-H., & Shiau, S.-Y. (2015). Rheological, antioxidative and sensory properties of dough and Mantou (steamed bread) enriched with lemon fiber. *LWT-Food Science and Technology*, *61*(1), 56–62. doi:10.1016/j.lwt.2014.11.034

Garti, N., & Reichman, D. (1993). Hydrocolloids as food emulsifiers and stabilizers. *Food Structure*, *12*(4), 3.

Glicksman, M. (1991). Hydrocolloids and the search for the oily grail. *Food Technology*, *45*(10), 94–103.

Gujral, H. S., Sharma, A., & Singh, N. (2002). Effect of hydrocolloids, storage temperature, and duration on the consistency of tomato ketchup. *International Journal of Food Properties*, *5*(1), 179–191. doi:10.1081/JFP-120015600

Haley, T. A., & Smith, R. S. (2003). Evaluation of in-line absorption photometry to predict consistency of concentrated tomato products. *LWT-Food Science and Technology*, *36*(2), 159–164. doi:10.1016/S0023-6438(02)00218-9

Halter, R. J., Hartov, A., Heaney, J. A., Paulsen, K. D., & Schned, A. R. (2007). Electrical impedance spectroscopy of the human prostate. *Biomedical Engineering. IEEE Transactions on*, *54*(7), 1321–1327.

Hanim, M., Chin, N., & Yusof, Y. (2015). Effects of grinding time on rheological, textural and physical properties of natural peanut butter stored at different temperatures. *Journal of Texture Studies*.

Hayes, W., Smith, P., & Morris, A. (1998). The production and quality of tomato concentrates. *Critical Reviews in Food Science and Nutrition*, *38*(7), 537–564. doi:10.1080/10408699891274309 PMID:9813734

Hong, Y., Wu, Y., Liu, G., & Gu, Z. (2015). Effect of amylose on pasting and rheological properties of corn starch/xanthan blends. *Stärke*, *67*(1-2), 98–106. doi:10.1002/star.201400126

Huang, Y., Zheng, M., Lin, Z., Zhao, B., Zhang, S., Yang, J., & Shi, Y. et al. (2015). Flexible cathodes and multifunctional interlayers based on carbonized bacterial cellulose for high-performance lithium–sulfur batteries. *Journal of Materials Chemistry A*, *3*(20), 10910–10918. doi:10.1039/C5TA01515D

Jin Choi, Y., Milczarek, R. R., Fleck, C., Casey Garvey, T., McCarthy, K. L., & McCarthy, M. J. (2006). In-line monitoring of tomato concentrate physical properties during evaporation. *Journal of Food Process Engineering*, *29*(6), 615–632. doi:10.1111/j.1745-4530.2006.00085.x

Juszczak, L., Oczadły, Z., & Gałkowska, D. (2013). Effect of modified starches on rheological properties of ketchup. *Food and Bioprocess Technology*, *6*(5), 1251–1260. doi:10.1007/s11947-012-0813-x

Kaaki, D., Baghdadi, O. K., Najm, N., & Olabi, A. (2012). Preference mapping of commercial Labneh (strained yogurt) products in the Lebanese market. *Journal of Dairy Science*, *95*(2), 521–532. doi:10.3168/jds.2011-4409 PMID:22281316

Kadam, Y., Pochat-Bohatier, C., Sanchez, J., & El Ghzaoui, A. (2015). Modulating Viscoelastic Properties of Physically Crosslinked Self-Assembled Gelatin Hydrogels through Optimized Solvent Conditions. *Journal of Dispersion Science and Technology*, *36*(9), 1349–1356. doi:10.1080/01932691.2014.984721

Koocheki, A., Ghandi, A., Razavi, S., Mortazavi, S. A., & Vasiljevic, T. (2009). The rheological properties of ketchup as a function of different hydrocolloids and temperature. *International Journal of Food Science & Technology*, *44*(3), 596–602. doi:10.1111/j.1365-2621.2008.01868.x

Ma, Z., & Boye, J. I. (2015). Novel Health Ingredients and Their Applications in Salad Dressings and Other Food Emulsions. *Nutraceutical and Functional Food Processing Technology*, 151-187.

Mac Kenna, N., Calvert, P., & Morrin, A. (2015). Impedimetric transduction of swelling in pH-responsive hydrogels. *Analyst (London)*, *140*(9), 3003–3011. doi:10.1039/C4AN02252A PMID:25768307

Mackley, M., Marshall, R., Smeulders, J., & Zhao, F. (1994). The rheological characterization of polymeric and colloidal fluids. *Chemical Engineering Science*, *49*(16), 2551–2565. doi:10.1016/0009-2509(94)E0082-2

Maria, H. J., Lyczko, N., Nzihou, A., Joseph, K., Mathew, C., & Thomas, S. (2014). Stress relaxation behavior of organically modified montmorillonite filled natural rubber/nitrile rubber nanocomposites. *Applied Clay Science*, *87*, 120–128. doi:10.1016/j.clay.2013.10.019

McCarthy, K. L., & McCarthy, M. J. (2009). Relationship between In-line Viscosity and Bostwick Measurement during Ketchup Production. *Journal of Food Science*, *74*(6), E291–E297. doi:10.1111/j.1750-3841.2009.01205.x PMID:19723191

McCarthy, K. L., Sacher, R. F., & Garvey, T. C. (2008). Relationship between rheological behavior and Bostwick measurement during manufacture of ketchup. *Journal of Texture Studies*, *39*(5), 480–495. doi:10.1111/j.1745-4603.2008.00155.x

Mert, B. (2012). Using high pressure microfluidization to improve physical properties and lycopene content of ketchup type products. *Journal of Food Engineering*, *109*(3), 579–587. doi:10.1016/j.jfoodeng.2011.10.021

Mohanty, M. E., & Mishra, A. K. (2013). Estimating viscosity and polarity in the microenvironment of polymeric gels—introducing a microviscosity parameter. *Journal of Polymer Research*, *20*(7), 1–8. doi:10.1007/s10965-013-0185-6

Myhan, R., Markowski, M., Daszkiewicz, T., Zapotoczny, P., & Sadowski, P. (2015). Non-linear stress relaxation model as a tool for evaluating the viscoelastic properties of meat products. *Journal of Food Engineering*, *146*, 107–115. doi:10.1016/j.jfoodeng.2014.09.006

Pethig, R. (1987). Dielectric properties of body tissues. *Clinical Physics and Physiological Measurement*, *8*(4A), 5–12. doi:10.1088/0143-0815/8/4A/002 PMID:3568571

Pliquett, U., & Prausnitz, M. R. (2000). Electrical impedance spectroscopy for rapid and noninvasive analysis of skin electroporation. *Electrochemotherapy, Electrogenetherapy, and Transdermal Drug Delivery: Electrically Mediated Delivery of Molecules to Cells*, 377-406.

Pokorný, J. (2009). Textural and flavour characteristics of commercial tomato ketchups. *Czech J. Food Sci.*, *27*(3), 165–170.

Sahin, H., & Ozdemir, F. (2004). Effect of some hydrocolloids on the rheological properties of different formulated ketchups. *Food Hydrocolloids*, *18*(6), 1015–1022. doi:10.1016/j.foodhyd.2004.04.006

Şahin, H., & Özdemir, F. (2007). Effect of some hydrocolloids on the serum separation of different formulated ketchups. *Journal of Food Engineering*, *81*(2), 437–446. doi:10.1016/j.jfoodeng.2006.11.022

Sánchez-Moreno, C., De Ancos, B., Plaza, L., Elez-Martínez, P., & Cano, M. P. (2008). Nutritional characterization of tomato juices. *Tomatoes and Tomato Products: Nutritional, Medicinal, and Therapeutic Properties*, 235-258.

Sharoba, A., Senge, B., El-Mansy, H., Bahlol, H. E., & Blochwitz, R. (2005). Chemical, sensory and rheological properties of some commercial German and Egyptian tomato ketchups. *European Food Research and Technology*, *220*(2), 142–151. doi:10.1007/s00217-004-0981-7

Sidhu, J. S., Bawa, A., & Singh, N. (1997). Studies on the effect of hydrocolloids on the consistency of tomato ketchup. *Journal of Food science and Technology, 34*(5), 423-424.

Stading, M., & Bohlin, L. (2001). Contraction flow measurements of extensional properties. *Annual Transactions-Nordic Rheology Society*, *8*, 181–186.

Tan, J., & Kerr, W. (2015). Rheological properties and microstructure of tomato puree subject to continuous high pressure homogenization. *Journal of Food Engineering*, *166*, 45–54. doi:10.1016/j.jfoodeng.2015.05.025

Valencia, C., Sanchez, M., Ciruelos, A., & Gallegos, C. (2004). Influence of tomato paste processing on the linear viscoelasticity of tomato ketchup. *Food Science & Technology International*, *10*(2), 95–100. doi:10.1177/1082013204043880

Valencia, C., Sanchez, M., Ciruelos, A., Latorre, A., Madiedo, J., & Gallegos, C. (2003). Non-linear viscoelasticity modeling of tomato paste products. *Food Research International*, *36*(9), 911–919. doi:10.1016/S0963-9969(03)00100-5

Varela, P., Gambaro, A., Giménez, A., Duran, I., & Lema, P. (2003). Sensory and instrumental texture measures on ketchup made with different thickeners. *Journal of Texture Studies*, *34*(3), 317–330. doi:10.1111/j.1745-4603.2003.tb01064.x

Villay, A., de Filippis, F. L., Picton, L., Le Cerf, D., Vial, C., & Michaud, P. (2012). Comparison of polysaccharide degradations by dynamic high-pressure homogenization. *Food Hydrocolloids*, *27*(2), 278–286. doi:10.1016/j.foodhyd.2011.10.003

Warren, H., Gately, R. D., OBrien, P., Gorkin, R. III, & in het Panhuis, M. (2014). Electrical conductivity, impedance, and percolation behavior of carbon nanofiber and carbon nanotube containing gellan gum hydrogels. *Journal of Polymer Science. Part B, Polymer Physics*, *52*(13), 864–871. doi:10.1002/polb.23497

Xu, L., Gong, H., Dong, M., & Li, Y. (2015). Rheological properties and thickening mechanism of aqueous diutan gum solution: Effects of temperature and salts. *Carbohydrate Polymers*, *132*, 620–629. doi:10.1016/j.carbpol.2015.06.083 PMID:26256389

Yılmaz, M. T., Karaman, S., Cankurt, H., Kayacier, A., & Sagdic, O. (2011). Steady and dynamic oscillatory shear rheological properties of ketchup–processed cheese mixtures: Effect of temperature and concentration. *Journal of Food Engineering*, *103*(2), 197–210. doi:10.1016/j.jfoodeng.2010.10.016

Yoo, B., & Rao, M. (1994). Effect of unimodal particle size and pulp content on rheological properties of tomato puree. *Journal of Texture Studies*, *25*(4), 421–436. doi:10.1111/j.1745-4603.1994.tb00772.x

KEY TERMS AND DEFINITIONS

Compression Cycle: It is the period during which the capacity of a material to withstand the loads which are tending to reduce its size.

Hydrocolloid: Substance which converts into the gel in the presence of water.

Impedance: It is the effective opposition to the electric current in an electrical circuit.

Ketchup: Ketchup is a processed product of tomato.

Spreadability: It is the property of a fluid which defines the ability of the fluid to spread.

Stress Relaxation: It can be defined as the gradual decrease in stress in response to a constant strain applied on a substance.

Viscoelasticity: It is defined as the property of an object that possesses both viscous and elastic nature.

Chapter 20
Prediction of Water Quality Indices by Using Artificial Neural Network Models:
Prediction of Water Quality Indices

Prakash Chandra Mishra
Fakir Mohan University, India

Anil Kumar Giri
Fakir Mohan University, India

ABSTRACT

Conventionally, fixed techniques are used for prediction of future time-series data. Subsequently adaptive techniques are used to forecast improved future data. The adaptive techniques are essentially based on ANN and fuzzy logic techniques. It is observed that these techniques also perform poorly when the input data set available is less and when there is abrupt change in the input data set. In this paper the proposed hybrid technique is based on data farming for intermediate data generation and the ANN model for better learning and forecasting. The performance of the proposed model has been tested with actual pertaining to water quality indices of various water samples collected from different sources.

1. INTRODUCTION

Time series forecasting finds extensive applications in many areas such as financial engineering, power system engineering, pollution control, river flow, stock market etc. Conventionally, various fixed techniques are being used for forecasting purpose. These fixed techniques suffer from accurate forecasting of time series data. To alleviate this problem adaptive techniques have been proposed in the literature. These are techniques based on artificial neural network (ANN) (Werbos, 1988, Patra et *al.*, 1997, Widro et *al*, 1998) and fuzzy logic (Dash et *al*, 1995, Dash et *al*, 1995). Peng et *al., 1992* has applied the ANN technique for forecasting of electric load. These adaptive techniques involve more computational load and large training time. Conventionally, these techniques perform two tasks. In the first phase, an adap-

DOI: 10.4018/978-1-5225-2440-3.ch020

tive model is developed using known input/output data sets. This phase is known as training or learning phase. The model is trained by a learning algorithm so as to reduce the output mean square error (MSE). The second phase is known as testing phase in which a new input data set is applied and the estimated output is obtained from the model as the future value of the time series.

In the present paper we have evolved a new idea of generating more data so that adequate data sets are available to train the ANN based forecasting model efficiently. This technique is termed as data farming. The combination of data farming and the ANN based model is known as data farming linked neural network (DFL-ANN), which has made the overall system attractive in the sense that the forecasting of future data becomes more accurate. The new model developed in this paper has been used to predict the pollution level of drinking water, defined by the water quality index (WQI) (Satya Mohan et *al*, 1986). Water available from different sources of Rourkela City was collected and different water quality parameters were analyzed for five consecutive years and are used in the present study. Rourkela is an important industrial city of Orissa occupying an area of 121.7 km^2, situated on the Howrah-Mumbai railway line. The main industry here is the Steel Plant and the associated industries of steel plant are cement, fertilizer, foundries, chemicals and explosive industries. In addition, more than three hundred small scale industries are also present in and around this city. Various gases, particulate matters and liquid effluents emitted by the industries enter the biosphere daily, which contaminate the drinking water in various ways. Besides, in certain places unhygienic surroundings, improper storage and mishandling of drinking water render it unfit for consumption. These consequences made it necessary to study the WQI.

The organization of the proposed paper is as follows.

2. EXPERIMENTAL

2.1. Water Quality Index

Accurate and timely information on the quality of water is necessary to shape a sound public policy and to implement the water quality improvement programmes efficiently. One of the most effective ways to communicate information on water quality trends are with indices. Different water quality parameters such as pH, Dissolved Oxygen, Turbidity, Electrical Conductivity, Total Dissolved Solids, Hardness, Calcium, Magnesium, Chloride, Bio-chemical Oxygen Demand, Iron and Sulphate, Total Solids, Total Suspended Solids, Residual Chlorine, Lead, Mercury, Fluoride, Arsenic and Chemical Oxygen Demand were analyzed by collecting samples of drinking water from ten different areas of Rourkela. To analyze the suitability of water for drinking purpose the WQI (Naik et *al*, 1996) is calculated using these parameters. Water quality index (WQI) is commonly used for the detection and evaluation of water pollution and may be defined as "a rating, reflecting the composite influence of different quality parameters on the overall quality of water."

The indices are broadly characterized into two parts: the physico-chemical indices and the biological indices. The physico-chemical indices are based on the values of various physico-chemical parameters in a water sample, while biological indices are derived from the biological information and are calculated using the species composition of the sample, the diversity of species, their distribution pattern, the presence or absence of the indicator species or groups etc. (Trivedy and Goel, 1984). Here attempt has been made to calculate the water quality index of Rourkela on the basis of Harkins (1974), Lohani (1981) and subsequently modified by Tiwari et *al.*, (1986) based on physico-chemical data.

For the purpose of present investigation, twelve water quality parameters have been selected. These twelve parameters are pH, Dissolved Oxygen, Turbidity, Electrical Conductivity, Total Dissolved Solids, Hardness, Calcium, Magnesium, Chloride, Bio-chemical Oxygen Demand, Iron and Sulphate.

2.2. Quality Rating and Weight Age

In the formulation of water quality index, the importance of various parameters depends on the intended use of water; here water quality parameters are studied from the point of view of suitability for human consumption. The 'standards' (permissible values of various pollutants) for the drinking water, recommended by the Indian Council of Medical Research (ICMR) and unit weights are given in Table 1. When the ICMR standards are not available, the standards of United States Public Health Services (USPHS), World Health Organization (WHO), Indian Standards Institution (ISI) and European Economic Community (EEC) have been quoted.

The quality rating q_i for the ith water quality parameters (i = 1, 2, 3------------12) was obtained from the relation

$$q_i = 100\,(v_i / S_i) \tag{1}$$

where v_i = value of the ith parameter at a given sampling station and S_i = Standard permissible value of ith parameter. This equation ensures that $q_i = 0$ when a pollutant (the ith parameter) is absent in the water, while $q_i = 100$ if the value of this parameter is just equal to its permissible value for drinking water. Thus the larger the value of q_i the more polluted is the water with the ith pollutant.

However, quality ratings for pH and DO require special handling. The permissible range of pH for the drinking water is 7.0 to 8.5. Therefore, the quality rating for pH may be

Table 1. Drinking water standards and unit weights

Water Quality Parameters	Standards	Recommending Agency	Unit Weights (W_i)
pH	7.0-8.5	ICMR	0.0354
Dissolved Oxygen	5.0 mg/l	EEC	0.0496
Turbidity	10 NTU	ISI	0.0248
Total Alkalinity	120 mg/l	USPHS	0.0020
Total Dissolved Solids	500 mg/l	ICMR	0.0004
Total Hardness	300 mg/l	ICMR	0.0008
Calcium	75 mg/l	ICMR	0.0033
Magnesium	50 mg/l	ICMR	0.0049
Chloride	250 mg/l	ISI	0.0009
Biochemical Oxygen Demand	5.0 mg/l	WHO	0.0496
Iron	0.3 mg/l	ISI	0.8276
Sulphate	200 mg/l	ICMR	0.0012

$$q_{pH} = 100\left[\left(v_{pH} - 7.0\right)/\left(8.5 - 7.0\right)\right] \tag{2}$$

where v_{pH} = value of pH ~ 7, it means the numerical difference between v_{pH} and 7.0, ignoring algebraic sign. Equation (2) ensures the $q_{pH} = 0$ for pH = 7.0.

In contrast to other pollutants, the case of DO is slightly complicated because the quality of water is enhanced if it contains more DO. Therefore, the quality rating q_{DO} has been calculated from the relation

$$q_{DO} = 100\left[\left(14.6 - v_{DO}\right) / \left(14.6 - 5\right)\right] \tag{3}$$

where v_{DO} = value of DO.

In equation (3), 14.6 is the solubility of oxygen (mg/l) in distilled water at 0°C and 5.0 mg/l is the standard for drinking water. Equation (3) gives $q_{DO} = 0$ when DO = 14.6 mg/l and $q_{DO} = 100$ when v_{DO} = 5.0 mg/l.

The more harmful a given pollutant is, the smaller is its permissible value for drinking water. So the 'weights' for various water quality parameters are assumed to be inversely proportional to the recommended standards for the corresponding parameters i.e.

$$W_i = \frac{k}{S_i} \tag{4}$$

where W_i = unit weight for the ith parameter (i = 1, 2, 3 ---------------12), k = constant of proportionality which is determined from the condition and k =1 for sake of simplicity.

$$\sum_{i\ 1}^{12} W_i = 1 \tag{5}$$

The unit weights W_i calculated from equation (4) and (5) are listed in Table 2.

2.3. Calculation of WQI

To calculate the Water Quality Index, first the sub index $(SI)_i$ corresponding the ith parameter was calculated. These are given by the product of the quality rating q_i and the unit weight W_i of the ith parameter i.e.

$$(SI)_i = q_i W_i \tag{6}$$

The over all Water Quality Index was then calculated by aggregating these sub indices (SI) linearly. Thus Water Quality Index could be written as

$$WQI = \left[\sum_{i\ 1}^{12} q_i W_i \Big/ \sum_{i\ 1}^{12} W_i\right] \tag{7}$$

which gives

$$WQI = \sum_{i=1}^{12} qiWi \tag{8}$$

since $\sum W_i = 1$.

2.4. Data Generation and Proposed Model (DFL-ANN)

Data generation is an important issue in developing any adaptive forecasting model, particularly when the available data for training the model is scanty. This leads to improper training of the model and consequently inaccurate prediction. The problem has been circumvented in the proposed paper by a novel data farming technique as depicted below. We have computed 12 numbers of WQI for each month in a year. In this way for five years 60 values are available with us. Out of these 48 are used as the core data for farming purpose and remaining 12 data of the 5th year are used for testing the model. In the first case 48 data sets are used for training which seems to be inadequate. In the second case 36 more relevant data are generated from 48 WQI values by interpolation method. The intermediate data obtained for two consecutive years are based on averaging principle as shown in Figure 1 for two consecutive years.

The DFL-ANN model used for the prediction purpose is shown in Figure 2. The first block takes N_1 data points and generates N_2 data points based on interpolation which generates N2 number of input patterns, each of N_3 values.

Figure 1. Generation of additional data by data farming

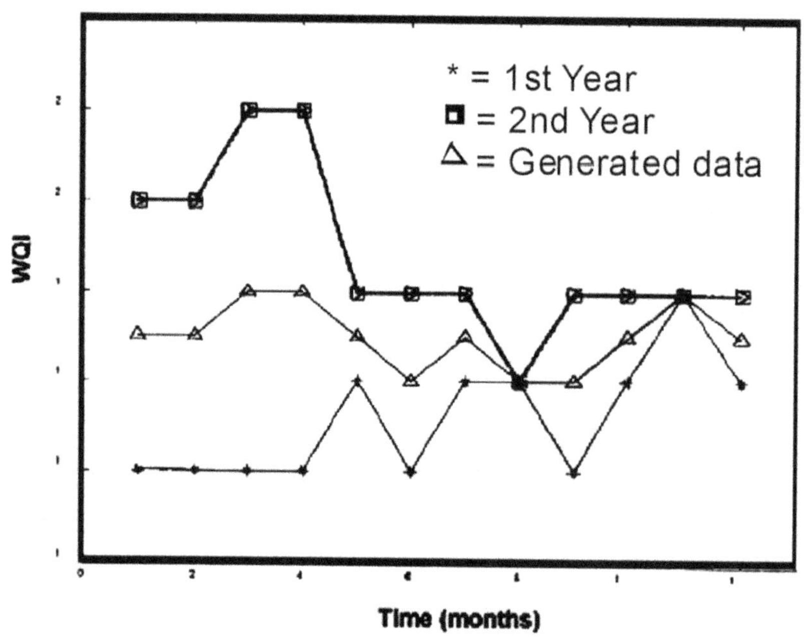

Figure 2. Model for Data Farming and Data prediction

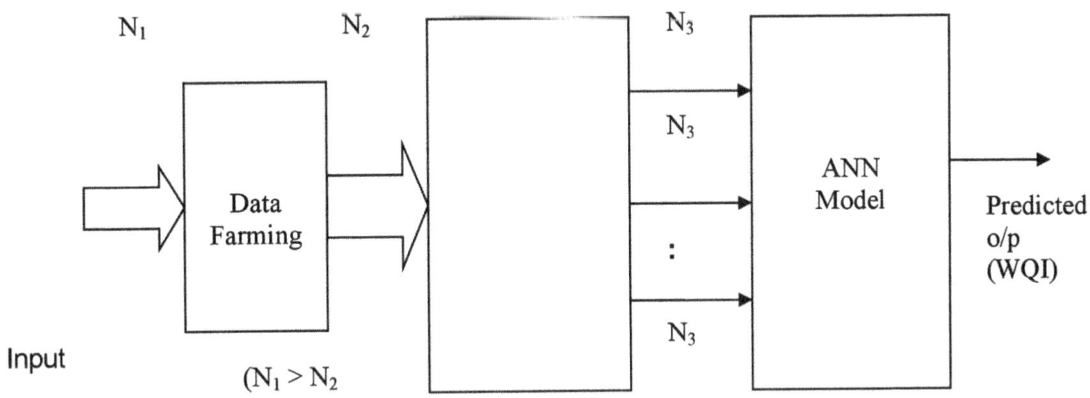

In the present case N_1= 48, N_2 = 84 and N_3 = 12. As an illustration, for June 2000 the input pattern (X_i, i= 1, 2,..12) is generated as follows. X_1 = WQI of June 1999, X_2 = WQI of June 1998, X_3 = Generated WQI for June, X_4 = Abs ($X_1 - X_2$), X_5 = Abs (X_2 - X_3), X_6 = Abs (X_1 - X_3), X_7= Average of WQI of previous 6 month i.e. from Jan 1999 to May 2000, X_8 = Root mean square of (X_1,X_2,X_3,), X_9 = sin(X_4), X_{10} = sin (X_5), X_{11} = sin (X_6), X_{12}=K, where K = 0 if there is a seasonal change in the same month, otherwise K = WQI of the previous month (May, 2000). Similar input patterns are obtained for other months.

The ANN model consists of N3 input points each of which is connected to M1 number of neurons in the 1st hidden layer (Lh1) through connecting weights. The output of each neuron of Lh1 is similarly

Figure 3. Structure of Multi-Layer Artificial Neural Network (ML - ANN)

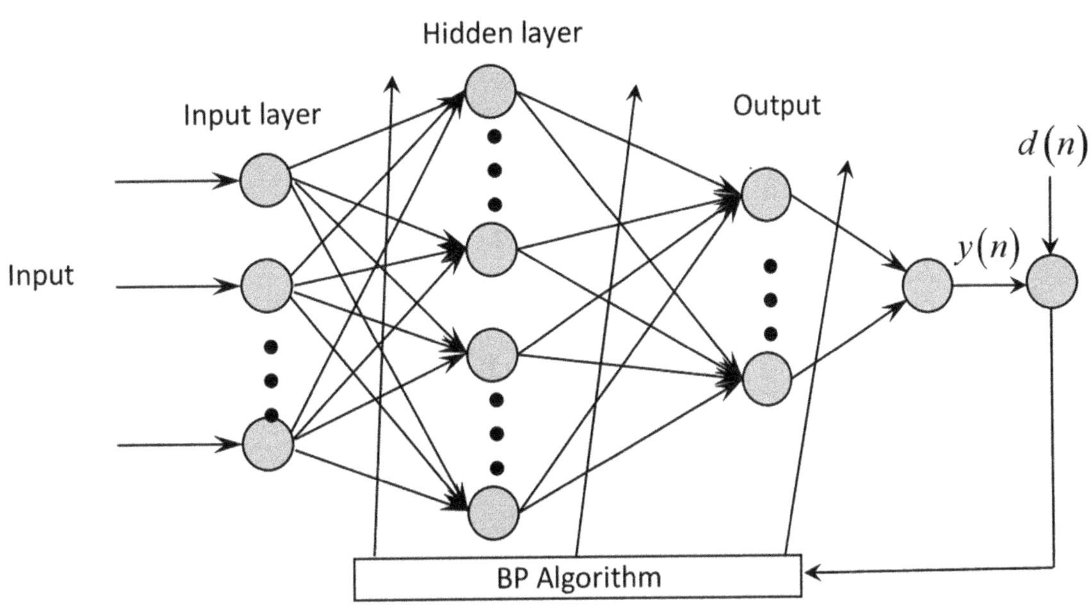

connected to each neuron of the 2nd hidden layer (Lh2) containing M2 number of neurons. Similarly the outputs of Lh2 are connected to a single output neuron through a set of connecting weights. In the present case we have taken M1 = 24, M2 = 9, M3 = 1. The connecting weights are updated in every iteration based on Back Propagation Algorithm as discussed in the next section.

2.4. Training Algorithm

The scheme of multi-layer neural network or multilayer perception (MLP) using three layers and the learning mechanism is depicted in Figure 3. $x_i(n)$ represent the input to the network, $f_j^{(1)}$ and $f_k^{(2)}$ represent the output of the input and hidden layers respectively and $y_l(n)$ represents the output of the final layer of the neural network. The connecting weights between the input to the first layer, first to second layer and the second layer to the output layers are represented by $w_{ij}^{(1)}$, $w_{jk}^{(2)}$ and $w_{kl}^{(3)}$ respectively. If P_1 is the number of neurons in the first layer, each element of the output vector may be calculated as,

$$f_j^{(1)} = g_j\left[\sum_{i=1}^{N} w_{ij}^{(1)} x_i + \theta_j^{(1)}\right] \tag{9}$$

$j=1, 2, 3,\ldots, P_1$

where $\theta_j^{(1)}$ is the threshold to the neurons at the first layer, N is the no. of inputs and $g_j(.)$ is the nonlinear activation function,

$$g(x) = \frac{1-e^{-\phi x}}{1+e^{-\phi x}} \tag{10}$$

Let P_2 be the number of neurons in the second layer. The output of this layer is represented as

$$f_k^{(2)} = g_k\left[\sum_{j=1}^{P_1} w_{jk}^{(2)} f_j^{(1)} + \theta_k^{(2)}\right] \tag{11}$$

$k=1, 2, 3, \ldots, P_2$

where, $\theta_k^{(2)}$ is the threshold to the neurons at the second layer. The output of the final layer can be calculated as

$$y_l = g_l\left[\sum_{k=1}^{P_2} w_{kl}^{(3)} f_k^{(2)} + \theta_l^{(3)}\right] \tag{12}$$

l=1, 2, 3, …, P_3

where, $\theta_l^{(3)}$ is the threshold to the neuron at the final layer and P_3 is the no. of neurons in the output layer. The output of the MLANN may be expressed as

$$y_l = g_n\left[\sum_{k=1}^{P_2} w_{kl}^{(3)} g_k \left(\sum_{j=1}^{P_1} w_{jk}^{(2)} g_j \left\{\sum_{i=1}^{N} w_{ij}^{(1)} x_i + \theta_j^{(1)}\right\} + \theta_k^{(2)}\right) + \theta_l^{(3)}\right] \tag{13}$$

The error signal can be computed as,

$$e(n) = d(n) - y_l(n) \tag{14}$$

where, $d(n)$=desired signal, and $y_l(n)$=estimated output signal, and n is the time index.

The instantaneous error energy is calculated by (7),

$$\xi(n) = \sum_{P_3} e^2(n), \tag{15}$$

where P_3 is the number of output nodes.

The cost function E=1/2. e^2 is minimized to obtain the update equation.

The connecting weights are trained according to the following set of equations.

Weight update: $w_{ij}^t(n+1) = w_{ij}^t(n) + \lambda e_j^t(n) f_i^{t-1}(n) + \alpha \Delta w_{ij}^t(n-1)$ (16)

For Thresholds: $\theta_j^t(n+1) = \theta_j^t(n) + \lambda e_j^t(n) + \alpha \Delta \theta_j^t(n-1)$ (17)

For output layers: $e_l^t(n) = \left(d_l(n) - y_l(n)\right)\left(1 - y_l(n)^2\right)/2$ (18)

For hidden layers: $e_m^t(n) = \left[1 - \left(f_k^t\right)^2\right]\left[\sum_m e_m^{t+1}(n) w_{km}^{t+1}(n)\right]/2$ (19)

3. RESULTS AND DISCUSSION

3.1. Computer Simulation Results

Computer simulation study has been carried out employing the generated patterns of four years as input and its true WQI value as the desired input. The ANN model has 12 inputs, 24 neurons in the 1st hidden layer, 9 neurons in the 2nd hidden layer and a single neuron in the output. By applying input patterns successively the connecting weights of different layers are trained according to the BP algorithm dealt in section III. The MSE is computed for each iteration until a steady state MSE is achieved. The same experiment is repeated for DFL-ANN also. The convergence characteristics obtained in both cases are shown in Figure 4.

In the experiment the λ has been set to 0.005. After the training is complete both the ANN and DFL-ANN models are tested with the actual data available for the 5th year. The comparison of the true results along with those obtained from ANN and DFL-ANN models are shown in Figure 5 for one area of Rourkela city.

It is observed that there is a close agreement between the true WQI value and its estimated one. The same trend is also observed for other areas of Rourkela. The forecasted Values of WQI for four different areas of Rourkela are shown in Table 2.

Figure 4. Comparison of convergence characteristics with λ = 0.005 and number of iterations = 200

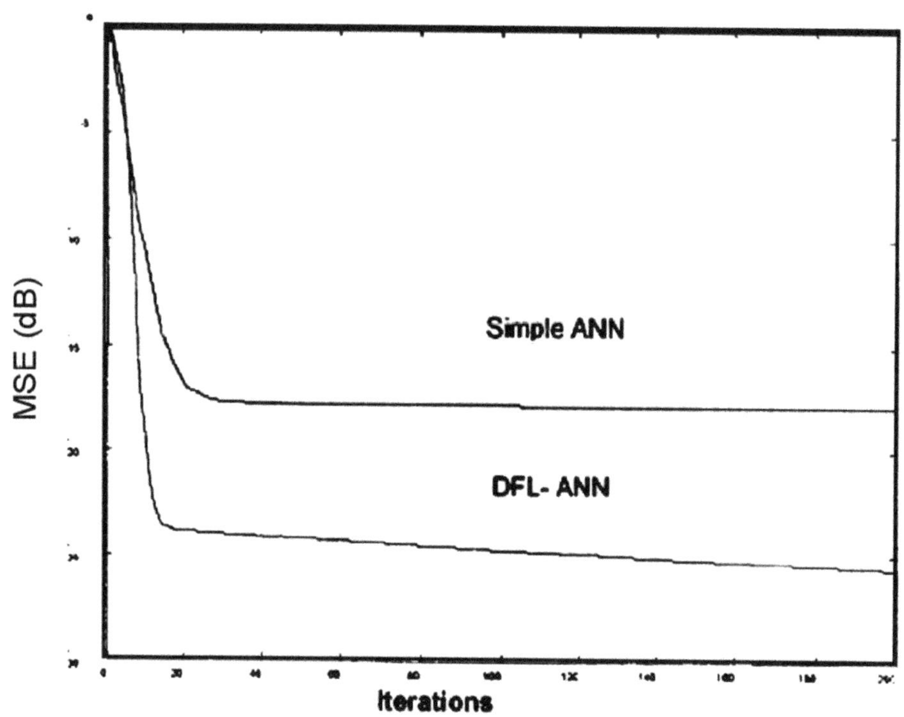

Figure 5. Comparison between actual WQI and estimated WQI

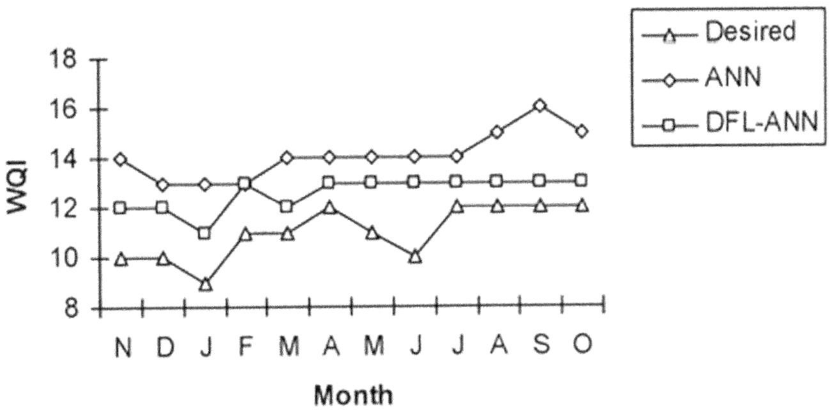

Table 2. Forecast WQIs of four different places of Rourklea

JHIRPANI			JAGDA		
Desired	ANN	DFL-ANN	Desired	ANN	DFL-ANN
17	19	17	17	18	17
18	20	18	17	19	18
18	19	18	17	19	18
19	19	18	18	19	18
20	18	19	17	18	17
18	18	19	17	18	18
19	17	19	19	18	17
19	17	18	19	18	17
18	17	18	15	19	16
18	17	18	15	19	17
20	17	18	15	19	17
20	18	19	15	19	17
JALDA			HAMIRPUR		
Desired	ANN	DFL-ANN	Desired	ANN	DFL-ANN
14	24	17	10	14	12
14	22	18	10	13	12
13	21	17	09	13	11
15	21	18	11	13	13
15	22	21	11	14	12
15	22	21	12	14	13
14	21	20	11	14	13
16	21	20	10	14	13
15	24	23	12	14	13
15	24	23	12	15	13
16	24	22	12	16	13
16	23	22	12	15	13

The simulation results reveal that the results provided by the simple ANN model deviate form true values. But the DFL-ANN model that includes data farming yields superior performance in predicting the desired Water Quality indices. Slight deviations in the results are observed which may be due to sudden changes in the measured WQI values or may be due to seasonal changes. This difference can be minimized by improved design of the input pattern, which is presently under study.

4. CONCLUSION

This paper deals with the development of an ANN model based on back propagation algorithm to forecast water quality indices of different samples of water. The performance of this model is poor because of less number of data available for training the model. To improve this performance data farming technique has been used in this paper to generate intermediate input data so that ANN model can learn better and can yield accurate output. This has been tested with actual data and it is shown that the hybrid model comprising of data farming and ANN can produce better estimated output compared to the previous model.

REFERENCES

Dash, P. K., Liew, A. C., Rahman, S., & Dash, S. (1995). Fuzzy and Neuro-fuzzy Computing models for electric load forecasting. *Engineering Applications of Artificial Intelligence*, *8*(4), 423–433. doi:10.1016/0952-1976(95)00023-T

Dash, P. K., Ramakrishna, G., Liew, A. C., & Rahman, S. 1995. Fuzzy Neural networks for time series forecasting of electric load. *IEE Proc-Gender. Transm. Distrib.*, *142*(5).

Harkins, R. D. (1974). An objective water quality index. *Journal - Water Pollution Control Federation*, *46*, 589. PMID:4846806

ICMR. (1975). Manual of Standards of Quality of Drinking Water Supplied (2nd ed.). Indian Council of Medical Research.

ISI. (1983). *Indian Standard Specification for Drinking Water.* IS: 10500. ISI.

Lohani, B. N. (1981). Water Quality Indices. In C. K. Varshney (Ed.), *Water Pollution and Management Reviews* (pp. 53–69). New Delhi: South Asian Publ.

Naik, S., & Purohit, K. M. (1996). Physico-chemical analysis of some community ponds of Rourkela. *Indian J. of Env. Protection*, *16*(9), 679–684.

Patra, J. C., Panda, G., Pal, R. N., & Chatterjee, B. N. (1997). Neural Networks for signal Processing Applications. *Journal of Computer Society of India*, *27*(3), 34–58.

Peng, T. M., Hubble, N. F., & Karady, G. G. (1992). Advancement in the application of neural networks for short term load forecasting. *IEEE Transactions on Power Systems*, *7*(1), 250–258. doi:10.1109/59.141711

Satya Mohan, K., & Zafar, A. R. (1986). Water Quality of two drinking water reservoirs of Hyderabad, India. *Pollution Research*, *5*(3-4), 111–112.

Tiwari, T. N., Das, S. C., & Bose, P. K. (1986). Water quality index for the river Jhelum in Kashmir and its seasonal variation. *Pollution Research*, *5*(1), 1–5.

Trivedy, R.K., & Goel, P.K. (1984). Chemcial and biological methods for water pollution studies. *Env. Publ. India*, 215.

Werbos, P. J. (1988). Generation of Back Propagation with application to a recurrent gas market model. *Neural Networks*, *1*(4), 339–356. doi:10.1016/0893-6080(88)90007-X

WHO. (1984a). International Drinking Water Supply and Sanitation Decade-Review of National Baseline Data. WHO off set publication No. 85, World Health Organisation.

WHO. (1984b). Guidelines for Drinking Water Quality Vol. 1, Recommendations. World Health Organization.

WHO. (1984c). Guidelines for Drinking Water Quality. Vol. 2. Health Criteria and other Supporting Information. World Health Organization.

Widrow, B., Winter, R. G., & Barxter, R. A. (1998). Layered Neural Nets for Pattern Recognition. *IEEE Transactions on Acoustics, Speech, and Signal Processing*, *36*(7), 1109–1118. doi:10.1109/29.1638

Chapter 21
Designing and Evaluation of Aluminium Thin-Film Electrochemical Sensors for Biomedical Analysis

Gaurav Dinesh Kulkarni
National Institute of Technology Rourkela, India

Suraj Kumar Nayak
National Institute of Technology Rourkela, India

Kailash Das
National Institute of Technology Rourkela, India

Jyoti Prakash Kar
National Institute of Technology Rourkela, India

Biswajit Mohapatra
Vesaj Patel Hospital, India

D. N. Tibarewala
Jadavpur University, India

Arfat Anis
King Saud University, Saudi Arabia

Kunal Pal
National Institute of Technology Rourkela, India

ABSTRACT

In recent years, Electrochemical Impedance Spectroscopy (EIS) has received special attention of the researchers due to its suitability for sensing/analysis of the cells and the biological molecules in a label-free manner. The current study proposes the fabrication of thin-film aluminium electrodes on a glass substrate by thermal evaporation method. The electrodes were fabricated having gaps of varied lengths. Different electrical models were used to determine the electrical stability of the system using an in-lab developed LabVIEW program. The electrode having the best stability was used for EIS analysis of distilled water and saline solutions of different strengths to validate its proper functioning.

DOI: 10.4018/978-1-5225-2440-3.ch021

INTRODUCTION

There has been an increase in the use of impedance-based sensing of cells and biological molecules (G.-H. Hui, Ji, Mi, & Deng, 2013; G. Hui, Mi, Chen, & Chen, 2014; Perfézou, Turner, & Merkoçi, 2012). These sensors are widely categorized as electrochemical biosensors. The main advantages of this type of biosensor are its accuracy and efficiency in detecting the biological molecules or cells. Apart from this, the impedance-based monitoring system provides the analysis of the cells and the biological molecules in a label-free environment (Benson, Cramer, & Galla, 2013; Ke, Nguyen, Irelan, & Abassi, 2015; Nguyen, Yin, Reyes, & Urban, 2013). This helps not only in the reduction of the analysis cost but also reduces the time required for the analysis. In this type of analysis, a low-amplitude sinusoidal signal of varied frequency is passed through the sample to be analyzed, and the corresponding impedance is measured. In the presence of a specific analyte and/or biological event or activity, the impedance profile of the sample changes. Some of the commonly studied biological activities using impedance-based method include interactions amongst the target molecules and the receptors (e.g. antigen-antibody reaction) (Mashazi, Tetyana, Vilakazi, & Nyokong, 2013; Özcan, Demirbakan, Yeşiller, & Sezgintürk, 2014; Ramanathan, et al., 2016), change in the number of the bacterial/mammalian cells and ionic concentration of the media associated with some metabolic activity (Gómez, Bashir, & Bhunia, 2002; Varshney & Li, 2009; Yang & Bashir, 2008). In this paper, we have fabricated thin-film aluminium electrodes on a glass substrate by thermal evaporation method. The electrodes were fabricated having gaps of varying lengths. Different electrical models were used to determine the electrical stability of the system using a LabVIEW program, which was developed in-lab. The electrode having the best stability was used for further analysis of saline solutions with different strengths.

BACKGROUND

The sensing and monitoring of biomedical processes are very difficult mainly due to the rapid variations in their physical or chemical properties with time. Till now, various methods have been proposed by scientists to solve such problems. In recent years, many attempts have been made in solving many complex problems using electrochemical impedance spectroscopy (EIS). Especially, when it comes to the multidisciplinary fields like biomedical engineering, researchers are more focused on the development of impedance-based cheap, portable and fast instruments and sensing devices (Feldman, 1998; T. C. Johnson, Balbierz, & Pearson, 2005; Krüger, et al., 2002; Solly, Wang, Xu, Strulovici, & Zheng, 2004). Although certain major limitations still exist, the sensing efficiency and sensitivity of these devices are increasing rapidly. This section provides a brief discussion on a few major biomedical problems and their proposed solutions using EIS.

One of the major areas of application of EIS is the diagnosis of various types of diseases. One such application is the differentiation between benign lesion and melanoma of the skin (Rigel, Russak, & Friedman, 2010). Melanoma has been reported to be the primary cause of death resulting from skin cancer and its early detection can play an important role in saving the life of a patient (Baade, Meng, Youlden, Aitken, & Youl, 2012; Tsao, Atkins, & Sober, 2004). Although progress has been made in the detection of melanoma, visual detection of early melanoma still remains a challenging task. Therefore, numerous techniques have been proposed by researchers to assist the medical practitioners (Åberg,

Birgersson, Elsner, Mohr, & Ollmar, 2011; Rigel, et al., 2010). Mohr *et al.* (2013) have reported the development of an EIS based classification algorithm that can distinguish between the benign and the malignant (melanocytic and non-melanocytic) stages of skin cancer with a sensitivity of more than 98% (Mohr, et al., 2013).

According to Ferlay *et al.* (2010), oral cancers are the ninth most common cancers across the globe with 127,000 deaths reported in the year 2008. The global incidence of oral cancers is increasing day by day and it has a very deprived prognosis having less than 55% overall 5 year survival rate (N. W. Johnson, Jayasekara, & Amarasinghe, 2011; Napier & Speight, 2008; Warnakulasuriya, Johnson, & Van der Waal, 2007). Potentially malignant lesions (PML), appearing as white (leukoplakia) or red (erythroplakia) oral lesions, can lead to different types of oral cancers (N. W. Johnson, et al., 2011; Warnakulasuriya, et al., 2007). But, it has been reported that only a small percentage (5%-15%) of PML actually grows into cancer (Lee, et al., 2000; Napier & Speight, 2008). Clinical signs like heterogeneous surface or a stippled appearance also hint for the high possibility of malignant transformation. However, no trustworthy clinical indicators are available which can accurately predict whether a lesion will lead to cancer or not. The architecture and cellular composition of tissues have an important role in defining their electrical properties. Murdoch *et al.* (2014) have reported that the changes in the impedance can be utilized to distinguish between the various stages of cervical dysplasia and cervical cancer (e.g. benign, potentially malignant lesions (PML) and oral cancer) and have suggested that it is more convenient than the traditional scalpel based method (Murdoch, et al., 2014).

Currently, the visual diagnosis of bladder is performed using cystoscopy. Its definitive diagnosis is possible only by biopsies, which are collected from the suspected areas. This is a relatively difficult procedure and is associated with the discomfort of the patient and morbidity along with high cost. Keshtkar *et al.* (2006) have studied the applicability of EIS technology for diagnosis of bladder pathology. In this investigation, both *ex vivo* and *in vivo* studies were carried out. A total of 145 points from 16 patients (for *ex vivo* study) and 106 points from 38 patients (for *in vivo* study) were used to take impedance and biopsy samples. A significant difference in impedance values between the malignant and benign patients was observed with the malignant bladder tissue, exhibiting higher impedance values at lower frequencies (p=0.001). Also, the receiver operating characteristics (ROC) curve for the measured impedance values suggested the suitability of EIS technique for divulging diagnostic information. Based on the results, the authors have proposed that the electrical impedance measurement of bladder tissue can be regarded as a new and minimally invasive method for characterizing bladder urothelium and can act as a complementary technique for cystoscopy and biopsy evaluation of bladder pathology (Keshtkar, Keshtkar, & Smallwood, 2006).

For the last several decades, the enzymatic methods have been used in medical practice for blood glucose measurement. But, these techniques are not appropriate for online monitoring of blood glucose level because of their invasive nature. In recent years, the EIS based methods have shown great promises to this goal. Shervedani *et al.* (2006) have reported the development of a novel EIS-based method for quantitative determination of glucose concentration using immobilization of glucose oxidase (GOx) on modified gold electrode (Shervedani, Mehrjardi, & Zamiri, 2006). Pradhan *et al.* (2013) have reported the design and fabrication of different electrode geometries for EIS-based quantitative measurement of blood glucose level and have proposed the relation between the electrical impedance, glucose concentration and different electrode geometries (Pradhan, Mitra, & Das, 2013).

Dhurjaty *et al.* (2015) have reported the feasibility of an EIS-based device for the measurement and real-time monitoring of glucose metabolism levels with the help of a phantom study. This study involved the use of a multi-frequency electrical signal for scanning of the subject that resulted in resonance (Dhurjaty, Qiu, Tan, Liu, & Zheng, 2015).

MAIN FOCUS OF THE CHAPTER

The main focus of this study is to develop thin-film aluminium electrodes for EIS analysis of materials having biomedical importance. The proposed electrodes were fabricated on a glass substrate (having gaps of varying lengths) by thermal evaporation method. Various electrical models were employed to verify the electrical stability of the system using an in-lab developed LabVIEW program. The electrode having the best stability was used for further analysis of distilled water and saline solutions of different strengths.

MATERIALS

Copper clad board, glass slides, Aluminium block (99.99% pure), op-amp IC (OP-07, Analog Devices, USA), resistances and connecting wires were procured from the local suppliers. Dynaxact function generator (Model: DX 4050) was purchased from Dynalog Limited, India. Digital storage oscilloscope (TBS-1022, Tektronix Inc, USA) was purchased from Peridot Technologies, India. LabVIEW 2010 (National Instruments, USA) and Excel 2013 (Microsoft, USA) were used for the analysis. PDMS (Polydimethylsiloxane) and micro-tips were provided as a gift by Dr. Indranil Banerjee, Assistant Professor, Department of Biotechnology & Medical Engineering, National Institute of Technology Rourkela, India.

METHODS

Fabrication of Electrodes

Thin films of aluminium of 200 μm thickness were deposited on microscopic glass slides by vacuum thermal evaporation technique as reported by Amusan *et al.* (2010) (Amusan, et al., 2010). The electrode systems were fabricated such that the distance between the successive electrodes was 40 μm, 80 μm, 150 μm and 300 μm. The electrodes were regarded as E40, E80, E150 and E300, respectively.

Determination of Stability of the Electrode by Electrical Modeling

Wells were prepared by cutting the bases of the micropipette tips. The wells were fixed over the electrodes using PDMS substrate. 200 μl of the distilled water (DW) was put in the wells and the impedance was determined at the frequency range of 50 Hz – 10 kHz using the in-house developed impedance analyzer. A sinusoidal waveform having an amplitude of 1 $V_{(p-p)}$ was used for the analysis. The impedance data obtained were fitted to the mathematical models representing (RQ)Q, (RQ)(RQ) and (RQ)(RQ)Re electrical models using solver add-in of Excel 2013. The stability of the equivalent electrical circuit was predicted and the most stable model was used for the analysis of saline solutions of different strengths. A correlation coefficient amongst the experimental and the calculated values (from models) was determined to get the best-fit model.

Development of LabVIEW-Based Software

A LabVIEW-based software was developed. The software was capable of determining impulse response, Bode and Nyquist plots of the selected models. The software was made such that when a particular model was selected, a dialogue box was prompted to the user for providing input values of the electrical components of the model. The input of the relevant values resulted in the generation of the above-mentioned plots. The algorithm of the LabVIEW program has been shown in Figure 1 and the user interface of the program has been shown in Figure 2.

RESULTS AND DISCUSSIONS

Fabrication of Electrode

A thin layer of aluminium was deposited on the microscopic glass slides by vacuum vapor deposition technique. The electrodes were prepared such that the distance between the two electrodes was 40 μm, 80 μm, 150 μm and 300 μm. The pictographs of the designed electrodes have been shown in Figure 3a-d. The distance between the electrode edges was calculated using ImageJ software (National Institute of Health, US). The distance was found to be 40.13 μm, 82.26 μm, 150.93 μm and 316.53 μm. The micrographs of the electrode graphs have been shown in Figure 3e-h.

Figure 1. Algorithm of the LabVIEW program

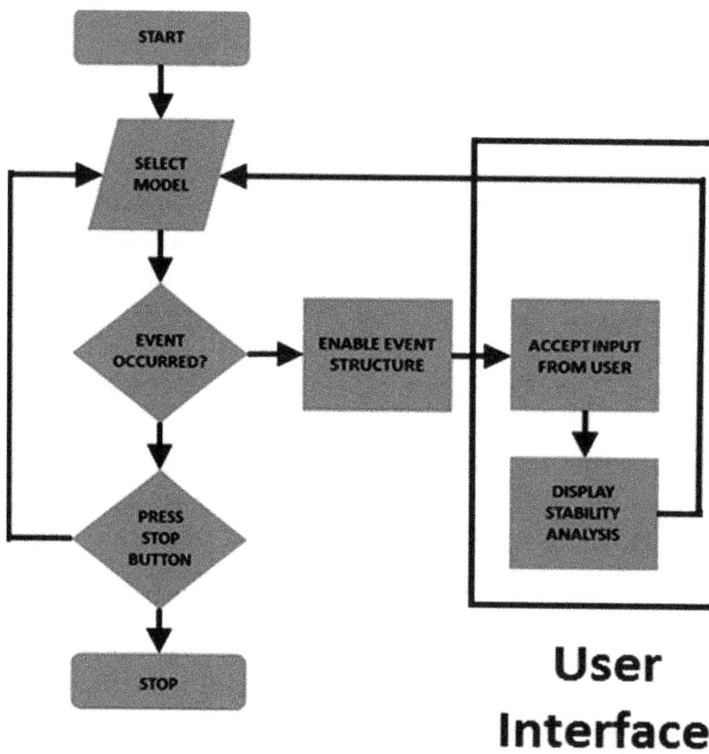

Figure 2. The user interface of the LabVIEW program

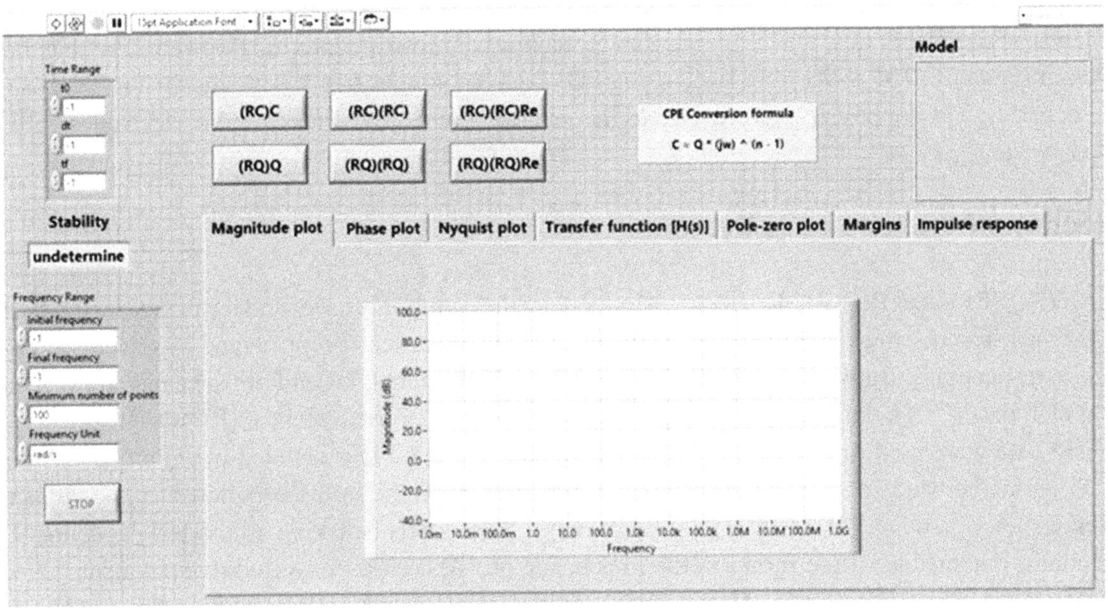

Figure 3. The pictograph of the electrodes: (a) E40, (b) E80, (c) E150, and (d) E300; the micrographs of electrodes used for the calculation of the electrode spacing: (e) E40, (f) E80, (g) E150, and (h) E300; and (i) the pictograph set-up used for the impedance analysis

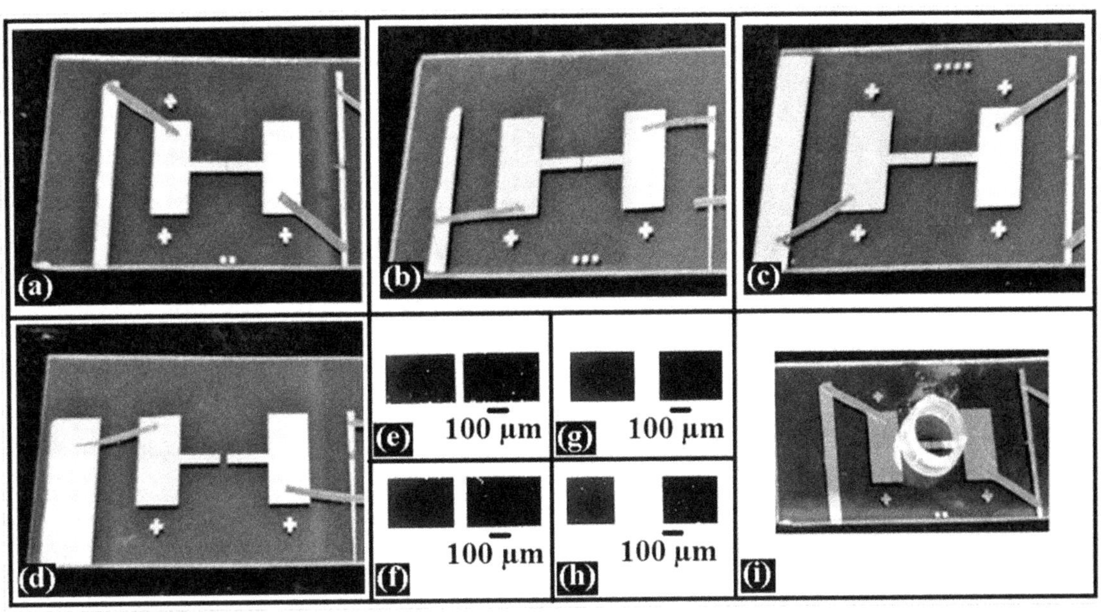

After the characterization of the electrodes, wells made using micropipette tips were fixed onto the glass slides such that the mid-point of the gap line (perpendicular line running through mid-point of the edges of the electrodes which are parallel to each other) forms center of the well. This has been shown in Figure 3i. The wells were fixed using PDMS. Thin copper strips were attached to the base of the electrodes. The copper strips were used as the junction points for connecting the electrodes with an impedance analyzer.

Selection of the Electrode System

(RC)C, (RC)(RC) and (RC)(RC)Re electrical models are the commonly used models for the electrochemical analysis. The transfer functions of these models are mathematically represented by the equations 1, 2 and 3, respectively. But for the analysis of the biological elements, instead of true capacitor constant phase elements (CPE), defined as a non-ideal capacitor, are often used. Hence, (RQ)Q, (RQ)(RQ) and (RQ)(RQ)Re were used for the analysis of the saline solutions. CPE is denoted by 'Q'. The reactance of CPE is mathematically represented by Equation 4. The 'n' value in the equation 4 is known as the inhomogeneity constant. As the 'n' value approaches 1, CPE starts behaving as an ideal capacitor. The 'Q' values, obtained from the fitting of the impedance profiles, were converted to equivalent 'C' (capacitor) values using equation 6. The calculated equivalent 'C' values were used for further analysis.

$$H\left(s\right)=\frac{1+\left(RC_1+RC_2\right)s}{C_2s+RC_1C_2s^2} \tag{1}$$

$$H\left(s\right)=\frac{\left(R_1+R_2\right)+\left(R_1R_2C_1+R_1R_2C_2\right)s}{1+\left(R_1C_1+R_2C_2\right)s+R_1R_2C_1C_2s^2} \tag{2}$$

$$H\left(s\right)=\frac{R_e\left(R_1+R_2\right)+\left(R_1R_2R_eC_1+R_1R_2R_eC_2\right)s}{\left(R_1+R_2+R_e\right)+\left(R_1R_2C_1+R_1R_2C_2+R_1R_eC_1+R_2R_eC_2\right)s+R_1R_2R_eC_1C_2s^2} \tag{3}$$

$$X_Q=\frac{1}{(j\omega)^nQ} \tag{4}$$

$$C=(j\omega)^{n-1}Q \tag{5}$$

The values of these elements were obtained from the fitting of the experimental impedance profiles. The fitting was done by non-linear least square method using Solver add-in in Excel 2013. The impedance analysis was done in the frequency range of 50 Hz-10 kHz. For the selection of the electrodes, 200 μl of distilled water was used for the analysis. The impedance profiles, their fit and the values of the calculated elements using different electrodes have been shown in Figures 4, 5, and 6.

The values of the electrical circuit elements obtained from the modeling and the equivalent 'C' values were used as the input for the developed program. Impulse response, Bode and Nyquist plots of the equivalent electrical circuit models were obtained from the program. The time constants of the equivalent circuits were calculated from the fitting of the impulse response profile using equation 6. The correlation coefficients amongst the impulse response profile and the fitting parameters models were equal to 1.

$$Y = A + C_1 e^{-t/T_1} + C_2 e^{-t/T_2} \tag{6}$$

where, A= Residual value after decay, C_1 and C_2= System constants and T_1 and T_2= Time constants.

The impulse responses of the electrode systems obtained from (RQ)Q, (RQ)(RQ) and (RQ)(RQ)Re models have been shown in Figure 7, 8, and 9. The (RQ)Q model showed a non-zero residual value, whereas, the (RQ)(RQ) and (RQ)(RQ)Re models showed zero residual amplitude values for all the electrode systems. Hence, the (RQ)Q model was not considered for further analysis. Except for E300, the final time constant was lower for the (RQ)(RQ) model as compared to the (RQ)(RQ)Re model. From the electrical system point of view, the system which shows lower time constants is regarded as compara-

Figure 4. Fitting of impedance profile using (RQ)Q model: (a) E40, (b) E80, (c) E150, and (d) E300

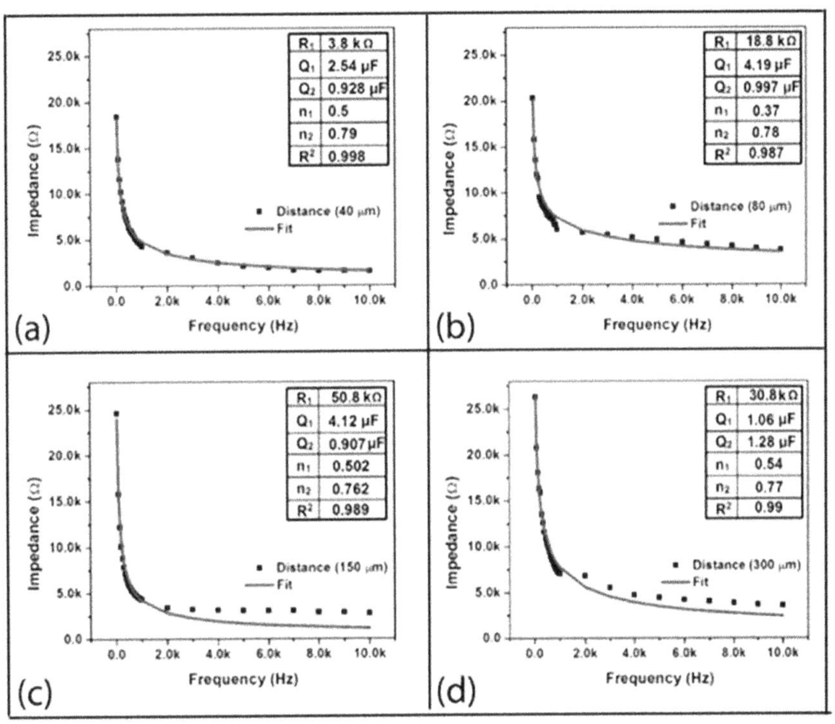

Figure 5. Fitting of impedance profile using (RQ)(RQ) model: (a) E40, (b) E80, (c) E150, and (d) E300

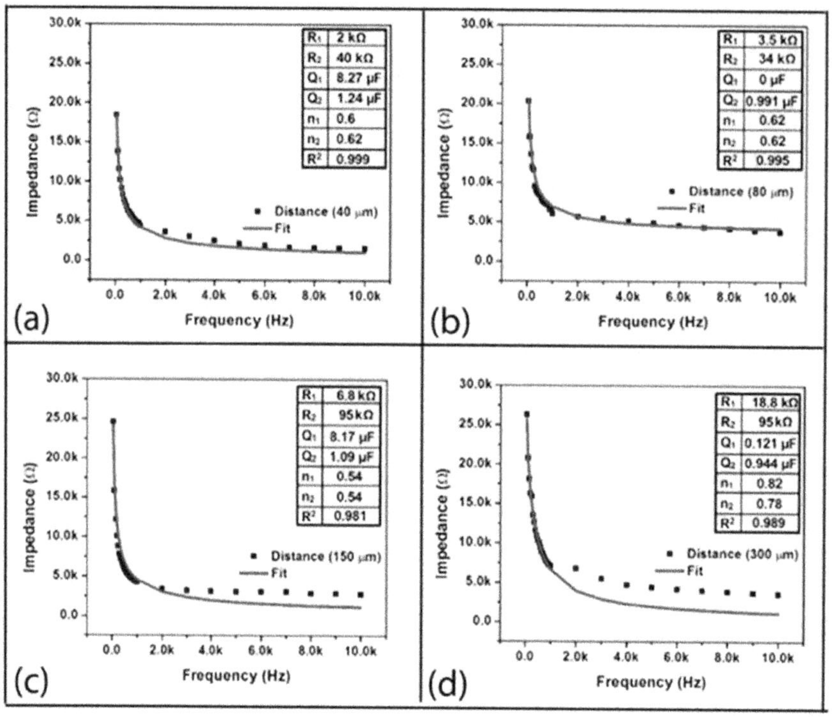

Figure 6. Fitting of impedance profile using (RQ)(RQ)Re model: (a) E40, (b) E80, (c) E150, and (d) E300

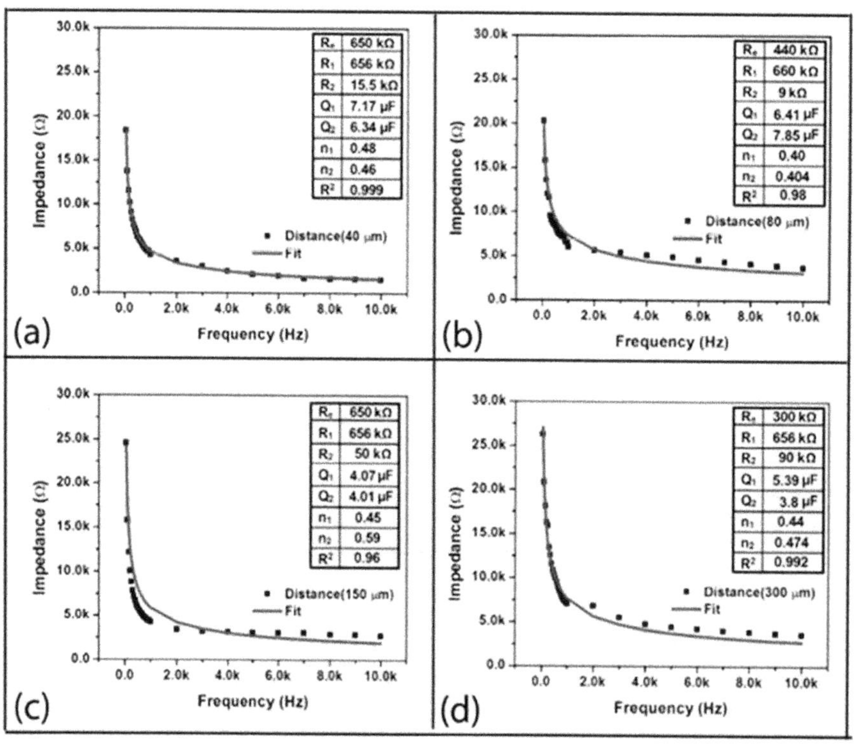

Figure 7. Fitting of impulse response profile of (RQ)Q model: (a) E40, (b) E80, (c) E150, and (d) E300

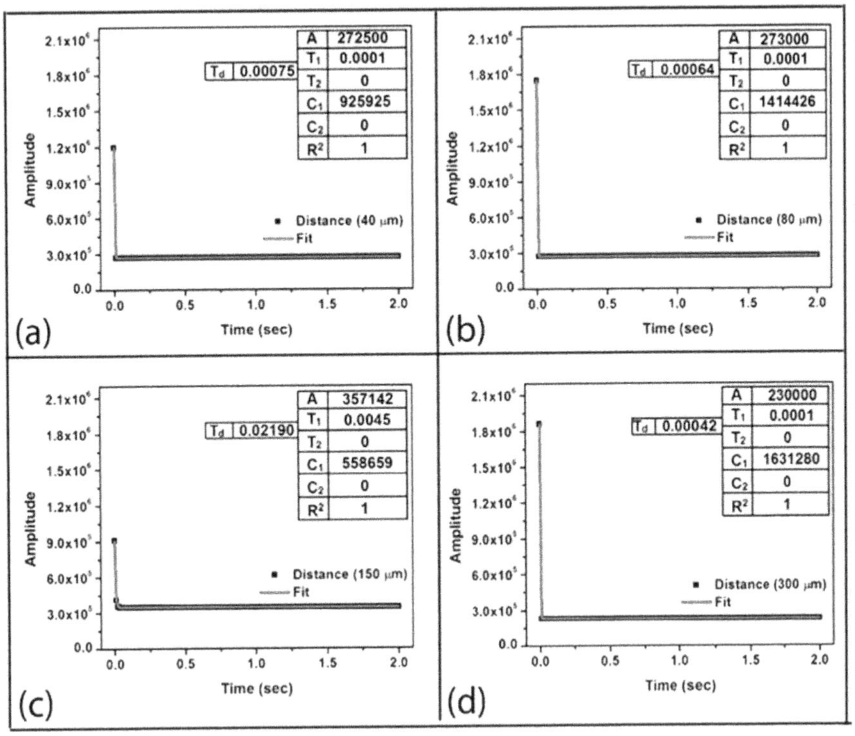

Figure 8. Fitting of impulse response profile of (RQ)(RQ) model: (a) E40, (b) E80, (c) E150, and (d) E300

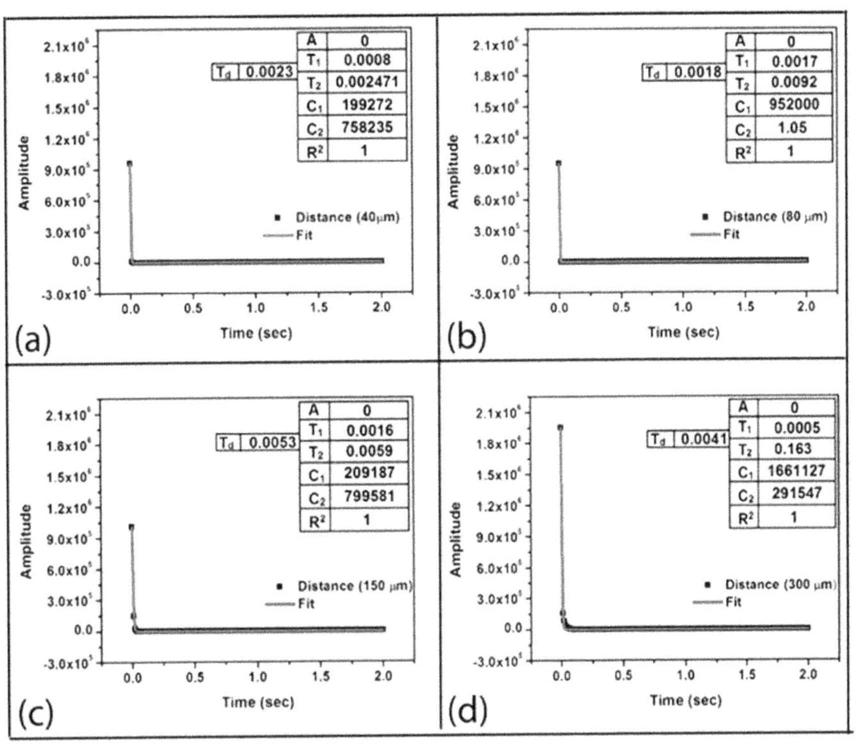

Figure 9. Fitting of impulse response profile of (RQ)(RQ)Re model: (a) E40, (b) E80, (c) E150, and (d) E300

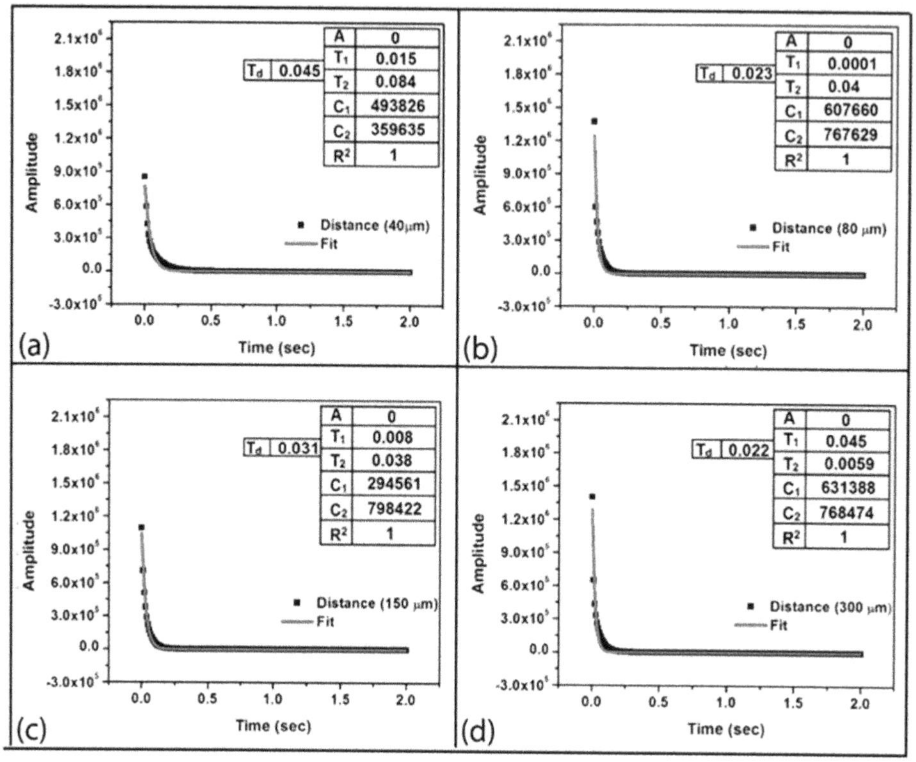

tively more stable. Taking a note of this, (RQ)(RQ) model was selected for further analysis. Also, from a careful examination of the time constants of the different electrode systems using (RQ)(RQ) model it was found that the final time constant was lowest for E40 electrode system. From the above, it can be summarized that the E40 electrode system was most stable amongst all the electrode systems studied. Hence, E40 electrode system was used for the analysis of the saline solutions.

Analysis of Distilled Water

The Nyquist plots of all the electrode systems using distilled water (DW) as an analyte have been shown in Figure 10, Figure 11, and Figure 12 for (RQ)Q, (RQ)(RQ) and (RQ)(RQ)Re models, respectively. The Nyquist plot obtained from (RQ)Q model showed an initiation of a semicircular region. In neither of the cases, the semicircle was completed. Instead, there was a spike having slopes of different magnitudes. The occurrence of the spikes has been related to the imperfections present in the electrodes, which result in the altered electrical phenomenon (faradic current) at the sample electrode interface. This type of phenomenon is undesirable for any electrochemical impedance analysis.

The Nyquist plot for the (RQ)(RQ) model showed the formation of semicircles when E40, E80 and E150 electrodes were used, even though, there are two constant phase elements (CPEs) present in the circuit. On the contrary, when E300 electrode was used, a combination of two semicircles was seen. This

Figure 10. Nyquist plot of (RQ)Q model: (a) E40, (b) E80, (c) E150, and (d) E300

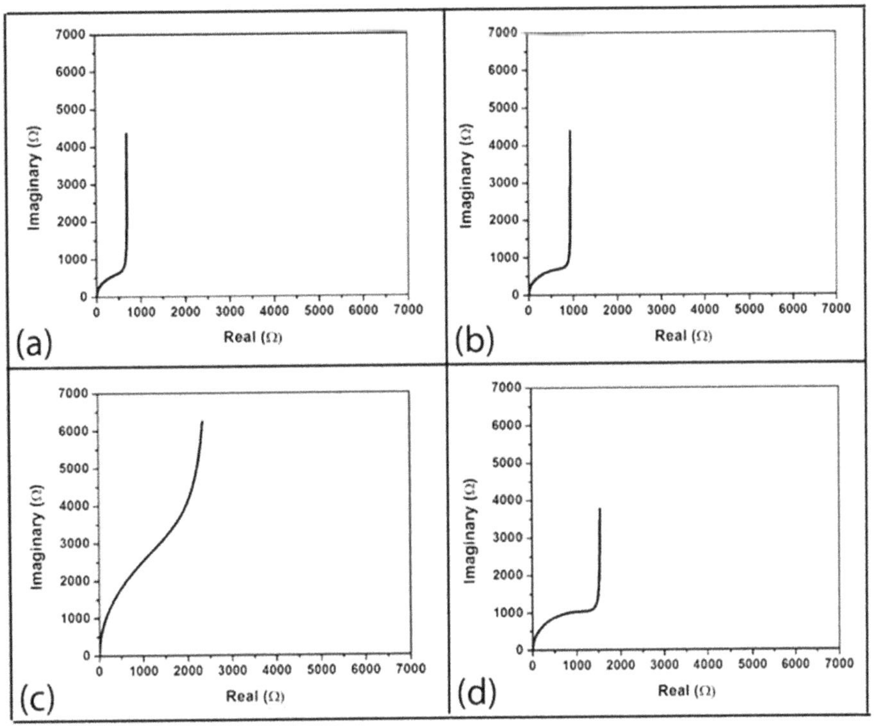

Figure 11. Nyquist plot of (RQ)(RQ) model: (a) E40, (b) E80, (c) E150, and (d) E300

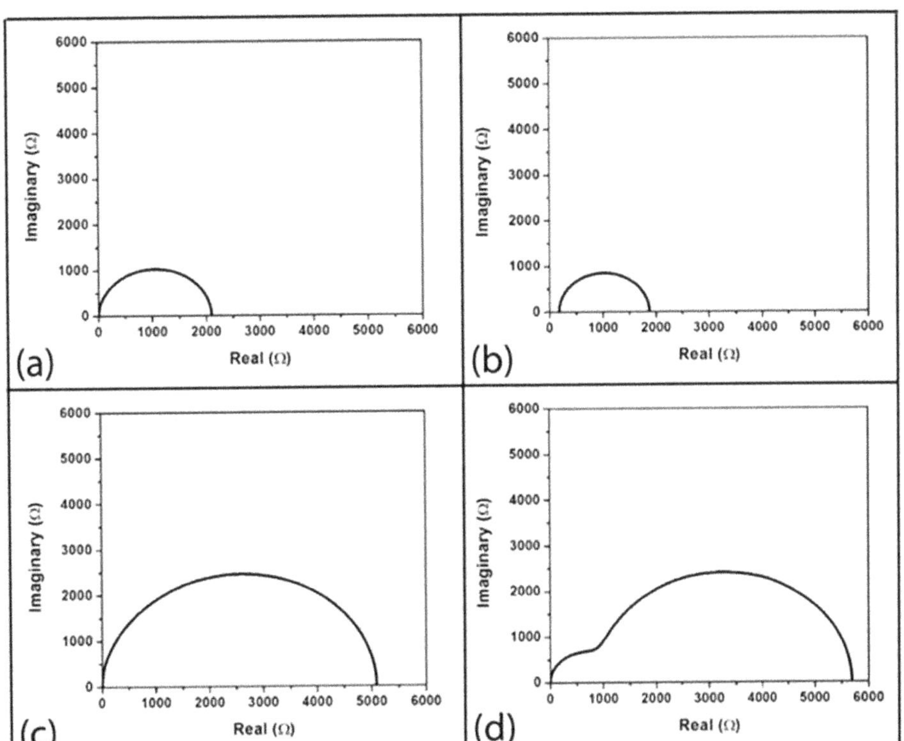

Figure 12. Nyquist plot of (RQ)(RQ)Re model: (a) E40, (b) E80, (c) E150, and (d) E300

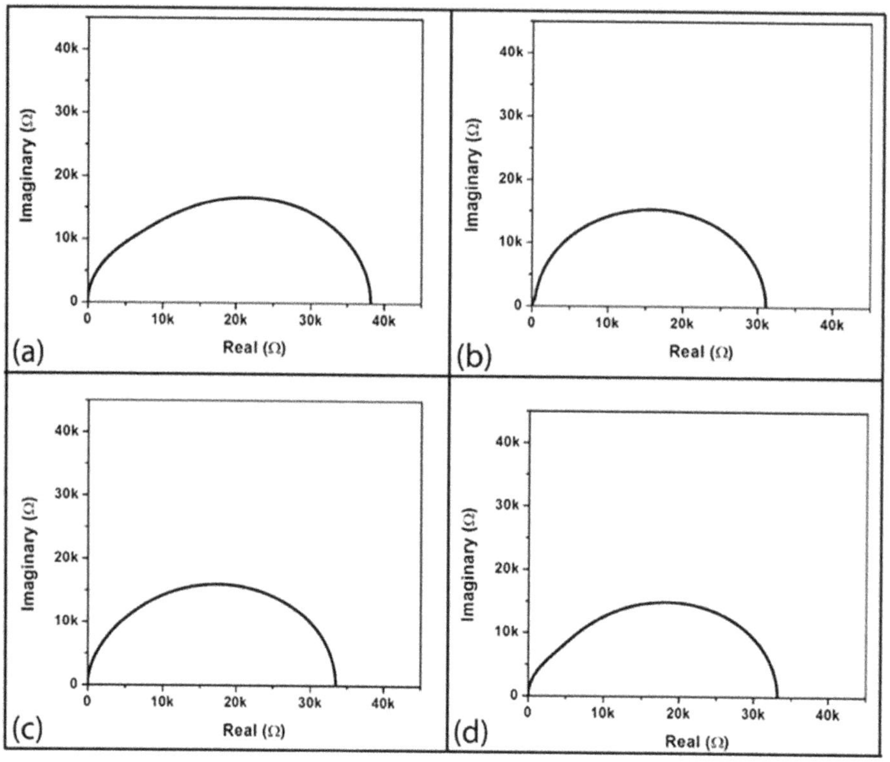

indicated that the time constants of the two CPEs might be significantly different. The bulk resistance was found to be 2100 Ω, 1875 Ω and 5090 Ω respectively for E40, E80 and E150 electrode systems. The bulk resistance of E300 electrode system could not be calculated for the plots due to bulk property, and electrical double layer was merged. The combination of the bulk and the faradic resistances was ~5690 Ω. The sample resistance (R_∞) (resistance at high frequency) was found to be zero when E40, E150 and E300 electrode systems were used. R_∞ for the E80 system was 175 Ω. This suggested that the resistance of the distilled water was ~175Ω.

The Nyquist plots of the (RQ)(RQ)Re model also showed profiles that could be due to the presence of CPEs of distinctly different time constants. Unlike in the Nyquist plot of (RQ)(RQ) model of E300, the two semicircular regions were not distinctly visible. The initial semicircular regions appeared as bulges to the more prominent semicircular plots (marked by arrows). Similar to the (RQ)(RQ) model, the combination of the bulk and the faradic resistances was noted from the Nyquist plot for (RQ)(RQ) Re model. The combination of the bulk and faradic resistances was 38.2 kΩ, 31.1kΩ, 33.5 kΩ and 33.2 kΩ for E40, E80, E150 and E300 systems respectively.

The Bode plots of different models have been shown in Figure 13, Figure 14, and Figure 15 for (RQ) Q, (RQ)(RQ) and (RQ)(RQ)Re models, respectively. The amplitude profiles of the (RQ)Q model was similar for all the electrode systems. The phase profiles for all the electrode systems showed Gaussian profile with a distinct peak. The phase profiles of the E40, E80 and E300 electrode systems showed a peak maxima at ~75 Hz, whereas, the peak maxima for E150 electrode system was at ~30 Hz. The phase margin for the E40, E80, E150 and E300 electrode systems were found to be 90.049, 90.0434, 90.0085 and 90.0286, respectively. The electrode systems showed three distinct regions. In the low-frequency region, there was a decrease in the magnitude response. Thereafter, the decrease in the amplitude response was slow in the mid-frequency region due to the presence of a capacitive element, confirmed from the phase plot, which shows a phase of -90°. Subsequently, the magnitude plot again decreased monotonically with the steeper slope at the high-frequency region. The magnitude plot of E150 showed only two distinct regions. In the low-frequency region, the steepness of the magnitude plot was slower as compared to the steepness in the high-frequency region.

The Bode plot of the electrode system fitted with (RQ)(RQ) model has been shown in Figure 14. The magnitude response of the E40 and E150 electrode system showed two regions. The initial part of the magnitude response was a constant region. This region was followed by a monotonically decreasing amplitude region in the higher frequency range. The amplitude plot of E80 electrode system showed three distinct regions. In the low and the high-frequency regions, amplitude response was constant and was parallel to the frequency axis. In the mid frequency region, there was a decrease in the amplitude response. The E300 electrode system also showed three distinct regions but, its profile was completely different than E80 electrode system. The amplitude response was steeply decreasing in the low and the high-frequency region as compared to the mid-frequency region. The phase response of the E40 and E300 systems showed the tendency of attaining a peak in the flat amplitude response region. This indicated the probability of the presence of a capacitive element in the circuit. This was confirmed from the attainment of a -90° phase at higher frequencies. The phase response of E80 electrode system showed a valley in the mid frequency region where there was a decrease in the amplitude profile. This indicated

Figure 13. Bode plots of (RQ)Q model for (a) E40, (b) E80, (c) E150, and (d) E300

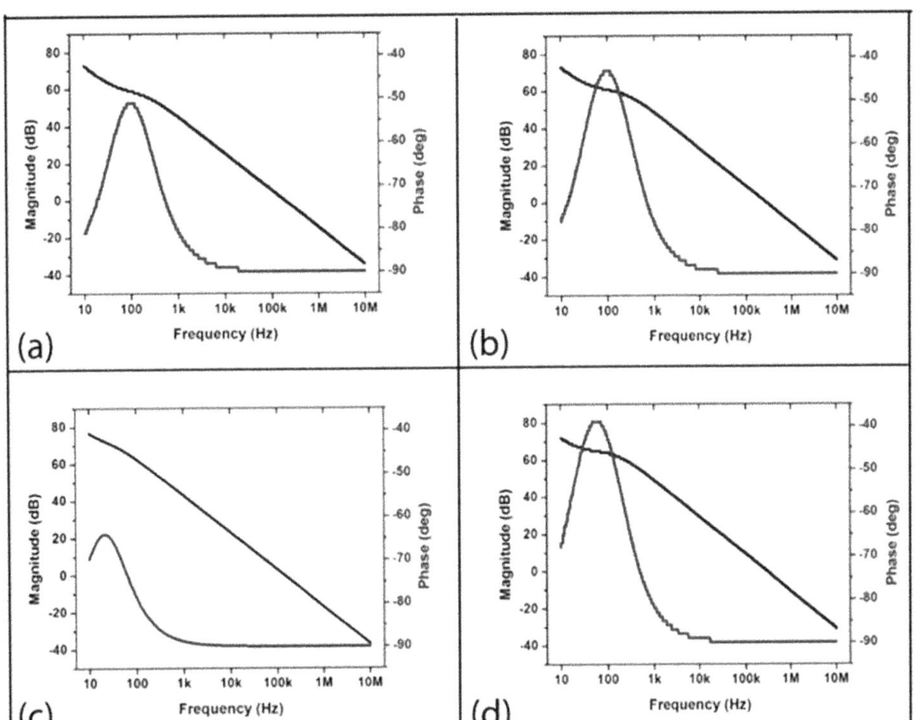

Figure 14. Bode plots of (RQ)(RQ) mode for (a) E40, (b) E80, (c) E150, and (d) E300

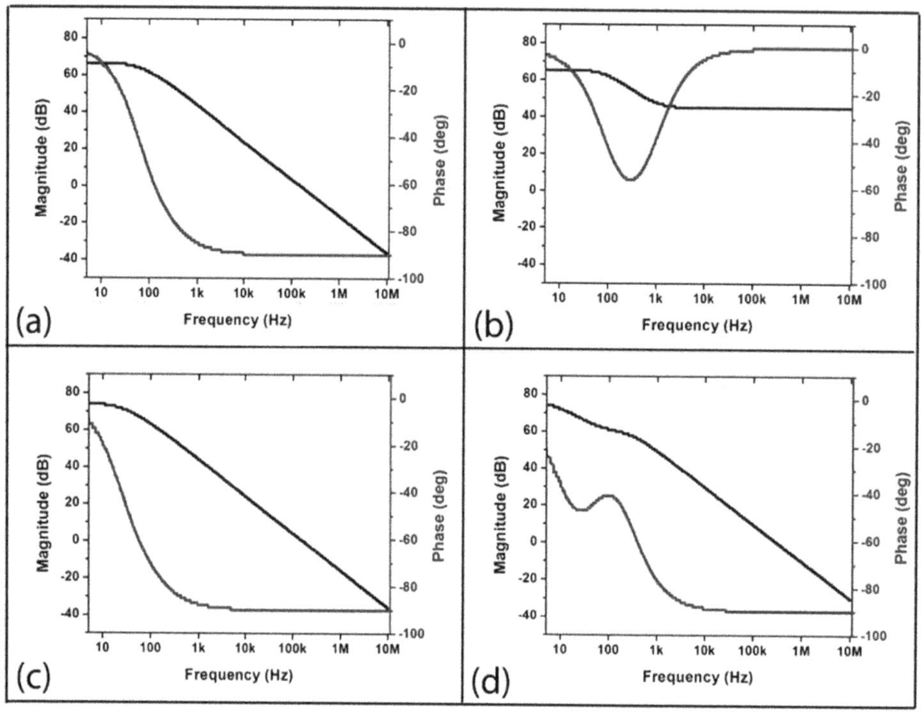

the presence of a capacitive element in the circuit. It would be wise to mention over here that during modeling of impedance profile, Q_2 was found to be zero. Due to this reason, this type of plot was obtained. In the E300 electrode system, a peak in mid frequency region was observed. The phase profile reached a -90° phase at higher frequencies.

The Bode plots of the (RQ)(RQ)Re model have been shown in Figure 15. E40, E150 and E300 electrode system showed two regions. In the initial region, there was a slow decrease in the amplitude response. Thereafter, there was a steep decrease in the amplitude at in the mid and the high-frequency region. In all the three systems, the phase at lower frequencies was higher which decreased to a residual value at the high-frequency region. The amplitude response of E80 electrode system showed three distinct regions. In the low and the high-frequency regions, there was a steep decrease in the amplitude response. In the mid frequency region, the steepness of decrease in the amplitude response was slower. The phase response of the system showed a peak in the mid-frequency region, which subsequently attains a residual value at the high-frequency region.

Analysis of Saline Solution

Saline solutions of different strengths were analyzed using the (RQ)(RQ) model using E40 electrode system. The impedance profiles of the samples have been shown in Figure 16. The experimental data was fitted using (RQ)(RQ) mathematical model. The values of the electrical elements were calculated. R, R_1 and Q_1 were found to be lowest in S1. An increase in the strength of saline solutions in S2, S3 and S4 did not alter the R, R_1 and Q values of the samples. An increase in the strengths from S2 to S3 resulted in

Figure 15. Bode plots of (RQ)(RQ)Re model for (a) E40, (b) E80, (c) E150, and (d) E300

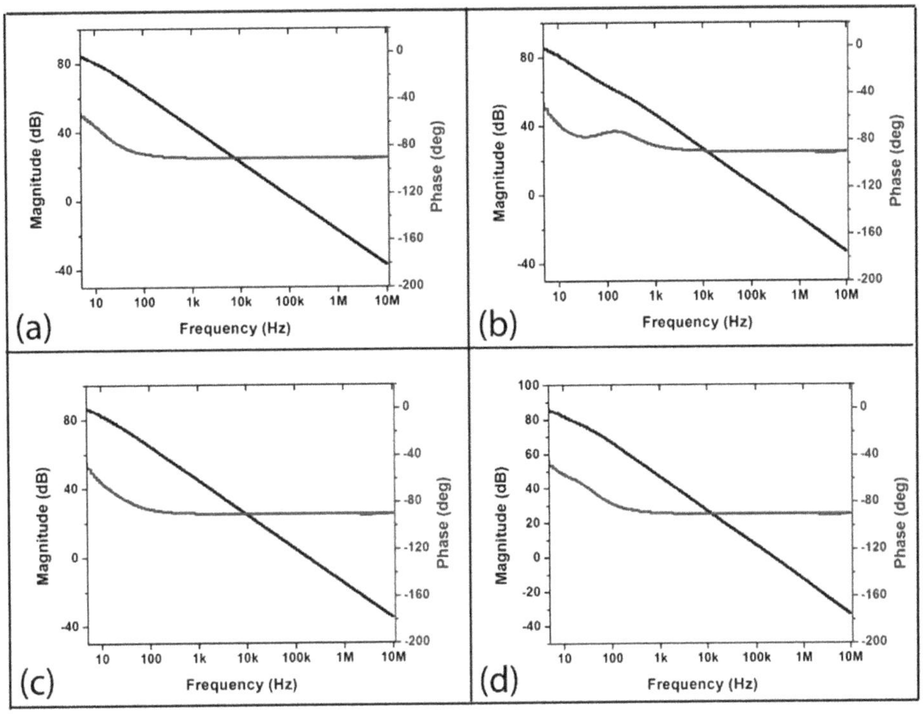

Figure 16. Impedance profiles of E40 electrode system: (a) S1, (b) S2, (c) S3, and (d) S4

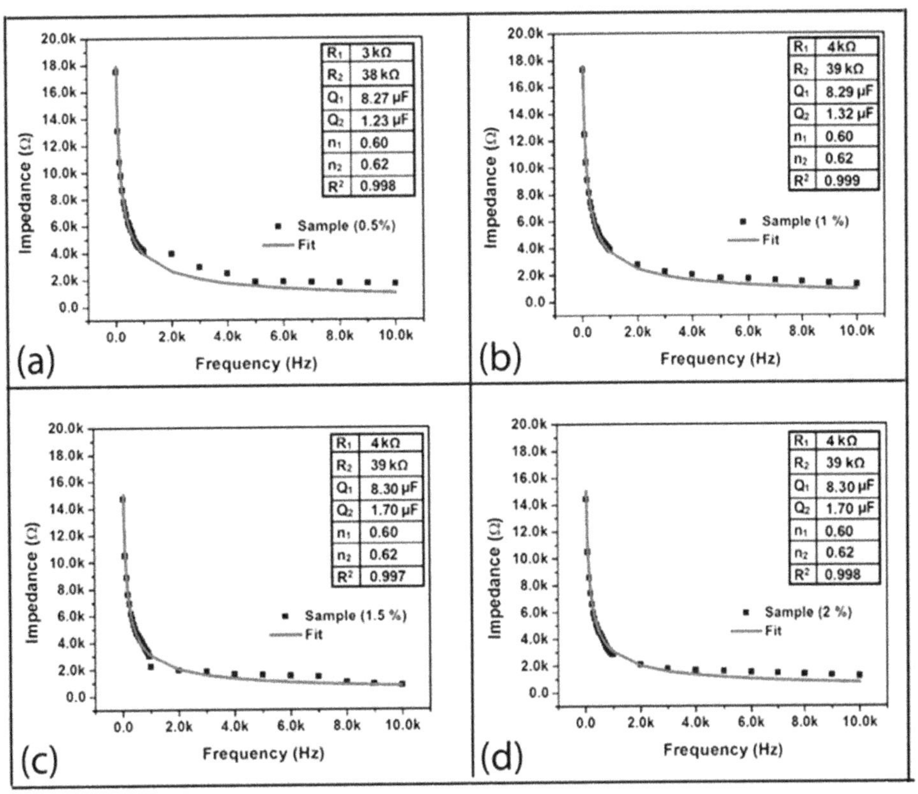

the increase Q_1 value. A further increase in the saline strengths in S4 did not show any alteration in Q1 values as compared to S3. The CPE inhomogeneous constants (n and n_1) were equal in all the samples. The results over here suggested that the capacitive elements (due to both sample and electrical double layer) were behaving as non-ideal capacitors.

The impulse response profiles of the samples have been shown in Figure 17. The impulse response profile was fitted using the previously mentioned mathematical model (Equation 6). The time constants of the circuits were calculated. The differences amongst the time constants within the samples and amongst the samples were not distinctly different. This suggested that the electrical properties of the samples would behave in a similar manner.

The Nyquist plot of the samples has been shown in Figure 18. All the samples showed similar semi-circular profiles. From the graph, it can be observed that the semicircular profile intersected at 0 Ω. This suggested that the resistance of the solution at high frequency is 0 Ω. Even though there are two capacitive elements present in the circuit, additional semicircle or bulges were not observed. This can be explained by the relaxation times of the capacitive elements. The changes in the relaxation times of the capacitive elements were insignificant. The presence of a similar type of Nyquist profile of the sample can be explained by the absence of the activation controlled process. Since there are no changes in the

Figure 17. Impulse response profiles of E40 electrode system for (a) S1, (b) S2, (c) S3, and (d) S4

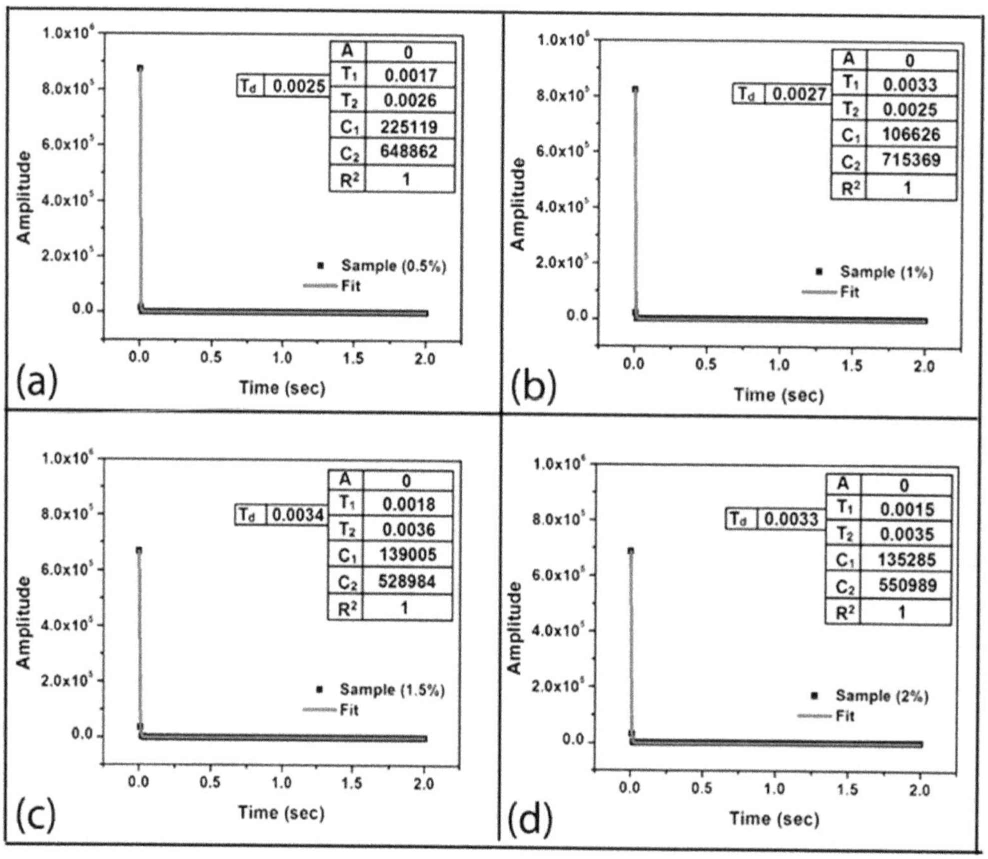

Figure 18. Nyquist plots of E40 electrode system for (a) S1, (b) S2, (c) S3, and (d) S4

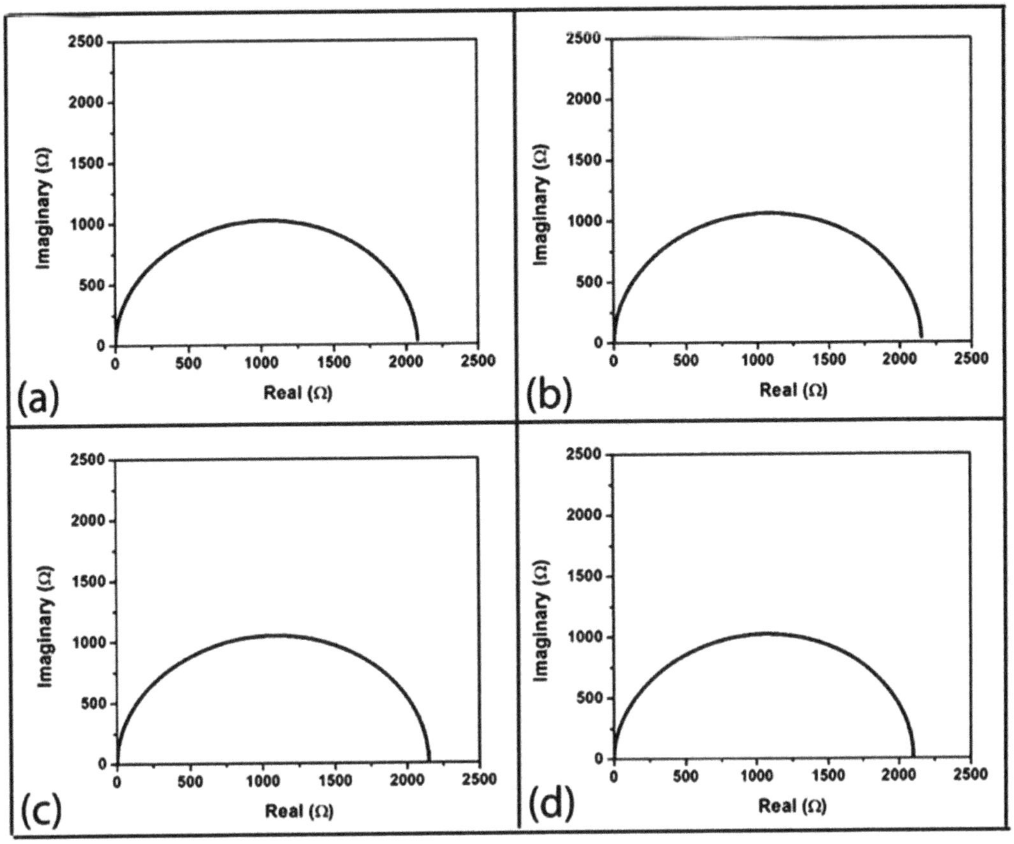

rate of reaction of the samples during the analysis, the potential changes in the presence of the samples were mainly due to the formation of the electrical double layer across the electrode surfaces.

The Bode plots of the sample have been shown in Figure 19. Similar to the Nyquist plot, Bode plots of all the samples were found to be similar. The amplitude profiles of all the samples showed two distinct regions. In the low-frequency region, the amplitude response was constant. An increase in the frequency resulted in the monotonical decrease in the amplitude plot. At lower frequencies, the phase component was highest, which rapidly decreased to a constant value of -90° at higher frequencies. This indicated the presence of capacitive element within the circuit. The occurrence of similar plots in all the cases also suggested that the variation in the Bode plot was not due to the variation in the sample concentration, but due to a constant parameter of the electrode system. This confirmed that, when E40 electrode system was used, the electrode polarization effect hampered the sample analysis.

FUTURE RESEARCH DIRECTIONS

The current study was devoted to the development of aluminium electrodes for EIS applications. The testing of the electrodes was achieved by EIS studies of distilled water and saline solutions of four different concentrations. Future studies can be performed to explore the suitability of the developed electrodes in

Figure 19. Bode plots of E40 electrode system for (a) S1, (b) S2, (c) S3, and (d) S4

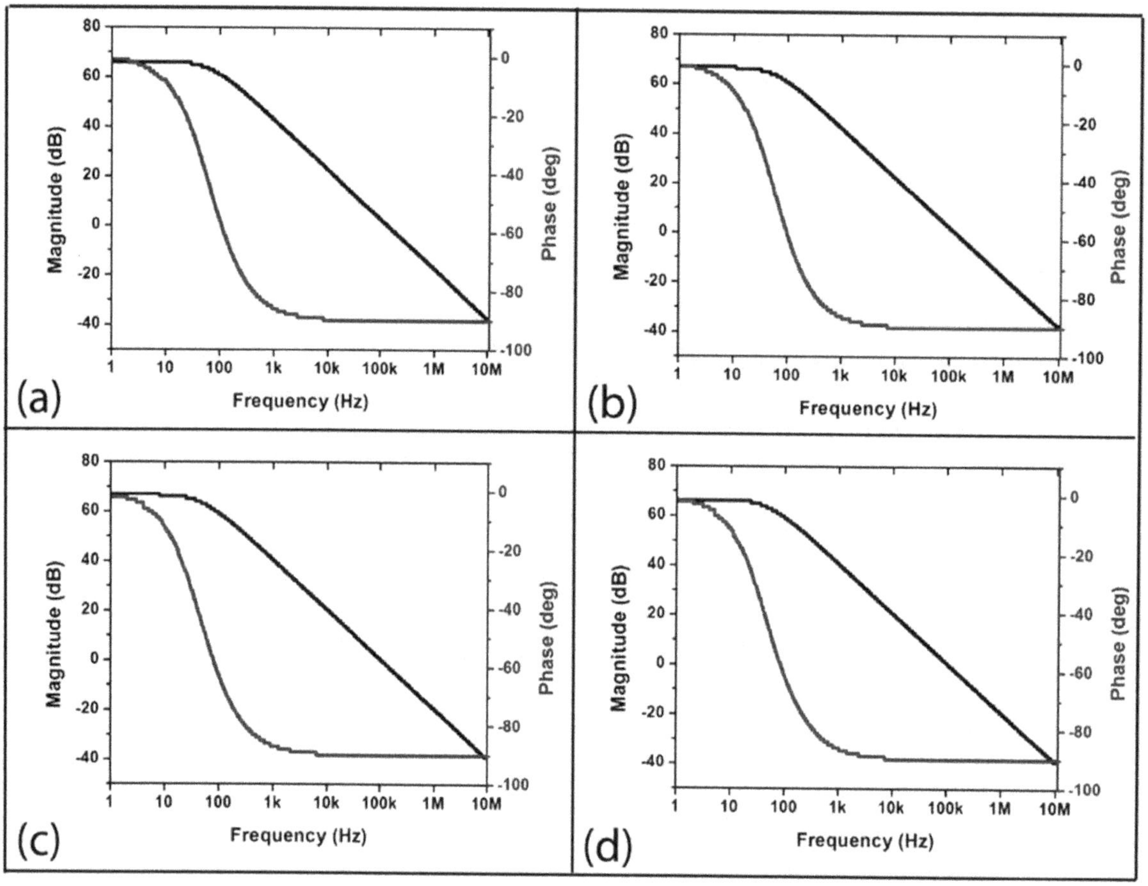

biomedical applications like blood glucose concentration monitoring, diagnosis of bladder pathologies, identification of different types of cancer (e.g. oral and skin), estimation of blood clotting time and antigen-antibody presence in blood. Studies can also be performed to implement different geometries of aluminium electrodes in EIS studies to establish the most suitable geometry for biomedical applications. Suitability of electrodes made from different metals can also be explored in future studies.

CONCLUSION

A thin layer of aluminium was deposited on the microscopic glass slides by vacuum vapor deposition technique to act as electrodes. The distance between two successive electrodes was 40 μm, 80 μm, 150 μm and 300 μm, respectively. The distances between the electrode edges were measured using ImageJ software. After the characterization of the electrodes, wells made using micropipette tips were fixed onto the glass slides using PDMS. Thin copper strips were attached to the base of the electrodes to act as the junction points for connecting the electrodes with an impedance analyzer. The impedance analysis was done in the frequency range of 50 Hz-10 kHz. For the selection of the electrodes, 200 μl of distilled water was used for the analysis. The experimental impedance profiles were fitted to the popularly used models for electrochemical analysis (i.e., (RQ)Q, (RQ)(RQ) and (RQ)(RQ)Re) by a non-linear least square method using Solver add-in in Excel 2013. The values of the electrical circuit elements obtained from the modeling and the equivalent 'C' values were used as the input for the in-lab developed LabVIEW program. The program provided the Impulse response, Bode and Nyquist plots of the equivalent electrical circuit models. The time constants of the equivalent circuits were calculated. The correlation coefficients amongst the impulse response profile and the fitting parameters models were equal to 1. A non-zero residual value was obtained from the impulse response profile for the (RQ)Q model. Hence, the (RQ)Q model was not considered for further analysis. Except for E300, the final time constant was lower for the (RQ)(RQ) model as compared to the (RQ)(RQ)Re model which suggested its better electrical stability. Taking a note of this, (RQ)(RQ) model was selected for further analysis. Also, a careful examination of the time constants of the different electrode systems using (RQ)(RQ) model showed the lowest final time constant value for E40 electrode system. Hence, E40 electrode system was used for the analysis of the saline solutions. Finally, EIS study of distilled water and saline solution was done using the developed setup to validate its performance.

REFERENCES

Åberg, P., Birgersson, U., Elsner, P., Mohr, P., & Ollmar, S. (2011). Electrical impedance spectroscopy and the diagnostic accuracy for malignant melanoma. *Experimental Dermatology*, *20*(8), 648–652. doi:10.1111/j.1600-0625.2011.01285.x PMID:21539620

Amusan, L., Olayinka, A., Nwambo, Y., Alayande, S., Ojuh, O., James, C., et al. (2010). Electrical Characterisation of Vacuum Thermally Deposited Aluminium Thin Film. *Research Journal of Applied Sciences, 5*(2), 96-100.

Baade, P., Meng, X., Youlden, D., Aitken, J., & Youl, P. (2012). Time trends and latitudinal differences in melanoma thickness distribution in Australia, 1990–2006. *International Journal of Cancer*, *130*(1), 170–178. doi:10.1002/ijc.25996 PMID:21344376

Benson, K., Cramer, S., & Galla, H.-J. (2013). Impedance-based cell monitoring: Barrier properties and beyond. *Fluids and Barriers of the CNS*, *10*(1), 1. doi:10.1186/2045-8118-10-5 PMID:23305242

Dhurjaty, S., Qiu, Y., Tan, M., Liu, H., & Zheng, B. (2015). *A new application of electrical impedance spectroscopy for measuring glucose metabolism: a phantom study.* Paper presented at the SPIE Medical Imaging.

Feldman, C. L. (1998). *Process for monitoring patients with chronic congestive heart failure*. Google Patents.

Gómez, R., Bashir, R., & Bhunia, A. K. (2002). Microscale electronic detection of bacterial metabolism. *Sensors and Actuators. B, Chemical*, *86*(2), 198–208. doi:10.1016/S0925-4005(02)00175-2

Hui, G., Mi, S., Chen, Q., & Chen, X. (2014). Sweet and bitter tastant discrimination from complex chemical mixtures using taste cell-based sensor. *Sensors and Actuators. B, Chemical*, *192*, 361–368. doi:10.1016/j.snb.2013.10.119

Hui, G.-H., Ji, P., Mi, S.-S., & Deng, S.-P. (2013). Electrochemical impedance spectrum frequency optimization of bitter taste cell-based sensors. *Biosensors & Bioelectronics*, *47*, 164–170. doi:10.1016/j.bios.2013.03.018 PMID:23578970

Johnson, N. W., Jayasekara, P., & Amarasinghe, A. (2011). Squamous cell carcinoma and precursor lesions of the oral cavity: Epidemiology and aetiology. *Periodontology 2000*, *57*(1), 19–37. doi:10.1111/j.1600-0757.2011.00401.x PMID:21781177

Johnson, T. C., Balbierz, D. J., & Pearson, R. (2005). *Method for detecting and treating tumors using localized impedance measurement*. Google Patents.

Ke, N., Nguyen, K., Irelan, J., & Abassi, Y. A. (2015). Multidimensional GPCR profiling and screening using impedance-based label-free and real-time assay. *G Protein-Coupled Receptor Screening Assays: Methods and Protocols*, 215-226.

Keshtkar, A., Keshtkar, A., & Smallwood, R. H. (2006). Electrical impedance spectroscopy and the diagnosis of bladder pathology. *Physiological Measurement*, *27*(7), 585–596. doi:10.1088/0967-3334/27/7/003 PMID:16705257

Krüger, J., Singh, K., ONeill, A., Jackson, C., Morrison, A., & OBrien, P. (2002). Development of a microfluidic device for fluorescence activated cell sorting. *Journal of Micromechanics and Microengineering*, *12*(4), 486. doi:10.1088/0960-1317/12/4/324

Lee, J. J., Hong, W. K., Hittelman, W. N., Mao, L., Lotan, R., & Shin, D. M. et al.. (2000). Predicting cancer development in oral leukoplakia: Ten years of translational research. *Clinical Cancer Research*, *6*(5), 1702–1710. PMID:10815888

Mashazi, P., Tetyana, P., Vilakazi, S., & Nyokong, T. (2013). Electrochemical impedimetric immunosensor for the detection of measles-specific IgG antibodies after measles infections. *Biosensors & Bioelectronics*, *49*, 32–38. doi:10.1016/j.bios.2013.04.028 PMID:23708815

Mohr, P., Birgersson, U., Berking, C., Henderson, C., Trefzer, U., Kemeny, L., & Paoli, J. et al. (2013). Electrical impedance spectroscopy as a potential adjunct diagnostic tool for cutaneous melanoma. *Skin Research and Technology*, *19*(2), 75–83. doi:10.1111/srt.12008 PMID:23350668

Murdoch, C., Brown, B. H., Hearnden, V., Speight, P. M., DApice, K., & Hegarty, A. M. et al.. (2014). Use of electrical impedance spectroscopy to detect malignant and potentially malignant oral lesions. *International Journal of Nanomedicine*, *9*, 4521. doi:10.2147/IJN.S64087 PMID:25285005

Napier, S. S., & Speight, P. M. (2008). Natural history of potentially malignant oral lesions and conditions: An overview of the literature. *Journal of Oral Pathology & Medicine*, *37*(1), 1–10. doi:10.1111/j.1600-0714.2007.00579.x PMID:18154571

Nguyen, T. A., Yin, T.-I., Reyes, D., & Urban, G. A. (2013). Microfluidic chip with integrated electrical cell-impedance sensing for monitoring single cancer cell migration in three-dimensional matrixes. *Analytical Chemistry*, *85*(22), 11068–11076. doi:10.1021/ac402761s PMID:24117341

Özcan, B., Demirbakan, B., Yeşiller, G., & Sezgintürk, M. K. (2014). Introducing a new method for evaluation of the interaction between an antigen and an antibody: Single frequency impedance analysis for biosensing systems. *Talanta*, *125*, 7–13. doi:10.1016/j.talanta.2014.02.067 PMID:24840408

Perfézou, M., Turner, A., & Merkoçi, A. (2012). Cancer detection using nanoparticle-based sensors. *Chemical Society Reviews*, *41*(7), 2606–2622. doi:10.1039/C1CS15134G PMID:21796315

Pradhan, R., Mitra, A., & Das, S. (2013). Quantitative evaluation of blood glucose concentration using impedance sensing devices. *Journal of Electrical Bioimpedance*, *4*(1), 73–77. doi:10.5617/jeb.657

Ramanathan, M., Patil, M., Epur, R., Yun, Y., Shanov, V., Schulz, M., & Kumta, P. N. et al. (2016). Gold-coated carbon nanotube electrode arrays: Immunosensors for impedimetric detection of bone biomarkers. *Biosensors & Bioelectronics*, *77*, 580–588. doi:10.1016/j.bios.2015.10.014 PMID:26476598

Rigel, D. S., Russak, J., & Friedman, R. (2010). The evolution of melanoma diagnosis: 25 years beyond the ABCDs. *CA: a Cancer Journal for Clinicians*, *60*(5), 301–316. doi:10.3322/caac.20074 PMID:20671054

Shervedani, R. K., Mehrjardi, A. H., & Zamiri, N. (2006). A novel method for glucose determination based on electrochemical impedance spectroscopy using glucose oxidase self-assembled biosensor. *Bioelectrochemistry (Amsterdam, Netherlands)*, *69*(2), 201–208. doi:10.1016/j.bioelechem.2006.01.003 PMID:16580891

Solly, K., Wang, X., Xu, X., Strulovici, B., & Zheng, W. (2004). Application of real-time cell electronic sensing (RT-CES) technology to cell-based assays. *Assay and Drug Development Technologies*, *2*(4), 363–372. doi:10.1089/adt.2004.2.363 PMID:15357917

Tsao, H., Atkins, M. B., & Sober, A. J. (2004). Management of cutaneous melanoma. *The New England Journal of Medicine*, *351*(10), 998–1012. doi:10.1056/NEJMra041245 PMID:15342808

Varshney, M., & Li, Y. (2009). Interdigitated array microelectrodes based impedance biosensors for detection of bacterial cells. *Biosensors & Bioelectronics*, *24*(10), 2951–2960. doi:10.1016/j.bios.2008.10.001 PMID:19041235

Warnakulasuriya, S., Johnson, N., & Van der Waal, I. (2007). Nomenclature and classification of potentially malignant disorders of the oral mucosa. *Journal of Oral Pathology & Medicine*, *36*(10), 575–580. doi:10.1111/j.1600-0714.2007.00582.x PMID:17944749

Yang, L., & Bashir, R. (2008). Electrical/electrochemical impedance for rapid detection of foodborne pathogenic bacteria. *Biotechnology Advances*, *26*(2), 135–150. doi:10.1016/j.biotechadv.2007.10.003 PMID:18155870

ADDITIONAL READING

Gimsa, J., & Wachner, D. (1998). A unified resistor-capacitor model for impedance, dielectrophoresis, electrorotation, and induced transmembrane potential. *Biophysical Journal*, *75*(2), 1107–1116. doi:10.1016/S0006-3495(98)77600-3 PMID:9675212

Grimnes, S., & Martinsen, O. (2005). Cole electrical impedance model-a critique and an alternative. *IEEE Transactions on Bio-Medical Engineering*, *52*(1), 132–135. doi:10.1109/TBME.2004.836499 PMID:15651574

Gudivaka, R., Schoeller, D., Kushner, R., & Bolt, M. (1999). Single-and multifrequency models for bioelectrical impedance analysis of body water compartments. *Journal of Applied Physiology*, *87*(3), 1087–1096. PMID:10484581

Hui, G., Mi, S., Chen, Q., & Chen, X. (2014). Sweet and bitter tastant discrimination from complex chemical mixtures using taste cell-based sensor. *Sensors and Actuators. B, Chemical*, *192*, 361–368. doi:10.1016/j.snb.2013.10.119

Hui, G.-H., Ji, P., Mi, S.-S., & Deng, S.-P. (2013). Electrochemical impedance spectrum frequency optimization of bitter taste cell-based sensors. *Biosensors & Bioelectronics*, *47*, 164–170. doi:10.1016/j.bios.2013.03.018 PMID:23578970

Johnson, N. W., Jayasekara, P., & Amarasinghe, A. (2011). Squamous cell carcinoma and precursor lesions of the oral cavity: Epidemiology and aetiology. *Periodontology 2000*, *57*(1), 19–37. doi:10.1111/j.1600-0757.2011.00401.x PMID:21781177

Jorcin, J.-B., Orazem, M. E., Pébère, N., & Tribollet, B. (2006). CPE analysis by local electrochemical impedance spectroscopy. *Electrochimica Acta*, *51*(8), 1473–1479. doi:10.1016/j.electacta.2005.02.128

Keshtkar, A., Keshtkar, A., & Smallwood, R. H. (2006). Electrical impedance spectroscopy and the diagnosis of bladder pathology. *Physiological Measurement*, *27*(7), 585–596. doi:10.1088/0967-3334/27/7/003 PMID:16705257

Lee, J. J., Hong, W. K., Hittelman, W. N., Mao, L., Lotan, R., & Shin, D. M. et al.. (2000). Predicting cancer development in oral leukoplakia: Ten years of translational research. *Clinical Cancer Research*, *6*(5), 1702–1710. PMID:10815888

Mohr, P., Birgersson, U., Berking, C., Henderson, C., Trefzer, U., Kemeny, L., & Paoli, J. et al. (2013). Electrical impedance spectroscopy as a potential adjunct diagnostic tool for cutaneous melanoma. *Skin Research and Technology*, *19*(2), 75–83. doi:10.1111/srt.12008 PMID:23350668

Murdoch, C., Brown, B. H., Hearnden, V., Speight, P. M., DApicc, K., & Hegarty, A. M. et al.. (2014). Use of electrical impedance spectroscopy to detect malignant and potentially malignant oral lesions. *International Journal of Nanomedicine*, *9*, 4521. doi:10.2147/IJN.S64087 PMID:25285005

Napier, S. S., & Speight, P. M. (2008). Natural history of potentially malignant oral lesions and conditions: An overview of the literature. *Journal of Oral Pathology & Medicine*, *37*(1), 1–10. doi:10.1111/j.1600-0714.2007.00579.x PMID:18154571

Perfézou, M., Turner, A., & Merkoçi, A. (2012). Cancer detection using nanoparticle-based sensors. *Chemical Society Reviews*, *41*(7), 2606–2622. doi:10.1039/C1CS15134G PMID:21796315

Pradhan, R., Mitra, A., & Das, S. (2013). Quantitative evaluation of blood glucose concentration using impedance sensing devices. *Journal of Electrical Bioimpedance*, *4*(1), 73–77. doi:10.5617/jeb.657

Rigel, D. S., Russak, J., & Friedman, R. (2010). The evolution of melanoma diagnosis: 25 years beyond the ABCDs. *CA: a Cancer Journal for Clinicians*, *60*(5), 301–316. doi:10.3322/caac.20074 PMID:20671054

Schiller, C., & Strunz, W. (2001). The evaluation of experimental dielectric data of barrier coatings by means of different models. *Electrochimica Acta*, *46*(24), 3619–3625. doi:10.1016/S0013-4686(01)00644-2

Shervedani, R. K., Mehrjardi, A. H., & Zamiri, N. (2006). A novel method for glucose determination based on electrochemical impedance spectroscopy using glucose oxidase self-assembled biosensor. *Bioelectrochemistry (Amsterdam, Netherlands)*, *69*(2), 201–208. doi:10.1016/j.bioelechem.2006.01.003 PMID:16580891

Tsao, H., Atkins, M. B., & Sober, A. J. (2004). Management of cutaneous melanoma. *The New England Journal of Medicine*, *351*(10), 998–1012. doi:10.1056/NEJMra041245 PMID:15342808

Wang, Q., Moser, J.-E., & Grätzel, M. (2005). Electrochemical impedance spectroscopic analysis of dye-sensitized solar cells. *The Journal of Physical Chemistry B*, *109*(31), 14945–14953. doi:10.1021/jp052768h PMID:16852893

Warnakulasuriya, S., Johnson, N., & Van der Waal, I. (2007). Nomenclature and classification of potentially malignant disorders of the oral mucosa. *Journal of Oral Pathology & Medicine*, *36*(10), 575–580. doi:10.1111/j.1600-0714.2007.00582.x PMID:17944749

KEY TERMS AND DEFINITIONS

Bode Plot: Bode plot refers to a standard method of graphical representation of the frequency response of a system. It consists of a magnitude plot and phase plot.

Electrical Stability: Electrical stability of a linear system refers to the ability of the system to produce a finite output for a finite input.

Electrochemical Impedance Spectroscopy (EIS): EIS is a technique to characterize the dielectric properties of an electrochemical system as a function of frequency.

Electrode: An electrode is a conductor used to set up electrical contact with the non-metallic part of an electric circuit.

Impulse Response: Impulse response of a system refers to the output of the system for an impulse input.

Nyquist Plot: Nyquist Plot is a method of representing the frequency response of a linear system.

Sensor: A sensor is a device which responds to a stimulus and provides an output signal.

Chapter 22
Artificial Intelligence:
Current Issues and Applications

Kijpokin Kasemsap
Suan Sunandha Rajabhat University, Thailand

ABSTRACT

This chapter explains the Artificial Intelligence (AI) techniques in terms of Artificial Neural Networks (ANNs), fuzzy logic, expert systems, machine learning, Genetic Programming (GP), Evolutionary Polynomial Regression (EPR), and Support Vector Machine (SVM); the AI applications in modern education; the AI applications in software engineering development; the AI applications in Content-Based Image Retrieval (CBIR); and the multifaceted applications of AI in the digital age. AI is a branch of science which deals with helping machines find the suitable solutions to complex problems in a more human-like manner. AI technologies bring more complex data-analysis features to the existing applications in various industries and greatly contribute to management's organization, planning, and controlling operations, and will continue to do so with more frequency as programs are refined.

INTRODUCTION

Artificial intelligence (AI) techniques become useful as the alternate approaches to conventional techniques or as thc components of integrated systems (Kalogirou, Metaxiotis, & Mellit, 2010). The accuracy of AI methods is superior to that of traditional statistical methods in dealing with financial problems, especially regarding nonlinear patterns (Bahrammirzaee, 2010). AI methods are employed throughout the information retrieval process toward value-added services (Mandl, 2009). Information retrieval has considerably changed with the expansion of the Internet and the advent of inexpensive graphical user interfaces and mass storage devices (Kasemsap, 2017a). AI methods outperform the conventional methods in terms of response time and controllability (Kow, Wong, Rajkumar, & Rajkumar, 2016), and are utilized in various applications, such as automation and process control (Ali, Hussain, Tade, & Zhang, 2015).

The utilization of AI has increased since the middle of the 20th century as recognized in its application in a wide range of engineering and science problems (Yaseen, El-shafie, Jaafar, Afan, & Sayl, 2015). Since the idea of AI with knowledge had been introduced, so many thoughts, theories, and ideas in

DOI: 10.4018/978-1-5225-2440-3.ch022

various fields of engineering, science, geology, social study, economics, and management methods have been proposed (Sugiyama, 2013). Examples of the available AI techniques are artificial neural networks (ANNs), genetic programming (GP), evolutionary polynomial regression (EPR), support vector machine (SVM), and k-nearest neighbor algorithm (Elshorbagy, Corzo, Srinivasulu, & Solomatine, 2010).

This chapter is based on a literature review of AI. The extensive literature of AI provides a contribution to practitioners and researchers by explaining the current issues and applications of AI in order to maximize the impact of AI in the digital age.

BACKGROUND

The beginning of modern AI can be traced to the classical philosophers' attempts to describe human thinking as a symbolic system. But the field of AI was not formally founded until 1956, at a conference at Dartmouth College, in Hanover, New Hampshire, where the term "artificial intelligence" was coined. The organizers included John McCarthy, Marvin Minsky, Claude Shannon, Nathaniel Rochester, all of whom went on to greatly contribute to the field. In the years following the Dartmouth Conference, impressive advances were made in AI. Machines were built that could solve school mathematics problems, and a program called Eliza became the world's first chatbot, occasionally fooling users into thinking that it was conscious.

The first "AI winter" lasted from 1974 until around 1980. It was followed in the 1980s by another boom, thanks to the advent of expert systems, and the Japanese fifth generation computer initiative, which adopted massively parallel programming. Expert systems limit themselves to solving narrowly defined problems from single domains of expertise (for instance, litigation) using vast databases. They avoid the messy complications of everyday life, and do not tackle the perennial problem of trying to inculcate common sense. The funding dried up again in the late 1980s because the difficulties of the tasks being addressed was once again underestimated, and also because desktop computers overtook mainframes in speed and power, rendering very expensive legacy machines redundant.

AI has crossed the threshold for the simple reason that it works. AI has provided effective services that make a huge difference in people's lives, toward enabling companies to make a lot of money. A central goal of AI is the design of automated systems that can accomplish a task despite uncertainty (Poupart, 2012). Such systems can be viewed as taking inputs from the environment and producing outputs toward the realization of some goals (Poupart, 2012). Modern intelligent agents approaches should combine methodologies, techniques, and architectures from many areas of computer science, cognitive science, operation research, and cybernetics (Marinagi, Panayiotopoulos, & Spyropoulos, 2005). AI planning is an essential function of intelligence that is necessary in intelligent agents applications (Marinagi et al., 2005).

SIGNIFICANT ASPECTS OF ARTIFICIAL INTELLIGENCE

This section provides the AI techniques in terms of ANNs, fuzzy logic, expert systems, machine learning, GP, EVR, and SVM; the AI applications in modern education; the AI applications in software engineering development; the AI applications in content-based image retrieval (CBIR); and the multifaceted applications of AI in the digital age.

Artificial Intelligence Techniques

Artificial Neural Networks

Artificial neural network (ANN) is a form of AI that attempts to imitate the function of the human brain and nervous systems (Shahin, 2016). Neurons in adjacent layers of ANNs are connected by weighted links passing signals from one layer to the next adjacent layer. Those links have numerical weights that are applied to the inputs of a neuron (Ali et al., 2015). Although the concept of ANNs was first introduced in 1943 (McCulloch & Pitts, 1943), research into applications of ANNs has expanded since the introduction of the back-propagation training algorithm for the multilayer perceptrons (MLPs) in 1986 (Rumelhart, Hinton, & Williams, 1986).

The architecture of ANNs consists of an array of processing elements (PEs), or nodes, that are arranged in layers: an input layer, an output layer, and one or more hidden layers (Shahin, 2016). Based on the ability of self-designing to solve a problem by learning the solution from data (Mutihac, 2009), ANNs have been used in power system applications including the classification of different power quality (PQ) disturbances (Khokhar, Zin, Mokhtar, & Pesaran, 2015). ANNs can perform tasks, such as pattern recognition, classification, function approximation, optimization, and data clustering (Uyar, Yildirim, & Gencoglu, 2008). ANN training is the learning process to extract the inherent complex relationship among input variables by which network can calculate the difference between actual and desired output (Raza & Khosravi, 2015).

Fuzzy Logic

Fuzzy logic is the modern technology that allows the realistic models of the real world to be defined with the fuzzy variables and fuzzy rules (Sadkhan Al Maliky & Jawad, 2015) and is an extension of the traditional binary logic that allows individuals to achieve the multi-evaluated logic by describing domains in a much more detailed manner and by classifying better through searches in a more extensive area (Castro & Arcay, 2009). Fuzzy-logic-based multi-criteria decision making techniques have become very popular in doing the optimization of different manufacturing processes (Das, Roy, Rai, & Saha, 2016). Optimization is the investigation of an alternative with the most cost effective or highest achievable performance under the given constraints, by maximizing desired factors and minimizing undesired ones (Kasemsap, 2017b).

Fuzzy logic is one of the possibilities to imitate the working of a human brain and to try to turn AI into real intelligence (Alsina Pagès, Mateo Segura, & Socoró Carriém, 2009) and has proved to be a powerful tool for decision-making systems, such as expert systems and pattern classification systems (Ghosal, Haldar, & Sutradhar, 2015). Fuzzy control is a type of intelligent control whose main feature is that a control knowledge base is available within the controller and control actions are generated by applying the existing conditions or data to the knowledge base, making use of an inference mechanism (Kanthi, 2015).

Recent developments in the robotic age have made robots and robotics more intelligent, affordable, and user-friendly in modern operations, ranging from manufacturing to health care (Kasemsap, 2017c). Fuzzy system technology has emerged as an effective methodology to solve many problems, ranging from control engineering, robotics, and automation to system identification, medical image, and signal processing (Mehran, Zahawi, & Giaouris, 2011). Fuzzy sets and fuzzy logic provide a logical framework for description of the varied amount of ambiguity, uncertainty, and imprecision exhibited in real-world data under consideration (Bhattacharyya & Dutta, 2012). Fuzzy logic due to its proficiency in handling uncertainty has gained its importance in a variety of applications in combination with the use of data mining techniques (Bhattacharya & Bhatnagar, 2015). Data mining is the process of applying these computational methods in showing unknown data formats in large data sets (Kasemsap, 2015a).

Fuzzy logic works on fuzzy sets that indicate the maximum limits of an element (Ali et al., 2015). Fuzzy logic is a type of logic or multi-valued logic that distinguishes more than just TRUE and FALSE values and can be represented with the degrees of truthfulness and falseness in the range of [0,1] where 0 is absolute FALSE and 1 is absolute TRUE (Ali et al., 2015). Despite commercial success of fuzzy logic, a type-1 fuzzy set (T1FS) does not capture uncertainty in its manifestations when it arises from vagueness in the shape of the membership function (Azar, 2012). Such uncertainties need to be depicted by fuzzy sets that have blur boundaries. The imprecise boundaries of a type-2 fuzzy set (T2FS) give rise to truth/membership values that are fuzzy sets in [0], [1], instead of a crisp number. Type-2 fuzzy logic systems (T2FLSs) offer opportunity to model levels of uncertainty which traditional fuzzy logic type1 struggles (Azar, 2012).

Expert Systems

Expert systems have been applied to many areas of research to handle various problems (Manjunath, 2005), and requires specific knowledge and skills (Atanasova, 2015). Because the main function of the expert systems is to imitate the expertise and distribute the expert knowledge to nonexperts, these benefits can be greatly enhanced with the emergence of the Internet (Duan, 2009). Originally, expert systems comprised the idea that computer programs can be managed to solve the complex problems of decision making and can be better than human experts (Svensson, 2005).

Expert systems are the programs with a set of rules encapsulating knowledge about a particular problem domain (Atanasova, 2015) and can be defined as computer systems that can solve a problem for which an answer is being sought like an expert and present solution recommendations (Mutlu-Bayraktar & Esgin, 2015). Knowledge is receiving recognition as a strategic force in organizations (Wagner, 2003). Correspondingly, one form of knowledge capture and maintenance organizations are tempted to apply is the expert system utilized by end users (Wagner, 2003). Expert systems differ from conventional programs both in the way problem knowledge is stored and used.

Machine Learning

As the size of multimedia databases grows, retrieval becomes a key challenge in multimedia database management (Wei & Li, 2011). Accordingly, it is necessary to apply machine learning techniques to automatically tune the mechanism of multimedia retrieval systems (Wei & Li, 2011). Machine learning is an emerging area of computer science that deals with the design and development of new algorithms based on various types of data (Kulkarni, 2012). Machine learning builds a computational model from past experience of what has been observed and studies the automated acquisition of domain knowledge looking for the improvement of systems performance as result of experience (Gama & de Carvalho, 2009).

The most common machine learning algorithms can be either supervised or unsupervised. Supervised learning algorithms generate a function that maps inputs to desired outputs, based on a set of examples with known output (labeled examples). Unsupervised learning algorithms find the patterns and relationships over a given set of inputs (unlabeled examples) (Tzanis, Berberidis, & Vlahavas, 2009). The automatic classification of web pages using machine learning techniques is the most efficient way used by search engines to provide accurate results to the users (Selvakuberan, Devi, & Rajaram, 2009). Machine learning classifiers can be trained to preserve the personal details from unauthenticated users and for privacy preserving data mining (Selvakuberan et al., 2009).

Genetic Programming

Genetic programming (GP) is an extension of genetic algorithms (GA), which are evolutionary computing search (optimization) methods that are based on the principles of genetics and natural selection (Shahin, 2016). GP is a method for producing computer programs through adaptation according to a user-defined fitness criterion (Hsu, 2009) and evolves programs based on symbolic regression regarding artificial evolution (Collet, 2007), whereas GA can be used as a tool for the design and generation of fuzzy rules for a fuzzy logic system (Mohammadian, 2009).

GP is a type of evolutionary algorithm (i.e., a search algorithm inspired by the Darwinian process of natural selection) that evolves computer programs or executable structures (Freitas & Pappa, 2009) and tends to have better consistency in terms of goodness of fit and accuracy across majority of data sets (Afzal, Torkar, Feldt, & Gorschek, 2010). GP offers solutions to many drawbacks inherent to the classic identification approaches and relates to evolutionary algorithm designed to evolve the graph-based individuals (Ferariu & Patelli, 2012). GP manipulates and optimizes a population of computer models proposed to solve a specific problem, so that the model that fits the problem is obtained.

Evolutionary Polynomial Regression

Evolutionary polynomial regression (EPR) is a hybrid regression technique based on evolutionary computing that was developed by Giustolisi and Savic (2006). EPR constructs the symbolic models by integrating the optimum features of numerical regression, with GP and symbolic regression (Koza, 1992). EPR is suitable for modeling the physical phenomena, based on two features: the introduction of previous knowledge about the physical process, to be modeled at three different times, namely before, during, and after EPR modeling calibration; and the production of symbolic patterns, enabling data mining to discover the patterns that describe the desired parameters. The following two steps describe the features of the EPR technique, aimed to search for polynomial structures representing a system (Shahin, 2016). In the first step of EPR, the selection of exponents for polynomial expressions is carried out, employing an evolutionary searching strategy by means of GA (Goldberg, 1989). In the second step, numerical regression using the least square method is conducted, aiming to compute the coefficients of the previously selected polynomial terms (Shahin, 2016).

Support Vector Machine

Support vector machine (SVM) is the recent revolutionary development for classification analysis (Sun, 2014) and is the machine learning technique that performs in solving common data mining problems (Musicant, 2005). SVM introduced by Vapnik (1995) is a supervised learning machine tool applied for pattern recognition and classification. SVM uses the nonlinear mapping to transform the original training data into a higher dimension (Jagan, Gundlapalli, & Samui, 2016). SVM has been applied in many pattern recognition and regression problems, such as for dependency estimation, forecasting, and constructing intelligent machines.

SVM is used in the large classification problems because the training of SVM can handle the large vector dimensions more effectively as compared to conventional classifiers (Khokhar et al., 2015), based on the theory of statistical learning theory (Das, Samui, & Sabat, 2011). Using the SVM that is trained on multiple demand series has produced the most accurate forecasts (Carbonneau, Vahidov, & Laframboise, 2007). The effect of the imbalance of the distribution of target classes in databases can be analyzed with respect to the performance of AI-related standard classifiers, such as decision trees and SVM (Vannucci, Colla, Cateni, & Sgarbi, 2012).

Artificial Intelligence Applications in Modern Education

Information and communication technologies have led to the new developments in education (Işik & Aslan, 2015). Type, goal, and specialty of intelligent systems programmed to satisfy the educational need are being developed with each passing day (Mutlu-Bayraktar, 2015). The expert systems gather all the knowledge about a subject which needs expertise from experts, printed sources, and online sources (e.g., journals, books, and articles) and store them in the database of the system (Mutlu-Bayraktar & Esgin, 2015).

Through AI, information is properly represented in real time for the effective study of knowledge inference mechanism (Mutlu-Bayraktar, 2015). The fuzzy logic technique has been used to classify student knowledge levels and topic difficulty levels (Deperlioglu, Sonugur, & Suzme, 2015). In order to adaptively determine next questions, the stored questions have been classified with division clustering methods, and the most suitable questions for the related student knowledge level have been found by using the nearest neighbor algorithm (Deperlioglu et al., 2015). Employing AI-based approaches is one of the most popular ways to improve the e-learning process and rapidly obtain the desired objectives (Kose, 2015).

Artificial Intelligence Applications in Software Engineering Development

The demands of software requirement imply a growing need for using AI to support all aspects of the software development process (Ramachandran, 2010). The software engineering development process usually involves many participants from multiple perspectives, who may be in the geographically dispersed locations (Liu, Khudkhudia, Wen, & Sajja, 2010). AI techniques (e.g., knowledge-based systems, neural networks, fuzzy logic, and data mining) have been advocated by many researchers and developers as the way to improve the software development activities (Meziane & Vadera, 2010). Existing argumentation-based software design can support the software development-related decision making by establishing the design explanation for critical decisions (Liu et al., 2010).

Software design can be recognized as the bridge between requirements and implementation (Soria, Diaz-Pace, Bass, Bachmann, & Campo, 2010). The tasks applied to find out, analyze, document, and check the needs and constraints placed on a software product, have become the category of a subject called requirement engineering (del Sagrado Martinez & del Aguila Cano, 2010). Requirement engineering, the first phase of any software development project, is the crucial part of the whole development process, as requirements documents are often inconsistent and incomplete (Kof, 2010).

AI techniques are successfully applied in various phases of software development life cycle (Kosmatov, 2010). One of the most innovative AI applications is using constraint-based techniques for automation of software testing (Kosmatov, 2010). Software testing is a technique for achieving the degree of software quality toward obtaining consumer confidence (Lam, 2010). A myriad of cost estimation models (e.g., multi-objective particle swarm optimization) have been proposed to predict the software engineering development costs early in the life cycle with the expectation of effectively managing the project within time and budget (Gonsalves, Yamagishi, Kawabata, & Itoh, 2010).

AI-related cost estimation methods are fundamental to help project managers allocate resources more adequately, thus supporting projects to be finished on time and within budget (Mendez, 2010). Web effort models and estimation techniques (e.g., AI-related Bayesian networks) provide the methods for the Web companies to formalize the way they estimate the effort for their software engineering development projects, and help in obtaining more accurate estimates (Mendez, 2010). The primary reason for the lack of high accuracy in prediction may be because most estimation models are linear in the parameters and are not sufficiently flexible (Sarcia, Cantone, & Basili, 2010).

Artificial Intelligence Applications in Content-Based Image Retrieval

The increasing trend of multimedia data utilization is likely to accelerate creating an urgent need of providing a clear method of capturing, storing, indexing, retrieving, analyzing, and summarizing data through image data (Ma, 2009). Content-based image retrieval (CBIR), which was proposed in the early 1990s, has attracted a broad range of research interests from many computer communities in the past decade (Chen & Chen, 2009). CBIR aims to search the images that are similar to the query based on the visual content of the images without the support of annotations (Zhang & Alhajj, 2009) and must be able to interact with users and discover the current user's information needs (Wei & Li, 2009).

With the increasing demand of multimedia information retrieval, such as image and video retrieval from the Web, there is a need to find ways to train a classifier when the training dataset is combined with a small number of labeled data and a large number of unlabeled one (Jiang & Ip, 2009). AI technologies allow for a more semantically complex metadata image representation toward handling the digital images (Avola, Ferri, & Grifoni, 2009). After the generation of multimedia data turning digital, an explosion of interest in their data storage, retrieval, and processing, has drastically increased in the database and data mining community. Retrieval is performed by image example where a query image is given as input by the user and an appropriate similarity measure is used to find the best matches in the corresponding feature space (Gondra, 2009).

The multimedia index makes it possible to group data according to similarity criteria through utilizing the k-nearest neighbor algorithm for retrieving the databases (Hacid & Zighed, 2009). Many different solutions regarding CBIR have been proposed to improve performance of retrieval, but the large part of these works have focused on sub-parts of the retrieval problem, providing the targeted solutions only for individual aspects, such as feature extraction, similarity measures, and indexing (García Pérez, Mosquera, Berretti, & del Bimbo, 2009). In hospitals and medical institutes, a large number of medical images are being produced in ever increasing quantities and used for diagnostics and therapy (Wei, Li, & Li, 2009). The need for effective methods to manage and retrieve those image resources has been actively pursued in the medical community (Wei et al., 2009). The images are searched by using the surrounding text, captions, and keywords (Zhang & Zhang, 2009).

Texture feature extraction is the important research content in content-based medical image retrieval and can be executed through AI-related co-occurrence matrix and perceptual texture features (Zhang, Ma, & Yan, 2009). Texture features are extracted, represented, and normalized in an attribute vector, followed by a generation of rough set dependency rules from the real value attribute vector (Ali, 2009). The rough set reduction technique is applied along with the minimal subset of attributes associated with a class label for classification (Ali, 2009).

Multifaceted Applications of Artificial Intelligence in the Digital Age

Automation with inherent AI is a growing field with diverse applications in various industries (Hengstler, Enkel, & Duelli, 2016). The enduring innovations in AI and robotics offer the promised capacity of computer consciousness, sentience, and rationality (Ashrafian, 2015). Expert systems and machine learning are the two AI techniques that have been widely used in telecommunications, while machine learning and distributed artificial intelligence are the two AI techniques which are most promising for the future (Qi, Wu, Li, & Shu, 2007). AI techniques play an important role in modeling, analysis, and prediction of the performance and control of renewable energy (Belu, 2014). In less democratic countries, government-affiliated AI-enabled national search engines can be introduced and these engines can control the access to the information at the individual level (Karaman, 2012).

In civil engineering, AI methods have been extensively used in the fields of civil engineering applications, such as construction management, building materials, hydraulic, optimization, and transportation engineering (Martínez-Barrera, Gencel, Beycioglu, Subaşı, & González-Rivas, 2015). The discipline of investing requires data identification, asset valuation, and risk management (Rada, 2009). AI techniques apply to many aspects of financial investing, and published work has shown an emphasis on the application of knowledge-based techniques for credit risk assessment and machine learning techniques for stock valuation (Rada, 2009). Econometric modeling and forecasting techniques regarding AI techniques are highly exploited to the understanding of international tourism demand (Goh, Mok, & Law, 2009).

Computational tools to solve the three challenges of natural resources management (i.e., data management and communication; data analysis; and optimization and control) can be implemented with AI techniques, because AI techniques have flexibility to treat the intrinsic dynamics of natural resources (Adamatti & de Aguiar, 2012). With their high performance and simplicity, the developed AI models can be effectively utilized to evaluate the effects of future management scenarios on groundwater nitrate concentration, leading to more reasonable groundwater resources management and decision making (Alagha, Said, & Mogheir, 2014).

Recent developments in information technology and AI tools, particularly in neural networks, fuzzy logic, and SVM, have provided the necessary support to develop the highly efficient automated diagnostic systems in health care settings (Begg, 2009). The ability to interact with the system at any place and any time significantly changes the coordinates of time and place and offers the opportunities to the health care delivery and management which lead to many opportunities as well as many challenges arising from the availability and accessibility of modern tools to access the information (Pombo, Garcia, Bousson, & Felizardo, 2015).

FUTURE RESEARCH DIRECTIONS

The classification of the extensive literature in the domains of AI will provide the potential opportunities for future research. AI can be effectively utilized in executing the repetitive and time-consuming tasks. Cloud computing includes network access to storage, processing power, development platforms, and software (Kasemsap, 2015b). Business intelligence involves creating any type of data visualization that provides insight into a business for the purpose of making a decision or taking an action (Kasemsap, 2016). The purpose of business intelligence is to support the business-related decision making via the interactive access, and the analysis of important corporate information (Kasemsap, 2017d). Critical thinking is one of the most vital parts of the problem-solving and decision-making process, as it is the act of clearly thinking through options that will lead to a final choice (Kasemsap, 2017e).

Intelligent decision support systems are the decision support systems that make extensive use of artificial intelligence techniques (Kasemsap, 2017f). Semantic Web is the extension of the World Wide Web that catalogs information on a web page and reprocesses it so that other machines including computers can understand the information. (Kasemsap, 2017g). Mobile ad-hoc network is a network that has many autonomous nodes, often composed of mobile devices or other mobile pieces, that can arrange themselves in various ways and operate without strict top-down network administration (Kasemsap, 2017h). An examination of linkages among AI, cloud computing, business intelligence, critical thinking, intelligent decision support systems, Semantic Web, and mobile ad-hoc network would seem to be viable for future research efforts.

CONCLUSION

This chapter explained the AI techniques in terms of ANNs, fuzzy logic, expert systems, machine learning, GP, EVR, and SVM; the AI applications in modern education; the AI applications in software engineering development; the AI applications in CBIR; and the multifaceted applications of AI in the digital age. There are many AI techniques (e.g., ANNs, fuzzy logic, expert systems, machine learning, GP, EVR, and SVM) that can be utilized in various perspectives, such as business, finance, supply chain, manufacturing, robotics, health care, education, software engineering development, and CBIR. AI is a branch of science which deals with helping machines find the suitable solutions to complex problems in a more human-like manner.

Modern AI has many practical developments and delivers the real benefits to users and encompasses a diverse number of computer applications, or components within applications, that use the sets of rules and knowledge to make inferences. The effective approach of AI can be taken depending on the requirements established, which influences how artificial the intelligent behavior appears. AI technologies bring more complex data-analysis features to the existing applications in various industries and greatly contribute to management's organization, planning, and controlling operations, and will continue to do so with more frequency as programs are refined. With AI, the chances of error are almost zero and greater precision is achieved.

AI technologies increase the world's economic prosperity, and enhance individual's ability to make progress on many important problems. In particular, any area where progress depends on analyzing and using huge amounts of data can be solved by AI techniques. Enterprises that utilize the AI-enhanced applications are expected to become more diverse, as the needs for the ability to analyze data across multiple variables, fraud detection, and customer relationship management emerge as the key business drivers to gain competitive advantage. In business applications, AI capabilities are integrated with systems that serve the day-to-day needs of the enterprise, such as inventory tracking, manufacturing process controls, and customer service databases.

REFERENCES

Adamatti, D. F., & de Aguiar, M. S. (2012). Artificial intelligence applied to natural resources management. In *Machine learning: Concepts, methodologies, tools and applications* (pp. 1566–1582). Hershey, PA: IGI Global. doi:10.4018/978-1-60960-818-7.ch604

Afzal, W., Torkar, R., Feldt, R., & Gorschek, T. (2010). Genetic programming for cross-release fault count predictions in large and complex software projects. In M. Chis (Ed.), *Evolutionary computation and optimization algorithms in software engineering: Applications and techniques* (pp. 94–126). Hershey, PA: IGI Global. doi:10.4018/978-1-61520-809-8.ch006

Alagha, J. S., Said, M. A. M., & Mogheir, Y. (2014). Modeling of nitrate concentration in groundwater using artificial intelligence approach: A case study of Gaza coastal aquifer. *Environmental Monitoring and Assessment*, *186*(1), 35–45. doi:10.1007/s10661-013-3353-6 PMID:23974533

Ali, J. M. (2009). Content-based image classification and retrieval: A rule-based system using rough sets framework. In Z. Ma (Ed.), *Artificial intelligence for maximizing content based image retrieval* (pp. 68–83). Hershey, PA: IGI Global. doi:10.4018/978-1-60566-174-2.ch004

Ali, J. M., Hussain, M. A., Tade, M. O., & Zhang, J. (2015). Artificial intelligence techniques applied as estimator in chemical process systems: A literature survey. *Expert Systems with Applications: An International Journal*, *42*(14), 5915–5931. doi:10.1016/j.eswa.2015.03.023

Alsina Pagès, R. M., Mateo Segura, C., & Socoró Carrié, J. (2009). Fuzzy logic estimator for variant SNR environments. In J. Rabuñal Dopico, J. Dorado, & A. Pazos (Eds.), *Encyclopedia of artificial intelligence* (pp. 719–727). Hershey, PA: IGI Global. doi:10.4018/978-1-59904-849-9.ch107

Ashrafian, H. (2015). Artificial intelligence and robot responsibilities: Innovating beyond rights. *Science and Engineering Ethics*, *21*(2), 317–326. doi:10.1007/s11948-014-9541-0 PMID:24737482

Atanasova, I. (2015). On design and development of QLIFEX: An expert system for social area. *International Journal of Applied Behavioral Economics*, *4*(2), 16–35. doi:10.4018/ijabe.2015040102

Avola, D., Ferri, F., & Grifoni, P. (2009). Genetic algorithms and other approaches in image feature extraction and representation. In Z. Ma (Ed.), *Artificial intelligence for maximizing content based image retrieval* (pp. 1–19). Hershey, PA: IGI Global. doi:10.4018/978-1-60566-174-2.ch001

Azar, A. T. (2012). Overview of type-2 fuzzy logic systems. *International Journal of Fuzzy System Applications*, *2*(4), 1–28. doi:10.4018/ijfsa.2012100101

Bahrammirzaee, A. (2010). A comparative survey of artificial intelligence applications in finance: Artificial neural networks, expert system and hybrid intelligent systems. *Neural Computing & Applications*, *19*(8), 1165–1195. doi:10.1007/s00521-010-0362-z

Begg, R. (2009). Artificial intelligence techniques in medicine and healthcare. In J. Tan (Ed.), *Medical informatics: Concepts, methodologies, tools, and applications* (pp. 784–791). Hershey, PA: IGI Global. doi:10.4018/978-1-60566-050-9.ch057

Belu, R. (2014). Artificial intelligence techniques for solar energy and photovoltaic applications. In *Robotics: Concepts, methodologies, tools, and applications* (pp. 1662–1720). Hershey, PA: IGI Global. doi:10.4018/978-1-4666-4607-0.ch081

Bhattacharya, S., & Bhatnagar, V. (2015). Critical parameters for fuzzy data mining. In *Research methods: Concepts, methodologies, tools, and applications* (pp. 1–18). Hershey, PA: IGI Global.

Bhattacharyya, S., & Dutta, P. (2012). Fuzzy logic: Concepts, system design, and applications to industrial informatics. In M. Khan & A. Ansari (Eds.), *Handbook of research on industrial informatics and manufacturing intelligence: Innovations and solutions* (pp. 33–71). Hershey, PA: IGI Global. doi:10.4018/978-1-4666-0294-6.ch003

Carbonneau, R., Vahidov, R., & Laframboise, K. (2007). Machine learning-based demand forecasting in supply chains. *International Journal of Intelligent Information Technologies*, *3*(4), 40–57. doi:10.4018/jiit.2007100103

Castro, A., & Arcay, B. (2009). Fuzzy logic applied to biomedical image analysis. In J. Rabuñal Dopico, J. Dorado, & A. Pazos (Eds.), *Encyclopedia of artificial intelligence* (pp. 710–718). Hershey, PA: IGI Global. doi:10.4018/978-1-59904-849-9.ch106

Chen, M., & Chen, S. (2009). MMIR: An advanced content-based image retrieval system using a hierarchical learning framework. In Z. Ma (Ed.), *Artificial intelligence for maximizing content based image retrieval* (pp. 371–387). Hershey, PA: IGI Global. doi:10.4018/978-1-60566-174-2.ch016

Collet, P. (2007). Genetic programming. In J. Rennard (Ed.), *Handbook of research on nature-inspired computing for economics and management* (pp. 59–73). Hershey, PA: IGI Global. doi:10.4018/978-1-59140-984-7.ch005

Das, B., Roy, S., Rai, R. N., & Saha, S. C. (2016). Application of grey fuzzy logic for the optimization of CNC milling parameters for Al–4.5%Cu–TiC MMCs with multi-performance characteristics. *Engineering Science and Technology: An International Journal*, *19*(2), 857–865.

Das, S. K., Samui, P., & Sabat, A. K. (2011). Application of artificial intelligence to maximum dry density and unconfined compressive strength of cement stabilized soil. *Geotechnical and Geological Engineering*, *29*(3), 329–342. doi:10.1007/s10706-010-9379-4

del Sagrado Martinez, J., & del Aguila Cano, I. M. (2010). A Bayesian network for predicting the need for a requirements review. In F. Meziane & S. Vadera (Eds.), *Artificial intelligence applications for improved software engineering development: New prospects* (pp. 106–128). Hershey, PA: IGI Global. doi:10.4018/978-1-60566-758-4.ch006

Deperlioglu, O., Sonugur, G., & Suzme, K. (2015). Intelligent questioning system based on fuzzy logic. In U. Kose & D. Koc (Eds.), *Artificial intelligence applications in distance education* (pp. 72–94). Hershey, PA: IGI Global. doi:10.4018/978-1-4666-6276-6.ch005

Duan, Y. (2009). Web-based expert systems. In M. Khosrow-Pour (Ed.), *Encyclopedia of information science and technology* (2nd ed.; pp. 4105–4118). Hershey, PA: IGI Global. doi:10.4018/978-1-60566-026-4.ch656

Elshorbagy, A., Corzo, G., Srinivasulu, S., & Solomatine, D. P. (2010). Experimental investigation of the predictive capabilities of data driven modeling techniques in hydrology-Part 1: Concepts and methodology. *Hydrology and Earth System Sciences, 14*(10), 1931–1941. doi:10.5194/hess-14-1931-2010

Ferariu, L., & Patelli, A. (2012). Genetic programming for system identification. In P. Cong-Vinh (Ed.), *Formal and practical aspects of autonomic computing and networking: Specification, development, and verification* (pp. 135–168). Hershey, PA: IGI Global. doi:10.4018/978-1-60960-845-3.ch006

Freitas, A. A., & Pappa, G. L. (2009). Genetic programming for automatically constructing data mining algorithms. In J. Wang (Ed.), *Encyclopedia of data warehousing and mining* (2nd ed.; pp. 932–936). Hershey, PA: IGI Global. doi:10.4018/978-1-60566-010-3.ch144

Gama, J., & de Carvalho, A. C. (2009). Machine learning. In M. Khosrow-Pour (Ed.), *Encyclopedia of information science and technology* (2nd ed.; pp. 2462–2468). Hershey, PA: IGI Global. doi:10.4018/978-1-60566-026-4.ch392

García Pérez, D., Mosquera, A., Berretti, S., & del Bimbo, A. (2009). Content based image retrieval using active-nets. In Z. Ma (Ed.), *Artificial intelligence for maximizing content based image retrieval* (pp. 85–114). Hershey, PA: IGI Global. doi:10.4018/978-1-60566-174-2.ch005

Ghosal, K., Haldar, P., & Sutradhar, G. (2015). Application of fuzzy expert system in medical treatment. In A. Kumar (Ed.), *Fuzzy expert systems for disease diagnosis* (pp. 200–246). Hershey, PA: IGI Global. doi:10.4018/978-1-4666-7240-6.ch008

Giustolisi, O., & Savic, D. A. (2006). A symbolic data-driven technique based on evolutionary polynomial regression. *Journal of Hydroinformatics, 8*(3), 207–222.

Goh, C., Mok, H. M., & Law, R. (2009). Artificial intelligence applications in tourism. In M. Khosrow-Pour (Ed.), *Encyclopedia of information science and technology* (2nd ed.; pp. 241–247). Hershey, PA: IGI Global. doi:10.4018/978-1-60566-026-4.ch042

Goldberg, D. E. (1989). *Genetic algorithms in search optimization and machine learning*. Reading, MA: Addison–Wesley.

Gondra, I. (2009). Personalized content-based image retrieval. In Z. Ma (Ed.), *Artificial intelligence for maximizing content based image retrieval* (pp. 261–287). Hershey, PA: IGI Global. doi:10.4018/978-1-60566-174-2.ch012

Gonsalves, T., Yamagishi, K., Kawabata, R., & Itoh, K. (2010). Optimizing software development cost estimates using multi-objective particle swarm optimization. In F. Meziane & S. Vadera (Eds.), *Artificial intelligence applications for improved software engineering development: New prospects* (pp. 46–65). Hershey, PA: IGI Global. doi:10.4018/978-1-60566-758-4.ch003

Hacid, H., & Zighed, A. D. (2009). A machine learning-based model for content-based image retrieval. In Z. Ma (Ed.), *Artificial intelligence for maximizing content based image retrieval* (pp. 171–190). Hershey, PA: IGI Global. doi:10.4018/978-1-60566-174-2.ch008

Hengstler, M., Enkel, E., & Duelli, S. (2016). Applied artificial intelligence and trust: The case of autonomous vehicles and medical assistance devices. *Technological Forecasting and Social Change*, *105*, 105–120. doi:10.1016/j.techfore.2015.12.014

Hsu, W. H. (2009). Genetic programming. In J. Wang (Ed.), *Encyclopedia of data warehousing and mining* (2nd ed.; pp. 926–931). Hershey, PA: IGI Global. doi:10.4018/978-1-60566-010-3.ch143

Işik, A. H., & Aslan, G. (2015). Review and proposal for intelligent distance education. In U. Kose & D. Koc (Eds.), *Artificial intelligence applications in distance education* (pp. 95–109). Hershey, PA: IGI Global. doi:10.4018/978-1-4666-6276-6.ch006

Jagan, J., Gundlapalli, P., & Samui, P. (2016). Utilization of classification techniques for the determination of liquefaction susceptibility of soils. In S. Bhattacharyya, P. Banerjee, D. Majumdar, & P. Dutta (Eds.), *Handbook of research on advanced hybrid intelligent techniques and applications* (pp. 124–160). Hershey, PA: IGI Global. doi:10.4018/978-1-4666-9474-3.ch005

Jiang, J., & Ip, H. H. (2009). Active learning with SVM. In J. Rabuñal Dopico, J. Dorado, & A. Pazos (Eds.), *Encyclopedia of artificial intelligence* (pp. 1–7). Hershey, PA: IGI Global. doi:10.4018/978-1-59904-849-9.ch001

Kalogirou, S., Metaxiotis, K., & Mellit, A. (2010). Artificial intelligence techniques for modern energy applications. In K. Metaxiotis (Ed.), *Intelligent information systems and knowledge management for energy: Applications for decision support, usage, and environmental protection* (pp. 1–39). Hershey, PA: IGI Global. doi:10.4018/978-1-60566-737-9.ch001

Kanthi, M. (2015). Fuzzy logic-based intelligent control system for active ankle foot orthosis. In A. Kumar (Ed.), *Fuzzy expert systems for disease diagnosis* (pp. 248–279). Hershey, PA: IGI Global. doi:10.4018/978-1-4666-7240-6.ch009

Karaman, F. (2012). Artificial intelligence enabled search engines (AIESE) and the implications. In C. Jouis, I. Biskri, J. Ganascia, & M. Roux (Eds.), *Next generation search engines: Advanced models for information retrieval* (pp. 438–455). Hershey, PA: IGI Global. doi:10.4018/978-1-4666-0330-1.ch019

Kasemsap, K. (2015a). The role of data mining for business intelligence in knowledge management. In A. Azevedo & M. Santos (Eds.), *Integration of data mining in business intelligence systems* (pp. 12–33). Hershey, PA: IGI Global. doi:10.4018/978-1-4666-6477-7.ch002

Kasemsap, K. (2015b). The role of cloud computing adoption in global business. In V. Chang, R. Walters, & G. Wills (Eds.), *Delivery and adoption of cloud computing services in contemporary organizations* (pp. 26–55). Hershey, PA: IGI Global. doi:10.4018/978-1-4666-8210-8.ch002

Kasemsap, K. (2016). The fundamentals of business intelligence. *International Journal of Organizational and Collective Intelligence*, *6*(2), 12–25. doi:10.4018/IJOCI.2016040102

Kasemsap, K. (2017a). Mastering web mining and information retrieval in the digital age. In A. Kumar (Ed.), *Web usage mining techniques and applications across industries* (pp. 1–28). Hershey, PA: IGI Global. doi:10.4018/978-1-5225-0613-3.ch001

Kasemsap, K. (2017b). Mastering customer service, customer experience, and customer orientation in the hospitality and tourism industry. In P. Vasant & K. M. (Eds.), Handbook of research on holistic optimization techniques in the hospitality, tourism, and travel industry (pp. 115–140). Hershey, PA: IGI Global. doi:10.4018/978-1-5225-1054-3.ch005

Kasemsap, K. (2017c). Robotics: Theory and applications. In M. Moore (Ed.), *Cybersecurity breaches and issues surrounding online threat protection* (pp. 311–345). Hershey, PA: IGI Global. doi:10.4018/978-1-5225-1941-6.ch013

Kasemsap, K. (2017d). Mastering business process management and business intelligence in global business. In M. Tavana, K. Szabat, & K. Puranam (Eds.), *Organizational productivity and performance measurements using predictive modeling and analytics* (pp. 192–212). Hershey, PA: IGI Global. doi:10.4018/978-1-5225-0654-6.ch010

Kasemsap, K. (2017e). Promoting critical thinking in the modern learning environments. In R. Batko & A. Szopa (Eds.), *Strategic imperatives and core competencies in the era of robotics and artificial intelligence* (pp. 50–80). Hershey, PA: IGI Global. doi:10.4018/978-1-5225-1656-9.ch004

Kasemsap, K. (2017f). Mastering intelligent decision support systems in enterprise information management. In G. Sreedhar (Ed.), *Web data mining and the development of knowledge-based decision support systems* (pp. 35–56). Hershey, PA: IGI Global. doi:10.4018/978-1-5225-1877-8.ch004

Kasemsap, K. (2017g). Software as a service, Semantic Web, and big data: Theories and applications. In A. Turuk, B. Sahoo, & S. Addya (Eds.), *Resource management and efficiency in cloud computing environments* (pp. 264–285). Hershey, PA: IGI Global. doi:10.4018/978-1-5225-1721-4.ch011

Kasemsap, K. (2017h). Radio frequency identification and mobile ad-hoc network: Theories and applications. In S. Bhattacharyya, N. Das, D. Bhattacharjee, & A. Mukherjee (Eds.), *Handbook of research on recent developments in intelligent communication application* (pp. 63–95). Hershey, PA: IGI Global. doi:10.4018/978-1-5225-1785-6.ch003

Khokhar, S., Zin, A. A. B. M., Mokhtar, A. S. B., & Pesaran, M. (2015). A comprehensive overview on signal processing and artificial intelligence techniques applications in classification of power quality disturbances. *Renewable & Sustainable Energy Reviews*, *51*, 1650–1663. doi:10.1016/j.rser.2015.07.068

Kof, L. (2010). From textual scenarios to message sequence charts. In F. Meziane & S. Vadera (Eds.), *Artificial intelligence applications for improved software engineering development: New prospects* (pp. 83–105). Hershey, PA: IGI Global. doi:10.4018/978-1-60566-758-4.ch005

Kose, U. (2015). For an intelligent e-learning: A managerial model suggestion for artificial intelligence supported e-learning content flow. In U. Kose & D. Koc (Eds.), *Artificial intelligence applications in distance education* (pp. 149–160). Hershey, PA: IGI Global. doi:10.4018/978-1-4666-6276-6.ch009

Kosmatov, N. (2010). Constraint-based techniques for software testing. In F. Meziane & S. Vadera (Eds.), *Artificial intelligence applications for improved software engineering development: New prospects* (pp. 218–232). Hershey, PA: IGI Global. doi:10.4018/978-1-60566-758-4.ch011

Kow, K. W., Wong, Y. W., Rajkumar, R. K., & Rajkumar, R. K. (2016). A review on performance of artificial intelligence and conventional method in mitigating PV grid-tied related power quality events. *Renewable & Sustainable Energy Reviews*, *56*, 334–346. doi:10.1016/j.rser.2015.11.064

Koza, J. R. (1992). *Genetic programming: On the programming of computers by natural selection*. Cambridge, MA: MIT Press.

Kulkarni, S. (2012). *Machine learning algorithms for problem solving in computational applications: Intelligent techniques* (pp. 1–464). Hershey, PA: IGI Global. doi:10.4018/978-1-4666-1833-6.ch001

Lam, C. P. (2010). Computational intelligence for functional testing. In F. Meziane & S. Vadera (Eds.), *Artificial intelligence applications for improved software engineering development: New prospects* (pp. 233–258). Hershey, PA: IGI Global. doi:10.4018/978-1-60566-758-4.ch012

Liu, X., Khudkhudia, E., Wen, L., & Sajja, V. (2010). An intelligent computational argumentation system for supporting collaborative software development decision making. In F. Meziane & S. Vadera (Eds.), *Artificial intelligence applications for improved software engineering development: New prospects* (pp. 167–180). Hershey, PA: IGI Global. doi:10.4018/978-1-60566-758-4.ch009

Ma, Z. (2009). *Artificial intelligence for maximizing content based image retrieval* (pp. 1–450). Hershey, PA: IGI Global. doi:10.4018/978-1-60566-174-2

Mandl, T. (2009). Artificial intelligence for information retrieval. In J. Rabuñal Dopico, J. Dorado, & A. Pazos (Eds.), *Encyclopedia of artificial intelligence* (pp. 151–156). Hershey, PA: IGI Global. doi:10.4018/978-1-59904-849-9.ch023

Manjunath, R. (2005). Hierarchical architecture of expert systems for database management. In L. Rivero, J. Doorn, & V. Ferraggine (Eds.), *Encyclopedia of database technologies and applications* (pp. 271–275). Hershey, PA: IGI Global. doi:10.4018/978-1-59140-560-3.ch046

Marinagi, C. C., Panayiotopoulos, T., & Spyropoulos, C. D. (2005). AI planning and intelligent agents. In I. Vlahavas & D. Vrakas (Eds.), *Intelligent techniques for planning* (pp. 225–258). Hershey, PA: Idea Group Publishing. doi:10.4018/978-1-59140-450-7.ch007

Martínez-Barrera, G., Gencel, O., Beycioglu, A., Subaşı, S., & González-Rivas, N. (2015). Artificial intelligence methods and their applications in civil engineering. In H. Lim (Ed.), *Handbook of research on recent developments in materials science and corrosion engineering education* (pp. 166–189). Hershey, PA: IGI Global. doi:10.4018/978-1-4666-8183-5.ch009

McCulloch, W. S., & Pitts, W. (1943). A logical calculus of ideas imminent in nervous activity. *Bulletin of Mathematical Biology*, *5*(4), 115–133.

Mehran, K., Zahawi, B., & Giaouris, D. (2011). Fuzzy logic for non-smooth dynamical systems. In Y. Dai, B. Chakraborty, & M. Shi (Eds.), *Kansei engineering and soft computing: Theory and practice* (pp. 147–168). Hershey, PA: IGI Global. doi:10.4018/978-1-61692-797-4.ch008

Mendez, E. (2010). Using Bayesian networks for Web effort estimation. In F. Meziane & S. Vadera (Eds.), *Artificial intelligence applications for improved software engineering development: New prospects* (pp. 26–44). Hershey, PA: IGI Global. doi:10.4018/978-1-60566-758-4.ch002

Meziane, F., & Vadera, S. (2010). Artificial intelligence in software engineering: Current developments and future prospects. In F. Meziane & S. Vadera (Eds.), *Artificial intelligence applications for improved software engineering development: New prospects* (pp. 278–299). Hershey, PA: IGI Global. doi:10.4018/978-1-60566-758-4.ch014

Mohammadian, M. (2009). Supervised learning of fuzzy logic systems. In J. Rabuñal Dopico, J. Dorado, & A. Pazos (Eds.), *Encyclopedia of artificial intelligence* (pp. 1510–1517). Hershey, PA: IGI Global. doi:10.4018/978-1-59904-849-9.ch221

Musicant, D. R. (2005). Support vector machines illuminated. In J. Wang (Ed.), *Encyclopedia of data warehousing and mining* (pp. 1071–1076). Hershey, PA: IGI Global. doi:10.4018/978-1-59140-557-3.ch201

Mutihac, R. (2009). Adaptive neural algorithms for PCA and ICA. In J. Rabuñal Dopico, J. Dorado, & A. Pazos (Eds.), *Encyclopedia of artificial intelligence* (pp. 22–30). Hershey, PA: IGI Global. doi:10.4018/978-1-59904-849-9.ch004

Mutlu-Bayraktar, D. (2015). Intelligent educational support system. In U. Kose & D. Koc (Eds.), *Artificial intelligence applications in distance education* (pp. 32–46). Hershey, PA: IGI Global. doi:10.4018/978-1-4666-6276-6.ch003

Mutlu-Bayraktar, D., & Esgin, E. (2015). Expert systems in distance education. In U. Kose & D. Koc (Eds.), *Artificial intelligence applications in distance education* (pp. 136–148). Hershey, PA: IGI Global. doi:10.4018/978-1-4666-6276-6.ch008

Pombo, N., Garcia, N. M., Bousson, K., & Felizardo, V. (2015). Artificial neural learning based on Big Data process for eHealth applications. In T. Issa & P. Isaías (Eds.), *Artificial intelligence technologies and the evolution of Web 3.0* (pp. 291–306). Hershey, PA: IGI Global. doi:10.4018/978-1-4666-8147-7.ch014

Poupart, P. (2012). An introduction to fully and partially observable Markov decision processes. In L. Sucar, E. Morales, & J. Hoey (Eds.), *Decision theory models for applications in artificial intelligence: Concepts and solutions* (pp. 33–62). Hershey, PA: IGI Global. doi:10.4018/978-1-60960-165-2.ch003

Qi, J., Wu, F., Li, L., & Shu, H. (2007). Artificial intelligence applications in the telecommunications industry. *Expert Systems: International Journal of Knowledge Engineering and Neural Networks*, *24*(4), 271–291. doi:10.1111/j.1468-0394.2007.00433.x

Rada, R. (2009). Artificial intelligence and investing. In M. Khosrow-Pour (Ed.), *Encyclopedia of information science and technology* (2nd ed.; pp. 237–240). Hershey, PA: IGI Global. doi:10.4018/978-1-60566-026-4.ch041

Ramachandran, M. (2010). Knowledge engineering support for software requirements, architectures and components. In F. Meziane & S. Vadera (Eds.), *Artificial intelligence applications for improved software engineering development: New prospects* (pp. 129–145). Hershey, PA: IGI Global. doi:10.4018/978-1-60566-758-4.ch007

Raza, M. Q., & Khosravi, A. (2015). A review on artificial intelligence based load demand forecasting techniques for smart grid and buildings. *Renewable & Sustainable Energy Reviews*, *50*, 1352–1372. doi:10.1016/j.rser.2015.04.065

Rumelhart, D. E., Hinton, G. E., & Williams, R. J. (1986). Learning internal representation by error propagation. In D. Rumelhart & J. McClelland (Eds.), *Parallel distributed processing* (pp. 318–362). Cambridge, MA: MIT Press.

Sadkhan Al Maliky, S. B., & Jawad, S. F. (2015). Fuzzy logic-based security evaluation of stream cipher. In *Research methods: Concepts, methodologies, tools, and applications* (pp. 277–298). Hershey, PA: IGI Global. doi:10.4018/978-1-4666-7456-1.ch013

Sarcia, S. A., Cantone, G., & Basili, V. R. (2010). Auto-associative neural networks to improve the accuracy of estimation models. In F. Meziane & S. Vadera (Eds.), *Artificial intelligence applications for improved software engineering development: New prospects* (pp. 66–81). Hershey, PA: IGI Global. doi:10.4018/978-1-60566-758-4.ch004

Shahin, M. A. (2016). State-of-the-art review of some artificial intelligence applications in pile foundations. *Geoscience Frontiers*, *7*(1), 33–44. doi:10.1016/j.gsf.2014.10.002

Soria, A., Diaz-Pace, J. A., Bass, L., Bachmann, F., & Campo, M. (2010). Supporting quality-driven software design through intellectual assistants. In F. Meziane & S. Vadera (Eds.), *Artificial intelligence applications for improved software engineering development: New prospects* (pp. 181–216). Hershey, PA: IGI Global. doi:10.4018/978-1-60566-758-4.ch010

Sugiyama, S. (2013). Dynamics in knowledge. In *Industrial engineering: Concepts, methodologies, tools, and applications* (pp. 1595–1612). Hershey, PA: IGI Global. doi:10.4018/978-1-4666-1945-6.ch085

Sun, M. (2014). Support vector machine models for classification. In J. Wang (Ed.), *Encyclopedia of business analytics and optimization* (pp. 2395–2409). Hershey, PA: IGI Global. doi:10.4018/978-1-4666-5202-6.ch215

Svensson, J. S. (2005). Legal expert systems in administrative organizations. In M. Khosrow-Pour (Ed.), *Encyclopedia of information science and technology* (1st ed.; pp. 1827–1830). Hershey, PA: IGI Global. doi:10.4018/978-1-59140-553-5.ch321

Tzanis, G., Berberidis, C., & Vlahavas, I. (2009). Machine learning and data mining in bioinformatics. In V. Ferraggine, J. Doorn, & L. Rivero (Eds.), *Handbook of research on innovations in database technologies and applications: Current and future trends* (pp. 612–621). Hershey, PA: IGI Global. doi:10.4018/978-1-60566-242-8.ch066

Uyar, M., Yildirim, S., & Gencoglu, M. T. (2008). An effective wavelet-based feature extraction method for classification of power quality disturbance signals. *Electric Power Systems Research*, *78*(10), 1747–1755. doi:10.1016/j.epsr.2008.03.002

Vannucci, M., Colla, V., Cateni, S., & Sgarbi, M. (2012). Artificial intelligence techniques for unbalanced datasets in real world classification tasks. In *Machine learning: Concepts, methodologies, tools and applications* (pp. 414–427). Hershey, PA: IGI Global. doi:10.4018/978-1-60960-818-7.ch304

Vapnik, V. N. (1995). *The nature of statistical learning theory*. New York, NY: Springer–Verlag. doi:10.1007/978-1-4757-2440-0

Wagner, C. (2003). Knowledge management through end user developed expert systems: Potential and limitations. In M. Mahmood (Ed.), *Advanced topics in end user computing* (Vol. 2, pp. 148–172). Hershey, PA: Idea Group Publishing. doi:10.4018/978-1-59140-065-3.ch008

Wei, C., & Li, C. (2009). Content analysis from user's relevance feedback for content-based image retrieval. In Z. Ma (Ed.), *Artificial intelligence for maximizing content based image retrieval* (pp. 216–234). Hershey, PA: IGI Global. doi:10.4018/978-1-60566-174-2.ch010

Wei, C., Li, C., & Li, Y. (2009). Content-based retrieval for mammograms. In Z. Ma (Ed.), *Artificial intelligence for maximizing content based image retrieval* (pp. 315–341). Hershey, PA: IGI Global. doi:10.4018/978-1-60566-174-2.ch014

Wei, C., & Li, Y. (2011). *Machine learning techniques for adaptive multimedia retrieval: Technologies applications and perspectives* (pp. 1–408). Hershey, PA: IGI Global. doi:10.4018/978-1-61692-859-9.ch001

Yaseen, Z. M., El-shafie, A., Jaafar, O., Afan, H. A., & Sayl, K. N. (2015). Artificial intelligence based models for stream-flow forecasting: 2000–2015. *Journal of Hydrology (Amsterdam)*, *530*, 829–844. doi:10.1016/j.jhydrol.2015.10.038

Zhang, G., Ma, Z. M., & Yan, L. (2009). Review on texture feature extraction and description methods in content-based medical image retrieval. In Z. Ma (Ed.), *Artificial intelligence for maximizing content based image retrieval* (pp. 44–67). Hershey, PA: IGI Global. doi:10.4018/978-1-60566-174-2.ch003

Zhang, M., & Alhajj, R. (2009). Content-based image retrieval: From the object detection/recognition point of view. In Z. Ma (Ed.), *Artificial intelligence for maximizing content based image retrieval* (pp. 115–144). Hershey, PA: IGI Global. doi:10.4018/978-1-60566-174-2.ch006

Zhang, R., & Zhang, Z. M. (2009). Solving the small and asymmetric sampling problem in the context of image retrieval. In Z. Ma (Ed.), *Artificial intelligence for maximizing content based image retrieval* (pp. 192–215). Hershey, PA: IGI Global. doi:10.4018/978-1-60566-174-2.ch009

ADDITIONAL READING

Armstrong, S., Bostrom, N., & Shulman, C. (2016). Racing to the precipice: A model of artificial intelligence development. *AI & Society*, *31*(2), 201–206. doi:10.1007/s00146-015-0590-y

Buyukyildiz, M., Tezel, G., & Yilmaz, V. (2014). Estimation of the change in lake water level by artificial intelligence methods. *Water Resources Management*, *28*(13), 4747–4763. doi:10.1007/s11269-014-0773-1

Christodoulou, C. A., Perantzakis, G., Spanakis, G. E., & Karampelas, P. (2012). Evaluation of lightning performance of transmission lines protected by metal oxide surge arresters using artificial intelligence techniques. *Energy Systems*, *3*(4), 379–399. doi:10.1007/s12667-012-0061-1

Duer, S. (2013). Applications of an artificial intelligence for servicing of a technical object. *Neural Computing & Applications*, *22*(5), 955–968. doi:10.1007/s00521-011-0788-y

Erdal, H. I., & Ekinci, A. (2013). A comparison of various artificial intelligence methods in the prediction of bank failures. *Computational Economics*, *42*(2), 199–215. doi:10.1007/s10614-012-9332-0

Fabrikantova, E. F. (2012). Modeling systems by means of artificial intelligence. *Automatic Documentation and Mathematical Linguistics*, *46*(6), 229–234. doi:10.3103/S0005105512060027

Fayala, F., Alibi, H., Jemni, A., & Zeng, X. (2014). Study the effect of operating parameters and intrinsic features of yarn and fabric on thermal conductivity of stretch knitted fabrics using artificial intelligence system. *Fibers and Polymers*, *15*(4), 855–864. doi:10.1007/s12221-014-0855-y

Ghasemi, E., Amini, H., Ataei, M., & Khalokakaei, R. (2014). Application of artificial intelligence techniques for predicting the flyrock distance caused by blasting operation. *Arabian Journal of Geosciences*, *7*(1), 193–202. doi:10.1007/s12517-012-0703-6

Hadas, Z., Kurfurst, J., Ondrusek, C., & Singule, V. (2012). Artificial intelligence based optimization for vibration energy harvesting applications. *Microsystem Technologies*, *18*(7), 1003–1014. doi:10.1007/s00542-012-1432-1

Homami, R. M., Tehrani, A. F., Mirzadeh, H., Movahedi, B., & Azimifar, F. (2014). Optimization of turning process using artificial intelligence technology. *International Journal of Advanced Manufacturing Technology*, *70*(5), 1205–1217. doi:10.1007/s00170-013-5361-7

Issa, T., & Isaías, P. (2015). *Artificial intelligence technologies and the evolution of Web 3.0* (pp. 1–422). Hershey, PA: IGI Global. doi:10.4018/978-1-4666-8147-7

Jayachandran, J., & Sachithanandam, R. M. (2015). Performance investigation of artificial intelligence based controller for three phase four leg shunt active filter. *Frontiers in Energy*, *9*(4), 446–460. doi:10.1007/s11708-015-0378-2

Khaki, M., Yusoff, I., & Islami, N. (2015). Simulation of groundwater level through artificial intelligence system. *Environmental Earth Sciences*, *73*(12), 8357–8367. doi:10.1007/s12665-014-3997-8

Kolchugina, E. A. (2015). Self-organizing software systems with distributed artificial intelligence. *Automatic Control and Computer Sciences*, *49*(4), 216–220. doi:10.3103/S0146411615040069

Kuznetsov, O. P. (2013). Cognitive semantics and artificial intelligence. *Scientific and Technical Information Processing*, *40*(5), 269–276. doi:10.3103/S0147688213050067

Lim, J., & Kim, S. (2011). Analysis of woven fabric structure using image analysis and artificial intelligence. *Fibers and Polymers*, *12*(8), 1062–1068. doi:10.1007/s12221-011-1062-8

Liu, X., Guo, C., & Zhang, L. (2014). Scholar metadata and knowledge generation with human and artificial intelligence. *Journal of the Association for Information Science and Technology*, *65*(6), 1187–1201. doi:10.1002/asi.23013

Mazinan, A. H., & Sheikhan, M. (2012). On the practice of artificial intelligence based predictive control scheme: A case study. *Applied Intelligence*, *36*(1), 178–189. doi:10.1007/s10489-010-0253-0

Megill, J. (2014). Emotion, cognition and artificial intelligence. *Minds and Machines*, *24*(2), 189–199. doi:10.1007/s11023-013-9320-8

Moret-Bonillo, V. (2015). Can artificial intelligence benefit from quantum computing? *Progress in Artificial Intelligence*, *3*(2), 89–105. doi:10.1007/s13748-014-0059-0

Nanni, L., Brahnam, S., & Lumini, A. (2011). Artificial intelligence systems based on texture descriptors for vaccine development. *Amino Acids*, *40*(2), 443–451. doi:10.1007/s00726-010-0654-8 PMID:20552381

Nasiri, M. R., Mahjoob, M. J., & Aghakasiri, A. (2011). Damage detection in a composite plate using modal analysis and artificial intelligence. *Applied Composite Materials*, *18*(6), 513–520. doi:10.1007/s10443-011-9231-x

Okamoto, T. (2011). An artificial intelligence membrane to detect network intrusion. *Artificial Life and Robotics*, *16*(1), 44–47. doi:10.1007/s10015-011-0880-5

Ratrout, N. T., Rahman, S. M., & Reza, I. (2015). Calibration of PARAMICS model: Application of artificial intelligence-based approach. *Arabian Journal of Science and Engineering*, *40*(12), 3459–3468. doi:10.1007/s13369-015-1816-5

Sambath, S., Nagaraj, P., & Selvakumar, N. (2011). Automatic defect classification in ultrasonic NDT using artificial intelligence. *Journal of Nondestructive Evaluation*, *30*(1), 20–28. doi:10.1007/s10921-010-0086-0

Samui, P., & Kim, D. (2014). Applicability of artificial intelligence to reservoir induced earthquakes. *Acta Geophysica*, *62*(3), 608–619. doi:10.2478/s11600-014-0201-1

Tecuci, G. (2012). Artificial intelligence. *Wiley Interdisciplinary Reviews: Computational Statistics*, *4*(2), 168–180. doi:10.1002/wics.200

Thomas, M. S. C., Forrester, N. A., & Ronald, A. (2016). Multiscale modeling of gene–behavior associations in an artificial neural network model of cognitive development. *Cognitive Science*, *40*(1), 51–99. doi:10.1111/cogs.12230 PMID:25845802

Velichenko, V. V., & Pritykin, D. A. (2011). Using artificial intelligence and computer technologies for developing treatment programs for complex immune diseases. *Journal of Mathematical Sciences*, *172*(5), 635–649. doi:10.1007/s10958-011-0209-x

Vijayaraghavan, V., Garg, A., Wong, C. H., & Tai, K. (2014). Estimation of mechanical properties of nanomaterials using artificial intelligence methods. *Applied Physics. A, Materials Science & Processing*, *116*(3), 1099–1107. doi:10.1007/s00339-013-8192-3

Wauters, M., & Vanhoucke, M. (2016). A comparative study of artificial intelligence methods for project duration forecasting. *Expert Systems with Applications: An International Journal*, *46*, 249–261. doi:10.1016/j.eswa.2015.10.008

KEY TERMS AND DEFINITIONS

Artificial Intelligence: The area of computer science that emphasizes the creation of intelligent machines that work and react like humans.

Artificial Neural Network: The computational model based on the structure and function of biological neural networks.

Data Mining: The process of analyzing the hidden patterns of data according to the different perspectives for the categorization into the useful information.

Evolutionary Polynomial Regression: The evolutionary data-driven modeling technique for the identification and construction of models.

Expert System: The computer program that is designed to imitate human intelligence, skills, and behavior.

Fuzzy Logic: The system of theories utilized in mathematics, computing, and philosophy to deal with the statements that are neither true nor false.

Genetic Programming: The model of programming, which utilizes the ideas of biological evolution to handle the complex problem.

Machine Learning: The artificial intelligence discipline geared toward the technological development of human knowledge.

Optimization: The act of making something as good as possible.

Support Vector Machine: The supervised machine learning algorithm, which can be used for the classification and regression methods.

Compilation of References

Afzal, W., Torkar, R., Feldt, R., & Gorschek, T. (2010). Genetic programming for cross-release fault count predictions in large and complex software projects. In M. Chis (Ed.), *Evolutionary computation and optimization algorithms in software engineering: Applications and techniques* (pp. 94–126). Hershey, PA: IGI Global. doi:10.4018/978-1-61520-809-8.ch006

Alagha, J. S., Said, M. A. M., & Mogheir, Y. (2014). Modeling of nitrate concentration in groundwater using artificial intelligence approach: A case study of Gaza coastal aquifer. *Environmental Monitoring and Assessment*, *186*(1), 35–45. doi:10.1007/s10661-013-3353-6 PMID:23974533

Ali, J. M. (2009). Content-based image classification and retrieval: A rule-based system using rough sets framework. In Z. Ma (Ed.), *Artificial intelligence for maximizing content based image retrieval* (pp. 68–83). Hershey, PA: IGI Global. doi:10.4018/978-1-60566-174-2.ch004

Ashrafian, H. (2015). Artificial intelligence and robot responsibilities: Innovating beyond rights. *Science and Engineering Ethics*, *21*(2), 317–326. doi:10.1007/s11948-014-9541-0 PMID:24737482

Atanasova, I. (2015). On design and development of QLIFEX: An expert system for social area. *International Journal of Applied Behavioral Economics*, *4*(2), 16–35. doi:10.4018/ijabe.2015040102

Avola, D., Ferri, F., & Grifoni, P. (2009). Genetic algorithms and other approaches in image feature extraction and representation. In Z. Ma (Ed.), *Artificial intelligence for maximizing content based image retrieval* (pp. 1–19). Hershey, PA: IGI Global. doi:10.4018/978-1-60566-174-2.ch001

Azar, A. T. (2012). Overview of type-2 fuzzy logic systems. *International Journal of Fuzzy System Applications*, *2*(4), 1–28. doi:10.4018/ijfsa.2012100101

Bahrammirzaee, A. (2010). A comparative survey of artificial intelligence applications in finance: Artificial neural networks, expert system and hybrid intelligent systems. *Neural Computing & Applications*, *19*(8), 1165–1195. doi:10.1007/s00521-010-0362-z

Begg, R. (2009). Artificial intelligence techniques in medicine and healthcare. In J. Tan (Ed.), *Medical informatics: Concepts, methodologies, tools, and applications* (pp. 784–791). Hershey, PA: IGI Global. doi:10.4018/978-1-60566-050-9.ch057

Bhattacharyya, S., & Dutta, P. (2012). Fuzzy logic: Concepts, system design, and applications to industrial informatics. In M. Khan & A. Ansari (Eds.), *Handbook of research on industrial informatics and manufacturing intelligence: Innovations and solutions* (pp. 33–71). Hershey, PA: IGI Global. doi:10.4018/978-1-4666-0294-6.ch003

Carbonneau, R., Vahidov, R., & Laframboise, K. (2007). Machine learning-based demand forecasting in supply chains. *International Journal of Intelligent Information Technologies*, *3*(4), 40–57. doi:10.4018/jiit.2007100103

Castro, A., & Arcay, B. (2009). Fuzzy logic applied to biomedical image analysis. In J. Rabuñal Dopico, J. Dorado, & A. Pazos (Eds.), *Encyclopedia of artificial intelligence* (pp. 710–718). Hershey, PA: IGI Global. doi:10.4018/978-1-59904-849-9.ch106

Chen, M., & Chen, S. (2009). MMIR: An advanced content-based image retrieval system using a hierarchical learning framework. In Z. Ma (Ed.), *Artificial intelligence for maximizing content based image retrieval* (pp. 371–387). Hershey, PA: IGI Global. doi:10.4018/978-1-60566-174-2.ch016

Collet, P. (2007). Genetic programming. In J. Rennard (Ed.), *Handbook of research on nature-inspired computing for economics and management* (pp. 59–73). Hershey, PA: IGI Global. doi:10.4018/978-1-59140-984-7.ch005

Das, S. K., Samui, P., & Sabat, A. K. (2011). Application of artificial intelligence to maximum dry density and unconfined compressive strength of cement stabilized soil. *Geotechnical and Geological Engineering*, *29*(3), 329–342. doi:10.1007/s10706-010-9379-4

del Sagrado Martinez, J., & del Aguila Cano, I. M. (2010). A Bayesian network for predicting the need for a requirements review. In F. Meziane & S. Vadera (Eds.), *Artificial intelligence applications for improved software engineering development: New prospects* (pp. 106–128). Hershey, PA: IGI Global. doi:10.4018/978-1-60566-758-4.ch006

Deperlioglu, O., Sonugur, G., & Suzme, K. (2015). Intelligent questioning system based on fuzzy logic. In U. Kose & D. Koc (Eds.), *Artificial intelligence applications in distance education* (pp. 72–94). Hershey, PA: IGI Global. doi:10.4018/978-1-4666-6276-6.ch005

Duan, Y. (2009). Web-based expert systems. In M. Khosrow-Pour (Ed.), *Encyclopedia of information science and technology* (2nd ed.; pp. 4105–4118). Hershey, PA: IGI Global. doi:10.4018/978-1-60566-026-4.ch656

Elshorbagy, A., Corzo, G., Srinivasulu, S., & Solomatine, D. P. (2010). Experimental investigation of the predictive capabilities of data driven modeling techniques in hydrology-Part 1: Concepts and methodology. *Hydrology and Earth System Sciences*, *14*(10), 1931–1941. doi:10.5194/hess-14-1931-2010

Ferariu, L., & Patelli, A. (2012). Genetic programming for system identification. In P. Cong-Vinh (Ed.), *Formal and practical aspects of autonomic computing and networking: Specification, development, and verification* (pp. 135–168). Hershey, PA: IGI Global. doi:10.4018/978-1-60960-845-3.ch006

Freitas, A. A., & Pappa, G. L. (2009). Genetic programming for automatically constructing data mining algorithms. In J. Wang (Ed.), *Encyclopedia of data warehousing and mining* (2nd ed.; pp. 932–936). Hershey, PA: IGI Global. doi:10.4018/978-1-60566-010-3.ch144

Gama, J., & de Carvalho, A. C. (2009). Machine learning. In M. Khosrow-Pour (Ed.), *Encyclopedia of information science and technology* (2nd ed.; pp. 2462–2468). Hershey, PA: IGI Global. doi:10.4018/978-1-60566-026-4.ch392

García Pérez, D., Mosquera, A., Berretti, S., & del Bimbo, A. (2009). Content based image retrieval using active-nets. In Z. Ma (Ed.), *Artificial intelligence for maximizing content based image retrieval* (pp. 85–114). Hershey, PA: IGI Global. doi:10.4018/978-1-60566-174-2.ch005

Ghosal, K., Haldar, P., & Sutradhar, G. (2015). Application of fuzzy expert system in medical treatment. In A. Kumar (Ed.), *Fuzzy expert systems for disease diagnosis* (pp. 200–246). Hershey, PA: IGI Global. doi:10.4018/978-1-4666-7240-6.ch008

Goh, C., Mok, H. M., & Law, R. (2009). Artificial intelligence applications in tourism. In M. Khosrow-Pour (Ed.), *Encyclopedia of information science and technology* (2nd ed.; pp. 241–247). Hershey, PA: IGI Global. doi:10.4018/978-1-60566-026-4.ch042

Goldberg, D. E. (1989). *Genetic algorithms in search optimization and machine learning*. Reading, MA: Addison–Wesley.

Gondra, I. (2009). Personalized content-based image retrieval. In Z. Ma (Ed.), *Artificial intelligence for maximizing content based image retrieval* (pp. 261–287). Hershey, PA: IGI Global. doi:10.4018/978-1-60566-174-2.ch012

Gonsalves, T., Yamagishi, K., Kawabata, R., & Itoh, K. (2010). Optimizing software development cost estimates using multi-objective particle swarm optimization. In F. Meziane & S. Vadera (Eds.), *Artificial intelligence applications for improved software engineering development: New prospects* (pp. 46–65). Hershey, PA: IGI Global. doi:10.4018/978-1-60566-758-4.ch003

Hacid, H., & Zighed, A. D. (2009). A machine learning-based model for content-based image retrieval. In Z. Ma (Ed.), *Artificial intelligence for maximizing content based image retrieval* (pp. 171–190). Hershey, PA: IGI Global. doi:10.4018/978-1-60566-174-2.ch008

Hengstler, M., Enkel, E., & Duelli, S. (2016). Applied artificial intelligence and trust: The case of autonomous vehicles and medical assistance devices. *Technological Forecasting and Social Change*, *105*, 105–120. doi:10.1016/j.techfore.2015.12.014

Hsu, W. H. (2009). Genetic programming. In J. Wang (Ed.), *Encyclopedia of data warehousing and mining* (2nd ed.; pp. 926–931). Hershey, PA: IGI Global. doi:10.4018/978-1-60566-010-3.ch143

Işik, A. H., & Aslan, G. (2015). Review and proposal for intelligent distance education. In U. Kose & D. Koc (Eds.), *Artificial intelligence applications in distance education* (pp. 95–109). Hershey, PA: IGI Global. doi:10.4018/978-1-4666-6276-6.ch006

Jagan, J., Gundlapalli, P., & Samui, P. (2016). Utilization of classification techniques for the determination of liquefaction susceptibility of soils. In S. Bhattacharyya, P. Banerjee, D. Majumdar, & P. Dutta (Eds.), *Handbook of research on advanced hybrid intelligent techniques and applications* (pp. 124–160). Hershey, PA: IGI Global. doi:10.4018/978-1-4666-9474-3.ch005

Jeong, I. J., & Kim, K. J. (2009). An interactive desirability function method to multi-response optimization. *European Journal of Operational Research*, *195*(2), 412–426. doi:10.1016/j.ejor.2008.02.018

Jiang, J., & Ip, H. H. (2009). Active learning with SVM. In J. Rabuñal Dopico, J. Dorado, & A. Pazos (Eds.), *Encyclopedia of artificial intelligence* (pp. 1–7). Hershey, PA: IGI Global. doi:10.4018/978-1-59904-849-9.ch001

Kalogirou, S., Metaxiotis, K., & Mellit, A. (2010). Artificial intelligence techniques for modern energy applications. In K. Metaxiotis (Ed.), *Intelligent information systems and knowledge management for energy: Applications for decision support, usage, and environmental protection* (pp. 1–39). Hershey, PA: IGI Global. doi:10.4018/978-1-60566-737-9.ch001

Kanthi, M. (2015). Fuzzy logic-based intelligent control system for active ankle foot orthosis. In A. Kumar (Ed.), *Fuzzy expert systems for disease diagnosis* (pp. 248–279). Hershey, PA: IGI Global. doi:10.4018/978-1-4666-7240-6.ch009

Karaman, F. (2012). Artificial intelligence enabled search engines (AIESE) and the implications. In C. Jouis, I. Biskri, J. Ganascia, & M. Roux (Eds.), *Next generation search engines: Advanced models for information retrieval* (pp. 438–455). Hershey, PA: IGI Global. doi:10.4018/978-1-4666-0330-1.ch019

Kasemsap, K. (2015a). The role of data mining for business intelligence in knowledge management. In A. Azevedo & M. Santos (Eds.), *Integration of data mining in business intelligence systems* (pp. 12–33). Hershey, PA: IGI Global. doi:10.4018/978-1-4666-6477-7.ch002

Kasemsap, K. (2015b). The role of cloud computing adoption in global business. In V. Chang, R. Walters, & G. Wills (Eds.), *Delivery and adoption of cloud computing services in contemporary organizations* (pp. 26–55). Hershey, PA: IGI Global. doi:10.4018/978-1-4666-8210-8.ch002

Kasemsap, K. (2016). The fundamentals of business intelligence. *International Journal of Organizational and Collective Intelligence*, *6*(2), 12–25. doi:10.4018/IJOCI.2016040102

Kasemsap, K. (2017a). Mastering web mining and information retrieval in the digital age. In A. Kumar (Ed.), *Web usage mining techniques and applications across industries* (pp. 1–28). Hershey, PA: IGI Global. doi:10.4018/978-1-5225-0613-3.ch001

Kasemsap, K. (2017c). Robotics: Theory and applications. In M. Moore (Ed.), *Cybersecurity breaches and issues surrounding online threat protection* (pp. 311–345). Hershey, PA: IGI Global. doi:10.4018/978-1-5225-1941-6.ch013

Kasemsap, K. (2017d). Mastering business process management and business intelligence in global business. In M. Tavana, K. Szabat, & K. Puranam (Eds.), *Organizational productivity and performance measurements using predictive modeling and analytics* (pp. 192–212). Hershey, PA: IGI Global. doi:10.4018/978-1-5225-0654-6.ch010

Kasemsap, K. (2017e). Promoting critical thinking in the modern learning environments. In R. Batko & A. Szopa (Eds.), *Strategic imperatives and core competencies in the era of robotics and artificial intelligence* (pp. 50–80). Hershey, PA: IGI Global. doi:10.4018/978-1-5225-1656-9.ch004

Kasemsap, K. (2017f). Mastering intelligent decision support systems in enterprise information management. In G. Sreedhar (Ed.), *Web data mining and the development of knowledge-based decision support systems* (pp. 35–56). Hershey, PA: IGI Global. doi:10.4018/978-1-5225-1877-8.ch004

Kasemsap, K. (2017g). Software as a service, Semantic Web, and big data: Theories and applications. In A. Turuk, B. Sahoo, & S. Addya (Eds.), *Resource management and efficiency in cloud computing environments* (pp. 264–285). Hershey, PA: IGI Global. doi:10.4018/978-1-5225-1721-4.ch011

Kasemsap, K. (2017h). Radio frequency identification and mobile ad-hoc network: Theories and applications. In S. Bhattacharyya, N. Das, D. Bhattacharjee, & A. Mukherjee (Eds.), *Handbook of research on recent developments in intelligent communication application* (pp. 63–95). Hershey, PA: IGI Global. doi:10.4018/978-1-5225-1785-6.ch003

Kof, L. (2010). From textual scenarios to message sequence charts. In F. Meziane & S. Vadera (Eds.), *Artificial intelligence applications for improved software engineering development: New prospects* (pp. 83–105). Hershey, PA: IGI Global. doi:10.4018/978-1-60566-758-4.ch005

Kose, U. (2015). For an intelligent e-learning: A managerial model suggestion for artificial intelligence supported e-learning content flow. In U. Kose & D. Koc (Eds.), *Artificial intelligence applications in distance education* (pp. 149–160). Hershey, PA: IGI Global. doi:10.4018/978-1-4666-6276-6.ch009

Kosmatov, N. (2010). Constraint-based techniques for software testing. In F. Meziane & S. Vadera (Eds.), *Artificial intelligence applications for improved software engineering development: New prospects* (pp. 218–232). Hershey, PA: IGI Global. doi:10.4018/978-1-60566-758-4.ch011

Kow, K. W., Wong, Y. W., Rajkumar, R. K., & Rajkumar, R. K. (2016). A review on performance of artificial intelligence and conventional method in mitigating PV grid-tied related power quality events. *Renewable & Sustainable Energy Reviews*, *56*, 334–346. doi:10.1016/j.rser.2015.11.064

Koza, J. R. (1992). *Genetic programming: On the programming of computers by natural selection*. Cambridge, MA: MIT Press.

Lam, C. P. (2010). Computational intelligence for functional testing. In F. Meziane & S. Vadera (Eds.), *Artificial intelligence applications for improved software engineering development: New prospects* (pp. 233–258). Hershey, PA: IGI Global. doi:10.4018/978-1-60566-758-4.ch012

Liu, X., Khudkhudia, E., Wen, L., & Sajja, V. (2010). An intelligent computational argumentation system for supporting collaborative software development decision making. In F. Meziane & S. Vadera (Eds.), *Artificial intelligence applications for improved software engineering development: New prospects* (pp. 167–180). Hershey, PA: IGI Global. doi:10.4018/978-1-60566-758-4.ch009

Mandl, T. (2009). Artificial intelligence for information retrieval. In J. Rabuñal Dopico, J. Dorado, & A. Pazos (Eds.), *Encyclopedia of artificial intelligence* (pp. 151–156). Hershey, PA: IGI Global. doi:10.4018/978-1-59904-849-9.ch023

Manjunath, R. (2005). Hierarchical architecture of expert systems for database management. In L. Rivero, J. Doorn, & V. Ferraggine (Eds.), *Encyclopedia of database technologies and applications* (pp. 271–275). Hershey, PA: IGI Global. doi:10.4018/978-1-59140-560-3.ch046

Marinagi, C. C., Panayiotopoulos, T., & Spyropoulos, C. D. (2005). AI planning and intelligent agents. In I. Vlahavas & D. Vrakas (Eds.), *Intelligent techniques for planning* (pp. 225–258). Hershey, PA: Idea Group Publishing. doi:10.4018/978-1-59140-450-7.ch007

Martínez-Barrera, G., Gencel, O., Beycioglu, A., Subaşı, S., & González-Rivas, N. (2015). Artificial intelligence methods and their applications in civil engineering. In H. Lim (Ed.), *Handbook of research on recent developments in materials science and corrosion engineering education* (pp. 166–189). Hershey, PA: IGI Global. doi:10.4018/978-1-4666-8183-5.ch009

McCulloch, W. S., & Pitts, W. (1943). A logical calculus of ideas imminent in nervous activity. *Bulletin of Mathematical Biology*, *5*(4), 115–133.

Mehran, K., Zahawi, B., & Giaouris, D. (2011). Fuzzy logic for non-smooth dynamical systems. In Y. Dai, B. Chakraborty, & M. Shi (Eds.), *Kansei engineering and soft computing: Theory and practice* (pp. 147–168). Hershey, PA: IGI Global. doi:10.4018/978-1-61692-797-4.ch008

Mendez, E. (2010). Using Bayesian networks for Web effort estimation. In F. Meziane & S. Vadera (Eds.), *Artificial intelligence applications for improved software engineering development: New prospects* (pp. 26–44). Hershey, PA: IGI Global. doi:10.4018/978-1-60566-758-4.ch002

Meziane, F., & Vadera, S. (2010). Artificial intelligence in software engineering: Current developments and future prospects. In F. Meziane & S. Vadera (Eds.), *Artificial intelligence applications for improved software engineering development: New prospects* (pp. 278–299). Hershey, PA: IGI Global. doi:10.4018/978-1-60566-758-4.ch014

Mohammadian, M. (2009). Supervised learning of fuzzy logic systems. In J. Rabuñal Dopico, J. Dorado, & A. Pazos (Eds.), *Encyclopedia of artificial intelligence* (pp. 1510–1517). Hershey, PA: IGI Global. doi:10.4018/978-1-59904-849-9.ch221

Montogomery, D. C., Peck, E. A., & Vining, G. G. (2003). *Introduction to linear regression analysis* (3rd ed.). New York: Wiley.

Musicant, D. R. (2005). Support vector machines illuminated. In J. Wang (Ed.), *Encyclopedia of data warehousing and mining* (pp. 1071–1076). Hershey, PA: IGI Global. doi:10.4018/978-1-59140-557-3.ch201

Mutihac, R. (2009). Adaptive neural algorithms for PCA and ICA. In J. Rabuñal Dopico, J. Dorado, & A. Pazos (Eds.), *Encyclopedia of artificial intelligence* (pp. 22–30). Hershey, PA: IGI Global. doi:10.4018/978-1-59904-849-9.ch004

Mutlu-Bayraktar, D., & Esgin, E. (2015). Expert systems in distance education. In U. Kose & D. Koc (Eds.), *Artificial intelligence applications in distance education* (pp. 136–148). Hershey, PA: IGI Global. doi:10.4018/978-1-4666-6276-6.ch008

Mutlu-Bayraktar, D. (2015). Intelligent educational support system. In U. Kose & D. Koc (Eds.), *Artificial intelligence applications in distance education* (pp. 32–46). Hershey, PA: IGI Global. doi:10.4018/978-1-4666-6276-6.ch003

Pombo, N., Garcia, N. M., Bousson, K., & Felizardo, V. (2015). Artificial neural learning based on Big Data process for eHealth applications. In T. Issa & P. Isaías (Eds.), *Artificial intelligence technologies and the evolution of Web 3.0* (pp. 291–306). Hershey, PA: IGI Global. doi:10.4018/978-1-4666-8147-7.ch014

Poupart, P. (2012). An introduction to fully and partially observable Markov decision processes. In L. Sucar, E. Morales, & J. Hoey (Eds.), *Decision theory models for applications in artificial intelligence: Concepts and solutions* (pp. 33–62). Hershey, PA: IGI Global. doi:10.4018/978-1-60960-165-2.ch003

Puri, A. B., & Banerjee, S. (2013). Multiple-response optimization of electrochemical grinding characteristics through response surface methodology. *International Journal of Advanced Manufacturing Technology*, *64*(5), 715–725.

Qi, J., Wu, F., Li, L., & Shu, H. (2007). Artificial intelligence applications in the telecommunications industry. *Expert Systems: International Journal of Knowledge Engineering and Neural Networks*, *24*(4), 271–291. doi:10.1111/j.1468-0394.2007.00433.x

Qu, N. S., Zhang, Q. L., Fang, X. L., Ye, E. K., & Zhu, D. (2015). Experimental Investigation on Electrochemical Grinding of Inconel 718. *15th Machining Innovations Conference for aerospace Industry*, *35*, 16-19. doi:10.1016/j.procir.2015.08.055

Rada, R. (2009). Artificial intelligence and investing. In M. Khosrow-Pour (Ed.), *Encyclopedia of information science and technology* (2nd ed.; pp. 237–240). Hershey, PA: IGI Global. doi:10.4018/978-1-60566-026-4.ch041

Ramachandran, M. (2010). Knowledge engineering support for software requirements, architectures and components. In F. Meziane & S. Vadera (Eds.), *Artificial intelligence applications for improved software engineering development: New prospects* (pp. 129–145). Hershey, PA: IGI Global. doi:10.4018/978-1-60566-758-4.ch007

Raza, M. Q., & Khosravi, A. (2015). A review on artificial intelligence based load demand forecasting techniques for smart grid and buildings. *Renewable & Sustainable Energy Reviews*, *50*, 1352–1372. doi:10.1016/j.rser.2015.04.065

Reddy, H. P., Ramamoorthy, K. B., & Nair, K. P. (1998). Estimation of Optimum Grinding Conditions to improve the quality of the Ground Surface by Applying Taguchi Techniques.*International Conference on CAD/CAM, Robotics & Factories of Future, India*, 473-479.

Reddy, H. P., Ramamoorthy, K. B., & Nair, K. P. (1999). Investigation on the influence of grinding parameters on the Microstrain of Medium Carbon Steel.*ASME Mechanics and Materials Conference.*

Rumelhart, D. E., Hinton, G. E., & Williams, R. J. (1986). Learning internal representation by error propagation. In D. Rumelhart & J. McClelland (Eds.), *Parallel distributed processing* (pp. 318–362). Cambridge, MA: MIT Press.

Sarcia, S. A., Cantone, G., & Basili, V. R. (2010). Auto-associative neural networks to improve the accuracy of estimation models. In F. Meziane & S. Vadera (Eds.), *Artificial intelligence applications for improved software engineering development: New prospects* (pp. 66–81). Hershey, PA: IGI Global. doi:10.4018/978-1-60566-758-4.ch004

Shahin, M. A. (2016). State-of-the-art review of some artificial intelligence applications in pile foundations. *Geoscience Frontiers*, *7*(1), 33–44. doi:10.1016/j.gsf.2014.10.002

Soria, A., Diaz-Pace, J. A., Bass, L., Bachmann, F., & Campo, M. (2010). Supporting quality-driven software design through intellectual assistants. In F. Meziane & S. Vadera (Eds.), *Artificial intelligence applications for improved software engineering development: New prospects* (pp. 181–216). Hershey, PA: IGI Global. doi:10.4018/978-1-60566-758-4.ch010

Sun, M. (2014). Support vector machine models for classification. In J. Wang (Ed.), *Encyclopedia of business analytics and optimization* (pp. 2395–2409). Hershey, PA: IGI Global. doi:10.4018/978-1-4666-5202-6.ch215

Svensson, J. S. (2005). Legal expert systems in administrative organizations. In M. Khosrow-Pour (Ed.), *Encyclopedia of information science and technology* (1st ed.; pp. 1827–1830). Hershey, PA: IGI Global. doi:10.4018/978-1-59140-553-5.ch321

Tzanis, G., Berberidis, C., & Vlahavas, I. (2009). Machine learning and data mining in bioinformatics. In V. Ferraggine, J. Doorn, & L. Rivero (Eds.), *Handbook of research on innovations in database technologies and applications: Current and future trends* (pp. 612–621). Hershey, PA: IGI Global. doi:10.4018/978-1-60566-242-8.ch066

Uyar, M., Yildirim, S., & Gencoglu, M. T. (2008). An effective wavelet-based feature extraction method for classification of power quality disturbance signals. *Electric Power Systems Research*, *78*(10), 1747–1755. doi:10.1016/j.epsr.2008.03.002

Vapnik, V. N. (1995). *The nature of statistical learning theory*. New York, NY: Springer–Verlag. doi:10.1007/978-1-4757-2440-0

Vasudevan, R. (2007). Utilization of Waste Polymers for Flexible Pavement and Easy Disposal of Waste Polymers. *International Conference on Sustainable Solid Waste Management*, 105-111.

Wagner, C. (2003). Knowledge management through end user developed expert systems: Potential and limitations. In M. Mahmood (Ed.), *Advanced topics in end user computing* (Vol. 2, pp. 148–172). Hershey, PA: Idea Group Publishing. doi:10.4018/978-1-59140-065-3.ch008

Wei, C., Li, C., & Li, Y. (2009). Content-based retrieval for mammograms. In Z. Ma (Ed.), *Artificial intelligence for maximizing content based image retrieval* (pp. 315–341). Hershey, PA: IGI Global. doi:10.4018/978-1-60566-174-2.ch014

Wei, C., & Li, C. (2009). Content analysis from user's relevance feedback for content-based image retrieval. In Z. Ma (Ed.), *Artificial intelligence for maximizing content based image retrieval* (pp. 216–234). Hershey, PA: IGI Global. doi:10.4018/978-1-60566-174-2.ch010

Yaseen, Z. M., El-shafie, A., Jaafar, O., Afan, H. A., & Sayl, K. N. (2015). Artificial intelligence based models for stream-flow forecasting: 2000–2015. *Journal of Hydrology (Amsterdam)*, *530*, 829–844. doi:10.1016/j.jhydrol.2015.10.038

Zhang, G., Ma, Z. M., & Yan, L. (2009). Review on texture feature extraction and description methods in content-based medical image retrieval. In Z. Ma (Ed.), *Artificial intelligence for maximizing content based image retrieval* (pp. 44–67). Hershey, PA: IGI Global. doi:10.4018/978-1-60566-174-2.ch003

Zhang, M., & Alhajj, R. (2009). Content-based image retrieval: From the object detection/recognition point of view. In Z. Ma (Ed.), *Artificial intelligence for maximizing content based image retrieval* (pp. 115–144). Hershey, PA: IGI Global. doi:10.4018/978-1-60566-174-2.ch006

Zhang, R., & Zhang, Z. M. (2009). Solving the small and asymmetric sampling problem in the context of image retrieval. In Z. Ma (Ed.), *Artificial intelligence for maximizing content based image retrieval* (pp. 192–215). Hershey, PA: IGI Global. doi:10.4018/978-1-60566-174-2.ch009

Adamatti, D. F., & de Aguiar, M. S. (2012). Artificial intelligence applied to natural resources management. In *Machine learning: Concepts, methodologies, tools and applications* (pp. 1566–1582). Hershey, PA: IGI Global. doi:10.4018/978-1-60960-818-7.ch604

Alsina Pagès, R. M., Mateo Segura, C., & Socoró Carrié, J. (2009). Fuzzy logic estimator for variant SNR environments. In J. Rabuñal Dopico, J. Dorado, & A. Pazos (Eds.), *Encyclopedia of artificial intelligence* (pp. 719–727). Hershey, PA: IGI Global. doi:10.4018/978-1-59904-849-9.ch107

Belu, R. (2014). Artificial intelligence techniques for solar energy and photovoltaic applications. In *Robotics: Concepts, methodologies, tools, and applications* (pp. 1662–1720). Hershey, PA: IGI Global. doi:10.4018/978-1-4666-4607-0.ch081

Bhattacharya, S., & Bhatnagar, V. (2015). Critical parameters for fuzzy data mining. In *Research methods: Concepts, methodologies, tools, and applications* (pp. 1–18). Hershey, PA: IGI Global.

Chatterjee, A. (2015). A Survey on Different Digital Media Based Steganography Techniques Using Genetic Algorithms. *International Journal of Innovative Research in Computer and Communication Engineering*, *3*(12), 12990–12998.

Das, B., Roy, S., Rai, R. N., & Saha, S. C. (2016). Application of grey fuzzy logic for the optimization of CNC milling parameters for Al–4.5%Cu–TiC MMCs with multi-performance characteristics. *Engineering Science and Technology: An International Journal*, *19*(2), 857–865.

Giustolisi, O., & Savic, D. A. (2006). A symbolic data-driven technique based on evolutionary polynomial regression. *Journal of Hydroinformatics*, *8*(3), 207–222.

Kaczmarek, J., & Zachwieja, T. (1966). Investigation on the material removal rate by electrochemical grinding of cutting tool materials in dependence on the properties of the grinding wheel. *International Journal of Machining Tool Design and Research*, *6*(1), 1–13. doi:10.1016/0020-7357(66)90002-3

Kaldos, E. (1974). A study of electrochemical grinding of small diameter specimens made of tungsten carbide. *Periodica Polytechnica*, *8*, 21–32.

Khokhar, S., Zin, A. A. B. M., Mokhtar, A. S. B., & Pesaran, M. (2015). A comprehensive overview on signal processing and artificial intelligence techniques applications in classification of power quality disturbances. *Renewable & Sustainable Energy Reviews*, *51*, 1650–1663. doi:10.1016/j.rser.2015.07.068

Kulkarni, S. (2012). *Machine learning algorithms for problem solving in computational applications: Intelligent techniques* (pp. 1–464). Hershey, PA: IGI Global. doi:10.4018/978-1-4666-1833-6.ch001

Langdon, W. B., & Barrett, S. J. (2004). *Genetic Programming in Data Mining for Drug Discovery. In Evolutionary Computing in Data Mining* (pp. 211–234). Springer.

Ma, Z. (2009). *Artificial intelligence for maximizing content based image retrieval* (pp. 1–450). Hershey, PA: IGI Global. doi:10.4018/978-1-60566-174-2

Ozgan, E. (2011). Artificial neural network based modelling of the Marshall Stability of asphalt concrete. *Expert Systems with Applications*, *38*(5), 6025–6030. doi:10.1016/j.eswa.2010.11.018

Park, S. H. (1996). *Robust Design and Analysis for Quality Engineering*. Chapman & Hall.

Roy, S., Bhattacharrya, A., & Banerjee, S. (2007). Analysis of effect of voltage on surface texture in electrochemical grinding by autocorrelation function. *Tribology International*, *40*(9), 1387–1393. doi:10.1016/j.triboint.2007.03.008

Sadkhan Al Maliky, S. B., & Jawad, S. F. (2015). Fuzzy logic-based security evaluation of stream cipher. In *Research methods: Concepts, methodologies, tools, and applications* (pp. 277–298). Hershey, PA: IGI Global. doi:10.4018/978-1-4666-7456-1.ch013

Stefanowski, J. (2009). *Data Mining- Clustering. Presented at SE Master Course*. Poznan, Poland: Poznan University of Technology.

Sugiyama, S. (2013). Dynamics in knowledge. In *Industrial engineering: Concepts, methodologies, tools, and applications* (pp. 1595–1612). Hershey, PA: IGI Global. doi:10.4018/978-1-4666-1945-6.ch085

Vannucci, M., Colla, V., Cateni, S., & Sgarbi, M. (2012). Artificial intelligence techniques for unbalanced datasets in real world classification tasks. In *Machine learning: Concepts, methodologies, tools and applications* (pp. 414–427). Hershey, PA: IGI Global. doi:10.4018/978-1-60960-818-7.ch304

Wei, C., & Li, Y. (2011). *Machine learning techniques for adaptive multimedia retrieval: Technologies applications and perspectives* (pp. 1–408). Hershey, PA: IGI Global. doi:10.4018/978-1-61692-859-9.ch001

Ali, J. M., Hussain, M. A., Tade, M. O., & Zhang, J. (2015). Artificial intelligence techniques applied as estimator in chemical process systems: A literature survey. *Expert Systems with Applications: An International Journal*, *42*(14), 5915–5931. doi:10.1016/j.eswa.2015.03.023

El-Hoffy, H. (2007). *Fundamentals of Machining Process*. CRC Press Taylor & Francis.

Goswami, R. N., Mitra, S., & Sarkar, S. (2009). Experimental investigation on electrochemical grinding (ECG) of alumina-aluminum interpenetrating phase composite. *International Journal of Advanced Manufacturing Technology*, *40*(7), 729–741. doi:10.1007/s00170-008-1382-z

Gupta, R. (2011). A novel approach to improve road quality by utilizing plastic waste in road construction. *Journal of Environmental Research and Development*, *5*(4), 1036–1042.

Hole, K. R., Gulhane, V. S., & Shellokar, N. D. (2013). Application of Genetic Algorithm for Image Enhancement and Segmentation. *International Journal of Advanced Research in Computer Engineering & Technology*, *2*(4), 1342–1346.

Lee, P., & Park, K. S. (2015). An Interactive Approach to Multiple Response Optimization. *Journal of the Korean Operations Research and Management Science Society*, *40*(3), 49–61. doi:10.7737/JKORMS.2015.40.3.049

Levinger, R., & Malkin, S. (1979). Electrochemical Grinding of WC-Co Cemented Carbides. *Journal of Engineering for Industry*, *101*(3), 285. doi:10.1115/1.3439509

Mat Sah, S. B., Ciesielski, V., D'Souza, D., & Berry, M. (2008). Comparison between Genetic Algorithm and Genetic Programming for Photo mosaic Generation. In *LNCS 5361* (pp. 259–268). Springer.

Ramakalyani, K., & Umadevi, D. (2012). Fraud Detection of Credit card Payment System by Genetic Algorithm. *International Journal of Scientific & Engineering Research*, *3*(7), 1–6.

Sarda, V., Sakaria, P., & Deulkar, K. (2015). Football Team Selection Using Genetic Algorithm. *International Journal of Engineering and Technical Research*, *3*(2), 153–156.

Soleimanpour, M., Talebi, S., & Azadi Motlagh, H. (2013). A Novel Technique for Steganography Method Based on Improved Genetic Algorithm optimization in Spatial Domain. *Iranian Journal of Electrical & Electronics Engineering*, *9*(2), 67–75.

Swami, V. (2012). Use of waste plastic in construction of bituminous roads. *International Journal of Engineering Science and Technology*, *4*(5), 2351–235.

Tamjidyamcholo, A., & Al-Dabbagh, R. D. (2012). Genetic Algorithm Approach for Risk Reduction of Information Security. *International Journal of Cyber-Security and Digital Forensics*, *1*(1), 59–66.

Tapkın, S., Çevik, A., & Uşar, Ü. (2010). Prediction of Marshall test results for polypropylene modified dense bituminous mixtures using neural networks. *Expert Systems with Applications*, *37*(6), 4660–4670. doi:10.1016/j.eswa.2009.12.042

Yilmaz, M. (2011). Effects of using asphaltite as filler on mechanical properties of hot mix asphalt. *Construction & Building Materials*, 1–8.

Al-Hadidy, A. I., & Yi-qiu, T. (2009). Effect of polyethylene on life of flexible pavements. *Construction & Building Materials*, *23*(3), 1456–1464. doi:10.1016/j.conbuildmat.2008.07.004

Habib, N. Z. (2011). Effect of Mixing Process on Polypropylene Modified Bituminous Concrete Mix Properties. *International Journal of Civil, Environmental, Structural, Construction and Architectural Engineering Vol*, *5*(:10), 477–482.

Harding,, S., & Leitner,, J., & Schmidhuber. (2013). Cartesian Genetic Programming for Image Processing. *Genetic and Evolutionary Computation: Genetic Programming Theory and Practice*, *10*, 1–17.

Paulians,, M., & Usinskas, A. (2007). A Survey of Genetic Algorithms Applications for Image Enhancement and Segmentation. *Information Technology and Control*, *36*(3), 278–284.

Sumithra,, A., & Ramraj, E. (2012). A Genetic Algorithmic Approach to the Identification of the Crucial Human Factors Involved in the Development of Secure Software. *International Journal of Emerging Trends & Technology in Computer Science*, *1*(2), 256–258.

Zaiane, O. R. (1999). Introduction to Data mining. In *CMPUT690, Principles of Knowledge Discovery in Databases* (pp. 1–15). University of Alberta Publishing.

Chatterjee, A. (2015). A Comparative Analysis of Three Important Data Hiding Schemes in Steganography. *International Journal of Innovative Trends In Engineering*, *11*(2), 61–65.

Cho, Y. H., & Park, S. H. (2006). Optimization in Multi Response Model with Modification Desirability Function. *Asian Journal on Quality*, *7*(3), 46–57. doi:10.1108/15982688200600026

Menaria, Y., & Sankhla, R. (2015). Use of Waste Plastic in Flexible Pavements-Green Roads. *Open Journal of Civil Engineering*, *5*(03), 299–311. doi:10.4236/ojce.2015.53030

Zaki, & Meira, Jr. (2014). Data mining and Analysis. Fundamental Concepts and Algorithms. New York: Cambridge University Press.

Afroz Sultana, S. K. (2012). Utilization of Waste Plastic as a Strength Modifier in Surface Course of Flexible and Rigid Pavements. *International Journal of Engineering Research and Applications, 2*(4), 1185-1191.

Dhodapkar, A. N. (2008). Use of waste plastics in road construction. Indian Highways Technical Paper Journal, 31-32.

Han & Kamber. (2006). Data Mining: Concepts and Techniques (2nd ed.). Urbana, IL: Elsevier

Takac, A. (2004). Cellular Genetic Programming Algorithm applied to classification task. Comenius University.

Anam, S., Islam, M. S., Kashem, M.A., Islam, M.N., Islam, M.R., & Islam, M.S. (2009). *Face Recognition Using Genetic Algorithm and Back Propagation Neural Network*. Paper presented at International Multi Conference of Engineers and Computer Scientists, IMECS 2009, Hong Kong, China

Chavan. (2013). Use of plastic waste in flexible pavements. *International Journal of Application and Innovation in Engineering and Management, 2*(4), 540-552.

Elkaffas, S. M., & Toony, A. A. (2006). *Applications of Genetic Programming in Data Mining*. Paper presented at World Academy of Science, Engineering and Technology.

Geva, M., Lenz, E., & Nadiv, S. (1976). Peripheral electrochemical grinding of sintered carbides -Effect on surface finish. Technion — Israel Institute of Technology, 38(2), 325-339.

Goldberg, A. B. (2003). *Highly Constrained Sports Scheduling with Genetic Algorithm* (Bachelor's dissertation). Amherst College.

Harding, S., & Banzhaf, W. (2008). *Genetic Programming on GPUs for Image Processing*. Paper Presented in First International Workshop on Parallel and Bioinspired Algorithms. doi:10.1504/IJHPSA.2008.024207

Jafar. (2015). Utilisation of Waste plastic in bituminous mix for improved performance of roads. *KSCE Journal of Civil Engineering*, 1-7.

Kasemsap, K. (2017b). Mastering customer service, customer experience, and customer orientation in the hospitality and tourism industry. In P. Vasant & K. M. (Eds.), Handbook of research on holistic optimization techniques in the hospitality, tourism, and travel industry (pp. 115–140). Hershey, PA: IGI Global. doi:10.4018/978-1-5225-1054-3.ch005

Khamrui, A., & Mandal, J.K. (2013). *A Genetic Algorithm Based Steganography using Discrete Cosine Transformation (GASDCT)*. Paper presented at International Conference on Computational Intelligence: Modeling Techniques and Applications (CIMTA 2013). doi:10.1016/j.protcy.2013.12.342

Kowaliw, T., Banzhaf, W., Kharma, N., & Harding, S. (2009). *Evolving novel image features using genetic programming based image transforms*. Paper presented at IEEE conference on Commerce and Enterprise on Evolutionary Computation. doi:10.1109/CEC.2009.4983255

Lin, L., Cao, L., Wang, J., & Zhang, C. (2007). *The applications of Genetic algorithms in Stock Market Data Mining Optimisation*. Paper presented at NSW Australia on Capital Market CRC, Sydney, Australia.

Marmelstein, R. E. (1997). *Application of Genetic Algorithms to Data Mining*. Paper presented at Modern Artificial Intelligence and Cognitive Science Conference.

Muharam, M. A. Y. (2005). *The Application of Genetic Programming for Feature Construction in Classification* (Unpublished doctoral dissertation). School of Computer Science, University of East Anglia.

Rajmane, P. B. (2007). Effective utilization of waste plastic in construction of flexible pavement for improving their performance. *IOSR Journal of Mechanical and Civil Engineering*, 27-30.

Raymer, M. L., Punch, W. F., Goodman, E. D., & Kuhn, L. A. (1996). *Genetic Programming for Improved Data Mining: application to the biochemistry of protein interactions*. Paper presented at 1st annual conference on genetic programming.

Salmasnia, A., Bashiri, M., & Salehi, M. (2013). *A robust interactive approach to optimize correlated multiple responses*. Academic Press.

Schoreels, C., Logan, B., & Garibaldi, J. M. (2014). *Agent based genetic Algorithm employing financial Technical analysis for making trading decisions using historical equity market data*. Paper presented at International Conference on Intelligent agent Technology.

Vats, S., Dubey, S.K., & Dubey, N.K. (2013). *Genetic Algorithms for credit card fraud detection*. Paper presented at International Conference on Education and Educational Technologies.

Atkinson, J. (2007). Work piece Surface Hardness as an Indicator of Process Regime in Peripheral Electrochemical Grinding. In *Proceedings of the 35th International MATADOR Conference* (pp. 89-94). London: Springer London. doi:10.1007/978-1-84628-988-0_20

Bhandari, S., & Shukla, N. (2015). Parametric Optimization of Electrochemical Grinding Operation by Particle Swarm Optimization Technique. *SSRG International Journal of Mechanical Engineering, 2*(4), 1-6.

Molla, K. Z. & Manna, A. (2013). Optimization of Electrochemical Grinding Parameters for Effective Finishing of Hybrid Al/(Al2O3+ZrO2) MMC. *International Journal of Surface Engineering and Interdisciplinary Material Science*, *1*(2), 35-45.

Senthilkumar, C., Ganesan, G., & Karthikeyan, R. (2011). *Parametric optimization of electrochemical machining of Al/15%SiCp composite using NSGA-II.* Academic Press.

Abed, S., Knapp, W., Traub, M., Hoffmann, D., Poprawe, R., & Loosen, P. (2004). Development of simultaneous laser welding process applied to thermoplastic polymers. *Proceedings of the 23rd International Congress on Applications of Lasers & Electro-Optics*, 97.

Åberg, P., Birgersson, U., Elsner, P., Mohr, P., & Ollmar, S. (2011). Electrical impedance spectroscopy and the diagnostic accuracy for malignant melanoma. *Experimental Dermatology*, *20*(8), 648–652. doi:10.1111/j.1600-0625.2011.01285.x PMID:21539620

Abliz, D., Duan, Y., Zhao, X., & Li, D. (2014). Low-energy electron beam cured tape placement for out-of-autoclave fabrication of advanced polymer composites. *Composites. Part A, Applied Science and Manufacturing*, *65*(0), 73–82. doi:10.1016/j.compositesa.2014.06.004

Acherjee, B., Kuar, A. S., Mitra, S., & Misra, D. (2015). Empirical modeling and multi-response optimization of laser transmission welding of polycarbonate to ABS. *Lasers in Manufacturing and Materials Processing*, *2*(3), 103–123. doi:10.1007/s40516-015-0009-0

Acherjee, B., Kuar, A. S., Mitra, S., Misra, D., & Acharyya, S. (2012). Experimental investigation on laser transmission welding of PMMA to ABS via response surface modeling. *Optics & Laser Technology*, *44*(5), 1372–1383. doi:10.1016/j.optlastec.2011.12.029

Aggarwal, A., Singh, H., Kumar, P., & Singh, M. (2008). Optimizing power consumption for CNC turned parts using response surface methodology and Taguchis technique—a comparative analysis. *Journal of Materials Processing Technology*, *200*(1), 373–384. doi:10.1016/j.jmatprotec.2007.09.041

Akovali, G. (2001). Handbook of composite fabrication. Shropshire, UK: iSmithers Rapra Publishing.

Aksoy, A., & Öztürk, N. (2011). Supplier selection and performance evaluation in just-in-time production environments. *Expert Systems with Applications*, *38*(5), 6351–6359. doi:10.1016/j.eswa.2010.11.104

Alaneme, K. K., & Adewale, T. (2013). Influence of rice husk ash–silicon carbide weight ratios on the mechanical behaviour of Al-Mg-Si alloy matrix hybrid composites. Tribology in Industry, 1(35), 163–172.

Alaneme, K. K., & Sanusi, K. O. (2015). Microstructural characteristics, mechanical and wear behaviour of aluminium matrix hybrid composites reinforced with alumina, rice husk ash and graphite. Engineering Science and Technology, an International Journal, 18(3), 416–422.

Alaneme, K. K., Akintunde, I. B., Olubambi, P. A., & Adewale, T. M. (2013). Fabrication characteristics and mechanical behaviour of rice husk ash–alumina reinforced al-mg-si alloy matrix hybrid composites. Journal of Materials Research and Technology, 2(1), 60–67.

Alaneme, K. K., & Olubambi, P. A. (2013). Corrosion and wear behaviour of rice husk ashalumina reinforced al–mg–si alloy matrix hybrid composites. *Journal of Materials Research and Technology*, *2*(2), 188–194. doi:10.1016/j.jmrt.2013.02.005

Albano, K. M., Franco, C. M., & Telis, V. R. (2014). Rheological behavior of Peruvian carrot starch gels as affected by temperature and concentration. *Food Hydrocolloids*, *40*, 30–43. doi:10.1016/j.foodhyd.2014.02.003

Aleboyeh, A., Kasiri, M. B., Olya, M. E., & Aleboyeh, H. (2008). Prediction of azo dye decolorization by UV/H2O2 using artificial neural networks. *Dyes Pigm.*, *77*(2), 288–294. doi:10.1016/j.dyepig.2007.05.014

Alfredson, R. J., & Mathew, J. (1985). Frequency domain methods for monitoring the condition of rolling element bearings. Transactions of the Institution of Engineers, Australia. *Mechanical Engineering (New York, N.Y.)*, *10*(2), 108–112.

Altintas, Y. (2010). *Manufacturing automation, metal cutting mechanics, machine tool vibration and CNC design*. Cambridge University Press.

Amasaka, K. (2014). New JIT, New Management Technology Principle: Surpassing JIT. *Procedia Technology*, *16*, 1135–1145. doi:10.1016/j.protcy.2014.10.128

Amusan, L., Olayinka, A., Nwambo, Y., Alayande, S., Ojuh, O., James, C., et al. (2010). Electrical Characterisation of Vacuum Thermally Deposited Aluminium Thin Film. *Research Journal of Applied Sciences*, *5*(2), 96-100.

Anthon, G. E., Diaz, J. V., & Barrett, D. M. (2008). Changes in pectins and product consistency during the concentration of tomato juice to paste. *Journal of Agricultural and Food Chemistry*, *56*(16), 7100–7105. doi:10.1021/jf8008525 PMID:18605731

Antoni, J., & Randall, R. B. (2006). The spectral kurtosis: Application to the vibratory surveillance and diagnostics of rotating machines. *Mechanical Systems and Signal Processing*, *20*(2), 308–331. doi:10.1016/j.ymssp.2004.09.002

Aouici, H., & Yallese, M. A. (2013). Experimental Investigation of Cutting Parameters influence on Surface Roughness and Cutting forces in Hard Turning of X38CrMoV5-1 with CBN Tool. *Sadhana*, *384*, 29–445.

Asadi, H. H., Sansoleimani, A., Fatehi, M., & Carranza, E. J. M. (2016). An AHP–TOPSIS Predictive Model for District-Scale Mapping of Porphyry Cu–Au Potential: A Case Study from Salafchegan Area (Central Iran). *Natural Resources Research*, 1–13.

Ashby, M. F., & Easterling, K. E. (2004). The transformations hardening of steel surface by laser beams. *Acta Metallurgica*, *32*(11), 1935–1948. doi:10.1016/0001-6160(84)90175-5

Ashley, S. (1995). High speed machining goes mainstream. *Mechanical Engineering (New York, N.Y.)*, *117*, 56–61.

Ashtakov, V. P. (2007). Effects of the cutting feed, depth of cut, and workpiece (bore) diameter on the tool wear rate. *International Journal of Advanced Manufacturing Technology*, *34*(7-8), 631–640. doi:10.1007/s00170-006-0635-y

Assambo, C., & Burke, M. J. (2012). *Low-Frequency Response and the Skin-Electrode Interface in Dry-Electrode Electrocardiography*. INTECH Open Access Publisher. doi:10.5772/22777

Assari, A., Mahesh, T. M., & Assari, E. (2012). Role of public participation in sustainability of historical city: Usage of TOPSIS method. *Indian Journal of Science and Technology*, *5*(3), 2289–2294.

Athavale, S. M., & Strenkowski, J. S. (2013). Material damage-based model for predicting chipbreakability. *Journal of Manufacturing Science and Engineering*, *119*(11), 675–680.

Augusto, P. E., Falguera, V., Cristianini, M., & Ibarz, A. (2011). Viscoelastic properties of tomato juice. *Procedia Food Science*, *1*, 589–593. doi:10.1016/j.profoo.2011.09.089

Baade, P., Meng, X., Youlden, D., Aitken, J., & Youl, P. (2012). Time trends and latitudinal differences in melanoma thickness distribution in Australia, 1990–2006. *International Journal of Cancer*, *130*(1), 170–178. doi:10.1002/ijc.25996 PMID:21344376

Babu, G. D., Babu, K. S., & Gowd, B. (2013). Optimization of machin ing parameters in drilling hemp fiber reinforced composites to maximi ze the tensile strength using design experiments. *Indian Journals of Engineering and Materials Science*, *20*, 385–390.

Bachmann, F. G., & Russek, U. A. (2002). Laser welding of polymers using high power diode lasers. *Proceedings of Photonics West Conference*.

Bahram, A. (2009). Optimising the automated plasma cutting process by design of experiments. *International Journal of Rapid Manufacturing*, *1*(1), 19–40. doi:10.1504/IJRAPIDM.2009.028930

Bajer, K., & Braun, U. (2014). Different aspects of the accelerated oxidation of polypropylene at increased pressure in an autoclave with regard to temperature, pretreatment and exposure media. *Polymer Testing*, *37*(0), 102–111. doi:10.1016/j.polymertesting.2014.05.006

Baker, M., Chen, Y., Lague, C., Landry, ¨. H., Peng, Q., Zhong, W., & Wang, J. (2010). Hemp fibre decortications using a planetary ball mill. *Canadian Biosystems Engineering*, *52*(2), 2–7.

Baker, M., Rosler, J., & Siemers, C. (2012). A finite element model of high speed metal cutting with adiabatic shearing. *Computers & Structures*, *80*(5), 495–513.

Barringer, S. A., Azam, A. T. M. S., Heskitt, B., & Sastry, S. (1998). On-line prediction of bostwick consistency from pressure differential in pipe flow for ketchup and related tomato products1. *Journal of Food Processing and Preservation*, *22*(3), 211–220. doi:10.1111/j.1745-4549.1998.tb00346.x

Barthel, K. (1978, February). The Shock Pulse Method for Determining the Condition of Anti-Friction Bearings. *Proceedings of the Machinery Vibration Monitoring Analysis Seminar*.

Bartolozzi, G., Baldanzini, N., & Pierini, M. (2014). Equivalent properties for corrugated cores of sandwich structures: A general analytical method. *Composite Structures*, *108*(0), 736–746. doi:10.1016/j.compstruct.2013.10.012

Bass, I. (2007). *Six sigma statistics with Excel and Minitab New York*. McGraw Hill.

Batish, A., Bhattacharya, A., Kaur, M., & Cheema, M. S. (2014). Hard turning: Parametric optimization using genetic algorithm for rough/finish machining and study of surface morphology. *Journal of Mechanical Science and Technology*, *8*(5), 1629–1640. doi:10.1007/s12206-014-0308-y

Baylis, B. (2002). Welding thermoplastic elastomers to polypropylene with a diode laser. *Proceedings of the 21st International Congress on Applications of Lasers & Electro-Optics*, 94.

Bayod, E., Willers, E. P., & Tornberg, E. (2008). Rheological and structural characterization of tomato paste and its influence on the quality of ketchup. *LWT-Food Science and Technology*, *41*(7), 1289–1300. doi:10.1016/j.lwt.2007.08.011

Bayoumi, L. (2001). Cold drawing of Regular Polygonal Tubular Sections from Round Tubes. *International Journal of Mechanical Sciences*, *43*(11), 2541–2553. doi:10.1016/S0020-7403(01)00056-X

Bayoumi, L., & Attia, A. S. (2009). Determination of the Forming Tool Load in Plastic Shaping of a Round Tube into a Square Tubular Section. *Journal of Materials Processing Technology*, *209*(4), 1835–1842. doi:10.1016/j.jmatprotec.2008.04.047

Beckwith, S. W., & Hyland, C. R. (1998). Resin Transfer Molding: A decade of technology advances. *SAMPE Journal*, *34*(6), 7–19.

Belanche, L., Valdes, J. J., Comas, J., Roda, I. R., & Poch, M. (2000). Prediction of the bulking phenomenon in wastewater treatment plants. *Artificial Intelligence in Engineering*, *14*(4), 307–317. doi:10.1016/S0954-1810(00)00012-1

Béland, J. F., Fafard, M., Rahem, A., DAmours, G., & Côté, T. (2011). Optimization on the Cold Drawing Process of 6063 Aluminium Tubes. *Applied Mathematical Modelling*, *35*(11), 5302–5313. doi:10.1016/j.apm.2011.04.025

Ben Ammar, I., Karra, C., El Mahi, A., El Guerjouma, R., & Haddar, M. (2014). Mechanical behavior and acoustic emission technique for detecting damage in sandwich structures. *Applied Acoustics*, *86*(0), 106–117. doi:10.1016/j.apacoust.2014.04.016

Benardos, P. G., & Vosniakos, G. C. (2003). Predicting surface roughness in machining: A review. *International Journal of Machine Tools & Manufacture*, *43*(8), 833–844. doi:10.1016/S0890-6955(03)00059-2

Benson, K., Cramer, S., & Galla, H.-J. (2013). Impedance-based cell monitoring: Barrier properties and beyond. *Fluids and Barriers of the CNS*, *10*(1), 1. doi:10.1186/2045-8118-10-5 PMID:23305242

Bently, D. (1989). *Predictive maintenance through the monitoring and diagnostics of rolling element bearings*. Bently Nevada Co. *Applications Note*, *ANO44*, 2–8.

Bergenwall, A. L., Chen, C., & White, R. E. (2012). TPSs process design in American automotive plants and its effects on the triple bottom line and sustainability. *International Journal of Production Economics*, *140*(1), 374–384. doi:10.1016/j.ijpe.2012.04.016

Berta, M., Wiklund, J., Kotzé, R., & Stading, M. (n.d.). Correlation between in-line measurements of tomato ketchup shear viscosity and extensional viscosity. *Journal of Food Engineering*.

Bhuvenesh, R., Norizaman, M. H., & Manan, M. S. A. (2012). Surface roughness and MRR effect on manual plasma arc cutting machining. *International Journal of Mechanical Engineering*, *6*, 131–134.

Bi, W., Li, X., Zhao, Y., Zhao, W., He, S., Ge, W., et al. (2015). Imitation cheese manufacture using Rapid Visco-Analyzer and its optimization. *International Journal of Food Properties*.

Bihamta, R., Bui, Q. H., Guillot, M., DAmours, G., Rahem, A., & Fafard, M. (2011). Application of a New Procedure for the Optimization of Variable Thickness Drawing of Aluminium Tubes. *Journal of Materials Processing Technology*, *211*, 578–589. doi:10.1016/j.jmatprotec.2010.11.012

Bil, H., Kilic, S. E., & Tekkaya, A. E. (2014). A comparison of orthogonal cutting data from experiments with three different finite element models. *International Journal of Machine Tools & Manufacture*, *44*(9), 933–944. doi:10.1016/j.ijmachtools.2004.01.016

Bini, R., Colosimo, B. M., Kutlu, A. E., & Monno, M. (2007). Experimental study of the features of the kerf generated by a 200 A high tolerance plasma cutting system. *Journal of Materials Processing Technology*, *196*(1-3), 345–355. doi:10.1016/j.jmatprotec.2007.05.061

Bisio, I., Braccini, C., Delucchi, S., Lavagetto, F., & Marchese, M. (2014, June). Dynamic multi-attribute network selection algorithm for vertical handover procedures over mobile ad hoc networks. In *2014 IEEE International Conference on Communications (ICC)* (pp. 342-347). IEEE. doi:10.1109/ICC.2014.6883342

Biswal, D., Anupriya, B., Uvanesh, K., Anis, A., Banerjee, I., & Pal, K. (2016). Effect of mechanical and electrical behavior of gelatin hydrogels on drug release and cell proliferation. *Journal of the Mechanical Behavior of Biomedical Materials*, *53*, 174–186. doi:10.1016/j.jmbbm.2015.08.017 PMID:26327452

Biswas, S., & Satapathy, A. (2010). A comparative study on erosion characteristics of red mud filled bamboo–epoxy and glass–epoxy composites. *Materials & Design*, *31*(4), 1752–1767. doi:10.1016/j.matdes.2009.11.021

Boey, F. Y. C. (1990). Reducing the void content and its variability in polymeric fibre reinforced composite test specimens using a vacuum injection moulding process. *Polymer Testing*, *9*(6), 363–377. doi:10.1016/0142-9418(90)90029-D

Boey, F. Y. C., & Lye, S. W. (1992). Void reduction in autoclave processing of thermoset composites: Part 1: High pressure effects on void reduction. *Composites*, *23*(4), 261–265. doi:10.1016/0010-4361(92)90186-X

Bonavia, T., & Marin-Garcia, J. A. (2011). Integrating human resource management into lean production and their impact on organizational performance. *International Journal of Manpower*, *32*(8), 923–938. doi:10.1108/01437721111181679

Bonten, C., & Tüchert, C. (2002). Welding of plastics-Introduction into heating by radiation. *Journal of Reinforced Plastics and Composites*, *21*(8), 699–710. doi:10.1177/073168402128988436

Bottiglieri, P., Sio, F., Fasanaro, G., Mojoli, G., Impembo, M., & Castaldo, D. (1991). Rheological characterization of ketchup. *Journal of Food Quality*, *14*(6), 497–512. doi:10.1111/j.1745-4557.1991.tb00089.x

Bouacha, K., Yallese, M. A., Mabrouki, T., & Rigal, J. F. (2010). Statistical analysis ofsurface roughness and cutting forces using response surface methodology in hard turning of AISI 52100 bearing steel with CBN tool. *International Journal of Refractory Metals & Hard Materials*, *28*(3), 349–361. doi:10.1016/j.ijrmhm.2009.11.011

Bourithis, E., Tazedakis, A., & Papadimitriou, G. (2012). A study on the surface treatment of Calmax tool steel by a plasma transferred arc (PTA) process. *Journal of Materials Processing Technology*, *128*(1), 169–177.

Broderick, J. J., Burchill, R. F., & Clark, H. L. (1972). *Design and fabrication of prototype system for early warning of impending bearing failure*. Academic Press.

Budak, E., Altintas, Y., & Armarego, E. J. A. (1996). Prediction of milling force coefficient from orthogonal cutting data. Cambridge University Press.

Budak, E., & Altintas, Y. (2012). Flexible milling force model for surface error prediction. *Journal of Engineering System Design and Analysis*, *12*(1), 89–94.

Bui, Q. H., Bihamta, R., Guillot, M., DAmours, G., Rahem, A., & Fafarda, M. (2011, March). Investigation of the Formability Limit of Aluminium Tubes Drawn with Variable Wall Thickness. *Journal of Materials Processing Technology*, *211*(3), 402–414. doi:10.1016/j.jmatprotec.2010.10.016

Bui, Q. H., Bihamta, R., Guillot, M., Rahem, A., & Fafard, M. (2011). Effect of Cross Section Reduction on the Mechanical Properties of Aluminium Tubes Drawn with Variable Wall Thickness. *Journal of Manufacturing Science and Engineering*, *133 /061004*, 1–10.

Burgess, P. F. (1988). Antifriction bearing fault detection using envelope detection. *Transactions of the Institution of Professional Engineers New Zealand: Electrical/Mechanical/Chemical Engineering Section, 15*(2), 77.

Butler, D. E. (1973). The shock-pulse method for the detection of damaged rolling bearings. *Non-Destructive Testing*, *6*(2), 92–95. doi:10.1016/0029-1021(73)90116-3

Butukuri, R., Bheemreddy, V., Chandrashekhara, K., Berkel, T., & Rupel, K. (2012). Evaluation of skin-core adhesion bond of out-of-autoclave honeycomb sandwich structures. *Journal of Reinforced Plastics and Composites*, *31*(5), 331–339. doi:10.1177/0731684412437267

Büyüközkan, G., Kayakutlu, G., & Karakadılar, İ. S. (2015). Assessment of lean manufacturing effect on business performance using Bayesian Belief Networks. *Expert Systems with Applications*, *42*(19), 6539–6551. doi:10.1016/j.eswa.2015.04.016

Byon, S. M., Lee, S. J., Lee, D. W., Lee, Y. H., & Lee, Y. (2011). Effect of Coating Material and Lubricant on Forming Force and Surface Defects in Wire Drawing Process. *Trans. Nonferrous Met. Soc. China*, *21*, S104–S110. doi:10.1016/S1003-6326(11)61071-6

CALFEM. (n.d.). *A Finite Element Toolbox, V3.3*. LUND University.

Callister, W. D. J., & Rethwich, D. G. (2010). Material Science and Engineering: An Introduction (8th ed.). Wiley.

Camacho, A. M., Gonzalez, C., Rubio, E. M., & Sebastian, M. A. (2006). Influence of Geometrical Conditions on Central Burst Appearance in Axisymmetrical Drawing Processes. *Journal of Materials Processing Technology*, *177*(1-3), 304–306. doi:10.1016/j.jmatprotec.2006.04.037

Carpinteri, A., Brighenti, R., & Vantadori, S. (2010). Influence of the Cold-Drawing Process on Fatigue Crack Growth of a V-notched Round Bar. *International Journal of Fatigue*, *32*(7), 1136–1145. doi:10.1016/j.ijfatigue.2009.12.014

Carrino, L., Giuliano, G., & Napolitano, G. (2013). Finite element simulation of orthogonal metal cutting. *Journal of Materials Processing Technology*, *12*(1), 138–157.

Caydas. (2010). Machinability Evaluation in Hard Turning of AISI 4340 Steel with Different Cutting Tools using Statistical Techniques. *Journal of Engineering Manufacture*, *224*, 1043–1055.

Celentano, D. J., Palacios, M. A., Rojas, E. L., Cruchag, M. A., Artigas, A. A., & Monsalve, A. E. (2009). Simulation and Experimental Validation of Multiple-Step Wire Drawing Processes. *Finite Elements in Analysis and Design*, *45*(3), 163–180. doi:10.1016/j.finel.2008.09.001

Centea, T., & Hubert, P. (2011). Measuring the impregnation of an out-of-autoclave prepreg by micro-CT. *Composites Science and Technology*, *71*(5), 593–599. doi:10.1016/j.compscitech.2010.12.009

Centea, T., & Hubert, P. (2012). Modelling the effect of material properties and process parameters on tow impregnation in out-of-autoclave prepregs. *Composites. Part A, Applied Science and Manufacturing*, *43*(9), 1505–1513. doi:10.1016/j.compositesa.2012.03.028

Ceretti, E., Fallbohmer, P., Wu, W. T., & Altan, T. (2015). Application of 2D FEM to chip formation in orthogonal cutting. *Journal of Materials Processing Technology*, *59*(1), 169–180.

Cetinarslan, C. S. (2012). A Study on Influences of some Process Parameters on Cold Drawing of Ferrous Wires. *Indian Journal of Engineering and Materials Sciences*, *19*, 221–228.

Chamodrakas, I., & Martakos, D. (2012). A utility-based fuzzy TOPSIS method for energy efficient network selection in heterogeneous wireless networks. *Applied Soft Computing*, *12*(7), 1929–1938. doi:10.1016/j.asoc.2012.04.016

Chandramohan, D., & Marimuthu, K. (2011). Drilling of natural fibe r particle reinforced polymer composite material. *International Journal of Advanced Engineering Research and Studies*, *1*(1), 134–145.

Chandrasekaran, M., Muralidhar, M., Murali Krishna, C., & Dixit, U. S. (2010). Application of soft computing techniques in machining performance prediction and optimization: A literature review. *International Journal of Advanced Manufacturing Technology*, *46*(5–8), 445–464. doi:10.1007/s00170-009-2104-x

Chang, F. J., Chen, P. A., Liu, C. W., Liao, V. H. C., & Lio, C. M. (2013). Regional estimation of groundwater arsenic concentrations through systematical dynamic-neural modeling. *Journal of Hydrology (Amsterdam)*, *499*, 265–274. doi:10.1016/j.jhydrol.2013.07.008

Chaturvedi, D. K. (2008). *Soft Computing: Techniques and its applications in electrical engineering*. Springer-Verlag Berlin and Heidelberg GmbH & Co. KG. doi:10.1007/978-3-540-77481-5

Chaturvedi, G. K., & Thomas, D. W. (1982). Bearing fault detection using adaptive noise cancelling. *Journal of Mechanical Design*, *104*(2), 280–289. doi:10.1115/1.3256337

Chen, D. C., & Huang, J. Y. (2007). Design of Brass Alloy Drawing Process using Taguchi Method. *Materials Science and Engineering*, *464*(1-2), 135–140. doi:10.1016/j.msea.2007.01.139

Cheng, P. J., & Lin, S. C. (2010). Using neural networks to predict bending angle of sheet metal formed by laser. *International Journal of Machine Tools & Manufacture*, *40*(1), 1185–1197.

Chen, W. (2000). Cutting forces and surface finish when machining medium hardness steel using CBN tools. *International Journal of Machine Tools & Manufacture*, *40*(3), 455–466. doi:10.1016/S0890-6955(99)00011-5

Chen, Y. L., He, P. L., & Zhang, Y. H. (2015). Combining penalty function with modified chicken swarm optimization for constrained optimization. *International Conference on Information Sciences, Machinery, Materials and Energy (ICISMME 2015)*, 1899-1907. doi:10.2991/icismme-15.2015.386

Chen, Z. X., & Tan, K. H. (2011). The perceived impact of JIT implementation on operations performance: Evidence from Chinese firms. *Journal of Advances in Management Research*, *8*(2), 213–235. doi:10.1108/09727981111175957

Chiang, Y. M., & Hsieh, H. H. (2009). The use of the Taguchi method with grey relational analysis to optimize the thin film sputtering process with multiple quality characteristics in color filter manufacturing. *Computers & Industrial Engineering*, *56*(2), 648–661. doi:10.1016/j.cie.2007.12.020

Cho, P. J., Kim, B. H., Choi, D. K., & Chu, C. N. (2008). Characteristics of MEDM for micro-holes with respect to circuit elements. *Journal of Mechanical Science and Technology*, *22*(2), 300–307. doi:10.1007/s12206-007-1039-0

Choudhury, A., & Tandon, N. (1998). A theoretical model to predict vibration response of rolling bearings to distributed defects under radial load. *Journal of Vibration and Acoustics*, *120*(1), 214–220. doi:10.1115/1.2893808

Chowdhury, S., & Saha, P. D. (2013). Artificial neural network (ANN) modeling of adsorption of methylene blue by NaOH-modified rice husk in a fixed-bed column system. *Environmental Science and Pollution Research International*, *20*(2), 1050–1058. doi:10.1007/s11356-012-0912-2 PMID:22562342

Chuiko, Savin, & Kalashnikov. (1973). Cold Drawing of Stainless Steel Tubes on Short Mandrel. *Metallurg*, (3), 32-33.

Chu, K. H. (2003). Prediction of Two-Metal Biosorption Equilibria Using a Neural Network. *European Journal of Mineral Processing and Environmental Protection*, *3*, 119–127.

Chu, P. K., Chen, J. Y., Wang, L. P., & Huang, N. (2012). Plasma surface modification of biomaterials. *Materials Science and Engineering*, *36*(1), 143–156.

Coelho, J. M. P., Abreu, M. A., & Pires, M. C. (2000). High-speed laser welding of plastic films. *Optics and Lasers in Engineering*, *34*(4-6), 385–395. doi:10.1016/S0143-8166(00)00071-3

Costa, M. L., & Almeida, S. F. M. (2005). Critical Void Content for Polymer Composite Laminates. *AIAA Journal*, *43*(6), 1336–1341. doi:10.2514/1.5830

Costa, M. L., Almeida, S. F. M., & Rezende, M. C. (2001). The influence of porosity on the interlaminar shear strength of carbon/epoxy and carbon/bismaleimide fabric laminates. *Composites Science and Technology*, *61*(14), 2101–2108. doi:10.1016/S0266-3538(01)00157-9

Črepinšek, M., Liu, S. H., & Mernik, L. (2012). A note on teaching–learning-based optimization algorithm. *Information Sciences*, *212*, 79–93. doi:10.1016/j.ins.2012.05.009

Cruz, D., & Pinho, F. (2003). Turbulent pipe flow predictions with a low Reynolds number k–ε model for drag reducing fluids. *Journal of Non-Newtonian Fluid Mechanics*, *114*(2), 109–148. doi:10.1016/S0377-0257(03)00119-8

Danese, P., Romano, P., & Bortolotti, T. (2012). JIT production, JIT supply and performance: Investigating the moderating effects. *Industrial Management & Data Systems*, *112*(3), 441–465. doi:10.1108/02635571211210068

Dash, P. K., Ramakrishna, G., Liew, A. C., & Rahman, S. 1995. Fuzzy Neural networks for time series forecasting of electric load. *IEE Proc-Gender. Transm. Distrib.*, *142*(5).

Dash, P. K., Liew, A. C., Rahman, S., & Dash, S. (1995). Fuzzy and Neuro-fuzzy Computing models for electric load forecasting. *Engineering Applications of Artificial Intelligence*, *8*(4), 423–433. doi:10.1016/0952-1976(95)00023-T

Davé, R. S., & Loos, A. C. (Eds.). (2000). *Processing of composites*. Munich, Germany: Hanser Verlag. doi:10.3139/9783446401778

Dean, D., Ramanathan, T., Machado, D., & Sundararajan, R. (2008). Electrical impedance spectroscopy study of biological tissues. *Journal of Electrostatics*, *66*(3), 165–177. doi:10.1016/j.elstat.2007.11.005 PMID:19255614

Debbaut, B., & Crochet, M. (1988). Extensional effects in complex flows. *Journal of Non-Newtonian Fluid Mechanics*, *30*(2-3), 169–184. doi:10.1016/0377-0257(88)85023-7

Debnath, A., Majumder, M., Pal, M., Das, N. S., Chattopadhyay, K. K., & Saha, B. (2016). Enhanced Adsorption of Hexavalent Chromium onto Magnetic Calcium Ferrite Nanoparticles: Kinetic, Isotherm, and Neural Network Modeling. *Journal of Dispersion Science and Technology.*, *37*(12), 1806–1818. doi:10.1080/01932691.2016.1141100

Dedeepya, Raju, & Kumar. (2012). Effect of alkali ne treatment on mechanical and thermal properties oftypha angustifolia fiber reinforced composites. *International Journal of Mechanical and Industrial Engineering*.

Dehghani Soufi, M., Ghobadian, B., Najafi, G., Sabzimaleki, M. R., & Yusaf, T. (2015). TOPSIS Multi-Criteria Decision Modeling Approach for Biolubricant Selection for Two-Stroke Petrol Engines. *Energies*, *8*(12), 13960–13970. doi:10.3390/en81212408

Demuth, H., & Beale, M. (2004). *Neural Netwotk Toolbox User's Guide. Version 4.0.* The Mathworks Inc.

Deng, J. (1982). Control problems of grey system. *Systems & Control Letters*, *5*, 288–294.

Deng, J. (1989). Introduction to grey system. *Journal of Grey System*, *1*(1), 1–24.

Deng, J. (1989). Introduction to Grey System. *Journal of Grey System*, *1*, 1–24.

Despagne, F., & Massart, D. (1998). Neural networks in multivariate calibration. *Analyst (London)*, *123*(11), 157R–178R. doi:10.1039/a805562i PMID:10396805

Dhurjaty, S., Qiu, Y., Tan, M., Liu, H., & Zheng, B. (2015). *A new application of electrical impedance spectroscopy for measuring glucose metabolism: a phantom study*. Paper presented at the SPIE Medical Imaging.

Dickinson, E. (2003). Hydrocolloids at interfaces and the influence on the properties of dispersed systems. *Food Hydrocolloids*, *17*(1), 25–39. doi:10.1016/S0268-005X(01)00120-5

Dimopoulos, I., Chronopoulos-Serelia, J. A., & Lek, S. (1999). Neural network models to study relationships between lead concentration in grasses and permanenturban descriptors in Athens city (Greece). *Ecological Modelling*, *120*(2-3), 157–165. doi:10.1016/S0304-3800(99)00099-X

Donea, J., Fasoli-Stella, D. P., & Giuliani, S. (2014). Lagrangian and eulerian finite element techniques for transient fluid structure interaction problems. *Transactions of the ASME*, *11*(2), 15-19.

DUrso, G., & Merla, C. (2014). Workpiece and electrode influence on micro-EDM drilling performance. *Precision Engineering*, *38*(4), 903–914. doi:10.1016/j.precisioneng.2014.05.007

Dwivedi, S., Huang, S. H., Shi, J., & VerDuin, W. H. (2008). Yield Prediction for Seamless Tubing Processes: A Computational Intelligence Approach. *International Journal of Advanced Manufacturing Technology*, *37*(3-4), 314–322. doi:10.1007/s00170-007-0964-5

Dwyer, R. (1983, April). Detection of non-Gaussian signals by frequency domain kurtosis estimation. In *Acoustics, Speech, and Signal Processing, IEEE International Conference on ICASSP'83* (Vol. 8, pp. 607-610). IEEE. doi:10.1109/ICASSP.1983.1172264

Dyer, D., & Stewart, R. M. (1978). Detection of rolling element bearing damage by statistical vibration analysis. *Journal of Mechanical Design*, *100*(2), 229–235. doi:10.1115/1.3453905

Eisenlohr, A., Gutierrez-Urrutia, I., & Raabe, D. (2012). Adiabatic temperature increase associated with deformation twinning and dislocation plasticity. *Acta Materialia*, *60*(9), 3994–4004. doi:10.1016/j.actamat.2012.03.008

El Dabee, F., Marian, R., & Amer, Y. (2013). An optimisation model for a simultaneous cost-risk reduction in just-in-time systems. *Australian Journal of Multi-Disciplinary Engineering*, *9*(2), 139–147. doi:10.7158/N13-GC03.2013.9.2

Enrione, J. I., Sáez, C., López, D., Skurtys, O., Acevedo, C., Osorio, F., & Hill, S. et al. (2012). Structural relaxation of salmon gelatin films in the glassy state. *Food and Bioprocess Technology*, *5*(6), 2446–2453. doi:10.1007/s11947-011-0618-3

Eroglu, A., Bayrambaş, K., Eroglu, Z., Toker, O. S., Yilmaz, M. T., & Karaman, S. (2015). Steady, dynamic, creep/recovery, and textural properties of yoghurt/molasses blends: Temperature sweep tests and applicability of Cox–Merz rule. *Food Science & Technology International*. PMID:25614154

Ezema, Menon, Obayi, & Omah. (2014). Effect of surface treatment and fiber orientation on the tensile and morphological properties of banana stem fiber reinforced natural rubber composite. *Journal of Minerals and Materials Characterization and Engineering*.

Ezugwu, E. O., Fadare, D. A., Bonney, J., Da Silva, R. B., & Sales, W. F. (2015). Modeling the correlation between cutting and process parameters in high-speed machining of Inconel alloy using an artificial neural network. *International Journal of Machine Tools & Manufacture*, *45*(1), 1375–1385.

Feldman, C. L. (1998). *Process for monitoring patients with chronic congestive heart failure*. Google Patents.

Feng, X.-B., & Wang, W. (2010). Materials and process research of 2 megawatt wind turbine blade. *Fiber Reinforced Plastics, 4*, 22.

Fenton, R. G., & Oxley, P.L.B. (2013). Mechanics of orthogonal machining - Predicting chip geometry and cutting forces from work. *Material Properties and Cutting Conditions, 18*(1), 927-942.

Fernandes, F. A. N., & Lona, L. M. F. (2005). Neural networks applications in polymerization processes. *Brazilian Journal of Chemical Engineering*, *22*(3), 401–418. doi:10.1590/S0104-66322005000300009

Fernlund, G., Rahman, N., Courdji, R., Bresslauer, M., Poursartip, A., Willden, K., & Nelson, K. (2002). Experimental and numerical study of the effect of cure cycle, tool surface, geometry, and lay-up on the dimensional fidelity of autoclave-processed composite parts. *Composites. Part A, Applied Science and Manufacturing*, *33*(3), 341–351. doi:10.1016/S1359-835X(01)00123-3

Findley, W. N., & Davis, F. A. (2013). *Creep and relaxation of nonlinear viscoelastic materials*. Courier Corporation.

Fisher, R. A. (1951). *Design of experiments Edinburgh*. Oliver and Boyd.

Foster, T. (2011). Natural structuring with cell wall materials. *Food Hydrocolloids*, *25*(8), 1828–1832. doi:10.1016/j.foodhyd.2011.05.016

Frank-Susich, D., Laananen, D. H., & Ruffner, D. (1993). Cure cycle simulation for thermoset composites. *Composites Manufacturing*, *4*(3), 139–146. doi:10.1016/0956-7143(93)90098-S

Fu, J.-T., Chang, Y.-H., & Shiau, S.-Y. (2015). Rheological, antioxidative and sensory properties of dough and Mantou (steamed bread) enriched with lemon fiber. *LWT-Food Science and Technology*, *61*(1), 56–62. doi:10.1016/j.lwt.2014.11.034

Fullerton, R. R., Kennedy, F. A., & Widener, S. K. (2014). Lean manufacturing and firm performance: The incremental contribution of lean management accounting practices. *Journal of Operations Management*, *32*(7–8), 414–428. doi:10.1016/j.jom.2014.09.002

Gaikwad & Jatti. (2016). Optimization of material removal rate during electrical discharge machining of cryo-treated niti alloys using taguchis method. *Journal of King Saud University-Engineering Sciences.*

Gaitonde, V. N., Karnik, S. R., Figueira, L., & Davim, J. P. (2009). Machinability investigations in hard turning of AISI D2 cold work tool steel with conventional and wiper ceramic inserts. *International Journal of Refractory Metals & Hard Materials*, *27*(4), 754–763. doi:10.1016/j.ijrmhm.2008.12.007

Gandy, H. T. N. (2012). *Adhesiveless honeycomb sandwich structure with carbon graphite prepreg for primary structural application: a comparative study to the use of adhesive film*. Wichita State University.

Gane, N., Rogozinski, M. W., Polivka, F., Doolette, A. G., & Ramakrishnan, S. (1994). *Quality of cut in air plasma cutting*. Paper presented at the Washington Technology Industry Association 42nd annual National Welding Conference, Melbourne, Australia.

García, J. L., Maldonado, A. A., Iniesta, A., Robles, G., & Hernández, G. (2014). A systematic review/survey for JIT implementation: Mexican maquiladoras as case study. *Computers in Industry*, *65*(4), 761–773. doi:10.1016/j.compind.2014.02.013

Gardea-Torresdey, J.L., Tiemann, K.J., & Armendariz, V. (2000). Characterization of Cr(VI) binding and reduction to Cr(III) by the agricultural byproducts of *Avena monida* (Oat) biomass. *J. Hazard. Mater., B 80*, 175-188.

Gariboldi, E., & Previtali, B. (2004). High tolerance plasma arc cutting of commercially pure titanium. *Journal of Materials Processing Technology*, *160*(1), 77–89. doi:10.1016/j.jmatprotec.2004.04.366

Garschke, C., Weimer, C., Parlevliet, P. P., & Fox, B. L. (2012). Out-of-autoclave cure cycle study of a resin film infusion process using in situ process monitoring. *Composites. Part A, Applied Science and Manufacturing*, *43*(6), 935–944. doi:10.1016/j.compositesa.2012.01.003

Garti, N., & Reichman, D. (1993). Hydrocolloids as food emulsifiers and stabilizers. *Food Structure*, *12*(4), 3.

George, P., Raghunath, B., Manocha, L., & Warrier, A. M. (2004). Edm machining of carbon–carbon compositea taguchi approach. *Journal of Materials Processing Technology*, *145*(1), 66–71. doi:10.1016/S0924-0136(03)00863-X

Glicksman, M. (1991). Hydrocolloids and the search for the oily grail. *Food Technology*, *45*(10), 94–103.

Gob, S., Oliveros, E., Bossmann, S. H., Braun, A. M., Guardani, R., & Nascimento, C. A. O. (1999). Modeling the kinetics of a photochemical water treatment process by means of artificial neural networks. *Chemical Engineering and Processing*, *38*(4-6), 373–382. doi:10.1016/S0255-2701(99)00028-8

Gómez, R., Bashir, R., & Bhunia, A. K. (2002). Microscale electronic detection of bacterial metabolism. *Sensors and Actuators. B, Chemical*, *86*(2), 198–208. doi:10.1016/S0925-4005(02)00175-2

Gonzalez-Alvarez, D. L., Vega-Rodrıguez, M. A., Gomez-Pulido, J. A., & Sanchez-Perez, J. M. (2012). Multiobjective teaching-learning- based optimization (mo-tlbo) for motif finding. *Computational Intelligence and Informatics (CINTI), 2012 IEEE 13th International Symposium on.*, 141–146. doi:10.1109/CINTI.2012.6496749

Gopal, A. K., Adali, S., & Verijenko, V. E. (2000). Optimal temperature profiles for minimum residual stress in the cure process of polymer composites. *Composite Structures*, *48*(1–3), 99–106. doi:10.1016/S0263-8223(99)00080-X

Graham, N. (1997). *Patent No. EP 0805746 A1*. Munich, Germany: E. P. Office.

Green Jr, K. W., Inman, R. A., Birou, L. M., and Whitten, D. (2014). Total JIT (T-JIT) and its impact on supply chain competency and organizational performance. *International Journal of Production Economics, 147, Part A*, 125-135. doi:10.1016/j.ijpe.2013.08.026

Grunenfelder, L., Centea, T., Hubert, P., & Nutt, S. (2013). Effect of room-temperature out-time on tow impregnation in an out-of-autoclave prepreg. *Composites. Part A, Applied Science and Manufacturing*, *45*, 119–126. doi:10.1016/j.compositesa.2012.10.001

Gujral, H. S., Sharma, A., & Singh, N. (2002). Effect of hydrocolloids, storage temperature, and duration on the consistency of tomato ketchup. *International Journal of Food Properties*, *5*(1), 179–191. doi:10.1081/JFP-120015600

Gullu, A., & Atici, U. (2006). Investigation of the effects of plasma arc parameters on the structure variation of AISI 304 and St 52 steels. *Materials & Design*, *27*(10), 1157–1162. doi:10.1016/j.matdes.2005.02.014

Gunasekera, J. S., & Hoshino, S. (1982). Analysis of Extrusion or Drawing of Polygonal Sections through Straightly Converging Dies. *Journal of Engineering for Industry*, *104*(1), 38–45. doi:10.1115/1.3185795

Guo, Z.-S., Liu, L., & Zhang, B.-M. (2009). Critical Void Content for Thermoset Composite Laminates. *Journal of Composite Materials*, *43*(17), 1775–1790. doi:10.1177/0021998306065289

Hagan, T. M., Demuth, H. B., & Beale, M. (1996). *Neural Network Design*. PWS Publishing Company.

Hair, J. F., Hult, G. T. M., Ringle, C., & Sarstedt, M. (2013). *A Primer on Partial Least Squares Structural Equation Modeling (PLS-SEM)*. SAGE Publications.

Haley, T. A., & Smith, R. S. (2003). Evaluation of in-line absorption photometry to predict consistency of concentrated tomato products. *LWT-Food Science and Technology*, *36*(2), 159–164. doi:10.1016/S0023-6438(02)00218-9

Halter, R. J., Hartov, A., Heaney, J. A., Paulsen, K. D., & Schned, A. R. (2007). Electrical impedance spectroscopy of the human prostate. *Biomedical Engineering. IEEE Transactions on*, *54*(7), 1321–1327.

Hamed, M. M., Khalafallah, M. G., & Hassanien, E. A. (2004). Prediction of wastewater treatment plant performance using artificial neural networks. *Environmental Modelling & Software*, *19*(10), 919–928. doi:10.1016/j.envsoft.2003.10.005

Hanim, M., Chin, N., & Yusof, Y. (2015). Effects of grinding time on rheological, textural and physical properties of natural peanut butter stored at different temperatures. *Journal of Texture Studies*.

Hansch, D., Haff, D., Putz, H., Treusch, H. G., Gillner, A., & Poprawe, R. (1998). Welding of plastics with diode laser. *Proceedings of the 17th International Congress on Applications of Lasers & Electro-Optics*, 85, 131-140.

Harikrishnan & Shivakumar. (n.d.). *Mode i fracture toughness of jute/glass fibre hybrid composite–an experimental and numeric al study*. Academic Press.

Harkins, R. D. (1974). An objective water quality index. *Journal - Water Pollution Control Federation*, *46*, 589. PMID:4846806

Hasda, R. K. (2013). *Optimization of EDM Process Parameters through Teaching Learning Based Optimization Algorithm* (Doctoral dissertation). National Institute of Technology, Rourkela, India.

Hayes, W., Smith, P., & Morris, A. (1998). The production and quality of tomato concentrates. *Critical Reviews in Food Science and Nutrition*, *38*(7), 537–564. doi:10.1080/10408699891274309 PMID:9813734

Herfurth, H., Ehlers, B., Heinemann, S., & Haensch, D. (1999). New approaches in plastic welding with diode lasers. *Proceedings of the 18th International Congress on Applications of Lasers & Electro-Optics*, *87b*, 48-56.

Herring, M., Mardel, J., & Fox, B. (2010). The effect of material selection and manufacturing process on the surface finish of carbon fibre composites. *Journal of Materials Processing Technology*, *210*(6), 926–940. doi:10.1016/j.jmatprotec.2010.02.005

He, X. (2012). Finite Element Analysis of Laser Welding: A State of Art Review. *Materials and Manufacturing Processes*, *27*(12), 1354–1365. doi:10.1080/10426914.2012.709345

He, X. (2014). Influence of boundary conditions on stress distributions in a single-lap adhesively bonded joint. *International Journal of Adhesion and Adhesives*, *53*, 34–43. doi:10.1016/j.ijadhadh.2014.01.009

He, X., Gu, F., & Ball, A. (2014). A review of numerical analysis of friction stir welding. *Progress in Materials Science*, *65*, 1–66. doi:10.1016/j.pmatsci.2014.03.003

He, X., Zhao, L., Deng, C., Xing, B., Gu, F., & Ball, A. (2015). Self-piercing riveting of similar and dissimilar metal sheets of aluminum alloy and copper alloy. *Materials & Design*, *65*, 923–933. doi:10.1016/j.matdes.2014.10.002

Ho, D., & Randall, R. B. (2000). Optimisation of bearing diagnostic techniques using simulated and actual bearing fault signals. *Mechanical Systems and Signal Processing*, *14*(5), 763–788. doi:10.1006/mssp.2000.1304

Holl, A., Pardo, R., & Rama, R. (2013). Spatial patterns of adoption of just-in-time manufacturing * Spatial patterns of adoption of just-in-time manufacturing. *Papers in Regional Science*, *92*(1), 51–67. doi:10.1111/j.1435-5957.2011.00406.x

Hong, Y., Wu, Y., Liu, G., & Gu, Z. (2015). Effect of amylose on pasting and rheological properties of corn starch/xanthan blends. *Stärke*, *67*(1-2), 98–106. doi:10.1002/star.201400126

Hossain, M. R., Islam, M. A., Van Vuurea, A., & Verpoest, I. (2013). Tensile behavior of environment friendly jute epoxy laminated composite. *Procedia Engineering*, *56*, 782–788. doi:10.1016/j.proeng.2013.03.196

Huang, Y., Zheng, M., Lin, Z., Zhao, B., Zhang, S., Yang, J., & Shi, Y. et al. (2015). Flexible cathodes and multifunctional interlayers based on carbonized bacterial cellulose for high-performance lithium–sulfur batteries. *Journal of Materials Chemistry A*, *3*(20), 10910–10918. doi:10.1039/C5TA01515D

Hui, G.-H., Ji, P., Mi, S.-S., & Deng, S.-P. (2013). Electrochemical impedance spectrum frequency optimization of bitter taste cell-based sensors. *Biosensors & Bioelectronics*, *47*, 164–170. doi:10.1016/j.bios.2013.03.018 PMID:23578970

Hui, G., Mi, S., Chen, Q., & Chen, X. (2014). Sweet and bitter tastant discrimination from complex chemical mixtures using taste cell-based sensor. *Sensors and Actuators. B, Chemical*, *192*, 361–368. doi:10.1016/j.snb.2013.10.119

Hwang, C. L., & Yoon, K. (2012). *Multiple attribute decision making: methods and applications a state-of-the-art survey* (Vol. 186). Springer Science & Business Media.

Hwang, J. R., & Fung, C. P. (2014). Effect of electron beam surface hardening on fatigue crack growth rate in AISI 4340 steel. *Surface and Coatings Technology*, *80*(1), 278–291.

ICMR. (1975). Manual of Standards of Quality of Drinking Water Supplied (2nd ed.). Indian Council of Medical Research.

Inman, R. A., Sale, R. S., Green, K. W. Jr, & Whitten, D. (2011). Agile manufacturing: Relation to JIT, operational performance and firm performance. *Journal of Operations Management*, *29*(4), 343–355. doi:10.1016/j.jom.2010.06.001

Ion, J. C., & Anisdahl, L. M. (2007). A PC based system for produce development in laser transformation hardening. *Journal of Materials Processing Technology*, *65*(1), 261–267.

ISI. (1983). *Indian Standard Specification for Drinking Water.* IS: 10500. ISI.

Ismail, M. I. S., & Taha, Z. (2011). Experimental design and performance analysis in plasma arc surface hardening, World Academy of Science. *Engineering and Technology*, *56*, 1052–1058.

Itoigawa, F., Childs, T. H. C., Nakamura, T., & Belluco, W. (2006). Effects and mechanisms in minimal quantity lubrication machining of aluminum alloy. *Journal of Wear*, *260*(3), 339–344. doi:10.1016/j.wear.2005.03.035

Iwata, K., Osakada, K., & Terasaka, Y. (2015). Process modeling of orthogonal cutting by the rigid plastic finite element method. *Transactions of the ASME*, *106*(2), 132-138.

Jabbar, A., & Militky, J. (2016). Modeling and analysis of the creep behavior of jute/green e poxy composites incorporated with chemically treated pulverized nano/micro jute fibers. *Industrial Crops and Products*, *84*, 230–240. doi:10.1016/j.indcrop.2015.12.052

Jangra, K., Grover, S., & Aggarwal, A. (2011). Simultaneous optimization of material removal rate and surface roughness for WEDM of WC-Co composite using grey relational analysis along with Taguchi method. *International Journal of Industrial Engineering Computations*, *2*(3), 479–490. doi:10.5267/j.ijiec.2011.04.005

Jangra, K., Grover, S., & Aggarwal, A. (2012). Optimization of multi machining characteristics in WEDM of WC-5.3%Co composite using integrated approach of Taguchi, GRA and entropy method. *Frontiers of Mechanical Engineering*, *7*(3), 288–299. doi:10.1007/s11465-012-0333-4

Jansson, A., Kouvo, S., & Kujanpää, V. (2005). Quasi-simultaneous laser welding of polymers - the process and applications for mass-production. *Proceedings of the 24th International Congress on Applications of Lasers & Electro-Optics*, 98, 381-387.

Jansson, A., Kouvo, S., Salminen, A., & Kujanpää, V. (2003). The effect of parameters on laser transmission welding of polymers. *Proceedings of the 22nd International Congress on Applications of Lasers & Electro-Optics*, 95, 124-133.

Jarek, B., & Kubik, A. (2015). The examination of the glass fiber reinforced polymer composite rods in terms of the application for concrete reinforcement. *Procedia Engineering*, *108*, 394–401. doi:10.1016/j.proeng.2015.06.163

Jayaram, J., Das, A., & Nicolae, M. (2010). Looking beyond the obvious: Unraveling the Toyota production system. *International Journal of Production Economics*, *128*(1), 280–291. doi:10.1016/j.ijpe.2010.07.024

Jin Choi, Y., Milczarek, R. R., Fleck, C., Casey Garvey, T., McCarthy, K. L., & McCarthy, M. J. (2006). In-line monitoring of tomato concentrate physical properties during evaporation. *Journal of Food Process Engineering*, *29*(6), 615–632. doi:10.1111/j.1745-4530.2006.00085.x

Jirarat, T., Ketlada, K., & Nuttapon, S. (2011). Relationship model and supporting activities of JIT, TQM and TPM. *Songklanakarin Journal of Science & Technology*, *33*(1), 101–106.

Johnson, T. C., Balbierz, D. J., & Pearson, R. (2005). *Method for detecting and treating tumors using localized impedance measurement*. Google Patents.

Johnson, N. W., Jayasekara, P., & Amarasinghe, A. (2011). Squamous cell carcinoma and precursor lesions of the oral cavity: Epidemiology and aetiology. *Periodontology 2000*, *57*(1), 19–37. doi:10.1111/j.1600-0757.2011.00401.x PMID:21781177

Joosten, M. W., David, M., Kelly, D., & Thomson, R. (2014). Static and dynamic evaluation of a multifunctional crashworthy pi-joint/sandwich energy absorbing structure. *Composite Structures*, *118*(0), 423–431. doi:10.1016/j.compstruct.2014.07.023

Joshi, S. C., Liu, X. L., & Lam, Y. C. (1999). A numerical approach to the modeling of polymer curing in fibre-reinforced composites. *Composites Science and Technology*, *59*(7), 1003–1013. doi:10.1016/S0266-3538(98)00138-9

Junsheng, C., Dejie, Y., & Yu, Y. (2007). Application of an impulse response wavelet to fault diagnosis of rolling bearings. *Mechanical Systems and Signal Processing*, *21*(2), 920–929. doi:10.1016/j.ymssp.2005.09.014

Juszczak, L., Oczadły, Z., & Gałkowska, D. (2013). Effect of modified starches on rheological properties of ketchup. *Food and Bioprocess Technology*, *6*(5), 1251–1260. doi:10.1007/s11947-012-0813-x

Kaaki, D., Baghdadi, O. K., Najm, N., & Olabi, A. (2012). Preference mapping of commercial Labneh (strained yogurt) products in the Lebanese market. *Journal of Dairy Science*, *95*(2), 521–532. doi:10.3168/jds.2011-4409 PMID:22281316

Kadam, Y., Pochat-Bohatier, C., Sanchez, J., & El Ghzaoui, A. (2015). Modulating Viscoelastic Properties of Physically Crosslinked Self-Assembled Gelatin Hydrogels through Optimized Solvent Conditions. *Journal of Dispersion Science and Technology*, *36*(9), 1349–1356. doi:10.1080/01932691.2014.984721

Kadirgama, K., Noor, M. M., Zuki, N. M., Rahman, M. M., Rejab, M. R. M., Daud, R. & Abou-El-Hossein, K. A. (2008). Optimization of surface roughness in end milling on mould aluminium alloys (AA6061-T6) using response surface method and radian basis function network. *Jourdan Journal of Mechanical and Industrial Engineering*, *2*(4), 209- 214.

Kadirgama, K., Noor, M. M., Harun, W. S. W., & Aboue-EI-Hossein, K. A. (2010). Optimization of heat affected zone by partial swarm optimization in air plasma cutting operation. *Journal of Scientific and Industrial Research*, *69*, 439–443.

Kagan, V. A., Bray, R. G., & Kuhn, W. P. (2002). Laser transmission welding of semi-crystalline thermoplastics: part I: Optical characterization of nylon based plastics. *Journal of Reinforced Plastics and Composites*, *21*(12), 1101–1122. doi:10.1177/073168402128987699

Kalia, S., Kaith, B., & Kaur, I. (2009). Pretreatments of natural fibers and their application as reinforcing material in polymer compositesa review. *Polymer Engineering and Science*, *49*(7), 1253–1272. doi:10.1002/pen.21328

Kallivayalil, J. A., & Zehnder, A. T. (1994). Measurement of the temperature field induced by dynamic crack growth in Beta-C titanium. *International Journal of Fracture*, *66*(2), 99–120. doi:10.1007/BF00020077

Kanagarajan, D., Karthikeyan, R., Palanikumar, K., & Davim, J. P. (2008). Optimization of electrical discharge machining characteristics of WC/Co composites using non-dominated sorting genetic algorithm (NSGA-II). *International Journal of Advanced Manufacturing Technology*, *36*(11-12), 1124–1132. doi:10.1007/s00170-006-0921-8

Kandpal, B. C., & Singh, H. (2015). Machining of aluminium metal matrix composites with electrical discharge machining-a review. *Materials Today: Proceedings*, *2*(4), 1665–1671. doi:10.1016/j.matpr.2015.07.094

Kaneko, J., & Nojiri, W. (2008). The logistics of Just-in-Time between parts suppliers and car assemblers in Japan. *Journal of Transport Geography*, *16*(3), 155–173. doi:10.1016/j.jtrangeo.2007.06.001

Kannan, S., & Ramanathan, K. (2014). Optimization of edm parameter of al/tic composite using taguchi methodology. *Middle-East Journal of Scientific Research*, *22*(1), 121–127.

Karaboga, D. (2005). *An idea based on honey bee swarm for numerical optimization, Technical Report-TR06*. Erciyes University, Engineering Faculty, Computer Engineering Department.

Karacay, T., & Akturk, N. (2009). Experimental diagnostics of ball bearings using statistical and spectral methods. *Tribology International*, *42*(6), 836–843. doi:10.1016/j.triboint.2008.11.003

Ke, N., Nguyen, K., Irelan, J., & Abassi, Y. A. (2015). Multidimensional GPCR profiling and screening using impedance-based label-free and real-time assay. *G Protein-Coupled Receptor Screening Assays: Methods and Protocols*, 215-226.

Keshtkar, A., Keshtkar, A., & Smallwood, R. H. (2006). Electrical impedance spectroscopy and the diagnosis of bladder pathology. *Physiological Measurement*, *27*(7), 585–596. doi:10.1088/0967-3334/27/7/003 PMID:16705257

Ketkar, M., & Vaidya, O. S. (2012). Study of Emerging Issues in Supply Risk Management in India. *Procedia: Social and Behavioral Sciences*, *37*(0), 57–66. doi:10.1016/j.sbspro.2012.03.275

Khalifa, M., & Chappar, S. S. (2014). Experimental investigati on of tensile properties of banana reinforced epoxy composites. *International Journal of Science and Research, 3*(6), 832–836.

Khanna, R., Kumar, A., Garg, M. P., Singh, A., & Sharma, N. (2015). Multiple performance characteristics optimization for Al 7075 on electric discharge drilling by Taguchi grey relational theory. *Journal of Industrial Engineering International*, *11*(4), 459–472. doi:10.1007/s40092-015-0112-z

Kim, D., Centea, T., & Nutt, S. R. (2014a). In-situ cure monitoring of an out-of-autoclave prepreg: Effects of out-time on viscosity, gelation and vitrification. *Composites Science and Technology*, *102*(0), 132–138. doi:10.1016/j.compscitech.2014.07.027

Kim, D., Centea, T., & Nutt, S. R. (2014b). Out-time effects on cure kinetics and viscosity for an out-of-autoclave (OOA) prepreg: Modelling and monitoring. *Composites Science and Technology*, *100*(0), 63–69. doi:10.1016/j.compscitech.2014.05.027

Kim, I. S., Son, J. K., & Yarlagadda, P. K. D. V. (2013). A study on the quality improvement of robotic GMA welding process. *Robotics and Computer-integrated Manufacturing*, *19*(1), 567–578.

Kim, J. W., Kim, H. G., & Lee, D. G. (2004). Compaction of thick carbon/phenolic fabric composites with autoclave method. *Composite Structures*, *66*(1-4), 467–477. doi:10.1016/j.compstruct.2004.04.070

Kim, S. W., Kwon, Y. N., Lee, Y. S., & Lee, J. H. (2007). Design of Mandrel in Tube Drawing Process for Automotive Steering Input Shaft. *Journal of Materials Processing Technology*, *187–188*, 182–186. doi:10.1016/j.jmatprotec.2006.11.134

Kitamura, T., Ito, K., Teramura, S., Zhang, Z., & Hamada, II. (2014). Application of multilayered paper processing to hybrid random natural fiber mat. *Energy Procedia*, *56*, 247–254. doi:10.1016/j.egypro.2014.07.155

Kock, N. (2010). Using WarpPLS in E-collaboration Studies: An Overview of Five Main Analysis Steps. *International Journal of e-Collaboration*, *6*(4), 1–11. doi:10.4018/jec.2010100101

Kock, N. (2013). Using WarpPLS in e-collaboration studies: What if I have only one group and one condition. *International Journal of e-Collaboration*, *9*(3), 12. doi:10.4018/jec.2013070101

Kock, N. (2015). *WarpPLS 5.0 User Manual*. USA: ScriptWarp System.

Ko, J. H., Yun, W. S., Cho, D. W., & Ehmann, K. F. (2002). Development of a virtual machining system, Part 1: Approximation of the size effect for cutting force prediction. *International Journal of Machine Tools & Manufacture*, *42*(15), 1595–1605. doi:10.1016/S0890-6955(02)00137-2

Koocheki, A., Ghandi, A., Razavi, S., Mortazavi, S. A., & Vasiljevic, T. (2009). The rheological properties of ketchup as a function of different hydrocolloids and temperature. *International Journal of Food Science & Technology*, *44*(3), 596–602. doi:10.1111/j.1365-2621.2008.01868.x

Kraposhin, V. S., Bobrov, A. V., & Gaponenko, O. S. (2009). Surface hardening of 9KhF steel by heating with a plasma gun. *Journal of Materials Processing Technology*, *11*(1), 13–17.

Kratz, J., & Hubert, P. (2011). Processing out-of-autoclave honeycomb structures: Internal core pressure measurements. *Composites. Part A, Applied Science and Manufacturing*, *42*(8), 1060–1065. doi:10.1016/j.compositesa.2011.04.009

Kratz, J., & Hubert, P. (2013). Anisotropic air permeability in out-of-autoclave prepregs: Effect on honeycomb panel evacuation prior to cure. *Composites. Part A, Applied Science and Manufacturing*, *49*(0), 179–191. doi:10.1016/j.compositesa.2013.02.013

Krishnan & Ramesh. (n.d.). *Synthesis and characterization of cellulose nanofibers from coconut coir fibers*. Academic Press.

Krohling, R. A., & Campanharo, V. C. (2011). Fuzzy TOPSIS for group decision making: A case study for accidents with oil spill in the sea. *Expert Systems with Applications*, *38*(4), 4190–4197. doi:10.1016/j.eswa.2010.09.081

Krüger, J., Singh, K., ONeill, A., Jackson, C., Morrison, A., & OBrien, P. (2002). Development of a microfluidic device for fluorescence activated cell sorting. *Journal of Micromechanics and Microengineering*, *12*(4), 486. doi:10.1088/0960-1317/12/4/324

Kumar, & Agnihotri. (2013). Cold Drawing Process–A Review. *Metal Forming, 3*(3).

Kumar, P., & Agnihotri, G. (2013). *Application of finite element method on seamless tube cold drawing* (Unpublished Master's thesis). M.A.N.I.T. Bhopal, India.

Kumar, V. R. (2013). *Teaching learning based optimization applied to mechanical constrained design problems* (Doctoral dissertation).

Kumar, N. M., Kumaran, S. S., & Kumaraswamidhas, L. (2015). An investigation of mechanical properties and material removal rate, tool wear rate in edm machining process of al2618 alloy reinforced with si 3 n 4, aln and zrb 2 composites. *Journal of Alloys and Compounds*, *650*, 318–327. doi:10.1016/j.jallcom.2015.07.292

Kumar, R., & Singh, M. (2013). Outer race defect width measurement in taper roller bearing using discrete wavelet transform of vibration signal. *Measurement*, *46*(1), 537–545. doi:10.1016/j.measurement.2012.08.012

Kuppan, P., Rajadurai, A., & Narayanan, S. (2008). Influence of EDM Process parameters in deep hole drilling of Inconel 718. *International Journal of Advanced Manufacturing Technology*, *38*(1-2), 74–84. doi:10.1007/s00170-007-1084-y

Lahby, M., Cherkaoui, L., & Adib, A. (2013, May). Hybrid network selection strategy by using M-AHP/E-TOPSIS for heterogeneous networks. In *Intelligent Systems: Theories and Applications (SITA), 2013 8th International Conference on* (pp. 1-6). IEEE.

Leao, F. N., & Pashby, I. R. (2005). Optimisation of EDM fast hole-drilling through Evaluation of dielectric and electrode materials. *18th International Congress of Mechanical Engineering*.

Lee, T.H., & Mathew, P. (2007). Experimental and Theoretical Investigation of AISI D2 hardened steel machining with varying nose radius CBN tools. *International Journal of Machining and Machinability of Materials, 2*, 254-269.

Lee, D., & Wilkening, W. W. (2012). Material modeling and high speed machining process. *Transactions of the ASME, 13*(2), 324-341.

Lee, E. H., & Schaffer, B. W. (2011). Theory of plasticity applied to a problem of machining. *Journal of Applied Mechanics*, *18*(1), 405–416.

Lee, J. J., Hong, W. K., Hittelman, W. N., Mao, L., Lotan, R., & Shin, D. M. et al.. (2000). Predicting cancer development in oral leukoplakia: Ten years of translational research. *Clinical Cancer Research*, *6*(5), 1702–1710. PMID:10815888

LeGrand, M., & Bellenger, V. (1998). The cure optimisation of carbon/epoxy prepregs. *Composites Science and Technology*, *58*(5), 639–644. doi:10.1016/S0266-3538(97)00136-X

Lei, S., Shin, Y. C., & Incropera, F. P. (2013). Thermo-mechanical modeling of orthogonal machining process by finite element analysis. *International Journal of Machine Tools & Manufacture*, *39*(5), 731–750. doi:10.1016/S0890-6955(98)00059-5

Lek, S., & Guegan, J. F. (1999). Artificial neural networks as a tool in ecological modeling:an introduction. *Ecological Modelling*, *120*(2-3), 65–73. doi:10.1016/S0304-3800(99)00092-7

Li, J., & Yang, Q. (2010). *The Analysis of Human Resource Allocation and Incentive Based on JIT System*. Paper presented at the Information Science and Management Engineering (ISME), 2010 International Conference of.

Li, C. H., & Tsai, M. J. (2009). Multi-objective optimization of laser cutting for flash memory modules with special shapes using grey relational analysis. *Optics & Laser Technology*, *41*(5), 634–642. doi:10.1016/j.optlastec.2008.09.009

Lima, J. G., Avila, R. F., & Abrao, A. M. (2007). Turning of hardened AISI 4340 steel using coated carbide inserts. *IMechE Part B: J. Engineering Manufacture*, *221*(8), 1359–1366. doi:10.1243/09544054JEM847

Lin, J. L., & Lin, C. L. (2002). The use of the orthogonal array with grey relational analysis to optimize the electrical discharge machining process with multiple performance characteristics. *International Journal of Machine Tools & Manufacture*, *42*(2), 237–244. doi:10.1016/S0890-6955(01)00107-9

Linnik, V. A., Onegina, A. K., Andrev, A. I., Aldarkin, K. K., Sinaiskii, V. M., & Grigorenko, L. P. (2015). Surface hardening of steel by plasma hardening. *Journal of Materials Processing Technology*, *5*(4), 2–23.

Liu, H. S., Yan, B. H., Huang, F. Y., & Qui, K. H. (2005). A study on the characterization of high nickel alloy micro-holes using micro-EDM and their applications. *Journal of Materials Processing Technology*, *169*(3), 418–426. doi:10.1016/j.jmatprotec.2005.04.084

Liu, L., Zhang, B.-M., Wang, D.-F., & Wu, Z.-J. (2006). Effects of cure cycles on void content and mechanical properties of composite laminates. *Composite Structures*, *73*(3), 303–309. doi:10.1016/j.compstruct.2005.02.001

Lodhi, B. K., & Agarwal, S. (2014). Optimization of machining parameters in wedm of aisi d3 steel using taguchi technique. *Procedia CIRP*, *14*, 194–199. doi:10.1016/j.procir.2014.03.080

Lohani, B. N. (1981). Water Quality Indices. In C. K. Varshney (Ed.), *Water Pollution and Management Reviews* (pp. 53–69). New Delhi: South Asian Publ.

Loharkar. (2015). Parametric analysis of industrial cold drawing process. *International Journal of Engineering Research and General Science*, *3*(5).

Lokare, Gholap, & Gavade. (n.d.). *To develop and evaluate impact behaviour of natural fibre reinforced composite-review*. Academic Press.

Low Sui, P., & Gao, S. (2011). The Application of the Just-in-Time Philosophy in the Chinese Construction Industry. *Journal of Construction in Developing Countries*, *16*(1), 91–111.

Low, S. P., & Show, M. Y. (2008). Facilities design incorporating just-in-time principles for ramp-up light factories in Singapore. *Facilities*, *26*(7/8), 321–342. doi:10.1108/02632770810877967

Lu, H. S., Chang, C. K., Hwang, N. C., & Chung, C. T. (2009). Grey relational analysis coupled with principle component analysis for optimization design of the cutting parameters in high-speed end milling. *Journal of Materials Processing Technology*, *209*(8), 3808–3817. doi:10.1016/j.jmatprotec.2008.08.030

Luo, W. (2013). The corrosion resistance of 0Cr19Ni9 stainless steel arc welding joints with and without arc surface melting. *Materials Science and Engineering A*, *345*(1), 7–19.

Ma, Z., & Boye, J. I. (2015). Novel Health Ingredients and Their Applications in Salad Dressings and Other Food Emulsions. *Nutraceutical and Functional Food Processing Technology*, 151-187.

Mac Kenna, N., Calvert, P., & Morrin, A. (2015). Impedimetric transduction of swelling in pH-responsive hydrogels. *Analyst (London)*, *140*(9), 3003–3011. doi:10.1039/C4AN02252A PMID:25768307

Mackley, M., Marshall, R., Smeulders, J., & Zhao, F. (1994). The rheological characterization of polymeric and colloidal fluids. *Chemical Engineering Science*, *49*(16), 2551–2565. doi:10.1016/0009-2509(94)E0082-2

Madhavan, V., Chandrasekar, S., & Farris, T. N. (2010). Machining as a wedge indentation. *Transactions ASME*, *67*(1), 128-139.

Madhukiran, J., Rao, S., & Madhusudan, S. (2013). Fabrication and testing of natural fiber reinforced hybrid composites banana/pineapple. *International Journal of Modern Engineering Research*, *3*(4), 2239–2243.

Mahapatra, S., & Patnaik, A. (2007). Optimization of wire electrical discharge machining (wedm) process parameters using taguchi method. *International Journal of Advanced Manufacturing Technology*, *34*(9-10), 911–925. doi:10.1007/s00170-006-0672-6

Maistros, G. M., & Partridge, I. K. (1998). Monitoring autoclave cure in commercial carbon fibre/epoxy composites. *Composites. Part B, Engineering*, *29*(3), 245–250. doi:10.1016/S1359-8368(97)00020-6

Maity, K. P., & Bagal, D. K. (2015). Effect of process parameters on cut quality of stainless steel of plasma arc cutting using hybrid approach. *International Journal of Advanced Manufacturing Technology*, *78*(1-4), 161–175. doi:10.1007/s00170-014-6552-6

Mamalis, A. G., Horvath, M., Branis, A. S., & Manolakos, D. E. (2012). Finite element simulation of chip formation in orthogonal metal cutting. *Journal of Materials Processing Technology*, *110*(1), 19–27. doi:10.1016/S0924-0136(00)00861-X

Maria, H. J., Lyczko, N., Nzihou, A., Joseph, K., Mathew, C., & Thomas, S. (2014). Stress relaxation behavior of organically modified montmorillonite filled natural rubber/nitrile rubber nanocomposites. *Applied Clay Science*, *87*, 120–128. doi:10.1016/j.clay.2013.10.019

Martelloti, M. E. (2011). An analysis of the milling process. *Transactions of the ASME*, *63*(1), 677-700.

Martinez, G. M. (1991). Fast cures for thick laminated organic matrix composites. *Chemical Engineering Science*, *46*(2), 439–450. doi:10.1016/0009-2509(91)80005-J

Martínez-Jurado, P. J., Moyano-Fuentes, J., & Jerez-Gómez, P. (2014). Human resource management in Lean Production adoption and implementation processes: Success factors in the aeronautics industry. *BRQ Business Research Quarterly*, *17*(1), 47–68. doi:10.1016/j.cede.2013.06.004

Martins, L. G., & Gerges, S. N. Y. (1984). Comparison of Signal Analyses for Detecting Incipient Bearing Damage. *Condition Monitoring*, *84*, 191-204.

Mary Mangaiyarkari, M. S., Vincent, S., Janarthanan, S., Subba Rao, T., & Tata, B. V. R. (2011). Bioreduction of Cr (VI) by alkaliphilic *Bacillus subtilis* and interaction of the membrane groups. *Saudi J. of Biol. Sci.*, *18*(2), 157–167. doi:10.1016/j.sjbs.2010.12.003 PMID:23961119

Masai, P., Parrend, P., & Zanni-Merk, C. (2015). Towards a Formal Model of the Lean Enterprise. *Procedia Computer Science*, *60*, 226–235. doi:10.1016/j.procs.2015.08.122

Mashazi, P., Tetyana, P., Vilakazi, S., & Nyokong, T. (2013). Electrochemical impedimetric immunosensor for the detection of measles-specific IgG antibodies after measles infections. *Biosensors & Bioelectronics*, *49*, 32–38. doi:10.1016/j.bios.2013.04.028 PMID:23708815

Mashitah, M. D., Zulfadhly, Z., & Bhatia, S. (1999). Binding mechanism of heavy metals biosorption by Pycnoporus sanguineus. J. Artificial Cells. *Blood Substitutes and Immobilization Biotechnology*, *27*(5/6), 441–445. doi:10.3109/10731199909117717

Mathew, J., & Alfredson, R. J. (1984). The condition monitoring of rolling element bearings using vibration analysis. *Journal of Vibration, Acoustics, Stress, and Reliability in Design*, *106*(3), 447-453.

MATLAB V7.0. (n.d.). MathWorks, Inc. Retrieved from www.mathworks.com

Matsuo, H. (2015). Implications of the Tohoku earthquake for Toyota's coordination mechanism: Supply chain disruption of automotive semiconductors. *International Journal of Production Economics*, *161*, 217–227. doi:10.1016/j.ijpe.2014.07.010

McCarthy, K. L., & McCarthy, M. J. (2009). Relationship between In-line Viscosity and Bostwick Measurement during Ketchup Production. *Journal of Food Science*, *74*(6), E291–E297. doi:10.1111/j.1750-3841.2009.01205.x PMID:19723191

McCarthy, K. L., Sacher, R. F., & Garvey, T. C. (2008). Relationship between rheological behavior and Bostwick measurement during manufacture of ketchup. *Journal of Texture Studies*, *39*(5), 480–495. doi:10.1111/j.1745-4603.2008.00155.x

McFadden, P. D., & Smith, J. D. (1984). Vibration monitoring of rolling element bearings by the high-frequency resonance technique—a review. *Tribology International*, *17*(1), 3–10. doi:10.1016/0301-679X(84)90076-8

Meng, X., Liu, Y., Gao, X., & Zhang, H. (2014). A new bio-inspired algorithm: Chicken swarm optimization. *Advances in Swarm Intelligence. Lecture Notes in Computer Science*, *8794*, 86–94. doi:10.1007/978-3-319-11857-4_10

Mert, B. (2012). Using high pressure microfluidization to improve physical properties and lycopene content of ketchup type products. *Journal of Food Engineering*, *109*(3), 579–587. doi:10.1016/j.jfoodeng.2011.10.021

Meyer, L. D., Ahlgren, F. F., & Weichbrodt, B. (1980). An analytic model for ball bearing vibrations to predict vibration response to distributed defects. *Journal of Mechanical Design*, *102*(2), 205–210. doi:10.1115/1.3254731

Mia, L., & Winata, L. (2014). Manufacturing strategy and organisational performance: The role of competition and MAS information. *Journal of Accounting & Organizational Change*, *10*(1), 83–115. doi:10.1108/JAOC-10-2011-0048

Miracle, D. B., Donaldson, S. L., & Vander Voort, G. F. (2001). ASM handbook: Vol. 21. *Asm International Materials Park*. Academic Press.

Miyachi, T., & Seki, K. (1986, September). An Investigation of the Early Detection of Defects in Ball Bearings Using Vibration Monitoring, Practical Limit of Detectability and Growth Speed of Defects. *Proceedings of the International Conference of Rotordynamics*, 403-408.

Mohandas, K. N., Ramesh, C. S., Prasad, K. E., Balashanmugam, N., & Shashikumar, P. V. (2012). Study of the flank wear during hard turning of hard chrome plated surfaces. *Int. Conf. on materials processing and characterization.*

Mohanty, A. R. (2014). *Machinery Condition Monitoring: Principles and Practices*. CRC Press.

Mohanty, M. E., & Mishra, A. K. (2013). Estimating viscosity and polarity in the microenvironment of polymeric gels—introducing a microviscosity parameter. *Journal of Polymer Research*, *20*(7), 1–8. doi:10.1007/s10965-013-0185-6

Mohr, P., Birgersson, U., Berking, C., Henderson, C., Trefzer, U., Kemeny, L., & Paoli, J. et al. (2013). Electrical impedance spectroscopy as a potential adjunct diagnostic tool for cutaneous melanoma. *Skin Research and Technology*, *19*(2), 75–83. doi:10.1111/srt.12008 PMID:23350668

Montgomery, D. C. (2001). *Design and Analysis of Experiments*. New York: Wiley.

Montgomery, D., & Altintas, Y. (2014). Mechanism of cutting force and surface generation in dynamic milling. *ASME Journal of Engineering for Industry*, *113*(1), 160–168.

Mori, K., Abe, Y., & Kato, T. (2014). Self-pierce riveting of multiple steel and aluminium alloy sheets. *Journal of Materials Processing Technology*, *214*(10), 2002–2008. doi:10.1016/j.jmatprotec.2013.09.007

Mori, K., Kasashima, N., Yoshioka, T., & Ueno, Y. (1996). Prediction of spalling on a ball bearing by applying the discrete wavelet transform to vibration signals. *Wear*, *195*(1), 162–168. doi:10.1016/0043-1648(95)06817-1

Moriwaki, T., Sugimura, N., & Luan, S. (2013). Combined stress, material flow and heat analysis of orthogonal micromachining of copper. *Transactions of the ASME*, *42*(1), 75-78, 1993.

Morteza, H., & Ghasempour, M. M. (2016). An analysis of tube drawing process used to produce squared sections from round tubes through FE simulation and response surface methodology. *International Journal of Advanced Manufacturing Technology*. doi:10.1007/s00170-016-8532-5

Mortimer, S., Smith, M. J., & Olk, U. E. (2010). *Product development for out-of-autoclave (OOA) manufacture of aerospace structures*. Paper presented at the SAMPE conference.

Movahhedy, M., Gadala, M. S., & Altintas, Y. (2010). Simulation of the orthogonal metal cutting process using an arbitrary Lagrangian-Eulerian finite-element method. *Journal of Materials Processing Technology*, *103*(2), 267–275. doi:10.1016/S0924-0136(00)00480-5

Mucha, J. (2011). The analysis of lock forming mechanism in the clinching joint. *Materials & Design*, *32*(10), 4943–4954. doi:10.1016/j.matdes.2011.05.045

Murdoch, C., Brown, B. H., Hearnden, V., Speight, P. M., DApice, K., & Hegarty, A. M. et al.. (2014). Use of electrical impedance spectroscopy to detect malignant and potentially malignant oral lesions. *International Journal of Nanomedicine*, *9*, 4521. doi:10.2147/IJN.S64087 PMID:25285005

Mustapha, S., & Ye, L. (2014). Leaky and non-leaky behaviours of guided waves in CF/EP sandwich structures. *Wave Motion*, *51*(6), 905–918. doi:10.1016/j.wavemoti.2014.03.004

Myhan, R., Markowski, M., Daszkiewicz, T., Zapotoczny, P., & Sadowski, P. (2015). Non-linear stress relaxation model as a tool for evaluating the viscoelastic properties of meat products. *Journal of Food Engineering*, *146*, 107–115. doi:10.1016/j.jfoodeng.2014.09.006

Nadarajan, S., Chandren, S., & Mohamed Elias, E. (2013). Hiccups in just–in–time practices for electrical & electronic manufacturing. *International Journal of Academic Research*, *5*(5), 269–271. doi:10.7813/2075-4124.2013/5-5/B.41

Nagarkar, M. P., Zaware, R. N., & Ghalme, S. G. (2012). Finite Element Simulation of Sink Pass Round Tubes using ANSYS. *APTEFF*, *43*(43), 179–188. doi:10.2298/APT1243179N

Naik, S., & Purohit, K. M. (1996). Physico-chemical analysis of some community ponds of Rourkela. *Indian J. of Env. Protection*, *16*(9), 679–684.

Nakayama, K., Arai, M., & Kanda, T. (1998). Machining characteristics of hard materials, CIRP. *Annals of CIRP*, *37*(1), 89–92. doi:10.1016/S0007-8506(07)61592-3

Napier, S. S., & Speight, P. M. (2008). Natural history of potentially malignant oral lesions and conditions: An overview of the literature. *Journal of Oral Pathology & Medicine*, *37*(1), 1–10. doi:10.1111/j.1600-0714.2007.00579.x PMID:18154571

Narasaraju, G., & Raju, D. L. (2015). Characterization of hybrid rice husk and fly ash-reinforced aluminium alloy (alsi10mg) composites. *Materials Today: Proceedings*, *2*(4), 3056–3064. doi:10.1016/j.matpr.2015.07.245

Narayanan, Idapalapati, & Subbiah. (2010). Effect of Cold Work on the Mechanical Response of Drawn Ultra-Fine Gold Wire. *Computational Materials Science, 12*(38).

Natarajan, R., & Jeelani, S. (2013). Residual stresses in machining using finite element method. *Computer Software and Applications*, *3*(1), 79–80.

Neves, Button, Caminaga, & Gentile. (2005). Numerical and Experimental Analysis of Tube Drawing with Fixed Plug. *J. of the Braz. Soc. of Mech. Sci. & Eng., 27*(4), 426-431.

Neves, P., Fernandes, A., Lantz, D., Carldén, L., & Andersson, K. (2014). *Bus BiW Weight Optimization Using Composite Sandwich Panels.* Paper presented at the Transport Research Arena (TRA) 5th Conference: Transport Solutions from Research to Deployment.

Nguyen, T. A., Yin, T.-I., Reyes, D., & Urban, G. A. (2013). Microfluidic chip with integrated electrical cell-impedance sensing for monitoring single cancer cell migration in three-dimensional matrixes. *Analytical Chemistry*, *85*(22), 11068–11076. doi:10.1021/ac402761s PMID:24117341

Niknam, T., Golestaneh, F., & Sadeghi, M. S. (2012). Multiobjective Teaching–Learning-Based Optimization for Dynamic Economic Emission Dispatch. *IEEE Systems Journal*, *6*(2), 341–352. doi:10.1109/JSYST.2012.2183276

Nikolaou, N. G., & Antoniadis, I. A. (2002). Demodulation of vibration signals generated by defects in rolling element bearings using complex shifted Morlet wavelets. *Mechanical Systems and Signal Processing*, *16*(4), 677–694. doi:10.1006/mssp.2001.1459

Nitkiewicz, Z., & Jeiorski, L. (2011). Plasma heat treatment of steel: Microstructures, properties and applications. *Metal Science and Engineering*, *140*(1), 474–493.

Nourbakhsh, M. N. (2002). Biosorption of Cr6+, Pb2+ and Cu2+ ions in industrial wtewater on Bacillus sp. *Chemical Engineering Journal*, *85*(2-3), 351–355. doi:10.1016/S1385-8947(01)00227-3

Obaid, M. H., & Tawfeeq, J. F. (n.d.). *Measurement the Distance between Points on the Map*. Academic Press.

Obi, A. I., & Oyinlola, A. K. (1996). Frictional Characteristics of Fatty-Based Oils in Wire Drawing. *Wear, 194*(1-2), 30–37. doi:10.1016/0043-1648(95)06664-0

Ocak, H., & Loparo, K. A. (2005). HMM-based fault detection and diagnosis scheme for rolling element bearings. *Journal of Vibration and Acoustics*, *127*(4), 299–306. doi:10.1115/1.1924636

Oh, J. H., & Lee, D. G. (2002). Cure Cycle for Thick Glass/Epoxy Composite Laminates. *Journal of Composite Materials*, *36*(1), 19–45. doi:10.1177/0021998302036001300

Oktem, H., Erzurumlu, T., & Erzincanli, F. (2006). Prediction of minimum surface roughness in end milling mold parts using neural network and genetic algorithm. *Materials & Design*, *27*(9), 735–744. doi:10.1016/j.matdes.2005.01.010

Olivier, P., Cottu, J. P., & Ferret, B. (1995). Effects of cure cycle pressure and voids on some mechanical properties of carbon/epoxy laminates. *Composites*, *26*(7), 509–515. doi:10.1016/0010-4361(95)96808-J

Osei-Antwi, M., de Castro, J., Vassilopoulos, A. P., & Keller, T. (2014). Fracture in complex balsa cores of fiber-reinforced polymer sandwich structures. *Construction & Building Materials*, *71*(0), 194–201. doi:10.1016/j.conbuildmat.2014.08.029

Osuagwu, C. C., & Thomas, D. W. (1982). Effect of inter-modulation and quasi-periodic instability in the diagnosis of rolling element incipient defect. *Journal of Mechanical Design*, *104*(2), 296–302. doi:10.1115/1.3256339

Özcan, B., Demirbakan, B., Yeşiller, G., & Sezgintürk, M. K. (2014). Introducing a new method for evaluation of the interaction between an antigen and an antibody: Single frequency impedance analysis for biosensing systems. *Talanta*, *125*, 7–13. doi:10.1016/j.talanta.2014.02.067 PMID:24840408

Ozek, C., Caydas, U., & Unal, E. (2012). A fuzzy model for predicting surface roughness in plasma arc cutting of AISI 4140 steel. *Materials and Manufacturing Processes*, *27*(1), 95–102. doi:10.1080/10426914.2011.551952

Ozel, T., & Altan, T. (2007). Process simulation using finite element method - prediction of cutting forces, tool stresses and temperatures in high-speed flat end milling. *International Journal of Machine Tools & Manufacture*, *40*(5), 713–738. doi:10.1016/S0890-6955(99)00080-2

Ozel, T., & Zeren, E. (2006). Finite element modeling the influence of edge roundness on the stress and temperature fields induced by high-speed machining. *International Journal of Advanced Manufacturing Technology*, *35*(3), 255–267.

Paliwal, D., Choudhury, A., & Tingarikar, G. (2014). Wavelet and scalar indicator based fault assessment approach for rolling element bearings. *Procedia Materials Science*, *5*, 2347–2355. doi:10.1016/j.mspro.2014.07.478

Pan, C. X., Chen, Y. W., & Male, A. T. (2012). Microstructural development in plasma jet forming of sheet steels. *Materials Science and Technology*, *18*(1), 1151–1168.

Pan, L. K., Wang, C. C., Wei, S. L., & Sher, H. F. (2007). Optimizing multiple quality characteristics via Taguchi method based grey relational analysis. *Journal of Materials Processing Technology*, *182*(1-3), 107–116. doi:10.1016/j.jmatprotec.2006.07.015

Pantale, O., Bacaria, J. L., Dalverny, O., Rakotomalala, R., & Caperaa, S. (2004). 2D and 3D numerical models of metal cutting with damage effects. *Computer Methods in Applied Mechanics and Engineering*, *193*(39), 4383–4399. doi:10.1016/j.cma.2003.12.062

Pan, W. X., Meng, X., Li, G., Fei, Q. X., & Wu, C. K. (2005). Feasibility of laminar plasma jet hardening of cast iron surface. *Surface and Coatings Technology*, *197*(1), 345–350. doi:10.1016/j.surfcoat.2004.06.043

Park, Y. S., Chon, T. S., Kwak, I. S., & Lek, S. (2004). Hierarchical community classification and assessment of aquatic ecosystems using artificial neuralnetworks. *The Science of the Total Environment*, *327*(1-3), 105–122. doi:10.1016/j.scitotenv.2004.01.014 PMID:15172575

Parshin, S. V. (2009). Multipass Drawing of Finned Tubes. *Russian Journal of Non-Ferrous Metals*, *50*(2), 128–132. doi:10.3103/S1067821209020102

Patra, J. C., Panda, G., Pal, R. N., & Chatterjee, B. N. (1997). Neural Networks for signal Processing Applications. *Journal of Computer Society of India*, *27*(3), 34–58.

Pawar, P. J., & Rao, R. V. (2013). Parameter optimization of machining processes using teaching–learning-based optimization algorithm. *International Journal of Advanced Manufacturing Technology*, *67*(5-8), 995–1006. doi:10.1007/s00170-012-4524-2

Peng, T., Shi, H., Liang, X., Luo, F., & Wu, X. (n.d.). Experimental investigation on sandwich structure ring-type ultrasonic motor. *Ultrasonics.*

Peng, T. M., Hubble, N. F., & Karady, G. G. (1992). Advancement in the application of neural networks for short term load forecasting. *IEEE Transactions on Power Systems*, *7*(1), 250–258. doi:10.1109/59.141711

Peng, Z. K., & Chu, F. L. (2004). Application of the wavelet transform in machine condition monitoring and fault diagnostics: A review with bibliography. *Mechanical Systems and Signal Processing*, *18*(2), 199–221. doi:10.1016/S0888-3270(03)00075-X

Pentakota, S. N., Prasad, V., Ramji, K., & Pusty, G. (2014). Synthe sis of bio–degradable banana nanofibers. *IJITR*, *2*(1), 730–734.

Perfézou, M., Turner, A., & Merkoçi, A. (2012). Cancer detection using nanoparticle-based sensors. *Chemical Society Reviews*, *41*(7), 2606–2622. doi:10.1039/C1CS15134G PMID:21796315

Pethig, R. (1987). Dielectric properties of body tissues. *Clinical Physics and Physiological Measurement*, *8*(4A), 5–12. doi:10.1088/0143-0815/8/4A/002 PMID:3568571

Petrucci, R., Nisini, E., Ghelli, D., Santulli, C., Puglia, D., & Sarasini, F. (2012). *Mechanical and impact characterisation of hybrid composite laminates based on flax, hemp, basalt and glass fibers produced by vacuum infusion.* Paper presented at the ECCM15 Conference, Venezia, Italy.

Phadke, S. M. (1989). *Quality Engineering Using Robust Design*. Englewood Cliffs, NJ: Prentice Hall.

Pham, D. T., Ivanov, A., Bigot, S., Popov, K., & Dimov, S. (2007). An investigation of tube and rod electrode wear in micro EDM drilling. *International Journal of Advanced Manufacturing Technology*, *33*(1–2), 103–109. doi:10.1007/s00170-006-0639-7

Pheng, L. S., Arain, F. M., & Fang, J. W. Y. (2011). Applying just-in-time principles in the delivery and management of airport terminal buildings. *Built Environment Project and Asset Management*, *1*(1), 104–121. doi:10.1108/20441241111143812

Pickering, K. L., Efendy, M. A., & Le, T. M. (2016). A review of recent developments in natural fibre composites and their mecha nical performance. *Composites. Part A, Applied Science and Manufacturing*, *83*, 98–112. doi:10.1016/j.compositesa.2015.08.038

Piispanen, V. (2014). Theory of formation of metal chips. *Journal of Applied Physics*, *19*(1), 876–881.

Pillai, V. K., Beris, A. N., & Dhurjati, P. S. (2002). Implementation of Model-Based Optimal Temperature Profiles for Autoclave Curing of Composites Using a Knowledge-Based System. *Industrial & Engineering Chemistry Research*, *33*(10), 2443–2452. doi:10.1021/ie00034a029

Pimenta, S., & Robinson, P. (2014). Wavy-ply sandwich with composite skins and crushable core for ductility and energy absorption. *Composite Structures*, *116*(0), 364–376. doi:10.1016/j.compstruct.2014.05.020

Plaza, S., Sanchez, J. A., Perez, E., Gil, R., Izquierdo, B., Ortega, N., & Pombo, I. (2014). Experimental study on micro EDM-drilling of Ti6Al4V using helical electrode. *Precision Engineering*, *38*(4), 821–827. doi:10.1016/j.precisioneng.2014.04.010

Pliquett, U., & Prausnitz, M. R. (2000). Electrical impedance spectroscopy for rapid and noninvasive analysis of skin electroporation. *Electrochemotherapy, Electrogenetherapy, and Transdermal Drug Delivery: Electrically Mediated Delivery of Molecules to Cells*, 377-406.

Podder, M. S., & Majumder, C. B. (2016). *Corynebacterium glutamicum* MTCC 2745 immobilized on granular activated carbon/$MnFe_2O_4$ composite: A novel biosorbent for removal of As(III) and As(V) ions. *Spectrochimica Acta. Part A: Molecular and Biomolecular Spectroscopy, 168*, 159–179. doi:10.1016/j.saa.2016.05.056 PMID:27289352

Podder, M. S., & Majumder, C. B. (2016). Predictive approach for simultaneous biosorption and bioaccumulation of arsenic by *Corynebacterium glutamicum* MTCC 2745 biofilm supported on NL/$MnFe_2O_4$ composite. *Journal of Water Process Engineering., 11*, 8–31. doi:10.1016/j.jwpe.2016.03.010

Pokorný, J. (2009). Textural and flavour characteristics of commercial tomato ketchups. *Czech J. Food Sci., 27*(3), 165–170.

Polmear, I. J. (1995). *Light Metals: Metallurgy of the light metals* (3rd ed.; Vol. 3). London: Halsted Pr.

Pospiech, J. (1998). Description of a Mathematical Model of Deformability for the Process of Drawing Tubes on a Fixed Mandrel. *Journal of Materials Engineering and Performance, 7*(1), 71–78. doi:10.1361/105994998770348061

Prabhakaran, R., Kontopoulou, M., Zak, G., Bates, P. J., & Baylis, B. K. (2006). Contour laser – Laser-transmission welding of glass reinforced nylon 6. *Journal of Thermoplastic Composite Materials, 19*(4), 427–439. doi:10.1177/0892705706062200

Prabhakar, S., Mohanty, A. R., & Sekhar, A. S. (2002). Application of discrete wavelet transform for detection of ball bearing race faults. *Tribology International, 35*(12), 793–800. doi:10.1016/S0301-679X(02)00063-4

Pradhan, R., Mitra, A., & Das, S. (2013). Quantitative evaluation of blood glucose concentration using impedance sensing devices. *Journal of Electrical Bioimpedance, 4*(1), 73–77. doi:10.5617/jeb.657

Pramanik, A., Basak, A., Islam, M. N., & Littlefair, G. (2015). Electrical discharge machining of 6061 aluminium alloy. *Transactions of Nonferrous Metals Society of China, 25*(9), 2866–2874. doi:10.1016/S1003-6326(15)63912-7

Prashad, H., Ghosh, M., & Biswas, S. (1985). Diagnostic monitoring of rolling-element bearings by high-frequency resonance technique. *ASLE Transactions, 28*(4), 439-448.

Pujari, Ramakrishna, & Padal. (n.d.). *Experimental investigation of mechanical properties of uni directional jute and banana fiber composites*. Academic Press.

Pujari, Ramakrishna, & Suresh Kumar. (2014). Comparison of jute and banana fiber composites: A review. *International Journal of Current Engineering and Technology*.

Purslow, D., & Childs, R. (1986). Autoclave moulding of carbon fibre-reinforced epoxies. *Composites, 17*(2), 127–136. doi:10.1016/0010-4361(86)90250-8

Py, L. K. (1983). A review of rolling element bearing health monitoring. *Proceedings of Machinery Vibration Monitoring and Analysis Meeting*, 145.

Qiu, H., Lee, J., Lin, J., & Yu, G. (2006). Wavelet filter-based weak signature detection method and its application on rolling element bearing prognostics. *Journal of Sound and Vibration, 289*(4), 1066–1090. doi:10.1016/j.jsv.2005.03.007

Radovanovic, M., & Madic, M. (2011). Modeling the plasma arc cutting using ANN. *Nonconventional Technologies Review, 4*, 43–48.

Rajan, K. M., & Narasimhan, K. (2001). An Investigation of the Development of Defects During Flow Forming of High Strength Thin Wall Steel Tubes. *Practical Failure Analysis, 1*(5), 69–76. doi:10.1007/BF02715366

Rajasekhar, A., Rani, R., Ramya, K., & Abraham, A. (2012, October). Elitist teaching learning opposition based algorithm for global optimization. In *2012 IEEE International Conference on Systems, Man, and Cybernetics (SMC)* (pp. 1124-1129). IEEE. doi:10.1109/ICSMC.2012.6377882

Raji, N. A., & Oluwole, O. O. (2011). Influence of Degree of Cold-Drawing on the Mechanical Properties of Low Carbon Steel. *Materials Sciences and Applications*, *2*(11), 1556–1563. doi:10.4236/msa.2011.211208

Rajkumar, K., Santosh, S., Ibrahim, S. J. S., & Gnanavelbabu, A. (2014). Effect of electrical discharge machining parameters on microwave heat treated aluminium-boron carbide-graphite composites. *Procedia Engineering*, *97*, 1543–1550. doi:10.1016/j.proeng.2014.12.438

Ramakrishnan, S., Shrinet, V., Polivka, F. B., Kearney, T. N., & Koltun, P. (2000). Influence of gas composition on plasma arc cutting of steel. *Journal of Physics. D, Applied Physics*, *33*(18), 2288–2299. doi:10.1088/0022-3727/33/18/313

Ramanand, K. R., Krishnanand, K. R., Panigrahi, B. K., & Mallick, M. K. (2012, December). Brain storming incorporated teaching–learning–based algorithm with application to electric power dispatch. In *International Conference on Swarm, Evolutionary, and Memetic Computing* (pp. 476-483). Springer Berlin Heidelberg. doi:10.1007/978-3-642-35380-2_56

Ramanathan, M., Patil, M., Epur, R., Yun, Y., Shanov, V., Schulz, M., & Kumta, P. N. et al. (2016). Gold-coated carbon nanotube electrode arrays: Immunosensors for impedimetric detection of bone biomarkers. *Biosensors & Bioelectronics*, *77*, 580–588. doi:10.1016/j.bios.2015.10.014 PMID:26476598

Ramesh, M., Atreya, T. S. A., Aswin, U., Eashwar, H., & Deepa, C. (2014). Processing and mechanical property evaluation of banana fiber reinforced polymer composites. *Procedia Engineering*, *97*, 563–572. doi:10.1016/j.proeng.2014.12.284

Ranganath, S., Campbell, A. B., & Gorkiewicz, D. W. (2007). A model to calibrate and predict forces in machining with honed cutting tools or inserts. *International Journal of Machine Tools & Manufacture*, *47*(5), 820–840. doi:10.1016/j.ijmachtools.2006.06.019

Rao, R. V., & Kalyankar, V. D. (n.d.). *Grinding parameters selection using TLBO method*. Academic Press.

Rao, R. V., & Kalyankar, V. D. (2011). Parameters optimization of advanced machining processes using TLBO algorithm. *EPPM, Singapore*, *20*, 21–31.

Rao, R. V., & Kalyankar, V. D. (2013). Parameter optimization of modern machining processes using teaching–learning-based optimization algorithm. *Engineering Applications of Artificial Intelligence*, *26*(1), 524–531. doi:10.1016/j.engappai.2012.06.007

Rao, R. V., & Patel, V. (2013). Multi-objective optimization of heat exchangers using a modified teaching-learning-based optimization algorithm. *Applied Mathematical Modelling*, *37*(3), 1147–1162. doi:10.1016/j.apm.2012.03.043

Rao, R. V., & Waghmare, G. G. (2014). A comparative study of a teaching–learning-based optimization algorithm on multi-objective unconstrained and constrained functions. *Journal of King Saud University-Computer and Information Sciences*, *26*(3), 332–346. doi:10.1016/j.jksuci.2013.12.004

Rao, R., & Patel, V. (2012). An elitist teaching-learning-based optimization algorithm for solving complex constrained optimization problems. *International Journal of Industrial Engineering Computations*, *3*(4), 535–560. doi:10.5267/j.ijiec.2012.03.007

Rashed, H., Islam, M., & Rizvi, F. (2006). Effects of process para meters on tensile strength of jute fiber reinforced thermoplastic co mposites. *Journal of Naval Architecture and Marine Engineering*, *3*(1), 1–6.

Rashid, W. B., Goel, S., Davim, P. J., Shrikrishna, N., & Joshi, S. N. (2016). Parametric design optimization of hard turning of AISI 4340 steel (69 HRC). *International Journal of Advanced Manufacturing Technology*, *82*(1), 451–462. doi:10.1007/s00170-015-7337-2

Ravichandran, G., Rosakis, A. J., Hodowany, J., & Rosakis, P. (2002). AIP Conf. Proc.: Vol. 620. *On the Conversion of Plastic Work into Heat During High-Strain-Rate Deformation*. doi:10.1063/1.1483600

Ray, A. (1980, September). Monitoring rolling contact bearings under adverse conditions. *2nd International Conference of Vibrations in rotating Machinery, Institution of Mechanical Engineers*, 1-4.

Razak, M., Abdul-Rani, A., Rao, T., Pedapati, S., & Kamal, S. (2016). Electrical discharge machining on biodegradable az31 magnesium alloy using taguchi method. *Procedia Engineering*, *148*, 916–922. doi:10.1016/j.proeng.2016.06.501

Rech, J., & Moisan, A. (2003). Surface integrity in finish hard turning of case-hardened steels. *International Journal of Machine Tools & Manufacture*, *43*(5), 543–550. doi:10.1016/S0890-6955(02)00141-4

Reif, Z., & Lai, M. S. (1989). *Detection of developing bearing failures by means of vibration*. New York: The American Society of Mechanical Engineers.

Rigel, D. S., Russak, J., & Friedman, R. (2010). The evolution of melanoma diagnosis: 25 years beyond the ABCDs. *CA: a Cancer Journal for Clinicians*, *60*(5), 301–316. doi:10.3322/caac.20074 PMID:20671054

Rion, J., Leterrier, Y., & Månson, J. A. E. (2008). Prediction of the adhesive fillet size for skin to honeycomb core bonding in ultra-light sandwich structures. *Composites. Part A, Applied Science and Manufacturing*, *39*(9), 1547–1555. doi:10.1016/j.compositesa.2008.05.022

Robinson, J. P., & Amirtharaj, E. H. (2013). A Strategic TOPSIS algorithm with correlation coefficient of interval vague sets. *International Journal of Computing Algorithm*, *2*, 314–331.

Rocha, Nunes, & Hirsch. (2012). Analysis by Design of Experiments of Distortion Potentials in Drawn and Induction Hardened Wire. *Materials Research, 15*(2), 266-276.

Rojas, Gonza´lez, Calvet, & Bubnovich. (2008). A New Analytical Solution for Prediction of Forward Tension in the Drawing Process. *Journal of Materials Processing Technology, 198*, 93–98.

Romoli, L., Rashed, C. A., & Fiaschi, M. (2014). Experimental characterization of the inner surface in micro-drilling of spray holes: A comparison between ultra short pulsed laser and EDM. *Optics & Laser Technology*, *56*, 35–42. doi:10.1016/j.optlastec.2013.07.010

Rosakis, P., Rosakis, A. J., Ravichandran, G., & Hodowany, J. (2000). Thermodynamic internal variable model for the partition of plastic work into heat and stored energy in metals. *Journal of the Mechanics and Physics of Solids*, *48*(3), 581–607. doi:10.1016/S0022-5096(99)00048-4

Rosa, P. A. R., Martins, P. A. F., & Atkins, A. G. (2007). Revisiting the fundamentals of metal cutting by means of finite elements and ductile fracture mechanics. *International Journal of Machine Tools & Manufacture*, *47*(3), 607–617. doi:10.1016/j.ijmachtools.2006.05.003

Ross, P. J. (1996). *Taguchi Techniques for Quality Engineering* (2nd ed.). New York: McGraw Hill.

Rout, A. K., & Satapathy, A. (2012). Study on mechanical and tribo-performance of rice-husk filled glass–epoxy hybrid composites. *Materials & Design*, *41*, 131–141. doi:10.1016/j.matdes.2012.05.002

Roy, P., Mondal, N. K., Das, K. (2014). Modeling of the adsorptive removal of arsenic: A statistical approach. *Journal of Environmental Chemical Engineering, 2*(1), 585-597.

Roy, R. K. (1990). *A primer on the Taguchi method Dearborn*. Michigan Society of Manufacturing Engineers.

Roy, R. K. (2001). *Design of Experiments using the Taguchi approach: 16 steps to Product and Process Improvement*. New York: Wiley-Interscience.

Rubio, E. M. (2006). Analytical methods application to the study of tube drawing processes with fixed conical inner plug: Slab and Upper Bound Methods. *Journal of Achievements in Materials and Manufacturing Engineering*, *14*(1-2), 383–386.

Rubio, E. M., Camacho, A. M., Sevilla, L., & Sebastian, M. A. (2005). Calculation of the Forward Tension in Drawing Processes. *Journal of Materials Processing Technology*, *162–163*, 551–557. doi:10.1016/j.jmatprotec.2005.02.122

Rubio, E. M., Gonzalez, C., Marcos, M., & Sebastian, M. A. (2006). Energetic Analysis of Tube Drawing Processes with Fixed Plug by Upper Bound Method. *Journal of Materials Processing Technology*, *177*(1-3), 175–178. doi:10.1016/j.jmatprotec.2006.03.193

Ruiz, J., Lopez, V., & Fernandez, B. J. (2013). Effect of surface laser treatment on the microstructure and wear behaviour of grey iron. *Materials & Design*, *17*(5/6), 273–287.

Russek, U.-A., Staub, H., Palmen, A., & Kind, H. (2003). Simultaneous laser beam welding of thermoplastics – innovations and challenges. *Proceedings of the 22nd International Congress on Applications of Lasers & Electro-Optics*, 95, 99-107.

Ryder, A. (2009). *The design and manufacturing of primary aircraft structures via out-of-autoclave techniques*. Paper presented at the ICMAC Conference, Belfast Waterfront, UK.

Sahin, H., & Ozdemir, F. (2004). Effect of some hydrocolloids on the rheological properties of different formulated ketchups. *Food Hydrocolloids*, *18*(6), 1015–1022. doi:10.1016/j.foodhyd.2004.04.006

Şahin, H., & Özdemir, F. (2007). Effect of some hydrocolloids on the serum separation of different formulated ketchups. *Journal of Food Engineering*, *81*(2), 437–446. doi:10.1016/j.jfoodeng.2006.11.022

Sahoo, A. K., & Sahoo, B. (2013). Performance Studies of Multilayer Hard Surface Coatings (TiN/TiCN/Al2O3/TiN) of Indexable Carbide Inserts in Hard Machining: Part-II (RSM, greyrelational and techno economical approach). *Measurement*, *46*(8), 2868–2884. doi:10.1016/j.measurement.2012.09.023

Sahu, R. K., Verma, S., & Mishra, S. K. (n.d.). *Optimization of parameters to minimize the skin friction coefficient in abrasive water suspension jet machining through TLBO*. Academic Press.

Sakthivel, M., & Ramesh, S. (2013). Mechanical properties of nat ural fiber (banana, coir, sisal) polymer composites. Science Park, 1(1), 1–6.

Salari, D., Daneshvar, N., Aghazadeh, F., & Khataee, A. R. (2005). Application of artificial neural networks for modeling of the treatment of wastewater contaminated with methyl tert-butyl ether (MTBE) by UV/H2O2 process. *J. Hazard. Mater. B*, *125*(1-3), 205–210. doi:10.1016/j.jhazmat.2005.05.030 PMID:15996818

Saliba, T. E., Hofmann, D., & Smolinski, P. (1994). Development of an in situ Hall-effect sensor for on-line monitoring of thickness and compaction during composite curing. *Composites Science and Technology*, *51*(1), 1–9. doi:10.1016/0266-3538(94)90151-1

Salonitis, K., & Vatousianos, S. (2012). Experimental investigation of the plasma arc cutting process. *Procedia CIRP*, *3*, 287–292. doi:10.1016/j.procir.2012.07.050

Samat, S. N. A., Azahari, B., & Rashid, A. A. (2013). Effect of ball milling time on the particle size and morphology of banana stem powder soaked in sodium hypochlorite solution. *Journal of Chemistry and Chemical Engineering*, *7*(3), 206.

Samotugin, S. S. (2008). Plasma treatment of tool steels. *Welding International*, *12*(3), 228–235.

Sánchez-Moreno, C., De Ancos, B., Plaza, L., Elez-Martínez, P., & Cano, M. P. (2008). Nutritional characterization of tomato juices. *Tomatoes and Tomato Products: Nutritional, Medicinal, and Therapeutic Properties*, 235-258.

Sanjay, M., Arpitha, G., & Yogesha, B. (2015). Study on mechanic al properties of natural-glass fibre reinforced polymer hybrid composites: A review. *Materials Today: Proceedings*, *2*(4), 2959–2967. doi:10.1016/j.matpr.2015.07.264

Saravanan, S., & Kumar, M. S. (2013). Effect of mechanical properties on rice husk ash reinforced aluminum alloy (AlSi10mg) matrix composites. *Procedia Engineering*, *64*, 1505–1513. doi:10.1016/j.proeng.2013.09.232

Sassi, S., Badri, B., & Thomas, M. (2006, October). "TALAF" and "THIKAT" as Innovative Time-Domain Indicators for Tracking Ball Bearings. *Proceedings of the 14th Seminar on Machinery Vibration*, 24-27.

Satasivam, S., & Bai, Y. (2014). Mechanical performance of bolted modular GFRP composite sandwich structures using standard and blind bolts. *Composite Structures*, *117*(0), 59–70. doi:10.1016/j.compstruct.2014.06.011

Satya Mohan, K., & Zafar, A. R. (1986). Water Quality of two drinking water reservoirs of Hyderabad, India. *Pollution Research*, *5*(3-4), 111–112.

Savino, M. M., & Mazza, A. (2015). Kanban-driven parts feeding within a semi-automated O-shaped assembly line: A case study in the automotive industry. *Assembly Automation*, *35*(1), 3–15. doi:10.1108/AA-07-2014-068

Schellbach, S. L., Monteiro, S. N., & Drelich, J. W. (2016). A novel method for contact angle measurements on natural fibers. *Materials Letters*, *164*, 599–604. doi:10.1016/j.matlet.2015.11.039

Schlimbach, J., & Ogale, A. (2012). 14 - Out-of-autoclave curing process in polymer matrix composites. In S. G. Advani & K.-T. Hsiao (Eds.), *Manufacturing Techniques for Polymer Matrix Composites (PMCs)* (pp. 435–480). Woodhead Publishing. doi:10.1533/9780857096258.3.435

Schubel, P. (2010). Technical cost modelling for a generic 45-m wind turbine blade producedby vacuum infusion (VI). *Renewable Energy*, *35*(1), 183–189. doi:10.1016/j.renene.2009.02.030

Seferis, J. C., Hillermeier, R. W., & Buehler, F. D. R. U. (2000). *2.20 Prepregging and Autoclaving of Thermoset Composites* (A. Kelly & C. Zweben, Eds.). Comprehensive Composite Materials.

Sekhon, G. S., & Chenot, J. L. (2013). Numerical simulation of continuous chip formation during non-steady orthogonal cutting. *Transactions of the ASME*, *10*(1), 31-48.

Seki, H., Suzuki, A., & Maruyama, H. (2005). Biosorption of chromium (VI) and arsenic (V) onto methylated yet biomass. *Journal of Colloid and Interface Science*, *281*(2), 261–266. doi:10.1016/j.jcis.2004.08.167 PMID:15571680

Selvan, J. S., Subramaniam, K., & Nath, A. K. (2009). Effect of laser surface hardening on En18 36–(AISI 5135) steel. *Journal of Materials Processing Technology*, *91*(1), 29–43.

Sepehr, A., & Zucca, C. (2012). Ranking desertification indicators using TOPSIS algorithm. *Natural Hazards*, *62*(3), 1137–1153. doi:10.1007/s11069-012-0139-z

Sharoba, A., Senge, B., El-Mansy, H., Bahlol, H. E., & Blochwitz, R. (2005). Chemical, sensory and rheological properties of some commercial German and Egyptian tomato ketchups. *European Food Research and Technology*, *220*(2), 142–151. doi:10.1007/s00217-004-0981-7

Shervedani, R. K., Mehrjardi, A. H., & Zamiri, N. (2006). A novel method for glucose determination based on electrochemical impedance spectroscopy using glucose oxidase self-assembled biosensor. *Bioelectrochemistry (Amsterdam, Netherlands)*, *69*(2), 201–208. doi:10.1016/j.bioelechem.2006.01.003 PMID:16580891

Shetty, G. R., & Chellam, S. (2003). Predicting membrane fouling during municipal drinkingwater nanofiltration using artificial neural Networks. *Journal of Membrane Science*, *217*(1-2), 69–86. doi:10.1016/S0376-7388(03)00075-9

Shi, D. F., Wang, W. J., & Qu, L. S. (2004). Defect detection for bearings using envelope spectra of wavelet transform. *Journal of Vibration and Acoustics*, *126*(4), 567–573. doi:10.1115/1.1804995

Shinohara, T., & Yoshida, K. (2005). Deformation Analysis of Surface Flaws in Stainless Steel Wire Drawing. *Journal of Materials Processing Technology*, *162–163*, 579–584. doi:10.1016/j.jmatprotec.2005.02.125

Siddika, S., Mansura, F., & Hasan, M. (2013). Physico-mechanic al properties of jute-coir fiber reinforced hybrid polypropylene composites. In Proceedings of World Academy of Science, Engineering and Technology. World Academy of Science, Engineering and Technology (WASET).

Sidhu, J. S., Bawa, A., & Singh, N. (1997). Studies on the effect of hydrocolloids on the consistency of tomato ketchup. *Journal of Food science and Technology, 34*(5), 423-424.

Sinha, K., Chowdhury, S., Saha, P. D., & Datta, S. (2013). Modeling of microwave-assisted extraction of natural dye from seeds of *Bixa orellana* (Annatto) using response surface methodology (RSM) and artificial neural network (ANN). *Industrial Crops and Products*, *41*, 165–171. doi:10.1016/j.indcrop.2012.04.004

Smith, J. D. (1982). Vibration monitoring of bearings at low speeds. *Tribology International*, *15*(3), 139–144. doi:10.1016/0301-679X(82)90130-X

Smith, S., & Tlusty, J. (2011). An overview of modelling and simulation of milling process. *ASME Journal of Engineering for Industry*, *113*(2), 169–175. doi:10.1115/1.2899674

Solly, K., Wang, X., Xu, X., Strulovici, B., & Zheng, W. (2004). Application of real-time cell electronic sensing (RT-CES) technology to cell-based assays. *Assay and Drug Development Technologies*, *2*(4), 363–372. doi:10.1089/adt.2004.2.363 PMID:15357917

Song, R. G., Zhang, K., & Chen, G. N. (2013). Electron beam surface treatment. Part I: Surface hardening of AISI D3 tool steel. *Journal of Materials Processing Technology*, *69*(1), 513–527.

Stading, M., & Bohlin, L. (2001). Contraction flow measurements of extensional properties. *Annual Transactions-Nordic Rheology Society*, *8*, 181–186.

Stark, W., Jaunich, M., & McHugh, J. (2013). Carbon-fibre epoxy prepreg (CFC) curing in an autoclave analogue process controlled by Dynamic Mechanical Analysis (DMA). *Polymer Testing*, *32*(8), 1487–1494. doi:10.1016/j.polymertesting.2013.09.014

Stewart, R. (n.d.). New prepreg materials offer versatility, top performance. *Reinforced Plastics, 53*(5), 28-33.

Stich, T. J., Spoerre, J. K., & Velasco, T. (2010). The application of artificial neural networks to monitoring and control of an induction hardening process. *Journal of Information Technology*, *16*(1), 1–11.

Strik, D. P. B. T. B., Domnanovich, A. M., Zani, L., Braun, R., & Holubar, P. (2005). Prediction of trace compounds in biogas from anaerobic digestion using the MATLAB Neural Network Toolbox. *Environmental Modelling & Software*, *20*(6), 803–810. doi:10.1016/j.envsoft.2004.09.006

Sumaila, M., Amber, I., & Bawa, M. (2013). Effect of fiber length on the physical and mechanical properties of random oriented, nonwoven short banana (Musa balbisiana) fibre/epoxy composite. Asian Journal of Natural and Applied Sciences, 2(1), 39–49.

Summers, G. J., & Scherpereel, C. M. (2008). Decision making in product development: Are you outside-in or inside-out? *Management Decision*, *46*(9), 1299–1312. doi:10.1108/00251740810911957

Sun, C. C., & Lin, G. T. (2009). Using fuzzy TOPSIS method for evaluating the competitive advantages of shopping websites. *Expert Systems with Applications*, *36*(9), 11764–11771. doi:10.1016/j.eswa.2009.04.017

Sunnersjö, C. S. (1978). Varying compliance vibrations of rolling bearings. *Journal of Sound and Vibration*, *58*(3), 363–373. doi:10.1016/S0022-460X(78)80044-3

Suresh, R., Basavarajappa, S., Gaitonde, V. N., Samuel, G. L., & Davim, J. P. (2013). State-of-the-art research in machinability of hardened steels, *Proc. IMechE Part B: J. Engineering Manufacture*, *227*(2), 191–209. doi:10.1177/0954405412464589

Suresh, R., Basavarajappa, S., & Samuel, G. L. (2012). Some studies on hard turning of AISI4340 steel using multilayer coated carbide tool. *Measurement*, *45*(7), 1872–1884. doi:10.1016/j.measurement.2012.03.024

Tabachnick, B., & Fidell, L. (2013). *Using Multivariate Statistics*. Pearson.

Taguchi, G. (1990). *Introduction to Quality Engineering*. Tokyo: Asian Productivity Organization.

Talla, G., Sahoo, D. K., Gangopadhyay, S., & Biswas, C. (2015). Modeling and multi-objective optimization of powder mixed electric discharge machining process of aluminum/alumina metal matrix composite. Engineering Science and Technology, an International Journal, 18(3), 369–373.

Tallian, T. E., & Gustafsson, O. G. (1965). Progress in rolling bearing vibration research and control. *ASLE Transactions, 8*(3), 195-207.

Tandon, N., & Nakra, B. C. (1992). Vibration and acoustic monitoring techniques for the detection of defects in rolling element bearings—a review. *The Shock and Vibration Digest, 24*(3), 3-11.

Tandon, N. (1994). A comparison of some vibration parameters for the condition monitoring of rolling element bearings. *Measurement*, *12*(3), 285–289. doi:10.1016/0263-2241(94)90033-7

Tandon, N., & Choudhury, A. (1999). A review of vibration and acoustic measurement methods for the detection of defects in rolling element bearings. *Tribology International*, *32*(8), 469–480. doi:10.1016/S0301-679X(99)00077-8

Tandon, N., & Nakra, B. C. (1993). Detection of defects in rolling element bearings by vibration monitoring. *Indian Journal of Mechanical Engineering Division*, *73*, 271–282.

Tang, K. K., Li, Z. X., & Wang, J. (2011). Numerical Simulation of Damage Evolution in Multi-Pass Wire Drawing Process and its Applications. *Materials & Design*, *32*(6), 3299–3311. doi:10.1016/j.matdes.2011.02.026

Tan, J., & Kerr, W. (2015). Rheological properties and microstructure of tomato puree subject to continuous high pressure homogenization. *Journal of Food Engineering*, *166*, 45–54. doi:10.1016/j.jfoodeng.2015.05.025

Tarng, Y. S., Juang, S. C., & Chang, C. H. (2002). The use of grey based Taguchi methods to determine submerged arc welding process parameters in hard facing. *Journal of Materials Processing Technology*, *128*(1-3), 1–6. doi:10.1016/S0924-0136(01)01261-4

Tarsha-Kurdi, K. E., & Olivier, P. (2002). Thermoviscoelastic analysis of residual curing stresses and the influence of autoclave pressure on these stresses in carbon/epoxy laminates. *Composites Science and Technology*, *62*(4), 559–565. doi:10.1016/S0266-3538(01)00148-8

Tastle, W. J., & Wierman, M. J. (2007). Consensus and dissention: A measure of ordinal dispersion. *International Journal of Approximate Reasoning*, *45*(3), 531–545. doi:10.1016/j.ijar.2006.06.024

Taylor, J. I. (1994). *The vibration analysis handbook*. Vibration Consultants.

Teulet, Ph., Girard, L., Razafinimanana, M., Gleizes, A., Bertrand, Ph., Camy-Peyret, F., & Richard, F. et al. (2006). Experimental study of an oxygen plasma cutting torch: II. Arc material interaction, energy transfer and anode attachment. *Journal of Physics. D, Applied Physics*, *39*(8), 1557–1573. doi:10.1088/0022-3727/39/8/015

Tillema, S., & van der Steen, M. (2015). Co-existing concepts of management control: The containment of tensions due to the implementation of lean production. *Management Accounting Research*, *27*, 67–83. doi:10.1016/j.mar.2015.01.002

Tiwari, T. N., Das, S. C., & Bose, P. K. (1986). Water quality index for the river Jhelum in Kashmir and its seasonal variation. *Pollution Research*, *5*(1), 1–5.

Tlusty, J. (2000). *Manufacturing Process and equipment*. Prentice Hall.

Tonshoff, H. K., Arendt, C., & Ben, A. R. (2000). Cutting hardened steels. *Annals of CIRP*, *49*(2), 215–220. doi:10.1016/S0007-8506(07)63455-6

Toribio, J., Kharin, V., Lorenzo, M., & Vergara, D. (2011). Role of Drawing-Induced Residual Stresses and Strains in the Hydrogen Embrittlement Susceptibility of Prestressing Steels. *Corrosion Science*, *53*(10), 3346–3355. doi:10.1016/j.corsci.2011.06.012

Trivedy, R.K., & Goel, P.K. (1984). Chemcial and biological methods for water pollution studies. *Env. Publ. India*, 215.

Tsao, H., Atkins, M. B., & Sober, A. J. (2004). Management of cutaneous melanoma. *The New England Journal of Medicine*, *351*(10), 998–1012. doi:10.1056/NEJMra041245 PMID:15342808

Ugrasen, G., Ravindra, H., Prakash, G. N., & Prasad, Y. T. (2015). Optimization of process parameters in wire edm of hchcr material using taguchis technique. *Materials Today: Proceedings*, *2*(4), 2443–2452. doi:10.1016/j.matpr.2015.07.185

Uno, Y., & Okada, A. et al.. (1999). High efficiency fine boring of monocrystalline silicon ingot by electrical discharge machining. *Precision Engineering*, *23*(2), 126–133. doi:10.1016/S0141-6359(98)00029-4

Vaezi, M., Seitz, H., & Yang, S. (2013). A review on 3D micro-additive manufacturing technologies. *International Journal of Advanced Manufacturing Technology*, *67*(5-8), 1721–1754. doi:10.1007/s00170-012-4605-2

Valencia, C., Sanchez, M., Ciruelos, A., & Gallegos, C. (2004). Influence of tomato paste processing on the linear viscoelasticity of tomato ketchup. *Food Science & Technology International*, *10*(2), 95–100. doi:10.1177/1082013204043880

Valencia, C., Sanchez, M., Ciruelos, A., Latorre, A., Madiedo, J., & Gallegos, C. (2003). Non-linear viscoelasticity modeling of tomato paste products. *Food Research International*, *36*(9), 911–919. doi:10.1016/S0963-9969(03)00100-5

Varela, P., Gambaro, A., Giménez, A., Duran, I., & Lema, P. (2003). Sensory and instrumental texture measures on ketchup made with different thickeners. *Journal of Texture Studies*, *34*(3), 317–330. doi:10.1111/j.1745-4603.2003.tb01064.x

Varshney, M., & Li, Y. (2009). Interdigitated array microelectrodes based impedance biosensors for detection of bacterial cells. *Biosensors & Bioelectronics*, *24*(10), 2951–2960. doi:10.1016/j.bios.2008.10.001 PMID:19041235

Vasilev, K. V. (2003). Plasma-arc cutting – a promising method of thermal cutting. *Welding International*, *17*(2), 147–151. doi:10.1533/wint.2003.3096

Velmurugan, C., Subramanian, R., Thirugnanam, S., & Ananadavel, B. (2011). Experimental investigations on machining characteristics of al 6061 hybrid metal matrix composites processed by electrical discharge machining. *International Journal of Engineering Science and Technology*, *3*(8), 87–101.

Venkata Rao, R., & Patel, V. (2012). Multi-objective optimization of combined Brayton and inverse Brayton cycles using advanced optimization algorithms. *Engineering Optimization*, *44*(8), 965–983. doi:10.1080/0305215X.2011.624183

Vijayan, S., Raju, R., & Rao, S. R. K. (2010). Multi objective Optimization of Friction Stir Welding Process Parameters on Aluminium Alloy AA5083 using Taguchi Based Grey Relational Analysis. *Materials and Manufacturing Processes*, *25*(11), 1206–1212. doi:10.1080/10426910903536782

Vikas & Kumar. (2013). *Effect and optimization of machine process parameters on over-cut in edm for en41 material using taguchi*. Academic Press.

Vilay, V., Mariatti, M., Taib, R. M., & Todo, M. (2008). Effect of fiber surface treatment and fiber loading on the properties of bagasse fiber– reinforced unsaturated polyester composites. *Composites Science and Technology*, *68*(3), 631–638. doi:10.1016/j.compscitech.2007.10.005

Villa, A., & Taurino, T. (2013). From JIT to Seru, for a Production as Lean as Possible. *Procedia Engineering*, *63*, 956–965. doi:10.1016/j.proeng.2013.08.172

Villay, A., de Filippis, F. L., Picton, L., Le Cerf, D., Vial, C., & Michaud, P. (2012). Comparison of polysaccharide degradations by dynamic high-pressure homogenization. *Food Hydrocolloids*, *27*(2), 278–286. doi:10.1016/j.foodhyd.2011.10.003

Wang, C., & Gao, R. X. (2003). Wavelet transform with spectral post-processing for enhanced feature extraction . *IEEE Transactions on Instrumentation and Measurement*, *52*(4), 1296–1301. doi:10.1109/TIM.2003.816807

Wang, J. W., Cheng, C. H., & Huang, K. C. (2009). Fuzzy hierarchical TOPSIS for supplier selection. *Applied Soft Computing*, *9*(1), 377–386. doi:10.1016/j.asoc.2008.04.014

Wang, J. Y., Kusumoto, K., & Nezu, K. (2013). Modelling and prediction of cut shape for plasma arc cutting based on artificial neural network. *Science and Technology of Welding and Joining*, *4*(4), 195–200. doi:10.1179/136217199101537770

Wang, J., Kusumoto, K., & Nezu, K. (2011). Plasma arc cutting torch tracking control. *Technology of Welding and Joining*, *6*(3), 154–156. doi:10.1179/136217101101538695

Wang, J., Zhu, Z., He, C., & Yang, F. (2011). Effect of dual swirling plasma arc cutting parameters on kerf characteristics. *International Journal of Material*, *4*, 39–43.

Wang, K. L., & Argyropoulos, V. (2005). Design and Analysis of Direct Cold Drawing of Section Rods through a Single Die. *Journal of Materials Processing Technology*, *166*(3), 345–358. doi:10.1016/j.jmatprotec.2004.06.030

Wang, T. C., & Chang, T. H. (2007). Application of TOPSIS in evaluating initial training aircraft under a fuzzy environment. *Expert Systems with Applications*, *33*(4), 870–880. doi:10.1016/j.eswa.2006.07.003

Wansheng, Z., Zhenlong, W., Shichun, D., Guanxin, C., & Hongyu, W. (2002). Ultrasonic and electric discharge machining to deep and small hole on titanium alloy. *Journal of Materials Processing Technology*, *120*(1-3), 101–106. doi:10.1016/S0924-0136(01)01149-9

Wardle, F. P., & Poon, S. Y. (1983). Rolling bearing noise-cause and cure. *Chartered Mechanical Engineer, 30*, 36-40.

Warnakulasuriya, S., Johnson, N., & Van der Waal, I. (2007). Nomenclature and classification of potentially malignant disorders of the oral mucosa. *Journal of Oral Pathology & Medicine*, *36*(10), 575–580. doi:10.1111/j.1600-0714.2007.00582.x PMID:17944749

Warren, H., Gately, R. D., OBrien, P., Gorkin, R. III, & in het Panhuis, M. (2014). Electrical conductivity, impedance, and percolation behavior of carbon nanofiber and carbon nanotube containing gellan gum hydrogels. *Journal of Polymer Science. Part B, Polymer Physics*, *52*(13), 864–871. doi:10.1002/polb.23497

Weichert, R., & Schonert, K. (2011). Heat Generation at the tip of a moving crack. *Journal of the Mechanics and Physics of Solids*, *26*(3), 151–161. doi:10.1016/0022-5096(78)90006-6

Werbos, P. J. (1988). Generation of Back Propagation with application to a recurrent gas market model. *Neural Networks*, *1*(4), 339–356. doi:10.1016/0893-6080(88)90007-X

White, S. R., & Kim, Y. K. (1996). Staged curing of composite materials. *Composites. Part A, Applied Science and Manufacturing*, *27*(3), 219–227. doi:10.1016/1359-835X(95)00023-U

WHO. (1984a). International Drinking Water Supply and Sanitation Decade-Review of National Baseline Data. WHO off set publication No. 85, World Health Organisation.

WHO. (1984b). Guidelines for Drinking Water Quality Vol. 1, Recommendations. World Health Organization.

WHO. (1984c). Guidelines for Drinking Water Quality. Vol. 2. Health Criteria and other Supporting Information. World Health Organization.

WHO. (1993). *Guidelines for drinking-water quality*. Geneva, Switzerland: World Health Organization.

Widrow, B., Winter, R. G., & Barxter, R. A. (1998). Layered Neural Nets for Pattern Recognition. *IEEE Transactions on Acoustics, Speech, and Signal Processing*, *36*(7), 1109–1118. doi:10.1109/29.1638

Wong, A. K., & Kirby, G. C. III. (1990). A hybrid numerical/experimental technique for determining the heat dissipated during low cycle fatigue. *Engineering Fracture Mechanics*, *37*(3), 493–504. doi:10.1016/0013-7944(90)90375-Q

Woo, H. G., & Cho, H. S. (2008). Estimation of hardened layer dimensions in laser surface hardening processes with variations of coating thickness. *Surface and Coatings Technology*, *102*(1), 205–217.

Xu, L., Gong, H., Dong, M., & Li, Y. (2015). Rheological properties and thickening mechanism of aqueous diutan gum solution: Effects of temperature and salts. *Carbohydrate Polymers*, *132*, 620–629. doi:10.1016/j.carbpol.2015.06.083 PMID:26256389

Xu, W. J., Fang, J. C., & Lu, Y. S. (2002). Study on ceramic cutting by plasma arc. *Journal of Materials Processing Technology*, *129*(1-3), 152–156. doi:10.1016/S0924-0136(02)00600-3

Xu, W., Wang, K., Wang, P., & Zhou, J. (2011). A Newly Developed Plug in the Drawing Process for Achieving the High Accuracy of Aluminium Rectangular Tube. *International Journal of Advanced Manufacturing Technology*, *57*(1-4), 1–9. doi:10.1007/s00170-011-3292-8

Yan, B. H., Tsai, H. C., & Lin, Y. C. (1997). Study on EDM characteristics of cemented carbides. *Proceedings of the 14th National Conference on Mechanical Engineering*, 157 - 164.

Yan, B. H., Wang, A. C., Huang, C. Y., & Huang, F. Y. (2002). Study of precision micro-holes in borosilicate glass using micro EDM combined with micro ultrasonic vibration machining. *International Journal of Machine Tools & Manufacture*, *42*(10), 115–112. doi:10.1016/S0890-6955(02)00061-5

Yang, L. J. (2001). Plasma surface hardening of ASSAB 760 steel specimens with Taguchi optimization of the processing parameters. *Journal of Materials Processing Technology*, *113*(1-3), 521–526. doi:10.1016/S0924-0136(01)00583-0

Yang, L. J. (2007). The effect of specimen thickness on the hardness of plasma surface hardened ASSAB 760 steel specimens. *Journal of Materials Processing Technology*, *185*(1-3), 113–119. doi:10.1016/j.jmatprotec.2006.03.117

Yang, L. J. (2011). Plasma arc surface hardening of ASSAB 760 steel specimens with taguchi optimization of the processing parameters. *Journal of Materials Processing Technology*, *113*(1), 521–526.

Yang, L., & Bashir, R. (2008). Electrical/electrochemical impedance for rapid detection of foodborne pathogenic bacteria. *Biotechnology Advances*, *26*(2), 135–150. doi:10.1016/j.biotechadv.2007.10.003 PMID:18155870

Yan, M. (2003). Micro-beam plasma arc surface processing for ferrous and non-ferrous metals. *Journal of Materials Science*, *38*(1), 3219–3222. doi:10.1023/A:1025169517526

Yan, M., & Zhu, W. Z. (2007). Surface treatment of 45 steel by plasma arc melting. *Surface and Coatings Technology, 91*(1), 183–191.

Yan, M., & Zhu, W. Z. (2008). A new approach of surface treatment: Micro-beam plasma arc remelting. *Materials Letters, 34*(1), 222–225.

Yetilmezsoy, K. (2006). Determination of optimum body diameter of air cyclones using a new empirical model and a neural network approach. *Environmental Engineering Science, 23*(4), 680–690. doi:10.1089/ees.2006.23.680

Yetilmezsoy, K., & Demirel, S. (2008). Artificial neural network (ANN) approach for modeling of Pb(II) adsorption from aqueous solution by Antep pistachio (Pistacia vera L.) shells. *Journal of Hazardous Materials, 153*(3), 1288–1300. doi:10.1016/j.jhazmat.2007.09.092 PMID:17980484

Yetilmezsoy, K., & Saral, A. (2007). Stochastic modeling approaches based on neural network and linear–nonlinear regression techniques for the determination of single droplet collection efficiency of countercurrent spray towers. *Environmental Modeling and Assessment, 12*(1), 13–26. doi:10.1007/s10666-006-9048-4

Yılmaz, M. T., Karaman, S., Cankurt, H., Kayacier, A., & Sagdic, O. (2011). Steady and dynamic oscillatory shear rheological properties of ketchup–processed cheese mixtures: Effect of temperature and concentration. *Journal of Food Engineering, 103*(2), 197–210. doi:10.1016/j.jfoodeng.2010.10.016

Yilmaz, O., & Okka, M. A. (2010). Effect of single and multi-channel electrodes application on EDM fast hole drilling performance. *International Journal of Advanced Manufacturing Technology, 51*(1-4), 185–194. doi:10.1007/s00170-010-2625-3

Yoo, B., & Rao, M. (1994). Effect of unimodal particle size and pulp content on rheological properties of tomato puree. *Journal of Texture Studies, 25*(4), 421–436. doi:10.1111/j.1745-4603.1994.tb00772.x

Yoon, K. P., & Hwang, C. L. (1995). *Multiple attribute decision making: an introduction* (Vol. 104). Sage publications. doi:10.4135/9781412985161

Yoshizawa, M., Sumimoto, D., & Kazuhiro, K. (1984). *The Forming Technique and Quality Characteristics of ERW Rifled Boiler Tube*. 106th ISIJ Meeting, S1221 and S1394.

Young, W.-B. (1995). Compacting pressure and cure cycle for processing of thick composite laminates. *Composites Science and Technology, 54*(3), 299–306. doi:10.1016/0266-3538(95)00067-4

Yu, T., Zhou, J., Zhao, K., Wang, W., & Wang, W. (2008, October). Study on project experts' evaluation based on analytic hierarchy process and fuzzy comprehensive evaluation. In *Intelligent Computation Technology and Automation (ICICTA), 2008 International Conference on* (Vol. 1, pp. 941-945). IEEE.

Yuan, C., Li, M., Zhang, Z., & Gu, Y. (2008). Experimental investigation on the Co-cure processing of honeycomb structure with self-adhesive prepreg. *Applied Composite Materials, 15*(1), 47–59. doi:10.1007/s10443-008-9056-4

Yuangang, W., Fuling, Z., & Jin, W. (2009). Wear resist Electrodes for Micro-EDM. *Chinese Journal of Aeronautics, 22*(3), 339–342. doi:10.1016/S1000-9361(08)60108-9

Yuhazri, M., & Sihombing, H. (2010). A comparison process between vacuum infusion and hand lay-up method toward kenaf/polyester composite. *International Journal of Basic & Applied Sciences, 10.*

Yun, W. S., Hoon, J., Cho, D., & Ehman, K. (2012). Development of a virtual machining system. Part2: Prediction and analysis of a machined surface error. *International Journal of Machine Tools & Manufacture, 42*(2), 1607–1615.

Yu, Z. Y., Zhang, Y., Luan, L. J., Zhao, F., & Gua, D. (2009). High aspect ratio micro-hole drilling aided with ultrasonic vibration and planetary movement of electrode by micro-EDM. *CIRP Annals-Manufacturing Technology*, *58*(1), 213–216. doi:10.1016/j.cirp.2009.03.111

Zavadskas, E. K., Zakarevicius, A., & Antucheviciene, J. (2006). Evaluation of ranking accuracy in multi-criteria decisions. *Informatica*, *17*(4), 601–618.

Zehnder, A. T., & Rosakis, A. J. (1993). Temperature Rise at the Tip of Dynamically Propagating Cracks: Measurements Using High-speed Infrared Detectors. Experimental Techniques in Fracture III. Society for Experimental Mechanics.

Zehnder, A. T. (1991). A model for the heating due to plastic work. *Mechanics Research Communications*, *18*(1), 23–28. doi:10.1016/0093-6413(91)90023-P

Zehnder, A. T., Babinsky, E., & Palmer, T. (1998). Hybrid Method for Determining the Fraction of Plastic Work Converted to Heat. *Experimental Mechanics*, *38*(4), 295–302. doi:10.1007/BF02410392

Zehnder, A. T., & Rosakis, A. J. (1991). On the temperature distribution at the vicinity of dynamically propagating cracks in 4340 steel. *Journal of the Mechanics and Physics of Solids*, *39*(3), 385–415. doi:10.1016/0022-5096(91)90019-K

Zelbst, P. J., Kenneth, W., Green, J., Abshire, R. D., & Sower, V. E. (2010). Relationships among market orientation, JIT, TQM, and agility. *Industrial Management & Data Systems*, *110*(5), 637–658. doi:10.1108/02635571011044704

Zhai, Z., Li, S., & Liu, Y. (2015). Parameter Determination of Milling Process Using a Novel Teaching-Learning-Based Optimization Algorithm. *Mathematical Problems in Engineering*.

Zhang, Y. F., Lu, L., & Yap, S. M. (2009). Prediction of the amount of PCA for mechanical milling. *Journal of Materials Processing Technology*, *89*(1), 260–265.

Zhao, W., Yu, Q.-q., Li, H.-s., & Tian, Y.-z. (2014). *Study on the relationship between JIT practices and operational performance based on the cost leading strategy.* Paper presented at the Management Science & Engineering (ICMSE), 2014 International Conference on.

About the Contributors

Mohan Kumar Pradhan is Assistant Professor in the Department of Mechanical Engineering, and Head of Production Engineering Lab.and CAM lab. of the Maulana Azad National Institute of Technology, Bhopal, India. He received his M. Tech and PhD from National Institute of Technology, Rourkela, India. He has over 15 years of teaching and research experience in manufacturing and 5 years of postdoctoral research experience. He has advised over 50 graduates, 20 Post graduate and three PhD students. Dr. Pradhan's research interests include manufacturing, non-traditional machining, metrology, micro-machining, hybrid machining, and process modelling and optimization. Dr. Pradhan has more than 50 refereed publications and nearly 50 technically edited papers, which were published in conference proceedings, edited two books, Five Conference Proceedings and five journals as Guest editor authored seven book chapters. Dr. Pradhan is Charted Engineer, a life fellow of IIPE and life member of ISTE, IACSIT, IAENG and MIE (I).

* * *

Bappa Acherjee is presently working as an Assistant Professor in the Department of Production Engineering at BIT Mesra, Deoghar Campus. He received his Ph.D. in Engineering from Jadavpur University, Kolkata, India in 2012, in the field of laser beam welding. His research interest includes laser material processing and application of finite element method and soft computing techniques in manufacturing processes. He is the author of over 50 peer-reviewed publications including journal papers, conference papers and book chapters. Dr. Acherjee is a member of ASME and a life member of IIPE.

S. Ahmed, PhD in Electrical & Electronics Engineering from University of Surrey, UK, is the Advisor (Engineering Programs) to the Rector & President, International Islamic University (IIU), Islamabad, and the Principal Investigator of Islamic Development Bank's Funded Advanced Electronics Laboratories Project as well as Photovoltaic Energy Engineering Laboratory. Besides serving in EU/UK and USA during his professional career; he is also the former Dean of Faculty of Engineering & Technology at IIU, Islamabad. He is also the Trustee of Centre for Emerging Sciences, Engineering & Technology and Member Board of Governors, Indus University Karachi. As guest scientist and visiting professor, Dr. Shuja maintained very close collaborative linkages with some of the world's most renowned scientific facilities such as Lawrence Berkeley National Laboratory, Department of Energy, USA; Lucent-Bell Labs, USA and Nano-electronic fabrication engineering facilities at MIT, KSU and Michigan, USA. Dr. Shuja also co-founded two companies in 2006 and 2012, which are now having presence in UK, USA and Pakistan. Higher Education Commission of Pakistan has also focused and highlighted Dr. Shuja's contributions in its official newsletter in February 2011 and May 2015. Dr. Shuja instituted and chaired three international conferences and one national workshop in Pakistan. With a track record of winning competitive projects funded both on national and international level; Dr. Shuja has published numerous international research communications including the book chapters, impact journal papers as well as peer-reviewed international conference proceedings. Dr. Shuja is a member of various national committees, and has a commendable track record of working closely with national and international bodies for the initiation, approval and accreditation of institutions, academic programs, and devising the policies and strategies in Science and Technology arena.

Tapan Kr. Barman received his BME, MME and PhD from Jadavpur University, Kolkata, India. He is presently working as Assistant Professor in Mechanical Engineering at Jadavpur University, Kolkata. His areas of interests are characterisation of machined surface topography and coatings tribology. He has 50 journals and 40 conference publications to his credit.

Dibyajyoti Biswal has received her M.Tech in Biomedical Engineering from National Institute of Technology Rourkela, Odisha, India. Prior to that she has done B.Tech from Gandhi Institute for technological advancement, Odisha, India. Presently, she is pursuing her PhD degree in Biomedical Engineering at University of Kentucky. Her research interests include Biomedical Signal Processing, Biomedical Instrumentation and Drug Delivery.

Goutam Kumar Bose is currently working as Professor in Mechanical Engineering Department, Haldia Institute of Technology, Haldia, India. He obtained his PhD in Production Engineering from the Jadavpur University, Kolkata, India. He has obtained a Master's in Engineering in Mechanical Engineering from the Bengal Engineering & Science University, Shibpur, India. He has worked as an Assistant Professor in the Department of Mechanical Engineering at the College of Engineering & Management, Kolaghat for ten years. He was an Engineer in R & D Centre of M/s Hindustan Motors Ltd. West Bengal, India. His active areas of interest are Metal Cutting, Non- conventional machining and Industrial and Production Management. He has published research papers in journals of International repute. He has attended several international conferences in India and abroad.

Ayan Chatterjee, MCA, is an assistant professor of D.EL.ED section at Sarboday Public Academy, Purba Medinipur, India. He holds a Master's degree in Computer Application from Vidyasagar University, India. Before that, he did B.Sc (Hons.) degree in Mathematics from University of Calcutta, India. Mr. Chatterjee started his professional career since January, 2016. In his academic career, he published some research papers with good impact factors. His Master's thesis is on optimization technique of Travelling Salesman Problem (TSP) using Artificial Neural Network under supervision of Dr. Asit Kumar Das, IIEST, India. His areas of interests are Software Engineering, Computational Methodology, Soft Computing, Steganography and Cryptography.

Kailash Chandra Das completed his Master of Science in Physics at the Indian Institute of Technology, Kharagpur during the year 2010–2012. He has done one year MSc project in the field of organic thin film and memory devices entitled "fabrication and characterization of optimum organic memory devices" at Indian Institute of Technology, Kharagpur. Now, he continued his Doctor of Philosophy as a PhD Scholar in Department of Physics and Astronomy at National Institute of Rourkela from the year 2012.

Manas Das works in the area of advanced finishing and nano-finishing processes. He also concentrates on Machining of Advanced Engineering Materials, Micromanufacturing, Tribology. He has a number of journal papers and conferences papers. He is presently working as Assistant Professor at IIT Guwahati, India.

Milan Kumar Das received his BME, MME and PhD from Jadavpur University, Kolkata. Presently he is working as Lecturer in Mechanical Engineering at Bandwan Polytechnic College, Purulia, India. His area of research interest is roughness modelling and machining characteristics.

Jorge Luis Garcia-Alcaraz is a full-time researcher of the Department of Industrial Engineering at the Autonomous University of Ciudad Juarez. He received a PhD in Industrial Engineering from Ciudad Juarez Technology Institute (Mexico). His research areas are related to Multicriteria decision making applied to manufacturing, production process modeling and statistical inference. He is founding member of the Mexican Society of Operation Research and active member in the Mexican Academy of Industrial Engineering. Currently, Dr. Garcia is a National Researcher recognized by the National Council of Science & Technology of México.

Anil Giri, Ph.D. from NIT, Rourkela, and presently working as a Lecturer in the PG Dept. of Environmental Science, F M University, Balasore.

Ajith G. Joshi is an Assistant Professor in the Department of Mechanical Engineering of the Canara Engineering College, Bantwal, Mangalore, India. He received his B.E. in Mechanical Engineering from University B.D.T. College of Engineering, Davangere under Kuvempu University in 2009. He has completed M.Sc. (Engg.) by Research under VTU, Belagavi, Karnataka India. His research interests include wear, composites, machining of hard materials, and optimization techniques. He has authored more than 20 international/national journal and conference publications including one edited book chapter.

Jyoti Prakash Kar received his MTech and PhD from IIT Delhi in 2001 and 2007, respectively. He has worked as a Visiting Researcher, Department of Electrical and Computer Engineering, University of Oklahoma, Norman, USA. Thereafter, he joined as a Postdoctoral Researcher in the Department of Materials Science and Engineering, Yonsei University, Seoul, South Korea. He has also worked as a Postdoctoral Researcher in Department of Electronics Engineering, University of Tor Vergata, Rome, Italy. Currently, he is working as an Assistant Professor, in Department of Physics and Astronomy, National Institute of Technology, Rourkela, India. His major area of research is fabrication and characterizations of electronic materials and devices.

Kijpokin Kasemsap received his BEng degree in Mechanical Engineering from King Mongkut's University of Technology, Thonburi, his MBA degree from Ramkhamhaeng University, and his DBA degree in Human Resource Management from Suan Sunandha Rajabhat University. Dr. Kasemsap is a Special Lecturer in the Faculty of Management Sciences, Suan Sunandha Rajabhat University, based in Bangkok, Thailand. Dr. Kasemsap is a Member of the International Association of Engineers (IAENG), the International Association of Engineers and Scientists (IAEST), the International Economics Development and Research Center (IEDRC), the International Association of Computer Science and Information Technology (IACSIT), the International Foundation for Research and Development (IFRD), and the International Innovative Scientific and Research Organization (IISRO). Dr. Kasemsap also serves on the International Advisory Committee (IAC) for the International Association of Academicians and Researchers (INAAR). Dr. Kasemsap is the sole author of over 250 peer-reviewed international publications and book chapters on business, education, and information technology.

Laraib Alam Khan is PhD in Materials engineering and Sciences from Northwest Composites Centre (NWCC), School of Materials, University of Manchester, UK. He did MS (Mechanical) from EME College, NUST (Pakistan) and completed his Bachelor in Mechanical Engineering from NED University, Karachi (Pakistan). He has more than twelve years experience in composites product development and manufacturing and been involved in various R&D projects related to polymeric composite and nano-materials. He has hand on experience and expertise of composite manufacturing processes and several physical and mechanical characterizations for composite structures. He is author of more than 25 research papers published in International Journals and Conferences.

Arunanshu S. Kuar is Professor in the Department of Production Engineering at Jadavpur University, Kolkata, India. He has been awarded his PhD in the area of laser beam micro-machining. He has published more than 60 national and international journals and conference papers and guided a number of post graduate and doctoral research work.

Gaurav Kulkarni completed M.Tech in Biomedical engineering from NIT Rourkela in 2015. Currently working at Indian Institute of Technology, Kharagpur (IIT KGP) as PhD research scholar. Areas of research includes Bio-MEMS, impedance sensor, signal processing, 3D cell culture, cancer diagnostics, etc.

Kamal Kumar is an Assistant Professor in the Department of Mechanical Engineering, PEC University of Technology, Chandigarh, India since 2013. He completed his PhD in Mechanical Engineering in 2012. He has a teaching experience of 9 years and research experience of 6 years. He has published more than 25 research articles in international journals of repute. His current area of research is advance machining processes, Process optimization, Biodegradable metallic materials etc.

Kaushik Kumar, B.Tech (Mechanical Engineering, REC (Now NIT), Warangal), MBA (Marketing, IGNOU), and Ph.D (Engineering, Jadavpur University), is presently an Associate Professor in the Department of Mechanical Engineering, Birla Institute of Technology, Mesra, Ranchi, India. He has 14 years of Teaching & Research and over 11 years of industrial experience in a manufacturing unit of Global repute. His areas of teaching and research interest are Quality Management Systems, Optimization, Non-conventional machining, CAD / CAM, Rapid Prototyping and Composites. He has 9 Patents, 3 Book, 8 Book Chapters, 96 international Journal publications, 18 International and 8 National Conference publications to his credit. He is on the editorial board and review panel of 7 International and 1 National Journals of repute. He has been felicitated with many awards and honours.

Manoj Kumar received his PhD in Mechanical Engineering from the Indian Institute of Technology, Delhi, India. He is presently working as Director, Mother Parwati Education Services, India. He has authored or co-authored over 80 research papers in journals and conferences.

Praveen Loharkar is working (2013 onwards) as Assistant Professor in the Department of Mechanical Engineering at SVKM's NMIMS MPSTME Shirpur Campus. He has done (2011-2013) his M.Tech from MANIT Bhopal in Industrial design and has three years (2008-2011) of experience as a Maintenance Engineer in ISMT Ltd. (Pune), a seamless tube manufacturing industry. He is a qualified Internal Auditor for ISO TS 16949, ISO 14001 and OHSAS 18001.His areas of interest are computer aided design and analysis, mechanical vibrations, manufacturing process improvement through modern tools.

Debanjan Maity is a Graduate Engineer. He completed his graduation with first class honours from Jadavpur University, Kolkata, India in 2015, with Production Engineering specialization. He has carried out his undergraduate projects in the field of multi criteria decision making and gearbox design. He is presently working on optimization using nature inspired metaheuristic algorithms.

Valeria Martínez-Loya is a last-year Industrial and Systems Engineering student at Autonomous University of Ciudad Juarez, where she is an assistant professor in the Department of Industrial and Manufacturing Engineering.

Prakash Mishra, M. Sc. and PhD from NIT, Rourkela and presently working as Reader, PG ept. of Environmental Science, F M University, Balasore, Odisha, India. Having more than 15 years of teaching and research Experience.

Dipten Misra is the Director of School of Laser Science and Engineering, Jadavpur University. His principal research area includes heat transfer, finite element analysis and laser materials processing. He is involved in simulation and experimental validation of laser cutting, welding, forming and laser cleaning of archaeological objects and items of cultural heritage. He has supervised a number of post graduate and doctoral theses and research projects. He has several publications in international journals and conference proceedings.

Souren Mitra is a Professor in the Department of Production Engineering, Jadavpur University, Kolkata, India. His research areas include non-traditional machining, laser material processing and advanced manufacturing systems. He has published more than 100 research papers in international and national journals and conferences. Several doctoral and post graduate theses have been completed under his guidance. He is a recipient of the Research Award of UGC, Government of India, and has successfully completed five sponsored research projects as the principal investigator.

Teresita Molina received a master's degree in Industrial Engineering from the Autonomous University of Ciudad Juarez (Mexico) and a bachelor's degree in Industrial Engineering with a speciality in Chemistry from the Chihuahua Technology Institute (Mexico). She worked in the maquiladora sector for 12 years in the areas of manufacturing, processes, design, materials and quality.

Deepak Mylavarapu is an M.Tech. student from IIT Guwahati specializing in Computer-Aided Manufacturing.

Suraj Kumar Nayak received his B. Tech in Electrical Engineering from Government College of Engineering Keonjhar, Odisha, India and M. Tech in Biomedical Engineering from National Institute of Technology Rourkela, Odisha, India. Currently, he is pursuing his Ph.D. degree in Biomedical Engineering from National Institute of Technology Rourkela, Odisha, India., His research interests include Biomedical Signal and Image Processing, Biomedical Instrumentation, Artificial Neural Network and Drug Delivery.

Pritam Pain completed his B.Tech in Mechanical Engineering from West Bengal University of Technology, Kolkata, India. Currently, he is a final year M.Tech student in the Mechanical Engineering Department at Haldia Institute of Technology, Haldia. His areas of interest are nontraditional and modern machining.

Kunal Pal pursued his graduation in Pharmacy from University of Delhi, New Delhi, India in the year 2002 followed by post-graduation in Biomedical Engineering from Jadavpur University, Kolkata, India in the year 2004. Later, he did his Ph.D. in Materials Science from Indian Institute of Technology, Kharagpur, India in the year 2006. Subsequently, he joined Ryerson University, Toronto, Canada as a post-doctoral fellow in the year 2007. During his tenure at Ryerson University, he served as Vice-Chair (Elect), HQPA, AFMNet Network of Excellence, Canada. He joined National Institute of Technology, Rourkela, India in the year 2009 as an Assistant Professor in Biomedical Engineering. He is currently the Professor-in-Charge of Medical Electronics and Instrumentation Laboratory in the Department of Biotechnology and Medical Engineering, National Institute of technology, Rourkela, India. His major research interests revolve around Biomedical Signal Processing and Biomedical Equipment Design. He is also working in the fields of soft materials and controlled drug delivery. He has more than 140 publications in SCI cited journals of high repute, Book Chapters and peer-reviewed conferences. He is a very well cited researcher having more than 1700 citations.

Preeti Madhuri Pandey is an M. Tech in Biomedical Engineering from the Department of Biotechnology & Medical Engineering, National Institute of Technology, Rourkela, India.

Ganesh Narayanan R is currently Associate Professor at the Department of Mechanical Engineering at Indian Institute of Technology Guwahati, India. His areas of research include Material forming and Joining. He has contributed mainly in the areas of Forming of sheet materials including laser welded, friction stir welded sheets & adhesive bonded sheets, computational applications in material forming and joining, and end forming of tubes. He has published several articles in reputed journals and edited books including 'Advances in Material Forming and Joining' by Springer India and 'Metal Forming Technology and Process Modeling' by McGrawHill Education, India.

Mahendra Rong, M.Sc., Ph.D., is working as an Assistant Professor in Mathematics at Bangabasi Evening College under University of Calcutta, India. He is also handling a UGC (India) Minor Research Project. His fields of interest are optimization, operations research, fuzzy-random mathematics, etc. He has publications in international journals such as European Journal of Operational Research, Applied Mathematical Modelling, Central European Journal of Operation Research, etc.

Prasanta Sahoo is a Professor in Mechanical Engineering at Jadavpur University, Kolkata, India. He received his BME from Jadavpur University, Kolkata, India and MTech and PhD in Mechanical Engineering from Indian Institute of Technology, Kharagpur, India. He has over 20 years of experience in teaching and research with special emphasis on tribology. He has supervised 25 PhD theses and 30 Masters theses so far. He has authored two books, co-authored over 15 book chapters and over 400 technical articles in peer-reviewed international journals and conference proceedings. He is the Editor-in-Chief of one international journal, Associate Editor of two journals and on the editorial board of six international journals.

R. Suresh is an Associate Professor in the Department of Mechanical & Manufacturing Engineering of M.S. Ramaiah University of Applied Sciences, Bangalore; India. He received his B.E in Mechanical Engineering from Kuvempu University in 1995. He was awarded Master of Technology (MTech) and Doctoral degree (PhD) in 2002 and 2013, respectively from UBDT College of Engineering, Davangere, Kuvempu University, Karnataka, India. His research interests include machining of hard materials, micromachining, non-traditional machining and optimization techniques. He has authored more than 40 international/national journal and conference publications.

Nadia Sofia Tanino is an undergraduate student at Texas A&M University (USA).

D. N. Tibarewala, resently a Professor of Biomedical Engineering at Jadavpur University, Kolkata, India, since August 1996, has earlier served as Director in the School of Bioscience & Engineering of the same University during the period May 2004-November 2011. He did his BSc (Honours) in 1971 and B. Tech (Applied Physics) in 1974 from the Calcutta University, India. He was admitted to the Ph.D. (Tech) degree of the Calcutta University in 1980. Having professional, academic and research experience of more than 35 years, Dr. Tibarewala has contributed about 250 research papers in the areas of Rehabilitation Technology, Biomedical Instrumentation & Signal Processing, Human Computer Interfaces and, related branches of Biomedical Engineering.

Atul Tiwari is an M.Tech Scholar in Industrial Design, Department of Mechanical Engineering, Maulana Azad National Institute of Technology, Bhopal.

Srikant Tiwari is an M.Tech Scholar from Manit, Bhopal. His specialization is Engineering material. He is working on machining of Aluminum-RHA composites by using Electric Discharge Machining.

Diego Tlapa is a professor of industrial engineering in the Faculty of Engineering, Architecture and Design at the Autonomous University of Baja California, Mexico. He has participated in several research projects related to process improvement. Dr. Tlapa is member of the National System of Researches of the National Council of Science and Technology in Mexico and member of the optimization of industrial processes network (ROPRIN); his research interests include supply chain, six sigma and lean manufacturing. He is author/coauthor around 15 journal and conference papers in process improvement.

Indu Yadav is an assistant professor in the Department of Biotechnology, Madhav Institute of Technology and Science, Gwalior, India. She perused her B. Tech. in Biomedical Engineering from SGSITS Indore, India followed by M. Tech. in Biomedical Engineering from the Department of Biotechnology & Medical Engineering, National Institute of Technology, Rourkela, India. Her research interests include mathematical modeling of bio-films for drug delivery and tissue engineering applications.

Index

P

Q

R

S

T

V

W

www.ingramcontent.com/pod-product-compliance
Lightning Source LLC
LaVergne TN
LVHW080620300926
849162LV00005B/1216
* 9 7 8 1 5 2 2 5 2 4 4 0 3 *